Continuous Bivariate Distributions

N. Balakrishnan · Chin-Diew Lai

Continuous Bivariate Distributions

Second Edition

Springer

N. Balakrishnan
Department of Mathematics & Statistics
McMaster University
1280 Main St. W.
Hamilton ON L8S 4K1
Canada
bala@mcmaster.ca

Chin-Diew Lai
Institute of Fundamental Sciences
Massey University
11222 Private Bag
Palmerston North
New Zealand
c.lai@massey.ac.nz

ISBN 978-1-4419-1875-8 e-ISBN 978-0-387-09614-8
DOI 10.1007/b101765
Springer Dordrecht Heidelberg London New York

Printed on acid-free paper

Springer is part of Springer Science+Business Media (www.springer.com)

To my loving mother, Lakshmi,
and lovely memories of my father,
Narayanaswamy

N.B.

To Ai Ing, Joseph, Eugene, Serena,
my brother Chin-Yii, and my parents

C.D.L.

Preface

This volume, which is completely dedicated to continuous bivariate distributions, describes in detail their forms, properties, dependence structures, computation, and applications. It is a comprehensive and thorough revision of an earlier edition of "Continuous Bivariate Distributions, Emphasizing Applications" by T.P. Hutchinson and C.D. Lai, published in 1990 by Rumsby Scientific Publishing, Adelaide, Australia.

It has been nearly two decades since the publication of that book, and much has changed in this area of research during this period. Generalizations have been considered for many known standard bivariate distributions. Skewed versions of different bivariate distributions have been proposed and applied to model data with skewness departures. By specifying the two conditional distributions, rather than the simple specification of one marginal and one conditional distribution, several general families of conditionally specified bivariate distributions have been derived and studied at great length. Finally, bivariate distributions generated by a variety of copulas and their flexibility (in terms of accommodating association/correlation) and structural properties have received considerable attention. All these developments and advances necessitated the present volume and have thus resulted in a substantially different version than the last edition, both in terms of coverage and topics of discussion.

In a volume of this size and wide coverage, there will inevitably be some mistakes and omissions of some important published results. We have made a sincere effort to minimize these, and what are left and left out are accidental and are certainly not due to nonscientific antipathy. We welcome the readers to write to us about the contents of this volume and inform us of any errors, misrepresentations, and omissions that you find. If ever there is a next edition, we will take your comments into account and make the necessary changes (keep in mind that our expected residual lives are probably not large enough to guarantee the next edition!).

We express first and foremost our sincere thanks and gratitude to Paul Hutchinson for his generosity in permitting us to use good portions from the last edition that he was part of, and for his support and encouragement through out the course of this project. We also thank Ingram Olkin for proposing and initiating this revision through Springer-Verlag. Thanks are also due to John Kimmell (Editor, Springer-Verlag) for his interest in this book, and his support and immense patience during the long preparation period, and to Debbie Iscoe (McMaster University, Canada) for converting the not-so-presentable initial manuscript that we prepared into this fine-looking book that you hold in your hands. Our final special thanks go to our families who have endured all the countless hours we were away from them (it is quite possible, of course, that they enjoyed these times in our absence) just to make a bit of progress everytime.

We both enjoyed very much putting this book together and we sincerely hope that you, as reader, would enjoy it as much while using it!

N. Balakrishnan
Chin-Diew Lai

Hamilton, Canada
Palmerston North, New Zealand

November 2008

Contents

Chapter 0
Univariate Distributions

0.1 Introduction

A study of bivariate distributions cannot be complete without a sound background knowledge of the univariate distributions, which would naturally form the marginal or conditional distributions. The two encyclopedic volumes by Johnson et al. (1994, 1995) are the most comprehensive texts to date on continuous univariate distributions. Monographs by Ord (1972) and Hastings and Peacock (1975) are worth mentioning, with the latter being a convenient handbook presenting graphs of densities and various relationships between distributions. Another useful compendium is by Patel et al. (1976); Chapters 3 and 4 of Manoukian (1986) present many distributions and relations between them. Extensive collections of illustrations of probability density functions (denoted by p.d.f. hereafter) may be found in Hirano et al. (1983) (105 graphs, each with typically about five curves shown, grouped in 25 families of distributions) and in Patil et al. (1984). Useful bibliographies of univariate distributions, though dated now, have been given by Haight (1961) and Patel et al. (1976). A compact text on univariate distributions with a brief discussion of multivariate distributions at the end has been presented by Balakrishnan and Nevzorov (2003). Finally, it is of interest to mention here that most of the univariate distributions and related concepts discussed in this chapter are also present in the form of concise entries in the 16-volume set *Encyclopedia of Statistical Sciences* prepared by Kotz et al. (2006), which would serve as a valuable and useful general reference for readers of this volume.

In this chapter, we provide an elementary introduction and basic details on properties of various univariate distributions, and an understanding of these will be key to following the developments in subsequent chapters, as they will rely time and again on these univariate properties. In Section 0.2, we first introduce the pertinent notation and properties. In Section 0.3, we describe some of the useful measures that capture specific shape character-

N. Balakrishnan, C.D. Lai, *Continuous Bivariate Distributions*,
DOI 10.1007/b101765_1,

istics of univariate distributions. In Section 0.4, we present details on the normal distribution and its transformations. Section 0.5 discusses the beta distribution, while Section 0.6 handles the exponential, gamma, and Weibull and Stacy's generalized gamma distributions. A few important aging distributions are presented in Section 0.7. Some symmetric distributions, such as logistic, Laplace, and Cauchy distributions, are presented in Section 0.8. Next, in Sections 0.9 and 0.10, we describe the extreme-value and the Pareto distributions, respectively. The general broad families of Pearson and Burr distributions are presented in Sections 0.11 and 0.12. Section 0.13 discusses t- and F-distributions, while Section 0.14 presents the wrapped t-family of circular distributions. Some noncentral distributions are briefly mentioned in Section 0.15. Skew-families of distributions, which have seen a lot of activity recently in the literature, are described in Section 0.16. Jones' family of distributions is introduced in Section 0.17, and some lesser-known but useful distributions are described finally in Section 0.18.

0.2 Notation and Definitions

0.2.1 Notation

In the univariate case, the cumulative distribution function and the probability density function will be denoted by $F(x)$ and $f(x)$, respectively. The following is a list of terms and symbols that will be used in this chapter as well as all subsequent chapters.

Term	Symbols	Brief explanation
Moment generating function	$M(t)$	$E(e^{tX})$
Characteristic function	$\varphi(t)$	$E(e^{itX})$
Cumulant generating function	$K(t)$	$\log \varphi(t)$
rth moment (about the origin)	μ'_r	$E(X^r)$
rth central moment	μ_r	$E[(X-\mu)^r],\ \mu = \mu'_1$
rth cumulant	κ_r	The coefficient of $(it)^r/r!$ in the expression of $K(t)$
Variance	σ^2	μ_2
Coefficient of skewness	$\alpha_3 = \sqrt{\beta_1}$	μ_3/σ^3
Coefficient of kurtosis	$\alpha_4 = \beta_2$	μ_4/σ^4
Coefficient of variation		σ/μ
Survival function	$\bar{F}(x)$	$1 - F(x)$
Hazard (failure rate) function	$r(x)$	$f(x)/\{1 - F(x)\}$
Sample mean	$\bar{X}$	$\sum_{i=1}^{n} X_i/n$
Sample variance	s^2	$\sum_{i=1}^{n}(X_i - \bar{X})^2/(n-1)$

In the bivariate context, μ_1 and μ_2 will often be used for the means of the two variables. There is unlikely to be any confusion over this notation. Also, log simply means $\log_e$.

0.2.2 Explanations

Moment Generating Function

Let X be a random variable (denoted by r.v. hereafter) with cumulative distribution function (denoted by c.d.f. hereafter) $F(x)$ and p.d.f. $f(x)$. Then,

$$M(t) = E(e^{tX}) = \int_{-\infty}^{\infty} e^{tx} f(x)dx \tag{0.1}$$

is the moment generating function (denoted by m.g.f. hereafter) of X if the integral is convergent for all values of t belonging to an interval that contains the origin. The existence of the m.g.f. is not assured for all distributions; however, if it does exist, it will uniquely determine the distribution. When it exists, it may be written as

$$M(t) = \sum_{j=0}^{\infty} \mu'_j \frac{t^j}{j!}. \tag{0.2}$$

This readily implies that μ'_j is $M^{(j)}(0)$, i.e., the jth derivative of M, evaluated at 0. Note that it is possible to have μ'_j exist for all j and yet $M(t)$ not exist.

Let X_1 and X_2 be two independent r.v.'s with m.g.f.'s $M_1(t)$ and $M_2(t)$, respectively. It is easy to see that the m.g.f. of X_1+X_2 is $M_1(t)M_2(t)$. Hence, the m.g.f. is a convenient tool to study distributions of sums of independent r.v.'s.

For univariate distributions, the existence (finiteness) of a moment of some particular order implies the existence of all moments of lower order.[1]

0.2.3 Characteristic Function

The cumulative function (denoted by c.f. hereafter) φ of X is a complex-valued function defined as

---

[1] Is the same true for bivariate distributions? No. What we can say is that if moments of orders (κ, l), (k, λ), and (k, l) exist, then so do all the moments of order (m, n), where $\kappa \leq m \leq k$ and $\lambda \leq n \leq l$; see van der Vaart (1973).

$$\varphi(t) = E(e^{itX}) \tag{0.3}$$

$$= \int_{-\infty}^{\infty} e^{itx} f(x)dx \tag{0.4}$$

$$= \int_{-\infty}^{\infty} \cos tx\ f(x)dx + i \int_{-\infty}^{\infty} \sin tx\ f(x)dx, \tag{0.5}$$

where $i = \sqrt{-1}$, for all real t.

The c.f. uniquely determines the distribution. It has the following properties:

(i) $\varphi(0) = 1$,
(ii) $|\varphi(t)| \leq 1$ for all real t, and
(iii) $\varphi(-t) = \overline{\varphi(t)}$, where the bar denotes the complex conjugate.

Unlike the m.g.f., $\varphi(t)$ exists for all distributions.

Suppose that X has finite moments μ_j' up to order n. Then $\varphi^{(j)}(0) = i^j \mu_j'$ (for $1 \leq j \leq n$), where $\varphi^{(j)}$ is the jth derivative of φ.

The c.f. can be inverted to give the p.d.f. using the formula

$$f(x) = \frac{1}{2\pi} \int_{-\infty}^{\infty} e^{-itx} \varphi(t)dt. \tag{0.6}$$

If X_1 and X_2 are independent r.v.'s with c.f.'s φ_1 and φ_2, respectively, the c.f. of the sum $X_1 + X_2$ is simply the product of the c.f.'s $\varphi_1(t)\varphi_2(t)$.

An overview of the characteristic function and its various properties and applications is due to Laha (1982). The books by Lukacs (1970, 1983) are key references on this topic.

0.2.4 Cumulant Generating Function

Cumulant Generating Function

Let $K(t) = \log \varphi(t)$. Then, $K(t)$ is known as the *cumulant generating function.* Assuming again that the first n moments of X exist, we have

$$K(t) = \sum_{j=1}^{n} \frac{\kappa_j}{j!}(it)^j + o(|t|^n)$$

as $t \to 0$. The coefficients κ_j in this expression are called the *cumulants* (or semi-invariants) of X. Clearly,

$$\kappa_j = \frac{1}{i^j} K^{(j)}(0),$$

where $K^{(j)}(0)$ is the jth derivative of $K(t)$, evaluated at 0.

It is of interest to note here that the normal distribution has the unique characterizing property that all its cumulants of order 3 and higher are zero.

Interrelationships of Moments and Cumulants

Relationships between the lower moments about the origin μ'_j, central moments μ_j, and cumulants κ_j are as follows:

$$\begin{aligned}
\kappa_1 &= \mu'_1 = \mu = \text{(the mean)},\\
\kappa_2 &= \mu'_2 - \mu_1^2 = \sigma^2 \text{(the variance)},\\
\kappa_3 &= \mu'_3 - 3\mu'_1\mu'_2 + 2\mu_1'^3 = \mu_3,\\
\kappa_4 &= \mu'_4 - 3\mu_2'^2 - 4\mu'_1\mu'_3 + 12\mu_1'^2\mu_2 - 6\mu_1'^4,\\
\mu'_1 &= \kappa_1,\\
\mu'_2 &= \kappa_2 + \kappa_1^2,\\
\mu'_3 &= \kappa_3 + 3\kappa_2\kappa_1 + \kappa_1^3,\\
\mu'_4 &= \kappa_4 + 3\kappa_2^2 + 4\kappa_1\kappa_2 + 6\kappa_1^2\kappa_2 + \kappa_1^4.
\end{aligned}$$

0.3 Some Measures of Shape Characteristics

0.3.1 Location and Scale

If $F(x)$ is the cumulative distribution of a variable X, we may introduce a *location parameter* a and a *scale parameter* b into it by writing $F\left(\frac{x-a}{b}\right)$. These parameters a and b are often the mean and the standard deviation, respectively, but they need not be—(i) the mean and standard deviation may not be finite (in such a case, we might set a = median and b = semiquartile range), and (ii) it may be more convenient for distributions whose p.d.f. is zero for $X < x_0$ to set a as x_0 instead of as the mean (in which case a is often referred to as a *threshold parameter*).

0.3.2 Skewness and Kurtosis

The most common measure of skewness is the normalized third central moment,

$$\alpha_3 = \sqrt{\beta_1} = \frac{\mu_3}{\mu_2^{3/2}}. \tag{0.7}$$

For symmetric p.d.f.'s such as the normal, logistic, and Laplace, this is zero.

The normalized fourth moment,

$$\alpha_4 = \beta_2 = \frac{\mu_4}{\mu_2^2}, \tag{0.8}$$

is the usual measure of kurtosis. The normal distribution has $\beta_2 = 3$, and so sometimes $\gamma_2 = \beta_2 - 3$ is referred to as the "excess of kurtosis." There is some controversy as to what kurtosis actually means, but a distribution with $\beta_2 < 3$ ("platykurtic") usually is less sharply peaked in the center and has thinner tails than the normal distribution having the same standard deviation, whereas a distribution with $\beta_2 > 3$ ("leptokurtic") usually is more sharply peaked in the center and has heavier tails than the normal distribution having the same standard deviation. For all distributions, they satisfy the inequality $\beta_2 \geq \beta_1 + 1$.

The shape of a distribution is not completely determined by the values of β_1 and β_2. Nevertheless, these two quantities are helpful while evaluating the shape when we have decided on a particular family of distributions (such as Pearson or Johnson families) because we can plot them on a chart marked with what regions of (β_1, β_2) correspond to which member of the family and hence make the choice of a member suitable for modeling.

0.3.3 Tail Behavior

While considering this aspect, we are not concerned with tail behavior as affected by the standard deviation or any other measure of scale—we assume such effects have been taken care of by some process of standardization. Even when this has been done, it is still possible to classify distributions as short-, median-, or long-tailed; see, for example, Parzen (1979) and Schuster (1984).

0.3.4 Some Multiparameter Systems of Univariate Distributions

Among systems of univariate distributions having several parameters —typically, four, so that skewness and tail-heaviness can be captured properly while fitting to empirical data—are Pearson's, the transformed normal system of Johnson, the transformed logistic system of Tadikamalla and Johnson, the generalized lambda, and Tukey's g and h families. Mendoza and Iglewicz (1983) used these 5 to fit 12 of the symmetric distributions commonly used in simulation studies and compared them in terms of ease of fitting and goodness of fit at selected percentiles. Pearson et al. (1979) compared the percentage points of distributions chosen from the Pearson, Johnson, and Burr systems.

0.3.5 Reliability Classes

Patel (1973) classified 15 continuous distributions as to whether they have the increasing (or decreasing) failure rate on average property.

A table of formulas including the reliability function ($\bar{F}$), the hazard (failure rate) function, and the mean residual life function has been given by Sheikh et al. (1987); the distributions included are the normal, gamma, and Weibull, and also their reciprocals.

Lai and Xie (2006, Chapter 2) have discussed various concepts of aging for lifetime random variables.

0.4 Normal Distribution and Its Transformations

0.4.1 Normal Distribution

The normal (Gaussian) distribution is symmetric about μ and has a density function

$$f(x) = \frac{1}{\sqrt{2\pi}\sigma} \exp\left\{-\frac{(x-\mu)^2}{2\sigma^2}\right\}, \quad -\infty < x < \infty. \tag{0.9}$$

For the unit normal (the standard form), the density is conventionally denoted by ϕ with the argument as z rather than x; i.e.,

$$\phi(z) = \frac{1}{\sqrt{2\pi}} \exp\left\{-\frac{z^2}{2}\right\}, \quad -\infty < x < \infty. \tag{0.10}$$

The corresponding c.d.f. is conventionally denoted by Φ, and there is no explicit expression for it. It can be shown from (0.57) that

$$E(X) = \mu, \tag{0.11}$$

$$\mathrm{var}(X) = \sigma^2, \tag{0.12}$$

$$\kappa_r = 0 \quad \text{for } r > 2.$$

The mode and median are the same as the mean, μ.

0.4.2 Lognormal Distribution

The p.d.f. is given by

$$f(x) = \frac{1}{x\sqrt{2\pi}\sigma} \exp\left\{-\frac{(\log x - \xi)^2}{2\sigma^2}\right\}, \quad x > 0. \tag{0.13}$$

With Φ denoting the standard normal distribution function, we have

$$F(x) = \Phi\left(\frac{\log x - \xi}{\sigma}\right). \tag{0.14}$$

It can be shown from (0.13) that

$$\mu = e^{\xi + \frac{1}{2}\sigma^2}, \tag{0.15}$$

$$\mathrm{var}(X) = e^{2\xi} e^{\sigma^2}(e^{\sigma^2} - 1). \tag{0.16}$$

It should be noted that if $\log X$ has a normal distribution, then X is said to have a lognormal distribution.

0.4.3 Truncated Normal

A normal distribution can be singly or doubly truncated. Johnson et al. (1994, pp. 156–162) have provided a detailed description of these truncated forms. Barr and Sherrill (1999) have given simpler expressions for the mean and variance and their estimates. Castillo and Puig (1999) showed that the likelihood ratio test of exponentiality against singly truncated normal alternatives is the uniformly most powerful unbiased test and that it can be expressed in terms of the sampling coefficient of variation.

0.4.4 Johnson's System

Johnson's (1949) system of distributions is obtained by starting with a standard normal variate Z [with p.d.f. as in (0.10)] and applying one of several simple transformations to it,

$$Z = \gamma + \delta T(Y), \tag{0.17}$$

where

- $T(Y) = \log Y$ gives the lognormal family, denoted by S_L;

- $T(Y) = \sinh^{-1} Y$ gives the S_U system with unbounded range, $-\infty < Y < \infty$;
- $T(Y) = \log\left(\frac{Y}{1-Y}\right)$ gives the S_B family with bounded range, $0 < Y < 1$;
- the normal distribution may be considered within this family (by taking $T(Y) = Y$) and be denoted by S_N.

Making one of the choices above determines the shape of the distribution. Location and scale parameters may naturally be introduced by setting $Y = (X - a)/b$.

Detailed discussions may be found in Johnson et al. (1994, Section 4.3, Chapter 12) and Bowman and Shenton (1983). DeBrota et al. (1988) have provided software to help in the choice of an appropriate member of this system for fitting to practical data.

0.4.5 Box–Cox Power Transformations to Normality

If X is not normally distributed, a power function transformation may often bring it close to normality. One such transformation is the *Box–Cox transformation* given by

$$\begin{aligned} (X^\lambda - 1)/\lambda \quad &\text{for } \lambda \neq 0, \\ \log X \quad &\text{for } \lambda = 0. \end{aligned} \tag{0.18}$$

0.4.6 g and h Families of Distributions

These families of distributions are obtained by starting with a standard normal variable Z and then applying the transformation of the form

$$T_{g,h}(Z) = \frac{e^{gZ} - 1}{g} \exp(hZ^2/2), \tag{0.19}$$

where g and h are constants, with the former controlling asymmetry or skewness and the latter controlling elongation, or the extent to which the tails are stretched relative to the normal.

When $g = 0$, a symmetric distribution is obtained from $Z \exp(hZ^2/2)$. When $h = 0$, the lognormal distribution is obtained.

0.4.7 Efron's Transformation

Efron (1982) considered the question of whether there is a single transformation $Y = a(X)$ such that Y has nearly a normal distribution when the distribution of X comes from some one-parameter family of distributions. Efron developed a general theory to answer this question without considering a specific form of a and, in those cases where the answer is positive, he gave formulas for calculating a.

0.4.8 Distribution of a Ratio

Rogers and Tukey (1972) discussed distributions obtained from the ratio form X/V, where X has a normal distribution and V is a positive r.v. independent of X. Among the special cases of this form are:

- The normal distribution itself (the denominator being a constant).
- t-distribution (the denominator being the square root of a chi-squared variate divided by its degrees of freedom), including the special case of the Cauchy distribution (the denominator being half-normal).
- The so-called *contaminated distributions* (the denominator taking only two values).
- The *slash distribution* (the denominator being uniformly distributed).
- If V is another independent normal denoted by Y, then X/Y has a Cauchy distribution.
- Suppose Y has a punctured normal distribution with a small interval containing zero being removed [Lai et al. (2004)]. Then $E(X/Y)$ is well defined.

0.4.9 Compound Normal Distributions

Starting from a normal distribution for X, denoted as usual by $N(\mu, \sigma^2)$, we may now suppose that μ or σ^2 are themselves random variables.

- If μ has a normal distribution, $N(\xi, \sigma_\mu^2)$, then the distribution of X will also be normal and is given by $N(\xi, \sigma^2 + \sigma_\mu^2)$.
- If $X \sim N(\mu + \beta U, \sigma^2 U)$, with U being a random variable, the resulting distribution of X is called a *normal variance mean mixture*; see Barndorff-Nielsen et al. (1982). If $\beta = 0$, it is a *normal variance mixture*.

0.5 Beta Distribution

0.5.1 The First Kind

The density function is

$$f(x) = \frac{1}{B(p,q)} x^{p-1}(1-x)^{q-1}, \quad 0 \le x \le 1, \tag{0.20}$$

where p and q are shape parameters and $B(p,q)$ is the complete beta function.[2]

The distribution function (denoted by d.f. hereafter) cannot be expressed in a closed form other than as an incomplete beta function.

We shall use beta(α, β) to denote the beta distribution with shape parameters α and β.

From the p.d.f. in (0.20), it can be readily shown that

$$\mu'_r = \frac{B(p+r,q)}{B(p,q)}, \tag{0.21}$$

$$\mu = \frac{p}{p+q}, \tag{0.22}$$

$$\sigma^2 = \frac{pq}{(p+q)^2(p+q+1)}. \tag{0.23}$$

For $p > 1$, $q > 1$, the mode can be shown to be at $(p-1)/(p+q+2)$.

---

[2] The beta function with arguments α and β is defined as

$$B(\alpha,\beta) = \int_0^1 t^{\alpha-1}(1-t)^{\beta-1}dt$$

$(\alpha > 0, \beta > 0)$. The incomplete beta function is defined as

$$B_x(\alpha,\beta) = \int_0^x t^{\alpha-1}(1-t)^{\beta-1}dt.$$

We shall see that the beta function is related to the gamma function. With argument α, the gamma function is defined as

$$\Gamma(\alpha) = \int_0^\infty t^{\alpha-1}e^{-t}dt.$$

It satisfies the recurrence relation

$$\Gamma(\alpha+1) = \alpha\Gamma(\alpha).$$

Also, $\Gamma(\frac{1}{2}) = \sqrt{\pi}$, $\Gamma(1) = 1$, and if α is an integer, $\Gamma(\alpha+1) = \alpha!$. The incomplete gamma function is defined as

$$\Gamma_x(\alpha) = \int_0^x t^{\alpha-1}e^{-t}dt.$$

For methods for computing Γ_x, see DiDonato and Morris (1986) and Shea (1988).

The beta and gamma functions are connected by the relationship

$$B(\alpha,\beta) = \Gamma(\alpha)\Gamma(\beta)/\Gamma(\alpha+\beta).$$

0.5.2 Uniform Distribution

A special case of the beta distribution is the uniform distribution over the range $0 < x < 1$. The following expressions hold for the more general case of $a < x < b$:

$$f(x) = \frac{1}{b-a}, \tag{0.24}$$

$$F(x) = \frac{x-a}{b-a}. \tag{0.25}$$

(Outside this range, f and F are either 0 or 1.)

From the p.d.f. in (0.24), it can be readily shown that

$$\mu_r = \begin{cases} 0 & \text{for } r \text{ odd} \\ \left(\frac{b-a}{2}\right)^r/(r+1) & \text{for } r \text{ even} \end{cases}, \tag{0.26}$$

$$\mu = (b-a)/2, \tag{0.27}$$

$$\sigma^2 = (b-a)^2/12. \tag{0.28}$$

One reason why this distribution is so important is its role in generating random variates. Specifically, if U is uniformly distributed over $[0, 1]$, then $F^{-1}(U)$ has a distribution F, and thus random variates from any required distribution F can be generated through uniform variates.

0.5.3 Symmetric Beta Distribution

Let $p = q$ in (0.20), and further let $Y = 2X - 1$. Then, the density function of Y is given by

$$f(y) = \frac{1}{2^{2q-1}B(q,q)}(1-y^2)^{q-1}, \quad -1 < y < 1, \tag{0.29}$$

which is symmetric in y. This is the reason for the name *symmetric beta.* It is in fact the Pearson type II distribution.

0.5.4 Inverted Beta Distribution

This is commonly known as the *beta distribution of the second kind* and is in fact the Pearson type VI distribution. Its p.d.f. is given by

$$f(x) = \frac{1}{B(\alpha,\beta)}\frac{x^{\alpha-1}}{(1+x)^{\alpha+\beta}}, \quad x > 0, \tag{0.30}$$

where α and β are shape parameters. The c.d.f. F can be expressed once again in terms of an incomplete beta function.

From the p.d.f. in (0.30), it can be readily shown that

$$\mu'_r = B(\alpha + r, \beta + 2 - r). \tag{0.31}$$

This is a transformation of the beta distribution in (0.20). Suppose X has a beta distribution. Then, $X/(1-X)$ is distributed as (0.30). A convenient summary of the interrelationships between the beta, inverted beta, gamma, t-, F-, and Cauchy distributions has been given by Devroye (1986, p. 430).

When we take the logarithmic transformation of an inverted beta variate, the resulting distribution is sometimes termed the *Z-distribution.* For $-\infty < x < \infty$, $\lambda_1 > 0$, $\lambda_2 > 0$, we obtain the density

$$f(x) = \frac{1}{B(\lambda_1, \lambda_2)} \frac{e^{-\lambda_2 x}}{(1 + e^{-x})^{\lambda_1 + \lambda_2}}. \tag{0.32}$$

Its properties include

$$M(t) = \frac{\Gamma(\lambda_1 + t)\Gamma(\lambda_2 - t)}{\Gamma(\lambda_1)\Gamma(\lambda_2)}$$

and

$$\kappa_r = \psi^{r-1}(\lambda_1) + (-1)^r \psi^{(r-1)}(\lambda_2),$$

where $\psi^{(r)}(t) = \frac{d^r[\Gamma'(t)/\Gamma(t)]}{dt^r}$. When $\lambda_1 + \lambda_2 = 1$, this becomes an example of the Meixner hypergeometric distribution discussed briefly in Section 0.18.2.

0.6 Exponential, Gamma, Weibull, and Stacy Distributions

0.6.1 Exponential Distribution

For scale parameter $\lambda > 0$, the p.d.f. and c.d.f. are given by

$$f(x) = \lambda e^{-\lambda x}, \quad x \geq 0, \tag{0.33}$$

$$F(x) = 1 - e^{-\lambda x}, \quad x \geq 0. \tag{0.34}$$

From the p.d.f. in (0.33), it can be readily shown that

$$\mu'_r = r!/\lambda^r, \tag{0.35}$$
$$\mu = 1/\lambda, \tag{0.36}$$
$$\text{median} = \log 2/\lambda, \tag{0.37}$$
$$\text{mode} = 0, \tag{0.38}$$
$$\sigma^2 = 1/\lambda^2. \tag{0.39}$$

This distribution is characterized by the "lack of memory" property,

$$\Pr(X \le x + y | X > y) = \Pr(X \le x). \tag{0.40}$$

0.6.2 Gamma Distribution

For $\alpha > 0$, $\beta > 0$, the p.d.f. is given by

$$f(x) = \frac{x^{\alpha-1}\exp(x/\beta)}{\beta^\alpha \Gamma(\alpha)}, \quad x > 0, \tag{0.41}$$

where $\Gamma(\alpha)$ is the gamma function, defined earlier. An expression for F, with the use of the incomplete gamma function, is given by

$$F(x) = \Gamma_{x/\beta}(\alpha)/\Gamma(\alpha), \quad x > 0. \tag{0.42}$$

From the p.d.f. in (0.41), it can be readily shown that

$$\mu'_r = \beta^r \prod_{i=0}^{r-1}(\alpha + i), \tag{0.43}$$
$$\mu = \alpha\beta, \tag{0.44}$$
$$\sigma^2 = \alpha\beta^2. \tag{0.45}$$

We use gamma(α, β) to denote the gamma distribution with shape parameter α and scale parameter β. The *Erlang distribution* is simply a gamma distribution with α being a positive integer. When $\alpha \ge 1$, the mode of the distribution can be shown to be at $\beta(\alpha - 1)$.

0.6.3 Chi-Squared and Chi Distributions

The chi-squared distribution is the gamma distribution written in a slightly different form (and often thought of in different contexts). ν, effectively a shape parameter, is referred to in this case as the *degrees of freedom* of the distribution. For $\nu > 0$, the p.d.f. is

$$f(x) = \frac{1}{2^{\nu/2}\Gamma(\nu/2)} e^{-x/2} x^{\frac{\nu}{2}-1}, \quad x > 0. \tag{0.46}$$

The chi-squared variate may be obtained as the sum of ν squared independent standard normal variates.

As to the chi distribution, $\chi_\nu = \sqrt{\chi^2_\nu}$ has as its density function

$$f(x) = \frac{1}{2^{\frac{\nu-2}{2}}\Gamma(\frac{\nu}{2})} e^{-x^2/2} x^{\nu-1}, \quad x > 0, \tag{0.47}$$

and its moments are given by

$$\mu'_r = 2^{r/2}\Gamma[(\nu + r)/2]/\Gamma(\nu/2). \tag{0.48}$$

The case $\nu = 2$ is commonly known as the *Rayleigh distribution.*

0.6.4 Weibull Distribution

For positive α (a shape parameter) and λ (a scale parameter), the p.d.f. and c.d.f. are given by

$$f(x) = \alpha\lambda(\lambda x)^{\alpha-1} e^{-(\lambda x)^\alpha}, \quad x > 0, \tag{0.49}$$

$$F(x) = 1 - e^{-(\lambda x)^\alpha}, \quad x > 0. \tag{0.50}$$

From the p.d.f. in (0.49), it can be shown that

$$\mu'_r = \lambda^{-r}\Gamma[(\alpha + r)/\alpha], \tag{0.51}$$

$$\mu = \lambda^{-1}\Gamma[(\alpha + 1)/\alpha], \tag{0.52}$$

$$\sigma^2 = \lambda^{-2}\left\{\Gamma\left(\frac{\alpha+2}{\alpha}\right) - \left[\Gamma\left(\frac{\alpha+1}{\alpha}\right)\right]^2\right\}. \tag{0.53}$$

0.6.5 Stacy Distribution

Seeing the p.d.f.'s in (0.41) and (0.49), a general density can be easily thought of in the form

$$f(x) = \frac{1}{\beta^{c\alpha}\Gamma(\alpha)}\, c x^{c\alpha-1} e^{-(x/\beta)^c}, \quad x > 0, \tag{0.54}$$

where $\alpha, \beta, c > 0$. This is generally called the *Stacy distribution*, after Stacy (1962), but it dates back at least as far as Knibbs (1911). In the study of hydrology, in the former U.S.S.R., it was known as the Kritsky–Menkel distribution; see Sokolov et al. (1976, Section 2.3.3.1).

From the p.d.f. in (0.54), it can be easily shown that

$$\mu'_r = \beta^r \Gamma(\alpha + r/c)/\Gamma(\alpha). \tag{0.55}$$

When $\beta = 1$, the cumulant generating function becomes

$$K(t) = \log \Gamma(\alpha + t/i) - \log \Gamma(\alpha). \tag{0.56}$$

It is also easy to verify that if $X \sim \text{gamma}(\alpha, \beta^c)$, then $Y = X^{1/c}$ has a Stacy distribution in (0.54).

0.6.6 Comments on Skew Distributions

Basically, the shapes of the gamma, Weibull, and lognormal distributions are somewhat similar. If the starting point is a free parameter (so that the p.d.f. is nonzero for $X > a$, instead of $X > 0$), they all have three parameters. In such a three-parameter form, methods of estimating the parameters have been compared by Kappenman (1985).

0.6.7 Compound Exponential Distributions

Because of its lack-of-memory property, the exponential distribution is often considered to be the embodiment of true randomness. However, in the life-testing context, it can easily be imagined that the specimens tested differ in their quality, and hence their lifetimes do not have an exponential distribution. This is the compounding model; i.e., the parameter λ of the exponential distribution is itself a random variable with some distribution.

If λ has a gamma distribution, the resulting compound distribution is a Pareto distribution.

Bhattacharya and Kumar (1986) considered the case of $1/\lambda$ having an inverse Gaussian distribution. They then obtained a p.d.f. that involves a modified Bessel function of the third kind, and this distribution has a decreasing failure rate. Earlier, Bhattacharya and Holla (1965) and Bhattacharya (1966) had considered $1/\lambda$ having various elementary distributions.

0.7 Aging Distributions

Section 2.3 of Lai and Xie (2006) discusses ten commonly used aging distributions, which are exponential, gamma, truncated normal, Weibull, lognormal, Birnbaum–Saunders, inverse Gaussian, Gompertz, Makeham, linear failure rate, Lomax, log-logistic, Burr XII, and the exponential-geometric (EG) distributions. Details of these distributions can also be found in the two volumes by Johnson et al. (1994, 1995). The exponential-geometric is a special case of Marshall and Olkin's family described below.

0.7.1 Marshall and Olkin's Family of Distributions

Let $\bar{G}$ be the survival function of a lifetime variable X. Marshall and Olkin's (1997) family of life distributions is obtained by adding a parameter β to the original survival function $\bar{G}$ resulting in the form

$$\bar{F}(x) = \frac{\beta \bar{G}(x)}{1-(1-\beta)\bar{G}(x)}, \quad 0 < x < \infty,\ \beta > 0. \tag{0.57}$$

Note that, in their original paper, $x \in (-\infty, \infty)$ is taken to be the support of the random variable X.

The special case where $\bar{G}(x) = \exp(-\lambda x)$ was discussed in detail, and it was shown in this case that

$$E(X) = \frac{\beta \log \beta}{\lambda(1-\beta)}$$

and

$$\text{mode}(X) = \begin{cases} 0, & \beta \le 2 \\ \lambda^{-1}\log(\beta-1), & \beta \ge 2 \end{cases}.$$

The failure (hazard) rate function is given by

$$h(x) = \frac{\lambda e^{\lambda x}}{e^{\lambda x} - (1-\beta)},$$

which is decreasing in t for $0 < \beta < 1$ and increasing for $\beta > 1$.

For $\beta = 1 - p < 1$, the model reduces to the EG (exponential-geometric) distribution mentioned above. If $\beta = 1$, it becomes the exponential distribution.

0.7.2 Families of Generalized Weibull Distributions

The Weibull distribution is by far the most popular lifetime model in the area of reliability. There are several reasons for this, and the two most important ones are: (i) it has a simple survival function, and (ii) the model is flexible, and its parameters are easy to estimate. Despite its popularity, many researchers still find the original Weibull model to be inadequate while modeling for one reason or another. During the last decade, many modifications and generalizations of the Weibull distribution have been proposed. A key motivation behind this development is the desire to produce a generalized Weibull distribution that yields a more meaningful failure rate shape than merely decreasing or increasing as in the case of the original Weibull.

From (0.50), we have

$$\bar{F}(x) = \exp\{-(\lambda x)^\alpha\}, \quad \alpha, \lambda > 0, \ x > 0. \tag{0.58}$$

For any lifetime distribution, the survival function can be expressed as

$$\bar{F}(x) = \exp\{-H(x)\}, \tag{0.59}$$

where H is the cumulative hazard function defined as $H(x) = \int_0^x h(t)dt$. Loosely speaking, any $H(x)$ that generalizes $(\lambda x)^\alpha$ would thus constitute a generalized Weibull. We now select four such families as listed below:

- *Additive Weibull* [Xie and Lai (1995)],

$$\bar{F}(x) = \exp\{-(x/\beta_1)^{\alpha_1} - (x/\beta_2)^{\alpha_2}\}, \quad \alpha_1, \alpha_2, \beta_1, \beta_2 > 0, \ x > 0;$$

- *Modified Weibull* [Lai et al. (2003)],

$$\bar{F}(t) = \exp\{-ax^\alpha e^{\lambda x}\}, \quad a, \alpha, \lambda > 0, \ x > 0;$$

- *Flexible Weibull* [Bebbington et al. (2007)],

$$\bar{F}(x) = \exp\{-(e^{\alpha x - \beta/x})\}, \quad \alpha, \beta, \ x > 0;$$

- Weibull family of Marshall and Olkin (1997),

$$\bar{F}(x) = \frac{\beta e^{-(\lambda x)^\alpha}}{1-(1-\beta)e^{-(\lambda x)^\alpha}}, \quad \alpha, \beta > 0, \ 0 < x < \infty.$$

For other Weibull related distributions and details, we refer the reader to Murthy et al. (2003) and Lai and Xie (2006, Chapter 5).

0.8 Logistic, Laplace, and Cauchy Distributions

These three distributions are grouped together since they are symmetric and have their support as $-\infty < x < \infty$ and so may be seen as competitors for the normal distribution.

0.8.1 Logistic Distribution

For the scale parameter $\beta > 0$ and location parameter α,

$$f(x) = \frac{1}{\beta} \frac{e^{-(x-\alpha)/\beta}}{(1 + e^{-(x-\alpha)/\beta})^2}, \tag{0.60}$$

$$F(x) = \frac{1}{1 + e^{-(x-\alpha)/\beta}} \tag{0.61}$$

$$= \frac{1}{2}\left[1 + \tanh\left(\frac{x-\alpha}{2\beta}\right)\right]. \tag{0.62}$$

The mean, median, and mode are all equal to α, and the variance is $\beta^2\pi^2/3$.

Johnson's system of transformations can be applied to a logistic variate instead of starting with a normal variate; see Tadikamalla and Johnson (1982).

Tukey's lambda distribution may be regarded as a generalization of the logistic. In this case, instead of a friendly form for F in terms of x, there is a simple expression for x in terms of F,

$$x = [F^\lambda - (1-F)^\lambda]/\lambda. \tag{0.63}$$

On letting $\lambda \to 0$, we find $F = (1 + e^{-x})^{-1}$. An extended Tukey family may be written as

$$x = \lambda_1 + [F^{\lambda_3} - (1-F)^{\lambda_4}]/\lambda_2. \tag{0.64}$$

0.8.2 Laplace Distribution

This is also known as the *double exponential distribution*, and its p.d.f. and c.d.f. are

$$f(x) = \frac{1}{2\phi}\exp(-|x-\theta|/\phi), \quad -\infty < x < \infty,\ \phi > 0, \tag{0.65}$$

$$F(x) = \begin{cases} \frac{1}{2}\exp[-(\theta - x)/\phi] & \text{for } x \le \theta \\ 1 - \frac{1}{2}\exp[-(x-\theta)/\phi] & \text{for } x \ge \theta \end{cases}. \tag{0.66}$$

The mean, median and mode all equal θ, and the variance is $2\phi^2$.

Johnson's system of transformations can once again be applied to a Laplace variate instead of starting with a normal variate; see Johnson (1954).

0.8.3 The Generalized Error Distribution

To subsume the normal and Laplace distributions within one family, we can consider the generalized error distribution with p.d.f.

$$f(x) = \left[2^{(\delta+2)/2}\Gamma\left(\frac{\delta}{2}+1\right)\right]^{-1} \exp\left(-\frac{1}{2}\left|\frac{x-\theta}{\phi}\right|^{2/\delta}\right), \quad -\infty < x < \infty. \tag{0.67}$$

0.8.4 Cauchy Distribution

For scale parameter $\lambda > 0$ and location parameter θ, the p.d.f. and c.d.f. are given by

$$f(x) = \frac{1}{\pi\lambda}\frac{1}{1+(\frac{x-\theta}{\lambda})^2}, \quad -\infty < x < \infty, \tag{0.68}$$

$$F(x) = \frac{1}{2} + \frac{1}{\pi}\tan^{-1}\left(\frac{x-\theta)}{\lambda}\right). \tag{0.69}$$

The moments do not exist. However, θ and λ are location and scale parameters, respectively. Both the median and mode are at θ.

The distribution, like the normal, is *stable*, meaning that the distribution of the sample mean is of the same form as the parent distribution. In contrast to the normal distribution, the distribution of the sample mean has the same scale parameter as the parent distribution.

0.9 Extreme-Value Distributions

0.9.1 Type 1

This is also known as the *Gumbel distribution*, and its c.d.f. and p.d.f. are

$$F(x) = \exp(-e^{-x}), \quad -\infty < x < \infty, \tag{0.70}$$
$$f(x) = e^{-x}\exp(-e^{-x}), \quad -\infty < x < \infty, \tag{0.71}$$

respectively.

0.9.2 Type 2

This is also known as the *Fréchet distribution.* For $\alpha > 0$, the c.d.f. is given by

$$F(x) = \exp(-x^{-\alpha}), \quad x \geq 0. \tag{0.72}$$

Note that if X has the Fréchet distribution in (0.72), then $Y = X^{-\alpha}$ has an exponential distribution.

0.9.3 Type 3

This is related to the Weibull distribution, and its c.d.f. is given by

$$F(x) = \exp\left\{-(-x)^{\alpha}\right\}, \quad \alpha > 0,\ x < 0. \tag{0.73}$$

It is then evident that $-X$ has a Weibull distribution.

Distributions (0.72) and (0.73) can be transformed readily to type 1 by the simple transformations

$$Y = \log X, \qquad Y = -\log(-X).$$

A book-length account on extreme-value distributions is Kotz and Nadarajah (2000).

0.10 Pareto Distribution

For $x \geq k > 0$ and $a > 0$, we have as the p.d.f. and c.d.f.

$$f(x) = \frac{ak^a}{x^{a+1}}, \tag{0.74}$$
$$F(x) = 1 - \left(\frac{k}{x}\right), \tag{0.75}$$

respectively. From the p.d.f. in (0.74), it can be readily shown that

$$\mu_r' = \frac{ak^r}{a-r}, \quad \text{if } a > r, \tag{0.76}$$

$$\mu = \frac{ak}{a-1}, \quad \text{if } a > 1, \tag{0.77}$$

$$\sigma^2 = \frac{ak^2}{(a-1)^2(a-2)}, \quad \text{if } a > 2. \tag{0.78}$$

This is sometimes referred to as the *Pareto distribution of the first kind.*

Another form of this distribution, known as the *Pareto distribution of the second kind* (sometimes also called the Lomax distribution), is given by

$$F(x) = 1 - c^a/(x+c)^a, \quad c > 0,\ x \geq 0, \tag{0.79}$$

$$f(x) = ac^a/(x+c)^{(a+1)}; \tag{0.80}$$

see Chapter 20 of Johnson et al. (1994) for details.

A monograph devoted to Pareto distributions is Arnold (1983). The so-called *Pareto IV distribution* in that monograph has been termed the *generalized Pareto distribution* in Arnold et al. (1999) and has a survival function of the form

$$\bar{F}(x) = \left[1 + \left(\frac{x}{\sigma}\right)^\delta\right]^{-\alpha}, \quad x > 0, \tag{0.81}$$

where σ, δ, and α are all positive parameters.

0.11 Pearson System

All members of Karl Pearson's system of continuous densities satisfy the differential equation

$$\frac{df}{dx} = \frac{(x-a)f(x)}{b_0 + b_1 x + b_2 x^2}. \tag{0.82}$$

For $b_1 = b_2 = 0$, the density f is normal. There are 12 other types, many of which are better known under other names, as presented in the following table.

Common name	Type	Density	Support
Beta (shifted)	I	$(1+x)^{m_1}(1-x)^{m_2}$	-1 to 1
Symmetric beta	II	$(1-x^2)^m$	-1 to 1
Gamma	III	$x^m e^{-x}$	0 to ∞
	IV	$(1+x^2)^{-m}\exp(-v\tan^{-1}x)$	$-\infty$ to ∞
Reciprocal of gamma	V	$x^{-m}\exp(-x^{-1})$	0 to ∞
Inverted beta (F)	VI	$x^{m_2}(1+x)^{-m_1}$	0 to ∞
t	VII	$(1+x^2)^{-m}$	$-\infty$ to ∞
	VIII	$(1+x)^{-m}$	0 to 1
	IX	$(1+x)^m$	0 to 1
Exponential	X	e^{-x}	1 ∞
Pareto	XI	x^{-m}	1 to ∞
	XII	$[(1+x)/(1-x)]^m$	-1 to 1

0.12 Burr System

There are 12 types of Burr distributions. The two most important ones are presented below. In both cases, the parameters c and k are positive and, as usual, location and scale parameters can be introduced if required.

Type XII:

$$F(x) = 1 - (1 + x^c)^{-k}, \quad x > 0; \tag{0.83}$$

Type III:

$$F(x) = (1 + x^{-c})^{-k}, \quad x > 0. \tag{0.84}$$

If X has a Burr type XII distribution, then $Y = X^c$ has a Lomax distribution. Equation (0.83) is a special case of (0.81).

0.13 t- and F-Distributions

These distributions are not models that describe the variability of some directly observed quantity such as length or time but are usually obtained as the theoretical distribution of some statistics of interest.

0.13.1 t-Distribution

With ν being the degrees of freedom (effectively a shape parameter), the p.d.f. is given by

$$f(x) = \frac{1}{\sqrt{\nu}B(\frac{1}{2}, \frac{\nu}{2})}\left(1 + \frac{x^2}{\nu}\right)^{-(\nu+1)/2}, \quad -\infty < x < \infty. \tag{0.85}$$

Simple expressions for F can be given for the cases when $\nu = 1, 2, 3$. The mean is zero for $\nu > 1$, while the variance is $\nu/(\nu - 2)$ for $\nu > 2$. When $\nu = 1$, X has a Cauchy distribution.

The ratio $Z/\sqrt{X/\nu}$ has a t-distribution in (0.85) when Z has a standard normal distribution, X has a chi-squared distribution with ν degrees of freedom, and Z and X are independent random variables.

0.13.2 F-Distribution

This distribution is effectively the inverted beta introduced earlier written in a slightly different way. The pair ν_1 and ν_2, effectively two shape parameters, is referred to as the degrees of freedom of the distribution. The p.d.f. is given by

$$f(x) = \frac{\Gamma[(\nu_1 + \nu_2)/2]}{\Gamma(\nu_1/2)\Gamma(\nu_2/2)}\left(\frac{\nu_1}{\nu_2}\right)^{\nu_1/2} x^{(\nu_1+\nu_2)/2}\left(1 + \frac{\nu_1}{\nu_2}x\right)^{-(\nu_1+\nu_2)/2}, \quad x > 0. \tag{0.86}$$

The c.d.f. $F(x)$ cannot be expressed in an elementary form.

For $\nu_2 > 2$, the mean is $\nu_2/(\nu_2 - 2)$. For $\nu_2 > 4$, the variance is $2\nu_2^2(\nu_1 + \nu_2 - 2)/[\nu_1(\nu_2 - 2)^2(\nu_2 - 4)]$. For $\nu_1 > 1$, the mode is $\nu_2(\nu_1 - 2)/[\nu_1(\nu_2 + 2)]$.

The ratio $(X_1/\nu_1)/(X_2/\nu_2)$ has a F-distribution if X_1 and X_2 are independent chi-squared variates with ν_1 and ν_2 degrees of freedom, respectively. The chi-squared—i.e., the gamma—is not the only distribution for which this is true; see Section 9.14 of Springer (1979).

0.14 The Wrapped *t* Family of Circular Distributions

Pewsey et al. (2007) considered the three-parameter family of symmetric unimodal distributions obtained by wrapping the location-scale extension of Student's t distribution onto the unit circle. The family contains the wrapped normal and wrapped Cauchy distributions as special cases, and can closely approximate the von Mises distributions as special cases.

Let X have a t-distribution with ν degrees of freedom, and let $Y = \mu + \lambda X$, where μ is a real number and $\lambda > 0$. Wrapping Y onto the unit circle $\theta = Y(\text{mod } 2\pi)$, we obtain a circular random variable having probability density function

$$f(\theta;\mu_0,\lambda,\nu)=\frac{c}{\lambda}\sum_{p=-\infty}^{\infty}\left(1+\frac{(\theta+2\pi p-\mu_0)^2}{\lambda^2\nu}\right)^{-\frac{\nu+1}{2}},\quad 0\le\theta<2\pi,$$

with $\mu_0=\mu(\text{mod}2\pi)$.

0.15 Noncentral Distributions

The noncentral chi-squared variate, with ν degrees of freedom and noncentrality parameter λ, arises as the distribution of $\sum_{i=1}^{\nu}(Z_i+a_i)^2$, where the Z_i's are independent standard normal variates and $\lambda=\sum_{i=1}^{\nu}a_i^2$.

The noncentral F-variate is obtained from the ratio of a noncentral chi-squared variate to an independent chi-squared variate of the form

$$\frac{\nu_2\sum_{i=1}^{\nu_1}(Z_i+a_i)^2}{\nu_1\sum_{i=\nu_1+1}^{\nu_1+\nu_2}Z_i^2}.$$

The doubly noncentral F-variate is similarly obtained from the ratio of two independent noncentral chi-squared variates.

The noncentral t-variate with ν degrees of freedom and noncentrality parameter δ arises as the distribution of $(Z+\delta)/\sqrt{X/\nu}$, where Z is a standard normal variate and X is an independent chi-squared variate with ν degrees of freedom. The doubly noncentral t-variate is similarly obtained if X has a noncentral chi-squared distribution.

The noncentral beta variate is obtained as $X/(X+Y)$, where Y and X are independent chi-squared and noncentral chi-squared variates, respectively. If they are both noncentral chi-squared variates, $X/(X+Y)$ has the doubly noncentral beta distribution.

These distributions do not have elementary expressions for either their densities or their distribution functions.

0.16 Skew Distributions

There are various ways to skew a distribution, and some important developments in this direction are described in this section.

0.16.1 Skew-Normal Distribution

A random variable X is said to be *skew-normal* with parameter λ if its density function can be written as

$$f(x;\lambda) = 2\phi(x)\Phi(\lambda x), \quad -\infty < x < \infty, \tag{0.87}$$

where $\phi(x)$ and $\Phi(x)$ denote the density and distribution function, respectively, of the standard normal. The parameter λ, which regulates the skewness, varies in $(-\infty, \infty)$, and $\lambda = 0$ corresponds to the standard normal density. For detailed properties, see Azzalini (1985, 1986) and Henze (1986). The distribution has been used by Arnold et al. (1993) in the analysis of screening procedures.

An alternative skew extension of normal is considered in Mudholkar and Hutson (2000) by splitting two half-normal distributions and introducing an explicit skewness parameter so that the new p.d.f. can be expressed as

$$f(x,\varepsilon) = \phi\Big(\frac{\varepsilon}{1+\varepsilon}\Big)I_{(x<0)} + \phi\Big(\frac{\varepsilon}{1-\varepsilon}\Big)I_{(x\geq 0)}.$$

The distribution above is called the *epsilon-skew-normal distribution.*

Log-Skew-Normal Distribution

Following the same connection as between the normal and the lognormal distributions, Azzalini et al. (2003) obtained the log-skew-normal distribution.

0.16.2 Skew t-Distributions

There are several types of skew t-distributions, and we present here a brief review of these forms.

General Type

A general method of skewing a symmetric density function $g(x)$ with distribution function $G(x)$ is to define

$$f(x;\lambda) = 2g(x)G(\lambda x). \tag{0.88}$$

This family of skew distributions obviously includes the skew-normal in (0.87). An equivalent definition of X is to regard it as a scale mixture of skew-normal variates.

If $g(x)$ is the t-density with ν degrees of freedom, (0.88) becomes a skew t-distribution. The resulting distribution function is relatively intractable; see some comments by Jones and Faddy (2003).

Skew *t*-distribution of Azzalini and Capitanio

Suppose Y is skew-normal with density as given in (0.87). Azzalini and Capitanio (2003) defined a skew t-distribution through the transformation

$$X = \xi + V^{-1/2}Y, \tag{0.89}$$

where $V \sim \chi^2_\nu/\nu$, independent of Y. The density function of X has the form $t_\nu(x)T(w(x))$. Here, $w(x)$ is not a linear function of x, and thus it differs from the previous skew t-distribution.

Log-Skew *t*-Distributions

The log-skew t-distribution was obtained by Azzalini et al. (2003) in the same manner as for the log-skew-normal. They found it to fit the American family income data satisfactorily.

Skew *t*-Distribution of Jones and Faddy

Jones and Faddy (2003) derived a skew t-distribution having density

$$\begin{aligned} f(x) &= f(x;a,b) \\ &= C_{a,b}^{-1}\left\{1+\frac{t}{(a+b+t^2)^{1/2}}\right\}^{a+1/2}\left\{1-\frac{t}{(a+b+t^2)^{1/2}}\right\}^{b+1/2}, \end{aligned} \tag{0.90}$$

where $C_{a,b} = 2^{a+b-1}B(a,b)(a+b)^{1/2}, a > 0, b > 0$. When $a = b$, $f(x)$ in (0.90) reduces to the t-distribution with $2a$ degrees of freedom. When $a < b$ or $a > b$, f is negatively or positively skewed, respectively. Furthermore, it should be noted that $f(x;b,a) = f(-x;a,b)$.

0.16.3 Skew-Cauchy Distribution

Arnold and Beaver (2000) introduced a skew-Cauchy distribution with density function

$$f(x) = \psi(x)\Psi(\lambda_0+\lambda_1 x)/\Psi\left(\frac{\lambda_0}{1+\lambda_1}\right), \quad -\infty < x < \infty, \tag{0.91}$$

where

$$\psi(x) = \frac{1}{\pi(1+x^2)}, \quad -\infty < x < \infty,$$

and

$$\Psi(x) = \frac{1}{2} + \frac{1}{\pi}\tan^{-1} x$$

are, respectively, the density and distribution function of the standard Cauchy distribution.

If $\lambda_0 = 0$, (0.91) reduces to

$$f(x) = 2\psi(x)\Psi(\lambda_1 x) \tag{0.92}$$

which has the same form as (0.88).

0.17 Jones' Family of Distributions

Jones (2004) constructed a family of distributions arising from distributions of order statistics, and it has a p.d.f. to be of the form

$$f(x) = \frac{\Gamma(a+b)}{\Gamma(a)\Gamma(b)} g(x)\{G(x)\}^{a-1}\{1-G(x)\}^{b-1}, \quad a > 0,\ b > 0, \tag{0.93}$$

where G is a symmetric distribution with density g, i.e., $G' = g$.

Starting from a symmetric f with $a = b = 1$, a large family of distributions can be generated with the parameters a and b controlling skewness and tail weight. In particular, if $a = b$, the corresponding distributions remain symmetric; if a and b become large, tail weights are decreased, with normality being the limiting case as $a, b \to \infty$; if a and b are small, tail weights are increased; if a and b differ, skewness is introduced, with the sign of skewness depending on the sign of $a - b$; and if only one of a or b tends to infinity, a standard extreme-value type distribution arises.

0.18 Some Lesser-Known Distributions

0.18.1 Inverse Gaussian Distribution

This is also sometimes called the *Wald distribution*. For $\phi > 0$, the p.d.f. and c.d.f. are given by

$$f(x) = \sqrt{\frac{\phi}{2\pi}} e^{\phi} x^{-3/2} \exp\left[-\frac{1}{2}\phi(x + x^{-1})\right], \quad x > 0, \tag{0.94}$$

$$F(x) = \Phi\left[(x-1)\sqrt{\phi/x}\right] + e^{2\phi}\Phi\left[-(x+1)\sqrt{\phi/x}\right], \quad x > 0, \tag{0.95}$$

where Φ denotes the distribution function of a standard normal. It can be shown that $\mu = 1$ and $\sigma^2 = \phi^{-1}$. When the mean μ is other than 1, the Wald distribution is generally known as the *inverse Gaussian distribution* (because of the inverse relationship between the cumulant generating function of this distribution and that of the normal (Gaussian) distribution). In this case, the density becomes

$$f(x) = \sqrt{\frac{\phi}{2\pi}} x^{-3/2} \exp\left[-\frac{\phi(x-\mu)^2}{2\mu^2 x}\right], \tag{0.96}$$

and the variance is now $\phi^{-1}\mu^3$.

0.18.2 Meixner Hypergeometric Distribution

The p.d.f. is given by

$$f(x) = [\pi\Gamma(a)]^{-1} 2^{a-2} \left|\Gamma\left(\frac{a}{2} + \frac{ix}{2}\right)\right|^2 e^{\gamma x} (\cos\gamma)^a \tag{0.97}$$

(in which $|\gamma| < \frac{\pi}{2}$ and $a > 0$). This is called the *generalized hyperbolic secant distribution* if $\gamma = 0$ [Harkness and Harkness (1968)]. If, in addition, $a = 1$, it is known as the *hyperbolic secant distribution.* The distribution function $F(x)$ can be expressed through an incomplete beta function.

From the density in (0.97), it can be shown that

$$\mu = a \tan\gamma, \tag{0.98}$$
$$\sigma^2 = a[1 + (\tan\gamma)^2]. \tag{0.99}$$

0.18.3 Hyperbolic Distributions

The logarithm of the p.d.f. is a hyperbola, and omitting the location and scale parameters, we have the p.d.f.

$$f(x) \propto \exp\left[-\varsigma\left(\sqrt{(1+\eta^2)(1+x^2)} - \eta x\right)\right]. \tag{0.100}$$

0.18.4 Stable Distributions

If X's are i.i.d. r.v.'s and there exist constants $a_n > 0$ and b_n such that $a_n^{-1} \sum_{i=1}^n X_i - b_n$ has the same distribution as the X's, then this distribution is said to be *stable.* $a_n = n^{1/\alpha}$, where α is known as the characteristic exponent ($0 < \alpha \leq 2$); $\alpha = 2$ for the normal distribution and $\alpha = 1$ for the Cauchy distribution. In addition to α and scaling and centering constants, a skew parameter β is involved. The expression for the characteristic function is reasonably simple, but not so for the p.d.f. (except for some special cases). The main area of application of stable distributions is in modeling certain economic phenomena that seem to possess very heavy-tailed distributions.

References

1. Arnold, B.C.: Pareto Distributions. International Co-operative Publishing House, Fairland, Maryland (1983)
2. Arnold, B.C., Beaver, R.J.: The skew-Cauchy distribution. Statistics and Probability Letters **49**, 285-290 (2000)
3. Arnold, B.C., Beaver, R.J., Groenveld, R.A., Meeker, W.Q.: The nontruncated marginal of a truncated bivariate normal distribution. Psychometrika **58**, 471–488 (1993)
4. Arnold, B.C., Castillo, E., Sarabia, J.M.: Conditional Specification of Statistical Models. Springer-Verlag, New York (1999)
5. Azzalini, A.: A class of distributions which includes normal ones. Scandinavian Journal of Statistics **12**, 171–178 (1985)
6. Azzalini, A.: Further results on a class of distributions which includes the normal ones. Statistica **4**, 199–208 (1986)
7. Azzalini, A., Capitanio, A.: Distributions generated by perturbation of symmetry with emphasis on multivariate skew t distribution. Journal of the Royal Statistical Society, Series B **65**, 367–390 (2003)
8. Azzalini, A., Dal Cappello, T., Kotz, S.: Log-skew normal and log-skew-t distributions as models for family income data. Journal of Income Distribution **11**, 12–20 (2003)
9. Balakrishnan, N., Nevzorov, V.: A Primer on Statistical Distributions, John Wiley and Sons, Hoboken, New Jersey (2003)
10. Barndorff-Nielsen, O., Kent, J., Sørensen, M.: Normal variance-mean mixtures and z distributions. International Statistical Review **50**, 145–159 (1982)
11. Barr, R., Sherrill, E.T.: Mean and variance of truncated normal. The American Statistician **53**, 357–361 (1999)
12. Bebbington, M., Lai, C.D., Zitikis, R.: A flexible Weibull extension. Reliability Engineering and System Safety **92**, 719–726 (2007)
13. Bhattacharya, S.K.: A modified Bessel function model in life testing. Metrika **11**, 131–144 (1966)
14. Bhattacharya, S.K., Holla, M.S.: On a life distribution with stochastic deviations in the mean. Annals of the Institute of Statistical Mathematics **17**, 97–104 (1965)
15. Bhattacharya, S.K., Kumar, S.: E-IG model in life testing. Calcutta Statistical Association Bulletin **35**, 85–90 (1986)
16. Bowman, K.O., Shenton, L.R.: Johnson's system of distributions. In: Encyclopedia of Statistical Sciences, Volume 4, S. Kotz and N.L. Johnson (eds.), pp. 303–314. John Wiley and Sons, New York (1983)
17. Castillo, J.D., Puig, P.: The best test of exponentiality against singly truncated normal alternatives. Journal of the American Statistical Association **94**, 529–532 (1999)
18. DeBrota, D.J., Dittus, R.S., Roberts, S.D., Wilson, J.R., Swain, J.J., Venkatraman, S.: Input modeling with the Johnson system of distributions. In: 1988 Winter Simulation Conference Proceedings, M.A. Abrams, P.L. Haigh, and J.C. Comfort (eds.),

pp. 165–179. Institute of Electrical and Electronics Engineers, Piscataway, New Jersey (1988)
19. Devroye, L.: Non-uniform Random Variate Generation. Springer-Verlag, New York (1986)
20. DiDonato, A.R., Morris, A.H.: Computation of the incomplete gamma function ratios and their inverse. ACM (Association for Computing Machinery) Transactions on Mathematical Software **12**, 377–393; **13**, 318–319 (1986)
21. Efron, B.: Transformation theory: How normal is a family of distributions? Annals of Statistics **10**, 323–339 (1982)
22. Haight, F.A.: Index to the distributions of mathematical statistics. Journal of Research of the National Bureau of Standards, Series B. Mathematics and Mathematical Physics **65B**, 2360 (1961)
23. Harkness, W.L., Harkness, M.L.: Generalized hyperbolic secant distributions. Journal of the American Statistical Association **63**, 329–337 (1968)
24. Hastings, N.A.J., Peacock, J.B.: Statistical Distributions. Butterworths, London (1975)
25. Henze, N.: A probabilstic representation of the "skew-normal" distribution. Scandinavian Journal of Statistics **13**, 271–275 (1986)
26. Hirano, K., Kuboki, H., Aki, S., Kuribayashi, A.: Figures of probability density functions in statistics. I-Continuous univariate case. Computer Science Monograph No. 19, Institute of Statistical Mathematics, Tokyo (1983)
27. Johnson, N.L.: Bivariate distributions based on simple translation systems. Biometrika **36**, 297–304 (1949)
28. Johnson, N.L.: Systems of frequency curves derived from the first law of Laplace. Trabajos de Estadistica **5**, 283–291 (1954)
29. Johnson, N.L., Kotz, S., Balakrishnan, N.: Continuous Univariate Distributions, Volume 1, 2nd edition. John Wiley and Sons, New York (1994)
30. Johnson, N.L., Kotz, S., Balakrishnan, N.: Continuous Univariate Distributions, Volume 2, 2nd edition. John Wiley and Sons, New York (1995)
31. Jones, M.C.: Families of distributions arising from distributions of order statistics (with discussion). Test **13**, 1–43 (2004)
32. Jones, M.C., Faddy, M.J.: A skew extension of the t-distribution, with applications. Journal of the Royal Statistical Society, Series B **65**, 159–174 (2003)
33. Kappenman, R.F.: Estimation for the three-parameter Weibull, lognormal, and gamma distributions. Computational Statistics and Data Analysis **3**, 11–23 (1985)
34. Knibbs, G.H.: Studies in statistical representation: On the nature of the curve $y = Ax^m e^{nx^p}$. Journal of the Royal Society of New South Wales **44**, 341–367 (1911)
35. Kotz, S., Balakrishnan, N., Read, C.B., Vidakovic, B. (eds.): Encyclopedia of Statistical Sciences, Volumes 1-16, 2nd edition. John Wiley and Sons, Hoboken, New Jersey (2006)
36. Kotz, S., Nadarajah, S.: Extreme Value Distributions: Theory and Applications. Imperial College Press, London, England (2000)
37. Laha, R.G.: Characteristic functions. In: Encyclopedia of Statistical Sciences, Volume 1, S. Kotz and N.L. Johnson (eds.), pp. 415–422. John Wiley and Sons, New York (1982)
38. Lai, C.D., Wood, G.R., Qiao, C.G.: The mean of the inverse of a truncated normal distribution. Biometrical Journal **46**, 420–429 (2004)
39. Lai, C.D., Xie, M.: Stochastic aging and Dependence for Reliability. Springer-Verlag, New York (2006)
40. Lai, C.D., Xie, M., Murthy, D.N.P.: A modified Weibull distribution. IEEE Transactions in Reliability **52**, 33–37 (2003)
41. Lukacs, E.: Characteristic Functions, 2nd edition. Griffin, London (1970)
42. Lukacs, E.: Developments in Characteristic Function Theory. Griffin, High Wycombe (1983)

43. Manoukian, E.B.: Modern Concepts and Theorems of Mathematical Statistics. Springer-Verlag, New York (1986)
44. Marshall, A.W., Olkin, I.: A new method for adding a parameter to a family of distributions with application to the exponential and Weibull families. Biometrika **84**, 641–652 (1997)
45. Mendoza, G.A., Iglewicz, B.: A comparative study of systems of univariate frequency distributions. American Statistical Association, 1983 Proceedings of the Statistical Computing Section, pp. 249–254 (1983)
46. Mudholkar, G.S., Hutson, A.D.: The exponentiated Weibull family: Some properties and a flood data application. Communications in Statistics: Theory and Methods **25**, 3059–3083 (1996)
47. Mudholkar, G.S., Hutson, A.D.: The epsilon-skew-normal distribution for analyzing near-normal data. Journal Statistical Planning and Inference **83** 291–309 (2000)
48. Murthy, D.N.P., Xie, M., Jiang, R.: Weibull Models. John Wiley and Sons, New York (2003)
49. Ord, J.K.: Families of Frequency Distributions. Griffin, London (1972)
50. Parzen, E.: Nonparametric statistical data modeling. Journal of the American Statistical Association **74**, 105–121 (1979)
51. Patel, J.K., Kapadia, C.H., Owen, D.B.: Handbook of Statistical Distributions. Marcel Dekker, New York (1976)
52. Patel, J.K.: A catalog of failure distributions. Communications in Statistics: Theory and Methods, **1**, 281–284 (1973)
53. Patil, G.P., Boswell, M.T., Ratnaparkhi, M.V.: Dictionary and Classified Bibliography of Statistical Distributions in Scientific Work, Volume 2: Continuous Univariate Models. International Co-operative Publishing House, Fairland, Maryland (1984)
54. Pearson, E.S., Johnson, N.L., Burr, I.W.: Comparisons of the percentage points of distributions with the same first four moments, chosen from eight different systems of frequency curves. Communications in Statistics: Simulation and Computation **8**, 191–229 (1979)
55. Pewsey, A., Lewis, T., Jones, M.C.: The wrapped t family of circular distributions. Australian and New Zealand Journal of Statistics **49**, 79–91 (2007)
56. Rogers, W.H., Tukey, J.W.: Understanding some longtailed symmetrical distributions. Statistica Neerlandica **26**, 211–226 (1972)
57. Schuster, E.F.: Classification of probability laws by tail behavior. Journal of the American Statistical Association **79**, 936–939 (1984)
58. Shea, B.L.: Algorithm AS 239: Chi-squared and incomplete gamma integral. Applied Statistics **37**, 466–473 (1988)
59. Sheikh, A.K., Ahmed, M., Zulfiqar, A.: Some remarks on the hazard functions of the inverted distributions. Reliability Engineering **19**, 255–261 (1987)
60. Sokolov, A.A., Rantz, S.E., Roche, M.: Floodflow Computation: Methods Compiled from World Experience. UNESCO Press, Paris (1976)
61. Springer, M.D.: The Algebra of Random Variables. John Wiley and Sons, New York (1979)
62. Stacy, E.W.: A generalization of gamma distribution. Annals of Mathematical Statistics **33**, 1187–1192 (1962)
63. Tadikamalla, P.R., Johnson, N.L.: Systems of frequency curves generated by transformations of logistic variables. Biometrika **69**, 461–465 (1982)
64. van der Vaart, H.R.: On the existence of bivariate moments of lower order given the existence of moments of higher order. Statistica Neerlandica **27**, 97–102 (1973)
65. Xie, M., Lai, C.D.: Reliability analysis using additive Weibull model with bathtub-shaped failure rate function. Reliability Engineering and System Safety **52**, 87–93 (1995)

Chapter 1
Bivariate Copulas

1.1 Introduction

The study of copulas is a growing field. The construction and properties of copulas have been studied rather extensively during the last 15 years or so. Hutchinson and Lai (1990) were among the early authors who popularized the study of copulas. Nelsen (1999) presented a comprehensive treatment of bivariate copulas, while Joe (1997) devoted a chapter of his book to multivariate copulas. Further authoritative updates on copulas are given in Nelsen (2006). Copula methods have many important applications in insurance and finance [Cherubini et al. (2004) and Embrechts et al. (2003)].

What are copulas? Briefly speaking, copulas are functions that join or "couple" multivariate distributions to their one-dimensional marginal distribution functions. Equivalently, copulas are multivariate distributions whose marginals are uniform on the interval $(0,1)$. In this chapter, we restrict our attention to bivariate copulas.

Fisher (1997) gave two major reasons as to why copulas are of interest to statisticians: "Firstly, as a way of studying scale-free measures of dependence; and secondly, as a starting point for constructing families of bivariate distributions." Specifically, copulas are an important part of the study of dependence between two variables since they allow us to separate the effect of dependence from the effects of the marginal distributions. This feature is analogous to the bivariate normal distribution where the mean vectors are unlinked to the covariance matrix and jointly determine the distribution. Many authors have studied constructions of bivariate distributions with given marginals: This may be viewed as constructing a copula.

In this chapter, we present an overview of the properties of a copula as well as a brief sketch on constructions and simulation of copulas. Following this introduction, we describe the basic properties of bivariate copulas in Section 1.2. Some further properties of copulas are presented in Section 1.3. Next, in Sections 1.4–1.6, the survival, Archimedean, extreme-value, and Archimax

N. Balakrishnan, C.D. Lai, *Continuous Bivariate Distributions*,
DOI 10.1007/b101765_2,

copulas are discussed, respectively. In Sections 1.8 and 1.9, the Gaussian, t, and copulas of the elliptical distribution in general and the order statistics copulas are described. In Section 1.10, the polynomial copulas and their use in approximating a copula are discussed. In Section 1.11, we describe some measures of dependence between two variables with a given copula such as Kendall's tau, Spearman's rho, and the geometry of correlation under a copula. We also present in this section some measures based on Gini's coefficient, tail dependence, and local dependence measures. The distribution of the variable $Z = C(U, V)$ is discussed in Section 1.12. The simulation of copulas and different methods of constructing copulas are presented in Sections 1.13 and 1.14, respectively. Section 1.15 details some important applications of copulas in different fields of study. Finally, the chapter closes with some criticisms levied against copulas in Section 1.16 and brief concluding remarks in Section 1.17.

1.2 Basic Properties

Let $C(u, v)$ denote a bivariate copula. Then:

- For every $u, v \in (0, 1)$,

$$C(u,0) = 0 = C(0,v),\ C(u,1) = u,\ C(1,v) = v.$$

- $C(u_2, v_2) - C(u_2, v_1) - C(u_1, v_2) + C(u_1, v_1) \geq 0.$
- A copula is continuous in u and v; actually, it satisfies the stronger Lipschitz condition [see Schweizer and Sklar (1983)]

$$|C(u_2, v_2) - C(u_1, v_1)| \leq |u_2 - u_1| + |v_2 - v_1|;$$

- For $0 \leq u_1 < u_2 \leq 1$ and $0 \leq v_1 < v_2 \leq 1$.

$$\begin{aligned} &\Pr(u_1 \leq U \leq u_2, v_1 \leq V \leq v_2) \\ &= C(u_2, v_2) - C(u_1, v_2) - C(u_2, v_1) + C(u_1, v_1) > 0. \end{aligned}$$

It is easy to verify that the following are valid copulas:

$$C^+(u,v) = \min(u,v),\ C^-(u,v) = \max(u+v-1,0),\ \text{and}\ C^0(u,v) = uv.$$

Sklar's theorem below elucidates the role that copulas play in the relationship between bivariate distribution functions and their univariate marginals [see Sklar (1959)].

Theorem 1.1. *Let H be a joint distribution function with marginals F and G. Then, there exists a copula C such that, for all $x, y \in [-\infty, \infty]$,*

$$H(x, y) = C(F(x), G(y)). \tag{1.1}$$

If F and G are continuous, then the copula C is unique; otherwise, C is uniquely determined on (Range of $F\times$ Range of G). Conversely, if C is a copula and F and G are univariate distribution functions, then H is a joint distribution function with marginals F and G.

It follows from the representation in (1.1) that if F and G are uniform, then $H(x, y) = C(x, y)$, which indicates that the copula is in the form of a bivariate distribution with its marginals transformed to be uniform over the range (0, 1). In other words, a bivariate copula is simply the uniform representation of the bivariate distribution in question. The dictionary definition of copula is "something that connects," and the word is used here to indicate that it is what interconnects the marginal distributions to produce a joint distribution.

Let h, f, g, and c be the density functions of H, F, G, and C, respectively. Then, the relation (1.1) yields

$$h(x, y) = c(F(x), G(y))f(x)g(y). \tag{1.2}$$

1.3 Further Properties of Copulas

- For every copula C and every $(u, v) \in [0, 1] \times [0, 1]$,

  $$C^-(u, v) \leq C(u, v) \leq C^+(u, v),$$

  where $C^+(u, v) = \min(u, v)$ and $C^-(u, v) = \max(u + v - 1, 0)$ are the Fréchet upper and lower bounds, respectively.
- For every $v \in [0, 1]$, the partial derivative $\partial C/\partial u$ exists for almost all u and $0 \leq \frac{\partial}{\partial u}C(u, v) \leq 1$. Similarly, $0 \leq \frac{\partial}{\partial v}C(u, v) \leq 1$.
- $C(u, v) = uv$ is the copula associated with a pair (U, V) of independent random variables.
- A convex combination of two copulas C_1 and C_2 is a copula as well. For example,

  $$C(u, v) = \alpha C^+(u, v) + (1 - \alpha)C^-(u, v), \qquad 0 \leq \alpha \leq 1,$$

  is also a copula. Generalizing this, we can conclude that any convex linear combination of copulas is a copula, i.e., $\sum_{i=1}^{n} \alpha_i C_i$ is a copula for $\alpha_i > 0$ and $\sum \alpha_i = 1$. A family of copulas that includes C^+, C^0, and C^- is said to be *comprehensive*. The two-parameter comprehensive copula given below is due to Fréchet (1958):

  $$C_{\alpha,\beta} = \alpha C^+(u, v) + \beta C^-(u, v) + (1 - \alpha - \beta)C^0(u, v),$$

  commonly known as the Fréchet copula.

A one-parameter comprehensive family due to Mardia (1970) is

$$C_\theta(u,v) = \frac{\theta^2(1+\theta)}{2}C^+(u,v) + (1-\theta^2)C^0 + \frac{\theta^2(1-\theta)}{2}C^-(u,v);$$

- Strictly increasing transformations of the underlying random variables result in the transformed variables having the same copula. See Nelsen (2006, Theorem 2.4.3), for example, for a proof.
- The copula associated with the standard bivariate normal density (i.e., the marginals are standard normal with zero mean and standard deviation 1) has a density

$$c(u,v) = \frac{1}{\sqrt{(1-\rho^2)}} \exp\Big[-\frac{\rho^2}{2(1-\rho^2)}\{(\Phi^{-1}(u))^2 + (\Phi^{-1}(v))^2\} + \frac{\rho}{1-\rho^2}\Phi^{-1}(u)\Phi^{-1}(v)\Big]. \tag{1.3}$$

Note. The copula that corresponds to (1.3) is an important one. It is known as the Gaussian copula in finance and extreme-value study. We will discuss this further in Section 1.8.

1.4 Survival Copula

If one replaces C by $\hat{C}$, u by $1-u$, and v by $1-v$ in the copula formula, one is effectively moving the origin of the coordinate system from (0,0) to (1,1) and results in measuring the variates in the reverse direction. Although this is such a trivial procedure, the two distributions are perhaps best regarded as distinct, as the results of fitting them to data are different (unless there is symmetry).

The copula $\hat{C}$ obtained in this way is called the *survival copula* [Nelsen (2006, p. 33)] or *complementary copula* [Drouet-Mari and Kotz (2001, p. 85)], satisfying

$$\hat{C}(u,v) = u + v - 1 + C(1-u, 1-v) \tag{1.4}$$

and

$$\bar{H}(x,y) = \hat{C}(\bar{F}(x), \bar{G}(y)). \tag{1.5}$$

It is clear that $\hat{C}$ is a copula that "couples" the joint survival function $\bar{H}$ to the univariate marginal survival functions in a manner completely analogous to the way in which a copula connects the joint distribution to its margins. The term survival copula is a bit misleading, in our opinion, as $\hat{C}$ is not a survival function.

Let $\bar{C}$ be the joint survival function of two uniform variables whose joint distribution is the copula C. Then we have the relationship

$$\bar{C}(u,v) = 1 - u - v + C(u,v) = \hat{C}(1-u, 1-v). \tag{1.6}$$

Example 1.2. Consider the bivariate Pareto distribution considered in Hutchinson and Lai (1990). Let X and Y be a pair of random variables whose joint survival function is given by

$$\bar{H}_\theta(x,y) = \begin{cases} (1+x+y)^{-\theta}, & x \geq 0, y \geq 0 \\ (1+x)^{-\theta}, & x \geq 0, y < 0 \\ (1+y)^{-\theta}, & x < 0, y \geq 0 \\ 1, & x < 0, y < 0 \end{cases},$$

where $\theta > 0$. The marginal survival functions are $\bar{F}(x) = (1+x)^{-\theta}$ and $\bar{G}(y) = (1+y)^{-\theta}$. It can be shown that the survival copula is

$$\hat{C}_\theta(u,v) = \left(u^{-1/\theta} + v^{-1/\theta} - 1\right)^{-\theta}.$$

1.5 Archimedean Copula

In some situations, there exists a function φ such that

$$\varphi(C(u,v)) = \varphi(u) + \varphi(v). \tag{1.7}$$

Copulas of the form above are called *Archimedean copulas* [Genest and MacKay (1986a)]. Equivalently, we have

$$\varphi(H(x,y)) = \varphi(F(x)) + \varphi(G(y)); \tag{1.8}$$

i.e., we can write $H(x,y)$ as a sum of functions of marginals F and G. Since we are interested in expressions that we can use for the construction of copulas, we want to solve the relation $\varphi(C(u,v)) = \varphi(u) + \varphi(v)$. We thus need to find an appropriately defined "inverse" $\varphi^{[-1]}$ so that

$$C(u,v) = \varphi^{[-1]}(\varphi(u) + \varphi(v)). \tag{1.9}$$

Definition 1.3. [Nelsen (2006, p. 110)] Let φ be a continuous, strictly decreasing function from [0, 1] to [0, ∞] such that $\varphi(1) = 0$. The pseudoinverse of φ is the function $\varphi^{[-1]}$, with domain [0, ∞] and range [0, 1], given by

$$\varphi^{[-1]}(t) = \begin{cases} \varphi^{-1}(t), & 0 \leq t \leq \varphi(0) \\ 0, & \varphi(0) \leq t \leq \infty \end{cases}.$$

Note that if $\varphi(0) = \infty$, then $\varphi^{[-1]}(t) = \varphi^{-1}(t)$ and

$$C(u,v) = \varphi^{-1}(\varphi(u) + \varphi(v)). \tag{1.10}$$

C is a copula if and only if the pseudoinverse (or inverse if $\varphi(0) = \infty$) is a convex decreasing function; see Nelsen (2006, p. 111) for a proof.

The function φ is called a *generator of the copula.* If $\varphi(0) = \infty$, we then say that φ is a strict generator and $C(u,v) = \varphi^{-1}(\varphi(u)+\varphi(v))$ is said to be a strict Archimedean copula. Nelsen (2006) and Drouet-Mari and Kotz (2001) have given several examples of Archimedean copulas.

Example 1.4 (Bivariate Pareto copula). In this case, $\varphi(t) = t^{-1/\alpha} - 1$ and

$$\hat{C}(u,v) = (u^{-1/\alpha} + v^{-1/\alpha} - 1)^{-\alpha}. \tag{1.11}$$

Example 1.5 (Gumbel–Hougaard copula). In this case, $\varphi(t) = (-\log t)^{\alpha}$ and

$$C_{\alpha}(u,v) = \exp\left(-[(-\log u)^{\alpha} + (-\log v)^{\alpha}]^{1/\alpha}\right). \tag{1.12}$$

Example 1.6 (Frank's copula). In this case, $\varphi(t) = \log(\frac{1-\alpha}{1-\alpha^t})$, $0 < \alpha < 1$, and

$$C(u,v) = \log_{\alpha}\left(1 + \frac{(\alpha^u - 1)(\alpha^v - 1)}{(\alpha - 1)}\right). \tag{1.13}$$

The survival copula of Frank's distribution is also Archimedean. In fact, this is the only family that satisfies $C(u,v) = \hat{C}(u,v)$.

It is shown by Drouet-Mari and Kotz (2001, pp. 78–79) that the frailty models are also Archimedean.

These authors have further considered the following aspects of Archimedean copulas:

- characterization of Archimedean copulas;
- limit of a sequence of Archimedean copulas;
- archimedean copulas with two parameters; and
- fitting an observed distribution with an Archimedean copula.

1.6 Extreme-Value Copulas

Let $(X_1,Y_1),(X_2,Y_2),\ldots,(X_n,Y_n)$ be independent and identically distributed pairs of random variables with a common copula C, and also let $C_{(n)}$ denote the copula of componentwise maxima $X_{(n)} = \max X_i$ and $Y_{(n)} = \max Y_i$. From Theorem 3.3.1 of Nelsen (2006), we know that

$$C_{(n)}(u,v) = C^n(u^{1/n}, v^{1/n}), \quad 0 \le u, v \le 1.$$

The limit of the sequence $\{C_{(n)}\}$ leads to the following notion of an extreme-value copula.

Definition 1.7. A copula C_* is an extreme value copula if there exists a copula C such that

$$C_*(u,v) = \lim_{n\to\infty} C^n(u^{1/n}, v^{1/n}), \, 0 \le u, v \le 1. \tag{1.14}$$

Furthermore, C is said to belong to the *domain of attraction* of C_*. It is easy to verify that C_* satisfies the relationship

$$C_*(u^k, v^k) = C_*^k(u,v), \quad k > 0.$$

Example 1.8 (Gumbel–Hougaard copula).

$$C(u,v) = \exp\Big(- \Big[(-\log u)^\alpha + (-\log v)^\alpha\Big]^{1/\alpha}\Big),$$

see Section 2.6 for a discussion.

The Gumbel–Hougaard copula is also an Archimedean copula; in fact, there is no other Archimedean copula that is also an extreme-value copula [Genest and Rivest (1989)].

Example 1.9 (Marshall and Olkin copula).

$$C(u,v) = uv\min(u^{-\alpha}, v^{-\beta}) = \min(uv^{1-\beta}, u^{1-\alpha}v),$$

see Section 2.5.1 for details.

1.7 Archimax Copulas

Capéraà et al. (2000) have defined a new family of copulas for which Archimedean copulas and extreme-value copulas are particular cases.

Recall that the extreme-value copula associated with the extreme-value distribution of a copula C is

$$C_{\max}(u,v) = \lim_{n\to\infty} C^n\left(u^{\frac{1}{n}}, v^{\frac{1}{n}}\right).$$

Following the work of Pickands (1981), Capéraà et al. (2000) obtained as a general form of a bivariate extreme-value copula

$$C_A(u,v) \equiv \exp\left[\log(uv)\ A\left\{\frac{\log(u)}{\log(uv)}\right\}\right], \tag{1.15}$$

where A is a convex function $[0,1] \to [1/2,1]$ such that $\max(t,1-t) \leq A(t) \leq 1$ for all $0 \leq t \leq 1$.

Let φ be the generator of a copula and A be defined as before. A bivariate distribution is said to be an *Archimax copula* [Capéraà et al. (2000)] if it can be expressed in the form

$$C_{\varphi,A}(u,v) = \varphi^{-1}\left[\{\varphi(u)+\varphi(v)\}\ A\left\{\frac{\varphi(u)}{\varphi(u)+\varphi(v)}\right\}\right]. \tag{1.16}$$

If $A \equiv 1$, we retrieve the Archimedean copula, and if $\varphi(t) = \log(t)$, we retrieve the extreme-value copula.

Note. This procedure to generate a bivariate copula is a particular case of Marshall and Olkin's generalization (Section 1.10.2), where the function K is the bivariate extreme-value copula $C_A(u,v)$ given in (1.15), and the mixture distribution has the Laplace transform $\phi = \varphi^{-1}$ and the generator $\varphi(t) = \log t$.

1.8 Gaussian, *t*-, and Other Copulas of the Elliptical Distributions

Gaussian Copula

The Gaussian copula is perhaps the most popular distribution in applications. Let Φ denote the standard univariate normal distribution function and Ψ denote the standard bivariate normal distribution function. Then the bivariate Gaussian (normal) copula is given by

$$\begin{aligned} C_\rho(u,v) &= \Psi(\Phi^{-1}(u),\Phi^{-1}(v)) \\ &= \frac{1}{2\pi\sqrt{1-\rho^2}} \int_{-\infty}^{\Phi^{-1}(u)} \int_{-\infty}^{\Phi^{-1}(v)} \exp\left[\frac{-(s^2-2\rho st+t^2)}{2(1-\rho^2)}\right] ds\,dt, \end{aligned} \tag{1.17}$$

where $\rho \in (0,1)$ is the correlation coefficient such that $\rho \neq 0$. The density of the Gaussian copula is simpler, as given in (1.3).

The bivariate Gaussian copula can be used to generate bivariate dispersion models [Song (2000)]. There are numerous applications of Gaussian copulas, particularly in hydrology and finance.

t-Copula

The t-copula is simply the copula that represents the dependence structure of the bivariate t-distribution discussed in Section 7.2. Its properties are studied in Embrechts et al. (2002), Fang et al. (2002), and Demarta and McNeil (2005). The model has received much attention recently, particularly in the context of modeling multivariate financial data (e.g., daily relative or logarithmic price changes on a number of stocks). Marshall et al. (2003) and Breymann et al. (2003) have shown that the empirical fit of the t-copula is often good and is almost always superior to that of the Gaussian copula. One reason for the success of the t-copula is its ability to capture the phenomenon of dependent extreme values, which is often observed in the financial return data.

The Gaussian and t- copulas are copulas of elliptical distributions (see Chapter 15); they are not elliptical distributions themselves.

The dependence in elliptical distributions is essentially determined by covariances. Covariances are considered by some as being poor tools for describing dependence for non-normal distributions, in particular for their extremal dependence; see Embrechts et al. (2002) for a critique in risk modeling and Glasserman (2004) for advocating t-distributions for risk management.

1.9 Order Statistics Copula

Let $X_{r:n}$ be the rth order statistic $(1 \le r \le n)$ from a sequence of independent and identically distributed variables $\{X_1, X_2, \ldots, X_n\}$. Nelsen (2003) showed that the copula $C_{1,n}$ of $X_{1:n}$ and $X_{n:n}$ is given by

$$C_{1,n} = v - \left[\max\{(1-u)^{\frac{1}{n}} + v^{\frac{1}{n}} - 1, 0\}\right]^n; \tag{1.18}$$

see also Schmitz (2004).

1.10 Polynomial Copulas

Drouet-Mari and Kotz (2001) utilized the Rüschendorf method to construct a polynomial copula. To begin with, let $f = u^k v^q$ and obtain

$$\begin{aligned} f^1(u,v) &= f - \int_0^1 f(u,v)dv - \int_0^1 f(u,v)du + \int_0^1\int_0^1 f(u,v)du\,dv \\ &= \left(u^k - \frac{1}{k+1}\right)\left(v^q - \frac{1}{q+1}\right), \qquad k \ge 1,\ q \ge 1. \end{aligned}$$

Therefore, the function

$$c(u,v) = 1 + \theta \left(u^k - \frac{1}{k+1}\right)\left(v^q - \frac{1}{q+1}\right) \tag{1.19}$$

with the constraint

$$0 < \theta \le \min\left(\frac{(k+1)(q+1)}{q}, \frac{(k+1)(q+1)}{k}\right)$$

is the density of a polynomial copula. Repeating the process above for all k and q ($k \ge 1,\ q \ge 1$), we obtain a general formula

$$\frac{\partial^2 C}{\partial u \partial v} = 1 + \sum_{k\ge 1, q\ge 1} \theta_{kq}\left(u^k - \frac{1}{k+1}\right)\left(v^k - \frac{1}{q+1}\right)$$

with the same constraints

$$0 \le \min\left(\sum_{k\ge 1, q\ge 1} \theta_{k}q\frac{q}{(k+1)(q+1)}, \sum_{k\ge 1, q\ge 1} \theta_{k}q\frac{k}{(q+1)(k+1)}\right) \le 1.$$

A polynomial copula of power m can now be obtained as

$$C(u,v) = uv\left[1 + \sum_{k\ge 1, q\ge 1}^{k+q\le m-2} \frac{\theta_{kq}}{(k+1)(q+1)}(u^k-1)(v^q-1)\right]. \tag{1.20}$$

Example 1.10 (Polynomial copula of order 5). The polynomial copula of the fifth power from (1.20) then becomes

$$\begin{aligned} C(u,v) = uv \Big[1 &+ \frac{\theta_{11}}{4}(u-1)(v-1) + \frac{\theta_{12}}{6}(u-1)(v^2-1) \\ &+ \frac{\theta_{21}}{6}(u^2-1)(v-1)\Big], \end{aligned} \tag{1.21}$$

which coincides with the expression given by Wei et al. (1998).

Example 1.11 (Iterated F-G-M family). Johnson and Kotz (1977) presented the iterated Farlie–Gumbel–Morgenstern (F-G-M) family with the copula

$$C(u,v) = uv\{1 + \alpha(1-u)(1-v) + \beta uv(1-u)(1-v)\}.$$

Example 1.12 (Woodworth's polynomial copula). The uniform representation of the Woodworth (1966) family of distributions is given by

$$c(u,v) = 1 + \theta[1-(m+1)u^m][1-(m+1)v^m], \quad 0 \le \theta \le 1/m^2, m \ge 1.$$

For $m = 1$, the equation above clearly coincides with the F-G-M distribution.

Example 1.13 (Nelsen's polynomial copula). In this case, the copula is given by

$$C(u,v) = uv + 2\theta uv(1-u)(1-v)(1+u+v-2uv), \quad 0 \le \theta \le 1/4$$

[Nelsen (1999, pp. 168–169)].

1.10.1 Approximation of a Copula by a Polynomial Copula

Suppose a copula $C_\theta(u,v)$, indexed by a parameter θ, has a continuous nth derivative. We can then express it by means of the Taylor expansion in the neighborhood of θ_0 as

$$C_\theta(u,v) \approx C_{\theta_0}(u,v) + \sum_{i=1}^{n} \frac{C_{\theta_0}^{(i)}(u,v)(\theta-\theta_0)^i}{i!}.$$

Choosing θ_0 corresponding to independence [i.e., with $C_{\theta_0}(u,v) = uv$], and if the derivatives of C_θ with respect to θ are powers in uv, we then obtain an approximation of C_θ by means of a polynomial copula.

Example 1.14 (The F-G-M family). The F-G-M family corresponds to its first-order expansion in Taylor's series around $\theta = 0$.

Example 1.15 (The Ali–Mikhail–Haq family). In this case,

$$C(u,v) = \frac{uv}{1-\theta(1-u)(1-v)} = uv\left[1 + \sum_{i\ge 1}(\theta(1-u)(1-v))^i\right], \quad (1.22)$$

where $|\theta| \le 1$. If we consider only the first order in (1.22), we obtain the F-G-M family, and with the second-order approximation, we arrive at the iterated F-G-M of Lin (1987). For an approximation of any order, we have a polynomial copula.

Example 1.16 (The Plackett family). Nelsen (1999) proved that the F-G-M family is a first-order approximation to the Plackett family by expanding it in Taylor's series around $\theta = 1$.

1.11 Measures of Dependence Between Two Variables with a Given Copula

Many measures of dependence are "scale-invariant"; i.e., they remain unchanged under strictly increasing transformations of random variables. Since the copula C of a pair of random variables X and Y is invariant under strictly increasing transformations of X and Y, several scale-invariant measures of dependence are expressible in terms of the copulas. Two such "scale-invariant" measures are Kendall's tau and Spearman's rho.

1.11.1 Kendall's Tau

Let (x_i, y_i) and (x_j, y_j) be two observations from (X, Y) of continuous random variables. The two pairs (x_i, y_i) and (x_j, y_j) are said to be *concordant* if $(x_i - x_j)(y_i - y_j) > 0$ and *discordant* if $(x_i - x_j)(y_i - y_j) < 0$.

Kendall's tau is defined as the probability of concordance minus the probability of discordance,

$$\tau = P[(X - X')(Y - Y') \geq 0] - P[(X - X')(Y - Y') \leq 0], \tag{1.23}$$

where (X', Y') is independent of (X, Y) and is distributed as (X, Y).

The sample version of Kendall's τ is defined as

$$t = \frac{c - d}{c + d} = (c - d)/n, \tag{1.24}$$

where c denotes the number of concordant pairs and d the number of discordant pairs from a sample of n observations from (X, Y). Just as H can be expressed as a function of copula C, Kendall's τ can be expressed in terms of the copula [see, for example, Nelsen (2006, p. 101)] as

$$\tau = 4 \int_0^1 \int_0^1 C(u, v) c(u, v) du dv - 1 = 4E(C(U, V)) - 1. \tag{1.25}$$

Let C be an Archimedean copula generated by φ. Then, Genest and MacKay (1986a,b) have shown that

$$\tau = 4E(C(U, V)) - 1 = 4 \int_0^1 \frac{\varphi(t)}{\varphi'(t)} dt. \tag{1.26}$$

Example 1.17 (Bivariate Pareto copula). In this case, $\varphi(t) = t^{-1/\alpha} - 1$ and so

$$\frac{\varphi(t)}{\varphi'(t)} = \alpha(t^{1+\frac{1}{\alpha}} - t)$$

and, consequently, $\tau = \frac{1}{2\alpha+1} - 1$.

1.11.2 Spearman's Rho

Like Kendall's tau, the population version of the measure of association known as *Spearman's rho* (denoted by ρ_S) is based on concordance and discordance. Let $(X_1, Y_1), (X_2, Y_2)$, and (X_3, Y_3) be three independent pairs of random variables with a common distribution function H. Then, ρ_S is defined to be proportional to the probability of concordance minus the probability of discordance for the two pairs (X_1, Y_1) and (X_2, Y_3); i.e.,

$$\rho_S = 3\Big(P[(X_1 - X_2)(Y_1 - Y_3) > 0] - P[(X_1 - X_2)(Y_1 - Y_3) < 0]\Big). \quad (1.27)$$

Equation (1.27) is really the grade correlation and can be expressed in terms of the copula as

$$\rho_S = 12 \int_0^1 \int_0^1 C(u, v) du dv - 3 = 12E(UV) - 3. \quad (1.28)$$

Rewriting the equation above as

$$\rho_S = \frac{E(UV) - \frac{1}{4}}{\frac{1}{12}}, \quad (1.29)$$

we simply observe that Spearman's rho between X and Y is simply Pearson's product-moment correlation coefficient between the uniform variates U and V.

1.11.3 Geometry of Correlation Under a Copula

Long and Krzysztofowicz (1996) provided a novel way of deriving and interpreting the correlation coefficient ρ under a copula.

The sample space of U and V can be partitioned into four polygons (equilateral triangles) by drawing two diagonal lines, $l_1 : v = u$ and $l_2 : v = 1 - u$. From a fixed point (u, v), the distance to l_1 is $d_1 = |u - v|/\sqrt{2}$ and the distance to l_2 is $d_2 = |u + v - 1|/\sqrt{2}$. Let

$$\lambda = d_2^2 - d_1^2 = [4uv - 2(u + v) + 1]/2, \quad (1.30)$$

which measures the relative closeness of the point (u, v) to the diagonals. Then, the function λ has the following behavior:

- $\lambda > 0$ when a point is closer to l_1 than to l_2;
- $\lambda = 0$ when either $u = \frac{1}{2}$ or $v = \frac{1}{2}$.
- Its minimum, $\lambda = -\frac{1}{2}$, is attained at (0,1) or (1,0).
- $\lambda = \frac{1}{2}$ is attained at (0,0) or (1,1).

Long and Krzysztofowicz (1996) showed that, as a continuous function of a random vector (U, V), the random distance Λ has an expectation that is determined by the density c of the copula as

$$\begin{aligned} E(\Lambda) &= \int_0^1 \int_0^1 \lambda(u, v) c(u, v) du dv \\ &= E[d_2^2(U, V) - d_1^2(U, V)] \\ &= 2E(U\ V) - \frac{1}{2}\ . \end{aligned} \tag{1.31}$$

Upon comparing (1.31) with (1.29), we readily find that $\rho_S = 6E(\Lambda)$. In other words, Spearman's ρ_S under the copula is proportional to the expected difference of the quadratic distance from a random point (U, V) to the diagonal lines l_1 and l_2 of the unit square.

1.11.4 Measure Based on Gini's Coefficient

The measure of concordance between X and Y known as Gini's γ can be expressed as

$$\gamma_C = 2 \int_0^1 \int_0^1 (|u + v - 1| - |u - v|)\, dC(u, v).$$

This is equivalent to

$$\gamma_C = 2E\left(|U + V - 1| - |U - V|\right), \tag{1.32}$$

which can be interpreted as the expected distance between (U, V) and the diagonal of $[0, 1] \times [0, 1]$. For further discussion, see Nelsen (2006, p. 212).

1.11.5 Tail Dependence Coefficients

The dependence concepts introduced so far are designed to show how large (or small) values of one random variable appear with large (or small) values of the other. The following tail dependence concepts measure the dependence between the variables in the upper-right quadrant and the lower quadrant of

$[0, 1] \times [0, 1]$. In practice, the concept of tail dependence represents the current standard to describe the amount of extremal dependence.

Definition 1.18. The upper tail dependence coefficient (parameter) λ_U is the limit (if it exists) of the conditional probability that Y is greater than the 100αth percentile of G given that X is greater then the 100αth percentile F as α approaches 1,

$$\lambda_U = \lim_{\alpha \uparrow 1} \Pr\left[Y > G^{-1}(\alpha) \middle| X > F^{-1}(\alpha)\right]. \tag{1.33}$$

If $\lambda_U > 0$, then X and Y are upper tail dependent and asymptotically independent otherwise.

Similarly, the lower tail dependence coefficient is defined as

$$\lambda_L = \lim_{\alpha \downarrow 0} \Pr\left[Y \le G^{-1}(\alpha) \middle| X \le F^{-1}(\alpha)\right]. \tag{1.34}$$

Let C be the copula of X and Y. It can be shown that

$$\lambda_U = \lim_{u \uparrow 1} \frac{\bar{C}(u, u)}{1 - u}, \quad \lambda_L = \lim_{u \downarrow 0} \frac{C(u, u)}{u},$$

where $\bar{C}(u, v) = \Pr(U > u, V > v)$.

Expressions for the coefficients of tail dependence for a wide range of bivariate distributions, as presented in Table 1.1, may be found in Heffernan (2001).

Table 1.1 Tail dependence of some of the families of copulas

Family	λ_U	λ_L
Fréchet	α	α
Cuadras and Augé	0	θ
Marshall and Olkin	0	$\min(\alpha, \beta)$
Placket	0	0

For explicit expressions for both the Cuadras and Augé copula and the Marshall and Olkin copula, see Section 4.5.

The tail dependence coefficient has become very popular for those interested in extreme-value techniques [Kolev et al. (2006, Section 4)]. However, Mikosch (2006a) did not think it very informative with regard to the joint extreme behavior of the vector (X, Y). For nonparametric estimation of tail dependence, see Schmidt and Stadmuller (2006).

1.11.6 A Local Dependence Measure

A local dependence measure defined as a correlation between X and Y given $X = x$, $Y = y$ was proposed by Kotz and Nadarajah (2002):

$$\gamma(x,y) = \frac{E\left([X - E(X|Y=y)][Y - E(Y|X=x)]\right)}{\sqrt{E[-E(X|Y=y)]^2 E[Y - E(Y|X=x)]^2}}, \ -\infty < x, y < \infty. \tag{1.35}$$

A copula analogue of (1.35) has been defined by Kolev et al. (2006) as

$$\gamma_S(u,v) = \frac{E\left([U - E(U|V=v)][V - E(V|U=u)]\right)}{\sqrt{E[-E(U|V=v)]^2 E[V - E(V|U=u)]^2}}, \quad 0 \le u, \ v \le 1. \tag{1.36}$$

The measure γ_S may be interpreted as a "conditional" Spearman ρ.

1.11.7 Tests of Dependence and Inferences

Genest and Favre (2007) presented an introduction to inference for copula models based on rank methods. In particular, they considered empirical estimates for measures of dependence and dependence parameters. Simple graphical tools and numerical techniques were presented for selecting an appropriate model, parameter estimation, and checking the model's goodness of fit.

Shih and Louis (1995) presented both parametric and nonparametric estimation procedures for the association (dependence) parameter in copula models.

1.11.8 "Concepts of Dependence" of Copulas

For "concepts of dependence" that are expressed in terms of various notions of positive dependence for copulas, see Section 5.7 of Nelsen (2006) and Chapter 3 of this volume.

1.12 Distribution Function of $Z = C(U, V)$

In Section 1.7.1, we presented the expression

$$\tau = 4E(C(U,V)) - 1 = 4E(Z) - 1, \tag{1.37}$$

where $Z = C(U,V)$ and $E(Z) = \int_0^1 \{1 - K(z)\}dz$, with K being the distribution function of Z. It is well known that for any random variable X with continuous distribution function F, $F(X)$ is uniformly distributed on $[0,1]$. However, it is not generally true that the distribution K of Z is uniform on $[0,1]$. The fact that K is related to Kendall's tau via (1.37) has encouraged several authors [see, e.g., Genest and Rivest (1993) and Wang and Wells (2000)] to develop estimation and goodness-of-fit procedures for different classes of copulas using the empirical version of K, whose asymptotic behavior as a process was first studied by Barbe et al. (1996).

For Archimedean copulas, Genest and Rivest (1993) showed that

$$K(z) = z - \frac{\varphi(z)}{\varphi'(z)}, \tag{1.38}$$

where φ is the generator of the copula C. The key results on K when C is an Archimedean copula given by Genest and MacKay (1986a) and Genest and Rivest (1993) are as follows:

(1) The function $K(z) = z - \frac{\varphi(z)}{\varphi'(z)}$ is the cumulative distribution function of the variable $Z = C(U,V)$. Hence, with a knowledge of $K(z)$, we can in principle retrieve the function $\varphi(z)$ and hence the Archimedean copula.
(2) The function $K(z)$ can be estimated by means of empirical distribution functions $K_n(z_i)$, where z_i is the proportion of pairs (X_j, Y_j) in the sample that are less than or equal to the pair (X_i, Y_i) componentwise.
(3) The empirical function $K_n(z)$ can be fitted by the distribution function $K_{\tilde{\theta}}$ of any family of Archimedean copulas, where the parameter θ is estimated in such a manner that the fitted distribution has a coefficient of concordance (τ) equal to the corresponding empirical coefficient (τ_n).
(4) Z and W are independent, with the latter given by the expression $W = \frac{\varphi(U)}{\varphi(U)+\varphi(V)}$, which is uniformly distributed on [0, 1].

For copulas not necessarily Archimedean, Chakak and Ezzerg (2000) have shown that K can be expressed in terms of the quantile function associated with the bivariate copula C. Genest and Rivest (2001) have also given a general formula for computing K.

1.13 Simulation of Copulas

The following method of simulation is described in Drouet-Mari and Kotz (2001).

1.13.1 The General Case

To generate a sample (U_i, V_i), $i = 1, 2, \ldots, n$, from a copula $C(u, v)$, we use the fact that the conditional copula $C_u(v) = C(V|U = v)$ is a distribution function and that $Z = C_u(V)$ follows a uniform distribution on $[0,1]$. Since U has a uniform distribution, its density is 1 over $[0,1]$ and thus $C_u(v) = \frac{\partial C(u,v)}{\partial v}$. Hence, the simulation procedure is as follows:

Step 1: Generate two variables U and Z independent and uniform over $[0,1]$.
Step 2: Calculate $V = C_u^{-1}(Z)$. Then, the pair (U, V) has the desired copula.

This procedure works well but requires an analytical expression for $V = C_u^{-1}(Z)$.

1.13.2 Archimedean Copulas

For Archimedean copulas, we can modify the procedure above. The method described below is due to Genest and MacKay (1986a). Since $\varphi(C) = \varphi(U) + \varphi(V)$, it follows that $\varphi'\left(C\frac{\partial C}{\partial u}\right) = \varphi'(u)$. An auxiliary variable $W = C(U, V)$ is calculated as

$$W = (\varphi')^{-1}\left(\frac{\varphi'(u)}{\frac{\partial C}{\partial u}}\right),$$

where $(\varphi')^{-1}$ is the inverse of the derivative of φ. The simulation procedure is then as follows:

Step 1. Generate two uniform and independent random variables U and Z on $[0,1]$.
Step 2: Calculate W using the formula above.
Step 3: Calculate $V = \varphi^{-1}[\varphi(W) - \varphi(V)]$.

This procedure works well for Clayton and Frank's families (see Section 2.4). However, for the Gumbel–Hougaard family, there is no analytical expression for $(\varphi')^{-1}$.

1.14 Construction of a Copula

1.14.1 Rüschendorf's Method

We shall now describe a general method of constructing a copula developed by Rüschendorf (1985).

Suppose $f^1(u,v)$ has integral zero on the unit square and its two marginals integrate to zero; i.e.,

$$\int_0^1 \int_0^1 f^1(u,v)du\,dv = 0 \tag{1.39}$$

and

$$\int_0^1 f^1(u,v)du = 0 \qquad \text{and} \qquad \int_0^1 f^1(u,v)dv = 0. \tag{1.40}$$

Equation (1.39) implies (1.40). In that case, $1+f^1(u,v)$ is a density of a copula. However, there is the constraint that $1+f^1(u,v)$ must be non-negative. If it is not the case, but f^1 is bounded, we can then find a constant α such that $1+\alpha f^1$ is positive.

A function of the type described above can be constructed quite easily. One needs to start with an arbitrary real integrable function f on the unit square and compute

$$V = \int_0^1 \int_0^1 f(u,v)du\,dv, \;\; f_1(u) = \int_0^1 f(u,v)dv, \;\; f_2(v) = \int_0^1 f(u,v)du.$$

Then set $f^1 = f - f_1 - f_2 + V$.

If we have two functions f^1 and g^1 possessing the properties stipulated above, then $1 + f^1 + g^1$ is the density of a copula, and more generally, $1 + \sum_{i=1}^n f_i^1$ is a density with f_i^1 satisfying the conditions in (1.39) and (1.40).

Example 1.19. Long and Krzysztofowicz (1995) utilized a particular case of the Rüschendorf method of construction. Let $f^1(u,v) = c_1(u,v) + c_2(u,v) - 2K(1)$, where

$$\begin{aligned} c_1(u,v) &= \kappa(u-v) \;\text{ if } v \le u \\ &= \kappa(v-u) \;\text{ if } v \ge u \end{aligned} \tag{1.41}$$

and

$$c_2(u,v) = \begin{cases} \kappa(u+v) & \text{if } u \le u - v \\ \kappa(v-u) & \text{if } u \ge 1-v, \end{cases}$$

and $K(1) = \int_0^1 \kappa(t)\,dt$, where $\kappa(t)$ is a continuous and monotonic function on [0,1].

Example 1.20. [Lai and Xie's extension of F-G-M] Lai and Xie (2000) extended the Farlie–Gumbel–Morgenstern family by considering

$$\begin{aligned} C(u,v) = uv + w(u,v) &= uv + \alpha u^b v^b (1-u)^a (1-v)^a, \\ & a, b, 0 \le \alpha \le 1. \end{aligned} \tag{1.42}$$

1.14.2 Generation of Copulas by Mixture

Marshall and Olkin (1988) and Joe (1993) considered a general method in generating bivariate distributions by mixture. Set

$$H(u,v) = \int\int K(F^{\theta_1}, G^{\theta_2})\, d\Lambda(\theta_1, \theta_2), \tag{1.43}$$

where K is a copula and Λ is a mixing distribution, ϕ_i being the Laplace transform of the marginal Λ_i of Λ. Thus, different selections of G and K lead to a variety of distributions with marginals as parameters. Note that F and G here are not necessarily the marginals of H.

If K is an independent bivariate distribution and the two marginals of Λ are equal such that it is the Fréchet bound [i.e., $\Lambda(\theta_1,\theta_2) = \min(\Lambda_1(\theta_1), \Lambda_2(\theta_2))$], then $H(u,v) = \int_0^\infty F^\theta(u) G^\theta(v) d\Lambda_1(\theta)$ with $\theta_1 = \theta$. Now, let $F(u) = \exp[-\phi^{-1}(u)]$ and $G(u) = \exp[-\phi^{-1}(u)]$, where $\phi(t)$ is the Laplace transform of Λ_1, i.e., $\phi(-t)$ is the moment generating function of Λ_1. It follows that

$$H(u,v) = \int_0^\infty \exp\left[-\theta\left(\phi^{-1}(u) + \phi^{-1}(v)\right)\right] d\Lambda_1(\theta). \tag{1.44}$$

From (1.44), it is clear that the marginals of H are uniform and so H is a copula. In other words, when ϕ is the Laplace transform of a distribution, then the function defined on the unit square by

$$C(u,v) = \phi\left(\phi^{-1}(u) + \phi^{-1}(v)\right) \tag{1.45}$$

is indeed a copula. However, the right-hand side of (1.45) is a copula for a broader class of functions than the Laplace transforms, and these copulas are called Archimedean copulas, mentioned earlier in Section 1.5.

Example 1.21. If the mixing distribution $\Lambda_1(\theta)$ has a negative binomial distribution with the Laplace transform $\phi(t) = \left(\frac{pe^{-t}}{1-qe^{-t}}\right)^\alpha, \alpha > 0,\ 0 < p < 1,$ $q = 1-p$, and the inverse function $\varphi(t) = \log\left(\frac{t^{1/\alpha}}{p+qt^{1/\alpha}}\right)$, then

$$C(u,v) = \frac{uv}{\left[1 - q(1-u^{1/\alpha})(1-v^{1\alpha})\right]^\alpha}, \tag{1.46}$$

which is the survival copula of the bivariate Lomax distribution (see Section 2.8).

1.14.3 Convex Sums

In Section 1.3, it was shown that if $\{C_\theta\}$ is a finite collection of copulas, then any convex combination of the copulas in $\{C_\theta\}$ is also a copula. Convex sums are an extension of this idea to an infinite collection of copulas indexed by a continuous parameter θ.

Suppose now that θ is an observation of a random variable with distribution function Λ. If we set

$$C(u,v) = \int_{-\infty}^{\infty} C_\theta \, d\Lambda(\theta), \tag{1.47}$$

then it is easy to verify that C is a copula, which was termed by Nelsen (1999) as the *convex sum* of $\{C_\theta\}$ with respect to Λ. In fact, Λ is simply a mixing distribution of the family $\{C_\theta\}$.

Consider a special case of Marshall and Olkin's method discussed earlier, in which K is an independent copula (i.e., $K(u,v) = uv$) and Λ is a univariate distribution so that

$$C(u,v) = \int_0^{\infty} F^u(\theta) G^v(\theta) \; d\Lambda. \tag{1.48}$$

The expression in (1.48) can clearly be considered as a convex sum of the family of copulas $\{(FG)^\theta\}$.

1.14.4 Univariate Function Method

Durante (2007) constructed a family of symmetric copulas from a univariate function $f : [0,1] \to [0,1]$ that is continuous, differentiable except at finitely many points. Define

$$C_f(x,y) = \min(x,y) f\big(\max(x,y)\big).$$

Then C_f is a copula if and only if

(i) $f(1) = 1$;
(ii) f is increasing; and
(iii) the function $t \mapsto f(t)/t$ is decreasing on $(0,1]$.

Example 1.22. $f(t) = \alpha t + (1-\alpha)$, $\alpha \in [0,1]$. Then $C_\alpha(u,v) = \alpha uv + (1-\alpha)\min(u,v)$ is a member of the Fréchet family of copulas (see Section 3.2).

Example 1.23. Let $f_\alpha(t) = t^\alpha$. Then

$$C_\alpha(u,v) = \begin{cases} uv^\alpha, & \text{if } u \le v \\ u^\alpha v, & \text{if } u \ge v \end{cases},$$

which is the Cuadras-Augé copula given in (2.25).

1.14.5 Some Other Methods

Nelsen (2006) presented several other methods for constructing of copulas, including the following.

The Inversion Method

This is simply the so-called marginal transformation method through inverse probability integral transforms of the marginals $F^{-1}(u) = x$ and $G^{-1}(v) = y$. If either one of the two inverses does not exist, we simply modify our definition so that $F^{-}(u) = \inf\{x : F(x) \geq u\}$, for example. Then, given a bivariate distribution function H with continuous marginals F and G, we obtain a copula

$$C(u, v) = H\left(F^{-1}(u), G^{-1}(v)\right). \tag{1.49}$$

Nelsen (2006) illustrated this procedure with two examples:

(1) The procedure above is used to find Marshall and Olkin's family of copulas (also known as the generalized Cuadras and Augé family) from Marshall and Olkin's system of bivariate exponential distributions.
(2) A copula is obtained from the circular uniform distribution with X and Y being the coordinates of a point chosen at random on the unit circle.

Geometric Methods

Several schemes are given by Nelsen (2006), including:

- singular copulas with prescribed support;
- ordinal sums;
- shuffles of Min [Mikusiński et al. (1992)];
- copulas with prescribed horizontal or vertical sections; and
- copulas with prescribed diagonal sections.

A particular copula of interest generated by geometry is the symmetric copula constructed by Ferguson (1995). In this copula, $C(u, v) = \hat{C}(u, v)$.

Algebraic Methods

Two well-known families of copulas, the Plackett and Ali–Mikhail–Haq families, were constructed using an algebraic relationship between the joint distribution function and its univariate marginals. In both cases, the algebraic relationship concerns an "odds ratio." In the first case, we generalize 2×2 contingency tables, and in the second case we work with a survival odds ratio.

1.15 Applications of Copulas

There is a fast-growing industry for copulas. They have useful applications in econometrics, risk management, finance, insurance, etc. The commercial statistics software SPLUS provides a module in *FinMetrics* that include copula fitting written by Carmona (2004). One can also get copula modules in other major software packages such as R, Mathematica, Matlab, etc. The International Actuarial Association (2004) in a paper on Solvency II,[1] recommends using copulas for modeling dependence in insurance portfolios. Moody's uses a Gaussian copula for modeling credit risk and provides software for it that is used by many financial institutions. Basle II[2] copulas are now standard tools in credit risk management.

There are many other applications of copulas, especially the Gaussian copula, the extreme-value copulas, and the Archimedean copula. We now classify these applications into several categories.

1.15.1 Insurance, Finance, Economics, and Risk Management

One of the driving forces for the popularity of copulas is their application in the context of financial risk management. Mikosch (2006a, Section 3) explains the reasons why the finance researchers are attracted to copulas.

- Risk modeling—van der Hoek and Sherris (2006)
- Daily equity return in Spanish stock market—Roch and Alegre (2006)
- Jump-driven financial asset model—Luciano and Schoutens (2006)
- Default correlation and pricing of collateralized obligation—P. Li et al. (2006)
- Credit derivatives—Charpentier and Juri (2006)
- Modeling asymmetric exchange rate dependence—Patton (2006)
- Credibility for aggregate loss—Frees and Wang (2006)
- Decomposition of bivariate inequality by attributes – Naga and Geoffard (2006)

[1] Solvency II is a treaty for insurances.

[2] Basle I and II are treaties for banks.

- Group aspects of regulatory reform in insurance sector—Darlap and Mayr (2006)
- Financial risk calculation with applications to Chinese stock markets—Li et al. (2005)
- Measurement of aggregate risk—Junker and May (2005)
- Interdependence in emerging markets—Mendes (2005)
- Application to financial data—Dobric and Schmid (2005)
- Tail dependence in Asian markets—Caillault and Guegan (2005)
- Modeling heterogeneity in dependent data—Laeven (2005)
- Bivariate option pricing—van den Goorbergh et al. (2005)
- Worst VaR scenarios—Embrechts et al. (2005)
- Correlated default with incomplete information—Giesecke (2004)
- Value-at-risk-efficient portfolios—Malevergne and Sornette (2004)
- Fitting bivariate cumulative returns—Hürlimann (2004)
- General cash flows—Goovaerts et al. (2003)
- Modeling in actuarial science—Purcaru (2003)
- Financial asset dependence—Malevergne and Sornette (2003)
- High-frequency data in finance—Breymann et al. (2003)
- Dependence between the risks of an insurance portfolio in the individual risk model—Cossette et al. (2002)
- Portfolio allocations—Hennessy and Lapan (2002)
- Relationship between survivorship and persistency of insurance policy holders—Valdez (2001)
- Loss and allocated loss adjustment expenses on a single claim—Klugman and Parsa (1999)
- Sum of dependent risks—Denuit et al. (1999)

1.15.2 Hydrology and Environment

- On the use of copulas in hydrology: Theory and practice—Salvadori and De Michele (2007).
- Case studies in hydrology—Renard and Lang (2007)
- Bivariate rainfall frequency—Zhang and Singh (2007)
- Bivariate frequency analysis of floods—Shiau (2006)
- Groundwater quality—Bardossy (2006)
- Flood frequency analysis—Grimaldi and Serinaldi (2006)
- Drought duration and severity—Shiau (2006)
- Temporal structure of storms—Salvadori and De Michele (2006)
- Successive wave heights and successive wave periods—Wist et al. (2004)
- Ozone concentration—Dupuis (2005)
- Phosphorus discharge to a lake—Reichert and Borsuk (2005)
- Frequency analysis of hydrological events—Salvadori and De Michele (2004)

- Adequacy of dam spillway—De Michele et al. (2005)
- Hydrological frequency analysis—Favre et al. (2004)
- Storm rainfall—De Michele and Salvadori (2003)

1.15.3 Management Science and Operations Research

- Decision and risk analysis—Clemen and Reilly (1999)
- Entropy methods for joint distributions in decision analysis—Abbas (2006)
- Field development decision process—Acciolya and Chiyoshi (2004)
- Uncertainty analysis—van Dorp (2004)
- Schedulability analysis—Burns et al. (2003)
- Database management—Sarathy et al. (2002)
- Decision and risk analysis—Clemen et al. (2000)
- Beneficial changes insurance—Tibiletti (1995)

1.15.4 Reliability and Survival Analysis

- Bivariate failure time data—Chen and Fan (2007)
- Competing risk survival analysis—Bond and Shaw (2006)
- Interdependence in networked systems—Singpurwalla and Kong (2004)
- Competing risk—Bandeen-Roche and Liang (2002)
- Time to wound excision and time to wound infection in a population of burn victims—van der Laan et al. (2002)
- Survival times on blindness for each eye of diabetic patients with adult onset diabetes—Viswanathan and Manatunga (2001)
- Bivariate current status data—Wang and Ding (2000)

1.15.5 Engineering and Medical Sciences

- Poliomyelitis incidence—Escarela et al. (2006)
- Modeling of vehicle axle weights—Srinivas et al. (2006)
- Plant-specific dynamic failure assessment—Meel and Seider (2006)
- Trait linkage analysis—M.Y. Li et al. (2006)
- Unsupervised signal restoration—Brunel and Pieczynski (2005)
- Real option valuation of oil projects—Armstrong et al. (2004)

- Probabilistic dependence among binary events—Keefer (2004)
- QTL mapping—Basrak et al. (2004)
- Modeling the dependence between the times to international adoption of two related technologies—Meade and Islam (2003)
- Signal processing—Davy and Doucet (2003)
- Interaction between toxic compounds—Haas et al. (1997)
- Removing cancer when it is correlated with other causes of death—Carriere (1995)

1.15.6 Miscellaneous

- Expert opinions—Jouini and Clemen (1996)
- Accident precursor analysis—Yi and Bier (1998)
- Generations of dispersion models—Song (2000)
- Health care demand—Zimmer and Trivedi (2005)
- Biometric data studies—Rukhin and Osmoukhina (2005)
- Uncertainty measures in expert systems—Goodman et al. (1991)

1.16 Criticisms about Copulas

Despite their immense popularity, copulas have their critics. In a critical article entitled "Copulas: Tales or Facts" published in *Extremes*, Mikosch (2006a,b) gave several far-reaching criticisms to caution readers about the problems associated with copulas. Below are his verbatim remarks that summarize his opinion about copulas.

- There are no particular advantages of using copulas when dealing with multivariate distributions. Instead one can and should use any multivariate distribution which is suited to the problem at hand and which can be treated by statistical techniques.
- The marginal distributions and the copula of a multivariate distribution are inextricably linked. The main selling point of the copula technology—separation of the copula (dependence function) from the marginal distributions—leads to a biased view of stochastic dependence, in particular when one fits a model to the data.
- Various copula models (Archimedean, t-, Gaussian, elliptical, extreme value) are mostly chosen because they are mathematically convenient; the rationale for their applications is murky.
- Copulas are considered as an alternative to Gaussian models in a non-Gaussian world. Since copulas generate any distribution, the class is too big to be understood and to be useful.

- There is little statistical theoretical theory for copulas. Sensitivity studies of estimation procedures and goodness-of-fit tests for copulas are unknown. It is unclear whether a good fit of the copula of the data yields a good fit to the distribution of the data.
- Copulas do not contribute to a better understanding of multivariate extremes.
- Copulas do not fit into the existing framework of stochastic processes and time series; they are essentially static models and are not useful for modeling dependence through time.

There were several discussants [de Haan (2006), de Vries (2006), Genest and Rémillard (2006), Joe (2006), Linder (2006), Embrechts (2006), Peng (2006), and Segers (2006)] of the paper, and some did agree on certain aspects, but others did not agree at all with the issues raised. A rejoinder is given by Mikosch (2006b).

1.17 Conclusions

Over the last decade, there has been significant and rapid development of the theory of copulas. Much of the work has been motivated by their applications to stochastic processes, economics, risk management, finance, insurance, the environment (hydrology, climate, etc.), survival analysis, and medical sciences.

In many statistical models, the assumption of independence between two or more variables is often due to convenience rather than to the problem at hand. In some situations, neglecting dependence effects may lead to an erroneous conclusion. However, fitting a bivariate or multivariate distribution to a dataset has often proved to be difficult. The copula approach is a way to solve the difficult problem of finding the whole bivariate or multivariate distribution by a two-stage statistical procedure; i.e., estimating the marginal distributions and the copula function separately from each other. A weakness of the copula approach is that it is difficult to select or find an appropriate copula for the problem at hand. Often, the only alternative is to commence with some educated guess by selecting a parametric family of copulas and then try to fit the parameters. As a result, the model obtained may suffer a certain degree of arbitrariness. Indeed, there are some authors who have strong misgivings about the copula approach. Nevertheless, judging from the amount of interest generated, the copulas certainly have secured themselves an important place in the world.

References

1. Abbas, A.E.: Entropy methods for joint distributions in decision analysis. IEEE Transactions on Engineering Management **53**, 146–159 (2006)
2. Acciolya, R.D.E., Chiyoshi, F.Y.: Modeling dependence with copulas: A useful tool for field development decision process. Journal of Petroleum Science and Engineering **44**, 83–91 (2004)
3. Armstrong, M., Galli, A., Bailey, W., Coue, B.: Incorporating technical uncertainty in real option valuation of oil projects. Journal of Petroleum Science and Engineering **44**, 67–82 (2004)
4. Bandeen-Roche, K., Liang, K.Y.: Modelling multivariate failure time associations in the presence of a competing risk. Biometrika **89**, 299–314 (2002)
5. Barbe, P., Genest, C., Ghoudi, K., Rémillard, B.: On Kendall's process. Journal of Multivariate Analysis **58**, 197–229 (1996)
6. Bardossy, A.: Copula-based geostatistical models for groundwater quality parameters. Water Resources Research **42**, Art. No. W11416 (2006)
7. Basrak, B., Klaassen, C.A.J., Beekman, M., Martin, N.G., Boomsma, D.I.: Copulas in QTL mapping. Behavior Genetics **34**, 161–171 (2004)
8. Bond, S.J., Shaw J.E.H.: Bounds on the covariate-time transformation for competing-risks survival analysis. Lifetime Data Analysis **12**, 285–303 (2006)
9. Breymann, W., Dias, A., Embrechts, P.: Dependence structures for multivariate high-frequency data in finance. Quantitative Finance **3**, 1–14 (2003)
10. Brunel, N., Pieczynski, W.: Unsupervised signal restoration using hidden Markov chains with copulas. Signal Processing **85**, 2304–2315 (2005)
11. Burns, A., Bernat, G.,Broster, I.: A probabilistic framework for schedulability analysis. In: Proceedings of the Third International Conference on Embedded Software (EMSOFT 2003), pp. 1–15 (2003)
12. Caillault, C., Guegan, D.: Empirical estimation of tail dependence using copulas: application to Asian markets. Quantitative Finance **5**, 489–501 (2005)
13. Capéraà, P., Fougères, A.L., Genest, C.: Bivariate distributions with given extreme value attractor. Journal of Multivariate Analysis **72**, 30–40 (2000)
14. Carmona, R.A.: Statistical Analysis of Financial Data in S-PLUS. Springer-Verlag, New York (2004)
15. Carriere, J.F.: Removing cancer when it is correlated with other causes of death. Biometrical Journal **37**, 339–350 (1995)
16. Chakak, A., Ezzerg, M.: Bivariate contours of copula. Communications in Statistics—Simulation and Computation **29**, 175–185 (2000)
17. Charpentier, A., Juri, A.: Limiting dependence structures for tail events, with applications to credit derivatives. Journal of Applied Probability **43**, 563–586 (2006)
18. Chen, X.H., Fan, Y.Q.: A model selection test for bivariate failure-time data. Econometric Theory **23**, 414–439 (2007)
19. Cherubini, U., Luciano, E., Vecchiato, W.: Copula Methods in Finance. John Wiley and Sons, Chichester (2004)
20. Clemen, R.T., Fischer, G.W., Winkler, R.L.: Assessing dependence: Some experimental results. Management Science **46**, 1100–1115 (2000)
21. Clemen, R.T., Reilly, T.: Correlations and copulas for decisions and risk analysis. Management Science **45**, 208–224 (1999)
22. Cossette, H., Gaillardetz, P., Marceau, E., Rioux, J.: On two dependent individual risk models. Insurance Mathematics and Economics **30**, 153–166 (2002)
23. Darlap, P., Mayr, B.: Group aspects of regulatory reform in the insurance sector. Geneva Papers on Risk and Insurance: Issues and Practice **31**, 96–123 (2006)
24. Davy, M., Doucet, A.: Copulas: A new insight into positive time-frequency distributions. IEEE Signal Processing Letters **10**, 215–218 (2003)

25. de Haan, L.: "Copulas: tales and facts," by Thomas Mikosch. Extremes **9**, 21–22 (2006)
26. De Michele, C., Salvadori, G.: A Generalized Pareto intensity-duration model of storm rainfall exploiting 2-Copulas. Journal of Geophysical Research: Atmospheres **108**, Art. No. 4067 (2003)
27. De Michele, C., Salvadori, G., Canossi, M., Petaccia, A., Rosso, R.: Bivariate statistical approach to check adequacy of dam spillway. Journal of Hydrologic Engineering **10**, 50–57 (2005)
28. de Vries, C.G., Zhou, C.: "Copulas: tales and facts," by Thomas Mikosch. Extremes **9**, 23–25 (2006)
29. Demarta, S., McNeil, A.J.: The *t* Copula and Related Copulas. International Statistical Review **73**, 111–129 (2005)
30. Denuit, M., Genest, C., Marceau, E.: Stochastic bounds on sums of dependent risks. Insurance Mathematics and Economics **25**, 85–104 (1999)
31. Dobric, J., Schmid, F.: Testing goodness of fit for parametric families of copulas: Application to financial data. Communications in Statistics—Simulation and Computation **34**, 1053–1068 (2005)
32. Drouet-Mari, D., Kotz, S.: Correlation and Dependence. Imperial College Press, London (2001)
33. Durante, F.: A new class of bivariate copulas. Comptes Rendus Mathematique **344**, 195–198 (2007)
34. Dupuis, D.J.: Ozone concentrations: A robust analysis of multivariate extremes. Technometrics **47**, 191–201 (2005)
35. Embrechts, P.: "Copulas: tales and facts," by Thomas Mikosch. Extremes **9**, 45–47 (2006)
36. Embrechts, P., Lindskog, F., McNeil, A.: Modelling dependence with copulas and applications to risk management. In: Handbook of Heavey Tailed Distributions in Finance, S.T. Rachev (ed.). Elsevier, Amsterdam (2003)
37. Embrechts, P., McNeil, A., Straumann, D.: Correlation and dependence in risk management: Properties and pitfalls. In: Risk Management: Value at Risk and Beyond, M.A.H. Dempster (ed.), pp. 176–223. Cambridge University Press, Cambridge (2002)
38. Embrechts, P., Hoing, A., Puccetti, G.: Worst VaR scenarios. Insurance Mathematics and Economics **37**, 115–134 (2005)
39. Escarela, G., Mena, R. H., Castillo-Morales, A.: A flexible class of parametric transition regression models based on copulas: Application to poliomyelitis incidence. Statistical Methods in Medical Research **15**, 593–609 (2006)
40. Fang, H.B., Fang K.T., Kotz, S.: The meta-elliptical distributions with given marginals. Journal of Multivariate Analysis **82**, 1–16 (2002)
41. Favre, A.C., El Adlouni, S., Perreault, L., Thiemonge, N., Bobee, B.: Multivariate hydrological frequency analysis using copulas. Water Resources Research **40**, Art. No. W01101 (2004)
42. Ferguson, T.S.: A class of symmetric bivariate uniform distributions. Statistics Papers **36**, 31–40 (1995)
43. Fisher, N.I.: Copulas. In: Encyclopedia of Statistical Sciences, Updated Volume 1, S. Kotz, C. B. Read, D. L. Banks (eds.), pp. 159–164. John Wiley and Sons, New York (1997)
44. Fréchet. M.: Remarque au sujet de la note précédente. Comptes Rendus de l'Académie des Sciences, Série I. Mathématique **246**, 2719–2720 (1958)
45. Frees, E.W., Wang, P.: Copula credibility for aggregate loss models. Insurance Mathematics and Economics **38**, 360–373 (2006)
46. Genest, C., Favre, A.-C.: Everything you always wanted to know about copula modelling but were afraid to ask. Journal of Hydrologic Engineering **12**, 347–368 (2007)
47. Genest, C., Rémillard, B.: "Copulas: tales and facts," by Thomas Mikosch. Extremes **9**, 27–36 (2006)

48. Genest, C., MacKay, J.: Copules archimédiennes et familles de lois bidimensionnelles dont les marges sont données. Canadian Journal of Statistics **14**, 145–159 (1986a)
49. Genest, C., MacKay, J.: The joy of copula: Bivariate distributions with uniform marginals. The American Statistician **40**, 280–285 (1986b)
50. Genest, C., Rivest, L.P.: A characterization of Gumbel's family of extreme-value distributions. Statistics and Probability Letters **8**, 207–211 (1989)
51. Genest, C., Rivest, L.P.: Statistical inference procedures for bivariate Archimedaen copulas. Journal of the American Statistical Association **88**, 1034–1043 (1993)
52. Genest, C., Rivest, L.P. On the multivariate probability integral transformation. Statistics and Probability Letters **53**, 391–399 (2001)
53. Glasserman, P.: Monte-Carlo Methods in Financial Engineering. Springer-Verlag, New York (2004)
54. Giesecke, K.: Correlated default with incomplete information. Journal of Banking and Finance **28**, 1521–1545 (2004)
55. Goodman, I.R., Nguyen, H.T., Rogers, G.S.: Admissibility of uncertainty measures in expert systems. Journal of Mathematical Analysis and Application **159**, 550–594 (1991)
56. Goovaerts, M.J., De Schepper, A., Hua, Y.: A new approximation for the distribution of general cash flows. Insurance Mathematics and Economics **33**, 419 (2003)
57. Grimaldi, S., Serinaldi, F.: Asymmetric copula in multivariate flood frequency analysis. Advances in Water Resources **29**, 1155–1167 (2006)
58. Haas, C.N., Kersten, S.P., Wright K., Frank, M.J., Cidambi, K.: Generalization of independent response model for toxic mixtures. Chemosphere **34**, 699–710 (1997)
59. Heffernan, J.E.: A directory of coefficients of tail dependence. Extremes **3**, 279–290 (2001)
60. Hennessy, D.A., Lapan, H.E.: The use of Archimedean copulas to model portfolio allocations. Mathematical Finance **12**, 143–154 (2002)
61. Hürlimann, W.: Fitting bivariate cumulative returns with copulas. Computational Statistics and Data Analysis **45**, 355–372 (2004)
62. Hutchinson, T.P., Lai, C.D.: Continuous Bivariate Distributions: Emphasising Applications. Rumsby Scientific Publishing, Adelaide (1990)
63. The International Actuarial Association, A global framework for insurer solvency assessment, Research Report. www.actuaries.org (2004)
64. Joe, H.: Parametric familes of multivariate distributions with given margins. Journal of Multivariate Analysis **46**, 262–282 (1993)
65. Joe, H.: Multivariate Models and Dependence Concepts. Chapman and Hall, London (1997)
66. Joe, H.: "Copulas: tales and facts," by Thomas Mikosch. Extremes **9**, 37–41 (2006)
67. Johnson, N.L., Kotz, S.: On some generalized Farlie–Gumbel–Morgenstern distributions. II. Regression, correlation and further generalizations. Communications in Statistics: Theory and Methods **6**, 485–496 (1977)
68. Jouini, M.N., Clemen, R.T.: Copula models for aggregating expert opinions. Operations Research **44**, 444–457 (1996)
69. Junker, M., May, A.: Measurement of aggregate risk with copulas. Econometrics Journal **8**, 428–454 (2005)
70. Keefer, D.L.: The underlying event model for approximating probabilistic dependence among binary events. IEEE Transactions of Engineering Management **51**, 173–182 (2004)
71. Klugman, S.A., Parsa, R.: Fitting bivariate loss distributions with copulas. Insurance Mathematics and Economics **24**, 139–148 (1999)
72. Kolev, N., Anjos, U., Mendes, B.: Copulas: A review and recent developments. Stochastic Models **22**, 617–660 (2006)
73. Kotz, K., Nadarajah, D.: Some local dependence functions for the elliptically symmetric distributions. Sankhyā, Series A **65**, 207–223 (2002)

74. Laeven, R.J.: Families of Archimedean copulas for modelling heterogeneity in dependent data. Insurance Mathematics and Economics **37**, 375 (2005)
75. Lai, C.D., Xie, M.: A new family of positive dependence bivariate distributions. Statistics and Probability Letters **46**, 359–364 (2000)
76. Li, M.Y., Boehnke, M., Abecasis, G.R., Song, P.X-K.: Quantitative trait linkage analysis using Gaussian copulas. Genetics **173**, 2317–2327 (2006)
77. Li, P., Chen, H.S., Deng, X.T., Zhang, S.M.: On default correlation and pricing of collateralized debt obligation by copula functions. International Journal of Information Technology and Decision Making **5**, 483–493 (2006)
78. Li, P., Shi, P., Huang, G.D.: A new algorithm based on copulas for financial risk calculation with applications to Chinese stock markets. In: Lecture Notes in Computer Science, Volume 3828, pp. 481–490. Springer, Berlin (2005)
79. Lin, G.D.: Relationships between two extensions of Farlie–Gumbel–Morgenstern distribution. Annals of the Institute of Statistical Mathematics **39**, 129–140 (1987)
80. Linder, A.: "Copulas: tales and facts," by Thomas Mikosch. Extremes **9**, 43–47 (2006)
81. Long, D., Krzysztofowicz, R.: A family of bivariate densities constructed from marginals. Journal of the American Statistical Association **90**, 739–746 (1995)
82. Long, D., Krzysztofowicz, R.: Geometry of a correlation coefficient under a copula. Communications in Statistics: Theory and Methods **25**, 1397–1406 (1996)
83. Luciano, E., Schoutens, W.: A multivariate jump-driven financial asset model. Quantitative Finance **6**, 385–402 (2006)
84. Malevergne, Y., Sornette, D.: Testing the Gaussian copula hypothesis for financial assets dependences. Quantitative Finance **3**(4), 231–250 (2003)
85. Malevergne, Y., Sornette, D.: Value-at-risk-efficient portfolios for a class of super- and sub-exponentially decaying asset return distributions. Quantitative Finance **4**, 17–36 (2004)
86. Mardia, K. V.: Families of Bivariate Distributions. Griffin, London (1970)
87. Marshall, A.W., Olkin, I.: Families of multivariate distributions. Journal of the American Statistical Association **83**, 834–841 (1988)
88. Marshall, R., Naldi, M., Zeevi, A.: On the dependence of equity and asset returns. Risk **16**, 83–87 (2003)
89. Meade, N., Islam, T.: Modelling the dependence between the times to international adoption of two related technologies. Technological Forecasting and Social Change **70**, 759–778 (2003)
90. Meel, A., Seider, W.D.: Plant-specific dynamic failure assessment using Bayesian theory. Chemical Engineering Science **61**, 7036–7056 (2006)
91. Mendes, B.V.D.: Asymmetric extreme interdependence in emerging equity markets. Applied Stochastic Models in Business and Industry **21**, 483–498 (2005)
92. Mikosch, T.: "Copulas: tales and facts," by Thomas Mikosch. Extremes **9**, 3–20 (2006a)
93. Mikosch, T.: "Copulas: tales and facts, rejoinder," by Thomas Mikosch. Extremes **9**, 55–66 (2006b)
94. Mikusiński, P., Sherwood, H., Taylor, M.D.: Shuffles of Min. Stochastica **13**, 61–74 (1992)
95. Naga, R.H.A., Geoffard, P.Y.: Decomposition of bivariate inequality indices by attributes. Economics Letters **90**, 362–367 (2006)
96. Nelsen, R.B.: An Introduction to Copulas, 2nd edition. Springer-Verlag, New York (2006)
97. Nelsen, R.B.: An Introduction to Copulas. Springer-Verlag, New York (1999)
98. Patton, A.J.: Modelling asymmetric exchange rate dependence. International Economic Review **47**, 527–556 (2006)
99. Peng, L.: "Copulas: Tales and facts," by Thomas Mikosch. Extremes **9**, 49–56 (2006)

100. Pickands, J.: Multivariate extreme value distributions. In: Proceedings of the 43rd Session of the International Statistical Institute, Buenos Aires, pp. 859–878. Amsterdam: International Statistical Institute (1981)
101. Purcaru, O.: Semi-parametric Archimedean copula modelling in actuarial science. Insurance Mathematics and Economics **33**, 419–420 (2003)
102. Reichert, P., Borsuk, M.E.: Does high forecast uncertainty preclude effective decision support? Environmental Modelling and Software **20**, 991–1001 (2005)
103. Renard, B., Lang, M.: Use of a Gaussian copula for multivariate extreme value analysis: Some case studies in hydrology. Advances in Water Resources **30**, 897–912 (2007)
104. Roch, O., Alegre, A.: Testing the bivariate distribution of daily equity returns using copulas: An application to the Spanish stock market. Computaional Statistics and Data Analysis **51**, 1312–1329 (2006)
105. Rukhin, A.L., Osmoukhina, A.: Nonparametric measures of dependence for biometric data studies. Journal of Statistical Planning and Inference **131**, 1–18 (2005)
106. Rüschendorf, L.: Construction of multivariate distributions with given marginals. Annals of the Institute of Statistical Mathematics **37**, 225–233 (1985)
107. Salvadori, G., De Michele, C.: Frequency analysis via copulas: Theoretical aspects and applications to hydrological events. Water Resources Research **40**, Art. No. W12511 (2004)
108. Salvadori, G., De Michele, C.: Statistical characterization of temporal structure of storms. Advances in Water Resources **29**, 827–842 (2006)
109. Salvadori, G., De Michele, C.: On the use of copulas in hydrology: Theory and practice. Journal of Hydologic Engineering **12**, 369–380 (2007)
110. Sarathy, R., Muralidhar, K., Parsa, R.: Perturbing non-normal confidential attributes: The copula approach. Management Science **48**, 1613–1627 (2002)
111. Schmidt, R., Stadmuller, U.: Non-parametric estimation of tail dependence. Scandinavian Journal of Statistics **33**, 307–335 (2006)
112. Schmitz, V.: Revealing the dependence structure between $X_{(1)}$ and $X_{(n)}$. Journal of Statistical Planning and Inference **123**, 41–47 (2004)
113. Schweizer, B. and Sklar, A.: Probabilistic Metric Spaces. North-Holland, New York (1983)
114. Segers, J.: “Copulas: tales and facts”, by Thomas Mikosch. Extremes **9**, 51–53 (2006)
115. Shiau, J.T.: Fitting drought duration and severity with two-dimensional copulas. Water Resources Management **20**, 795–815 (2006)
116. Shih, J.H., Louis, T.M.: Inferences on the association parameter in copula models for bivariate survival data. Biometrics **51**, 1384–1399 (1995)
117. Singpurwalla, N.D., Kong, C.W.: Specifying interdependence in networked systems. IEEE Transactions on Reliability **53**, 401–405 (2004)
118. Sklar, A.: Fonctions de repartition et leurs marges. Publications of the Institute of Statistics, Université de Paris **8**, 229–231 (1959)
119. Song, P.X-K.: Multivariate dispersion models generated from Gaussian copulas. Scandinavian Journal of Statistics **27**, 305–320 (2000)
120. Srinivas, S., Menon, D., Prasad, A.M.: Multivariate simulation and multimodal dependence modeling of vehicle axle weights with copulas. Journal of Transportation Engineering **132**, 945–955 (2006)
121. Tibiletti, L.: Beneficial changes in random variables via copulas: An application to insurance. Geneva Papers on Risk and Insurance Theory **20**, 191–202 (1995)
122. Valdez, E.A.: Bivariate analysis of survivorship and persistency. Insurance Mathematics and Economics **29**, 357–373 (2001)
123. van den Goorbergh, R.W.J., Genest, C., Werker, B.J.M.: Bivariate option pricing using dynamic copula models. Insurance Mathematics and and Economics **37**, 101–114 (2005)
124. van der Hoek, J., Sherris, M.: A flexible approach to multivariate risk modelling with a new class of copulas. Insurance Mathematics and Economics **39**, 398–399 (2006)

125. van der Laan, M.J., Hubbard, A.E., Robins, J.M.: Locally efficient estimation of a multivariate survival function in longitudinal studies. Journal of the American Statistical Association **97**, 494–507 (2002)
126. van Dorp J.R.: Statistical dependence through common risk factors: With applications in uncertainty analysis. European Journal of Operations Research **161**, 240–255 (2004)
127. Viswanathan, B., Manatunga, A.K.: Diagnostic plots for assessing the frailty distribution in multivariate survival data. Lifetime Data Analysis **7**, 143–155 (2001)
128. Wang, W., Ding, A.A.: On assessing the association for bivariate current status data. Biometrika **87**, 879–893 (2000)
129. Wang, W., Wells, M.T.: Model selection and semiparametric inference for bivariate failure-time data. Journal of the American Statistical Association **95**, 62–72 (2000)
130. Wei, G., Fang, H.B., Fang, K.T.: The dependence patterns of random variables: elementary algebraic and geometric properties of copulas. Technical Report, Hong Kong Baptist University, Hong Kong (1998)
131. Wist, H.T., Myrhaug, D., Rue, H.: Statistical properties of successive wave heights and successive wave periods. Applied Ocean Research **26**, 114–136 (2004)
132. Woodworth, G.G.: On the Asymptotic Theory of Tests of Independence Based on Bivariate Layer Ranks. Technical Report No. 75, Department of Statistics, University of Minnesota, Minneapolis (1966)
133. Yi, W-J., Bier, V.M.: An application of copulas to accident precursor analysis. Management Science, **44**, S257–S270 (1998)
134. Zhang, L., Singh, V.P.: Bivariate rainfall frequency distributions using Archimedean copulas. Journal of Hydrology **332**, 93–109 (2007)
135. Zimmer, D.M., Trivedi, P.K.: Using trivariate copulas to model sample selection and treatment effects: Application to family health care demand. Journal of Business and and Economic Statistics **24**, 63–76 (2006)

Chapter 2
Distributions Expressed as Copulas

2.1 Introduction

A feature common to all the distributions in this chapter is that $H(x, y)$ is a simple function of the uniform marginals $F(x)$ and $G(y)$. These types of joint distributions are known as copulas, as mentioned in the last chapter, and will be denoted by $C(u, v)$; the corresponding random variables will be denoted by U and V, respectively.

When the marginals are uniform, independence of U and V implies a flat p.d.f., and any deviation from this will indicate some form of dependence.

Most of the copulas presented in this chapter are of simple forms although in some cases [e.g., the distribution of Kimeldorf and Sampson (1975a) discussed in Section 2.12] they have a rather complicated expression. Some are obtained through marginal transformations, while several others already have uniform marginals and need no transformations to bring them to that form.

The great majority of the copulas described in this chapter have a single parameter that reflects the strength of mutual dependence between U and V. To emphasize its role, we could have chosen to use the same symbol in all these cases. We have not done this, however, since for some distributions it is customary to find α used, others θ, and yet others c.

Throughout this chapter, we assume that U and V are uniform with $C(u, v)$ as their joint distribution function and $c(u, v)$ as the corresponding density function. Thus, the supports of the bivariate distributions are unit squares. For each case, we state some simple properties such as the correlation coefficient and conditional properties. Also, we should note that for bivariate copulas, Pearson's product moment correlation coefficient is the same as the grade coefficient (Spearman's coefficient), as mentioned in Section 1.7.

Unless otherwise specified, the supports of all the distributions are over the unit square. Also, the distribution functions are in fact the cumulative distribution functions. Following this introduction, we discuss the Farlie–Gumbel–Morgenstern (F-G-M) copula and its generalization in Section 2.2.

N. Balakrishnan, C.D. Lai, *Continuous Bivariate Distributions*,
DOI 10.1007/b101765_3,

Next, in Sections 2.3 and 2.4, we discuss the Ali–Mikhail–Haq and Frank distributions. The distribution of Cuadras and Augé and its generalization are presented in Section 2.5. In Section 2.6, the Gumbel–Hougaard copula and its properties are detailed. Next, the Plackett and bivariate Lomax distributions are described in Sections 2.7 and 2.8, respectively. The Lomax copula is presented in Section 2.9. In Sections 2.10 and 2.12, the Gumbel type I bivariate exponential and Kimeldorf and Sampson's distributions are discussed, respectively. The Gumbel–Barnett copula and some other copulas of interest are described in Sections 2.11 and 2.14, respectively. In Section 2.13, the Rodríguez-Lallena and Úbeda-Flores families of bivariate copulas are discussed. Finally, in Section 2.15, some references to illustrations are presented for the benefit of readers.

2.2 Farlie–Gumbel–Morgenstern (F-G-M) Copula and Its Generalization

Formula for Distribution Function

$$C(u,v) = uv[1+\alpha(1-u)(1-v)], \qquad -1 \le \alpha \le 1. \tag{2.1}$$

Formula for Density Function

$$c(u,v) = 1+\alpha(1-2u)(1-2v). \tag{2.2}$$

Correlation Coefficient

The correlation coefficient is $\rho = \frac{\alpha}{3}$, which clearly ranges from $-\frac{1}{3}$ to $\frac{1}{3}$. After the marginals have been transformed to distributions other than uniform, Gumbel (1960a) and Schucany et al. (1978) showed that (i) ρ cannot exceed $\frac{1}{3}$ and (ii) determined it for some well-known distributions—for example, $\frac{\alpha}{\pi}$ for normal marginals and $\frac{\alpha}{4}$ for exponential ones.

Conditional Properties

The regression $E(V|U=u)$ is linear in u.

Dependence Properties

- Lai (1978) has shown that, for $0 \leq \alpha \leq 1$, U and V are positively quadrant dependent (PQD) and positively regression dependent (PRD).
- For $0 \leq \alpha \leq 1$, U and V are likelihood ratio dependent (LRD) (TP_2) [Drouet-Mari and Kotz (2001)].
- For $-1 \leq \alpha \leq 0$, its density is RR_2; see Drouet-Mari and Kotz (2001).

Remarks

- This copula is not Archimedean [Genest and MacKay (1986)].
- The p.d.f. is symmetric about the point $(\frac{1}{2}, \frac{1}{2})$, i.e., it is the same as at $(1-u, 1-v)$ as it is at (u, v), and so the survival (complementary) copula is the same as the original copula.
- Among the results established by Mikhail, Chasnov, and Wooldridge (1987) are the regression curves when the marginals are exponential. Drouet-Mari and Kotz (2001, pp. 115–116) have also provided expressions for the conditional mean and conditional variance when the marginal distributions are F and G.
- Mukherjee and Sasmal (1977) have worked out some properties of a two-component system whose components' lifetimes have the F-G-M distribution, with standard exponential marginals, such as the densities, m.g.f.'s, and tail probabilities of $\min(X, Y)$, $\max(X, Y)$, and $X + Y$, these being of relevance to series, parallel, and standby systems, respectively. Mukherjee and Sasmal (1977) have compared the densities and means of $\min(X, Y)$ and $\max(X, Y)$ with those of Downton (1970) and Marshall and Olkin (1967) distributions.
- Tolley and Norman (1979) obtained some results relevant to epidemiological applications with the marginals being exponential.
- Lingappaiah (1984) was also concerned with properties of the F-G-M distribution with gamma marginals in the context of reliability.
- Building a paper by Phillips (1981), Kotz and Johnson (1984) considered a model in which components 1 and 2 were subjected to "revealed" and "unrevealed" faults, respectively, with (Y, Z) having an F-G-M distribution, where Y is the time between unrevealed faults and Z is the time from an unrevealed fault to a revealed fault.
- In the context of sample selection, Ray et al. (1980) have presented results for the distributions having logistic marginals, with the copula being the F-G-M or the Pareto.

2.2.1 Applications

- Cook and Johnson (1986) used this distribution (with lognormal marginals) for fitting data on the joint occurrence of certain trace elements in water.
- Halperin et al. (1979) used this distribution, with exponential marginals, as a starting point when considering how a population p.d.f. $h(x, y)$ is altered in the surviving and nonsurviving groups by a risk function $a(x, y)$. (X and Y were blood pressure and cigarette smoking, respectively, in this study.).
- Durling (1974) utilized this distribution with logistic marginals for y, reanalyzing seven previously published datasets on the effects of mixtures of poisons.
- Chinchilli and Breen (1985) used a six-variate version of this distribution with logistic marginals to analyze multivariate binary response data arising in toxicological experiments—specifically, tumor incidence at six different organ sites of mice exposed to one of five dosages of a possible carcinogen [data from Brown and Fears (1981)].
- Thinking now of "lifetimes" in the context of component reliability, Teichmann (1986) used this distribution for (U_1, U_2), with U_i being a measure of association between an external factor and the failure of the ith unit—specifically, it was the ratio of how much the external factor increases the probability of failure compared with how much an always fatal factor would increase the probability of failure.
- With exponential marginals, Lai (1978) used the F-G-M distribution to model the joint distribution of two adjacent intervals in a Markov-dependent point process.
- In the context of hydrology, Long and Krzysztofowicz (1992) also noted that the F-G-M model is limited to describing weak dependence since $|\rho| \leq 1/3$.

2.2.2 Univariate Transformations

The following cases have been considered in the literature: the case of exponential marginals by Gumbel (1960a,b); of normal marginals by Gumbel (1958, 1960b); of logistic marginals by Gumbel (1961, Section 6); of Weibull marginals by Johnson and Kotz (1977) and Lee (1979); of Burr type III marginals by Rodriguez (1980); of gamma marginals by D'Este (1981); of Pareto marginals by Arnold (1983, Section 6.2.5), who cites Conway (1979); of "inverse Rayleigh" marginals (i.e., $F = \exp(-\theta/x^2)$) by Mukherjee and Saran (1984); and of Burr type XII marginals by Bagchi and Samanta (1985).

Drouet-Mari and Kotz (2001, pp. 122–124) have presented a detailed discussion on the bivariate F-G-M distribution with Weibull marginals. Kotz and Van Dorp (2002) have studied the F-G-M family with marginals as a two-sided power distribution.

2.2.3 A Switch-Source Model

For general marginals, the density is $f(x)g(y)\{1+\alpha[1-2F(x)][1-2G(y)]\}$. The density

$$a(x)a(y)[1+\alpha b(x)b(y)] \tag{2.3}$$

arises from a mixture model governed by a Markov process. Imagine a source producing observations from a density f_1, another source producing observations from a density f_2, a switch connecting one or the other of these sources to the output, a Markov process governing the operation of the switch, and X and Y being observations at two points in time; see Willett and Thomas (1985, 1987).

2.2.4 Ordinal Contingency Tables

The nonidentical marginal case of (2.3) is $a(x)b(y)[1+\alpha b(x)d(y)]$. This looks very much like the "rank-2 canonical correlation model" used to describe structure in ordinary contingency tables; see Gilula (1984), Gilula et al. (1988), and Goodman (1986).

Now, instead of generalizing (2.1) and comparing it with contingency table models, we shall explicitly write (2.1) in the contingency form and see what sort of restrictions are effectively being imposed on the parameters of a contingency table model. The probability within a rectangle $\{x_0 < X < x_1, y_0 < Y < y_1\}$ is $H_{11} - H_{01} - H_{10} + H_{00}$ (in an obvious notation), which equals

$$\begin{aligned}(x_1 - x_0)(y_1 - y_0) + \alpha[x_1(1-x_1) - x_0(1-x_0)][y_1(1-y_1) - y_0(1-y_0)] \\ = (x_1 - x_0)(y_1 - y_0)[1 + \alpha(1 - x_1 - x_0)(1 - y_1 - y_0)].\end{aligned}$$

Comparing this with equation (2.2) of Goodman (1986), we see that $(1 - x_1 - x_0)$ and $(1 - y_1 - y_0)$ play the role of row scores and column scores—in effect, Goodman's model U.

2.2.5 Iterated F-G-M Distributions

For the singly iterated case, the distribution function C and p.d.f. c are, respectively, given by

$$C(u,v) = uv[1 + \alpha(1-u)(1-v) + \beta uv(1-u)(1-v)], \tag{2.4}$$

$$c(u,v) = [1 + \alpha(1-2u)(1-2v) + \beta uv(2-3u)(2-3v)], \tag{2.5}$$

where the valid combinations of α and β are $-1 \leq \alpha \leq 1$ and $-1 - \alpha \leq \beta \leq \left(3 - \alpha + \sqrt{9 - 6\alpha - 3\alpha^2}\right)/2$. This distribution is obtained [Johnson and Kotz (1977) and Kotz and Johnson (1977)] by realizing that (2.1) may alternatively be written in terms of the survival function $\bar{C}$ as

$$\bar{C} = (1 - u)(1 - v)(1 + \alpha uv). \tag{2.6}$$

Now replacing the independent survival function $(1 - u)(1 - v)$ in (2.1) by this survival function of an F-G-M distribution, having a possibly different associated parameter, β/α (say) instead of α, we obtain the result in (2.4). This process can be repeated, of course. The correlation coefficient is $\mathrm{corr}(U, V) = \frac{\alpha}{3} + \frac{\beta}{12}$.

Note

For normal marginals, $\mathrm{corr}(X, Y) = \frac{\alpha}{\pi} + \frac{\beta}{4\pi}$. The first iteration increases the maximum attainable correlation to over 0.4. However, very little increase of the maximum correlation is achievable with further iterations, as noted by Kotz and Johnson (1977).

Lin (1987) suggested another way of iterating the F-G-M distribution: Start with (2.6), and replace uv by (2.1). After substituting for $\bar{C}$ in terms of C, we obtain

$$C(u, v) = uv[1 + \alpha(1 - u)(1 - v) + \beta(1 - u)^2(1 - v)^2]$$

at the first step.

Zheng and Klein (1994) studied an iterated F-G-M distribution of the form

$$C(u, v) = uv + \sum_j \alpha_j (uv)^{1/2}[(1 - u)(1 - v)]^{(j+1)/2}, \qquad -1 \leq \alpha_j \leq 1.$$

2.2.6 Extensions of the F-G-M Distribution

We shall discuss here a number of extensions of F-G-M copulas developed primarily to increase the maximal value of the correlation coefficient. Most of these are polynomial-type copulas (copulas that are expressed in terms of polynomials in u and v).

Huang and Kotz Extension

Huang and Kotz (1999) considered

$$C(u,v) = uv[1 + \alpha(1 - u^p)(1 - v^p)]. \tag{2.7}$$

The corresponding p.d.f. is

$$c(u,v) = 1 + \alpha\Big(1 - (1+p)u^p\Big)\Big(1 - (1+p)v^p\Big). \tag{2.8}$$

The admissible range for α is given by

$$-(\max\{1, p^2\})^{-2} \leq \alpha \leq p^{-1}.$$

The range for $\rho = \text{corr}(U,V) = 3\alpha(\frac{p}{p+2})^2$ is

$$-3(p+2)^{-2}\min\{1, p^2\} \leq \rho \leq \frac{3p}{(p+2)^2}.$$

Thus, for $p = 2$, $\rho_{\text{max}} = \frac{3}{8}$, and for $p = 1, \rho_{\text{min}} = \frac{-3}{16}$.

It is clear that the introduction of the parameter p has enabled us to increase the maximal correlation for the F-G-M copula.

Another extension of the bivariate F-G-M copula is given by

$$C(u,v) = uv[1 + \alpha(1-u)^p(1-v)^p], \qquad p > 0, \tag{2.9}$$

with p.d.f.

$$c(u,v) = 1 + \alpha(1-u)^{p-1}(1-v)^{p-1}\Big(1 - (1+p)u\Big)\Big(1 - (1+p)v\Big). \tag{2.10}$$

The admissible range of α is (for $p > 1$)

$$-1 \leq \alpha \leq \Big(\frac{p+1}{p-1}\Big)^{p-1}.$$

The range is empty for $p < 1$. The correlation

$$\rho = \text{corr}(U,V) = 12\alpha\Big(\frac{1}{(p+1)(p+2)}\Big)^2$$

in this case has the range

$$-12\Big(\frac{1}{(p+1)(p+2)}\Big)^2 \leq \rho \leq 12\frac{(p-1)^{1-p}(p+1)^{p-3}}{(p+2)^2}.$$

Thus, for $p = 1.877$, $\rho_{\text{max}} = 0.3912$ and $\rho_{\text{min}} = \frac{-1}{3}$, showing that the maximal correlation is even higher than the one attained by the first extension in (2.7).

Sarmanov's Extension

Sarmanov (1974) considered the following copula:

$$C(u,v) = uv\left\{1 + 3\alpha(1-u)(1-v) + 5\alpha^2(1-u)(1-2u)(1-v)(1-2v)\right\}. \tag{2.11}$$

The corresponding density function is

$$c(u,v) = 1 + 3\alpha(2u-1)(2v-1) + \frac{5}{4}\alpha^2[3(2u-1)^2 - 1][3(2v-1)^2 - 1].$$

Equation (2.11) is a probability distribution when $|\alpha| \le \frac{\sqrt{7}}{5} \simeq 0.55$.

Bairamov–Kotz Extension

Bairamov and Kotz (2000a) considered a two-parameter extension of the F-G-M copula given by

$$C(u,v) = uv[1 + \alpha(1-u^a)^b(1-v^a)^b], \quad a > 0, b > 0, \tag{2.12}$$

with the corresponding p.d.f.

$$c(u,v) = 1 + \alpha(1-x^a)^{b-1}(1-v^a)^{b-1}[1-u^a(1+ab)][1-v^a(1+ab)]. \tag{2.13}$$

The admissible range of α is as follows: For $b > 1$,

$$-\min\left\{1, \left[\frac{1}{a^b}\left(\frac{ab+1}{b-1}\right)^{b-1}\right]^2\right\} \le \alpha \le \left[\frac{1}{a^b}\left(\frac{ab+1}{b-1}\right)^{b-1}\right],$$

and for $b = 1$, the quantity inside the square bracket is taken to be 1. It can be shown in this case that $\mathrm{corr}(U,V) = 12\alpha\left[\frac{b}{ab+2}\frac{\Gamma(b)\Gamma(\alpha/2)}{\Gamma(b+\frac{2}{a})}\right]^2$. For $a = 2.8968$ and $b = 1.4908$, we have $\rho_{\max} = 0.5015$. For $a = 2$ and $b = 1.5$, $\rho_{\min} = -0.48$.

Another extension that does not give rise to a copula is

$$C(u,v) = u^p v^p[1 + \alpha(1-u^q)^n(1-v^q)^n], \quad p, q \ge 0,\ n > 1, \tag{2.14}$$

with marginals u^p and v^p, respectively.

Lai and Xie Extension

Lai and Xie (2000) considered the copula

$$C(u,v) = uv + \alpha u^b v^b (1-u)^a (1-v)^a, \quad a, b \ge 1, \tag{2.15}$$

and showed that it is PQD for $0 \leq \alpha \leq 1$. The corresponding p.d.f. is

$$c(u,v) = 1 + \alpha(uv)^{b-1}[(1-u)(1-v)]^{a-1}[b-(a+b)u][b-(a+b)v]. \quad (2.16)$$

The correlation coefficient is given by $\text{corr}(U,V) = 12\alpha[B(b+1,a+1)]^2$. Bairamov and Kotz (2000b) observed that (2.15) is a bivariate copula for α over a wider range satisfying

$$\min\left\{\frac{1}{[B^+(a,b)]^2}, \frac{1}{[B^-(a,b)]^2}\right\} \leq \alpha \leq \frac{1}{B^+(a,b)B^-(a,b)},$$

where B^+ and B^- are functions of a and b.

Bairamov–Kotz–Bekci Generalization

Bairamov et al. (2001) presented a four-parameter extension of the F-G-M copula as

$$C(u,v) = uv\Big\{1+\alpha(1-u^{p_1})^{q_1}(1-v^{p_2})^{q_2}\Big\}, \qquad p_1,p_2 \geq 1,\ q_1,q_2 \geq 1. \quad (2.17)$$

2.2.7 Other Related Distributions

- Farlie (1960) introduced the more general expression
$$H(x,y) = F(x)G(y)\{1+\alpha A[F(x)]B[G(y)]\}.$$
- Rodriguez (1980, p. 48), in the context of Burr type III marginals, made passing references to $H = FG[1+\alpha(1-F^a)(1-G^b)]$.
- Cook and Johnson (1986) discussed a compound F-G-M distribution.
- Regarding a distribution obtained by a Khintchine mixture using the F-G-M distribution as the bivariate F-G-M copula, see Johnson (1987, pp. 157–159).
- Cambanis (1977) has mentioned $C(u,v) = uv[1+\beta(1-u)+\beta(1-v)+\alpha(1-u)(1-v)]$, which arises as the conditional distribution in a multivariate F-G-M distribution.
- The following distribution was denoted u_8 in Kimeldorf and Sampson (1975b):
$$C(u,v) = uv[1+\alpha(1-u)(1-v)+\beta(1-u^2)(1-v^2)], \quad (2.18)$$
$$c(u,v) = 1+\alpha(1-2u)(1-2v)+\beta(1-3u^2)(1-3v^2), \quad (2.19)$$

with correlations $\tau = \frac{2\alpha}{9} + \frac{\beta}{2} + \frac{\alpha\beta}{450}$ and $\rho_S = \frac{\alpha}{3} + \frac{3\beta}{4}$.

2.3 Ali–Mikhail–Haq Distribution

$$C(u, v) = \frac{uv}{1 - \alpha(1-u)(1-v)} \tag{2.20}$$

and

$$c(u, v) = \frac{1 - \alpha + 2\alpha\frac{uv}{1-\alpha(1-u)(1-v)}}{[1 - \alpha(1-u)(1-v)]^2}. \tag{2.21}$$

Correlation Coefficients

The range of product-moment correlation is (−0.271, 0.478) for uniform marginals, (−0.227, 0.290) for exponential marginals, and approximately (−0.300, 0.600) for normal marginals; see Johnson (1987, pp. 202–203), crediting these results to Conway (1979).

Derivation

This distribution was introduced by Ali et al. (1978). They proposed searching for copulas for which the survival odds ratio satisfies

$$\frac{1 - C_\alpha(u, v)}{C_\alpha(u, v)} = \frac{1-u}{u} + \frac{1-v}{v} + (1-\alpha)\frac{1-u}{u} \times \frac{1-v}{v}.$$

Solving $C_\alpha(u, v)$ yields the Ali–Mikhail–Haq family given in (2.20).

Remarks

- This distribution is an example of an Archimedean copula:

$$\log\left[\frac{1 + \alpha(C-1)}{C}\right] = \log\left[\frac{1 + \alpha(u-1)}{u}\right] + \log\left[\frac{1 + \alpha(v-1)}{v}\right];$$

  i.e., the generator is $\varphi = \log\frac{1+\alpha(u-1)}{u}$.
- The distribution may be written as

$$C(u, v) = uv[1 + \alpha(1-u)(1-v)] + \sum_{i=2}^{\infty} \alpha^i (1-u)^i (1-v)^i,$$

with the first term being the F-G-M copula.

- Ali et al. (1978) showed that the copula is PQD, LTD, and PRD.
- Mikhail et al. (1987a) presented some further results, including the (mean) regression curves when the marginals are logistic. They also corrected errors in the calculations of the median regression by Ali et al. (1978).

Genest and MacKay (1986) showed that

$$\tau = \frac{3\alpha - 2}{3\alpha} - \frac{2(1-\alpha)^2}{3\alpha^2}\log(1-\alpha).$$

To obtain ρ_S, the second integration requires finding $\int_0^1 (1-u)^{-1}\log(1-\alpha+\alpha u)du$. By substituting $x = \alpha(1-u)$, it becomes $\int_0^\alpha x^{-1}\log(1-x)dx$, which is $\mathrm{diln}(1-\alpha)$, diln being the dilogarithm function.

The final expression for ρ_S is then

$$\rho_S = -\frac{12(1+\alpha)}{\alpha^2}\mathrm{diln}(1-\alpha) - \frac{3(12+\alpha)}{\alpha} - \frac{24(1-\alpha)}{\alpha^2}\log(1-\alpha).$$

2.3.1 Bivariate Logistic Distributions

A bivariate distribution that corresponds to (2.20),

$$C(u,v) = \frac{uv}{1-\alpha(1-u)(1-v)}$$

is

$$H(x,y) = \left[1 + e^{-x} + e^{-y} + (1-\alpha)e^{-x-y}\right]^{-1}, \quad -1 \le \alpha \le 1, \qquad (2.22)$$

[Ali et al. (1978)].

Properties

- The marginals are standard logistic distributions.
- When $\alpha = 0$, X and Y are independent.
- When $\alpha = 1$, we have Gumbel's bivariate logistic distribution discussed in Section 11.17:
$$H(x,y) = \left(1 + e^{-x} + e^{-y}\right)^{-1}.$$
- Gumbel's logistic lacks a parameter which limits its usefulness in applications. The generalized bivariate logistic (2.22) makes up for this lack.

2.3.2 Bivariate Exponential Distribution

The copula in (2.20) with $\alpha = 1$ also corresponds to the survival copula of a bivariate exponential distribution whose survival function is given by

$$\bar{H}(x,y) = \left(e^x + e^y - 1\right)^{-1}.$$

Clearly, X and Y are standard exponential random variables.

2.4 Frank's Distribution

$$C(u,v) = \log_\alpha \left[1 + \frac{(\alpha^u - 1)(\alpha^v - 1)}{\alpha - 1}\right] \tag{2.23}$$

and

$$c(u,v) = \frac{(\alpha - 1)\log_\alpha \alpha^{u+v}}{[\alpha - 1 + (\alpha^u - 1)(\alpha^v - 1)]^2}. \tag{2.24}$$

Correlation and Dependence

(i) For $0 < \alpha < 1$, we have (positive) association.
(ii) As $\alpha \to 1$, we have independence.
(iii) For $\alpha > 1$, we have negative association.

Nelsen (1986) has given an expression for Blomqvist's medial correlation coefficient. Nelsen (1986) and Genest (1987) have shown that

$$\tau = 1 + 4[D_1(\alpha^*) - 1]/\alpha,$$

$$\rho_S = 1 + 12[D_2(\alpha^*) - D_1(\alpha^*)]/\alpha^*,$$

where $\alpha^* = -\log(\alpha)$ and D_1 and D_2 are Debye functions defined by

$$D_k(\beta) = \frac{k}{\beta^k}\int_0^\beta \frac{t^k}{e^t - 1}dt.$$

Derivation

This is the distribution such that both C and $\hat{C} = u + v - C$ are associative, meaning $C[C(u,v),w] = C[u, C(v,w)]$ and similarly for $\hat{C}$ [Frank (1979)].

There does not seem to be a probabilistic interpretation of this associative property.

Remarks

- This distribution is an example of an Archimedean copula [Genest and MacKay (1986)],
$$\log\left(\frac{1-\alpha^C}{1-\alpha}\right)=\log\left(\frac{1-\alpha^u}{1-\alpha}\right)+\log\left(\frac{1-\alpha^v}{1-\alpha}\right),$$
so that $\varphi(t)=\log(\frac{1-\alpha^t}{1-\alpha})$.
- The p.d.f. is symmetric about $(\frac{1}{2},\frac{1}{2})$, and consequently the copula and the survival (complementary) copula are the same. In fact, this family is the only copula that satisfies the functional equation $\hat{C}(u,v)=C(u,v)$.
- When $0<\alpha<1$, this distribution is positive likelihood ratio dependent [Genest (1987)].
- This distribution has the "monotone regression dependence" property [Bilodeau (1989)].

2.5 Distribution of Cuadras and Augé and Its Generalization

This distribution, put forward by Cuadras and Augé (1981), is given by

$$C(u,v)=uv[\max(u,v)]^{-c}=uv[\min(u^{-c},v^{-c})], \tag{2.25}$$

with c being between 0 and 1. It is usually met with identical exponential marginals in the form of Marshall and Olkin given by

$$\bar{H}(x,y)=\exp(-\lambda x-\lambda y-\lambda_{12}\max(x,y)).$$

2.5.1 Generalized Cuadras and Augé Family (Marshall and Olkin's Family)

The Marshall and Olkin bivariate exponential distribution in the original form is
$$\bar{H}(x,y)=\exp\big(-\lambda_1 x-\lambda_2 y-\lambda_{12}\max(x,y)\big).$$

Nelsen (2006, p. 53) considered the uniform representation of the survival function above. In order to obtain it, we rewrite the preceding equation in the form

$$\begin{aligned}\bar{H}(x,y) &= \exp(-(\lambda_1+\lambda_{12})x-(\lambda_2+\lambda_{12})y+\lambda_{12}\min(x,y)) \\ &= \bar{F}(x)\bar{G}(y)\min\{\exp(\lambda_{12}x),\exp(\lambda_{12}y)\}.\end{aligned} \quad (2.26)$$

Set $u=\bar{F}(x)$ and $v=\bar{G}(y)$, and let $\alpha=\frac{\lambda_{12}}{(\lambda_1+\lambda_{12})}$, and $\beta=\frac{\lambda_{12}}{(\lambda_2+\lambda_{12})}$. Then, $\exp(\lambda_{12}x)=u^{-\alpha}$ and $\exp(\lambda_{12}y)=v^{-\beta}$, with the survival copula (complementary copula) $\hat{C}$ given by

$$\hat{C}(u,v)=uv\min(u^{-\alpha},v^{-\beta})=\min(uv^{1-\beta},u^{1-\alpha}v). \quad (2.27)$$

Since the λ's are all positive, it follows that α and β satisfy $0<\alpha,\beta<1$. Hence, the survival copula for the Marshall and Olkin bivariate exponential distribution yields a two-parameter family of copulas given by

$$C_{\alpha,\beta}(u,v)=\min(u^{1-\alpha},v^{1-\beta})=\begin{cases}u^{1-\alpha}v, & u^\alpha\geq v^\beta\\ uv^{1-\beta}, & u^\alpha\leq v^\beta\end{cases}. \quad (2.28)$$

This family is known as the *Marshall and Olkin family* and the *generalized Cuadras and Augé family*. When $\alpha=\beta=c$, (2.28) reduces to the Cuadras and Augé family in (2.25). Hanagal (1996) studied the distribution above with Pareto distributions of the first kind as marginals.

A slight complicating factor with this is that the p.d.f. has a singularity along $y=x$. For $\alpha=\beta=c$, Cuadras and Augé determined Pearson's correlation to be $3c/(4-c)$. Since the marginals are uniform, ρ_S is the same value. It may also be shown that $\tau=c/(2-c)$, and so $\rho_S=3\tau/(2+\tau)$.

Nelsen (2006, Chapter 5) showed that $\rho_S=3\tau/(2+\tau)$ also holds for the asymmetric case $H=\min(xy^{1-\beta},x^{1-\alpha}y)$, but $\tau=\frac{\alpha\beta}{\alpha-\alpha\beta+\beta}$.

2.6 Gumbel–Hougaard Copula

The copula satisfies the equation

$$[-\log C(u,v)]^\alpha=(-\log u)^\alpha+(-\log v)^\alpha. \quad (2.29)$$

Rewriting it in a different form gives

$$C(u,v)=\exp\left(-\left[(-\log u)^\alpha+(-\log v)^\alpha\right]^{1/\alpha}\right). \quad (2.30)$$

Letting $-\log u=e^{-x}$, $-\log v=e^{-y}$ in (2.30), we can verify that the joint distribution of X and Y is

$$H(x,y) = \exp\left[- \left(e^{-\alpha x} + e^{-\alpha y} \right)^{1/\alpha} \right], \tag{2.31}$$

which is the type B bivariate extreme-value distribution with type 1 extreme-value marginals, see Kotz et al. (2000, p. 628) and Nelsen (2006, p. 28).

Correlation Coefficient

Kendall's τ is $(\alpha-1)/\alpha$ [Genest and MacKay (1986)]. The correlation between $\log U$ and $\log V$ is $1-\alpha^2$.

Derivation

Perhaps surprisingly, the survival copula corresponding to (2.30) can be derived by compounding [Hougaard (1986)].

Suppose there are two independent components having failure rate functions given by $\theta\lambda(x)$ and $\theta\lambda(y)$. Then the joint survival probability is $e^{-\theta[\Lambda(x)+\Lambda(y)]}$. Now assuming θ has a stable distribution with the Laplace transform $E(e^{-\theta s}) = e^{-s^\gamma}$, then $E(e^{-\theta[\Lambda(x)+\Lambda(y)]}) = e^{-[\Lambda(x)+\Lambda(y)]^\gamma}$. Finally, we might suppose that $\lambda(u)$ is of the Weibull form $\varepsilon\alpha u^{\alpha-1}$, in which case $\Lambda(t) = \varepsilon t^\alpha$, so that

$$\bar{H}(x,y) = \exp[-(\varepsilon x^\alpha + \varepsilon y^\alpha)^\gamma], \quad x,y>0. \tag{2.32}$$

Set $\gamma = 1/\alpha$, and it follows that

$$\bar{H}(x,y) = \exp[-(\varepsilon x^\alpha + \varepsilon y^\alpha)^{1/\alpha}], \quad x,y>0.$$

Clearly, $\bar{H}(x,y) = C(\bar{F}(x), \bar{G}(y))$ where C is the Gumbel–Hougaard copula and $\bar{F}(x) = e^{-\varepsilon^{1/\alpha}x}$ and $\bar{G}(y) = e^{-\varepsilon^{1/\alpha}y}$.

It now follows from (1.4) that the Gumbel–Hougaard copula is the survival copula of the bivariate exponential distribution given by (2.31).

The Pareto distribution is obtained in a similar manner, but with θ having a gamma distribution. Hougaard (1986, p. 676) has mentioned the possibility of using a distribution that subsumes both gamma and positive stable distributions in order to arrive at a bivariate distribution that subsumes both the Gumbel–Hougaard and Pareto copulas.

Independently, Crowder (1989) had the same idea but added a new wrinkle to it. His distribution, in the bivariate form, is

$$\bar{H}(x,y) = \exp[\kappa^\alpha - (\kappa + \varepsilon x^\gamma + \varepsilon y^\gamma)^\alpha], \tag{2.33}$$

where we see an extra parameter κ; also, note that ε's and γ's are allowed to be different for X and Y. An interpretation of κ is in terms of selection based on $Z > z_0$ from a population having trivariate survival distribution $\exp[-(\varepsilon x^\gamma + \varepsilon y^\gamma + \varepsilon z^\gamma)^\alpha]$. Crowder has discussed further the dependence and association properties, hazard functions and failure rates, the marginal distributions, the density functions, the distribution of minima, and the fitting of the model to data.

Remarks

- We have called this a Gumbel–Hougaard copula since it appeared in the works of Gumbel (1960a, 1961) and a derivation of it has been given by Hougaard (1986).
- Clearly, from the form of (2.29), it is an Archimedean copula [Genest and MacKay (1986)].

Fields of Applications

- Gumbel and Mustafi (1967) fitted this distribution, in the extreme value form, to data on the sizes of annual floods (1918–1950) of the Fox River (Wisconsin) at two points.
- Hougaard (1986) used a trivariate version of this distribution to analyze data on tumor appearance in rats with 50 liters of a drug treated and two control animals.
- Hougaard (1986) analyzed insulation failure data using a trivariate form of the Weibull version of this distribution.
- Crowder (1989) fitted (2.33) to data on the sensitivity of rats to tactile stimulation of rats that did or did not receive an analgesic drug.

2.7 Plackett's Distribution

The distribution function is derived from the functional equation

$$\frac{C(1-u-v+C)}{(u-C)(v-C)} = \psi. \tag{2.34}$$

The equation above can be interpreted as (having the support divided into four regions by dichotomizing U and V)

$$\frac{\text{Probability in lower-left region} \times \text{Probability in upper-right region}}{\text{Probability in upper-left region} \times \text{Probability in lower-right region}}$$
$$= \text{a constant}$$

independent of where the variates are dichotomized. Expressed alternatively,

$$C = \frac{[1 + (\psi - 1)(u + v)] - \sqrt{[1 + (\psi - 1)(u + v)]^2 - 4\psi(\psi - 1)uv}}{2(\psi - 1)}. \quad (2.35)$$

It needs to be noted that the other root is not a proper distribution function, not falling within the Fréchet bounds.

The probability density function is

$$c = \frac{\psi[(\psi - 1)(u + -2uv) + 1]}{\{[1 + (\psi_1)(u + v)]^2 - 4\psi(\psi - 1)uv\}^{3/2}}.$$

Correlation Coefficient

Spearman's correlation is $\rho_S = \frac{\psi+1}{\psi-1} - \frac{2\psi}{(\psi-1)^2} \log \psi$. Kendall's τ does not seem to be known as a function of ψ. For the product-moment correlation when the marginals are normal, see Mardia (1967).

Conditional Properties

The regression of V on U is linear. After the marginals have been transformed to be normal, the conditional densities are skew and the regression of Y on X is nonlinear [Pearson (1913)].

Remarks

- Interest in this distribution was stimulated by the papers of Plackett (1965) and Mardia (1967), but in fact it can be traced in the contingency table literature back to the days of Yule and Karl Pearson [see Goodman (1981)].
- As compared with the bivariate normal distribution, the outer contours of the p.d.f. of this distribution with normal marginals are more nearly circular—their ellipticity is less than that of the inner ones [Pearson (1913) and Anscombe (1981, pp. 306–310)].
- For low correlation, this distribution is equivalent to the F-G-M in the sense that, if we set $\psi = 1 - \alpha$ in (2.33), expand in terms of α, and then let α be small so that we can neglect α^2 and higher terms, we arrive at (2.1).
- Arnold (1983, Section 6.2.5) has made brief mention of the Pareto-marginals version of this distribution, citing Conway (1979).
- Another account of this distribution is by Conway (1986).

Fields of Application

- This distribution has received considerable attention in the contingency table literature, where it is known as the *constant global cross ratio model.* Suppose one has a square table of frequencies, the categories of the dimensions being ordinals. Then, if the model of independence fails and a degree of positive (or negative) association is evident, one model that has a single degree of freedom to describe the association is the bivariate normal. But this is inconvenient to handle computationally with most of the present-day packages for modeling tables of frequencies. Another model consisting of a single association model is Plackett's distribution, which is much easier computationally. Work in this direction has been carried out by Mardia (1970a, Example 8.1), Wahrendorf (1980), Anscombe (1981, Chapter 12), Goodman (1981), and Dale (1983, 1984, 1985, 1986).
- In the context of bivariate probit models, Amemiya (1985, p. 319) has mentioned that Lee (1982) applied Plackett's distribution with logistic marginals to the data of Ashford and Sowden (1970) and Morimune (1979).
- Mardia (1970b) fitted the S_U-marginals version of this distribution to Johansen's bean data.

2.8 Bivariate Lomax Distribution

The joint survival function of the bivariate Lomax distribution (Durling–Pareto distribution) is given by

$$\bar{H}(x,y) = (1+ax+by+\theta xy)^{-c}, \quad 0 \le \theta \le (c+1)ab, \; a,b,c>0, \tag{2.36}$$

with probability density function

$$h(x,y) = \frac{c[c(b+\theta x)(a+\theta y)+ab-\theta]}{(1+ax+by+\theta xy)^{c+2}}. \tag{2.37}$$

Marginal Properties

It has Lomax (Pareto of the second kind) marginals with

$$E[X] = \frac{1}{a(c-1)}, \quad E[Y] = \frac{1}{b(c-1)}, \quad c>1$$

(the mean exists only if $c > 1$) and

$$\text{var}(X) = \frac{c}{(c-1)^2(c-2)a^2}, \quad \text{var}(Y) = \frac{c}{(c-1)^2(c-2)b^2}, \quad c > 2$$

(the variance exists only if $c > 2$).

Derivations

- Begin with two exponential random variables X and Y with parameters θ_1 and θ_2, respectively. Conditional on (θ_1, θ_2), X and Y are independent. We now assume that $\theta_1, \theta_2)$ has Kibble's bivariate gamma distribution with density $h(\theta_1, \theta_2)$ (see Section 8.2). Then

$$\Pr(X > x, Y > y) = \int_0^\infty \int_0^\infty \exp(-\theta_1 x, \theta_2 y) h(\theta_1, \theta_2) d\theta_1 \theta_2$$

  will have the same form as (2.36).
- Begin with Gumbel's bivariate distribution of the type

$$\bar{F}(x, y) = \exp\Big(- \eta(\alpha x + \beta y + \lambda xy)\Big).$$

  Assuming that η has a gamma distribution with scale parameter m and shape parameter c, then (2.36) will be obtained by letting $a = \alpha/m, b = \beta/m$, and $\theta = \lambda/m$; see Sankaran and Nair (1993).

Properties of Bivariate Dependence

Lai et al. (2001) established the following properties:

- For the bivariate Lomax survival function, X and Y are positively (negatively) quadrant dependent if $0 \le \theta \le ab$ ($ab < \theta \le (c+1)ab$).
- The Lomax distribution is RTI if $\theta \le ab$ and RTD if $\theta \ge ab$.
- X and Y are associated if $\theta \le ab$.

Correlation Coefficients

- Lai et al. (2001) have shown that

$$\rho = \frac{(1-\theta)(c-2)}{c^2} F\left[1, 2; c+1; (1-\theta)\right], \; 0 \le \theta \le (c+1), \; a = b = 1,$$

  where $F(a, b; c; z)$ is Gauss' hypergeometric function; see, for example, Chapter 15 of Abramowitz and Stegun (1964).
- For $a \ne 1, b \ne 1$, the correlation is

$$\rho = \frac{(ab-\theta)(c-2)}{abc^2} F\left[1,2;c+1;(1-\theta/ab)\right], \quad 0 \le \theta \le (c+1)ab.$$

- For $c = n$ an integer and $ab = 1$,

$$\begin{aligned}
&\mathrm{corr}(X,Y) \\
&= \frac{\frac{\theta^{n-2}}{(n-1)(\theta-1)^{n-1}}\log\theta - \sum_{i=2}^{n-1}\frac{\theta^{n-1-i}}{n(i-1)(\theta-1)^{n-i}} - \frac{1}{(n-1)^2}}{\frac{n}{(n-1)^2(n-2)}} \\
&= \frac{\frac{\theta^{n-2}}{(\theta-1)^{n-1}}\log\theta - \sum_{i=2}^{n-1}\frac{\theta^{n-1-i}}{(i-1)(\theta-1)^{n-i}} - \frac{1}{n-1}}{\frac{n}{(n-1)(n-2)}}, \qquad n \ge 3.
\end{aligned}$$

(i) For $c = n = 2$, and $ab = 1$, in particular,

$$\mathrm{cov}(X,Y) = \frac{\log\theta}{\theta-1} - 1.$$

Thus, the covariance exists for $c = 2$ even though the correlation does not exist since the marginal variance does not exist for $c = 2$.
(ii) For $c = n = 3$, and $ab = 1$, in particular,

$$\rho = \mathrm{corr}(X,Y) = \left[\frac{2}{3(\theta-1)^2}\theta\log\theta - \frac{2}{3(\theta-1)} - \frac{1}{3}\right].$$

- For a given c and $ab = 1$, the correlation ρ decreases as θ increases. However, it does not decrease uniformly over c.
- For a given c and $ab = 1$, the value of ρ lies in the interval

$$-\frac{(c-2)}{c}F(1,2;c+1;-c) \le \rho \le 1/c.$$

Thus, the admissible range for ρ is $(-0.403, 0.5)$.
- This reasonably wide admissible range compares well with the well-known Farlie–Gumbel–Morgenstern bivariate distribution having the ranges of correlation (i) $-\frac{1}{3}$ to $\frac{1}{3}$ for uniform marginals, (ii) $-\frac{1}{4}$ to $\frac{1}{4}$ for exponential marginals, and (iii) $-\frac{1}{\pi}$ to $\frac{1}{\pi}$ for normal marginals, as mentioned earlier.

Remarks

- In order to have a well-defined bivariate Lomax distribution, we need to restrict ourselves to the case $c > 2$ so that the second moments exist.
- The bivariate Lomax distribution is also known as the Durling-Pareto distribution.
- Durling (1975) actually proposed an extra term in the Takahasi–Burr distribution rather than in the simpler Pareto form. Some properties of Durling's distribution were established by Bagchi and Samanta (1985).

- Durling has given the (product-moment) correlation coefficient for the general case in which x and y are each raised to some power.
- An application of this distribution in the special case where $c = 1$, considered in the literature, is in modeling the severity of injuries to vehicle drivers in head-on collisions between two vehicles of equal mass.
- Several reliability properties have been discussed by Sankaran and Nair (1993). Lai et al. (2001) have discussed some additional properties pertaining to reliability analysis.
- Rodriguez (1980) introduced a similar term into the bivariate Burr type III distribution, resulting in $H = (1+x^{-a}+y^{-b}+kx^{-a}y^{-b})^{-c}$. He included a number of plots of probability density surfaces of this distribution in the report. This distribution (with location and scale parameters present) was used by Rodriguez and Taniguchi (1980) to describe the joint distribution of customers' and expert raters' assessments of octane requirements of cars.
- The special case

$$\bar{H}(x,y) = \frac{1}{(1+ax+by)^c}, \quad c > 0, \tag{2.38}$$

  is also known as the bivariate Pareto and has been studied in detail by several authors, including Lindley and Singpurwalla (1986).
- Sums, products, and ratios for the special case given in (2.38) are derived in Nadarajah (2005).
- Shoukri et al. (2005) studied inference procedures for $\gamma = \Pr(Y < X)$ of the special case above; in particular, the properties of the maximum likelihood estimate $\hat{\gamma}$ are derived.

2.8.1 The Special Case of c = 1

Suppose now that we have a number of 2×2 contingency tables, each of which corresponds to some particular x and some particular y, and we want to fit the distribution $\bar{H} = (1+ax+by+kabxy)^{-1}$ to them. Notice that the parameter θ depends on a and b. This special case with $c = 1$ is very convenient in these circumstances because we have $p_{11} = (1+ax+by+kabxy)^{-1}$, $p_{10}+p_{11} = (1+ax)^{-1}$, and $p_{01}+p_{11} = (1+by)^{-1}$. We can then estimate a and b from the marginals by

$$\frac{1-(p_{10}+p_{11})}{p_{10}+p_{11}} = ax \quad \text{and} \quad \frac{1-(p_{01}+p_{11})}{p_{01}+p_{11}} = by,$$

and k can be estimated by

$$\frac{\frac{1-p_{11}}{p_{11}} - \frac{1-(p_{10}+p_{11})}{p_{10}+p_{11}} - \frac{1-(p_{01}+p_{11})}{p_{01}+p_{11}}}{\left[\frac{1-(p_{19}+p_{11})}{p_{10}+p_{11}}\right] + \left[\frac{1-(p_{01}+p_{11})}{p_{01}+p_{11}}\right]}.$$

Applications of this distribution in transformed form have been discussed by Morimune (1979) and Amemiya (1975).

2.8.2 Bivariate Pareto Distribution

In this case, we have

$$\bar{H}(x,y) = (1+x+y)^{-c}. \tag{2.39}$$

The marginal is known as the *Pareto distribution of the second kind* (sometimes the Lomax distribution). The p.d.f. is

$$h(x,y) = c(c+1)(1+x+y)^{-(c+2)}. \tag{2.40}$$

Correlation Coefficients and Conditional Properties

Pearson's product-moment correlation is $1/c$ for $c > 2$. The regression of Y on X is linear, $E(Y|X=x) = (x+1)/c$, and the conditional variance is quadratic, $\text{var}(Y|X=x) = \frac{c+1}{(c-1)c^2}(x+1)^2$ for $c > 1$. In fact, $Y|X=x$ is also Pareto.

Derivation

Starting with X and Y having independent exponential distributions with the same scale parameter and then taking the scale parameter to have a gamma distribution, this distribution is obtained by compounding. More generally, starting with $\Pr(X > x) = [1 - A(x)]^\theta$ and $\Pr(Y > y) = [1 - B(y)]^\theta$, where A and B are distribution functions, and then taking θ to have a gamma distribution, the distribution (2.39) is obtained by compounding, with the only difference being that monotone transformations have been applied to X and Y.

If compounding of the scale parameter is applied to an F-G-M distribution that has exponential marginals instead of an independent distribution with exponential marginals, a distribution proposed and used by Cook and Johnson (1986) results.

Remarks

- Barnett (1979, 1983b) has considered testing for the presence of an outlier in a dataset assumed to come from this distribution; see also Barnett and Lewis (1984, Section 9.3.3). An alternative proposal given by Barnett (1983a) involves transformations to independent normal variates.
- The bivariate failure rate is decreasing [Nayak (1987)].
- The product moment is $E(X^r Y^s) = \Gamma(c-r-s)\Gamma(r+1)\Gamma(s+1)/\Gamma(c)$ if $r+s<c$ and ∞ otherwise.
- Mardia (1962) wrote this distribution in the form $h \propto (bx+ay-ab)^{-(c+2)}$, with $x>a>0$, $y>b>0$. In this case, Malik and Trudel (1985) have derived the distributions of XY and X/Y.

Univariate Transformation

In the bivariate Burr type XII (Takahasi–Burr) distribution, x and y in the distribution function are replaced by their powers; see Takahasi (1965). Further results, oriented toward the repeated measurements experimental paradigm, for this case have been given by Crowder (1985). For generation of random variates following the method of the distribution's derivation (scale mixture), see Devroye (1986, pp. 557–558). Arnold (1983, p. 249) has referred to this as a type IV Pareto distribution.

Rodriguez (1980) has discussed the bivariate Burr distribution, $H(x,y) = (1+x^{-a}+y^{-b})^{-c}$. In that report, there is a derivation (by compounding an extreme-value distribution with a gamma), algebraic expressions for the conditional density, conditional distributions, conditional moments, and correlation, and a number of illustrations of probability density surfaces. Satterthwaite and Hutchinson (1978) replaced x and y in the distribution function by e^{-x} and e^{-y}. Gumbel (1961) had previously done this in the special case $c=1$, thus getting a distribution whose marginals are logistic; however, it lacks an association parameter.

Cook and Johnson (1981) and Johnson (1987, Chapter 9) have treated this copula [whether in Takahasi (1965) form, or Satterthwaite–Hutchinson (1978) form] systematically and have also provided several plots of densities. Cook and Johnson (1986) and Johnson (1987, Section 9.2) have generalized the distribution further.

2.9 Lomax Copula

Consider the bivariate Lomax distribution with the survival function given by (2.36). As $\bar{H}(x,y) = \hat{C}(\bar{F}(x), \bar{G}(x))$, we observe that (2.36) can be obtained from the survival copula

$$\hat{C}(u,v) = uv\left\{1 - \alpha(1-u^{\frac{1}{c}})(1-v^{\frac{1}{c}})\right\}^{-c}, \qquad -c \le \alpha \le 1, \tag{2.41}$$

by taking $\alpha = 1 - \frac{\theta}{ab}$. Recall that the survival function of C is related to the survival copula through $\bar{C}(u,v) = 1 - u - v + C(u,v) = \hat{C}(1-u, 1-v)$, and so the copula that corresponds to (2.41) is

$$C(u,v) = \frac{(1-u)(1-v)}{\left\{1 - \alpha[1-(1-u)^{\frac{1}{c}}][1-(1-v)^{\frac{1}{c}}]\right\}^{\frac{1}{c}}} + u + v - 1. \tag{2.42}$$

- Case $\theta = 0$ ($\alpha = 1$), so $\hat{C}(u,v) = (u^{-1/c} + v^{-1/c} - 1)^c$ is known as Clayton's copula.
- The case $\alpha = 0$ (i.e., $\theta = ab$) corresponds to the case of independence. Fang et al. (2000) have also shown that U and V are also independent as $c \to \infty$.
- When $c = 1$, the survival copula (2.41) becomes

  $$\hat{C}(u,v) = \frac{uv}{1 - \alpha(1-u)(1-v)}, \qquad -1 < \alpha < 1,$$

  which is nothing but the Ali–Mikhail–Haq family of an Archimedean copula with generator $\log \frac{1-\alpha(1-t)}{t}$. Thus, the survival copula in (2.41) can be considered to be a generalization of the Ali–Mikhail–Haq family.
- Fang et al. (2000) have shown that the correlation coefficient of the copula is

  $$\rho = 3\{{}_3F_2(1, 1, c : 1 + 2c, 1 + 2c; \alpha) - 1\}, \qquad 0 \le \alpha \le 1,\ c > 0,$$

  where

  $${}_3F_2(a, b, c; d, e; x) = \sum_{k=0}^{\infty} \frac{(a)_k (b)_k (c)_k}{(d)_k (e)_k} \frac{x^k}{k!}.$$

- It is noted in Fang et al. (2000) that the copula is LRD if $\alpha > 0$.
- For $\alpha = 1$, the survival copula is also known as the Pareto copula, which is discussed next.

2.9.1 Pareto Copula (Clayton Copula)

$$\hat{C}(u,v) = (u^{-1/c} + v^{-1/c} - 1)^{-c}, \qquad c > 0. \tag{2.43}$$

This is the survival copula that corresponds to the bivariate Pareto distribution in (2.39). This is not symmetric about $(\frac{1}{2}, \frac{1}{2})$. Equation (2.43) is also called the Clayton copula by Genest and Rivest (1993). Clearly, this is a special case of the Lomax copula in (2.41).

The survival function of the copula that corresponds to the bivariate Pareto distribution in (2.43) is given by

$$\bar{C}(u,v) = [(1-u)^{-1/c} + (1-v)^{-1/c} - 1]^{-c}, \quad c > 0,$$

which has been discussed by many authors, including Oakes (1982, 1986) and Cox and Oakes (1984, Section 10.3). Note that the copula that corresponds to the bivariate Pareto distribution is given by

$$C(u,v) = [(1-u)^{-1/c} + (1-v)^{-1/c} - 1]^{-c} + u + v - 1.$$

Remarks

- Johnson (1987, Section 9.1) has given a detailed account of this distribution and has paid more attention to the marginals than we have done here. Johnson has referred to this as the Burr–Pareto–logistic family.
- This distribution is an example of an Archimedean copula [Genest and MacKay (1986)] with generator $\varphi(t) = t^{-1/c} - 1$.
- Ray et al. (1980) have presented results relevant in the context of sample selection.
- This distribution has the "monotone regression dependence" property [Bilodeau (1989)].
- It is possible [see Drouet-Mari and Kotz (2001, p. 86)] to extend the Pareto copula in (2.43) to have negative dependence by allowing $c < 0$. In that case, $\hat{C}(u,v) = \max(u^{-1/c} + v^{-1/c} - 1, 0)^{-c}$, $c < 0$. As $c \to -1$, this distribution then tends to the lower Fréchet bound.

Fields of Applications

- Cook and Johnson (1981, 1986) fitted this distribution, among others, with lognormal marginals to data on the joint distribution of certain trace elements (e.g., cesium and scandium) in water.
- Concerning association in bivariate life tables, Clayton (1978) deduced that a bivariate survival function must be of the form $\bar{H}(x,y) = [1 + a(x) + b(y)]^{-c}$ if

$$h\bar{H}(x,y) = c\int_x^\infty h(u,y)du \int_y^\infty h(x,v)dv.$$

  Clayton's context is in terms of deaths of fathers and sons from some chronic disease, with association stemming from common environmental or genetic influences. The equation above arises as follows:

- Consider the ratio of the age-specific death rate for sons given that the father died at age y to the age-specific death rate for sons given that the father survived beyond age y. This ratio is assumed to be independent of the son's age.
- As a symmetric form of association is being considered, an analogous assumption holds for the ratio of fathers' age-specific death rates.
- The proportionality property $\frac{h}{\bar{H}} = c\frac{\partial}{\partial x}(-\log \bar{H})\frac{\partial}{\partial y}(-\log \bar{H})$ then holds. (The left-hand side of this equation is the bivariate failure rate, and the right-hand side is c times the product of the hazard function for sons of fathers who survive until y and the hazard function for sons of fathers who survive until x.) See also Oakes (1982, 1986) and Clayton and Cuzick (1985a,b).

- Klein and Moeschberger (1988) have used this form of association in the "competing risks" context.
- The bivariate Burr distribution, both with and without the extra association term introduced by Durling (1975), was used by Durling (1974) in reanalyzing seven previously published datasets on the effects of mixtures of poisons.
- The Takahasi–Burr distribution, in its quadrivariate form, was applied by Crowder (1985) in a repeated measurements context—specifically, in analyzing response times of rats to pain stimuli at four intervals after receiving a dose of an analgesic drug.

2.9.2 Summary of the Relationship Between Various Copulas

For ease of reference, we summarize the relationship between the Lomax copula and its special cases.

The Lomax copula (α, c) is given in (2.41):

(i) $\alpha = 1 \Rightarrow$ Pareto copula (Clayton copula) as given in (2.43).
(ii) $c = 1 \Rightarrow$ Ali–Mikhail–Haq copula as given in (2.20).

2.10 Gumbel's Type I Bivariate Exponential Distribution

Again, we depart from our usual pattern by describing this distribution, with exponential marginals, before the copula.

Formula for Cumulative Distribution Function

$$H(x,y) = 1 - e^{-x} - e^{-y} + e^{-(x+y+\theta xy)}, \qquad 0 \le \theta \le 1. \tag{2.44}$$

Formula for Density Function

$$h(x,y) = e^{-(x+y+\theta xy)}[(1+\theta x)(1+\theta y) - \theta]. \tag{2.45}$$

Univariate Properties

Both marginals are exponential.

Correlation Coefficients and Conditional Properties

$$\rho = -1 + \int_0^\infty \frac{e^{-y}}{1+\theta y} dy.$$

Gumbel (1960a) has plotted ρ as a function of θ. A compact expression may be obtained in terms of the exponential integral (but care is always necessary with this function, as the nomenclature and notation are not standardized). For $\theta = 0$, X and Y are independent and $\rho = 0$. As θ increases, ρ increases, reaching -0.404 at $\theta = 1$. Thus, this distribution is unusual in being oriented towards negative correlation. (Of course, positive correlation can be obtained by changing X to $-X$ or Y to $-Y$.)

Barnett (1983a) has discussed the maximum likelihood method for estimating θ as well as a method based on the product-moment correlation.

Gumbel (1960a) has further given the following expressions:

$$\begin{aligned} g(y|x) &= e^{-y(1+\theta x)}[(1+\theta x)(1+\theta y) - \theta], \\ E(Y|X=x) &= (1+\theta+\theta x)(1+\theta x)^{-2}, \\ \mathrm{var}(Y|X=x) &= \frac{(1+\theta+\theta x)^2 - 2\theta^2}{(1+\theta x)^4}. \end{aligned}$$

Remarks

- This distribution is characterized by

$$\begin{aligned} E(X-x|X>x \text{ and } Y>y) &= E(X|Y>y), \\ E(Y-y|X>x \text{ and } Y>y) &= E(Y|X>x), \end{aligned} \tag{2.46}$$

which is a form of the lack-of-memory property; see K.R.M. Nair and N.U. Nair (1988) and N.U. Nair and V.K.R. Nair (1988).

- Barnett (1979, 1983b) and Barnett and Lewis (1984, Section 9.3.2) have discussed testing for the presence of an outlier in a dataset assumed to come from this distribution. An alternative proposal by Barnett (1983a) involves transformation to independent normal variates.
- In the context of structural reliability, Der Kiureghian and Liu (1986) utilized this distribution (with $\theta = 1$) in the course of demonstrating a procedure to approximate multivariate integrals by transforming the marginals to normality and assuming multivariate normality; see also Grigoriu (1983, Example 2).

An Application

In describing this, let us quote the opening words of the paper by Moore and Clarke (1981):"The rainfall runoff models referred to in the title of this paper are (1) those that attempt to describe explicitly both the storage of precipitated water within a river basin and the translation or routing of water that is in temporary storage to the basin outfall, and (2) those in which the parameters of the model are estimated from existing records of mean areal rainfall, Penman potential evaporation E_T, or some similar measure of evaporation demand, and stream flow." On pp. 1373–1374 of the paper is a section entitled "A Bivariate Exponential Storage-Translation Model." This introduces distribution (2.44), the justification being that it has exponential marginals and that the correlation is negative ("a basin with thin soils in the higher altitude areas that are furthest from the basin outfall is likely to have s and t negatively correlated"). The variables s and t are, respectively, the depth of a (hypothesized) storage element and the time taken for runoff to reach the catchment outfall.

Moore and Clarke did not present in detail the results using (2.44), saying, "Application of storage-translation models using more complex distribution functions ... did not lead to any appreciable improvement in model performance ... One exception ... gives a correlation of −0.37 between s and t."

2.11 Gumbel–Barnett Copula

Gumbel (1960a,b) suggested the exponential-marginals form of this copula; many authors refer to this copula as another Gumbel family. We call it the *Gumbel–Barnett copula* since Barnett (1980) first discussed it in terms of the uniform marginals among the distributions he considered. The survival function of the copula $C(u,v)$ that corresponds to Gumbel's type 1 bivariate exponential distribution (2.44) is given by

$$\bar{C}(u,v) = (1-u)(1-v)e^{-\theta \log(1-u)\log(1-v)},$$

so that

$$C(u,v) = u + v - 1 + (1-u)(1-v)e^{-\theta \log(1-u)\log(1-v)} \tag{2.47}$$

because of the relationship $C(u,v) = \bar{C}(u,v) + u + v - 1$. The density of the copula is

$$c(u,v) = \{-\theta + [1-\theta\log(1-u)][1-\theta\log(1-v)]\}\, e^{-\theta \log(1-u)\log(1-v)}. \tag{2.48}$$

The survival copula that corresponds to (2.47) is

$$\hat{C}(u,v) = \bar{C}(1-u, 1-v) = uve^{-\theta \log u \log v}. \tag{2.49}$$

2.12 Kimeldorf and Sampson's Distribution

Kimeldorf and Sampson (1975a) studied a bivariate distribution on the unit square, with uniform marginals and p.d.f. as follows:

- β on each of $[\beta]$ squares of side $1/\beta$ arranged corner to corner up to the diagonal from $(0,0)$ towards $(1,1)$, $[\beta]$ being the largest integer not exceeding β;
- $\frac{\beta}{\beta-[\beta]}$ on one smaller square side of $1-[\beta]/\beta$ in the top-right corner of the unit square (unless β is an integer);
- and 0 elsewhere.

For this distribution, Johnson and Tenenbein (1979) showed that

$$\rho_S = \frac{[\beta]}{\beta}\frac{3\beta^2 - 3\beta[\beta] + [\beta^2] - 1}{\beta^2},$$

and Nelsen (in a private communication) showed that

$$\tau = \frac{[\beta]}{\beta}\frac{2\beta - [\beta] - 1}{\beta}.$$

Hence, if $1 \le \beta < 2$, $\rho_S = 3\tau/2$; and if β is an integer, $\rho_S = 2\tau - \tau^2$.

Remarks

- Clearly (2.49) is an Archimedean copula.
- If C_α and C_β are both Gumbel–Barnett copulas given by (2.49), then their geometric mean is again a Gumbel–Barnett copula given by $C_{(\alpha+\beta)/2}$; see Nelsen (2006, p. 133).

2.13 Rodríguez-Lallena and Úbeda-Flores' Family of Bivariate Copulas

Rodríguez-Lallena and Úbeda-Flores (2004) defined a new class of copulas of the form

$$C(u,v) = uv + f(u)g(v), \tag{2.50}$$

where f and g are two real functions defined on $[0,1]$ such that

(i) $f(0) = f(1) = g(0) = g(1)$;
(ii) f and g are absolutely continuous;
(iii) $\min\{\alpha\delta, \beta\gamma\} \geq -1$, where $\alpha = \inf\{f'(u), u \in A\} < 0$, $\beta = \sup\{f'(u), u \in A\} > 0$ $\gamma = \inf\{g'(v), v \in B\} < 0$, and $\delta = \sup\{g'(v), v \in B\} > 0$, with $A = \{u \in [0,1] : f'(u) \, \text{exists}\}$ and $B = \{v \in [0,1] : g'(v) \, \text{exists}\}$.

Example 2.1. The family studied by Lai and Xie (2000), $C(u,v) = uv + \lambda u^a v^b (1-u)^c (1-v)^d$, $u, v \in [0,1]$, $0 \leq \lambda \leq 1$, $a, b, c, d \geq 1$, is a special case of Rodríguez-Lallena and Úbeda-Flores' family.

Properties

- $\tau = 8 \int_0^1 f(t)\, dt \int_0^1 g(r)\, dt$, $\rho_S = 3\tau/2$.
- Let (X, Y) be a continuous random pair whose associated copula is a member of Rodríguez-Lallena and Úbeda-Flores' family. Then X and Y are positively quadrant dependent if and only if either $f \geq 0$ and $g \geq 0$ or $f \leq 0$ and $g \leq 0$.

2.14 Other Copulas

Table 4.1 of Nelsen (2006) presents some important one-parameter families of Archimedean copulas, along with their generators, the range of the parameter, and some special cases and limiting cases. We have discussed some of these here, and for the rest we refer the reader to this reference. Many other copulas are discussed throughout the book of Nelsen (2006), wherein we can find a comprehensive treatment of copulas.

2.15 References to Illustrations

We will now outline five important references that contain illustrations of distributions discussed in this chapter as well as some others to follow.

Conway (1981). Conway's graphs are contours of bivariate distributions; that is, for uniform marginals $F(x) = x$ and $G(y) = y$, y as a function of x has been plotted such that a contour of the (cumulative) distribution is the result (i.e., $H(x, y) = c$) a constant. The paper (i) presents such contours for various c for three reference distributions (upper and lower Fréchet bounds, and the independence), (ii) gives the $c = 0.2$ contour for distributions having various strengths of correlations drawn from the Farlie–Gumbel–Morgenstern, Ali–Mikhail–Haq, Plackett, Marshall–Olkin, and Gumbel–Hougaard families, and (iii) presents some geometric interpretations of properties of bivariate distributions.

Barnett (1980). The contours in this paper are of probability density functions. The distributions are again transformed to have uniform marginals; the bivariate normal, Farlie–Gumbel–Morgenstern, Plackett, Cauchy, and Gumbel–Barnett are the ones included.

Johnson et al. (1981). This contains both contours and three-dimensional plots of the p.d.f.'s of a number of distributions after their marginals have been transformed to be either normal or exponential. The well-known distributions included are the Farlie–Gumbel–Morgenstern, Plackett, Cauchy, and Gumbel's type I exponentials, plus a bivariate normal transformed to exponential marginals. However, the main purpose of this work is to give similar plots for distributions obtained by a trivariate reduction technique and by the Khintchin mixture.

Johnson et al. (1984). In this, there are 18 small contour plots of the p.d.f.'s of distributions after their marginals have been transformed to be normal. The well-known distributions included are the bivariate normal, Farlie–Gumbel–Morgenstern, Ali–Mikhail–Haq, Plackett, Gumbel's type I exponential, and the bivariate Pareto.

Johnson (1987). Chapters 9 and 10 of this book presents contour and three-dimensional plots of the p.d.f.'s of the following distributions: Farlie–Gumbel–Morgenstern (uniform, normal, and exponential marginals), Ali–Mikhail–Haq (normal marginals), Plackett (contour plots only; uniform, normal, and exponential marginals), Gumbel's type I exponential (uniform, normal, and exponential marginals), bivariate Pareto (uniform and normal marginals; contour plots only for exponential marginals), and Cook and Johnson's generalized Pareto (contour plots only; uniform, normal, and exponential marginals; and one three-dimensional plot of normal marginals).

When thinking of contours of p.d.f.'s, the subject of unimodality (or otherwise) of multivariate distributions comes to mind. An excellent reference for this topic is the book by Dharmadhikari and Joag-Dev (1985), and we refer readers to this book for all pertinent details.

References

1. Abramowitz, M., Stegun, I.A.: Handbook of Mathematical Functions. Applied Mathematics Series No. 55, National Bureau of Standards. U.S. Government Printing Office, Washington, D.C. [Republished subsequently by Dover, New York (1964)]
2. Ali, M.M., Mikhail, N.N., Haq, M.S.: A class of bivariate distributions including the bivariate logistic. Journal of Multivariate Analysis **8**, 405–412 (1978)
3. Amemiya, T.: Advanced Econometrics. Blackwell, Oxford (1985)
4. Anscombe, F.J.: Computing in Statistical Science Through APL. Springer-Verlag, New York (1981)
5. Arnold, B.C.: Pareto distributions. International Co-operative Publishing House, Fairland, Maryland (1983)
6. Ashford, J.R., Sowden, R.R.: Multivariate probit analysis. Biometrics **26**, 535–546 (1970)
7. Bagchi, S.B., Samanta, K.C.: A study of some properties of bivariate Burr distributions. Bulletin of the Calcutta Mathematical Society **77**, 370–383 (1985)
8. Bairamov, I.G., Kotz, S.: On a new family of positive quadrant dependent bivariate distribution. Technical Report, The George Washington University, Washington, D.C. (2000a)
9. Bairamov, I.G., Kotz, S.: Private communication (2000b)
10. Bairamov, I.G., Kotz, S., Bekci, M.: New generalized Farlie–Gumbel–Morgenstern distributions and concomitants of order statistics. Journal of Applied Statistics **28**, 521–536 (2001)
11. Barnett, V.: Some outlier tests for multivariate samples. South African Statistical Journal **13**, 29–52 (1979)
12. Barnett, V.: Some bivariate uniform distributions. Communications in Statistics: Theory and Methods **9**, 453–461 (Correction, **10**, 1457) (1980)
13. Barnett, V.: Reduced distance measures and transformation in processing multivariate outliers. Australian Journal of Statistics **25**, 64–75 (1983a)
14. Barnett, V.: Principles and methods for handling outliers in data sets. In: Statistical Methods and the Improvement of Data Quality, T. Wright (ed.), pp. 131–166. Academic Press, New York (1983b)
15. Barnett, V., Lewis, T.: Outliers in statistical data, 2nd edition. John Wiley and Sons, Chichester, England (1984)
16. Bilodeau, M.: On the monotone regression dependence for Archimedian bivariate uniform. Communications in Statistics: Theory and Methods **18**, 981–988 (1989)
17. Brown, C.C., Fears, T.R.: Exact significance levels for multiple binomial testing with application to carcinogenicity screens. Biometrics **37**, 763–774 (1981)
18. Cambanis, S.: Some properties and generalizations of multivariate Eyraud–Gumbel–Morgenstern distributions. Journal of Multivariate Analysis **7**, 551–559 (1977)
19. Chinchilli, V.M., Breen, T.J.: Testing the independence of q binary random variables. Technical Report, Department of Biostatistics, Medical College of Virginia, Richmond. (Abstract in Biometrics, **41**, 578) (1985)
20. Clayton, D.G.: A model for association in bivariate life tables and its application in epidemiological studies of familial tendency in chronic disease incidence. Biometrika **65**, 141–151 (1978)
21. Clayton, D., Cuzick, J.: Multivariate generalizations of the proportional hazards model (with discussion). Journal of the Royal Statistical Society, Series A **148**, 82–117 (1985a)

22. Clayton, D., Cuzick, J.: The semi-parametric Pareto model for regression analysis of survival times. Bulletin of the International Statistical Institute **51**, Book 4, Paper 23.3 (Discussion, Book 5, 175–180) (1985b)
23. Conway, D.: Multivariate distributions with specified marginals. Technical Report No. 145, Department of Statistics, Stanford University, Stanford, California (1979)
24. Conway, D.: Bivariate distribution contours. Proceedings of the Business and Economic Statistics Section, American Statistical Association, pp. 475–480 (1981)
25. Conway, D.: , Plackett family of distributions. In: Encyclopedia of Statistical Sciences, Volume 7, S. Kotz and N.L. Johnson (eds.), pp. 1–5. John Wiley and Sons, New York (1986)
26. Cook, R.D., Johnson, M.E.: A family of distributions for modelling nonelliptically symmetric multivariate data. Journal of the Royal Statistical Society, Series B **43**, 210–218 (1981)
27. Cook, R.D., Johnson, M.E.: Generalized Burr-Paretologistic distribution with applications to a uranium exploration data set. Technometrics **28**, 123–131 (1986)
28. Cox, D.R., Oakes, D.: Analysis of Survival Data. Chapman and Hall, London (1984)
29. Crowder, M.: A distributional model for repeated failure time measurements. Journal of the Royal Statistical Society, Series B **47**, 447–452 (1985)
30. Crowder, M.: A multivariate distribution with Weibull connections. Journal of the Royal Statistical Society, Series B **51**, 93–107 (1989)
31. Cuadras, C.M., Augé, J.: A continuous general multivariate distribution and its properties. Communications in Statistics: Theory and Methods **10**, 339–353 (1981)
32. Dale, J.R.: Statistical methods for ordered categorical responses. Report No. 114, Applied Mathematics Division, Department of Scientific and Industrial Research, Wellington, New Zealand (1983)
33. Dale, J.R.: Local versus global association for bivariate ordered responses. Biometrika **71**, 507–514 (1984)
34. Dale, J.R.: A bivariate discrete model of changing colour in blackbirds. In: Statistics in Ornithology, B.J.T. Morgan and P. M. North (eds.), pp. 25–36. Springer-Verlag, Berlin (1985)
35. Dale, J.R.: Global cross-ratio models for bivariate, discrete, ordered responses. Biometrics **42**, 909–917 (1986)
36. Der Kiureghian, A., Liu, P-L.: Structural reliability under incomplete probability information. Journal of Engineering Mechanics **112**, 85–104 (1986)
37. D'Este, G.M.: A Morgenstern-type bivariate gamma distribution. Biometrika **68**, 339–340 (1981)
38. Devroye, L.: Nonuniform Random Variate Generation. Springer-Verlag, New York (1986)
39. Dharmadhikari, S.W., Joag-Dev, K.: Multivariate unimodality. In: Encyclopedia of Statistical Sciences, Volume 6, S. Kotz and N.L. Johnson (eds.), pp. 130–132. John Wiley and Sons, New York (1985)
40. Downton, F.: Bivariate exponential distributions in reliability theory. Journal of the Royal Statistical Society, Series B **32**, 408–417 (1970)
41. Drouet-Mari, D., Kotz, S.: Correlation and Dependence. Imperial College Press, London (2001)
42. Durling, F.C.: Bivariate normit, logit, and Burrit analysis. Research Report No. 24, Department of Mathematics, University of Waikato, Hamilton, New Zealand (1974)
43. Durling, F.C.: The bivariate Burr distribution. In: A Modern Course on Statistical Distributions in Scientific Work. Volume I: Models and Structures, G.P. Patil, S. Kotz, J.K. Ord (eds.), pp. 329–335. Reidel, Dordrecht (1975)
44. Fang, K.T., Fang, H.B., von Rosen, D.: A family of bivariate distributions with non-elliptical contours. Communications in Statistics: Simulation and Computation **29**, 1885–1898 (2000)
45. Farlie, D.J.G.: The performance of some correlation coefficients for a general bivariate distribution. Biometrika **47**, 307–323 (1960)

46. Frank, M.J.: On the simultaneous associativity of $F(x, y)$ and $x + y - F(x, y)$. Aequationes Mathematicae **19**, 194–226 (1979)
47. Genest, C.: Frank's family of bivariate distributions. Biometrika **74**, 549–555 (1987)
48. Genest, C., MacKay, R.J.: Copules archimédiennes et familles de lois bidimensionnelles dont les marges sont données. Canadian Journal of Statistics **14**, 145–159 (1986)
49. Genest, C., Rivest, L.P.: Statistical inference procedures for bivariate Archimedan copulas. Journal of the American Statistical Association **88**, 1034–1043 (1993)
50. Gilula, Z.: On some similarities between canonical correlation models and latent class models for two-way contingency tables. Biometrika **71**, 523–529 (1984)
51. Gilula, Z., Krieger, A.M., Ritov, Y.: Ordinal association in contingency tables: Some interpretive aspects. Journal of the American Statistical Association **83**, 540–545 (1988)
52. Goodman, L.A.: Association models and the bivariate normal for contingency tables with ordered categories. Biometrika **68**, 347–355 (1981)
53. Goodman, L.A.: Some useful extensions of the usual correspondence analysis approach and the usual log-linear models approach in the analysis of contingency tables (with discussion). International Statistical Review **54**, 243–309 (1986)
54. Grigoriu, M.: Approximate analysis of complex reliability problems. Structural Safety **1**, 277–288 (1983)
55. Gumbel, E.J.: Distributions à plusieurs variables dont les marges sont données. Comptes Rendus de l'Académie des Sciences, Paris **246**, 2717–2719 (1958)
56. Gumbel, E.J.: Bivariate exponential distributions. Journal of the American Statistical Association **55**, 698–707 (1960a)
57. Gumbel, E.J.: Multivariate distributions with given margins and analytical examples. Bulletin de l'Institut International de Statistique **37**, Book 3, 363–373 (Discussion, Book 1,114–115) (1960b)
58. Gumbel, E.J.: Bivariate logistic distributions. Journal of the American Statistical Association **56**, 335–349 (1961)
59. Gumbel, E.J., Mustafi, C.K.: Some analytical properties of bivariate extremal distributions. Journal of the American Statistical Association **62**, 569–588 (1967)
60. Halperin, M., Wu, M., Gordon, T.: Genesis and interpretation of differences in distribution of baseline characteristics between cases and noncases in cohort studies. Journal of Chronic Diseases **32**, 483–491 (1979)
61. Hanagal, D.D.: A multivariate Pareto distribution. Communications in Statistics: Theory and Methods **25**, 1471–1488 (1996)
62. Hougaard, P.: A class of multivariate failure time distributions. Biometrika **73**, 671–678 (Correction, **75**, 395) (1986)
63. Huang, J.S., Kotz, S.: Modifications of the Farlie–Gumbel–Morgenstern distributions: A tough hill to climb. Metrika **49**, 307–323 (1999)
64. Johnson, M.E.: Multivariate Statistical Simulation. John Wiley and Sons, New York (1987)
65. Johnson, M.E., Tenenbein, A.: Bivariate distributions with given marginals and fixed measures of dependence. Informal Report LA-7700-MS, Los Alamos Scientific Laboratory, Los Alamos, New Mexico (1979)
66. Johnson, M.E., Bryson, M.C., Mills, C.F.: Some new multivariate distributions with enhanced comparisons via contour and three-dimensional plots. Report LA-8903-MS, Los Alamos Scientific Laboratory, Los Alamos, New Mexico (1981)
67. Johnson, M.E., Wang, C., Ramberg, J.: Generation of continuous multivariate distributions for statistical applications. American Journal of Mathematical and Management Sciences **4**, 225–248 (1984)
68. Johnson, N.L., Kotz, S.: On some generalized Farlie–Gumbel–Morgenstern distributions. II. Regression, correlation and further generalizations. Communications in Statistics: Theory and Methods **6**, 485–496 (1977)
69. Kimeldorf, G., Sampson, A.: One-parameter families of bivariate distributions with fixed marginals. Communications in Statistics **4**, 293–301 (1975a)

70. Kimeldorf, G., Sampson, A.: Uniform representations of bivariate distributions. Communications in Statistics, **4**, 617–627 (1975b)
71. Klein, J.P., Moeschberger, M.L.: Bounds on net survival probabilities for dependent competing risks. Biometrics **44**, 529–538 (1988)
72. Kotz, S., Johnson, N.L.: Propriétés de dépendance des distributions itérées, géneralisées a deux variables Farlie–Gumbel–Morgenstern. Comptes Rendus de l'Académie des Sciences, Paris, Série A **285**, 277–280 (1977)
73. Kotz, S., Johnson, N.L.: Some replacement-times distributions in two-component systems. Reliability Engineering **7**, 151–157 (1984)
74. Kotz, S., Van Dorp, J.R.: A versatile bivariate distribution on a bounded domain: Another look at the product moment correlation. Journal of Applied Statistics **29**, 1165–1179 (2002)
75. Lai, C.D.: Morgenstern's bivariate distibution and its application to point process. Journal of Mathematical Analysis and Applications **65**, 247–256 (1978)
76. Lai, C.D., Xie, M.: A new family of positive quadrant dependent bivariate distributions. Statistics and Probability Letters **46**, 359–364 (2000)
77. Lai, C.D., Xie, M., Bairamov, I.G.: Dependence and ageing properties of bivariate Lomax distribution. In: System and Bayesian Reliability: Essays in Honor of Prof. R.E. Barlow on His 70th Birthday, Y. Hayakawa, T. Irony, and M. Xie (eds.) pp. 243–256. World Scientific, Singapore (2001)
78. Lee, L.: Multivariate distributions having Weibull properties. Journal of Multivariate Analysis **9**, 267–277 (1979)
79. Lee, L-F.: A bivariate logit model. Technical Report, Center for Econometrics and Decision Sciences, University of Florida, Gainesville, Florida (1982)
80. Lin, G.D.: Relationships between two extensions of Farlie–Gumbel–Morgenstern distribution. Annals of the Institute of Statistical Mathematics **39**, 129–140 (1987)
81. Lindley, D.V., Singpurwalla, N.D.: Multivariate distributions for the life lengths of components of a system sharing a common environment. Journal of Applied Probability **23**, 418–431 (1986)
82. Lingappaiah, G.S.: Bivariate gamma distribution as a life test model. Aplikace Matematiky **29**, 182–188 (1984)
83. Long, D., Krzysztofowicz, R.: Farlie–Gumbel–Morgenstern bivariate densities: Are they applicable in hydrology? Stochastic Hydrology and Hydraulics **6**, 47–54 (1992)
84. Malik, H.J., Trudel, R.: Distributions of the product and the quotient from bivariate t, F and Pareto distributions. Communications in Statistics: Theory and Methods **14**, 2951–2962 (1985)
85. Mardia, K.V.: Multivariate Pareto distributions. Annals of Mathematical Statistics **33**, 1008–1015 (Correction, **34**, 1603) (1962)
86. Mardia, K.V.: Some contributions to contingency-type bivariate distributions. Biometrika **54**, 235–249 (Correction, **55**, 597) (1967)
87. Mardia, K.V.: Families of Bivariate Distributions. Griffin, London (1970a)
88. Mardia, K.V.: Some problems of fitting for contingency-type bivariate distributions. Journal of the Royal Statistical Society, Series B **32**, 254–264 (1970b)
89. Marshall, A.W., Olkin, I.: A multivariate exponential distribution. Journal of the American Statistical Association **62**, 30–44 (1967)
90. Mikhail, N.N., Falwell, J.A., Bogue, A., Weaver, T. L.: Regression curves to a class of bivariate distributions including the bivariate logistic with application. In: Computer Science and Statistics: Proceedings of the 19th Symposium on the Interface, R.M. Heiberger (ed.), pp. 525–530. American Statistical Association, Alexandria, Virginia (1987a)
91. Mikhail, N.N., Chasnov, R., Wooldridge, T.S.: Regression curves for Farlie–Gumbel–Morgenstern class of bivariate distribution. In: Computer Science and Statistics: Proceedings of the 19th Symposium on the Interface, R.M. Heiberger (ed.), pp. 531–532. American Statistical Association, Alexandria, Virginia (1987b)

92. Moore, R.J., Clarke, R.T.: A distribution function approach to rainfall runoff modeling. Water Resources Research **17**, 1367–1382 (1981)
93. Morimune, K.: Comparisons of normal and logistic models in the bivariate dichotomous analysis. Econometrica **47**, 957–975 (1979)
94. Mukherjee, S.P., Saran, L.K.: Bivariate inverse Rayleigh distributions in reliability studies. Journal of the Indian Statistical Association **22**, 23–31 (1984)
95. Mukherjee, S.P., Sasmal, B.C.: Life distributions of coherent dependent systems. Calcutta Statistical Association Bulletin **26**, 39–52 (1977)
96. Nadarajah, S.: Sums, products, and ratios for the bivariate Lomax distribution. Computational Statistics and Data Analysis **49**, 109–129 (2005)
97. Nair, K.R.M., Nair, N.U.: On characterizing the bivariate exponential and geometric distributions. Annals of the Institute of Statistical Mathematics **40**, 267–271 (1988)
98. Nair, N.U., Nair, V.K.R.: A characterization of the bivariate exponential distribution. Biometrical Journal **30**, 107–112 (1988)
99. Nayak, T.K.: Multivariate Lomax distribution: Properties and usefulness in reliability theory. Journal of Applied Probability **24**, 170–177 (1987)
100. Nelsen, R.B.: Properties of a one-parameter family of bivariate distributions with specified marginals. Communications in Statistics: Theory and Methods **15**, 3277–3285 (1986)
101. Nelsen, R.B.: An Introduction to Copulas, 2nd edition. Springer-Verlag, New York (2006)
102. Oakes, D.: A model for association in bivariate survival data. Journal of the Royal Statistical Society, Series B **44**, 414–422 (1982)
103. Oakes, D.: Semiparametric inference in a model for association in bivariate survival data. Biometrika **73**, 353–361 (1986)
104. Pearson, K.: Note on the surface of constant association. Biometrika **9**, 534–537 (1913)
105. Phillips, M.J.: A preventive maintenance plan for a system subject to revealed and unrevealed faults. Reliability Engineering **2**, 221–231 (1981)
106. Plackett, R.L.: A class of bivariate distributions. Journal of the American Statistical Association **60**, 516–522 (1965)
107. Ray, S.C., Berk, R.A., Bielby, W.T.: Correcting sample selection bias for bivariate logistic distribution of disturbances. In: Proceedings of the Business and Economic Statistics Section, American Statistical Association, pp. 456–459. American Statistical Association, Alexandria, Virgina (1980)
108. Rodriguez, R.N.: Multivariate Burr III distributions, Part I: Theoretical properties. Research Publication GMR-3232, General Motors Research Laboratories, Warren, Michigan (1980)
109. Rodriguez, R.N., Taniguchi, B.Y.: A new statistical model for predicting customer octane satisfaction using trained rater observations, (with discussion). Paper No. 801356, Society of Automotive Engineers, Washington D.C. (1980)
110. Rodríguez-Lallena, J.A., Úbeda-Flores, M.: A new class of bivariate copulas. Statistics and Probability Letters **66**, 315–325 (2004)
111. Sankaran, P.G., Nair, U.N.: A bivariate Pareto model and its applications to reliability. Naval Research Logistics **40**, 1013–1020 (1993)
112. Sarmanov, I.O.: New forms of correlation relationships between positive quantities applied in hydrology. In: Mathematical Models in Hydrology Symposium, IAHS Publication No. 100, International Association of Hydrological Sciences, pp. 104–109 (1974)
113. Satterthwaite, S.P., Hutchinson, T.P.: A generalisation of Gumbel's bivariate logistic distribution. Metrika **25**, 163–170 (1978)
114. Schucany, W.R., Parr, W.C., Boyer, J.E.: Correlation structure in Farlie–Gumbel–Morgenstern distributions. Biometrika **65**, 650–653 (1978)
115. Shoukri, M.M., Chaudhary, M.A., Al-halees, A.: Estimating $\Pr(Y < X)$ when X and Y are paired exponential variables. Journal of Statistical Computation and Simulation **75**, 25–38 (2005)

116. Takahasi, K.: Note on the multivariate Burr's distribution. Annals of the Institute of Statistical Mathematics **17**, 257–260 (1965)
117. Teichmann, T.: Joint probabilities of partially coupled events. Reliability Engineering **14**, 133–148 (1986)
118. Tolley, H.D., Norman, J.E.: Time on trial estimates with bivariate risks. Biometrika **66**, 285–291 (1979)
119. Wahrendorf, J.: Inference in contingency tables with ordered categories using Plackett's coefficient of association for bivariate distributions. Biometrika **67**, 15–21 (1980)
120. Willett, P.K., Thomas, J.B.: A simple bivariate density representation. In: Proceedings of the 23rd Annual Allerton Conference on Communication, Control, and Computing, pp. 888–897. Coordinated Science Laboratory and Department of Electrical and Computer Engineering, University of Illinois, Urbana-Champaign (1985)
121. Willett, P.K., Thomas, J.B.: Mixture models for underwater burst noise and their relationship to a simple bivariate density representation. IEEE Journal of Oceanic Engineering **12**, 29–37 (1987)
122. Zheng, M., Klein, J.P.: A self-consistent estimator of marginal survival functions based on dependent competing risk data and an assumed copula. Communications in Statistics: Theory and Methods **23**, 2299–2311 (1994)

Chapter 3
Concepts of Stochastic Dependence

3.1 Introduction

Dependence relations between two variables are studied extensively in probability and statistics. No meaningful statistical models can be constructed without some assumptions regarding dependence although in many cases one may simply assume the variables are not dependent, i.e., they are independent.

Karl Pearson is often credited as the first to introduce the concept of dependence by defining the product-moment correlation, which measures the strength of the linear relationship between two variables under consideration.

Basically, positive dependence means that large values of Y tend to accompany large values of X, and similarly small values of Y tend to accompany small values of X. By the same principle, negative dependence between two variables means large values of Y tend to accompany small values of X and vice versa. The focus of this chapter is on different concepts of positive dependence.

Various notions of dependence are motivated by applications in statistical reliability; see, for example, Barlow and Proschan (1975, 1981). Although the starting point of reliability models is independent of the lifetimes of components, it is often more realistic to assume some form of positive dependence among the components.

In the 1960s, several different notions of positive dependence between two random variables and their interrelationships were discussed by a number of authors including Harris (1960, 1970), Lehmann (1966), Esary et al. (1967), Esary and Proschan (1972), and Kimeldorf and Sampson (1987). Yanagimoto (1972) unified some of these notions by introducing a family of positive dependence. Some further notions of positive dependence were introduced by Shaked (1977, 1979, 1982). Joe (1993) characterized the distributions for which dependence is concentrated at the lower and upper tails. These con-

N. Balakrishnan, C.D. Lai, *Continuous Bivariate Distributions*,
DOI 10.1007/b101765_4, © Springer Science+Business Media, LLC 2009

cepts, which were initially defined for two variables, have been extended to a multivariate random vector $(X_1, X_2, \ldots, X_n)$ with $n \geq 2$.

In the case of $n = 2$, the negative dependence is easily constructed by reversing the concepts of positive dependence, as was done by Lehmann (1966). However, for $n > 2$, negative dependence is no longer a simple mirror reflection of positive dependence; see, for example, Joag-Dev and Proschan (1983).

In Section 3.2, the concept of positive dependence is introduced and then some conditions for a family to be positively dependent are presented. In Section 3.3, some dependence concepts that are stronger and weaker than positive dependence are outlined. Next, in Sections 3.4 and 3.5, concepts of positive dependence stronger and weaker than the positive quadrant dependence (PQD) are discussed, respectively. In Section 3.6, some positively quadrant dependent bivariate distributions are presented. Some additional concepts of dependence are introduced in Section 3.7. In Section 3.8, the concept of negative dependence is discussed in detail, while results on positive dependence orderings are described in Section 3.9.

For reviews of implications among different dependence concepts, we refer the reader to Joe (1997), Müller and Stoyan (2002), or Lai and Xie (2006).

3.2 Concept of Positive Dependence and Its Conditions

A basic motivation of Lehmann (1966) for introducing the basic concept of positive dependence was to provide tests of independence between two variables that are not biased. As a matter of fact, in order to construct an unbiased test, we need to specify the alternative hypothesis. Lehmann identified subfamilies of bivariate distributions for which this property of unbiasedness is valid. Kimeldorf and Sampson (1987) presented seven conditions in all that a subfamily of distributions $\mathcal{F}^+$ with given marginals should satisfy to be positively dependent. Recall that $H^+(x, y) = \min(F(x), G(y))$ and $H^-(x, y) = \max(0, F(x) + G(y) - 1)$ are the upper and lower Fréchet bounds, where $F(x)$ and $G(y)$ are the marginal distributions of X and Y, respectively. Then, the conditions of Kimeldorf and Sampson (1987) are as follows:

1. $H \in \mathcal{F}^+ \Rightarrow H(x, y) \geq F(x)G(y)$ for all x and y.
2. If $H(x, y) \in \mathcal{F}^+$, so does $H^+(x, y)$.
3. If $H(x, y) \in \mathcal{F}^+$, so does $H^0(x, y) = F(x)G(y)$.
4. If $(X, Y) \in \mathcal{F}^+$, so does $(\phi(X), Y) \in \mathcal{F}^+$, where ϕ is any increasing function.
5. If $(X, Y) \in \mathcal{F}^+$, so does (Y, X).
6. If $(X, Y) \in \mathcal{F}^+$, so does $(-X, -Y)$.
7. If H_n converges to H in distribution, then $H \in \mathcal{F}^+$.

We note that condition 1 is equivalent to the *positive quadrant dependence* (PQD) concept, which is discussed in the next section.

3.3 Positive Dependence Concepts at a Glance

We list several concepts of positive dependence that exist in the literature in the form of two tables where the PQD is used as a benchmark, and so Table 3.1 lists the dependence concepts that are stronger than PQD, while Table 3.2 lists the dependence concepts that are weaker than PQD.

Table 3.1 Dependence concepts that are stronger than PQD

PQD	Positive quadrant dependence
ASSOC	Associated
LTD RTI	Left-tail decreasing Right-tail increasing
SI (alias PRD)	Stochastically increasing (Positively regression dependent)
RCSI LCSD	Right corner set increasing Left corner set decreasing
TP_2 (alias LRD)	Total positivity of order 2 (Likelihood ratio dependence)

Table 3.2 Dependence concepts that are weaker than PQD

PQD	Positive quadrant dependence
PQDE	Positive quadrant dependence in expectation
$\text{cov}(X, Y) \geq 0$	Positively correlated

According to Jogdeo (1982), positive correlation, positive quadrant dependence, association, and positive regression dependence are the four basic conditions that describe positive dependence, and these are in increasing order of stringency. For multivariate dependence concepts, one may refer to Joe (1997).

3.4 Concepts of Positive Dependence Stronger than PQD

We now formally define the concepts of positive dependence that are stronger than positive quadrant dependence listed in Table 3.1. Throughout this chapter, we assume that X and Y are continuous random variables with joint distribution function H.

3.4.1 Positive Quadrant Dependence

We say that (X, Y) is positive quadrant dependent (PQD) if

$$\Pr(X \geq x, Y \geq y) \geq \Pr(X \geq x) \Pr(Y \geq y) \tag{3.1}$$

or, equivalently, if

$$\Pr(X \leq x, Y \leq y) \geq \Pr(X \leq x) \Pr(Y \leq y). \tag{3.2}$$

Later, in Section 3.6, we will present many families of positive quadrant dependent distributions.

Lehmann (1966) showed the conditions above to be

$$\mathrm{cov}[a(X), b(Y)] \geq 0 \tag{3.3}$$

for every pair of increasing functions a and b defined on the real line R.

The proof is based on Hoeffding's (1940) well-known lemma [also see Shea (1983)], which states that

$$\mathrm{cov}(X, Y) = \int_{-\infty}^{\infty} \int_{-\infty}^{\infty} \{H(x, y) - F(x)G(y)\} dx dy. \tag{3.4}$$

We observe from (3.4) that (X, Y) being PQD implies $\mathrm{cov}(X, Y) \geq 0$, with equality holding only if X and Y are independent. Further, if a and b are two increasing real functions, then (X, Y) being PQD implies $(a(X), b(Y))$ is also PQD, and so $\mathrm{cov}[a(X), b(Y)] \geq 0$. Suppose now $\mathrm{cov}[a(X), b(Y)] \geq 0$ for all increasing functions a and b. Set $a(X) = I_{\{X \geq x\}}$ and $b(Y) = I_{\{Y \geq y\}}$. Now, $\mathrm{cov}[a(X), b(Y)] = \Pr(X \geq x, Y \geq y) - \Pr(X \geq x) \Pr(Y \geq y) \geq 0$, which means (X, Y) is PQD. Therefore, $\mathrm{cov}[a(X), b(Y)] \geq 0$ for all increasing functions a and b and the PQD conditions in (3.1) are indeed equivalent.

PUOD and PLOD

Unlike other bivariate dependence concepts, which can be readily extended to the corresponding multivariate dependence of n variables, this is not the case with PDQ. This is because (3.1) and (3.2) are equivalent only for $n = 2$. For $n > 2$, we say that $X_1, X_2, \ldots, X_n$ are positively upper orthant dependent (PUOD) if

$$\Pr(X_1 > x_1, X_2 > x_2, \ldots, X_n > x_n) \geq \prod_{i=i}^{n} \Pr(X_i > x_i)$$

and are positively lower orthant dependent (PLOD) if

$$\Pr(X_1 \leq x_1, X_2 \leq x_2, \ldots, X_n \leq x_n) \geq \prod_{i=i}^{n} \Pr(X_i \leq x_i).$$

3.4.2 Association of Random Variables

Esary et al. (1967) introduced the following condition, termed *association*. X and Y are said to be "associated" if for every pair of functions a and b, defined on R^2, that are increasing in each of their arguments (separately), we have

$$\operatorname{cov}[a(X,Y), b(X,Y)] \geq 0. \tag{3.5}$$

A direct verification of this dependence concept is difficult in general, but it is often easier to verify one of the alternative positive dependence notions that do imply association. For example, it is easy to see that the condition in (3.5) implies (3.3); that is, "association" $\Rightarrow$ PQD.

The concept of "association" is very useful in reliability, particularly in the context of multivariate (as distinct from just bivariate) dependence. Jogdeo (1982) defined an n-variate random vector $\mathbf{X} = (X_1, \ldots, X_n)$ or its distribution to be *associated* if for every pair of increasing real functions a and b, defined on R^n, $\operatorname{cov}[a(\mathbf{X}), b(\mathbf{Y})] \geq 0$.

The property of association has a number of consequences as listed by Jogdeo (1982). Some of them are trivial, at least in the bivariate case. We note here that (i) increasing (or decreasing) functions of associated random variables are also associated, and (ii) if $(Y_1, \ldots, Y_n)$ is also associated, and the X's and Y's are positive, then $(X_1Y_1, \ldots, X_nY_n)$ is associated. Clearly, the condition in (3.5) can be expressed alternatively as

$$E[a(X,Y)b(X,Y)] \geq E[a(X,Y)]E[b(X,Y)]. \tag{3.6}$$

Barlow and Proschan (1981, p. 29) considered the following practical reliability situations in which the lifetimes of the components are not independent but are associated:

a. Minimal path structures of a coherent system having components in common.
b. Components subject to the same set of stresses.
c. Structures in which components share the same load, so that the failure of one component results in an increased load on each of the remaining components.

Observe that, in all the situations listed above, the random variables of interest act in a similar manner. In fact, all the positive dependence concepts share this characteristic.

An important application of the concept of association is to obtain probability bounds for system reliability. Many such bounds are presented by Barlow and Proschan (1981).

For a relation between association and multivariate total positivity, see Kim and Proschan (1988). Similarly, with regard to the association of chi-squared, t-, and F-distributions, one may refer to Abdel-Hameed and Sampson (1978).

Example 3.1 (Marshall and Olkin's bivariate exponential distribution). X and Y are associated in this case since they have a variable in common in their construction.

Remarks

- It is easy to prove [see, e.g., Theorem 3.2, Chapter 2 of Barlow and Proschan (1981)] that "association" implies both PUOD and PLOD.
- $X_1, X_2, \ldots, X_n$ are weakly associated [Christofides and Vaggelatou (2004) and Hu et al. (2004)] if for every pair of disjoint subsets A_1 and A_2 of $1, 2, \ldots, n$
$$\text{cov} = [a(X_i, i \in A_1), b(X_j, j \in A_2)] \geq 0$$
whenever a and b are increasing. If the inequality sign is reversed, then the random variables $X_1, X_2, \ldots, X_n$ are said to be negatively associated, see Definition 3.22.

3.4.3 Left-Tail Decreasing (LTD) and Right-Tail Increasing (RTI)

Y is *right-tail increasing* in X, denoted by $\text{RTI}(Y|X)$, if

$$\Pr(Y > y|X > x) \text{ is increasing in } x, \text{ for all } y; \tag{3.7}$$

and X is *right-tail increasing* in Y, denoted by RTI($X|Y$), if

$$\Pr(X > x|Y > y) \text{ is increasing in } y, \text{ for all } x. \tag{3.8}$$

Similarly, Y is *left-tail decreasing* in X, denoted by LTD($Y|X$), if

$$\Pr(Y \le y|X \le x) \text{ is increasing in } x, \text{ for all } y; \tag{3.9}$$

and X is *left-tail decreasing* in Y, denoted by LTD($X|Y$), if

$$\Pr(X \le x|Y \le y) \text{ is decreasing in } y, \text{ for all } x. \tag{3.10}$$

When there is no ambiguity, we will simply use RTI or LTD, for example.

Remarks

- Both RTI and LTD imply PQD. For example, suppose Y is right tail increasing in X so $\Pr(Y > y|X > x)$ is increasing in x for all y. Thus $\Pr(Y > y|X > x_1) \le \Pr(Y > y|X > x)$, $x_1 < x$. By choosing $x_1 = -\infty$, we have $\Pr(Y > y) \le \Pr(Y > y|X > x)$, giving $\Pr(X > y, Y > y) \ge \Pr(Y > y)\Pr(X > x)$. Hence RTI($Y|X$) $\Rightarrow$ PQD. Similarly, RTI($X|Y$) $\Rightarrow$ PQD and both LTD($Y|X$) and LTD($X|Y$) imply PQD.
- The positive quadrant dependence does not imply any of the four tail dependence concepts above. Nelsen (2006, p. 204) gives a counterexample.
- Nelsen (2006, p. 192) showed that LTD($Y|X$) and LTD($X|Y$) if and only if, for all u, u', v, v' such that $0 < u \le u' \le 1$ and $0 < v \le v' \le 1$,
$$\frac{C(u,v)}{uv} \ge \frac{C(u',v')}{u'v'}.$$
Similarly, the joint distribution is RTI($Y|X$) if and only if $[v - C(u,v)]/(1-u)$ decreasing in u; RTI($X|Y$) if and only if $[u - C(u,v)]/(1-v)$ decreasing in v.
- Verifying that a given copula satisfies one or more of the dependence conditions above can be tedious. Nelsen (2006, pp. 192–193) gave the following criteria for tail monotonicity in terms of partial derivatives of C:

(1) LTD($Y|X$) $\Leftrightarrow$ for any $v \in [0,1]$, $\partial C(u,v)/\partial u \le C(u,v)/u$ for almost all u.
(2) LTD($X|Y$) $\Leftrightarrow$ for any $u \in [0,1]$, $\partial C(u,v)/\partial v \le C(u,v)/v$ for almost all v.
(3) RTI($Y|X$) $\Leftrightarrow$ for any $v \in [0,1]$, $\partial C(u,v)/\partial u \ge [v - C(u,v)]/(1-u)$ for almost all u.

(4) RTI($X|Y$) $\Leftrightarrow$ for any $u \in [0,1]$, $\partial C(u,v)/\partial v \geq [u - C(u,v)]/(1-v)$ for almost all v.

Example 3.2 (LTD copula). Nelsen (2006, p. 205) showed that the distribution with the copula

$$C(u,v) = \begin{cases} \min\left(\frac{u}{2}, v\right), & 0 \leq v \leq \frac{1}{2}, \\ \min\left(u, \frac{u}{2} + v - \frac{1}{2}\right), & \frac{1}{2} \leq v \leq 1 \end{cases}$$

is $LTD(Y|X)$ and $RTI(Y|X)$.

Example 3.3 (Durling–Pareto distribution). Lai et al. (2001) showed that X and Y are right-tail increasing if $k \leq 1$ and right-tail decreasing if $k \geq 1$. From the relationships listed in Section 3.5.4, it is known that right-tail increasing implies association. Hence, X and Y are associated if $k \leq 1$.

3.4.4 Positive Regression Dependent (Stochastically Increasing)

Y is said to be *stochastically increasing* in X, denoted by SI($Y|X$), if

$$\Pr(Y > y|X = x) \text{ is increasing in } x, \text{ for all } y; \tag{3.11}$$

and X is *stochastically increasing* in Y, denoted by SI($X|Y$), if

$$\Pr(X > x|Y = y) \text{ is increasing in } y, \text{ for all } x. \tag{3.12}$$

If there is no cause for confusion, SI($Y|X$) may simply be denoted by SI. Some authors refer to this relationship as Y being *positively regression dependent* on X (denoted by PRD) and similarly X being *positively regression dependent* on Y.

Shaked (1977) showed that SI($Y|X$) is equivalent to

$$R(y|X = x) \text{ is decreasing in } x, \text{ for all } y \geq 0, \tag{3.13}$$

where R is the conditional hazard function defined by

$$R(y|X \in A) = -\log \Pr(Y > y|X \in A). \tag{3.14}$$

(The hazard function here is the cumulative hazard rate.) It is now clear that RTI($Y|X$) is equivalent to $R(y|X > x)$ is decreasing in x for all y, and therefore SI($Y|X$) $\Rightarrow$ RTI($Y|X$). Similarly, we can show that SI($Y|X$) $\Rightarrow$ LTD($Y|X$).

Further, it can be shown that RTI$(Y|X) \Rightarrow$"association," but the proof is quite involved; see Esary and Proschan (1972). However, it is not difficult to show that SI$(Y|X) \Rightarrow X$ and Y are "associated." It can be shown that

$$E(Y|X=x) = -\int_{-\infty}^{0} \Pr(Y \leq y|X=x)dx + \int_{0}^{\infty} \Pr(Y > y|X=x)dy, \tag{3.15}$$

which implies that $E(Y|X=x)$ is increasing if the condition in (3.11) holds.

Consider now the identity

$$\text{cov}(X,Y) = \text{cov}[E(X|\mathbf{Z}), E(Y|\mathbf{Z})] + E\{\text{cov}[(X,Y)|\mathbf{Z}]\},$$

in which we have taken expectation over an arbitrary random vector $\mathbf{Z}$. Now, with a and b again being increasing functions, we have

$$\begin{aligned}&\text{cov}[a(X,Y), b(X,Y)] \\ &= \text{cov}\{E(a(X,Y)|X], E(b(X,Y)|X]\} + E\{\text{cov}[a(X,Y), b(X,Y)|X]\}.\end{aligned}$$

If (3.11) holds, the expected values in the first term on the right-hand side of the equation above are increasing[1] in X; this, taken with the result that $\text{cov}[a(X), b(X)] \geq 0$, which we established earlier, means that the first term is non-negative. Also, a and b being monotone functions means that the conditional covariance in the second term is non-negative, so its expected value must be non-negative as well. As a result, X and Y are "associated."

Example 3.4 (Marshall and Olkin's bivariate exponential distribution). In this case, we have

$$\Pr[Y > y|X=x] = \begin{cases} \frac{\lambda_1}{\lambda_1+\lambda_2} \exp\left(-\lambda_{12}(y-x) - \lambda_2 y\right), & x \leq y, \\ \exp(-\lambda_2 y) & x \geq y; \end{cases}$$

see, for example, Barlow and Proschan (1981, p. 132). Clearly, this conditional survival function is nondecreasing in x, and so X and Y are SI$(Y|X)$. This in turn implies that X and Y are associated.

Example 3.5 (F-G-M bivariate exponential distribution). Rödel (1987) showed that, for an F-G-M distribution, X and Y are SI (i.e., positively regression dependent) if $\alpha > 0$. For the case with exponential marginals with $\alpha > 0$, a direct and easy proof for this result is

[1] $\Pr(Y > y|X=x)$ in x implies that $\Pr[a(X,Y) > a(x,y)|X=x]$ increases in x for every increasing function a defined on R^2. By using (3.15), we now have

$$\begin{aligned}E[a(X,Y)|X=x] = &-\int_{-\infty}^{0} \Pr[a(X,Y) \leq a(x,y)|X=x]dy \\ &+\int_{0}^{\infty} \Pr[a(X,Y) > a(x,y)|X=x]dy,\end{aligned}$$

which is therefore increasing in x.

$$\begin{aligned}\Pr(Y \le y|X=x) &= \{1-\alpha(2e^{-x}-1)\}(1-e^{-y})+\alpha(2e^{-x}-1)(1-e^{-2y})\\ &= (1-e^{-y})+\alpha(2e^{-x}-1)(e^{-y}-e^{-2y}),\end{aligned}$$

and so

$$\Pr(Y > y|X=x) = e^{-y}-\alpha(2e^{-x}-1)(e^{-y}-e^{-2y})$$

which is clearly increasing in x, from which we readily conclude that X and Y are positively regression dependent if $\alpha > 0$.

Example 3.6 (Kibble's bivariate gamma distribution). Rödel (1987) showed that Kibble's bivariate gamma distribution (see, e.g., Section 3.6.1) is also SI (i.e., PRD).

Example 3.7 (Sarmanov's bivariate exponential distribution). The conditional distribution is [Lee (1996)]

$$\Pr(Y \le y|X=x) = G(y)+\omega\phi_1(x)\int_{-\infty}^{y}\phi_2(z)g(z)dz,$$

where $\phi_i(x) = e^{-x}-\frac{\lambda_i}{1+\lambda_i}, i=1,2$. It then follows that

$$\Pr(Y > y|X=x) = e^{-\lambda_2 y}-\omega\left(e^{-x}-\frac{\lambda_1}{1+\lambda_1}\right)\int_{-\infty}^{y}\phi_2(z)g(z)dz$$

is increasing in x since $\int_{-\infty}^{y}\phi_2(z)g(z)dz \ge 0$, and so Y is SI increasing in x if $0 \le \omega \le \frac{(1+\lambda_1)(1+\lambda_2)}{\max(\lambda_1,\lambda_2)}$. Further, it follows from Lee (1996) that (X,Y) is TP_2 since $\omega\phi'(x)\phi'(y) \ge 0$ for $\omega \ge 0$.

Example 3.8 (Bivariate exponential distribution). We have

$$H(x,y) = 1-e^{-x}-e^{-y}+(e^x+e^y-1)^{-1}.$$

In this case, it can be shown easily that $\Pr(Y \le y|X=x) = 1+\frac{1}{(e^x+e^y-1)^2}$ and hence $\Pr(Y > y|X=x) = \frac{-1}{(e^x+e^y-1)^2}$, which is increasing in x; hence, Y is SI in X.

3.4.5 Left Corner Set Decreasing and Right Corner Set Increasing

X and Y are said to be *left corner set decreasing* (denoted by LCSD) if, for all x_1 and y_1,

$$\Pr(X \le x_1, Y \le y_1|X \le x_2, Y \le y_2) \text{ is decreasing in } x_2 \text{ and } y_2. \quad (3.16)$$

Similarly, we say that X and Y are *right corner set decreasing* (denoted by RCSI) if, for all x_1 and y_1,

$$\Pr(X > x_1, Y > y_1 | X > x_2, Y > y_2) \text{ is decreasing in } x_2 \text{ and } y_2. \tag{3.17}$$

By choosing $x_1 = -\infty$ and $y_2 = -\infty$ in (3.17), we see that RCSI$(Y|X) \Rightarrow$ RTI$(Y|X)$. We note that RCSI (LCSD) is on the same hierarchical order of stringency of dependence as SI(X|Y) (SI(Y|X)) are, and yet they do not seem to be directly related to each other.

3.4.6 Total Positivity of Order 2

The notation of a "totally positive" function of order was defined by Karlin (1968).

Definition 3.9. A function $f(x, y)$ is totally positive of order 2 (TP$_2$) if $f(x, y) \geq 0$ such that

$$\begin{vmatrix} f(x,y) & f(x,y' \\ f(x',y) & f(x',y') \end{vmatrix} \geq 0$$

whenever $x \leq x'$ and $y \leq y'$.

Let X and Y have a joint distribution function H, joint survival function $\bar{H}$, and joint density function $h(x, y)$. Then, we can define three types of total positive dependence, depending on whether we are basing it on $H, \bar{H}$, or h. We assume that $x_1 < x_2$, and $y_1 < y_2$ in the following definitions.

(i) We say that H is *totally positive of order 2* (H-TP$_2$) if

$$H(x_1, y_1)H(x_2, y_2) \geq H(x_1, y_2)H(x_2, y_1). \tag{3.18}$$

(ii) Similarly, $\bar{H}$ is said to be *totally positive of order 2* ($\bar{H}$-TP$_2$) if

$$\bar{H}(x_1, y_1)\bar{H}(x_2, y_2) \geq \bar{H}(x_1, y_2)\bar{H}(x_2, y_1). \tag{3.19}$$

(iii) Finally, we say that h is *totally positive of order 2* (h-TP$_2$) if

$$h(x_1, y_1)h(x_2, y_2) \geq h(x_1, y_2)h(x_2, y_1). \tag{3.20}$$

Abdel-Hameed and Sampson (1978) have presented a sufficient condition for $h(x, y)$ to be totally positive of order 2. Some authors refer to this property as X and Y being (positive) *likelihood ratio dependent* (denoted by LRD) since the inequality in (3.20) is equivalent to the requirement that the conditional density of Y given x have a monotone likelihood ratio.

Example 3.10 (Bivariate normal distribution). The bivariate normal density is TP_2 if and only if the correlation coefficient $0 \le \rho < 1$; see, for example, Barlow and Proschan (1981, p. 149).

Example 3.11 (Bivariate absolute normal distribution). Abdel-Hameed and Sampson (1978) have shown that the bivariate density of the absolute normal distribution is TP_2.

It is easy to see that h-TP_2 implies that both H and $\bar{H}$ are TP_2. It can also be shown [see, e.g., Nelsen (2006, pp. 199–201)] that LCSD is equivalent to H being TP_2, while RCSI is equivalent to $\bar{H}$ being TP_2.

Example 3.12 (Marshall and Olkin's bivariate exponential distribution). X and Y have Marshall and Olkin's bivariate exponential distribution with joint survival function

$$\bar{H}(x,y) = \exp\left[-\lambda_1 x - \lambda_2 y - \lambda_{12}\max(x,y)\right],$$

and so

$$\begin{aligned}&\bar{H}(x,y)\bar{H}(x',y')\\ &= \exp\left[-\lambda_1(x+x') - \lambda_2(y+y') - \lambda_{12}\{\max(x,y)+\max(x',y')\}\right]\end{aligned}$$

and

$$\begin{aligned}&\bar{H}(x,y')\bar{H}(x',y)\\ &= \exp\left[-\lambda_1(x+x') - \lambda_2(y+y') - \lambda_{12}\{\max(x',y)+\max(x,y')\}\right].\end{aligned}$$

Now, if $0 \le x \le x'$ and $0 \le y \le y'$, then

$$\max(x,y) + \max(x',y') \le \max(x',y) + \max(x,y').$$

It then follows that $\bar{H}(x,y)\bar{H}(x',y') \le \bar{H}(x,y')\bar{H}(x',y)$, and so $\bar{H}$ is TP_2, which is equivalent to X and Y being RCSI.

Note that if h is TP_2, then X and Y are LCSD $\Leftrightarrow$ H is TP_2. If h is TP_2, then X and Y are RSCI $\Leftrightarrow$ $\bar{H}$ is TP_2, i.e., h is TP_2 implies that both H and $\bar{H}$ are TP_2. Thus, as pointed out by Shaked (1977), the notion of h being TP_2 (positively likelihood dependent) is stronger than any notion of dependence we have discussed so far. We thus have the following implications:

$$\begin{array}{ccccccccc}
\mathrm{LRD(TP_2)} & \Rightarrow & \mathrm{SI}(Y|X) & \Rightarrow & \mathrm{RTI}(Y|X) & \Leftarrow & \mathrm{RCSI} & \Leftrightarrow & \bar{H}-\mathrm{TP}_2\\
 & & \Downarrow & & \Downarrow & & \Downarrow & & \\
\Downarrow & & \mathrm{LTD}(Y|X) & \Rightarrow & \mathrm{PQD} & \Leftarrow & \mathrm{RTI}(X|Y) & & \Uparrow\\
 & & \Uparrow & & \Uparrow & & \Uparrow & & \\
H-\mathrm{TP}_2 & \Leftrightarrow & \mathrm{LCSD} & \Rightarrow & \mathrm{LTD}(X|Y) & \Leftarrow & \mathrm{SI}(X|Y) & \Leftarrow & \mathrm{LRD(TP_2)}
\end{array}$$

3.4.7 $DTP_2(m, n)$ and Positive Dependence by Mixture

Shaked (1977) used the classical theory of total positivity to construct a family of concepts of dependence called *dependent by total positivity of order 2 with degree (m,n)*, denoted by DTP(m, n). He then showed that DTP(0,0) is equivalent to positive likelihood ratio dependence (LRD) and that DTP(1,1) is equivalent to RCSI. In different applied situations, especially in reliability theory and genetic studies, positive dependence by mixture is often assumed.

If (X, Y) are any two random variables, independent conditionally with respect to a (latent) variable W with distribution function K, then their joint distribution function is

$$H(x, y) = \int F^w(x)G^w(y)dK(w), \tag{3.21}$$

where $F^w(x)$ and $G^w(y)$ are the distribution functions of X and Y, given W. Using the properties of TP_2 functions, it is easy to associate a concept of dependence with the pair (X, Y). More precisely, if the joint distributions of the pair (X, W) and (Y, W) are DTP$(m, 0)$ and DTP$(n, 0)$, respectively, then the pair (X, Y) is DTP(m, n); see, for example, Shaked (1977). In particular, (X, Y) is DTP(0,0) (i.e., X and Y are LRD) if (X, W) and (Y, W) have LRD.

3.5 Concepts of Positive Dependence Weaker than PQD

3.5.1 Positive Quadrant Dependence in Expectation

We now present a slightly less stringent dependence notion than PQD. For any real number x, let Y_x be the random variable with distribution function $\Pr(Y \leq y|X > x)$. It is easy to verify that the inequality in the conditional distribution $\Pr(Y \leq y|X > x) \leq \Pr(Y \leq y)$ implies an inequality in expectation $E(Y_x) \geq E(Y)$ if Y is a non-negative random variable. We then say that Y is *positive quadrant dependent in expectation* on X (PQDE) if this last inequality involving expectations holds. Similarly, we say that there is *negative quadrant dependence in expectation* if $E(Y_x) \leq E(Y)$.

It is easy to show that the PQD$\Rightarrow$PQDE by observing that PQD is equivalent to $\Pr(Y > y|X > x) \geq \Pr(Y > y)$, which in turn implies $E(Y_x) \geq E(Y)$, assuming $Y \geq 0$. This establishes the fact that PQDE is a weaker concept than PQD.

3.5.2 Positively Correlated Distributions

We say that X and Y are *positively correlated* if

$$\operatorname{cov}(X, Y) \geq 0. \tag{3.22}$$

Now,

$$\begin{aligned}\operatorname{cov}(X, Y) &= \int\int \left[\bar{H}(x, y) - \bar{F}(x)\bar{G}(y)\right] dx\, dy \\ &= \int \bar{F}(x) \left(\int \left[\Pr(Y > y | X > x) - \bar{G}(y)\right] dy \right) dx \\ &= \int \bar{F}(x) \left\{E(Y_x) - E(Y)\right\} dx,\end{aligned}$$

which is ≥ 0 if X and Y are PQDE. Thus, PQDE implies that $\operatorname{cov}(X, Y) \geq 0$. This means that PQDE lies between PQD and positive correlation. Many bivariate random variables are PQDE since all the PQD distributions with $Y \geq 0$ are also PQDE.

Positive correlation is the weakest notion of dependence between two random variables X and Y. It is indeed easy to construct a positively correlated bivariate distribution. For example, such a distribution may be obtained by simply adopting a well-known trivariate reduction technique as follows: Set $X = X_1 + X_3$, $Y = X_2 + X_3$, with X_i $(i = 1, 2, 3)$ being mutually independent random variables, so that the correlation coefficient between X and Y is

$$\rho = \frac{\operatorname{var}(X_3)}{\sqrt{\operatorname{var}(X_1 + X_3)\operatorname{var}(X_2 + X_3)}} > 0.$$

3.5.3 Monotonic Quadrant Dependence Function

As described above, PQDE is based on a comparison of $E(Y_x)$ with $E(Y)$. Kowalczyk and Pleszczyńska (1977) introduced the *monotonic quadrant dependence function* to quantify the difference between these two expectations.

Let B be the set of all bivariate random variables with finite marginal means, and let x_p and y_p denote the pth quantiles of X and Y, respectively $(0 < p < 1)$. For each $(X, Y) \in B$, we define a difference function

$$L_{Y,X}(p) = E(Y | X > x_p) - E(Y). \tag{3.23}$$

We may then define a function that can be used as a measure of the strength of the monotonic quadrant dependence between X and Y as follows. With

$$\mu^{+}_{Y,X}(p) = \frac{L_{Y,X}(p)}{E(Y|Y > y_p) - E(Y)} \tag{3.24}$$

and

$$\mu^{-}_{Y,X}(p) = \frac{L_{Y,X}(p)}{E(Y) - E(Y|Y < y_{1-p})}, \tag{3.25}$$

we define

$$\mu_{Y,X}(p) = \begin{cases} \mu^{+}_{Y,X} & \text{if } L_{Y,X}(p) \geq 0 \\ \mu^{-}_{Y,X} & \text{if } L_{Y,X}(p) \leq 0 \end{cases}. \tag{3.26}$$

The function $\mu_{Y,X}$ is called the *monotonic quadrant dependence function.* Described in words, it is a function that compares the improvement in prediction of Y from knowing that X is big to the improvement in prediction of Y from knowing that X is small.

Interpretation of $\mu_{Y,X}$

From the definition above, we see that $\mu_{Y,X}$ is a suitably normalized expected value of Y under the condition that X exceeds its pth quantile. It is a measure of the strength of the monotonic quadrant dependence between X and Y in the following sense. Let (X, Y) and (X', Y') be two pairs of random variables from B having identical marginal distributions; then, the positive quadrant dependence between X and Y is said to be stronger than X' and Y' if $\mu_{Y,X}(p) \geq \mu_{Y',X'}(p)$ for all p between 0 and 1. This is because $E(Y|X > x_p) > E(Y'|X' > x_p)$ is equivalent to $\mu_{Y,X}(p) \geq \mu_{Y',X'}(p)$. The PQD is strongest when $\mu_{Y,X}(p) = 1$ and weakest when $\mu_{Y,X}(p) = -1$. Instead of B, if we consider only distributions for which $E(Y_x) \geq E(Y)$, then PQD is weakest when $\mu_{Y,X}(p) = 0$.

Properties of $\mu_{Y,X}$

The monotonic quadrant dependence function $\mu_{Y,X}(p)$ introduced above has the following properties:

- $-1 \leq \mu_{Y,X}(p) \leq 1$.
- $\mu_{Y,X}(p) = 1 \Leftrightarrow \Pr(X < x_p \text{ and } Y > y_p) = \Pr(X > x_p \text{ and } Y < y_p) = 0$.
- $\mu_{Y,X}(p) = -1 \Leftrightarrow \Pr(X < x_p,\ Y < y_{1-p}) = \Pr(X > x_p,\ Y > y_{1-p}) = 0$.
- Let k and l be functions such that $F(a) < F(b) \Rightarrow k(a) < k(b)$ and $l(a) > l(b)$. Then, for any real a and b ($a \neq 0$),

$$\mu_{aY+b,k(X)}(p) = (\text{sgn } a)\mu_{Y,X}(p),$$
$$\mu_{aY+b,k(X)}(p) = (-\text{sgn } a)\mu_{Y,X}(1-p).$$

- $\mu_{Y,X}(p) = 0$ if and only if $E(Y|X) = E(X)$ almost everywhere (i.e., the probability that they are unequal is 0).
- $\mu_{Y,X}(p)$ is $\mu^{+}_{Y,X}(p)$ if X and Y are PQDE and is $\mu^{-}_{Y,X}(p)$ if X and Y are NQDE.
- If X and Y are PQD, then $\mu_{X,Y} \geq 0$ and $\mu_{Y,X} \geq 0$.
- If X and Y are either PQD or NQD, then $\mu_{X,Y}(p) = 0$ if and only if X and Y are independent.
- If the distributions of (X, Y) and (X', Y') are both in B they have the same marginals, then $\mu_{Y,X} = \mu_{Y',X'}$ if and only if $E(Y|X)$ and $E(Y'|X')$ have the same distribution.

Remarks

The following observations about the monotonic quadrant dependence function are worth making:

- $\mu_{Y,X}$ is a function of p and thus takes on different values for different choices of p.
- $\mu_{Y,X}$ is not symmetric in X and Y; thus, it is more similar to a prediction-improvement index than to a conventional measure of correlation.
- $\mu_{Y,X}$ is invariant under increasing transformation of X and linear increasing transformation of Y. Note that the product-moment correlation, in contrast, is invariant under linear increasing transformations of both X and Y.
- For sample counterparts of $\mu_{Y,X}$, see Kowalczyk (1977). Kowalczyk and Ledwina (1982) discussed the grade monotone dependence function $\mu_{G(Y),F(X)}$, while Kowalczyk (1982) provided some interpretations.

3.5.4 Summary of Interrelationships

The most common dependence property is actually a "lack of dependence" property; viz., independence. If X and Y are two continuous random variables with joint distribution function $H(x, y)$, independence of X and Y is a property of the joint distribution function; i.e., $H(x, y) = F(x)G(y)$.

Given that X and Y are not independent, TP_2 is the strongest positive dependence concept we have introduced so far. On the other end, positive correlation is the weakest positive dependence. The positive quadrant dependence (PQD) is a common one among the positive dependence concepts, and we have therefore used it as a benchmark for comparing the strength of dependence between X and Y. Thus, we have conveniently divided various concepts of dependence into two categories: one consisting of bivariate distributions with dependence stronger than PQD and the other consisting of bivariate distributions with dependence weaker than PQD.

We have summarized below interrelationships between different dependence concepts after removing equivalent concepts (in which Y is conditional on X whenever a conditioning is involved in the definition):

$$\begin{array}{ccccccccc} \text{RSCI} & \Rightarrow & \text{RTI} & \Rightarrow & \text{ASSOC} & \Rightarrow & \text{PQD} & \Rightarrow & \text{PQDE} \\ \Uparrow & & \Uparrow & & \Uparrow & & & & \Downarrow \\ \text{LRD(TP}_2) & \Rightarrow & \text{SI (PRD)} & \Rightarrow & \text{LTD} & & & & \text{cov} \geq 0 \end{array}$$

Another account of some of these interrelationships is due to Ohi and Nishida (1978).

3.6 Families of Bivariate PQD Distributions

Consider a system of two components that are arranged in series. By assuming that the two components are independent when they are in fact positively quadrant dependent, we will underestimate the system reliability. For systems in parallel, on the other hand, assuming independence when components are in fact positively quadrant dependent will lead to overestimation of the system reliability. This is because the other component will fail earlier knowing that the first has failed. This, from a practical point of view, reduces the effectiveness of adding parallel redundancy. Thus, a proper knowledge of the extent of dependence among the components in a system will enable us to obtain a more accurate estimate of the reliability characteristic of the system under study.

Since the PQD concept is important in reliability applications, it is imperative for a reliability practitioner to know what kinds of bivariate PQD distributions are available for reliability modeling. In this section, we list several well-known bivariate PQD distributions, some of which were originally derived from a reliability perspective. Most of these bivariate PQD distributions can be found, for example, in Hutchinson and Lai (1990).

As mentioned earlier, the concept of PQD is quite useful in reliability applications; see Barlow and Proschan (1981) and Lai (1986). Before presenting further applications of PQD, we need to state the following result due to Lehmann (1966). Let r and s be a pair of real functions on R^n that are monotone in each of their n arguments. The functions r and s are said to be *concordant* in the ith argument if the directions of the monotonicity for the ith argument are the same (i.e., both functions are either simultaneously increasing or simultaneously decreasing in the ith argument while all others are kept fixed) and *discordant* if the directions are opposite. Let (X_i, Y_i), $i = 1, 2, \ldots, n$, be n independent pairs each satisfying PQD. Suppose r and s are concordant in each of these arguments. Then

$$\text{cov}[r(X_1, \ldots, X_n), s(Y_1, \ldots, Y_n)] \geq 0. \tag{3.27}$$

The result has the following implications [see also Jogdeo (1982)]:

1. Let $r(X_1, X_2) = \text{sgn}(X_2 - X_1)$ and $s(Y_1, Y_2) = \text{sgn}(Y_2 - Y_1)$. Then, $\tau = \text{cov}[\text{sgn}(X_2 - X_1), \text{sgn}(Y_2 - Y_1)]$, where τ is Kendall's tau. From (3.27), the condition PQD implies $\tau \geq 0$.
2. Spearman's $\rho_S = \text{cov}[\text{sgn}(X_2 - X_1), \text{sgn}(Y_3 - Y_1)]$. On letting

$$r(X_1, X_2, X_3) = X_2 - X_1 \text{ and } s(Y_1, Y_2, Y_3) = Y_3 - Y_1,$$

we see $\rho_S \geq 0$ under PQD.
3. Blomqvist (1950) proposed $(2p_n - 1)$ as a measure of dependence, with p_n being the proportion of pairs (X_i, Y_i) that fall in either the positive or the negative quadrants formed by the lines $X = \tilde{x}, Y = \tilde{y}$, where $\tilde{x}$ and $\tilde{y}$ are the medians of X and Y, respectively. The expectation of this measure is given by

$$E(2p_n - 1) = 2[\text{cov}(I_{\{X_i \geq \tilde{x}\}}, I_{\{Y_i \geq \tilde{y}\}}) + \text{cov}(I_{\{X_i \leq \tilde{x}\}}, I_{\{Y_i \leq \tilde{y}\}})], \quad (3.28)$$

which is ≥ 0 under PQD.

The class of all PQD distributions with fixed marginals has been shown by Bhaskara Rao et al. (1987) to be *convex*; that is, if H_1 and H_2 are both PQD, then so is $\lambda H_1 + (1 - \lambda) H_2$, for $0 \leq \lambda \leq 1$.

3.6.1 Bivariate PQD Distributions with Simple Structures

Some of the bivariate distributions whose PQD property can be established easily are now presented.

Example 3.13 (Farlie–Gumbel–Morgenstern bivariate distribution). We have

$$H_\alpha(x, y) = F(x)G(y)\left[1 + \alpha\left(1 - F(x)\right)\left(1 - G(y)\right)\right], \qquad x, y \geq 0. \quad (3.29)$$

The family above, denoted by F-G-M, is a general system of bivariate distributions widely studied in the literature. It is easy to verify that X and Y are positively quadrant dependent if $\alpha > 0$.

Consider the special case of the F-G-M system where both marginals are exponential. The joint distribution function in (3.29) is then of the form [see, e.g., Kotz et al. (2000)]

$$H(x, y) = (1 - e^{-\lambda_1 x})(1 - e^{-\lambda_2 y})\left(1 + \alpha e^{-\lambda_1 x - \lambda_2 y}\right), \qquad x, y \geq 0.$$

Evidently,

$$\begin{aligned} w(x,y) &= H(x,y) - F(x)G(y) \\ &= \alpha e^{-\lambda_1 x - \lambda_2 y}(1 - e^{-\lambda_1 x})(1 - e^{-\lambda_2 y}), \qquad 0 < \alpha \leq 1 \\ &\geq 0, \end{aligned}$$

and X and Y are therefore PQD.

Mukerjee and Sasmal (1977) discussed several properties of a system of two exponential components having the F-G-M distribution, and these included the densities, means, moment generating functions, and tail probabilities of $\min(X,Y)$, $\max(X,Y)$, and $X+Y$, which are relevant to series, parallel, and standby systems, respectively. Lingappaiah (1983) was also concerned with properties of the F-G-M distribution with gamma marginals.

Based on an earlier work of Philips (1981), Kotz and Johnson (1984) considered a model in which components 1 and 2 were subject to "revealed" and "unrevealed" faults, respectively, with (X,Y) having an F-G-M distribution, where X is the time between unrevealed faults and Y is the time from an unrevealed fault to a revealed fault.

Example 3.14 (Bivariate exponential distribution). We have as the joint distribution function

$$H(x,y) = 1 - e^{-x} - e^{-y} + (e^x + e^y - 1)^{-1}, \qquad x, y \geq 0.$$

This distribution has both its marginals exponential. The joint distribution function above can be rewritten as

$$\begin{aligned} H(x,y) &= 1 - e^{-x} - e^{-y} + e^{-(x+y)} + (e^x + e^y - 1)^{-1} - e^{-(x+y)} \\ &= F(x)G(y) + (e^x + e^y - 1)^{-1} - e^{-(x+y)}. \end{aligned}$$

Now, $(e^x + e^y - 1)^{-1} - e^{-(x+y)} = \frac{(e^x-1)(e^y-1)}{(e^x+e^y-1)e^{(x+y)}} = \frac{(1-e^{-x})(1-e^{-y})}{(e^x+e^y-1)} \geq 0$, and H is therefore PQD.

Example 3.15 (Bivariate Pareto distribution). We have as the joint survival function

$$\begin{aligned} \bar{H}(x,y) &= 1 - F(x) - G(y) + H(x,y) \\ &= (1 + x + y)^{-a}, \qquad a > 0; \end{aligned}$$

see, for example, Mardia (1970) and Kotz et al. (2000). Consider a system of two independent exponential components that share a common environment factor η that can be described by a gamma distribution. Then, Lindley and Singpurwalla (1986) showed that the resulting joint distribution has the bivariate Pareto distribution above. It is very easy to verify that this joint distribution is PQD since $(1+x+y)^{-a} \geq (1+x)^{-a}(1+y)^{-a}$. For a generalization to the multivariate case, see Nayak (1987).

Example 3.16 (Durling–Pareto distribution). We have as the joint survival function

$$\bar{H}(x,y) = (1+x+y+kxy)^{-a}, \qquad 0 \le k \le a+1,\ x,y \ge 0. \tag{3.30}$$

Clearly, this is a generalization of the bivariate Pareto example above. Consider a system of two dependent exponential components having Gumbel's bivariate exponential distribution $H(x,y) = 1 - e^{-x} - e^{-y} + e^{-x-y-\theta xy}$, $x, y \ge 0$, $0 \le \theta \le 1$, and sharing a common environment that has a gamma distribution. Sankaran and Nair (1993) then showed that the resulting bivariate distribution is given by (3.30). It follows from (3.30) that

$$\begin{aligned}
&\bar{H}(x,y) - \bar{F}(x)\bar{G}(y)\\
&= \frac{1}{(1+x+y+kxy)^a} - \frac{1}{\{(1+x)(1+y)\}^a}, \quad 0 \le k \le (a+1)\\
&= \frac{1}{(1+x+y+kxy)^a} - \frac{1}{\{1+x+y+xy)\}^a} \ge 0, \quad 0 \le k \le 1.
\end{aligned}$$

H is therefore PQD if $0 \le k \le 1$.

Example 3.17 (Marshall and Olkin's bivariate exponential distribution). We have as the joint survival function

$$P(X > x, Y > y) = \exp\{-\lambda_1 x - \lambda_2 y - \lambda_{12}\max(x,y)\}, \ \lambda_1, \lambda_2, \lambda_{12} \ge 0. \tag{3.31}$$

This has become a widely used bivariate exponential distribution over the last four decades after being derived by Marshall and Olkin (1967) in the reliability context as follows. Suppose we have a two-component system subjected to shocks that are always fatal. These shocks are assumed to be governed by three independent Poisson processes with parameters λ_1, λ_2, and λ_{12}, according to whether the shock applies to component 1 only, component 2 only, or to both components, respectively. Then, the joint survival function for the two components is given by (3.13). Barlow and Proschan (1981, p. 129) showed that X and Y are PQD.

Example 3.18 (Block and Basu's bivariate exponential distribution). For θ, $x, y \ge 0$, the joint survival function is

$$\bar{H}(x,y) = \frac{2+\theta}{2}\exp\left[-x - y - \theta\max(x,y)\right] - \frac{\theta}{2}\exp\left[-(2+\theta)\max(x,y)\right].$$

This was constructed by Block and Basu (1976) to modify Marshall and Olkin's bivariate exponential distribution, which has a singular part. It is, in fact, a reparametrization of a special case of Freund's (1961) bivariate exponential distribution. The marginal survival function of X is $\bar{F}(x) = \frac{1+\theta}{2}\exp\left[-(1+\theta)x\right] - \frac{\theta}{2}\exp\left[(1+\theta)x\right]$, and a similar expression exists for $\bar{G}(y)$. It is then easy to show that this distribution is PQD.

Example 3.19 (Kibble's bivariate gamma distribution). The joint density function is, for $0 \le \rho < 1$ and $x, y, \alpha \ge 0$,

$$h_\rho(x, y; \alpha)$$
$$= f_\alpha(x)g_\alpha(y)\exp\left[-\frac{\rho(x+y)}{1-\rho}\right] \times \frac{\Gamma(\alpha)}{1-\rho}(xy\rho)^{-(\alpha-1)/2} I_{\alpha-1}\left(\frac{2\sqrt{xy\rho}}{1-\rho}\right),$$

where $I_\alpha(\cdot)$ is the modified Bessel function of the first kind of the αth order. Lai and Moore (1984) showed that the distribution function is given by

$$H(x, y; \rho) = F(x)G(y) + \alpha \int_0^\rho f_t(x, y; \alpha+1)dt \geq F(x)G(y)$$

since $\alpha \int_0^\rho f_t(x, y; \alpha+1)dt$ is clearly positive.

For the special case where $\alpha = 1$, Kibble's bivariate gamma distribution presented above becomes the well-known Moran–Downton bivariate exponential distribution; see Downton (1970). Thus, the Moran–Downton bivariate exponential distribution in particular and Kibble's bivariate gamma distribution in general are PQD.

Example 3.20 (Bivariate normal distribution). The bivariate normal distribution has as its density function

$$h(x, y) = \left(2\pi\sqrt{1-\rho^2}\right)^{-1} \exp\left[-\{1/2(1-\rho^2)\}(x^2 - 2\rho xy + y^2)\right]$$

for $-\infty < x, y < \infty$ and $-1 < \rho < 1$. In this case, X and $Y$are PQD for $0 \leq \rho < 1$, and NQD for $-1 < \rho \leq 0$. This result follows easily from the following lemma.

Lemma 3.21. *Let* (X_1, Y_1) *and* (X_2, Y_2) *follow standard bivariate normal distributions with correlation coefficients* ρ_1 *and* ρ_2*, respectively. If* $\rho_1 \geq \rho_2$*, then* $\Pr(X_1 > x, Y_1 > y) \geq \Pr(X_2 > x, Y_2 > y)$.

This is known as *Slepian's inequality* [see Gupta (1963, p. 805)]. By letting $\rho_2 = 0$ (thus, $\rho_1 \geq 0$), we establish that X and Y are PQD. On the other hand, letting $\rho_1 = 0$ (thus $\rho_2 \leq 0$), X and Y are then NQD.

3.6.2 Construction of Bivariate PQD Distributions

Let $H(x, y)$ denote the joint distribution function of (X, Y) having continuous marginal c.d.f.'s $F(x)$ and $G(y)$ and with marginal p.d.f.'s $f = F'$ and $g = G'$, respectively. For a bivariate PQD distribution, the joint distribution function may be written as

$$H(x, y) = F(x)F(y) + w(x, y),$$

with $w(x, y)$ satisfying the following conditions:

(i) $w(x, y) \geq 0$.

(ii) $w(x,\infty) \to 0$, $w(\infty,y) \to 0$, $w(x,-\infty) = 0$, $w(-\infty,y) = 0$.
(iii) $\frac{\partial^2 w(x,y)}{\partial x \partial y} + f(x)f(y) \geq 0$.

Note that if both $X \geq 0$ and $Y \geq 0$, then condition (ii) may be replaced by

$$w(x,\infty) \to 0, \quad w(\infty,y) \to 0, \quad w(x,0) = 0, \quad w(0,y) = 0.$$

Lai and Xie (2000) used these conditions to construct a family of bivariate PQD distributions with uniform marginals.

Example 3.22 (Ali–Mikhail–Haq family). Consider the bivariate family of distributions associated with the copula

$$C(u,v) = \frac{uv}{1-\theta(1-u)(1-v)}, \qquad \theta \in [0,1].$$

It is clear that the copula is PQD. In fact, it was shown in Section 2.9 that this is a special case of the Lomax copula (the survival copula that corresponds to the bivariate Lomax; viz., the Durling–Pareto distribution) given in Section 2.8.

Nelsen (2006, p. 188) has pointed out that if X and Y are PQD, then their copula C is also PQD. Nelsen (1999) has provided a comprehensive treatment on copulas and several examples of PQD copulas can be found therein.

3.6.3 Tests of Independence Against Positive Dependence

Let us consider the problem of testing the null hypothesis of independence,

$$H_0: \quad H(x,y) = F(x)G(y), \quad \text{for all } x, y,$$

against the alternative of positive quadrant dependence,

$$H_A: \quad H(x,y) \geq F(x)G(y), \quad \text{for all } x, y,$$

with strict inequality holding on a set of nonzero probability. This problem was first considered by Lehmann (1966), who proposed the Kendall's tau and Spearman's correlation tests. Since then, a large number of tests have been proposed in the literature for this hypothesis testing problem; see, for example, Joag-Dev (1984) and Schriever (1987b).

On the basis of a random sample $(X_1,Y_1),\ldots,(X_n,Y_n)$ from the distribution H, we wish to test H_0 against H_A. Let $k \geq 2$ be a fixed positive integer, and consider the following kernels:

$$\phi_{1k}\left((x_1, y_1), \ldots, (x_k, y_k)\right) = \begin{cases} 1 \text{ if } (\max_{1\le i\le k} x_i, \max_{1\le i\le k} y_i) \text{ belongs to the same pair} \\ 0 \text{ otherwise} \end{cases}$$

and

$$\phi_{2k}\left((x_1, y_1), \ldots, (x_k, y_k)\right) = \begin{cases} 1 \text{ if } (\min_{1\le i\le k} x_i, \min_{1\le i\le k} y_i) \text{ belongs to the same pair} \\ 0 \text{ otherwise.} \end{cases}$$

For skewed distributions, which arise particularly when the random variables are non-negative, as in the case of reliability applications, Kochar and Gupta (1987) proposed a class of distribution-free statistics based on U-statistics defined by

$$U_k = \frac{1}{\binom{n}{k}} \sum \phi_{1k}((x_{i_1}, y_{i_1}), \ldots, (x_{i_k}, y_{i_k})),$$

where the summation is over all combinations of k integers $(i_1, i_2, \ldots, i_k)$ chosen out of $(1, 2, \ldots, n)$. Large values of U_k are significant for testing H_0 against H_A. Evidently, U_2 is the well-known Kendall's tau statistic. Kochar and Gupta (1987) observed that these tests are quite efficient for skewed distributions.

Let $\phi_k = \phi_{1k} + \phi_{2k}$. Kochar and Gupta (1990) then proposed another class of distribution-free tests based on the U-statistics corresponding to the kernel ϕ_k, defined by

$$V_k = \frac{1}{\binom{n}{k}} \sum \phi_k((x_{i_1}, y_{i_1}), \ldots, (x_{i_k}, y_{i_k})),$$

where the summation is over all combinations of k integers $(i_1, i_2, \cdots, i_k)$ chosen out of $(1, 2, \ldots, n)$. Yet again, V_2 is the well-known Kendall's tau statistic. Large values of V_k are significant for testing H_0 against H_1. In this case, Kochar and Gupta (1990) found these tests to be quite efficient for symmetric distributions.

Ledwina (1986) also considered two rank tests for testing independence against positive quadrant dependence. These test statistics are closely related to the monotonic quadrant dependence function described in Section 3.5.3.

3.6.4 Geometric Interpretations of PQD and Other Positive Dependence Concepts

Geometric interpretations of positive dependence may be provided via copulas. Graphs and contour diagrams of Fréchet upper and lower bounds C^+ and C^- and the independence copula $C^0(u, v) = uv$, are given in Nelsen (1999,

p. 10). Nelsen (1999, p. 152) has also shown that X and Y are PQD if and only if $C(u,v) \geq uv$ from which it is concluded that if X and Y are PQD, then the graph of the copula of X and Y lies on or above the graph of the independence copula.

There are similar geometric interpretations of the graph of the copula when the random variables satisfy one or more of the tail monotonicity properties (LTD and RTI). These interpretations involve the shape of regions determined by the horizontal and vertical sections of the copula [Nelsen (1999, pp. 156–157)].

3.7 Additional Concepts of Dependence

Shaked (1979) introduced further ideas of positive dependence, applicable to exchangeable bivariate random vectors (i.e., random vectors with permutation-invariant distributions). These include the following concepts:

- Diagonal square dependent (denoted by DSD).
- Generalized diagonal square dependent (denoted by GDSD).
- Positive dependent by mixture (denoted by PDM). A bivariate distribution is PDM if it can be expressed as a mixture of the form given in (3.21).
- Positive dependent by expansion (denoted by PDE).
- Positive definite dependent (denoted by PDD).

Definitions of DSD and PDD are as follows:

- DSD means that $\Pr(X \in I \text{ and } Y \in I) \geq \Pr(X \in I)\Pr(Y \in I)$.
- PDD means that $\mathrm{cov}(a(X), a(Y)) \geq 0$ for every real function a for which the covariance exists.
- For definitions and explanations of the others, one may refer to Shaked (1979).

These dependence concepts have interrelationships that can be summarized as follows:

$$
\begin{array}{ccc}
 & \text{PDE} & \\
 & \Downarrow & \\
\text{PDM} \Rightarrow & \text{PDD} & \Rightarrow \text{GDSD} \Rightarrow \text{DSD} \\
 & \Downarrow & \\
 & \mathrm{cov}(X,Y) \geq 0 &
\end{array}
$$

3.8 Negative Dependence

Having defined several concepts of dependence for the bivariate case, we can easily obtain analogous concepts of negative dependence as follows. If (X, Y) has a positive dependence, then $(X, -Y)$ on R^2, or if we have a constraint of positivity $(X, 1-Y)$ on the unit square, it has a negative dependence. However, if we have more than two variables, reversing the definition of positive dependence does not allow us to retain the same appealing properties.

The negative dependence was first introduced by Lehmann (1966), and this concept was further developed by others. All of them can be obtained by negative analogues of positive dependence; viz., when the inequality signs in (3.1), (3.7), and (3.20) are reversed, we obtain negative dependence. For example, the negative analogue of PQD is *negative quadrant dependent* (denoted by NQD), and there are concepts of NRD (*negatively regression dependent*), RCSD (*right corner set decreasing*), and RTD (*right-tail decreasing*) . However, "association" has no negative analogue since the definition refers to *every* pair of functions a and b, and the choice $a = b$ will necessarily lead to $\text{cov}[a(X, Y), a(X, Y)] \geq 0$.

Negative association of $X_1, X_2, \ldots, X_k$ is defined in a different way than the positive association given in Section 3.4.2.

Definition 3.23 (Joag-Dev and Proschan (1983)). $X_1, X_2, \ldots, X_n$ are said to be *negatively associated* (denoted by NA) if, for every pair of disjoint subsets A_1 and A_2 of $\{1, 2, \ldots, n\}$,

$$\text{cov}[a(X_i, i \in A_1), b(X_j, j \in A_2)] \leq 0 \tag{3.32}$$

whenever a and b are increasing.

Joag-Dev and Proschan (1983) pointed out that for a pair of random variables X and Y, NA is equivalent to NQD. This definition of the concept also leads to several properties; most of them are in the multivariate setting. Among these are the following:

(1) A subset of two or more NA random variables is NA.
(2) A set of independent random variables is NA.
(3) Increasing functions of a set of NA random variables are NA.
(4) The union of independent sets of NA random variables is NA.

For a further generalization of this concept, see Kim and Seo (1995).

Block et al. (1982a,b, 1988), Ebrahimi and Ghosh (1981, 1982), Karlin and Rinott (1980), Lee (1985), and Kim and Seo (1995) have all introduced and studied some other concepts of multivariate negative dependence; see also the pertinent references in Block et al. (1985).

Lehmann (1966) defined the concept of *negative likelihood ratio dependence.* This was called *reverse regular of order 2* (denoted by RR_2) by Karlin and Rinott (1980) and Block et al. (1982a). The latter authors showed that under a condition that essentially requires the sum of three independent r.v.'s to be fixed, two of them satisfy the RR_2 condition. They also showed further that $RR_2 \Rightarrow$ NQD.

These concepts of negative dependence have interrelationships that can be summarized as follows:

$$\begin{array}{ccccccc} & & & \text{NA} & & \\ & & & \Updownarrow & & \\ \text{RCSD} & \Rightarrow & \text{RTD} \Rightarrow & \text{NQD} & \Rightarrow & \text{cov} \le 0 \\ & & & \Uparrow & & \\ & & & \text{RR}_2 & & \end{array}$$

3.8.1 Neutrality

It is important to mention one more context where negative dependence is more natural than positive dependence: when concerned with three proportion probabilities, X_1, X_2, and X_3, that add to one, and we focus our attention on only two of them. Then, (X_1, X_2) is distributed over a triangle. The percentage composition of different minerals in rocks is an example, and the percentage of household expenditures spent on different groups of commodities is another.

The two variables are often taken to have a bivariate beta distribution. The idea of *neutrality* was introduced by Connor and Mosimann (1969) as follows. X_1 and X_2 are said to be *neutral* if X_i and $X_j/(1 - X_i)$ are independent $(i \neq j)$. It is well known that if X_1 and X_2 have a bivariate beta distribution, then they are neutral, and the converse is also true [Fabius (1973)]. It was pointed out by Lehmann (1966) that the bivariate beta is RR_2; hence, it is also NQD. Negative covariance can also be observed quite easily in this case.

A thorough account of the concept of neutrality is by Mosimann (1988); see also Mosimann (1975) and Mosimann and Malley (1981). We also note here that quite often variables that sum to 1 are obtained by dividing more basic variables by their total, as in $X = X_1/(X_1 + X_2 + X_3), Y = X_2/(X_1 + X_2 + X_3)$, and the spurious correlation may arise through the division by the same quantity; see Pendleton (1986) and Prather (1988).

3.8.2 Examples of NQD

Several bivariate distributions discussed in Section 3.6, such as the bivariate normal, F-G-M family, Durling–Pareto distribution, and bivariate exponential of Sarmanov are all NQD when the range of the dependence parameter is suitably restrained. The two variables in the following example can only be negatively dependent.

Example 3.24 (Gumbel's bivariate exponential distribution). The joint survival function is

$$H(x, y) = 1 - e^{-x} - e^{-y} + e^{-(x+y+\theta xy)}, \qquad 0 \leq \theta \leq 1,$$

so that

$$H(x, y) - F(x)G(y) = e^{-(x+y+\theta xy)} - e^{-x} - e^{-y} \leq 0, \quad 0 \leq \theta \leq 1,$$

showing that F is NQD. In this case, it is known that $-0.40365 \leq \operatorname{corr}(X, Y) \leq 0$; see Kotz et al. (2000, p. 351).

Example 3.25. Lehmann (1966) presented the following situation in which negative quadrant dependence occurs naturally. Consider the rankings of n objects by m persons. Let X and Y denote the rank sum for the ith and jth objects, respectively. Then, X and Y are NQD.

3.9 Positive Dependence Orderings

Consider two bivariate distributions having the same pair of marginals F and G, and assume that both are positively dependent. Naturally, we would like to know which of the two bivariate distributions is more positively dependent. In other words, we wish to order the two given bivariate distributions by the extent of their positive dependence between the two marginal variables, with higher in ordering meaning more positively dependent. In this section, the concept of *positive dependence ordering* is introduced.

For a comprehensive treatment of dependence orderings, see Joe (1997). Section 3.6 of Drouet-Mari and Kotz (2001) also contains a good summary on this subject.

Throughout this section, we let H and H' denote the bivariate distribution functions of (X, Y) and (X', Y'), respectively, having common marginal distributions F and G. We shall now introduce some (partial) orderings that compare the strength of positive dependence of (X, Y) with that of (X', Y'). The following definition is the one given by Kimeldorf and Sampson (1987).

Definition 3.26. A relation $\ll$ on a family of all bivariate distributions is a positive dependence ordering (denoted by PDO) if it satisfies the following ten conditions:

(P0) $H \ll H' \Rightarrow H(x, \infty) = H'(x, \infty)$ and $H(\infty, y) = H'(\infty, y)$;
(P1) $H \ll H' \Rightarrow H(x, y) \leq H'(x, y)$ for all x, y;
(P2) $H \ll H'$ and $H' \ll H^* \Rightarrow H \ll H^*$;

(P3) $H \ll H$;
(P4) $H \ll H'$ and $H' \ll H \Rightarrow H = H'$;
(P5) $H^- \ll H \ll H^+$, where $H^+(x,y) = \min[H(x,\infty), H(\infty,y)]$ and $H^-(x,y) = \max[H(x,\infty) + H(\infty,y) - 1,\ 0]$;
(P6) $(X,Y) \ll (U,V) \Rightarrow (a(X),Y) \ll (a(U),V)$, where $(X,Y) \ll (U,V)$ means the relation $\ll$ holds between the corresponding bivariate distributions;
(P7) $(X,Y) \ll (U,V) \Rightarrow (-U,V) \ll (-X,Y)$;
(P8) $(X,Y) \ll (U,V) \Rightarrow (Y,X) \ll (V,U)$;
(P9) $H_n \ll H'_n, H_n \to H$ in distribution, $H'_n \to H'$ in distribution $\Rightarrow H \ll H'$, where H_n, H, H'_n, H' all have the same pair of marginals.

We now present several positive dependence orderings, and it is assumed that $(x,y) \in \mathbf{R}^2$:

- H is said to be *more PQDE* than H', denoted by $H' \overset{e}{\ll} H$, if $E(Y|X > x) \geq E(Y'|X' > x)$ [Kowalczyk and Pleszczyńka (1977)].
- H is said to be *more quadrant dependent* [Yanagimoto and Okamoto (1969)] or *more concordant dependent* [Cambanis et al. (1976) and Tchen (1980)] than H', denoted by $H' \overset{c}{\ll} H$, if $H(x,y) \geq H'(x,y)$.
- H is said to be *more (positively) regression dependent* than H', denoted by $H' \overset{r}{\ll} H$, if $\Pr(Y \leq y|X = x) \geq \Pr(Y' \leq y'|X' = x)$ implies $\Pr(Y \leq y|X = x) \geq \Pr(Y'|X' = x')$ for any $x' > x$ [Yanagimoto and Okamoto (1969)]. More (positively) regression dependent is also known as "more SI." The ordering can also be expressed in terms of quantiles of the conditional distributions. A slight modification of the definition above was given by Capéraà and Genest (1990).
- H is said to be *more associated* than H', denoted by $H' \overset{a}{\ll} H$, if there exist functions u and v that map $R(f) \times R(g)$ onto $R(f)$ and $R(g)$, respectively, such that
$$\left.\begin{matrix} x_1 \leq x_2 \\ y_1 \leq y_2 \end{matrix}\right\} \Rightarrow \left\{\begin{matrix} u(x_1,y_1) \leq u(x_2,y_2) \\ v(x_1,y_1) \leq v(x_2,y_2) \end{matrix}\right.$$
$$\left.\begin{matrix} u(x_1,y_1) < u(x_2,y_2) \\ v(x_1,y_1) > v(x_2,y_2) \end{matrix}\right\} \Rightarrow \left\{\begin{matrix} x_1 < x_2 \\ y_1 > y_2 \end{matrix}\right.$$
$$(X,Y) \sim (u(X',X'), v(X',Y'));$$
see Schriever (1987a,b). In the special case where $u(x,y) = x, H$ is *more regression dependent* than H', as defined above. We note also that if X' and Y' are independent, then H is "more associated" than H' is equivalent to X and Y are associated.
- Kimeldorf and Sampson (1987) defined a *TP_2 ordering* as follows. Let $I \times J$ be a rectangle and $H(I,J)$ and $H'(I,J)$ be the associated probabilities. We write $I_1 < I_2$ if, for all $x \in I_1$ and all $y \in I_2$, $x < y$. We say that

$H' \overset{T}{\ll} H$ if, for all $I_1 < I_2$ and for all $J_1 < J_2$, we have

$$\begin{aligned} &H'(I_1, J_1)H'(I_2, J_2)H(I_1, J_2)H(I_2, J_1) \\ &\quad \leq H'(I_1, J_2)H'(I_2, I_1)H(I_1, J_1)H(I_2, J_2). \end{aligned} \tag{3.33}$$

Capéraà and Genest (1990) also defined an ordering H "more LRD" than H'; see (3.34) for the definition. Although the dependence concepts LRD and TP_2 are the same when the joint density function exists, "more LRD" is not equivalent to "more TP_2."

Genest and Verret (2002) pointed out that all one-parameter systems of Archimedean copulas listed by Nelsen (2006) in Chapter 4 of his book fail to be ordered by TP_2, with the possible exception of Ali–Mikhail–Haq and Gumbel–Barnett copulas. Some counterexamples outside the Archimedean class are provided by the bivariate Cauchy, Cuadras–Augé, and Plackett distributions. It seems that this positive ordering may be of limited use.

Among these different positive dependence orderings, the following implications hold:

$$\overset{r}{\ll} \Rightarrow \overset{a}{\ll} \Rightarrow \overset{c}{\ll} \Rightarrow \overset{e}{\ll};$$

see Yanagimoto and Okamoto (1969) and Schriever (1978b). Kimeldorf and Sampson (1987) also showed that $\overset{T}{\ll} \Rightarrow \overset{c}{\ll}$. However, Capéraà and Genest (1990) showed that $\overset{T}{\ll} \not\Rightarrow \overset{r}{\ll}$. It is not known, however, whether $\overset{T}{\ll} \Rightarrow \overset{a}{\ll}$.

In the special case when $H' = FG$ (i.e., X' and Y' are independent), the following implications hold:

$FG \overset{a}{\ll} H \Rightarrow X$ and Y are associated;
$FG \overset{r}{\ll} H \Rightarrow X$ and Y are PRD;
$FG \overset{c}{\ll} H \Rightarrow X$ and Y are PQD;
$FG \overset{e}{\ll} H \Rightarrow X$ and Y are PQDE;
$FG \overset{T}{\ll} H \Rightarrow H$ is TP_2. If H has a density, then
$FG \overset{T}{\ll} H$ if and only if h is TP_2 (X and Y are LRD).

Fang and Joe (1992) linked the concepts of the "more associated" and "more regression dependent" orderings with families of continuous bivariate distributions. They presented several equivalent forms of these two orderings so that the orderings are more easily verifiable for some bivariate distributions. For several parametric bivariate families, the dependence orderings are shown to be equivalent to the orderings of the underlying parameters.

Example 3.27 (Bivariate normal distribution with positive ρ). The Slepian inequality mentioned in Section 3.4.2 states that

$$\Pr(X_1 > x, Y_1 > y) \geq \Pr(X_2 > x, Y_2 > y) \text{ if } \rho_1 \geq \rho_2.$$

Hence, a more PQD ordering can be defined in this case in terms of the positive correlation coefficient ρ.

Genest and Verret (2002) have shown that the bivariate normal with given means and variances can be ordered by their correlation coefficient in TP_2 ordering.

Example 3.28 (Ali–Mikhail–Haq family of distributions). In this case, the governing copula is

$$C_\theta(u,v) = \frac{uv}{1-\theta(1-u)(1-v)}, \quad \theta \in [0,1].$$

It is easy to see in this case that $C_\theta \gg C_{\theta'}$ if $\theta > \theta'$, i.e., C_θ is more PQD than $C_{\theta'}$.

Example 3.29. A special case of Marshall and Olkin's BVE is given by

$$\Pr(X > x, Y > y) = \exp\{-(1-\lambda)(x+y) - \lambda \max(x,y)\}, \quad x, y \geq 0, 0 \leq \lambda \leq 1. \tag{3.34}$$

Fang and Joe (1992) showed that the distribution is increasing with respect to "more associated" ordering as λ increases but not with respect to "more SI."

Example 3.30. Kimeldorf and Sampson (1987) showed that the F-G-M copula

$$C_\alpha(u,v) = uv + \alpha uv(1-u)(1-v), \ 0 \leq u,v \leq 1, \ -1 \leq \alpha \leq 1$$

can be ordered by the relation (3.33). Note, however, that this ordering holds for $-1 \leq \alpha \leq 0$ even though X and Y are RR_2 for $\alpha < 0$.

3.9.1 Some Other Positive Dependence Orderings

H is said to be *more positive definite dependent* (PDD) than H', denoted by $H' \overset{d}{\ll} H$ [Rinott and Pollack (1980)] if,

$$\text{cov}(a(X), a(Y)) \geq \text{cov}(a(X'), a(Y')).$$

Capéraà and Genest's Orderings

Capéraà and Genest (1990) presented the following definitions for some orderings.

Definition 3.31. If the conditional distribution $H_{Y|x}(y) = H(x,y)/F(x)$ is continuous and strictly increasing, then it has an inverse $H_{Y|x}^{-1}(u)$, and we can then define, without ambiguity, a cumulative distribution function $H_{x',x}(u)$ that maps [0, 1] to [0, 1] such that $H_{x',x}(u) = H_{Y|x} \circ H_{Y|x}^{-1}(u)$.

The PRD (SI) property is then equivalent to

$$H_{x',x}(u) \le u \quad \text{for all } x < x', \text{ for all } 0 \le u \le 1.$$

They also defined H is *more LRD* than H', denoted by $H' \overset{L}{\ll} H$, if, for all $x < x'$ and for all $0 \le u < v < t < 1$,

$$\frac{H_{x',x}(t) - H_{x',x}(u)}{H_{x',x}(v) - H_{x',x}(u)} \le \frac{H'_{x',x}(t) - H'_{x',x}(u)}{H'_{x',x}(v) - H'_{x',x}(u)}. \tag{3.35}$$

This ordering is different from the TP_2 ordering discussed earlier. Unlike the more TP_2 property, $H' \overset{L}{\ll} H \Rightarrow H' \overset{r}{\ll} H$ if H and H' are two distribution functions with the same marginals and such that the conditional distributions $H_{Y|x}$ and $H'_{Y|x}$ have supports independent of x.

3.9.2 Positive Dependent Ordering with Different Marginals

When the relation $\ll$ was defined earlier on the entire family of bivariate distributions, property (P0) of the positive dependence ordering expressed the condition that only bivariate distributions having the same pair of marginals are comparable. Kimeldorf and Sampson (1987) showed that this definition can be extended to allow for the comparison of bivariate distributions not having the same pair of marginals. This is done through the uniform representation, i.e., the ordering of two bivariate distributions is carried out through the ordering of their copulas. Thus, we can extend the definition $\ll$ to $\ll^*$, where the latter relation is defined by

$$H' \ll H \Leftrightarrow C'_H \ll^* C_H.$$

It is clear that the relation $\ll^*$ satisfies (P2)-P(3), (P5)-(P9), and

$$\text{(P4)}^* \qquad H' \ll H \Rightarrow C'_H = C_H.$$

There are several other types of positive dependence ordering in the literature, and we refer the interested reader to the book by Shaked and Shantikumar (2005), which gives a comprehensive treatment on stochastic orderings.

In concluding this section, we note that Joe (1997, p. 19) has mentioned that the concepts of PQD discussed in Section 3.6 and the concordance ordering (more PQD) defined above are basic for the parametric families of copulas in determining whether a multivariate parameter is a dependence parameter.

3.9.3 Bayesian Concepts of Dependence

Brady and Singurwalla (1996) introduced several concepts of dependence in the Bayesian framework. They argue that the notion of dependence between two or more variables is conditional on a known parameter (θ) or (latent) variable. For example, if X and Y have a bivariate normal distribution, then they are independent or dependent conditionally on their correlation coefficient ρ. Thus, if we can define a prior distribution $\tilde{P}$ on the parameter ρ, we shall be able to associate a certain probability for independence or positive dependence of the pair (X, Y).

More specifically, let ρ denote the correlation coefficient between two variables X and Y, and if a prior distribution on ρ can be defined, we can compute the probability

$$\Pi(\alpha) = \Pr(|\rho(X, Y)| \geq \alpha),$$

which is termed by Brady and Singpurwalla as a *correlation survival function.*

Definition 3.32. The pair (X, Y) is *stochastically more correlated* than the pair (X', Y') if

$$\Pr(|\rho(X, Y)| \geq \alpha) \geq \Pr(|\rho(X', Y')| \geq \alpha).$$

Definition 3.33. The pair (X, Y) is *stochastically more correlated in expectation* than the pair (X', Y') if

$$\int \Pi_{X,Y}(\alpha) d\alpha \geq \int \Pi_{X',Y'}(\alpha) d\alpha,$$

where $\int \Pi_{X,Y}(\alpha) d\alpha = \Pi(\alpha) = \Pr(|\rho(X, Y)| \geq \alpha)$.

We conclude this chapter by mentioning that orderings of bivariate random variables seem to be a fruitful and inexhaustible topic of research that attracts the attention of theoretical as well as applied researchers.

References

1. Abdel-Hameed, M., Sampson, A.R.: Positive dependence of the bivariate and trivariate absolute normal t, χ^2 and F distributions. Annals of Statistics **6**, 1360–1368

(1978)
2. Barlow, R.E., Proschan, F.: Statistical Theory of Reliability and Life Testing: Probability Models. Holt, Rinehart and Winston, New York (1975)
3. Barlow, R.E., Proschan, F.: Statistical Theory of Reliability and Life Testing: Probability Models, 2nd edition. To Begin With, Silver Spring, Maryland (1981)
4. Bhaskara Rao, M., Krishnaiah, P.R., Subramanyam, K.: A structure theorem on bivariate positive quadrant dependent distributions and tests for independence in two-way contingency tables. Journal of Multivariate Analysis **23**, 93–118 (1987)
5. Block, H.W., Basu, A.P.: A continuous bivariate exponential distribution. Journal of the American Statistical Association **64**, 1031–1037 (1976)
6. Block, H.W., Savits, T.H.: Multivariate nonparametric classes in reliability. In: Handbook of Statistics, Volume 7, Quality Control and Reliability, P.R. Krishnaiah and C.R. Rao (eds.), pp. 121–129. North-Holland, Amsterdam (1988)
7. Block, H.W., Savits, T.H., Shaked, M.: Some concepts of negative dependence. Annals of Probability **10**, 765–772 (1982a)
8. Block, H.W., Savits, T.H., Shaked, M.: Negative dependence. In: Survival Analysis, J. Crowley and R.A. Johnson (eds.), pp. 206–215. Institute of Mathematical Statistics, Hayward, California (1982b)
9. Block, H.W., Savits, T.H., Shaked, M.: A concept of negative dependence using stochastic ordering. Statistics and Probability Letters **3**, 81–86 (1985)
10. Blomqvist, N.: On a measure of dependence between two random variables. Annals of Mathematical Statistics **21**, 593–600 (1950)
11. Brady, B., Singpurwalla, N.D.: Stochastically monotone dependence. Technical Report, George Washington University, Washington, D.C. (1996)
12. Cambanis, S., Simons, G, and Stout, W.: Inequalities for $Ek(X, Y)$ when marginals are fixed. Zeitschrift für Wahrscheinlichkeitstheorie und verwandte Gebiete **36**, 285–294 (1976)
13. Capéraà, P., Genest, C.: Concepts de dépendence et orderes stochastiques pour des lois bidimensionelles. Canadian Journal of Statistics **18**, 315–326 (1990)
14. Christofides, T.C., Vaggelatou, E.:, A connection between supermodular ordering and positive/negative association. Journal of Multivariate Analysis **88**, 138–151 (2004)
15. Connor, R.J., Mosimann, J.E.: Concepts of independence for proportions with a generalization of the Dirichlet distribution. Journal of the American Statistical Association **64**, 194–206 (1969)
16. Downton, F.: Bivariate exponential distributions in reliability theory. Journal of the Royal Statistical Society, Series B **32**, 408–417 (1970)
17. Drouet-Mari, D., Kotz, S.: Correlation and Dependence. Imperial College Press, London (2001)
18. Ebrahimi, N., Ghosh, M.: Multivariate negative dependence. Communications in Statistics: Theory and Methods **11**, 307–337 (1981)
19. Ebrahimi, N., Ghosh, M.: The ordering of negative quadrant dependence. Communications in Statistics: Theory and Methods **11**, 2389–2399 (1982)
20. Esary, J.D., Proschan, F.: Relationships among some bivariate dependences. Annals of Mathematical Statistics **43**, 651–655 (1972)
21. Esary, J.D., Proschan, F., Walkup, D.W.: Association of random variables, with applications. Annals of Mathematical Statistics **38**, 1466–1474 (1967)
22. Fabius, J.: Two characterizations of the Dirichlet distribution. Annals of Statistics **1**, 583–587 (1973)
23. Fang, Z., Joe, H.: Further developments of some dependence orderings for continuous bivariate distributions. Annals of the Institute of Statistical Mathematics **44**, 501–517 (1992)
24. Freund, J.: A bivariate extension of the exponential distribution. Journal of the American Statistical Association **56**, 971–977 (1961)
25. Genest, C., Verret, F.: The TP_2 ordering of Kimeldorf and Sampson has the normal-agreeing property. Statistics and Probability Letters **57**, 387–391 (2002)

26. Gupta, S.S.: Probability integrals of multivariate normal and multivariate t. Annals of Mathematical Statistics **34**, 792–828 (1963)
27. Harris, R.: A lower bound for critical probability in a certain percolation model. Proceedings of the Cambridge Philosophical Society **56**, 13–20 (1960)
28. Harris, R.: A multivariate definition for increasing hazard rate distribution. Annals of Mathematical Statistics **41**, 713–717 (1970)
29. Hoeffding, W.: Masstabinvariante Korrelationstheorie. Schriften des Mathematischen Instituts und des Instituts für Angewandte Mathematik der Universität Berlin **5**, 179–233 (1940)
30. Hu, T., Müller, A., Scarsini, M.: Journal of Statistical Planning and Inference **124**, 153–158 (2004)
31. Hutchinson, T.P., Lai, C.D.: Continuous Bivariate Distributions: Emphasising Applications. Rumsby Scientific Publishing, Adelaide, Australia (1990)
32. Jogdeo, K.: Dependence concepts. In: Encyclopedia of Statistical Sciences, Volume 2, S. Kotz and N.L. Johnson (eds.), pp. 324–334. John Wiley and Sons, New York, (1982)
33. Joag-Dev, K.: Measures of dependence. In: Handbook of Statistics, Volume 4, Nonparametric Methods, P.R. Krishnaiah and P.K. Sen (eds.), pp. 79–88. North-Holland, Amsterdam (1984)
34. Joag-Dev, K., Proschan, F.: Negative association of random variables. Annals of Statistics **11**, 286–295 (1983)
35. Joe, H.: Multivariate Models and Dependence Concepts. Chapman and Hall, London (1997)
36. Joe, H.: Parametric families of multivariate distributions with given margins. Journal of Multivariate Analysis **46**, 262–282 (1993)
37. Karlin, S.: Total Positivity. Stanford University Press, Stanford, California (1968)
38. Karlin, S., Rinott, Y.: Classes of orderings of measures and related correlation inequalities, II. Multivariate reverse rule distributions. Journal of Multivariate Analysis **10**, 499–516 (1980)
39. Kim, J.S., Proschan, F.: Total positivity. In: Encyclopedia of Statistical Sciences, Volume **9**, S. Kotz and N.L. Johnson (eds.), pp. 289–297. John Wiley and Sons, New York (1988)
40. Kim, T.S., Seo, H.Y.: A note on some negative dependence notions. Communications in Statistics: Theory and Methods **24**, 845–858 (1995)
41. Kimeldorf, G., Sampson, A.R.: Positive dependence orderings. Annals of the Institute of Statistical Mathematics **39**, 113–128 (1987)
42. Kochar, S.C., Gupta, R.P.: Competitors of the Kendall-tau test for testing independence against positive quadrant dependence. Biometrika **74**, 664–666 (1987)
43. Kochar, S.C., Gupta, R.P.: Distribution-free tests based on sub-sample extrema for testing independence against positive dependence. Australian Journal of Statistics **32**, 45–51 (1990)
44. Kotz, S., Balakrishnan, N., Johnson, N.L.: Continuous Multivariate Distributions, Volume 1: Models and Applications. John Wiley and Sons, New York (2000)
45. Kotz, S., Johnson, N.L.: Some replacement-times distributions in two-component systems. Reliability Engineering **7**, 151–157 (1984)
46. Kowalczyk, T.: General definition and sample counterparts of monotonic dependence functions of bivariate distributions. Statistics **8**, 351–365 (1977)
47. Kowalczyk, T.: Shape of the monotone dependence function. Statistics **13**, 183–192 (1982)
48. Kowalczyk, T., Ledwina, T.: Some properties of chosen grade parameters and their rank counterparts. Statistics **13**, 547–553 (1982)
49. Kowalczyk, T., Pleszczynska, E.: Monotonic dependence functions of bivariate distributions. Annals of Statistics **5**, 1221–1227 (1977)
50. Lai, C.D.: Bounds on reliability of a coherent system with positively correlated components. IEEE Transactions on Reliability **35**, 508–511 (1986)

51. Lai, C.D., Moore, T.: Probability integrals of a bivariate gamma distribution. Journal of Statistical Computation and Simulation **19**, 205–213 (1984)
52. Lai, C.D., Xie, M.: A new family of positive dependence bivariate distributions. Statistics and Probability Letters **46**, 359–364 (2000)
53. Lai, C.D., Xie, M.: Stochastic Ageing and Dependence for Relaibility. Springer-Verlag, New York (2006)
54. Lai, C.D., Xie, M., Bairamov, I.: Dependence and ageing properties of bivariate Lomax distribution. In: A Volume in Honor of Professor R.E. Barlow on his 70th Birthday, Y. Hayakawa, T. Irony, and M. Xie (eds.), pp. 243–256. World Scientific Publishers, Singapore (2001)
55. Lee, M.L.T.: Properties and applications of the Sarmanov family of bivariate distributions. Communications in Statistics: Theory and Methods **25**, 1207–1222 (1996)
56. Lee, S.Y.: Maximum likelihood estimation of polychoric correlations in $r \times s \times t$ contingency tables. Journal of Statistical Computation and Simulation **23**, 53–67 (1985)
57. Ledwina, T.: Large deviations and Bahadur slopes of some rank tests of independence. Sankhyā, Series A **48**, 188–207 (1986)
58. Lehmann, E.L.: Some concepts of dependence. Annals of Mathematical Statistics **37**, 1137–1153 (1966)
59. Lindley, D.V., Singpurwalla, N.D.: Multivariate distributions for the life lengths of components of a system sharing a common environment. Journal of Applied Probability **23**, 418–431 (1986)
60. Lingappaiah, G.S.: Bivariate gamma distribution as a life test model. Aplikace Matematiky **29**, 182–188 (1983)
61. Mardia, K.V.: Families of Bivariate Distributions. Griffin, London (1970)
62. Marshall, A.W., Olkin, I.: A multivariate exponential distribution. Journal of the American Statistical Association **62**, 30–44 (1967)
63. Mosimann, J.E.: Statistical problems of size and shape, II. Characterizations of the lognormal, gamma and Dirichlet distributions. In: A Modern Course on Distributions in Scientific Work, Volume 2, Model Building and Model Selection, G.P. Patil, S. Kotz and J.K. Ord (eds.), Reidel, Dordrecht, pp. 219–239 (1975)
64. Mosimann, J.E.: Size and shape analysis. In: Encyclopedia of Statistical Sciences, Volume 8, S. Kotz and N.L. Johnson (eds.), pp. 497–507. John Wiley and Sons, New York (1988)
65. Mosimann, J.E., Malley, J.D.: The independence of size and shape before and after scale change. In: Statistical Distributions in Scientific Work, Volume 4, Models, Structures, and Characterizations, C. Taillie, G.P. Patil, and B.A. Baldessari (eds.), pp. 137–145. Reidel, Dordrecht (1981)
66. Mukerjee, S.P., Sasmal, B.C.: Life distributions of coherent dependent systems. Calcutta Statistical Association Bulletin **26**, 39–52 (1977)
67. Müller, A., Stoyan, D.: Comparison Methods for Stochastic Models and Risks. John Wiley and Sons, Chichester (2002)
68. Nayak, T.K.: Multivariate Lomax distribution: Properties and usefulness in reliability theory. Journal of Applied Probability **24**, 170–177 (1987)
69. Nelsen, R.B.: Introduction to Copulas. Springer-Verlag, New York (1999)
70. Nelsen, R.B.: An Introduction to Copulas, 2nd edition. Springer-Verlag, New York (2006)
71. Ohi, F., Nishida, T.: Bivariate shock models and its application to the system reliability analysis. Mathematica Japonica **23**, 109-122 (1978)
72. Pendleton, B.F.: Ratio correlation. In: Encyclopedia of Statistical Sciences, Volume 7, S. Kotz and N.L. Johnson (eds.), pp. 636–639. John Wiley and Sons, New York (1986)
73. Philips, M.J.: A preventive maintenance plan for a system subject to revealed and unrevealed faults. Reliability Engineering **2**, 221–231 (1981)

74. Prather, J.E.: Spurious correlation. In: Encyclopedia of Statistical Sciences, Volume 8, S. Kotz and N.L. Johnson (eds.), pp. 613–614. John Wiley and Sons, New York (1988)
75. Rinott, Y., Pollack, M.: A stochastic ordering induced by a concept of positive dependence and monotonicity of asymptotic test sizes. Annals of Statistics **8**, 190–198 (1980)
76. Rödel, E.: A necessary condition for positive dependence. Statistics **18**, 351–359 (1987)
77. Sankaran, P.G., Nair, N.U.: A bivariate Pareto model and its applications to reliability. Naval Research Logistics **40** 1013–1020 (1993)
78. Sarmanov, O.V.: Generalized normal correlation and two-dimensional Frechet classes. Doklady (Soviet Mathematics) **168**, 596–599 (1966)
79. Shaked, M.: A concept of positive dependence for exchangeable random variables. Annals of Statistics **5**, 505–515 (1977)
80. Schriever, B.F.: Monotonicity of rank statistics in some nonparametric testing problems. Statistica Neerlandica **41**, 99–109 (1987a)
81. Schriever, B.F.: An ordering for positive dependence. Annals of Statistics **15**, 1208–1214 (1987b)
82. Shaked, M.: Some concepts of positive dependence for bivariate interchangeable distributions. Annals of the Institute of Statistical Mathematics **31**, 67–84 (1979)
83. Shaked, M.: A general theory of some positive dependence notions. Journal of Multivariate Analysis **12**, 199–218 (1982)
84. Shaked, M., Shantikumar, J.G. (eds.): Stochastic Orders and Their Applications. Academic Press, New York (1994)
85. Shea, G.A.: Hoeffding's lemma. In: Encyclopedia of Statistical Sciences, Volume 3, S. Kotz and N.L. Johnson (eds.), pp. 648–649. John Wiley and Sons, New York (1983)
86. Tchen, A.: Inequalities for distributions with given marginals. Annals of Probability **8**, 814–827 (1980)
87. Yangimoto, T.: Families of positive random variables. Annals of the Institute of Statistical Mathematics **26**, 559–557 (1972)
88. Yanagimoto, T., Okamoto, M.: Partial orderings of permutations and monotonicity of a rank correlation statistic. Annals of the Institute of Statistical Mathematics **21**, 489–506 (1969)

Chapter 4
Measures of Dependence

4.1 Introduction

A measure of dependence indicates in some particular manner how closely the variables X and Y are related; one extreme will include a case of complete linear dependence, and the other extreme will be complete mutual independence. Although it is customary in bivariate data analysis to compute a correlation measure of some sort, one number (or index) alone can never fully reveal the nature of dependence; hence a variety of measures are needed.

In Section 4.2, we describe the idea of total dependence, and then we present some global measures of dependence in Section 4.3. Next, Pearson's product-moment correlation coefficient, the most commonly used measure of dependence, is detailed in Section 4.4. In Section 4.5, the concept of maximal correlation, which is based on Pearson's product-moment correlation, is presented. The monotone correlation and its properties are described in Section 4.6. The rank correlation measures and their properties and relationships are presented in Section 4.7. Next, in Section 4.8, three measures of dependence proposed by Schweizer and Wolff (1976, 1981), which are based on Spearman's rank correlation, are presented, and some related measures are also outlined. The matrix of correlation is explained in Section 4.9, and tetrachoric and polychoric correlations are introduced in Section 4.10. In Section 4.11, the idea of compatibility with perfect rank ordering is explained in the context of contingency tables. Some brief concluding remarks on measures of dependence are then made in Section 4.12. Some local measures of dependence that have been proposed in the literature are presented in Section 4.13. Finally, the concept of regional dependence and some related issues are described in Section 4.14.

N. Balakrishnan, C.D. Lai, *Continuous Bivariate Distributions*,
DOI 10.1007/b101765_5,

4.2 Total Dependence

Let us now examine the concept of *total dependence*.

4.2.1 Functions

Before presenting different definitions of total dependence, it is helpful to remind ourselves what a function is.

- By a function b from a set A to another set B, we mean a mapping (rule) that assigns to each x in A a unique element $b(x)$ in B. (Because of the uniqueness requirement, $\pm\sqrt{x}$, for instance, is not a function.)
- b is said to be *one-to-one* if $b(x) = b(y)$ only when $x = y$.
- b is called *onto* if $b(A) = B$; that is, for each y in B, there exists at least one x in A such that $b(x) = y$.
- A function b that is one-to-one and onto is said to be a *one-to-one correspondence*. Such a function has an inverse, which is denoted by b^{-1}.
- b is said to be Borel measurable if, for each α, the set $\{x : b(x) > \alpha\}$ is a Borel set, which is typically a countable union of open or closed sets or complements of these. (The reader need not get bogged down with this, as most functions we come across are indeed Borel measurable.)

In this chapter, we assume all the functions are Borel measurable and onto.

4.2.2 Mutual Complete Dependence

If each of two random variables X and Y can be predicted from the other, then, intuitively, X is a function of Y and Y is a function of X, and so X and Y are dependent on each other. In order to define this more formally, we first need the following definition.

Definition 4.1. A random variable Y is *completely dependent* on X if there exists a function b such that

$$\Pr[Y = b(X)] = 1. \tag{4.1}$$

This equation essentially means that $Y = b(X)$, except on events of zero probability.

Definition 4.2. X and Y are *mutually completely dependent* if the equation above holds for some one-to-one function b; see Lancaster (1963).

The concept of mutual complete dependence is an antithesis of stochastic independence in that mutual complete dependence entails complete predictability of either random variable from the other (i.e., X and Y are mutually determined), while stochastic independence entails X and Y being completely useless in predicting one another.

4.2.3 Monotone Dependence

Clearly, if a sequence $\{(X_n, Y_n)\}$ of pairs of independent random variables converges in distribution to (X, Y), then X and Y must be mutually independent. However, Kimeldorf and Sampson (1978) constructed a sequence of pairs of mutually completely dependent random variables, all having a uniform distribution on [0, 1], that converges to a pair of independent random variables each having a uniform distribution on [0, 1]. From this point of view, mutual complete dependence is not a perfect opposite of independence. This defect of mutual complete dependence motivated Kimeldorf and Sampson (1978) to present a new concept of total statistical dependence, called *monotone dependence.*

Definition 4.3. Let X and Y be continuous random variables. Then Y is *monotonically dependent* on X if there exists a strictly monotone function b for which $\Pr[Y = b(X)] = 1$.

It is clear that Y is monotonically dependent on X if and only if X is monotonically dependent on Y. We can then present the following additional definitions.

Definition 4.4. If the function b in the preceding definition is increasing, X and Y are said to be *increasing dependent*; if b is decreasing, X and Y are said to be *decreasing dependent.*

Note that a function b may be one-to-one and yet not monotone; for example,

$$b(x) = \begin{cases} x, & 0 \le x < 1, \\ 3 - x, & 1 \le x \le 2, \\ x, & 2 < x \le 3. \end{cases}$$

Hence, monotone dependence is stronger than mutual dependence.

Kimeldorf and Sampson (1978) showed that a necessary and sufficient condition that X and Y be increasing (decreasing) monotonically dependent is that the joint distribution function of (X, Y) be H^+ (H^-), which are the Fréchet bounds.

4.2.4 Functional and Implicit Dependence

These are some weaker definitions of total dependence.

Definition 4.5. X and Y are *functionally dependent* if either $X = a(Y)$ or $Y = b(X)$ for some functions a and b; see Rényi (1959) and Jogdeo (1982). X and Y are functionally dependent if either X is completely dependent on Y or vice versa. An example is $Y = X^2$.

Definition 4.6. X and Y are *implicitly dependent* if there exist two functions a and b such that $a(X) = b(Y)$ with $\text{var}[a(X)] > 0$; see Rényi (1970, p. 283). In other words, there may exist no function connecting X and Y and yet they are related. For example, consider the relation $X^2+Y^2 = 1$. If we set $a(X) = X^2$ and $b(Y) = 1-Y^2$, then $a(X) = b(Y)$. However, $Y = \pm\sqrt{1-X^2}$ is not a function, as it assigns one value of X to two values of Y.

4.2.5 Overview

The different notions of total dependence in decreasing order of strength are as follows:

- linear dependence,
- monotone dependence,
- mutual complete dependence,
- functional dependence,
- Implicit dependence.

4.3 Global Measures of Dependence

If X and Y are not totally dependent, then it may be helpful to find some quantities that can measure the strength or degree of dependence between them. If such a measure can be expressed as a scalar, it is often more convenient to refer to it as an *index*. We may then ask what conditions ought an index ought to satisfy or what desirable properties it should have in order to be useful. Such indices are called the *global measures* in Drouet-Mari and Kotz (2001).

Rényi (1959) proposed a set of seven conditions for this purpose and showed that the maximal correlation (discussed in Section 4.5) fulfills all of them. Lancaster (1982b) modified and enlarged Rényi's set of axioms to nine conditions, described below.

Let $\delta(X,Y)$ denote an index of dependence between X and Y. The following conditions, apart from the last one, represent Lancaster's version of

Rényi's conditions. Condition (9) is taken from Schweizer and Wolff (1981) instead of Lancaster (1982b), as the latter is expressed in highly technical terms.

(1) $\delta(X, Y)$ is defined for any pair of random variables, neither of them being constant, with probability 1. This is to avoid trivialities.
(2) $\delta(X, Y) = \delta(Y, X)$. But notice that while independence is a symmetric property, total dependence is not, as one variable may be determined by the other, but not vice versa.
(3) $0 \leq \delta(X, Y) \leq 1$. Lancaster says that this is an obvious choice, but not everyone may agree.
(4) $\delta(X, Y) = 0$ if and only if X and Y are mutually dependent. Notice how strong this condition is made by the "only if" part.
(5) If the functions a and b map the spaces of X and Y, respectively, onto themselves, in a one-to-one manner then $\delta(a(X), b(Y)) = \delta(X, Y)$. The condition means that the index remains invariant under one-to-one transformation of the marginal random variables.
(6) $\delta(X, Y) = 1$ if and only if X and Y are mutually completely dependent.
(7) If X and Y are jointly normal, with correlation coefficient ρ, then $\delta(X, Y) = |\rho|$.
(8) In any family of distributions defined by a vector parameter θ, $\delta(X, Y)$ must be a function of θ.
(9) If (X, Y) and (X_n, Y_n), $n = 1, 2, \ldots$, are pairs of random variables with joint distributions H and H_n, respectively, and if $\{H_n\}$ converges to H, then $\lim_{n\to\infty} \delta(X_n, Y_n) = \delta(X, Y)$.

Another version of Rényi's axioms for a symmetric nonparametric measure of dependence is given in Schweizer and Wolff (1981). A similar set of criteria for a good measure of association (dependence) is also given by Gibbons (1971, pp. 204–207). The nonparametric measures of dependence such as Kendall's and Spearman's rank correlations will be discussed in Section 4.7.

The following comments are worth making about the conditions given above:

- A curious feature of the list of conditions is its mixture of the trivial and/or unhelpful with the strong and/or deep. We would say that (1), (3), (7), and (8) fall into the first category (unless there are subtle consequences to them that elude us), whereas (2), (4), (5), (6), and (9) fall into the second category.
- Summarizing, conditions (2), (5), (4), and (6) say that we are looking for a measure that is symmetric in X and Y, is defined by the ranks of X and Y, attains 0 only in the case of independence, and attains 1 whenever there is mutual complete dependence.
- Condition (3) is too restrictive for correlations, as the range of these is traditionally from -1 to $+1$.

- Condition (6) is stronger than the original condition which says $\delta(X,Y) = 1$ if either $X = a(Y)$ or $Y = b(X)$ for some functions a and b, i.e., $\delta(X,Y) = 1$ if X and Y are functionally dependent. Rényi intentionally left out the converse implication, i.e., $\delta(X,Y) = 1$ only if X and Y are functionally dependent, as he felt it to be too restrictive. The strengthening from functional dependence to mutual complete dependence is possibly due to Lancaster himself.
- Condition (7) is not appropriate to rank correlations; it should be replaced by δ, being a strictly increasing function of $|\rho|$, as is done by Schweizer and Wolff (1981).
- Schweizer and Wolff (1981) claimed that at least for nonparametric measures, Rényi's original conditions are too strong.
- The main point about these axioms is not their virtues or demerits, either individually or as a set, but that they make us think about what we mean by dependence and what we require from a measure of it. They provide a yardstick against which the properties of different measures may be measured.

There are three prominent global measures of dependence: correlation coefficient, Kendall's tau, and Spearman's correlation coefficient.

4.4 Pearson's Product-Moment Correlation Coefficient

Pearson's product-moment correlation coefficient is a measure of the strength of the linear relationship between two random variables, and is defined by

$$\rho(X,Y) = \frac{\mathrm{cov}(X,Y)}{\sqrt{\mathrm{var}(X)\mathrm{var}(Y)}}\,, \tag{4.2}$$

where $\mathrm{cov}(X,Y) = E\{[X - E(X)][Y - E(Y)]\}$ is the covariance of X and Y, and $\mathrm{var}(X)$ and $\mathrm{var}(Y)$ are the variances of X and Y, respectively. If either of the two variables is a constant, the correlation is undefined. If either has an infinite variance, it may be possible to extend this definition, as done for *bivariate stable* distributions, for example. From the definition, it is clear that conditions (1) and (2) of Section 4.3 are satisfied.

From Cauchy-Schwarz inequality, it is also clear that $|\rho(X,Y)| \leq 1$; equality occurs only when X and Y are linearly dependent; ρ takes the same sign as the slope of the regression line. Suppose the marginals $F(x)$ and $G(y)$ are given. Then, ρ can take all values in the range -1 to $+1$ if and only if these exist constants α and β such that $\alpha X + \beta Y$ has the same distribution as Y, and the distributions are symmetrical about their means; see Moran (1967).

If X and Y are independent, then $\rho(X,Y) = 0$. But zero correlation does not imply independence and therefore condition (4) of Section 4.3 is not satisfied. [Between uncorrelatedness and independence lies *semi-independence*. This means that $E(Y|X) = E(Y)$ and $E(X|Y) = E(X)$; see Jensen (1988).] As is well known, adding constants to X and Y does not alter $\rho(X, Y)$, and neither does the multiplication of X and Y by constant factors with the same sign. As $\rho(X, Y)$ may be negative, condition (3) is clearly violated. Furthermore, $\rho(X, Y)$ is not invariant under monotone transformations of the marginals, and so condition (5) is not satisfied. Further, since $\rho(X, -X) = -1$, the "if" part of condition (6) is not satisfied. Conditions (7) and (8) are obviously satisfied. Condition (9) is satisfied, which can be established by using the continuity theorem for two-dimensional characteristic functions [Cramér (1954, p. 102)] and the expansions of such characteristic functions in terms of product moments [Bauer (1972, pp. 264–265)].

As to estimating the correlation coefficient ρ from a sample of n bivariate observations $(x_1, y_1), \ldots, (x_n, y_n)$, the sample correlation coefficient

$$r = \frac{\sum_{i=1}^{n}(x_i - \bar{x})(y_i - \bar{y})}{\sqrt{\sum_{i=1}^{n}(x_i - \bar{x})^2(y_i - \bar{y})^2}} \tag{4.3}$$

could be used, where $\bar{x}$ and $\bar{y}$ are the respective sample means.

If $(x_1, y_1), \ldots, (x_n, y_n)$ are n independent pairs of observations from a bivariate normal distribution, r is indeed the maximum likelihood estimator and also an approximate unbiased estimator of ρ. A disadvantage of r is that it is very sensitive to contamination of the sample by outliers. Devlin et al. (1975) compared r with various other estimators of ρ in terms of robustness; see Ruppert (1988) for ideas on multivariate "trimming" (i.e., removal of extreme values in the multivariate setting).

The value $\rho(X, Y)$ will be simply denoted as ρ whenever there is no ambiguity; furthermore, the symbols ρ' and ρ^* will be used for other types of correlations.

The distribution of $z = \frac{1}{2}\log\left(\frac{1+r}{1-r}\right) = \tanh^{-1} r$, called Fisher's *variance-stabilizing transformation* of r, approaches normality (as n increases) much faster than that of r, particularly when $\rho \neq 0$. For a detailed discussion, see Rodriguez (1982). Mudholkar (1983) has made some comments on the behavior of this transformation when the parent distribution is non-normal.

4.4.1 Robustness of Sample Correlation

The distribution of r has been discussed rather extensively in Chapter 32 of Johnson et al. (1995). While the properties of r for the bivariate normal are clearly understood, the same cannot be said about bivariate non-normal populations. Cook (1951), Gayen (1951), and Nakagawa and Niki (1992) obtained expressions for the first four moments of r in terms of the cumulants and cross-cumulants of the parent population. However, the size of the bias and the variance of r are still rather hazy for general bivariate non-normal populations when $\rho \neq 0$, since the cross-cumulants are difficult to quantify in general. Although several non-normal populations have been investigated, the messages regarding the robustness of r are somewhat conflicting; see Johnson et al. (1995, p. 580).

Hutchinson (1997) noted that the sample correlation is possibly a poor estimator. Using the bivariate lognormal as a case study on the robustness of r as an estimate of ρ, Lai et al. (1999) found that for smaller sample sizes, r has a large bias and large variance when $\rho \neq 0$ with skewed marginals, which supports the claim that r is not a robust estimator. It is therefore important to check for the underlying assumptions of the population before reporting the size of r.

4.4.2 Interpretation of Correlation

Rodriguez (1982) described the historical development of correlation, and in it he has stated that although Karl Pearson was aware that high correlation between two variables may be due to a third variable, this was not generally recognized until Yule's (1926) paper. One difficulty in interpreting correlation is that it is still all too easy to confuse it with causation.

Rodriguez has argued that, for interpreting a calculated correlation, an accompanying probability model for the chance variation in the data is necessary, with the two most common ones being as follows:

- The bivariate normal distribution: In this case, r estimates the parameter ρ; confidence intervals may be constructed for ρ, and hypothesis tests may be carried out as well.
- The simple regression model $y_i = \alpha + \beta x_i +$ random error: Here, r^2 represents the proportion of total variability (as measured by the sum of squares) in the y's that can be explained by the linear regression,

$$r^2 = \frac{\sum_{i=1}^{n}(\hat{y}_i - \bar{y})^2}{\sum_{i=1}^{n}(y_i - \bar{y})^2}, \tag{4.4}$$

  where $\hat{y}_i$ is the predicted value of y_i calculated from the estimated regression equation. In the regression context, the x's are often prefixed and not random, and so there is no underlying bivariate distribution in which r can be an estimate of a parameter.

Even so, says Elffers (1980), "It can be difficult (i) to decide when a particular value of ρ indicates association strong enough for a given purpose, and (ii) in a given situation, to weigh the losses involved in obtaining more strongly associated variables against the gains." Elffers therefore puts forward functions of the correlation that can be interpreted as the probability of making a wrong decision in certain situations.

Although they are elementary, the following points are perhaps worth emphasizing:

- For certain bivariate distributions, ρ may not even exist. For example, the bivariate Pareto distribution (see Section 2.8) ρ does not exist when $0 < c \leq 2$.
- The equation $r = 0$ does not mean that there is no relationship between the x's and y's. A scatterplot might reveal a clear (though nonlinear) relationship.
- And even if the correlation is close to 1, the relationship may be nonlinear, either to the eye when plotted directly or because a transformation reveals a relationship that is incompatible with linearity. For example, if X has a uniform distribution over the range 8 to 10 and Y is proportional to X^2, then the correlation between X and Y is approximately 0.999; see Blake (1979).
- Lots of different-looking sets of points can all produce the same value of r; for example, Chambers et al. (1983, Section 4.2) have presented eight scatterplots all having $r = 0.7$.
- The value of r calculated from a small sample may be totally misleading if not viewed in the context of its likely sampling error.

In view of the above, the computation of r should be accompanied by the use of such devices as scatterplots. When the data are not from a bivariate normal population, r provides only limited information about the observations. Barnett (1985), citing two scatterplots in Barnett (1979), has expressed the view that for highly skewed bivariate distributions, such as those with exponential marginals, the ordinary correlation coefficient is not a very useful measure of association.

History of Correlation Coefficients

Drouet-Mari and Kotz (2001) devoted their Chapter 2 to describing the historical development of "independent event" and the correlation coefficient, and they also conducted a brief tour of its early applications and misinterpretations. Readers should find this account of the early development of statistical dependence useful.

14 Faces of Correlation Coefficients

Thirteen ways to look at the correlation coefficient have been discussed by Rodgers and Nicewander (1988). A fourteenth way has been added to the list by Rovine and Von Eye (1997). These are the following:

1. Correlation as a function of raw scores and means.
2. Correlation as a standardized covariance.
3. Correlation as a standardized slope of the regression line.
4. Correlation as the geometric mean of the regression slopes.
5. Correlation as the square root of the ratio of two variances (proportion of variability accounted for).
6. Correlation as the mean cross-product of standardized variables.
7. Correlation as a function of the angle between the two standardized regression lines.
8. Correlation as a function of the angle between two variable vectors.
9. Correlation as a rescaled variance of the difference between two standardized scores.
10. Correlation estimated from the balloon rule.
11. Correlation in relation to the bivariate ellipses of isoconcentration.
12. Correlation as a function of the test statistic from designed experiments.
13. Correlation as the ratio of two means.
14. Correlation as the proportion of matches.

Cube of Correlation Coefficient

Falk and Well (1997) have also discussed many faces of the correlation coefficient. Dodge and Rousson (2000) have added up some new faces of the correlation coefficient. One of their representations of results, the cube of the correlation coefficient, is given as the ratio of skewness of the response variable (γ_Y) to that of the explanatory variable (γ_X),

$$\rho_{XY}^3 = \frac{\gamma_Y}{\gamma_X},$$

if $\gamma_X \neq 0$ and the distribution of the error term is symmetric. Muddapur (2003) gave an alternative proof for the same result. It was pointed out that the quantity $|\rho_{XY}^3|$ can be interpreted as the proportion of skewness "preserved" by the linear model.

Dodge and Rousson (2000) argued that $\left(\rho_{XY}^2\right)^3 = \frac{\gamma_Y^2}{\gamma_X^2}$ can be used to determine the direction of the regression line (whether Y is dependent on X or X is dependent on Y in a regression line) as follows. Since the left-hand side of the equation is always less than or equal to 1, $\gamma_Y^2 \leq \gamma_X^2$. Thus, Y is linearly dependent on X. A similar argument can be provided for the linear regression dependence of X on Y. To put it simply, for a given $\rho_{XY}, \gamma_X^2 \geq \gamma_Y^2$ implies Y is the response variable and $\gamma_X^2 \leq \gamma_Y^2$ implies X is the response variable. It has been pointed out by Sungur (2005) that this approach of "directional dependence" stems from the marginal behavior of the variables rather than the joint behavior. We note that in the case where X and Y are uniform variables, their coefficients of skewness are zero, so this approach to define directional dependence is inappropriate for copulas. Thus, it is clear that Dodge and Rousson's criterion only works for the skewed X and Y.

4.4.3 Correlation Ratio

The interpretation of r^2 given above in Section 4.4.2, which presumes that $\text{var}(Y|X)$ is a constant, suggests writing the theoretical correlation as $\rho^2 = 1 - \frac{\text{var}(Y|X)}{\text{var}(Y)}$. More generally (i.e., beyond the context of linear regression), the quantity $\eta = 1 - \frac{E[Y-E(Y|X)]^2}{\text{var}(Y)}$ is termed the *correlation ratio* of Y on X and was introduced by Pearson (1905). For further details on this, one may refer to Chapter 26 of Kendall and Stuart (1979).

4.4.4 Chebyshev's Inequality

For any univariate distribution with zero mean and unit standard deviation, Chebyshev's inequality states that $\Pr(|X| \leq a) \geq 1 - a^{-2}$, for all $a > 0$. In the general case, when μ is the mean and σ is the standard deviation, the left hand side of the inequality becomes $\Pr(|X - \mu| \leq a\sigma)$.

For any bivariate distribution with zero mean, unit standard deviation, and correlation ρ,

$$\Pr(|X| \leq a,\ |Y| \leq a) \geq 1 - \frac{1 + \sqrt{1 - \rho^2}}{a^2} .$$

More generally,

$$\Pr(|X| \leq a_1,\ |Y| \leq a_2) \geq 1 - \frac{\frac{a_1}{2a_2} + \frac{a_2}{2a_1} + \sqrt{(\frac{a_1}{2a_2} + \frac{a_2}{2a_1})^2 - \rho^2}}{a_1 a_2} ;$$

see Tong (1980, Section 7.2).

4.4.5 ρ and Concepts of Dependence

If X and Y satisfy any concept of positive dependence, for example, they are PQD. Then ρ will always be positive. Indeed in that case, $\text{cov}(X, Y) \geq 0$ (Hoeffding's lemma). If $\rho > 0$ and (X, Y) has a bivariate normal distribution, then X and Y satisfy a more stringent dependence condition of LRD; see Section 3.4 for pertinent details.

4.5 Maximal Correlation (Sup Correlation)

A frequently quoted measure of dependence between two random variables X and Y is that of *maximal correlation*, introduced by Gebelein (1941) and studied by, among others, Rényi (1959) and Sarmanov (1962, 1963), defined by

$$\rho'(X,Y) = \sup \rho[a(X), b(Y)],$$

where the supremum is taken over all Borel-measurable functions a and b for which $\text{var}[a(X)]$ and $\text{var}[b(Y)]$ are finite and nonzero and where ρ represents the ordinary (Pearson product-moment) correlation coefficient. The maximal correlation is also known as *sup correlation*. This measure satisfies the following:

1. $0 \le \rho'(X,Y) \le 1$.
2. $\rho'(X,Y) = \rho'(Y,X)$.
3. $\rho'(X,Y) = 0$ if and only if X and Y are independent. To see this, consider indicator functions of $X \le \xi$, $Y \le \eta$, where ξ, η are varied.
4. If X and Y are mutually dependent, then $\rho'(X,Y) = 1$, but the converse is not true; see Lancaster (1963) for counterexamples and for necessary and sufficient conditions for the complete mutual dependence of random variables. Hence, condition (6) of Section 4.3 fails in part.
5. Obviously, $|\rho(X,Y)| \le \rho'(X,Y)$.
6. $\rho'(X,Y) = |\rho(X,Y)| = |\rho|$ if (X,Y) is a bivariate normal random variable. This is because, in this particular case, $|\rho[a(X), b(Y)]| \le |\rho'(X,Y)|$, equality holding only when a and b are identity functions; see Kendall and Stuart (1979, p. 600). This result was rediscovered by Klaassen and Wellner (1997).
7. Condition (9) of Section 4.3 is not fulfilled; as mentioned in the beginning of Section 4.2.3, Kimeldorf and Sampson (1978) presented an example of a sequence of mutually completely dependent random variables $\{(X_n, Y_n)\}$ converging in distribution to a distribution in which X and Y are independent. Clearly, in this case, $\rho(X_n, Y_n) = 1$ but $\rho'(X,Y) = 0$.

Rényi (1970, p. 283) proved that even if X and Y are only implicitly dependent, then $\rho'(X,Y)$ is still equal to 1.

If the bivariate distribution is ϕ^2-bounded [Lancaster (1958)], then the maximal correlation equals ρ_1, the first canonical correlation coefficient.

This measure has many good properties. However, according to Hall (1970), it has a number of drawbacks, too. For instance, it equals 1 too often and is also generally not readily computable.

4.6 Monotone Correlations

4.6.1 Definitions and Properties

In the beginning of Section 4.2.3, we noted that mutual complete dependence is not compatible with independence, so they can hardly be opposites! For this reason, Kimeldorf and Sampson (1978) suggested the notion of *monotonically dependence*. X and Y are monotone dependent if there exists a perfect monotone relation between them. If the random variables are not perfectly monotonically related, it may be useful to measure numerically the degree of monotone dependence between them. One such measure, called *monotone correlation*, can be defined as

$$\rho^*(X,Y) = \sup \rho[a(X), b(Y)], \tag{4.5}$$

where the supremum is taken over all monotone functions a and b for which $\mathrm{var}[a(X)]$ and $\mathrm{var}[b(Y)]$ are finite and nonzero.

The monotone correlation possesses the following properties:

1. $0 \le \rho^*(X,Y) \le 1$.
2. $\rho^*(X,Y) = \rho^*(Y,X)$.
3. $\rho^*(X,Y) = 0$ if and only if X and Y are independent.[1]
4. $|\rho(X,Y)| \le \rho^*(X,Y) \le \rho'(X,Y)$, which is obviously true.
5. $|\rho(X,Y)| = \rho^*(X,Y) = \rho'(X,Y)$ if (X,Y) has a bivariate normal distribution.
6. If X and Y are monotonically dependent, then $\rho^*(X,Y) = 1$, but the converse is not true; see an example given in Kimeldorf and Sampson (1978, p. 899).
7. If (V,W) has the same uniform representation as (X,Y), then $\rho^*(X,Y) = \rho^*(V,W)$.
8. $\rho^*(X,Y) = \sup\{|\rho(V,W)| : (V,W)$ having the same uniform representation as $(X,Y)\}$.
9. $\rho_S(X,Y) \le \rho^*(X,Y) \le \rho'(X,Y)$, where ρ_S is Spearman's rank correlation, $\rho_S(X,Y) = \rho[G(X), H(Y)]$. Note that the grade correlation (Spearman's) is the ordinary correlation coefficient of the uniform representations.
10. ρ^* is invariant under all order-preserving or order-reversing transformations of X and Y, and hence it satisfies a weaker condition (5) of Section 4.3.

For a more detailed discussion, one may refer to Kimeldorf and Sampson (1978).

[1] Suppose $\rho^*(X,Y) = 0$. For any real t define $a_t(x)$ to be 1 if $x < t$, and 0 otherwise. We claim that $\rho[a_s(X), a_t(Y)] = 0$. If not, then either $\rho[a_s(X), a_t(Y)] > 0$ or $\rho[a_s(X), -a_t(Y)] > 0$, which contradicts the hypothesis. Now, $\rho[a_s(X), a_t(Y)] = 0$ implies that $\Pr(X \le s, Y \le t) = \Pr(X \le s)\Pr(Y \le t)$, which implies independence.

4.6.2 Concordant and Discordant Monotone Correlations

The concept of monotone correlation can be refined by measuring separately the strength of relationship between X and Y in a positive direction and the strength of the relationship in a negative direction, i.e., the strength of concordancy and discordancy between X and Y. The following definitions are due to Kimeldorf et al. (1982).

Definition 4.7. If a and b in (4.5) are both restricted to be increasing (or, equivalently, both decreasing), the resulting measure sup $\rho[a(X), b(Y)]$ is called the *concordant monotone correlation* (denoted by CMC).

Definition 4.8. If a and b in (4.5) are both restricted to be increasing, then $\inf \rho[a(X), b(Y)]$ is called the *discordant monotone correlation* (denoted by DMC).

Kimeldorf et al. (1982) have mentioned that CMC and DMC have natural interpretations as measures of positive and negative association, respectively, for ordinal random variables.

It is easy to observe that, for any pair of increasing functions a and b, we have

$$\text{DMC} \le \rho[a(X), b(Y)] \le \text{CMC}.$$

Suppose it is desired to impose numeric monotone scalings for a pair of psychological tests. If the CMC and DMC are close, then by the equation above, it makes little difference which monotone scales are used. If DMC = CMC = 0, then X and Y are independent; however, it is possible for DMC < CMC = 0 and yet X and Y not be independent. Note that if X and Y are increasing dependent (Section 4.2.3), then CMC = 1; and if X and Y are decreasing dependent, then DMC = 1.

In some situations, X and Y should have the same scaling—for example, scores on a single test before and after treatment. This leads to two further definitions.

Definition 4.9. If $a = b$ in (4.5), then the resulting measure is called the *isoconcordant monotone correlation* (denoted by ICMC).

Definition 4.10. If $a = b$ in the definition of DMC, then the resulting measure is called the *isodiscordant monotone correlation* (denoted by IDMC).

Note that isoscaling (i.e., assuming $a = b$) is not appropriate when X and Y have inherently different ranges of values. Kimeldorf et al. (1982) evaluated these measures of association by means of a nonlinear optimization algorithm. Kimeldorf et al. (1981) have also described an interactive FORTRAN program, called MONCOR, for computing the monotone correlations described above.

4.7 Rank Correlations

Kendall's tau (τ) and Spearman's rho (ρ_S) are the best-known rank correlation coefficients. Essentially, these are measures of correlation between rankings, rather than between actual values, of X and Y; as a result, they are unaffected by any increasing transformation of X and Y, whereas the Pearson product-moment correlation coefficient ρ is unaffected only by linear transformations.

4.7.1 Kendall's Tau

Let (x_i, y_i) and (x_j, y_j) be two observations from (X, Y) of continuous random variables. The two pairs (x_i, y_i) and (x_j, y_j) are said to be concordant if $(x_i - x_j)(y_i - y_j) > 0$ and discordant if $(x_i - x_j)(y_i - y_j) < 0$.

Kendall's tau is defined to be the difference between the probabilities of concordance and discordance:

$$\tau = P[(X - X')(Y - Y') \geq 0] - P[(X - X')(Y - Y') \leq 0]. \tag{4.6}$$

The definition above is equivalent to

$$\tau = \mathrm{cov}[\mathrm{sgn}(X' - X), \mathrm{sgn}(Y' - Y)].$$

τ may also be defined as

$$\tau = 4 \int\int H(x, y)h(x, y)dx\,dy - 1. \tag{4.7}$$

The sample version of τ is defined as

$$\hat{\tau} = \frac{c - d}{c + d} = \frac{c - d}{\binom{n}{2}}\,, \tag{4.8}$$

where c denotes the number of concordant pairs and d the number of discordant pairs from a sample of n observations from (X, Y). $\hat{\tau}$ is an unbiased estimator of τ.

Since τ is invariant under any increasing transformations, it may be defined via the copula C of X and Y

$$4 \int_0^1 \int_0^1 C(u, v)c(u, v)du\,dv - 1 = 4E(C(U, V)) - 1; \tag{4.9}$$

see Nelsen (2006, p. 162).

Nelsen (1992) proved that $\frac{\tau}{2}$ represents an average measure of total positivity for the density h defined by

$$T = \int_{-\infty}^{\infty}\int_{-\infty}^{y_2}\int_{-\infty}^{\infty}\int_{-\infty}^{x_2} [h(x_2, y_2)h(x_1, y_1) \\ - h(x_2, y_1)h(x_1, y_2)]\, dx_1\, dy_1\, dx_2\, dy_2.$$

4.7.2 Spearman's Rho

As with Kendall's tau, the population version of the measure of association known as *Spearman's rho* (denoted by ρ_S) is based on concordance and discordance. Let $(X_1, Y_1), (X_2, Y_2)$, and (X_3, Y_3) be three independent pairs of random variables with a common distribution function H. Then, ρ_S is defined to be proportional to the probability of concordance minus the probability of discordance for the two pairs (X_1, Y_1) and (X_2, Y_3),

$$\rho_S = 3\Big\{P[(X_1 - X_2)(Y_1 - Y_3) > 0] - P[(X_1 - X_2)(Y_1 - Y_3) < 0]\Big\}. \quad (4.10)$$

Equation (4.10) is really the grade correlation and can be expressed in terms of the copula as follows:

$$\rho_S = 12\int_0^1\int_0^1 C(u, v)\, du\, dv - 3 \quad (4.11)$$

$$= 12\int_0^1\int_0^1 uv\, dC(u, v) - 3 \quad (4.12)$$

$$= 12E(UV) - 3. \quad (4.13)$$

Rewriting the equation above as

$$\rho_S = \frac{E(UV) - \frac{1}{4}}{\frac{1}{12}}, \quad (4.14)$$

we observe that Spearman's rank correlation between X and Y is simply Pearson's product-moment correlation coefficient between the uniform variates U and V.

Quadrant Dependence and Spearman's ρ_S

The pair (X, Y) is said to be positively quadrant dependent (PQD) if $H(x, y) - F(x)G(y) \geq 0$ for all x and y, and negatively quadrant dependent (NQD) when the inequality is reversed, as defined in Section 3.3. Nelsen (1992) considers that the expression $H(x, y) - F(x)G(y)$ measures "local" quadrant dependence at each point of $(x, y) \in R^2$. Now, (4.11) gives

$$\rho_S = 12 \int_{-\infty}^{\infty} \int_{-\infty}^{\infty} [H(x,y) - F(x)G(y)]\, dF(x)\, dG(y). \qquad (4.15)$$

It follows from the equation above that $\frac{1}{12}\rho_S$ represents an average measure of quadrant dependence, where the average is taken with respect to the marginal distributions of X and Y. It is easy to see from (4.15) that when X and Y are PQD, then $\rho_S \geq 0$.

The sample Spearman correlation for a sample of size n is defined as

$$R = \frac{12}{n(n^2-1)} \sum_i \left(r_i - \frac{n+1}{2}\right)\left(s_i - \frac{n+1}{2}\right), \qquad (4.16)$$

where $r_i = \text{rank}(x_i)$ and $s_i = \text{rank}(y_i)$. Yet another common expression for R is

$$R = 1 - \frac{6\sum_i d_i^2}{n(n^2-1)}, \qquad (4.17)$$

where $d_i = r_i - s_i$. R is not an unbiased estimator of ρ_S, and the expectation of R in fact is $E(R) = \frac{(n-2)\rho_S + 3\tau}{(n+1)} \to \rho_S$ as $n \to \infty$. If the distribution of (X, Y) is bivariate normal with correlation ρ, then it can be shown that $\rho_S = \frac{6}{\pi} \sin^{-1} \frac{\rho}{2}$.

It is important to note the following points:

- Independence of X and Y implies that $\tau = \rho_S = 0$, but the converse implication does not hold.
- τ and ρ_S are both restricted to the range -1 to $+1$, attaining these limits for perfect negative and perfect positive relationships, respectively.
- If X and Y are positive quadrant dependent, then $\tau \geq 0$ and $\rho_S \geq 0$.
- If two distributions H and H' have the same marginals and H is more concordant than H' (i.e., $H \geq H'$), then τ and ρ_S are at least as great for H as for H' [see Tchen (1980)].
- It was mentioned that the sample correlation r is very sensitive to outliers; the sample counterparts of τ and ρ_S are less so, but Gideon and Hollister (1987) proposed a statistic that is even more resistant to the influence of outliers.

For a review of measures including rank correlations, one may refer to Nelsen (1999).

4.7.3 The Relationship Between Kendall's Tau and Spearman's Rho

While both Kendall's tau and Spearman's rho measure the probability of concordance between two variables with a given distribution, the values of

ρ_S and τ are often quite different. In this section, we will determine just how different ρ and τ can be.

We begin by giving explicit relationships between the two indices for some of the distributions we have considered; these are summarized in Table 4.1.

Table 4.1 Relationship between ρ_S and τ

Distribution	Relationship
Bivariate normal	$\rho_S = \frac{6}{\pi}\sin^{-1}(\frac{1}{2}\sin\frac{\pi\tau}{2})$
F-G-M	$\rho_S = 3\tau/2$
Marshall & Olkin	$\rho_S = 3\tau/(2+\tau)$
Raftery family	$\rho_S = 3\tau(8-5\tau)/(4-\tau)^2$

We may now ask what the relation is between τ and ρ_S for other distributions and whether this relation can be used to determine the shape of an empirical distribution. (By "bivariate shape," we mean the shape remaining once the univariate shape has been discarded by ranking.)

General Bounds Between τ and ρ_S

Various examples indicate that a precise relation between the two measures does not exist for every bivariate distribution, but bounds or inequalities can be established. We shall now summarize some general relationships [see Kruskal (1958)]:

- $-1 \le 3\tau - 2\rho \le 1$ (first set of universal inequalities).
- $\frac{1+\rho}{2} \ge \left(\frac{1+\tau}{2}\right)^2$; $\frac{1-\rho}{2} \ge \left(\frac{1-\tau}{2}\right)^2$ (second set of universal inequalities).

Combining the preceding two sets of inequalities yields a slightly improved set,

$$\frac{3\tau-1}{2} \le \rho_S \le \frac{1+2\tau-\tau^2}{2}, \tau \ge 0 \text{ and } \frac{\tau^2+2\tau-1}{2} \le \rho_S \le \frac{1+3\tau}{2}, \tau \le 0. \tag{4.18}$$

Another relationship worth noting [see, e.g., Nelsen (1992)] is

$$E(W) = \frac{1}{12}(3\tau - \rho_S),$$

where $W = H(X,Y) - F(X)G(Y)$, which corresponds to a measure of quadrant dependence. So $E(W)$ is the "expected" measure of quadrant dependence. This equation alludes that the relationship between the two rank correlations may be affected by the strength of the positive dependence discussed in the preceding chapter.

Some Empirical Evidence

A figure ρ_S as a function of τ can be plotted for which the pair (τ, ρ_S) lies within a shaded region bounded by four constraints given in the preceding set of inequalities. Such a figure with bounds for ρ_S and τ can be found in Nelsen (1999, p. 104).

These bounds are remarkably wide: For instance, when $\tau = 0, \rho_S$ can range between -0.5 and $+0.5$. Daniels (1950) comments that the assumption that τ and ρ_S describe more or less the same aspect of a bivariate population of ranks may be far from true and suggests circumstances in which the message conveyed by the two indices is quite different. ["The worse discrepancy...occurs when the individuals fall into two groups of about equal size, within which corresponding pairs of ranks are nearly all concordant, but between which they are nearly all discordant"; Daniels (1950, p. 190)]. But Fieller et al. (1957) do not think this would happen very often, saying that although, after transforming the margins to normality, the resulting bivariate distribution will not necessarily be the bivariate normal, "We think it likely that in practical situations it would not differ greatly from this norm," adding "This is a field in which further investigation would be of considerable interest."

For a given value of τ, how much do distributions differ in their values of ρ_S? Table 4.2 shows that although ρ_S could theoretically take on a very wide range of values, for the distributions considered, the values are all very similar. The distributions that are most different from the others are Marshall and Olkin's, with its singularity in the p.d.f. at $y = x$, and Kimeldorf and Sampson's, with its oddly shaped support. With these exceptions, at $\tau = 0.5$, ρ_S lies in the range .667 to .707, even though it could theoretically take any value between .250 and .875.

Table 4.2 shows us that the bounds of ρ_S in terms of τ appear to be much narrower than implied by (6.18). In fact, Capéraà and Genest (1993) point out that many of the bivariate distributions have their ρ_S and τ at the same sign, with $|\rho_S| \geq |\tau|$. Table 4.2 confirms this general finding.

Some Conjectures on the Influence of Dependence Concepts on the Closeness Between τ and ρ_S

The discussion above suggests the following question. Is there some class of bivariate distributions that includes nearly all of those that occur for which only a narrow range of ρ_S (for given τ) is possible? For instance, if every quantile of y for a given x decreases with x, and vice versa [i.e., X and Y are SI (PRD)], can bounds for ρ_S in terms of τ be found? Hutchinson and Lai (1991) posed two conjectures when X and Y are SI:

(i) $\rho_S \leq 3\tau/2$.
(ii) $-1 + \sqrt{1 + 3\tau} \leq \rho_S \leq 2\tau - \tau^2$.

Table 4.2 Comparisons of the values of ρ_S with corresponding values of τ

Distribution	$\tau = \frac{1}{5}$	$\tau = \frac{1}{3}$	$\tau = \frac{1}{2}$	$\tau = \frac{3}{4}$
Lower bound	−0.200	0.000	0.250	0.625
Upper bound	0.680	0.778	0.875	0.969
Normal	0.296	0.483	0.690	0.917
F-G-M	0.300[c]	—	—	—
Ali–Mikhail–Haq	0.297	0.478	—	—
Frank	0.297[d]	0.484[d]	0.695[d]	0.922[d]
Pareto	0.295	0.478	0.682[e]	?
Marshall and Olkin	0.273	0.429	0.600	0.818
Kimeldorf and Sampson	0.300	0.500	0.750	0.937
Weighted linear combination: exponential	0.289	0.467	0.667	0.900
Weighted linear combination: Laplace	0.293[f]	0.473[g]	0.674	0.904[f]
Weighted linear combination: uniform	0.298[f]	0.490	0.707	0.927
Part uniform[a]	0.298	0.486	0.707	0.919
Nelsen[b]	0.291	0.471	0.673	0.905
New lower bound	0.265	0.414	0.581	0.803
New upper bound	0.300	0.500	0.750	0.937

Notes:

[a] Part uniform distribution: $h(x, y) = (1 + c)/(1 - c)$, $x^{1/c} \leq y \leq x^c$, $0 < c < 1$ and is 0 elsewhere.

[b] Nelsen's distribution: $H(x, y) = \min[x, y, (xy)^{(2-c)/2}]$, $x^{(2-c)/c} < y < x^{c/(2-c)}$.

[c] For the iterated F-G-M with $\tau = 0.2$, ρ_S lies between .297 and .301, depending on what α and β are. The former corresponds to $\alpha = 0.446, \beta = 1.784$, the latter to $\alpha = 1, \beta = -0.385$.

[d] We are grateful to Professor R.B. Nelsen of Lewis and Clark College for calculating these values.

[e] One way of finding this is to use equation (1) of Lavoie (1986).

[f] We are grateful to M.E. Johnson of Los Alamos National Laboratory for calculating these values.

[g] This result is implicit in Table III of David and Fix (1961).

Combining the two conjectures, we have

$$-1 + \sqrt{1 + 3\tau} \leq \rho_S \leq \min\left\{3\tau/2, 2\tau - \tau^2\right\}.$$

Nelsen (1999, pp. 168–169) has constructed a polynomial copula

$$C(u, v) = uv + 2\theta uv(1 - u)(1 - v)(1 + u + v - 2uv),$$

for which $\rho_S > 3\tau/2$ if $\theta \in (0, 1/4)$. Hence, the first conjecture is false. Hürlimann (2003) has proved conjecture (ii) for the class of bivariate extreme-value copulas.

We note that U and V of the bivariate extreme-value copula are stochastically increasing (SI).

Positive Dependence Concepts as an Influential Factor on the Relationship Between τ and ρ_S

Earlier in this section, we saw that Spearman's rho (ρ_S) can be interpreted as a measure of "average" quadrant dependence and that Kendall's tau (τ) can be interpreted as a measure of TP_2 (totally positive of order 2) or the likelihood dependence ratio. Of the dependence properties (concepts) discussed in the preceding chapter, positive quadrant dependence is the weakest (cov $(X, Y) \geq 0$ is even weaker, but we hardly discussed this in that chapter) and totally positive of order 2 is the strongest. Thus, the two most commonly used measures of association are related to two rather different stochastic dependence concepts, a fact that may partially explain the difference between the values of ρ_S and τ that we observed in several of the examples in this chapter. (By the way, the Pearson correlation coefficient ρ is clearly related to the dependence concept $\text{cov}(X, Y) \geq 0$.)

We now wish to raise the question of identifying, by means of necessary and sufficient conditions on the joint distribution $H(x, y)$, the weakest possible type of stochastic dependence between X and Y that will guarantee either $\rho_S > \tau \geq 0$ or $\rho_S < \tau \leq 0$.

Capéraà and Genest (1993) have provided a partial answer to this question and we now summarize their results.

Let X and Y be two continuous random variables. Then

$$\rho_S \geq \tau \geq 0 \tag{4.19}$$

if Y is left-tail decreasing and X is right-tail increasing. The same inequality holds if X is left-tail decreasing and Y is right-tail increasing.

Also, $\rho_S \leq \tau \leq 0$ if Y is left-tail increasing and X is right-tail decreasing. The same inequality holds if X is left-tail increasing and Y is right-tail decreasing.

Note. Fredricks and Nelsen (2007) also provided an alternative proof to the results of Capéraà and Genest.

Nelsen (1992) and Nelsen (2006, p. 188) showed that if (X, Y) is PQD (positive quadrant dependent), then

$$3\tau \geq \rho_S \geq 0.$$

Note that PQD implies $\text{Cov}(X, Y) \geq 0$, which in turn implies $\rho_S \geq 0$. Now, it was shown in Section 3.4.3 that both left-tail decreasing and right-tail increasing imply PQD. It now follows from (6.19) that

$$3\tau \geq \rho_S \geq \tau \geq 0$$

if Y is simultaneously LTD and RTI in X or X is simultaneously LTD and RTI in Y. However, Nelsen (1999, p. 158) gives an example showing that positive quadrant depndence alone is not sufficient to guarantee $\rho_S \geq \tau$.

Relationship Between ρ_S and τ When the Joint Distribution Approaches That of Two Independent Variables

It has long been known that, for many joint distributions exhibiting weak dependence, the sample value of Spearman's rho is about 50% larger than the sample value of Kendall's tau. Fredricks and Nelsen (2007) explained this behavior by showing that for the population analogues of these statistics, the ratio of ρ to τ approaches 3/2 as the joint distribution approaches that of two independent random variables. They also found sufficient conditions for determining the direction of the inequality between 3τ and 2ρ when the underlying joint distribution is absolutely continuous.

Relationship Between ρ_S and τ for Sample Minimum and Maximum

Consider two extreme order statistics $X_{(1)} = \min\{X_1, X_2, \ldots, X_n\}$ and $X_{(n)} = \max\{X_1, X_2, \ldots, X_n\}$ of n independent and identically distributed random variables. Let ρ_n and τ_n denote Spearman's rho and Kendall's tau for $X_{(1)}$ and $X_{(n)}$, respectively.

Schmitz (2004) conjectured that $\lim_{n\to\infty} \rho_n/\tau_n = 3/2$. The conjecture has now been proved true by Li and Li (2007). Since $\tau_n = \frac{1}{2n-1}$, Li and Li noted that ρ_n is given by $3/(4n-2)$ for large n. Chen (2007) has established inequalities between ρ_n and τ_n.

4.7.4 Other Concordance Measures

Gini Index

The Gini measure of association may be defined through the copula C as

$$\gamma_C = 4\left\{\int_0^1 C(u, 1-u)du - \int_0^1 [u - C(u,u)]du\right\}; \tag{4.20}$$

see Nelsen (2006, p. 180).

Blomqvist's β

This coefficient β, also known as the quadrant test of Blomqvist (1950), evaluates the dependence at the "center" of a distribution where the "center" is given by $(\tilde{x}, \tilde{y})$, with $\tilde{x}$ and $\tilde{y}$ being the medians of the two marginals. For this reason, β is often called the *medial correlation coefficient.* Note that $F(\tilde{x}) = G(\tilde{y}) = \frac{1}{2}$.

Formally, β is defined as

$$\beta = 2\Pr[(X - \tilde{x})(Y - \tilde{y}) > 0] - 1 = 4H(\tilde{x}, \tilde{y}) - 1, \tag{4.21}$$

which shows that $\beta = 0$ if X and Y are independent. Also, since $H(\tilde{x}, \tilde{y}) = C(\frac{1}{2}, \frac{1}{2})$, we have $\beta = 4C(\frac{1}{2}, \frac{1}{2}) - 1$.

It was pointed out by Nelsen (2006, pp. 182–183) that although Blomqvist's β depends on the copula only through its value at the center of $[0,1] \times [0,1]$, it can nevertheless often provide an accurate approximation to Spearman's ρ_S and Kendall's τ, as the following example illustrates.

Example 4.11. Let $C(u,v) = \frac{uv}{1-\theta(1-u)(1-v)}, \theta \in [-1,1]$, be the copula for the Ali–Mikhail–Haq family. We note from Section 2.3 that the expressions for ρ_S and τ involve logarithm and dilogarithm functions. However, it is easy to verify that $\beta = \frac{\theta}{4-\beta}$. If we reparametrize the expressions for ρ_S and τ by replacing θ by $4\beta/(1+\beta)$ and expand each of the expressions in a Maclaurin series, we obtain $\rho_S = \frac{4}{3}\beta + \frac{44}{75}\beta^3 + \frac{8}{28}\beta^4 + \cdots$ and $\tau = \frac{8}{9}\beta + \frac{8}{15}\beta^3 + \frac{16}{45}\beta^4 + \cdots$. Thus, $\frac{4\beta}{3}$ and $\frac{8\beta}{9}$ are reasonable second-order approximations to ρ_S and τ, respectively.

4.8 Measures of Schweizer and Wolff and Related Measures

Schweizer and Wolff (1976, 1981) proposed three measures of dependence that are based on Spearman's rho, which can be defined through the copula of X and Y as $\rho_S(X,Y) = 12\int_0^1\int_0^1[C(u,v) - uv]du\,dv$. Observing that the integral in this expression is simply the signed volume between the surfaces $z = C(u,v)$ and $z = uv$, and that X and Y are independent if and only if $C(u,v) = uv$, these authors suggested that any suitably normalized measure of distance, such as L_p-distance, should yield a symmetric nonparametric measure of distance. By considering $p = 1$, $p = 2$, and $p \to \infty$, they obtained the following three measures of dependence:

$$\sigma(X,Y) = 12\int_0^1\int_0^1 |C(u,v) - uv|\,du\,dv, \tag{4.22}$$

$$\gamma(X,Y) = \sqrt{90 \int_0^1 \int_0^1 [C(u,v) - uv]^2 \, du \, dv}, \tag{4.23}$$

and

$$\kappa(X,Y) = 4 \sup_{u,v \in [0,1]} |C(u,v) - uv|. \tag{4.24}$$

Equation (4.23) is equivalent to the Cramér–von Mises index given by

$$\int_{-\infty}^{\infty} \int_{-\infty}^{\infty} [H(x,y) - F(x)G(y)]^2 \, dF(x) dG(y), \tag{4.25}$$

which is equivalent to Φ^2 of Hoeffding (1940); also see Lancaster (1982b). On the other hand, (4.24) is equivalent to the Kolmogorov–Smirnov measure given by

$$\sup_{x,y} |H(x,y) - F(x)G(y)|.$$

Schweizer and Wolff (1981) showed that, when evaluating by a suitably modified version of Rényi's condition, σ possesses many desirable properties, including, in particular, condition (9) of Section 4.3. Therefore, a comparison of σ with ρ_S may be desirable. Schweizer and Wolff (1981) measure the volume and the signed volume between the surfaces $C(u,v)$ and uv, respectively. They also noted the following properties:

- $|\rho_S(X,Y)| \leq \sigma(X,Y)$.
- Equality holds for the bivariate normal distribution.
- The difference can be large.

4.9 Matrix of Correlation

In this section, we present a summary of relevant aspects of the diagonal expansion method [Lancaster (1982a,b)]. Specifically, let $\{\xi_i\}$ and $\{\eta_i\}$ be complete orthonormal systems on F and G, respectively, with $\xi_0 = \eta_0$; that is, $E(\xi_i \xi_j) = \delta_{ij}$, where δ_{ij} is either 1 or 0 depending on whether $i = j$ or $i \neq j$ and similarly for η. Let $\rho_{ij} = E(\xi_i \eta_j)$ and $\mathbf{R} = (\rho_{ij})$, for all positive integers i and j, be an infinite matrix. For given F and G, $\mathbf{R}$ completely determines H [Lancaster (1963)], so that $\mathbf{R}$ can be said to be a *matrix measure of dependence*. In particular, $\mathbf{R} = \mathbf{0}$ if and only if X and Y are independent. $\mathbf{R}$ is orthogonal if and only if X and Y are mutually completely dependent [Lancaster (1963)]. Special interest arises when $\{\xi_i\}$ and $\{\eta_i\}$ possess the biorthogonal property (i.e., $E(\xi_i \eta_j) = \delta_{ij} \rho_{ij}$) in this case, $\mathbf{R}$ is diagonal.

The scalar $\phi^2 = \mathrm{tr}(\mathbf{R}\mathbf{R}') = \sum_{i=1}^{\infty} \sum_{i=1}^{\infty} \rho_{ij}^2$ is an index for measuring dependence of two random variables. ϕ^2 here is also referred to as the *mean square contingency*, and it is zero if and only if X and Y are independent. In the case of the bivariate normal distribution, $\phi^2 + 1 = (1 - \rho^2)^{-1}$. As we have just mentioned, X and Y being mutually completely dependent implies $\mathbf{R}$ is orthogonal, which in turn implies $\phi^2 = \infty$. However, ϕ^2 can be infinite without having X and Y be mutually completely dependent. Consider a monotone transformation of ϕ^2 defined by $\lambda(X, Y) = \phi^2/(1 + \phi^2)$. It is clear from the present discussion that λ does not satisfy the "necessary" part of condition (6) of Rényi's measures of dependence listed in Section 4.3. However, Rényi (1959) showed that λ satisfies conditions (2)–(5) and (7). If the distribution is absolutely continuous or discrete, condition (1) will also be satisfied.

4.10 Tetrachoric and Polychoric Correlations

It is common for data to be recorded on an ordinal scale with only a few steps to it. A typical case from the social sciences is where subjects (respondents) are asked to report whether they approve strongly, approve, are neutral toward, disapprove, or disapprove strongly of some proposal. When analyzing this kind of data, a common approach is to assign an integer value to each category and proceed with the analysis as if the results were on an interval scale, with convenient distributional properties. Although this approach may work satisfactorily in some cases, it may lead to erroneous results in some others; see Olsson (1980). The *polychoric correlation* is suggested in the literature as an appropriate measure of correlation for bivariate tables of such data; it is termed the *tetrachoric correlations* when applied to 2×2 tables. The idea behind these measures is now described.

Formally, we denote the observed ordinal variables by X and Y, having r and s distinct categories, respectively. We assume that X and Y have been generated from some unobserved (latent) variables Z_1 and Z_2 that have a bivariate normal distribution. The relation between X and Z_1 may be written as

$$\begin{aligned} X &= 1 \text{ if } Z_1 < s_1 \\ X &= 2 \text{ if } s_1 \leq Z_1 < s_2 \\ &\vdots \quad \vdots \\ X &= r \text{ if } s_{r-1} \leq Z_1; \end{aligned}$$

similarly, there is a relation between Y and Z_2 in terms of class limits $t_1, t_2, \ldots, t_{s-1}$ of Z_2. The s's and the t's are sometimes referred to as *thresholds.*

Interest is often primarily in estimating $\rho(Z_1, Z_2)$, the correlation between Z_1 and Z_2. Suppose we want to do this by means of the maximum likelihood method. Given this general aim, the problem may be solved in at least two different ways. One way is to estimate ρ and the thresholds simultaneously. Alternatively, the thresholds are first estimated as the inverse of the normal distribution function, evaluated at the cumulative marginal proportions of the contingency table, and the maximum likelihood estimate of ρ is then computed with the thresholds fixed at those estimates. This may be referred to as a *two-step procedure*. It has the advantage of greater ease of numerical calculationough the former is formally more correct. In most practical situations, the results are almost identical [Olsson (1979)]. For a generalization of these methods to three- and higher-dimensional polytomous ordinal variables, one may refer to Lee (1985) and Lee and Poon (1987a,b). Divgi (1979b) describes a FORTRAN program for calculating tetrachoric correlation and offers to provide a listing of it to any interested reader. Martinson and Hamdan (1975) have presented a computer program for calculating the polychoric correlation.

Other discussions on these correlations are by Drasgow (1986) and Harris (1988), with the latter presenting a number of references to methods of approximating the tetrachoric correlations.

4.11 Compatibility with Perfect Rank Ordering

Suppose we have a two-way ordinal contingency table, as described in Section 4.10, which we imagine to have arisen from grouping two continuous variables. For simplicity, suppose each variate has been reduced to a dichotomy, so that our table is only a 2×2 table. Suppose the frequencies are $\begin{smallmatrix} 0 & 1 \\ 1 & 2 \end{smallmatrix}$. How well are X and Y correlated?

- One approach is to calculate the tetrachoric correlation, implicitly thinking of the bivariate normal distribution, or to estimate the association parameter of some other bivariate distribution.
- There is an alternative approach, which is especially relevant if X and Y are two different measures of the same characteristic (e.g., the severity of disease as assessed by two doctors). The question here is to what extent the data are compatible with perfect agreement between the X-ordering and the Y-ordering. The set of frequencies $\begin{smallmatrix} 0 & 1 \\ 1 & 2 \end{smallmatrix}$ is compatible with perfect agreement between two orderings, as it may be that if finer discrimination was insisted upon, the table would become $\begin{smallmatrix} 0 & 0 & 0 & 1 \\ 0 & 0 & 1 & 0 \\ 0 & 1 & 0 & 0 \\ 1 & 0 & 0 & 0 \end{smallmatrix}$. (The original table is obtained by combining the last three rows and combining the last three columns.) In a sense, we are starting with perfect correlation, and not zero, as our null hypothesis, and then asking to what extent the data are incompatible. We feel that a formalization of this could be as follows: Calculate
$$\sup \rho_r[a(Z), b(Z)]$$

(where ρ_r is a rank correlation coefficient, such as ρ_S or τ) subject to X being a nondecreasing function of a, and Y being a nondecreasing function of b. (The point is that $X \to a$ and $Y \to b$ are one-to-many relationships and not functions.)

From the work of Guttman (1986), we observe some common ground between the second approach above and Guttman's suggestion that "weak" coefficients of monotonicity are sometimes more appropriate than "strong" ones.

4.12 Conclusions on Measures of Dependence

There is, we fear, no universal answer to, "What is the best measure of dependence?" According to Lancaster (1982b), for some defined classes of distributions, the absolute product-moment correlation $|\rho|$ is the index of choice—for the bivariate normal distribution, for example, it satisfies all the conditions presented in Section 4.3 except for condition (5); for the random elements in common model, it completely determines the joint distribution. In other classes, there may be other indices useful for some purposes and the user needs to think about what purposes have priority. There is inevitably some loss of information in condensing the matrix of correlations to a single index. The absence of an always best measure should not surprise us if we reflect on the persistence in the literature of two competing measures of rank correlation, Kendall's and Spearman's.

4.13 Local Measures of Dependence

We saw earlier that ρ_S is an average measure of the PQD dependence. However, Kotz et al. (1992) presented an example to show that a distribution with a high ρ_S may not be PQD. Drouet-Mari and Kotz (2001, p. 149) have given the following rationale for defining a local index (measure) of dependence: "These indices (global measures) are defined from the moments of the distribution on the whole plane and can be zero when X and Y are not independent. One needs therefore the indices which measure the dependence locally. In the case when X and Y are survival variables, one needs to identify the time of maximal correlation: for example, the delay before the first symptom of a genetic disease by members of the same family will appear. The pairs (X, Y) and (X', Y') can have the same global measure of dependence but may possess two different distributions H and H': a local index will allow us to compare their variation in time. The variations with x and y of some local indices allow us to characterize certain distributions and conversely choosing a shape of variation for an index allows us sometimes to choose an appropriate model."

4.13.1 Definition of Local Dependence

The following definitions of local dependence measures can be found in Drouet-Mari and Kotz (2001).

Definition 4.12. If $V(x_0, y_0)$ is an open neighborhood of (x_0, y_0), then a distribution $H(x, y)$ is *locally PQD* in the neighborhood $V(x_0, y_0)$ if

$$\bar{H}(x, y) \geq \bar{F}(x)\bar{G}(y) \quad \text{for all } (x, y) \in V(x_0, y_0).$$

If $V(x_0, y_0) = (x_0, \infty) \times (y_0, \infty)$, we then arrive at the *remaining PQD*. (We use the term *remaining* to indicate a part in R^2 beyond a certain point of (x, y).) In a similar way, we can define a *local* or *remaining LRD.*

4.13.2 Local Dependence Function of Holland and Wang

The following concepts were introduced by Holland and Wang (1987a,b), motivated by the contingency table for two discrete random variables. Consider an $r \times s$ contingency table with cell proportions p_{ij}. For any two pairs of indices (i, j) and (k, l), the cross-product ratio is

$$\alpha_{ij,kl} = \frac{p_{ij}p_{kl}}{p_{il}p_{kj}}, \quad 1 \leq i, k \leq (r-1), \ 1 \leq j, l \leq (s-1). \tag{4.26}$$

Yule and Kendall (1937, Section 5.15) and Goodman (1969) suggested considering the following set of cross-product ratios:

$$\alpha_{ij} = \frac{p_{ij}p_{i+1,j+1}}{p_{i,j+1}p_{i+1,j}}, \quad 1 \leq i \leq (r-1), \ 1 \leq j \leq (s-1). \tag{4.27}$$

Further, let $\gamma_{ij} = \log \alpha_{ij}$. Both α_{ij} and γ_{ij} measure the association in the 2×2 subtables formed at the intersection of pairs of adjacent rows and columns. They are, of course, invariant under multiplications of rows and columns.

Now, let us go back to the continuous case. Let $R(h) = \{(x, y) : h(x, y) > 0\}$ be the region of the nonzero p.d.f. that has been partitioned by a very fine rectangular grid. The probability content of a rectangle containing the point (x, y) with sides dx and dy is then approximately $h(x, y)dx\,dy$. This probability may be viewed as one cell probability of a large two-way table, and so the cross-product ratio in (4.26) may be expressed as

$$\alpha(x, y; u, v) = \frac{h(x, y)h(u, v)}{h(x, v)h(u, y)}, \quad x < u, \ y < v, \tag{4.28}$$

assuming that all four points are in $R(h)$. The function in (4.28) is called the *cross-product ratio function.*

A local LRD may be defined by having $\alpha(x, y; u, v) > 0$. The logarithm of $\alpha(x, y; u, v)$, denoted by

$$\theta(x, y; u, v) = \log \alpha(x, y; u, v), \tag{4.29}$$

has been used by Holland and Wang (1987a,b) to derive a local measure of LRD as well.

4.13.3 Local ρ_S and τ

We can restrict ρ_S and τ to an open neighborhood of (x_0, y_0) and then define local ρ_S and τ as [Drouet-Mari and Kotz (2001, p. 172)]

$$\rho_{S(x_0,y_0)} = 12 \int\int_{V(x_0,y_0)} (C(u, v) - uv)\, du\, dv \tag{4.30}$$

and

$$\tau_{(x_0,y_0)} = 4 \int\int_{V(x_0,y_0)} C(u, v) du\, dv - 1, \tag{4.31}$$

upon noting that $F(x) = u$, $G(y) = v$ for all $(x, y) \in V(x_0, y_0)$. We may now interpret $\rho_{S(x_0,y_0)}/12$ as the average on the local PQD property, while $\tau_{(x_0,y_0)}/2$ is the average on the local LRD (TP_2).

When $V(x_0, y_0) = (x_0, \infty) \times (y_0, \infty)$, it is easy to estimate $\tau_{(x_0,y_0)}$ by counting the remaining concordant and discordant pairs and to estimate the variance of this estimator from n_0, the number of observations remaining.

4.13.4 Local Measure of LRD

Holland and Wang (1987a,b) defined a local dependence index that can be used to measure a local LRD property as

$$\gamma(x, y) = \lim_{dx,dy \to 0} \frac{\theta(x, y; x + dx, y + dy)}{dxdy} = \frac{\partial^2}{\partial x \partial y} \log h(x, y) \tag{4.32}$$

assuming the partial derivative of the second order exists. The expression $\gamma(x, y)$ is the local index that can be used to measure a local LRD property.

It follows from the preceding equation that

$$\gamma(x, y) = \lim_{dx,dy \to 0} \left[\log \left(\frac{h(x, y)h(x + dx, y + dy)}{h(x + dx, y)h(x, y + dy)} \right) /dx\, dy \right]. \tag{4.33}$$

Thus we see that $\gamma(x,y) \geq 0, \forall x, \forall y$ is equivalent to $h(x,y)$ being TP_2 or X and Y are LRD. Hence $\gamma(x,y)$ is an appropriate index for measuring local LRD dependence.

4.13.5 Properties of $\gamma(x, y)$

We shall assume that $R(h)$ is a rectangle, and R^2 may also be regarded as a rectangle for this purpose. (If $R(h)$ is not a rectangle, then the shape of $R(h)$ can introduce dependence between X and Y of a different nature that local dependence—we will take up this issue in the next section.) Note also that Drouet-Mari and Kotz (2001, p. 189) regard $\gamma(x,y)$ as a local measure of LRD even though it was referred to as the local dependence function in Holland and Wang (1987a,b).

The following properties are satisfied by the measure $\gamma(x,y)$:

- $-\infty < \gamma(x,y) < \infty$.
- $\gamma(x,y) = 0$ for all $(x,y) \in R(h)$ if and only if X and Y are independent. $\gamma(x,y)$ reveals more information about the dependence than other indices. Recall, for example, that the product-moment correlation ρ may be zero without being independent.
- $\gamma(x,y)$ is symmetric.
- $\gamma(x,y)$ is marginal-free, and so changing the marginals does not change $\gamma(x,y)$; in particular, $\frac{\partial^2}{\partial x \partial y} \log c(u,v) = \gamma(x,y)$, $F(x) = u, G(y) = v$, where c is the density of the associated copula.
- Holland and Wang (1987b) mentioned that when $\gamma(x,y)$ is a constant, any monotone function of that constant will be a "good" measure of association. But, when $\gamma(x,y)$ changes sign in $R(h)$, most measures of association will be inadequate or even misleading.
- $\gamma(x,y)$ is a function only of the conditional distribution of Y given X or the conditional distribution of X given Y.
- If X and Y have a bivariate normal distribution with correlation coefficient ρ, then $\gamma(x,y) = \frac{\rho}{1-\rho^2}$, a constant. Conversely, if $\gamma(x,y)$ is a constant, Jones (1998) pointed out that the density function $h(x,y)$ should have the form $a(x;\theta)b(y;\theta)\exp(\theta xy)$.

Jones (1996) has shown, using a kernel method, that $\gamma(x_0, y_0)$ is indeed a local version of the linear correlation coefficient.

4.13.6 Local Correlation Coefficient

Suppose the standard deviations of X and Y are σ_X and σ_Y, respectively. Let $\mu(x) = E(Y|X = x)$, $\sigma^2(x) = \mathrm{var}(Y|X = x)$ and $\beta(x) = \frac{\partial \mu(x)}{\partial x}$. Then, the *local correlation coefficient* of Bjerve and Doksum (1993) is defined as

$$\rho(x) = \frac{\sigma_X \beta(x)}{\{\sigma_X \beta(x)\}^2 + \sigma^2(x)}. \tag{4.34}$$

If (X, Y) has a bivariate normal distribution, then $\beta(x) = \beta$, a constant. It is important to mention the following properties of the local correlation coefficient $\rho(x)$:

- $-1 \leq \rho \leq 1$.
- X and Y being independent implies $\rho(x) = 0 \ \forall x$.
- $\rho = \pm 1$ for almost all x is equivalent to Y being a function of X.
- In general, $\rho(x)$ is not symmetric, but it is possible to construct a symmetrized version.
- $\rho(x)$ is scale-free but not marginal-free, i.e., linear transformations of X and Y (viz., $X^* = aX + b$ and $Y^* = cY + d$, with c and d having the same sign) leave $\rho(x)$ unchanged, but the transformation $U = F(X)$ and $V = G(Y)$ results in $\rho(u)$, which is different from $\rho(x)$.

Note that if $\rho(x) \geq 0$ for all x, then H is PRD. We can therefore define a local PRD when $\rho(x)$ is positive in a neighborhood of (x_0, y_0).

4.13.7 Several Local Indices Applicable in Survival Analysis

In the field of survival analysis, there is a need for time-dependent measures of dependence; for example, to identify in medical studies the time of maximal association between the interval from remission to relapse and the next interval from relapse to death or to determine the genetic character of a disease by comparing the degree of association between the lifetimes of monozygotic twins [Hougaard (2000)].

The following indices may be found in this connection in Drouet-Mari and Kotz (2001):

- Covariance function of Prentice and Cai (1992).
- Conditional covariance rate of Dabrowska et al. (1999).

4.14 Regional Dependence

4.14.1 Preliminaries

In this section, we shall discuss the notion of regional dependence introduced by Holland and Wang (1987a). In addition to the notation of the previous section, we will write $R(f) = \{x : f(x) > 0\}$ and $R(g) = \{y : g(y) > 0\}$ for the support of the marginals. We assume that $R(h)$ is an open convex set of the plane, $R(f)$ and $R(g)$ are open intervals, and f, g, and h are continuous in their respective regions of support.

Clearly, $R(h)$ is contained in the Cartesian product of $R(f)$ and $R(g)$, denoted by $R(f) \times R(g)$. If $R(h)$ is not equal to $R(f) \times R(g)$, then there exists a point (x_0, y_0) in $R(f) \times R(g)$ that is not in $R(h)$, at which $h(x_0, y_0) = 0$. Yet, $f(x_0)g(y_0) > 0$. So, $h(x_0, y_0) \neq f(x_0)g(y_0)$. Therefore, X and Y cannot be independent if their region of support is not a rectangle. This situation is parallel to the effects caused by structural zeros in a two-way contingency table. We are concerned here with the type of statistical dependence that is "caused" by the region of support.

4.14.2 Quasi-Independence and Quasi-Independent Projection

Let us define the x-section, $R_y(x)$, and the y-section, $R_x(y)$, of $R(h)$ by $R_y(x) = \{x : h(x, y) > 0\}$ and $R_x(y) = \{y : h(x, y) > 0\}$. Clearly, $R_y(x) \subseteq R(f)$ and $R_x(y) \subseteq R(g)$. The following definition of quasi-independence is analogous to quasi-independence in a two-way contingency table.

Definition 4.13. X and Y, having a joint density function $h(x, y)$, are said to be *quasi-independent* if there exist positive functions $f_1(x)$ and $g_1(y)$ such that $h(x, y) = f_1(x)g_1(y)$ for all $(x, y) \in R(h)$. If $R(h) = R(f) \times R(g)$, then, as we have seen, X and Y cannot be independent.

Definition 4.14. A positive density function $P_h(x, y)$ on $R(h)$ is the *quasi-projection* of $h(x, y)$ on $R(h)$ if there exist positive functions $a(x)$ and $b(y)$ such that the following three equations hold:

$$P_h(x, y) = a(x)b(y) \quad \text{for all } (x, y) \in R(h),$$

$$\int_{R_x(y)} a(x)b(y)dy = f(x) \qquad \forall\, x \in R(f),$$

$$\int_{R_y(x)} a(x)b(y)dx = g(y) \qquad \forall\, y \in R(g).$$

The quasi-independent projection of h is a joint density that has the same marginals as those of X and Y, and has the functional form of the product of two independent distributions. The explicit form of $P_h(x, y)$ can be obtained by solving the two integral equations presented above. Holland and Wang (1987a) have shown that if $R(f) = (a, b)$ and $R(g) = (c, d)$ are both finite intervals, then the quasi-independent projection $P_h(x, y)$ exists uniquely over $R(h)$.

4.14.3 A Measure of Regional Dependence

The regional dependence measure $M(x, y) = M(f, g, R(h))$ is defined by

$$M(X, Y) = 1 - \frac{1}{c},$$

where $c = \int_{R(f)\times R(g)} P_h(x, y)\, dx\, dy$. For discrete random variables, $M(X, Y)$ can be computed from incomplete two-way contingency tables. The following properties of the measure M are reported by Holland and Wang (1987b):

- $0 \leq M(X, Y) \leq 1$.
- X and Y being independent implies $M(X, Y) = 0$ but the converse is not true.
- If X and Y are monotonically dependent, then $M(X, Y) = 1$.
- If $h(x, y)$ is a constant throughout $R(h)$, then

$$M(X, Y) = 1 - \frac{\text{Area of } R(h)}{\text{Area of } [R(f) \times R(g)]} \,.$$

- For fixed $R(f)$ and $R(g)$, let h_1 and h_2 be two constant densities defined inside $R(f) \times R(g)$, such that the marginal densities are positive in $R(f)$ and $R(g)$. Then, $R(h_1) \subseteq R(h_2)$ implies that $M(X_1, Y_1) \geq M(X_2, Y_2)$, where (X_1, Y_1) and (X_2, Y_2) have joint densities h_1 and h_2, respectively.
- $M(X, Y)$ is invariant under smooth monotone transformation of the marginals. (Of course, it changes sign if the transformation of one marginal is increasing and the other is decreasing.)

Just as the maximal correlation and monotone correlation are difficult to calculate, $M(X, Y)$ may not be easy to calculate as well, and especially so when $R(f)$ and $R(g)$ are not finite intervals.

References

1. Barnett, V.: Some outlier tests for multivariate samples. South African Statistical Journal **13**, 29–52 (1979)
2. Barnett, V.: The bivariate exponential distribution; a review and some new results. Statistica Neerlandica **39**, 343–357 (1985)
3. Bauer, H.: Probability Theory and Elements of Measure Theory. Holt, Rinehart and Winston, New York (1972)
4. Bjerve, S., Doksum, K.: Correlation curves: Measures of association as function of covariates values. Annals of Statistics **21**, 890–902 (1993)

5. Blake, I.F.: An Introduction to Applied Probability. John Wiley and Sons, New York (1979)
6. Blomqvist, N.: On a measure of dependence between two random variables. Annals of Mathematical Statistics **21**, 593–600 (1950)
7. Capéraà, P., Genest, C.: Spearman's ρ_S is larger than the Kendall's τ for positively dependent random variables. Nonparametric Statistics **2**, 183–194 (1993)
8. Chambers, J.M., Cleveland, W.S., Kleiner, B., Tukey, P.A.: Graphical Methods for Data Analysis, Wadsworth, Belmont, California (1983)
9. Chen, Y.-P.: A note on the relationship between Spearman's ρ and Kendall's τ for extreme-order statistics. Journal of Statistical Planning and Inference **137**, 2165–2171 (2007)
10. Cook, M.B.: Bivariate k-statistics and cumulants of their joint sampling distribution. Biometrika **38**, 179–195 (1951)
11. Cramér, H.: Mathematical Methods of Statistics. Princeton University Press, Princeton, New Jersey (1954)
12. Dabrowska, D.M., Duffy, D.L., Zhang, D.Z.: Hazard and density estimation from bivariate censored data. Journal of Nonparametric Statistics **10**, 67–93 (1999)
13. Daniels, H.E.: Rank correlation and population models. Journal of the Royal Statistical Society, Series B **12**, 171–181 (Discussion, 182–191) (1950)
14. David, F.N., Fix, E.: Rank correlation and regression in a nonnormal surface. In: Proceedings of the 4th Berkeley Symposium on Mathematical Statistics and Probability, Volume 1, J. Neyman (ed.), pp. 177–197. University of California Press, Berkeley (1961)
15. Devlin, S.J., Gnanadesikan, R., Kettenring, J.R.: Robust estimation and outlier detection with correlation coefficients. Biometrika **62**, 531–545 (1975)
16. Divgi, D.R.: Calculation of the tetrachoric correlation coefficient. Psychometrika **44**, 169–172 (1979b)
17. Dodge, Y., Rousson, V.: Direction dependence in a regression line. Communications in Statistics: Theory and Methods **29**, 1957–1972 (2000)
18. Drasgow, F.: Polychoric and polyserial correlations, In: Encyclopedia of Statistical Sciences, Volume 7, S. Kotz and N.L. Johnson (eds.), pp. 68–74. John Wiley and Sons, New York (1986)
19. Drouet-Mari, D., Kotz, S.: Correlation and Dependence. Imperial College Press, London (2001)
20. Elffers, H.: On interpreting the product moment correlation coefficient. Statistica Neerlandica **34**, 3–11 (1980)
21. Falk, R., Well, A.D.: Many faces of correlation coefficient. Journal of Statistical Education **5**, 1–16 (1997)
22. Fieller, E.C., Hartley, H.O., Pearson, E.S.: Tests for rank correlation coefficients. 1. Biometrika **44**, 470–481 (1957)
23. Fredricks, G.A., Nelsen, R.B.: On the relationship between Spearman's rho and Kendall's tau for pairs of continuous random variables. Journal of Statistical Planning and Inference **137**, 2143–2150 (2007)
24. Gayen, A.K.: The frequency distribution of the product-moment correlation coefficient in random samples of any size from non-normal universe. Biometrika **38**, 219–247 (1951)
25. Gebelein, H.: Das statistische Problem der Korrelation als Variations und Eigenwertproblem und sein Zusammenhang mit der Ausgleichsrechnung. Zeitschrift für Angewandte Mathematik und Mechanik **21**, 364–379 (1941)
26. Gibbons, J.D.: Nonparametric Statistical Inference, McGraw-Hill, New York (1971)
27. Gideon, R.A., Hollister, R.A.: A rank correlation coefficient resistant to outliers. Journal of the American Statistical Association **82**, 656–666 (1987)
28. Goodman, L.A.: How to ransack social mobility tables and other kinds of cross-classification tables. American Journal of Sociology **75**, 1–40 (1969)

29. Guttman, L.: Polytonicity and monotonicity, Coefficients of. In: Encyclopedia of Statistical Sciences, Volume 7, S. Kotz and N.L. Johnson (eds.), pp. 80–87. John Wiley and Sons, New York (1986)
30. Hall, W.J.: On characterizing dependence in joint distributions. In: Essays in Probability and Statistics, R.C. Bose, I.M. Chakravarti, P.C. Mahalanobis, C.R. Rao, and K.J.C. Smith (eds.), pp. 339–376. University of North Carolina Press, Chapel Hill (1970)
31. Harris, B.: Tetrachoric correlation coefficient. In: Encyclopedia of Statistical Sciences, Volume 9, S. Kotz and N.L. Johnson (eds.), pp. 223–225. John Wiley and Sons, New York (1988)
32. Hoeffding, W.: Masstabinvariante Korrelationstheorie. Schriften des Mathematischen Instituts und des Instituts für Angewandte Mathematik der Universität Berlin **5**, 179–233 (1940)
33. Holland, P.W., Wang, Y.J.: Regional dependence for continuous bivariate densities. Communications in Statistics: Theory and Methods **16**, 193–206 (1987a)
34. Holland, P.W., Wang, Y.J.: Dependence function for continuous bivariate densities. Communications in Statistics: Theory and Methods **16**, 863–876 (1987b)
35. Hougaard, P.: Analysis of Multivariate Survival Data. Springer-Verlag, New York (2000)
36. Hürlimann, W.: Hutchinson-Lai's Conjecture for bivariate extreme value copulas. Statistics and Probability Letters **61**, 191–198 (2003)
37. Hutchinson, T.P.: A comment on correlation in skewed distributions. The Journal of General Psychology **124**, 211–215 (1997)
38. Hutchinson, T.P., Lai, C.D.: The Engineering Statistician's Guide to Continuous Bivariate Distributions. Rumsby Scientific Publishing, Adelaide (1991)
39. Jensen, D.R.: Semi-independence. In: Encyclopedia of Statistical Sciences, Volume 8, S. Kotz and N.L. Johnson (eds.), pp. 358–359. John Wiley and Sons, New York (1988)
40. Jogdeo, K.: Dependence, Concepts of. In: Encyclopedia of Statistical Sciences, Volume 2, S. Kotz and N.L. Johnson (eds.), pp. 324–334. John Wiley and Sons, New York (1982)
41. Johnson, N.L., Kotz, S., Balakrishnan, N.: Continuous Univariate Distributions, Volume 2, 2nd edition. John Wiley and Sons, New York (1995)
42. Jones, M.C.: The local dependence function. Biometrika **83**, 899–904 (1996)
43. Jones, M.C.: Constant local dependence. Journal of Multivariate Analysis **64**, 148–155 (1998)
44. Kendall, M.G., Stuart, A.: The Advanced Theory of Statistics, Volume 2: Inference and Relationship, 4th edition. Griffin, London (1979)
45. Kimeldorf, G., Sampson, A.: Monotone dependence. Annals of Statistics **5**, 895–903 (1978)
46. Kimeldorf, G., May, J.H., and Sampson, A.R.: MONCOR: A program to compute concordant and other monotone correlations. In: Computer Science and Statistics: Proceedings of the 13th Symposium on the Interface, W.F. Eddy (ed.), pp. 348–351. Springer-Verlag, New York (1981)
47. Kimeldorf, G., May, J.H., Sampson, A.R.: Concordant and discordant monotone correlations and their evaluation by nonlinear optimization. In: Optimization in Statistics, With a View Towards Applications in Management Science and Operations Research, S.H. Zanakis and J.S. Rustagi (eds.), pp. 117–130. North Holland, Amsterdam (1982)
48. Klaassen, C.A., Wellner, J.A.: Efficient estimation in the bivariate normal copula model: Normal margins are least favourable. Bernoulli **3**, 55–77 (1997)
49. Kotz, S., Wang, Q.S., Hung, K.: Interrelations among various definitions of bivariate positive dependence. In: Topics in Statistical Dependence, H.W. Block, A.R. Sampson and T. Savits (eds.), pp. 333–350. Institute of Mathematical Statistics, Hayward, California (1992).

50. Kruskal, W.H.: Ordinal measures of association. Journal of the American Statistical Association **53**, 814–861 (1958)
51. Lai, C.D., Rayner, J.C.W., Hutchinson, T.P.: Robustness of the sample correlation: The bivariate lognormal case. Journal of Applied Mathematics and Decision Sciences **3**, 7–19 (1999)
52. Lancaster, H.O.: The structure of bivariate distributions. Annals of Mathematical Statistics **29**, 719–736 (1958)
53. Lancaster, H.O.: Correlation and complete dependence of random variables. Annals of Mathematical Statistics **34**, 1315–1321 (1963)
54. Lancaster, H.O.: Chi-square distribution. In: Encyclopedia of Statistical Sciences, Volume 1, S. Kotz and N.L. Johnson (eds.), pp. 439–442. John Wiley and Sons, New York (1982a)
55. Lancaster, H.O.: Dependence, Measures and indices of. In: Encyclopedia of Statistical Sciences, Volume 2, S. Kotz and N.L. Johnson (eds.), pp. 334–339. John Wiley and Sons, New York (1982b)
56. Lavoie, J.L.: Some evaluations for the generalized hypergeometric series. Mathematics of Computation **46**, 215–218 (1986)
57. Lee, S-Y.: Maximum likelihood estimation of polychoric correlations in $r \times s \times t$ contingency tables. Journal of Statistical Computation and Simulation **23**, 53–67 (1985)
58. Lee, S-Y., Poon, W-Y.: Two-step estimation of multivariate polychoric correlation. Communications in Statistics: Theory and Methods **16**, 307–320 (1987a)
59. Lee, S-Y., Poon, W-Y.: Some algorithms in computing GLS estimates of multivariate polychoric correlations. In: American Statistical Association, 1987 Proceedings of the Statistical Computing Section, pp. 444–447. American Statistical Association, Alexandria, Virgina (1987b)
60. Li, X.-H., Li, Z.-P.: Proof of a conjecture on Spearman's ρ and Kendall's τ for sample minimum and maximum. Journal of Statistical Planning and Inference **137**, 359–361 (2007)
61. Martinson, E.O., Hamdan, M.A.: Algorithm AS 87: Calculation of the polychoric estimate of correlation in contingency tables. Applied Statistics **24**, 272–278 (1975)
62. Moran, P.A.P.: Testing for correlation between non-negative variates. Biometrika **54**, 385–394 (1967)
63. Muddapur, M.V.: On directional dependence in a regression line. Communications in Statistics: Theory and Methods **32**, 2053–2057 (2003)
64. Mudholkar, G.S.: Fisher's z-transformation. In: Encyclopedia of Statistical Sciences, Volume 3, S. Kotz and N.L. Johnson (eds.), pp. 130–135. John Wiley and Sons, New York (1983)
65. Nakagawa, S., Niki, N.: Distribution of sample correlation coefficient for non-normal populations. Journal of Japanese Society of Computational Statistics **5**, 1–19 (1992)
66. Nelsen, R.B.: Measures of association as measures of positive dependence. Statistics & Probability Letters **14**, 269–274 (1992)
67. Nelsen, R.B.: An Introduction to Copulas. Springer-Verlag, New York (1999)
68. Nelsen, R.B.: An Introduction to Copulas, 2nd edition. Springer-Verlag, New York (2006)
69. Olsson, U.: Maximum likelihood estimation of the polychoric correlation coefficient. Psychometrika **44**, 443–460 (1979)
70. Olsson, U.: Measuring correlation in ordered two-way contingency tables. Journal of Marketing Research **17**, 391–394 (1980)
71. Pearson, K.: Mathematical contributions to the theory of evolution. XIV: On the general theory of skew correlation and nonlinear regression. Drapers' Company Research Memoirs, Biometric Series, II (1905). [Reprinted in Karl Pearson's Early Statistical Papers, pp. 477–528. Cambridge University Press, Cambridge (1948)]
72. Prentice, R.L., Cai, J.: Covariance and survival function estimation using censored multivariate failure time data. Biometrika **79**, 495–512 (1992)

73. Rényi, A.: On measures of dependence. Acta Mathematica Academia Scientia Hungarica **10**, 441–451 (1959)
74. Rényi, A.: Probability Theory. North-Holland, Amsterdam (1970)
75. Rodgers, J.L., Nicewander, W.A.: Thirteen ways to look at the correlation coefficient. The American Statistician **42**, 59–66 (1988)
76. Rodriguez, R.N.: Correlation. In: Encyclopedia of Statistical Sciences, Volume 2, S. Kotz and N.L. Johnson (eds.), pp. 193–204. John Wiley and Sons, New York (1982)
77. Rovine, M.J., Von Eye, A.C.: A 14th way to look at a correlation: Correlation as the proportion of matches. The American Statistician **51**, 42–46 (1997)
78. Ruppert, D.: Trimming and Winsorization. In: Encyclopedia of Statistical Sciences, Volume 9, S. Kotz and N.L. Johnson (eds.), pp. 348–353. John Wiley and Sons, New York (1988)
79. Sarmanov, O.V.: Maximum correlation coefficient (nonsymmetric case). Selected Translations in Mathematical Statistics and Probability **2**, 207–210 (1962) (Original Russian article was dated 1958)
80. Sarmanov, O.V.: Maximum correlation coefficient (symmetric case). Selected Translations in Mathematical Statistics and Probability **4**, 271–275 (1963) (Original Russian article was dated 1959)
81. Schmitz, V.: Revealing the dependence structure between $X_{(1)}$ and $X_{(n)}$. Journal of Statistical Planning and Inference **123**, 41–47 (2004)
82. Schweizer, B., Wolff, E.F.: Sur une mesure de dépendence pour les variables aléatories. Comptes Rendus de l'Académie des Sciences, Série A **283**, 659–661 (1976)
83. Schweizer, B., Wolff, E.F.: On nonparametric measure of dependence for random variables. Annals of Statistics **9**, 879–885 (1981)
84. Sungur, E.A.: A note on directional dependence in regression setting. Communications in Statistics: Theory and Methods **34**, 1957–1965 (2005)
85. Tchen, A.H.: Inequalities for distributions with given marginals. Annals of Probability **8**, 814–827 (1980)
86. Tong, Y.L.: Probability Inequalities in Multivariate Distributions. Academic Press, New York (1980)
87. Yule, G.U.: Why do we sometimes get nonsense-correlations between time-series? A study in sampling and the nature of time-series. Journal of the Royal Statistical Society **89**, 1–64 (1926) (Reprinted in A. Stuart and M.G. Kendall (selectors), Statistical Papers of George Udny Yule, pp. 325–388. Griffin, London)
88. Yule, G.U., Kendall, M.G.: An Introduction to the Theory of Statistics, 11th edition. Griffin, London (1937)

Chapter 5
Construction of Bivariate Distributions

5.1 Introduction

In this chapter, we review methods of constructing bivariate distributions. There is no satisfactory mathematical scheme for classifying the methods. Instead, we offer a classification that is based on loosely connected common structures, with the hope that a new bivariate distribution can be fitted into one of these schemes. We focus especially on application-oriented methods as well as those with mathematical nicety.

Sections 5.2–5.11 of this chapter deal with the first major group of methods which have been repeatedly rediscovered and reinvented by applied scientists seeking models for statistical dependence in numerous applied fields. Sections 5.12–5.16 deal with approaches that are more specific to particular applications.

In Section 5.2, we explain the marginal transformation method. In Sections 5.3 and 5.4, we describe different methods of constructing copulas and the mixing and compounding methods, respectively. In Section 5.5, we present the variables in common and trivariate reduction techniques for constructing bivariate distributions. In Section 5.6, we explain the construction of a joint distribution based on specified conditional distributions. Next, in Section 5.7 the marginal replacement method is outlined. In Section 5.8 bivariate ad multivariate skew distributions are referenced. Sections 5.9 and 5.10 outline density generators and geometric approaches. In Sections 5.11 and 5.12, some other simple construction methods and the weighted linear combination method are detailed. Data-guided methods are described in Section 5.13, while some special methods used in applied fields are presented in Section 5.14. Some bivariate distributions that are derived as limits of discrete distributions are explained in Section 5.15. After describing some other methods that could potentially be useful in constructing bivariate distributions but are not in vogue in Section 5.16, we com-

N. Balakrishnan, C.D. Lai, *Continuous Bivariate Distributions*,
DOI 10.1007/b101765_6,

plete the discussion in this chapter by making some concluding remarks in Section 5.17.

In the remainder of this section, we present some preliminary details and notation that are used throughout this chapter.

5.1.1 Fréchet Bounds

Let f and g be marginal probability density functions. For given marginal distribution functions F and G, what limits must a joint distribution function H satisfy so as to have its p.d.f. be non-negative everywhere? Hoeffding (1940) and Fréchet (1951) showed in this regard that

$$H^-(x,y) \leq H(x,y) \leq H^+(x,y), \tag{5.1}$$

where

$$H^+(x,y) = \min[F(x), G(y)] \tag{5.2}$$

and

$$H^-(x,y) = \max[F(x) + G(y) - 1, 0]. \tag{5.3}$$

It is easy to verify that the Fréchet bounds H^+ and H^- are themselves d.f.'s and that they have maximum and minimum correlations for the given marginals. Also, H^+ concentrates all the probability on the increasing curve $F(x) = G(y)$, and H^- concentrates all the probability on the decreasing curve $F(x)+G(y) = 1$. For any F and G, $[F^{-1}(U)G^{-1}(U)]$ has d.f. H^+ and $[F^{-1}(U), G^{-1}(1-U)]$ has d.f. H^-, where U denotes a standard uniform(0,1) random variable. For proofs and discussion on H^+ and H^-, one may refer to Whitt (1976), who also proved that convolution of identical bivariate distributions results in an increase of the Fréchet upper bound and a decrease of the Fréchet lower bound.

In order to have notation for the independent case, we further define

$$H^0(x,y) = F(x)G(y). \tag{5.4}$$

Devroye (1986, p. 581) uses the term *comprehensive* for any family of distributions that includes H^+, H^0, and H^-.

What if the distribution is restricted to the region $X \leq Y$? Smith (1983) showed that, in this case, the bounds on H become

$$G(y) - \max\{0, \min[G(y) - G(x), F(y) - F(x)]\} \leq H(x,y) \leq \min[F(x), G(y)]. \tag{5.5}$$

In (5.5), the lower bound need not necessarily be a distribution function. A sufficient condition for it to be one is that there exist an x_0 such that $g(x) > f(x)$ for $x > x_0$ and $g(x) < f(x)$ for $x < x_0$.

Regarding Fréchet bounds for multivariate distributions, one may refer to Kwerel (1983).

5.1.2 Transformations

Suppose we have a density $h(x, y)$ and we form two new variables $A(X, Y)$ and $B(X, Y)$. What is the joint density of A and B? To answer this, we first need to express X and Y in terms of A and B. Letting, as usual, the values of variates X, Y, A, and B be x, y, a, and b, the density of (A, B) is then $h[x(a, b), y(a, b)]\ |J|$, where J is the *Jacobian*, given by

$$J = \begin{vmatrix} \frac{\partial x}{\partial a} & \frac{\partial x}{\partial b} \\ \frac{\partial y}{\partial a} & \frac{\partial y}{\partial b} \end{vmatrix} = \frac{\partial x}{\partial a}\frac{\partial y}{\partial b} - \frac{\partial y}{\partial a}\frac{\partial x}{\partial b} = \frac{1}{\frac{\partial a}{\partial x}\frac{\partial b}{\partial y} - \frac{\partial a}{\partial y}\frac{\partial b}{\partial x}}\ . \tag{5.6}$$

For a more detailed explanation that includes pictures of a rectangle being transformed into a distorted rectangle, see Blake (1979, Section 7.2). Transformations often encountered include $X+Y$, $X-Y$, XY, X/Y, $X/(X+Y)$, $\sqrt{X^2+Y^2}$, and $\tan^{-1}(Y/X)$; see, for example, Blake (1979, Section 7.2).

Let us now consider the special case of transforming the marginals. Suppose X and Y are each uniformly distributed between 0 and 1, and we transform the marginals so that they become F and G (with densities f and g, respectively). In this case, $A \equiv F^{-1}$ and $B \equiv G^{-1}$, so that $X \equiv F$ and $Y \equiv G$. Hence, the density of (A, B) will be given by $h[F(a), G(b)]\frac{\partial F}{\partial a}\frac{\partial G}{\partial b} = h[F(a), G(b)]f(a)g(b)$.

Physicists have apparently found it helpful to put the conditions that a p.d.f. has to satisfy (non-negative and integrates to 1), along with what happens under transformation of the marginals, into the following form. Bivariate densities having $f(x)$ and $g(y)$ as their marginal densities and $F(x)$ and $G(y)$ as their marginal d.f.'s must be of the form $h = fg[1 + a(F, G)]$, where $a(u, v)$ is any function on the unit square that is bounded below by -1 and satisfies $\int_0^1 a(u, v)du = \int_0^1 a(u, v)dv = 0$; see Finch and Groblicki (1984) and Cohen and Zaparovanny (1980).

5.2 The Marginal Transformation Method

5.2.1 General Description

The basic idea here, usually attributed to Nataf (1962), is that if we start with a bivariate distribution $H(x, y)$ (with density $h(x, y)$) and apply monotone transformations $X \to X^*$ and $Y \to Y^*$, there is a sense in which the new distribution $H^*(x^*, y^*)$ has the same bivariate structure as the original H, and all that has changed is the marginals (viz., F becoming F^* and G becoming G^*). In the univariate situation, familiar examples include (i) transforming the normal distribution so that it becomes lognormal and (ii) transforming the exponential distribution so that it becomes Weibull.

The emphasis when transforming marginals may take either of two forms, which is easier to illustrate in the context of the bivariate normal distribution:

- Start with the bivariate normal distribution. Accept its description of how X and Y are interconnected as satisfactory, but suppose normal marginals are unsatisfactory for the purpose at hand. Transform the marginals so that they become normal.
- Start with an empirical or unfamiliar bivariate distribution. In order to compare its contours or other properties with the bivariate normal distribution, free from the influence of the forms of the marginals, transform its marginals to be normal.

Other distributions are sometimes used as standard—uniform and exponential are two examples. In Chapter 2, a number of distributions were written as their uniform representations, from which it was easy to transform to any other required marginals.

Sometimes, the purpose of a transformation is to change the region of support of a distribution. For example, suppose (X, Y) has a bivariate normal distribution. Then, (e^X, e^Y) has a bivariate lognormal distribution (the support of which is the positive quadrant), and $\left(\frac{e^X}{1+e^X+e^Y}, \frac{e^y}{1+e^X+e^Y}\right)$ has a bivariate logistic-normal distribution (the support of which is the simplex).

5.2.2 Johnson's Translation Method

The best-known set of distributions constructed by marginal transformation is that due to Johnson (1949), who started with the bivariate normal and transformed X and/or Y so that the marginals

- remain the same,
- become lognormal,
- become logit-normal, and
- become $\sinh^{-1}$-normal.

This has traditionally been referred to as a *translation method*, though we feel that *transformation* would be a better term. Including no transformation of the normal marginals as one of the possibilities, subscripts N, L, B, and U are used for the four models. There being four choices for X and similarly four choices for Y, a total of 16 possible bivariate distributions result, which are all listed in Kotz et al. (2000). For example, h_{NN} is the bivariate normal density, while h_{LL} is the bivariate lognormal density function.

As already mentioned in Chapter 4, it is well known that Pearson's product-moment correlation is not affected by linear transformations of X and Y. But what happens when applying nonlinear transformations? The answer is that if we start with the bivariate normal distribution and do this, the correlation becomes smaller (in absolute magnitude).

5.2.3 Uniform Representation: Copulas

The great innovation in the study of bivariate distributions over the last 30 years has been the desire to separate the *bivariate structure* from the *marginal distributions*. One manifestation of this has been the great interest in the study of *copulas* (also known as the *uniform representation*) of the distribution. This is the form the distribution takes when X and Y are transformed so that they each have a uniform distribution over the range 0 to 1.

As an example, suppose we start with

$$H = xy[1 + \alpha(1 - x)(1 - y)] \tag{5.7}$$

for x and y between 0 and 1, with $-1 \leq \alpha \leq 1$. Setting $y = 1$, we see that the distribution of X is uniform, $F = x$; similarly, setting $x = 1$, we see the distribution of Y is uniform, $G = y$. Now, suppose we require the new marginals to be exponential, $F = 1 - e^{-x}$ and $G = 1 - e^{-y}$. Replacing x by $1 - e^{-x}$ and y by $1 - e^{-y}$ in (5.7), we obtain

$$H = (1 - e^{-x})(1 - e^{-y})[1 + \alpha e^{-(x+y)}]. \tag{5.8}$$

Equation (5.8) is a bivariate exponential distribution considered by Gumbel (1960), and the copula in (5.7) is known as the *Farlie–Gumbel–Morgenstern copula.*

Some of the important advantages of considering distributions after their marginals have been made uniform are as follows:

- Independence does not usually have a clear geometric meaning, in that the graph of the joint p.d.f. of X and Y provides us no insight as to whether or not X and Y are independent. However, independence takes on a geometric meaning for variates U and V with uniform marginals, in that they are independent if and only if their joint p.d.f. is constant. Any variation in the value of the p.d.f. is indicative of dependence between U and V.
- The copula is the natural framework in which to discuss nonparametric measures of correlation, such as Kendall's τ and Spearman's rank correlation ρ_S.

- Simulations of X and Y may become easier via simulations of the associated copulas.

5.2.4 Some Properties Unaffected by Transformation

For any family H_θ ($-1 \leq \theta \leq 1$) of d.f.'s having absolutely continuous marginals F and G, consider the following five conditions:

(1) The upper Fréchet bound corresponds to $\theta = 1$, i.e.,

$$H_1(x, y) = \min[F(x), G(y)].$$

(2) At $\theta = 0$, X and Y are independent, i.e., $H_0(x, y) = F(x)G(y)$.
(3) The lower Fréchet bound corresponds to $\theta = -1$, i.e.,

$$H_{-1}(x, y) = \max[F(x) + G(y) - 1, 0].$$

(4) For fixed x, y, H_θ is continuous in $[-1, 1]$.
(5) For fixed θ in $(-1, 1)$, H_θ is an absolutely continuous d.f.

Then, Kimeldorf and Sampson (1975) have given the following result. Let $H = \{H_\theta : -1 \leq \theta \leq 1\}$ be a family of d.f.'s with fixed marginals F_1, G_1, and satisfying any subset of conditions (1)–(5). Let F_2 and G_2 be any two continuous d.f.'s. Then,

$$J = \{J_\theta(x, y) = H_\theta[F_1^{-1}F_2(x), G_1^{-1}G_2(y)], -1 \leq \theta \leq 1\} \tag{5.9}$$

is a family of d.f.'s with fixed marginals F_2 and G_2, that satisfies the same subset of conditions (1)–(5) as does H_θ.

Example 5.1. Suppose we pick one of the distributions whose d.f. is simple in form and whose marginals are uniform—for example, Frank's copula (see Section 2.4) given by

$$H_\alpha = \log_\alpha \left\{ 1 + \frac{(\alpha^x - 1)(\alpha^y - 1)}{\alpha - 1} \right\}, \quad 0 < \alpha \neq 1.$$

Then, if we require a joint distribution with marginals F and G, we can write

$$J_\alpha(x, y) = \log_\alpha \left\{ 1 + \frac{(\alpha^{F(x)} - 1)(\alpha^{G(y)} - 1)}{\alpha - 1} \right\}.$$

5.3 Methods of Constructing Copulas

Copulas can be considered as a starting point for constructing families of bivariate distributions because a bivariate distribution H with given marginals F and G can be generated via Sklar's theorem that $H(x, y) = C(F(x), H(y))$ after the copula C is determined. Thus, constructions of copulas play an important role in producing various families of bivariate distributions.

5.3.1 The Inversion Method

This is simply the marginal transformation method through inverse probability integral transforms of the marginals $F^{-1}(u) = x$ and $G^{-1}(v) = y$. If either one of the two inverses does not exist, we simply modify our definition so that $F^{-1}(u) = \inf\{x : F(x) \geq u\}$, for example. Then, for a given bivariate distribution function H with continuous marginals F and G, we obtain a copula

$$C(u, v) = H(F^{-1}(u), G^{-1}(v)). \tag{5.10}$$

With this copula, new bivariate distributions with arbitrary marginals, say F' and G', can be constructed using the formula $H'(x, y) = C(F', G')$.

Note also that a *survival copula* (complementary copula) can be obtained by using the survival functions $\bar{F}, \bar{G}$, and $\bar{H}$ (in place of F, G, and H) as

$$\hat{C}(u, v) = \bar{H}(\bar{F}^{-1}(u), \bar{G}^{-1}(v)). \tag{5.11}$$

5.3.2 Geometric Methods

Several geometric schemes have been given in Chapter 3 of Nelsen (2006):

- singular copulas with prescribed support;
- ordinal sums;
- shuffles of Min [Mikusiński et al. (1992)];
- copulas with prescribed horizontal or vertical sections;
- copulas with prescribed diagonal sections.

Wei et al. (1998) constructed copulas with discontinuity constraints. Their procedures may be considered as geometric methods, and they obtained the following three families of copulas:

(1) piecewise additive copulas with the unit square being partitioned into measurable closed sets A_i such that the copula is piecewise additive [i.e., on each partition set A_i, $C(u, v)|_{A_i} = C_1(u) + C_2(v)$, where $C_1(u)$ and $C_2(v)$ are some increasing functions];

(2) piecewise quadratic copulas whose densities are piecewise constant over the four rectangular regions of the unit square (so that they are locally independent);
(3) quadratic copulas with holes constructed by shifting the omitted mass of holes along one axis, next along the other axis, and again along the first axis, so as to ensure that the marginals are unaffected.

5.3.3 Algebraic Methods

Two well-known families of copulas, the Plackett and Ali–Mikhail–Haq families, were constructed using an algebraic relationship between the joint distribution function and its univariate marginals. In both cases, the algebraic relationship concerns an *odds ratio*. In the first case, we generalize 2×2 contingency tables, and in the second case we work with a survival odds ratio.

5.3.4 Rüschendorf's Method

Rüschendorf (1985) developed a general method of constructing a copula as follows:

Step 1. Find a function $f^1(u,v)$ such that

$$\int_0^1 \int_0^1 f^1(u,v) du\, dv = 0 \tag{5.12}$$

and

$$\int_0^1 f^1(u,v) du = 0 \text{ and } \int_0^1 f^1(u,v) dv = 0. \tag{5.13}$$

Clearly, (5.13) implies (5.12).

Step 2. (Construction of f_1) One starts with an arbitrary real integrable function f on the unit square and then computes

$$V = \int_0^1 \int_0^1 f(u,v) du\, dv,\ f_1(v) = \int_0^1 f(u,v) du\, du,\ f_2(v) = \int_0^1 f(u,v) du.$$

Then, $f^1 = f - f_1 - f_2 + V$.

Step 3. Then, $c(u,v) = 1 + f^1(u,v)$ is a density of a copula. However, there is a constraint that $1 + f^1(u,v)$ must be positive. If this is not the case but f^1 is bounded, we can then find a constant α such that $1 + \alpha f^1$ is positive.

In general, $1 + \sum_{i=1}^{n} f_i^1$ is a density with f_i^1 satisfying the conditions above in (5.12) and (5.13).

Example 5.2. Lai and Xie (2000) extended the F-G-M copula as

$$C(u,v) = uv + w(u,v) = uv + \alpha u^b v^b (1-u)^a (1-v)^a, \; a,b, 0 \le \alpha \le 1. \quad (5.14)$$

This method allows us to generate all polynomial copulas discussed earlier in Section 1.10.

5.3.5 Models Defined from a Distortion Function

In the field of insurance pricing, one often uses [see, e.g., Frees and Valdez (1998)] a distortion function ϕ that maps $[0,1]$ onto $[0,1]$, with $\phi(0) = 0, \phi(1) = 1$, and ϕ increasing.

Starting with $H(x,y) = C(F(x), G(y))$, one defines another distribution function via such a function ϕ as

$$H^*(x,y) = \phi[H(x,y)] \quad (5.15)$$

with marginals $F^*(x) = \phi(F(x))$ and $G^*(y) = \phi(G(y))$. The associated copula is then

$$C^*(u,v) = \phi[C(\phi^{-1}(u), \phi^{-1}(v))]. \quad (5.16)$$

Example 5.3 (Frank's copula). Let

$$\phi(t) = \frac{1 - e^{\alpha t}}{1 - e^{-\alpha}}, \quad \alpha > 0,$$

with independent copula $C(u,v) = uv$, yielding the copula

$$C^*(u,v) = \log_\alpha \left(1 + \frac{(\alpha^u - 1)(\alpha^v - 1)}{\alpha - 1}\right),$$

which is the well-known *Frank's copula*; see Section 2.4 for pertinent details.

5.3.6 Marshall and Olkin's Mixture Method

Marshall and Olkin (1988) considered a general method for generating bivariate distributions through mixture. Set

$$H(u,v) = \int\int K(F^{\theta_1}, G^{\theta_2})\, d\Lambda(\theta_1, \theta_2), \quad (5.17)$$

where K is a copula, Λ is a mixing distribution, and ϕ_i is the Laplace transform of the marginal Λ_i of Λ. Thus, selections of Λ and K lead to a variety of distributions with marginals as parameters. Note that F and G here are not the marginals of H.

If K is an independent bivariate distribution and the two marginals of Λ are equal to the Fréchet bound (i.e., $\Lambda(\theta_1, \theta_2) = \min(\Lambda_1(\theta_1), \Lambda_2(\theta_2)))$, then $H(u,v) = \int_0^\infty F^\theta(u) G^\theta(v) d\Lambda_1(\theta)$ with $\theta_1 = \theta$. Now, let $F(u) = \exp[-\phi^{-1}(u)]$ and $G(u) = \exp[-\phi^{-1}(u)]$, where $\phi(t)$ is the Laplace transform of Λ_1 and so $\phi(-t)$ is the moment generating function of Λ_1. It then follows that

$$H(u,v) = \int_0^\infty \exp\left[-\theta\left(\phi^{-1}(u) + \phi^{-1}(v)\right)\right] d\Lambda_1(\theta). \tag{5.18}$$

Because $\phi^{-1} = 0$, it is clear that the marginals of H are uniform and so H is a copula. In other words, when ϕ is the Laplace transform of a distribution, then the function defined on the unit square by

$$C(u,v) = \phi\left(\phi^{-1}(u) + \phi^{-1}(v)\right) \tag{5.19}$$

is indeed a copula. Marshall and Olkin (1988) have presented several examples.

Joe (1993) studied the properties of a group of eight families of copulas, three of which were given by Marshall and Olkin (1988). Joe and Hu (1996) derived a class of bivariate distributions that are mixtures of the positive powers of a max-infinitely divisible distribution. Their approach is based on a generalization of Marshall and Olkin's (1988) mixture method.

5.3.7 Archimedean Copulas

An important family of copulas are *Archimedean copulas*, which were discussed in Section 1.5.

Any function φ that has two continuous derivatives and that satisfies $\varphi(1) = 0, \varphi'(u) < 0$, and $\varphi''(u) > 0$ (naturally, u is between 0 and 1) generates a copula. These conditions are equivalent to saying that $1 - \varphi^{-1}(t)$ is the distribution of a unimodal r.v. with mode at 0 [Genest and Rivest (1989)].

We can define an inverse (or quasi-inverse if $\varphi(0) < \infty$) by

$$\varphi^{[-1]}(t) = \begin{cases} \varphi^{-1}(t), & 0 \le t \le \varphi(0) \\ 0, & \varphi(0) \le t \le \infty. \end{cases}$$

An Archimedean copula is then defined as

$$C(u,v) = \varphi^{[-1]}(\varphi(u) + \varphi(v)). \tag{5.20}$$

Here, the function φ is called a *generator* of an Archimedean copula. In other words, one can construct an Archimedean copula C by finding a generator having the above-mentioned properties. Several examples were presented in Section 1.5.

5.3.8 Archimax Copulas

An Archimax copula is generated by a bivariate extreme-value copula and a convex function defined on $[0, 1]$ that maps onto $[1/2, 1]$ as

$$C_{\varphi,A}(u, v) = \varphi^{-1}\left[\{\varphi(u) + \varphi(v)\}A\left\{\frac{\varphi(u)}{\varphi(u) + \varphi(v)}\right\}\right], \tag{5.21}$$

subject to $\max(t, 1 - t) \leq A(t) \leq 1$ for all $t \in [0, 1]$.

5.4 Mixing and Compounding

In the statistical literature, the terms *mixing* and *compounding* are often used synonymously, with the latter being used rarely these days. Here, we prefer to reserve the term *mixing* for a finite mixture of distributions while the rest of the mixtures involve *compounding*.

5.4.1 Mixing

One of the easiest ways to generate bivariate distributions is to use the method of mixing along with two distributions. Specifically, if H_1 and H_2 are two bivariate distribution functions, then

$$H(x, y) = \theta H_1(x, y) + (1 - \theta)H_2(x, y), \quad 0 \leq \theta \leq 1, \tag{5.22}$$

is a new bivariate distribution. Examples are readily found in Fréchet bounds:

- Fréchet (1951) himself suggested a one-parameter family of bivariate distributions that attained the Fréchet bounds at the limits of the parameter θ as

$$H(x, y) = \theta H^{-1}(x, y) + (1 - \theta)H^{+1}(x, y), \quad 0 \leq \theta < 1; \tag{5.23}$$

  however, this family does not include H^0 as a special case.
- A second example of a one-parameter family with a meaningful θ that includes H^+ and H^{-1} is the one given by Mardia (1970, p. 33) as

$$H(x,y) = \frac{1}{2}\theta^2(1+\theta)H^{+1} + (1-\theta^2)H^0(x,y) + \frac{1}{2}\theta^2(1-\theta)H^{-1}(x,y) \quad (5.24)$$

for $-1 \le \theta \le 1$. This family does include H^0 as a special case.

Kimeldorf and Sampson (1975) generalized the idea to propose

$$L_\theta = t(\theta)H_\theta + [1 - t(\theta)]K_\theta, \quad -1 \le \theta \le 1, \quad (5.25)$$

where $\{H_\theta\}$ and $\{K_\theta\}$ are two families of d.f.'s having the same marginals and satisfying the conditions given in Section 5.2.4 . Here, t is a continuous mapping of $[-1, 1]$ into $[0, 1]$. This generalization allows us to generate a wide range of bivariate distributions, though its usefulness is questionable. For mixtures of two bivariate normal distributions, one may refer to Johnson (1987, pp. 55–62).

The concepts of mixture above can be readily extended to three components—an example is (5.24) above, though usually two or more of the proportions will be free parameters, not merely one. Mixing infinitely many components is called *compounding*, which is described in the following section.

For a more detailed account of applications of mixture distributions, see Everitt (1985), McLachlan and Basford (1988), and Titterington et al. (1985).

5.4.2 Compounding

The idea of generating distributions by compounding has a long history, especially in the univariate setting. Motivation is often from survival time applications in biological or engineering sciences, and this does apply to the bivariate case as much as to the univariate case. Let X and Y be two random variables with parameters θ_1 and θ_2, respectively. For a given value of (θ_1, θ_2), X and Y are assumed to be independent. The basic idea of compounding is to say that θ_1 and θ_2 are themselves random variables, not constants, and the observed distribution of X and Y results from integrating over the (unobserved) distribution of θ_1 and θ_2. It is usual to assume that θ_1 and θ_2 are identically equal so that only a single integration is necessary, but sometimes they are assumed to be merely correlated, thus making an integration with respect to their bivariate distribution necessary. It should be noted that if θ_1 and θ_2 are identical and play the role of a scale parameter of F and G, then compounding is equivalent to a version of the trivariate reduction method, which is discussed in Section 5.5.

Bivariate Gamma Distribution as an Example

We now present an example due to Gaver (1970). This illustrates how the joint moment generating function of the compound distribution can be obtained by summing or integrating over the distribution of θ (the common value of θ_1 and θ_2).

Let X and Y have the same gamma distribution with shape parameter $\theta + k$ (θ is an integer and $k > 0$ need not be an integer). For a given value of θ, X and Y are independent with moment generating functions $(1-s)^{-(\theta+k)}$ and $(1-t)^{-(\theta+k)}$, respectively. Assuming now that θ has a negative binomial distribution with the probability generating function

$$G_k(z) = \sum_{n=0}^{\infty} b_n(k) z^n = \left(\frac{\alpha}{1+\alpha-z}\right) k, \tag{5.26}$$

where $b_n(k)$ is the probability that θ takes on the value n, and k and $\alpha > 0$ are the two parameters of the negative binomial distribution, we derive the joint moment generating function of X and Y as

$$\begin{aligned} M(s,t) &= E(e^{sX+tY}) \\ &= \sum_{n=0}^{\infty} E(e^{(sX+tY)}|\theta = n)\ \Pr(\theta = n) \\ &= \sum_{n=0} b_n(k)[(1-s)(1-t)]^{-n}[(1-s)(1-t)]^{-k} \\ &= G_k\{[(1-s)(1-t)]^{-1}\}\ [(1-s)(1-t)]^{-k} \\ &= \left(1 - \frac{\alpha+1}{\alpha}s - \frac{\alpha+1}{\alpha}t + \frac{\alpha+1}{\alpha}st\right)^{-k}. \end{aligned}$$

Integration May Be Over Two Parameters

Suppose that the parameters pertaining to X and Y are not identical but merely correlated. Specifically, suppose they have Kibble's bivariate gamma distribution, i.e., their marginal densities are of gamma form with shape parameter c and their joint p.d.f. is

$$\begin{aligned} &h(\theta_1,\theta_2) \\ &= \frac{(\theta_1\theta_2)^{(c-1)/2}}{b^{c+1}(1-k)^{(c-1)/2}k\Gamma(c)} \exp\left\{-\frac{\theta_1+\theta_2}{bk}\right\} I_{c-1}\left(\frac{2\sqrt{(1-k)\theta_1\theta_2}}{bk}\right) \end{aligned} \tag{5.27}$$

where $0 < k < 1$ and I_ν is the modified Bessel function of the first kind. Then, upon performing the integration

$$\Pr(X > x, Y > y) = \int_0^\infty \exp(-\theta_1 x - \theta_2 y)\, h(\theta_1, \theta_2)\, d\theta_1 d\theta_2 \tag{5.28}$$

by making use of Eq. (18) of Erdélyi (1954, p. 197), we find the bivariate survival function to be

$$(1 + bx + by + kb2xy)^{-c}. \tag{5.29}$$

Marshall and Olkin's Construction Scheme

Marshall and Olkin's (1988) method of constructing bivariate distributions is a generalization of constructing bivariate survival models induced by frailties. Frailty models have been defined and widely used in the field of survival analysis; see, for example, Hougaard (2000) and Oakes (1989).

The procedure for constructing a bivariate survival function from the marginal survival functions by the Laplace transform of a frailty variable can be easily applied to F and G to obtain another joint distribution H as

$$H(x, y) = \varphi^{-1}[\varphi(F(x)) + \varphi(G(y))]. \tag{5.30}$$

Marshall and Olkin (1988) have generalized themethod above to the case where the mixing distribution is also a bivariate distribution $\Omega(w_1, w_2)$ defined on $[0, \infty] \times [0, \infty]$ with the Laplace transform φ and its marginals $\Omega_i, i = 1, 2$, with the Laplace transforms φ_i, and K a bivariate distribution with uniform marginals over $[0, 1]$. F and G are defined using F_0 and G_0, the two univariate baseline distribution functions, so that $F = \varphi_1(\log F_0)$ and $G = \varphi_2(\log G_0)$. Then there exists a distribution function H such that

$$H(x, y) = \int\int K(F_0^{w_1}(x), G_0^{w_2}(y))\, d\Omega(w_1, w_2). \tag{5.31}$$

Marshall and Olkin (1988) and Oakes (1989) have shown that for any distribution obtained as $\int \exp[-\theta A(x)] \exp[-\theta B(y)] f(\theta) d\theta$, the copula is Archimedean. That is, there exists a function φ such that $\varphi(H) = \varphi(F) + \varphi(G)$. Writing $T(t) = \int_0^\infty \exp(-\theta t) f(\theta) d\theta$, we obtain $H = T(A(x) + B(y))$ with marginals $F = T(A(x))$ and $G = T(B(y))$. Hence, $T^{-1}(H) = T^{-1}(F) + T^{-1}(G)$. What this means is that if we know the function $\varphi(.)$ defining the Archimedean copula and we want to know the compounding density $f(\theta)$, we invert φ to get T and then apply the inverse Laplace transform to get f from T; see Table 5.1. But, not all Archimedean copulas give rise to valid densities $f(\theta)$. Three Archimedean copulas are summarized in the following discussion.

Table 5.1 Laplace transform and compounding density

Compounding density $f(\theta)$	$T(t)$	$\varphi(u) = T^{-1}(u)$
Gamma	$(1+t)^{-c}$	$u^{-1/c} - 1$
Positive stable	$\exp(-t^{\alpha})$	$(-\log u)^{1/\alpha}$
Inverse Gaussian	$\exp[-\eta(\sqrt{1+2t}-1)]$	$(\log u)[\log(u) - 2\eta]/(2\eta^2)$

Whitmore and Lee (1991, p. 41) argued for the case of the inverse Gaussian as the compounding density on the grounds that "the level of imperfection in the item may be proportional to the length of time the reaction continues before a critical condition is first satisfied. Based on this reasoning, we shall consider here a physical model in which the hazard rate equals the stopping time of a stochastic process. Furthermore, because of the prevalence of Wiener diffusion processes in chemical and molecular reactions and in physical systems, we select the first hitting time of a fixed barrier in such a process as a model ... [this] distribution is inverse Gaussian."

5.5 Variables in Common and Trivariate Reduction Techniques

5.5.1 Summary of the Method

The idea here is to create a pair of dependent random variables from three or more random variables. In many cases, these initial random variables are independent, but occasionally they may be dependent—an example of the latter is the construction of a bivariate t-distribution from two variates that have a standardized correlated bivariate normal distribution and one that has a chi-distribution. An important aspect of this method is that the functions connecting these random variables to the two dependent random variables are generally elementary ones; random realizations of the latter can therefore be generated as easily as these of the former. A broad definition of the variables-in-common (or trivariate reduction) technique is as follows. Set

$$\left.\begin{aligned} X &= \tau_1(X_1, X_2, X_3) \\ Y &= \tau_2(X_1, X_2, X_3) \end{aligned}\right\}, \tag{5.32}$$

where X_1, X_2, X_3 are not necessarily independent or identically distributed. A narrow definition is

$$\left.\begin{aligned} X &= X_1 + X_3 \\ Y &= X_2 + X_3 \end{aligned}\right\}, \tag{5.33}$$

with X_1, X_2, X_3 being i.i.d. Another possible definition is

$$\left.\begin{aligned} X &= \tau(X_1, X_3) \\ Y &= \tau(X_2, X_3) \end{aligned}\right\}, \tag{5.34}$$

with (i) the X_i being independently distributed and having d.f. $F_0(x_i; \lambda_i)$ and (ii) X and Y having distributions $F_0(x; \lambda_1 + \lambda_2)$ and $F_0(y; \lambda_1 + \lambda_3)$, respectively.

Three well-known distributions that can be obtained in this way are:

- the bivariate normal, from the additive model in (5.33), with the X_i's having normal distributions;
- Cherian's bivariate gamma distribution, also obtained from (5.33), but with the X_i's having gamma distributions; and
- Marshall and Olkin's bivariate exponential distribution with joint survival function

$$\begin{aligned} \bar{H}(x,y) &= \exp(-(\lambda_1 + \lambda_{12})x - (\lambda_2 + \lambda_{12})y + \lambda_{12}\min(x,y)) \\ &= \bar{F}(x)\bar{G}(y)\ \min\{\exp(\lambda_{12}x), \exp(\lambda_{12}y)\}, \end{aligned} \tag{5.35}$$

with the transformation τ being the minimum and the X_i's having exponential distributions.

5.5.2 Denominator-in-Common and Compounding

The denominator-in-common version of the trivariate reduction method of constructing bivariate distributions sets $X = X_1/X_3$ and $Y = X_2/X_3$. This may readily be seen to be equivalent to compounding a scale parameter if we instead write them as $X = X_1/\theta$ and $Y = X_2/\theta$. Then,

$$\begin{aligned} H(x,y) &= \Pr(X \le x, Y \le y) \\ &= \Pr(X_1 \le \theta x, X_2 \le \theta y) \\ &= \int \Pr(X_1 \le \theta x) \Pr(X_2 \le \theta y) f(\theta)\, d\theta \\ &= \int F_{X_1}(\theta x) F_{X_2}(\theta y) f(\theta)\, d\theta, \end{aligned}$$

where $f(\theta)$ is the p.d.f. of θ, which is the familiar equation for compounding a scale parameter; see Lai (1987).

5.5.3 Mathai and Moschopoulos' Methods

Mathai and Moschopoulos (1991) constructed a bivariate gamma distribution whose components are positively correlated and have three-parameter distributions. Denote the three-parameter (shape, scale, and location) gamma by $V_i \sim G(\alpha_i, \beta_i, \gamma_i)$, $i = 0, 1, 2$, and let

$$X = \frac{\beta_1}{\beta_0} V_0 + V_1, \qquad Y = \frac{\beta_2}{\beta_0} V_0 + V_2.$$

The X and Y so defined have a bivariate distribution with gamma marginals.

Mathai and Moschopoulos (1992) constructed another form of bivariate gamma distribution. Let V_i, $i = 1, 2$, be defined as above. Form

$$X = V_1, \qquad Y = V_1 + V_2.$$

Then X and Y clearly have a bivariate gamma distribution. Theconstruction above is only part of a multivariate setup motivated by the consideration of the joint distribution of the total waiting times of a renewal process.

5.5.4 Modified Structure Mixture Model

Lai (1994) proposed a method of constructing bivariate distributions by a generalized trivariate reduction technique that may be considered as a modified structure mixture model.

The proposed model has the form

$$\begin{aligned} X_1 &= Y_1 + I_1 Y_2, \\ X_2 &= Y_3 + I_2 Y_2, \end{aligned} \tag{5.36}$$

where Y_i are independent random variables and I_i $(i = 1, 2)$ are indicator random variables that are independent of Y_i but where (I_1, I_2) has a joint probability function with joint probabilities given by p_{ij}, $i, j = 0, 1$.

Thus, new bivariate distributions can be constructed by specifying p_{00} and p_{10}.

5.5.5 Khintchine Mixture

The following method of generating bivariate distributions may be found in Bryson and Johnson (1982) and Johnson (1987, Chapter 8). Let

$$\left.\begin{aligned} X &= Z_1 U_1 \\ Y &= Z_2 U_2 \end{aligned}\right\}, \tag{5.37}$$

where U's are uniform on $(0, 1)$ and both U's and Z's can be either identical or independent.

5.6 Conditionally Specified Distributions

5.6.1 A Conditional Distribution with a Marginal Given

A bivariate p.d.f. can be expressed as the product of a marginal p.d.f. and a conditional p.d.f., $h(x,y) = f(x)g(y|x)$. This is easily understood and has been a popular approach in the literature, especially when Y can be thought of as being caused by, or predictable from, X. We will give one simple and one complicated example. Conditionally specified distributions have been discussed rather extensively in the books by Arnold et al. (1992, 1999).

Example 5.4 (The Beta-Stacy Distribution). Mihram and Hultquist (1967) discussed the idea of a warning-time variable, X, for $Y =$ the failure time of a component being tested, where $0 < X < Y$. A bivariate distribution was proposed, with Y having Stacy's generalized gamma distribution and X, conditional on $Y = y$, having a beta distribution over the range 0 to y. The p.d.f. is thus given by

$$h = \frac{|c|}{a^{bc}\Gamma(b)B(p,q)} x^{p-1}(y-x)^{q-1}y^{bc-p-q}\exp[-(y/a)^c]. \qquad (5.38)$$

5.6.2 Specification of Both Sets of Conditional Distributions

Methods of Characterizing a Bivariate Distribution

Gelman and Speed (1993) have stated three possible ways to define a joint distribution of two random variables X and Y by using conditional and marginal specifications:

(1) The conditional distribution of X given Y and the marginal distribution of Y specify the joint distribution uniquely.

(2) The conditional distributions of X given Y, along with the single distribution of Y given $X = x_0$, for some x_0, uniquely determine the joint density as

$$h(x,y) \propto \frac{f(x|y)g(y|x_0)}{f(x_0|y)}. \qquad (5.39)$$

Normalization determines the constant of proportionality; the discrete analogue of these results is due to Patil (1965).

(3) The conditional distributions of X given Y and Y given X determine the joint distribution from the formula above for each x_0. The conditional specification thus overdetermines the joint distribution and is

self-consistent only if the derived joint distributions agree for all values of x_0. The last sentence is effectively equivalent to the compatibility condition discussed below.

Compatibility

Let $f(x|y)$ and $g(y|x)$ be given conditional density functions. There exists a body of work that derives a bivariate density from specifying that $f(x|y)$ takes a certain form, with parameters depending on y, and $g(y|x)$ takes a certain form (perhaps the same, perhaps different), with parameters depending on x. This work has been brought together in important books by Arnold et al. (1992, 1999), and we will therefore repeatedly refer to these, rather than the original source. A key feature of the systematic development of this topic is a theorem relating to functional equations. Details of this would be out of place here, but we will give a summary of results in Section 5.6.5 below. As a preliminary, we present the following theorem.

Compatibility Theorem. A bivariate density $h(x, y)$ with conditional densities $f(x|y)$ and $g(y|x)$ will exist if and only if [see Section 1.6 of Arnold et al. (1999)]

1. $\{(x, y) : f(x|y) > 0\} = \{(x, y) : g(y|x) > 0\}$.
2. There exist $a(x)$ and $b(y)$ such that the ratio $f(x|y)/g(y|x) = a(x)b(y)$, where $a(\cdot)$ and $b(\cdot)$ are non-negative integrable functions.
3. $\int a(x)dx < \infty$.

The condition $\int a(x)dx < \infty$ is equivalent to $\int [1/b(y)]dy < \infty$, and only one of these needs to be checked in practice. Note that the joint density obtained may not be unique; see Arnold and Press (1989). The compatibility conditions given above are essentially those given by Abrahams and Thomas (1984) except that these authors overlooked the possible lack of uniqueness.

If the necessary and sufficient conditions above are satisfied, we then say that the two conditional densities are *compatible*.

5.6.3 Conditionals in Exponential Families

An l_1-parameter *exponential family* of densities $\{f_1(x; \boldsymbol{\theta}) : \boldsymbol{\theta} \in \Theta\}$ has the form

$$f_1(x; \boldsymbol{\theta}) = r_1(x)\beta_1(\boldsymbol{\theta}) \exp\left\{\sum_{i=1}^{l_1} \theta_i q_{1i}(x)\right\}. \tag{5.40}$$

Another l_2-parameter exponential family of densities $\{f_2(y; \boldsymbol{\tau}) : \boldsymbol{\tau} \in T\}$ has the form

$$f_2(y;\boldsymbol{\tau}) = r_2(y)\beta_2(\boldsymbol{\tau}) \exp\left\{\sum_{j=1}^{l_2} \tau_j q_{2j}(y)\right\}. \tag{5.41}$$

Suppose the conditional density functions of $X|(Y = y)$ and $Y|(X = x)$ are specified by

$$f(x|y) = f(x;\boldsymbol{\theta}(y)) \tag{5.42}$$

and

$$g(y|x) = f_2(y;\boldsymbol{\tau}(x)), \tag{5.43}$$

where f_1 and f_2 are as defined in (5.40) and (5.41), respectively. Arnold and Strauss (1991) then showed that [see also Arnold et al. (1999, pp. 75–78)] the joint density $h(x, y)$ is of the form

$$h(x,y) = r_1(x)r_2(y)\exp\{\mathbf{q}^{(1)}(x)\mathbf{M}\mathbf{q}^{(2)}(y)\}, \tag{5.44}$$

where

$$\mathbf{q}^{(1)}(x) = (q_{10}(x), q_{11}(x), \ldots, q_{1l_1}(x)),$$
$$\mathbf{q}^{(2)}(y) = (q_{20}(y), q_{21}(y), \ldots, q_{2l_1}(y)),$$

with $q_{10}(x) = q_{20} \equiv 1$, and $\mathbf{M}$ is an $(l_1 + 1) \times (l_2 + 1)$ matrix of constant parameters. Of course, the density is subject to the requirement $\int\int f(x,y)dx\,dy = 1$. We note that the conditionals in the exponential families are compatible.

This is an important result, as one can generate a host of bivariate distributions by selecting appropriate constant parameters in the matrix $\mathbf{M}$.

Normal Conditionals

If both sets of conditional densities are normal, we let $l_1 = l_2 = 2, r_1(t) = r_2(t) = 1$, and

$$\mathbf{q}^{(1)}(t) = \mathbf{q}^{(2)}(t) = \begin{pmatrix} 1 \\ t \\ t^2 \end{pmatrix}.$$

The choice $m_{22} = m_{12} = m_{21} = 0$ yields the classical bivariate normal provided $m_{22} < 0, m_{02} < 0, m_{11}^2 < 4m_{02}m_{20}$. Several nonclassical normal conditional models can be constructed subject to the parametric constraints

$$m_{22} < 0,\ m_{02} < 0,\ 4m_{22}m_{02} > m_{22}^2,\ 4m_{22}m20 > m_{21}^2.$$

If the means of both normal conditionals are zero, then we have a bivariate centered model. Plots of a density curve and its contour are presented in Arnold et al. (1999, p. 67).

5.6.4 Conditions Implying Bivariate Normality

Various sets of conditions on the conditional distributions are sufficient to imply a bivariate normal distribution. Most of those below are given by Bhattacharyya (1943), and Castillo and Galambos (1987); see also Kendall and Stuart (1979) and Chapter 3 of Arnold et al. (1992):

- The distribution of Y given $X = x$ is normal and homoscedastic (i.e., $\text{var}(Y|X = x)$ is a constant), together with one of the following:
  - marginal normality of X, together with linearity of the regression of Y on X or X on Y;
  - conditional normality of X given $Y = y$;
  - conditional normality of X given $Y = y_0$, for some y_0, together with linearity of the regression of Y on X [Fraser and Streit (1980)];
  - marginal distributions of X and Y being identical, together with linearity of the regression of Y on X [Ahsanullah (1985)].
- Both conditional distributions, of Y given $X = x$ and X given $Y = y$, are normal, together with one of the following:
  - marginal normality of X;
  - one or both regressions are linear and nonconstant.
- Both regressions, of Y on X and X on Y, are linear and have the identical errors property (meaning only the mean of the dependent variable changes when the independent variable does) [Kendall and Stuart (1979, Paragraph 28.8)]. In this case, X and Y can be independent or functionally related as alternatives to being bivariate normal;
- The contours of probability density are similar concentric ellipses, together with one of the following:
  - normality of Y given $X = x$;
  - homoscedasticity of Y given $X = x$;
  - marginal normality of X.

5.6.5 Summary of Conditionally Specified Distributions

The rest of the conditionals in the exponential families are presented below in Table 5.2.

Some other conditionally specified families of bivariate distributions are summarized in Table 5.3 below. Details of some of these conditionals will be discussed in Section 6.4.

Table 5.2 Both conditionals in exponential families

$X\|Y$	$Y\|X$
Exponential	Exponential
Normal	Normal
Gamma	Gamma
Weibull†	Weibull
Gamma	Normal
Power-function	Power-function
Beta	Beta
Inverse Gaussian	Inverse Gaussian

† Weibull distribution is not a member of the exponential family but can be expressed as a positive power of an exponential random variable $W = X^c$.

Table 5.3 Conditionals not members of the exponential family of distributions

$X\|Y$	$Y\|X$
Pareto	Pareto
Beta of the second kind	Beta of the second kind
Pearson type VI	Pearson type VI
Generalized Pareto	Generalized Pareto
Cauchy	Cauchy
Student t	Student t
Uniform	Uniform
Possibly translated exponential	Possibly translated exponential
Scaled beta	Scaled beta
Weibull	Logistic

Conditional Distributions in Location-Scale Families with Specified Moments

Arnold et al. (1999) have considered conditionals in unspecified families with specified conditional moments, which are as follows:

(1) linear regressions with conditionals in location families;
(2) specified regressions with conditionals in scale families;
(3) conditionals in location-scale families with specified moments;
(4) given one family of conditional distributions and the other a regression function.

We now present a brief description of item (3) above. Most families of distributions considered so far have their marginals specified. Narumi (1923a,b) took a different approach. His attack on the problem of creating bivariate distributions was by specifying the regression and scedastic (conditional standard deviation) curves. An account of his work has been detailed in Chapter 6 of Mardia (1970). This approach does fall into the broad scheme formulated in Arnold et al. (1999). Consider bivariate distributions with conditional densities of the form

$$f(x|y) = g_1\left(\frac{x - a(y)}{c(y)}\right)\frac{1}{c(y)}, \tag{5.45}$$

$$g(y|x) = g_2\left(\frac{y - b(x)}{d(x)}\right)\frac{1}{d(x)}, \tag{5.46}$$

where a and b are the regression curves, and c and d are scedastic curves of X on Y and Y on X. This type of conditionally specified bivariate distribution has also been discussed by Arnold et al. (1999).

Some bivariate distributions that can be written in this form are summarized below in Table 5.4.

Table 5.4 Some bivariate distributions derived from conditional moments

$a(y)$	$c(y)$	Type of $h(x, y)$
linear	constant	normal
linear	linear	beta, Pareto, F
constant	linear	McKay
linear	parabolic	t, Cauchy, Pearson type II
r.h.*	r.h.*	gamma conditionals
		$h \propto (x + b_1)^{\gamma_1}(y + b_2)^{\gamma_2}$
		$\times \exp[\gamma(x + c_1)(y + c_2)]$

* r.h. denotes rectangular hyperbola, i.e., of the form $1/(x + a)$.

5.7 Marginal Replacement

A simple general scheme of constructing a new bivariate distribution is to replace a marginal of the existing bivariate distribution by a new marginal. This method of construction is called *marginal replacement* by Jones (2002). Consider a bivariate density $h(x, y)$ which can obviously be written as

$$h(x, y) = f(x)g(y|x). \tag{5.47}$$

With appropriate considerations for the support, we can obtain a new bivariate density function by replacing $f(x)$ above by $f_1(x)$, giving

$$h_1(x, y) = f_1(x)g(y|x). \tag{5.48}$$

The only condition on this approach is that the support of f_1 be contained in, or equal to, the support of f. Indeed, h_1 then has support contained in, or equal to, the support of h.

5.7.1 Example: Bivariate Non-normal Distribution

Tiku and Kambo (1992) obtained a new symmetric bivariate distribution by replacing one of the marginals of a bivariate normal distribution by a univariate t-distribution.

5.7.2 Marginal Replacement of a Spherically Symmetric Bivariate Distribution

Jones (2002) obtained a bivariate beta/symmetric beta distribution as well as a bivariate t/skew t distribution using this approach. More details of these distributions will be presented in Chapter 9.

5.8 Introducing Skewness

Over the last decade or so, many families of bivariate and multivariate skew distributions have been constructed by introducing one or more skewness parameters in the multivariate distributions. A Google search at the site azzalini.stat.unipd.it/SN/list-publ.ps (updated on March 17, 2007) found approximately 150 references on the skew-normal distribution and related ones. The major multivariate skew distributions are listed below:

1. skew-normal family—Azzalini (2005, 2006);
2. skew t—Azzalini and Capitanio (2003);
3. skew-Cauchy—Arnold and Beaver (2000);
4. skew-elliptical—Branco and Dey (2001);.
5. log-skew-normal and log-skew-t—Azzalini et al. (2003);
6. general class of multivariate skew distributions—Sahu et al. (2003).

5.9 Density Generators

A bivariate density function may be obtained through composition of a density generator g that is a function of a univariate density function with one or more parameters.

Example 5.5 (Bivariate Liouville distributions).

$$h(x,y) = \frac{x^{a-1}y^{b-1}}{\Gamma(a)\Gamma(b)} g\Big(x+y\Big),$$

where g is beta, inverted beta, gamma, or others satisfying the condition $\int_0^\infty \frac{t^{a+b-1}}{\Gamma(a+b)} g(t)dt = 1$; see, for example, Gupta and Richards (1987). Ma and Yue (1995) have extended the above to obtain the bivariate pth-order Liouville distribution

$$h(x;y) = c\theta^{a+b} \frac{x^{a-1}y^{b-1}}{\Gamma(a)\Gamma(b)} g\left(\frac{(x^p + y^p)^{1/p}}{\theta}\right),$$

where θ is a parameter and c is the normalizing constant.

Example 5.6 (Elliptical contoured distributions and extreme type elliptical distributions). (X, Y) is said to have an elliptically contoured distribution if its joint density takes the form

$$h(x;y) = \frac{1}{\sqrt{1-\rho^2}} g\left(\frac{(x^2 - 2\rho xy + y^2)^{1/p}}{1-\rho^2}\right),$$

where $-1 < \rho < 1$ and $g(\cdot)$ is a scale function referred to as the density generator.

By setting $g(x) = \frac{h(x)}{2\pi \int_0^\infty yh(y^2)dy}$, where $h(x)$ is the density function of (i) Weibull, (ii) Fréchet, and (iii) Gumbel, Kotz, and Nadarajah (2001) have obtained three extremal-type elliptical distributions.

5.10 Geometric Approach

In Stoyanov (1997, p. 77), an interesting nonbivariate normal distribution is given, of which two marginal distributions are normal. This is a classical counterexample that involves geometry. The basic idea is to punch four square holes symmetrically in the domain of a bivariate normal density function and to move the probability mass over the four holes to the four other symmetrical positions so as to ensure that the marginals are not affected.

Inspired by this counterexample, Wei et al. (1998) also constructed copulas with holes that are constrained within an admissible rectangle. They also provided a construction algorithm called the *squeeze algorithm*.

Nelsen (2006, pp. 59–88) has described various geometric methods of constructing copulas in the following manner:

(1) Singular copulas with prescribed support: Utilize some information of a geometric nature, such as a description of the support or the shape of the graphs of horizontal, vertical, or diagonal sections.
(2) Ordinal sum construction: Members of a set of copulas are scaled and translated in order to construct a new copula.
(3) Shuffles of M: These are constructed from the Fréchet upper bound.

Johnson and Kotz (1999) constructed what they called *square tray distributions* by simple, piecewise uniform modifications of a copula on the unit square. The resulting bivariate distributions may not be copulas, as their marginals may not be uniform.

5.11 Some Other Simple Methods

The transformation method outlined in Section 5.1.2 is pretty trivial. All that is done is to take one distribution and stretch or compress it in the X and/or the Y direction. Other methods that may be thought of as trivial and inelegant include the following:

- Let the formula for h take one form for some region of the (X, Y) plane and another form for the remaining region. (A particular example occurs when the p.d.f. of a unimodal distribution is reduced to c within the contour $h = c$ and then h is rescaled so that it becomes a p.d.f. again.) Another simple example in constructing a copula is given by Wei et al. (1998) as follows. Divide the rectangle formed by $0 \leq u \leq 1$ and $0 \leq v \leq 1$ into four rectangular areas by drawing $u = \alpha$ and $v = \alpha$. Assign probability mass $\lambda\alpha$, $(1-\lambda)\alpha$, $(1-\lambda)\alpha$, and $1-(2-\lambda)\alpha$ uniformly to the four regions with $0 < \lambda < 1$ and $0 \leq \alpha \leq \frac{1}{2-\lambda}$.
- Take an existing distribution and truncate it, singly or doubly, in one or both the variates; for example, a truncated bivariate normal [Kotz et al. (2000, pp. 311–320)]. Nadarajah and Kotz (2007) also gave truncated versions of several well-known bivariate distributions.
- Take a trivariate distribution of (X, Y, Z) and find the conditional distribution of (X, Y) given $Z = z$. In the previous situation, find the marginal distribution of (X, Y).
- Take an existing distribution and extend its region of support by reflecting the p.d.f. into the previously empty area.
- Take an existing p.d.f. $h(x, y)$ and multiply it by some function $a(x, y)$. Provided the volume under the surface remains finite, the result can be treated as proportional to a probability density. A special case of this method is where $a(x, y)$ is a risk function, so that the densities in the surviving and nonsurviving (or, more generally, selected and nonselected) populations are $(1 - a)h$ and ah. Epidemiological studies often find it necessary to make an assumption about the joint distribution of two (or more) variables considered to be possible risk factors for the disease under consideration. For instance, Halperin et al. (1979) were concerned with the probability of death (from any cause) being a function of systolic blood pressure and the number of cigarettes smoked per day. The interest of Halperin et al. (1979) was primarily methodological: They demonstrated that if X and Y have a bivariate normal distribution, with the risk of

death being a probit function $\Phi(\alpha + \beta_1 x + \beta_2 y)$, then not only do X and Y have different means in the group that dies and the group that survives, but also the variances and covariances in the groups differ also.

- Calculate two summary statistics from a univariate sample. The sample mean and variance are often uncorrelated, and hence their joint distribution is often uninteresting, but this is not so for the sample minimum and maximum or for the sample skewness and kurtosis. The sample mean and sample median from a symmetric distribution are often asymptotically bivariate normal [Stigler (1992)].
- Calculate a summary statistic for both X and Y, starting from a bivariate sample. For example:
  - The sampling distribution of $(\bar{X}, \bar{Y})$ is often bivariate normal.
  - The maxima of X and Y have limiting bivariate extreme-value distributions.
- A popular method that often lacks any further justification is to write down a formula and then check whether it satisfies the criteria for being a bivariate distribution. As an example of this, we may give the Farlie–Gumbel–Morgenstern distribution. In copula form, this is

$$H = xy[1 + \alpha(1-x)(1-y)] \tag{5.49}$$

for $-1 \leq \alpha \leq 1$, and the corresponding density is

$$h = 1 + \alpha(1-2x)(1-2y). \tag{5.50}$$

5.12 Weighted Linear Combination

In many simulation applications, it is required to generate dependent pairs of continuous random variables for which there is limited information on the joint distribution. The example that Johnson and Tenenbein (1981) presented is that of a portfolio analysis simulation in which a joint distribution of stock and bond returns may have to be specified. Because of a lack of data, it may be difficult to specify completely the joint distribution of stock and bond returns. However, it may be realistic (so state Johnson and Tenenbein) to specify the marginal distributions and some measures of dependence between the random variables.

The weighted linear combination (WLC) technique is as follows. Let

$$\left.\begin{aligned} X &= U_1 \\ Y &= cU_1 + (1-c)U_2 \end{aligned}\right\}, \tag{5.51}$$

where U_1 and U_2 are independent and identically distributed with common probability density function f and c is a constant ($0 \leq c \leq 1$).

Johnson and Tenenbein (1981) were then concerned with using WLC in the case where the specified measure of dependence was Kendall's τ or Spearman's rank correlation ρ_S. For some choice of f, they obtained equations connecting c to τ and ρ_S. They handled the problem of getting the appropriate marginals by means of the transformation method discussed in Section 5.2. A slightly more general model than WLC, in which Y is an arbitrary combination of U_1 and U_2, was considered earlier by Jogdeo (1964). A general model in which both X and Y are linear combinations of U_1 and U_2 has been treated by Mardia (1970, Chapter 5).

5.13 Data-Guided Methods

The study of bivariate distributions usually tends very much toward the modeling end of the statistical spectrum rather than toward the analysis end. In this section, however, we emphasize the data analysis side: If we want to follow passively, without preconceptions about the appropriate model, where bivariate data was leading us, how best can we do this?

5.13.1 Conditional Distributions

An elementary idea that is often useful when exploring bivariate data is to examine the conditional distributions. That is, given that X equals (or is within a narrow range of) x, what properties does the distribution of Y have? And, similarly, the distribution of X for a given Y may be examined. The methods that are common for univariate distributions are then applied; in particular, the conditional mean, the conditional standard deviation, and (for necessarily possible variates) the conditional coefficient of variation may each be plotted. Recall that the conditional means are linear and the conditional standard deviations are constant in the case of a bivariate normal distribution.

Mardia (1970, p. 81) suggested focusing attention on the regression and scedastic curves after the observations have been transformed to uniform marginals.

One might also consider conditioning of the form $X > x$. Further, one might think in terms of quantiles. Then one might decide that the statistic of prime interest is the mean. This leads to asking how useful it is to know that X is big compared with how useful it is to know that Y is big for the purpose of predicting Y. Hence, one will want to calculate the following

$$\frac{E(Y|X > x_p) - E(Y)}{E(Y|Y > y_p) - E(Y)} \tag{5.52}$$

(which is a function of p), where x_p and y_p are the pth quantiles of X and Y, respectively. Kowalczyk and Pleszczynska (1977) referred to this as the *monotonic quadrant dependence function*; see Section 3.5.3 for details. Clearly, many variations can be played on this theme.

5.13.2 Radii and Angles

The probability density h of the class of elliptically symmetric bivariate distributions is a function only of a positive definite quadratic form

$$(\mathbf{x}-\boldsymbol{\mu})'\boldsymbol{\Sigma}^{-1}(\mathbf{x}-\boldsymbol{\mu}).$$

Let $R^2 = (X_1^2 - 2\rho X_1 X_2 + X_2^2)/(1-\rho^2)$, where ρ is the off-diagonal entry in the scaling matrix $\boldsymbol{\Sigma}$.

Let $\mathbf{L}$ be the lower triangle (Choleski) decomposition of $\boldsymbol{\Sigma}$. Then, for this class of distributions, $\mathbf{X}$ may be represented as $(X_1, X_2)' = R\mathbf{L}\mathbf{U}^{(2)} + \boldsymbol{\mu}$, where $\mathbf{U}^{(2)}$ is uniformly distributed on the circumference of a unit circle and is independent of R. The distribution of R discriminates the members within the class.

For most practical purposes, the bivariate normal distributions would be the first to come to mind. The radii and angles method is specifically for assessing bivariate normality. It was discussed by Gnanadesikan (1977, Chapter 5). Let $(X_1, X_2)'$ denote the bivariate normal vector with the variance–covariance matrix $\boldsymbol{\Sigma}$. First, transform the original variates X_1 and X_2 to independent standard normal variates X and Y using

$$\begin{pmatrix} X \\ Y \end{pmatrix} = \boldsymbol{\Sigma}^{-1/2} \begin{pmatrix} X_1 - \mu_1 \\ X_2 - \mu_2 \end{pmatrix}. \tag{5.53}$$

Second, transform (X, Y) to polar coordinates (R, θ). Then, under the hypothesis of bivariate normality, R^2 has a χ_2^2-distribution (i.e., exponential with mean 2) and θ has a uniform distribution over the range 0 to 2π. These consequences may be tested graphically—by plotting sample quantiles of R^2 versus quantiles of the exponential distribution with mean 2 and by plotting sample quantiles of the angle θ versus those of the uniform distribution; for illustration, see Gnanadesikan (1977). If bivariate normality holds, the two plots should be approximately linear. However, if $\boldsymbol{\mu}' = (\mu_1, \mu_2)$ and $\boldsymbol{\Sigma}$ are estimated, the distributional properties of R and θ are only approximate. For $n \geq 25$, the approximation is good. It is important to mention that the radii and angles approach, though informal, is an informative graphical aid.

5.13.3 The Dependence Function in the Extreme-Value Sense

In Section 12.5, we will see that bivariate extreme-value distributions having exponential marginals can be expressed as $\bar{H} = \exp\left[-(x+y)A\left(\frac{y}{x+y}\right)\right]$. Pickands (1981) [see also Reiss (1989)] suggested estimating $A(w)$ from a sample of n observations by

$$\hat{A}_n(w) = n \Big/ \sum_{i=1}^{n} \min\left(\frac{x_i}{1-w}, \frac{y_i}{1-w}\right),$$

with w being between 0 and 1. This suggestion was made by using the fact that $\min\left(\frac{X}{1-w}, \frac{Y}{w}\right)$ has an exponential distribution with mean $1/A(w)$. This estimate was applied by Tawn (1988) to data on annual maximum sea levels at Lowestoft and Sheerness and by Smith (1990) to maximum temperatures at two places and to best performances in mile races in successive years.

There is currently interest in modifying the estimate of $A(\cdot)$ above in order to obtain a smoother one; see Smith (1985), Smith et al. (1990), Deheuvels and Tiago de Oliveira (1989), and Tiago de Oliveira (1989b). Exactly what method of estimating $A(\cdot)$ will eventually emerge as the preferred one seems uncertain at present. Due to the availability of these procedures, one may suggest transforming observations so that the marginals become exponential and then use them to estimate the function A.

$A(w)$ is interpretable in terms of $\Pr\left(\frac{Y}{X+Y} < w\right)$. This suggests direct consideration of the angle $\tan^{-1}(y_i/x_i)$ after X and Y have been transformed to exponential marginals—calculate the values observed in the sample, show them as a histogram, determine various summary statistics, and so on.

5.14 Special Methods Used in Applied Fields

There will be five specialist fields considered in this section: shock models, queueing theory, compositional data, extreme-value models, and time series.

5.14.1 Shock Models

Marshall and Olkin's Model

This is a distribution having exponential distributions as marginals. It is obtained by supposing that there is a two-component system subject to shocks that may knock out the first component, the second component, or both of them. If these shocks result from independent Poisson processes with parameters λ_1, λ_2 and λ_{12}, respectively, Marshall and Olkin's distribution arises. Equivalently, $X = \min(Z_1, Z_3)$ and $Y = \min(Z_2, Z_3)$, where the Z's are independent exponential variates.

The upper right volume under the probability density surface is [Marshall and Olkin (1967)]

$$\bar{H} = \exp[-\lambda_1 x - \lambda_2 y - \lambda_{12} \max(x, y)], \tag{5.54}$$

where the λ's are positive.

This model is widely used in reliability. Certainly, the idea of simultaneous failure of two components is far from being merely an academic plaything; Hagen (1980) has given a review in the context of nuclear power, wherein it is pointed out that redundancy in a system reduces random component failure to insignificance, leading to the common-mode/common-cause type being predominant among system failures.

Raftery's Model

In its general form, Raftery's (1984, 1985) scheme for obtaining a bivariate distribution with exponential marginals is

$$\left.\begin{aligned} X &= (1 - p_{10} - p_{11})U + I_1 W \\ Y &= (1 - p_{01} - p_{11})V + I_2 W \end{aligned}\right\}, \tag{5.55}$$

where U, V, W are independent and exponentially distributed and I_1 and I_2 are each either 0 or 1, with probabilities as set out below:

	$I_2 = 0$	$I_2 = 1$
$I_1 = 0$	p_{00}	p_{01}
$I_1 = 1$	p_{10}	p_{11}

Raftery obtained the correlation as $2p_{11} - (p_{01} + p_{11})(p_{10} + p_{11})$. There is also an extension of the model to permit negative correlation. The distribution arises from a shock model. This refers to a system that has two components, S_1 and S_2, each of which can be functioning normally, unsatisfactorily, or have failed. The system is subject to three kinds of shocks governed by independent Poisson processes. These kinds of shocks cause normal components to become unsatisfactory, an unsatisfactory S_1 to fail, and an unsatisfactory S_2 to fail, respectively.

A special case of this model sets $p_{01} = p_{10} = 0$ so that

$$\left.\begin{aligned} X &= (1 - p_{11})U + IW \\ Y &= (1 - p_{11})V + IW \end{aligned}\right\}, \tag{5.56}$$

and the distribution in this case is a mixture of independence and trivariate reduction.

Downton's Model

This distribution has exponential marginals and has the joint p.d.f.

$$h(x,y) = \frac{1}{1-\rho} \exp\left[-\frac{x+y}{1-\rho}\right] I_0\left(\frac{2\sqrt{xy\rho}}{1-\rho}\right), \tag{5.57}$$

where I_0 is the modified Bessel function of the first kind of order zero. It has become associated with the name of Downton, though his paper explicitly acknowledged that it was not new at that time. The paper of Downton (1970) was concerned with the context of reliability studies and used the following model to obtain (5.57). Consider a system of two components, each being subjected to shocks, the intervals between successive ones having exponential distributions. Suppose the numbers of shocks N_1 and N_2 required for the components to fail follow a bivariate geometric distribution with joint probability generating function

$$P(z_1, z_2) = \frac{z_1 z_2}{1+\alpha+\beta+\gamma-\alpha z_1-\beta z_2-\gamma z_1 z_2}. \tag{5.58}$$

Let

$$(X,Y) = \left(\sum_{i=1}^{N_1} X_i, \sum_{i=1}^{N2} Y_i\right), \tag{5.59}$$

where X_i and Y_i are the intershock intervals, mutually independent exponential variates. Then the component lifetimes (X,Y) have the joint density as in (5.57). Several different bivariate geometric distributions in (5.58) give rise to the density in (5.57); all that is required is that $\rho = \frac{\alpha\beta+\alpha\gamma+\beta\gamma+\gamma+\gamma^2}{(1+\alpha+\gamma)(1+\beta+\gamma)}$. In particular, the case in which N_1 and N_2 are identical corresponds to $\alpha = \beta = 0$. Gaver (1972) gave a slightly different motivation for this distribution.

Equation (5.59) may be termed the *random sums method* of constructing bivariate distributions. As far as we know, only the case in which the X_i and Y_i have exponential distributions and N_1 and N_2 have geometric distributions has received much attention.

5.14.2 Queueing Theory

Consider a single-server queueing system such that the interarrival time X and the service time Y have exponential distributions, as is a common assumption in this context. If it is desired to introduce positive correlation (arising from cooperative service) into the model, Downton's distribution is a suitable choice [Conolly and Choo (1979)]. Langaris (1986) applied it to a queueing system with infinitely many servers. Other relevant works include Mitchell and Paulson (1979) and Niu (1981). Naturally, one of the important issues in this context is how waiting time is affected by the presence of such a correlation.

5.14.3 Compositional Data

The distinctive feature of compositional data is that it consists of proportions, which must sum to unity (or to less than unity when considering just n of the $n+1$ components). A field where such data are particularly important is within the earth sciences when dealing with the composition of rocks. A bivariate distribution with support $0 \le x+y \le 1$ will be required when n is 2.

The univariate beta distribution has support $[0,1]$ and is therefore often used as a distribution of a proportion or probability. Its density is proportional to $x^{\theta_1-1}(1-x)^{\theta_2-1}$. Correspondingly, the bivariate beta distribution has the correct region of support for the joint distribution of two proportions. With support being that part of the unit square such that $x+y \le 1$, the bivariate beta distribution has density

$$h(x,y) = \frac{\Gamma(\theta_1+\theta_2+\theta_3)}{\Gamma(\theta_1)\Gamma(\theta_2)\Gamma(\theta_3)} x^{\theta_1-1} y^{\theta_2-1} (1-x-y)^{\theta_3-1}. \tag{5.60}$$

This distribution may be constructed by a form of trivariate reduction: If $X_i \sim \text{Gamma}(\theta_i, 1)$, then $X_1/(X_1+X_2+X_3)$ and $X_2/(X_1+X_2+X_3)$ jointly have a bivariate beta distribution; see, for example, Wilks (1963, p. 179). This distribution chiefly arises in the context of a trivariate distribution of three quantities that must sum to 1—for example, the probabilities of events or the proportions of substances in a mixture, which are mutually exclusive and exhaustive. When considering just two of these quantities, a bivariate beta distribution may be a natural model to adopt.

5.14.4 Extreme-Value Models

All types of extreme-value distributions can be transformed to the exponential distribution easily, and in what follows we will take the marginals to have this form.

With the support being the positive quadrant, the upper right volume under the probability density surface must take on the form

$$\bar{H} = \exp\left[-(x+y)A\left(\frac{y}{x+y}\right)\right], \tag{5.61}$$

where the function A satisfies

$$A(w) = \int_0^1 \max[(1-w)q, w(1-q)]\, \frac{dB}{dq} dq, \tag{5.62}$$

in which B is a positive increasing function on [0, 1].

A is often termed the *dependence function* of (X, Y) [Pickands (1981) and Tawn (1988)]. The following properties of A are worth noting:

(1) $A(0) = A(1) = 1$.
(2) $\max(w, 1-w) \leq A(w) \leq 1$, where $0 \leq w \leq 1$. Thus $A(w)$ lies within the triangle in the (w, A) plane bounded by $(0, 1), (\frac{1}{2}, \frac{1}{2})$, and $(1, 1)$.
(3) $A(w) = 1$ implies that X and Y are independent. $A(w) = \max(w, 1-w)$ implies that X and Y are equal; i.e., $\Pr(X = Y) = 1$.
(4) A is convex, i.e., $A[\lambda w_1 + (1-\lambda)w_2] \leq \lambda A(w_1) + (1-\lambda)A(w_2)$.
(5) If A_i are dependence functions, so is $\sum_{i=1}^n \alpha_i A_i$, where $\alpha_i \geq 0$ and $\sum_{i=1}^n \alpha_i = 1$.
(6) $\Pr\left(\frac{Y}{X+Y} < w\right) = w + w(1-w)\frac{A\prime(w)}{A(w)}$ [Tiago de Oliveira (1989a)].

A may or may not be differentiable. In the former case, H has a joint density everywhere; in the latter, H has a singular component, and is not differentiable in a certain region of its support. The dependence function $A(w)$ is analogous to the generator of an Archimedean copula discussed earlier in Section 1.5.

Some Special Cases of $A(w)$

The mixed model: Also known as Gumbel's type A bivariate extreme-value distribution, this sets $A(w) = \theta w^2 - \theta w + 1$ for $0 \leq \theta \leq 1$. Then,

$$\bar{H} = \exp\left[-(x+y) + \frac{\theta xy}{x+y}\right]. \tag{5.63}$$

The logistic model: This sets $A(w) = [(1-w)^r + w^r]^{1/r}$ for $r \geq 1$. Then,

$$\bar{H} = \exp[-(x^r + y^r)^{1/r}]. \tag{5.64}$$

The biextremal model: This sets $A(w) = \max(w, 1-\theta w)$ for $0 \leq \theta \leq 1$. Then,

$$\bar{H} = \exp\{-\max[x + (1-\theta)y, y]\}. \tag{5.65}$$

The Gumbel model: This sets $A(w) = \max[1 - \theta w, 1 - \theta(1 - w)](0 \le \theta \le 1)$. Then,

$$\bar{H} = \exp[-(1-\theta)(x+y) - \theta \max(x,y)]. \tag{5.66}$$

This is essentially the bivariate exponential distribution of Marshall and Olkin (1967a,b).

5.14.5 Time Series: Autoregressive Models

Joint Distribution of AR Models

Damsleth and El-Shaarawi (1989) considered autoregressive models in which the "noise" has either (i) a Laplace distribution or (ii) the more commonly assumed normal distribution. Most of their results are for the AR(1) model $X_t = \phi X_{t-1} + \varepsilon_t$ with ε_t having a Laplace or a normal distribution. Damsleth and El-Shaarawi obtained an expression (an infinite series) for the p.d.f. of X in the former case (notice that this is not a Laplace distribution). They then extended this to the joint distribution of X_t and X_{t-k} and presented six contour plots of the resulting p.d.f. (for $\phi = 0.25$ and 0.90 and $k = 1, 5$, and 10).

A Logistic Model

Developing the work of Yeh et al. (1988), Arnold and Robertson (1989) constructed a stationary Markov model with logistic marginals as follows. Let ε_t have a logistic distribution (mean $= \mu$ and scale parameter $= \sigma = (\sqrt{3}/\pi)$ s.d.), $X_0 = \varepsilon_0$, and

$$X_{t+1} = \begin{cases} X_t - \sigma \log \beta & \text{with probability } \beta \\ \min(X_t) - \sigma \log \beta & \text{with probability } 1 - \beta. \end{cases}$$

Then, all the X_t's have logistic distributions, and the joint survival function of $X = (X_t - \mu)/\sigma$ and $Y = (X_{t+1} - \mu)/\sigma$ is given by

$$\bar{H} = \frac{1 + \beta e^y}{(1 + e^y)[1 + \max(e^x, \beta e^y)]}. \tag{5.67}$$

A Pareto Model

Yeh et al. (1988) supposed ε_t to have the following Pareto distribution:

$$\Pr(\varepsilon_t > \varepsilon) = \left[1 + \left(\frac{\varepsilon}{\sigma}\right)^{1/\gamma}\right]^{-1} \quad \text{for } \varepsilon \geq 0.$$

They then set

$$X_{t+1} = \begin{cases} X_t & \text{with probability } \beta \\ \min(\beta^{-\gamma} X_t, \varepsilon_{t+1}) & \text{with probability } 1-\beta. \end{cases}$$

Then, all the X_t's have the same (Pareto) distribution as the ε_t's. The joint survival function of $X = X_t$ and $Y = X_{t+1}$ is

$$\bar{H} = \begin{cases} [1 + (y/\sigma)^{1/\gamma}]^{-1} & \text{for } 0 < x = b^\gamma y \\ \frac{1+\beta(y/\sigma)^{1/\gamma}}{[1+(x/\sigma)^{1/\gamma}][1+(y/\sigma)^{1/\gamma}]} & \text{for } 0 < b^\gamma y < x. \end{cases}$$

Exponential Models

Several models giving rise to exponential marginals for the X_t's were considered by Lawrance and Lewis (1980). The qualitative features of the bivariate distributions of (X_t, X_{t+1}) that are implied are clear from the methods of construction.

In the model they called EAR(1),

$$X_{t+1} = \begin{cases} \rho X_t & \text{with probability } \rho, \\ \rho X_t + \varepsilon_{t+1} & \text{with probability } 1-\rho, \end{cases}$$

with the ε's being exponentially distributed. For a discussion on this model, also see Gaver and Lewis (1980).

In the model Lawrance and Lewis called TEAR(1),

$$X_{t+1} = \begin{cases} (1-\alpha)\varepsilon_{t+1} + X_t & \text{with probability } \alpha, \\ (1-\alpha)\varepsilon_{t+1} & \text{with probability } 1-\alpha, \end{cases}$$

with the ε's, as before, being exponentially distributed.

In the model they called NEAR(1),

$$X_{t+1} = \begin{cases} \varepsilon_{t+1} + \beta X_t & \text{with probability } \alpha, \\ \varepsilon_{t+1} & \text{with probability } 1-\alpha, \end{cases}$$

with the ε's having a particular mixed exponential distribution that is necessary for getting an exponential distribution for the X_t's.

There have been further developments in this direction by Dewald et al. (1989) and Block et al. (1988).

5.15 Limits of Discrete Distributions

It is well known that many of the univariate distributions have their genesis in the Bernoulli distributions and are obtained as sums or limits. [1] Marshall and Olkin (1985a) extended these elementary probability ideas to two dimensions and obtained a number of bivariate distributions.

A random variable (X, Y) is said to have a *bivariate Bernoulli distribution* if it has only four possible values, $(1,1)$, $(1,0)$, $(0,1)$, and $(0,0)$, these occurring with probabilities p_{11}, p_{10}, p_{01}, and p_{00}, respectively. We also set $p_{1+} = p_{11} + p_{10} = 1 - p_{0+}$ and $p_{+1} = p_{11} + p_{01} = 1 - p_{+0}$ in the notation of Marshall and Olkin.

Many of the bivariate distributions obtained by Marshall and Olkin are discrete. As this book is concerned only with continuous distributions we only mention the construction of a bivariate exponential as the limit of a bivariate geometric distribution, and a bivariate gamma as the limit of a bivariate negative binomial distribution.

5.15.1 A Bivariate Exponential Distribution

If $(X_1, Y_1), (X_2, Y_2), \ldots$ is a sequence of i.i.d. bivariate Bernoulli variates and U and V are the number of 0's before the first 1 among the X's and among the Y's, respectively, then U and V each have a geometric distribution in general but not independent. The bivariate distribution function of U and V is given by

$$\Pr(U = u, V = v) = \begin{cases} p_{00}^u p_{01} p_{+0}^{v-u-1} p_{+1} & \text{if } 0 \le u < v \\ p_{00}^u p_{11} & \text{if } 0 \le u = v \end{cases} \tag{5.68}$$

and

$$\Pr(U \ge u, V \ge v) = p_{00}^u p_{+0}^{v-u} \quad \text{for } 0 \le u \le v. \tag{5.69}$$

Now, obtain a bivariate exponential distribution as a limit of this bivariate geometric distribution in (5.67): If independent Bernoulli variates $(X_1, Y_1), (X_2, Y_2), \ldots$ are observed at times $\frac{1}{n}, \frac{2}{n}, \ldots$, then

$$\Pr(U > t_1, V > t_2) = \begin{cases} p_{00}^{[nt_1]} p_{+0}^{[nt_2]-[nt_1]} & \text{if } t_1 < t_2 \\ p_{00}^{[nt_2]} p_{0+}^{[nt_1]-[nt_2]} & \text{if } t_1 > t_2 \end{cases} \tag{5.70}$$

provided nt_1 and nt_2 are not integers, where $[a]$ denotes the integer part of a. Writing $\lambda_{ij} = np_{ij}$ and passing to the limit, we find

[1] Examples are binomial, negative binomial, Poison, and gamma (integer shape parameter).

$$\lim_{n\to\infty} \Pr(U > t_1, V > t_2) = \exp\left[-\lambda_{10}t_1 - \lambda_{01}t_2 - \lambda_{11}\max(t_1, t_2)\right] \quad (5.71)$$

for $t_1, t_2 \geq 0$. This indeed is the bivariate exponential distribution of Marshall and Olkin (1967a).

5.15.2 A Bivariate Gamma Distribution

If $(X_1, Y_1), (X_2, Y_2), \ldots$ is a sequence of i.i.d. bivariate Bernoulli variates and (for positive integers r, s) U and V are the number of 0's before the rth 1 among the X's and before the sth 1 among the Y's, respectively, then U and V each have a negative binomial distribution in general but one that is not independent.

The negative binomial distribution obtained in this way has untidy expressions for its probability functions and for its cumulative distribution function, and we shall not present them here; see Marshall and Olkin (1985b). Proceeding as before, if independent bivariate Bernoulli variates $(X_1, Y_1), (X_2, Y_2), \ldots$ are observed at times $\frac{1}{n}, \frac{2}{n}, \ldots$, then, on setting $\lambda_{ij} = np_{ij}$, Marshall and Olkin (1985b) showed that

$$\begin{aligned}&\lim_{n\to\infty} \Pr(U > t_1, V > t_2)\\ &= \sum \frac{t_1^{a+l-i}\lambda_{11}^{i}\lambda_{01}^{a-i}}{i!(a-i)!(l-i)!} \exp\left[-(\lambda_{11} + \lambda_{01} + \lambda_{01})t_1 \frac{\lambda_{+1}^{m}}{m!}\right] \exp[-\lambda_{+1}(t_2 - t_1)^m]\end{aligned} \quad (5.72)$$

for $0 \leq t_1 \leq t_2$ and with nt_1 and nt_2 not being integers, where the summation is over those values of a, i, l, m such that $r-1 \geq l \geq i \geq 0$, $s-1 \geq a \geq 0$, and $s-1-a \geq m \geq 0$. This distribution has marginals to be gamma distributions with integer shape parameters r and s, respectively.

5.16 Potentially Useful Methods But Not in Vogue

The methods considered in this section are rather more heavily mathematical-based differential equation methods, diagonal expansion, and bivariate Edgeworth expansion, and they are potentially useful but are not in vogue.

5.16.1 Differential Equation Methods

Karl Pearson derived a family of univariate distributions through the differential equation

$$\frac{1}{f}\frac{df}{dx} = \frac{x-a}{b_0 + b_1 x + b_2 x^2}, \tag{5.73}$$

where a, b_0, b_1, and b_2 are constants. The univariate Pearson family of distributions has been discussed in detail in Chapter 12 of Johnson et al. (1994).

Early efforts to generalize this method to two dimensions were unsuccessful until van Uven (1947a,b; 1948a,b) succeeded. Note that the left-hand side of (5.73) is $\frac{d(\log f)}{dx}$. Van Uven started with the particular derivatives

$$\left.\begin{aligned} \frac{\partial \log h}{\partial x} &= \frac{L_1}{Q_1} \\ \frac{\partial \log h}{\partial y} &= \frac{L_2}{Q_2} \end{aligned}\right\}, \tag{5.74}$$

where h is the joint pdf of X and Y, L_1 and L_2 are linear functions of both x and y, and Q_1 and Q_2 are quadratic (or, possibly, linear) functions of both x and y. On fixing either x or y, it is clear that the conditional distributions of either variable, given the other, satisfying differential equations of the form (5.73), belong to the univariate Pearson family. A detailed discussion of the solutions to the differential equations above has been presented by Mardia (1970, pp. 5–9). We provide a condensed version of it as follows.

From (5.74), we obtain by a simple differentiation that

$$\frac{\partial^2 \log h}{\partial x \partial y} = \frac{\partial(\frac{L_1}{Q_1})}{\partial y} = \frac{\partial(\frac{L_2}{Q_2})}{\partial x}, \tag{5.75}$$

$$Q_2 \frac{\partial L_1}{\partial y} - Q_1 \frac{\partial L_2}{\partial x} = \frac{L_1 Q_2}{Q_1}\frac{\partial Q_1}{\partial y} - \frac{L_2 Q_1}{Q_2}\frac{\partial Q_2}{\partial x}. \tag{5.76}$$

The nature of the solution of (5.75) and (5.76) depends mainly on the structure of Q_1 and Q_2, and the usefulness of the solution will depend on their having common factors. If Q_1 and Q_2 do not have a common factor, then X and Y are independent, as shown by Mardia (1970, p. 8). Important cases are as follows:

Case 1. Q_1 and Q_2 have a common linear factor.
Case 2. Q_1 and Q_2 are identical.
Case 3. Q_2 is a linear factor of Q_1, i.e., $Q_1 = LQ_2$.

Case 1

The solution has the form

$$h(x,y) = k_0(ax+b)^{p_1}(cy+d)^{p_2}(a_1 x + b_1 y + c_1)^{p_3}. \tag{5.77}$$

This family of distributions includes the bivariate beta, Pareto, and F-distributions and the following two cases:

$$h(x, y) = \frac{\Gamma(-p_2) x^{p_1} y^{p_2} (-1 - x + y)^{p_3}}{\Gamma(1 + p_1)\Gamma(1 + p_2)\Gamma(-p_1 - p_2 - p_3 - 2)} \tag{5.78}$$

for $p_1, p_3 > -1$, $(p_1 + p_2 + p_3) < -3$, $y - 1 > x > 0$, and

$$h(x, y) = \frac{\Gamma(-p_1) x^{p_1} y^{p_2} (-1 - x + y)^{p_3}}{\Gamma(1 + p_2)\Gamma(1 + p_3)\Gamma(-p_1 - p_2 - p_3 - 2)} \tag{5.79}$$

for $p_2, p_3 > -1$, $(p_1 + p_2 + p_3) < -3$, $x - 1 > y > 0$.

The last two cases are effectively equivalent, though they are considered sometimes as distinct types.

Case 2

When $Q_1 = Q_2$, the solution is

$$h(x, y) = k_0(ax^2 + 2bxy + cy^2 + 2dx + 2ey + c_0)^p. \tag{5.80}$$

Examples include bivariate Cauchy, t-, and Pearson type II distributions. The bivariate normal distribution is a limit of Case 2, in which $a = c = -(1 - \rho^2)/2$, $b = \rho/(1 - \rho^2)$, $c_0 = 1$, $d = e = 0$, $k_0 = (2\pi\sqrt{(}1 - \rho^2)^{-1}$, and $p \to \infty$.

Case 3

When $Q_1 = LQ_2$, we get

$$h(x, y) = k_0(ax + b)^p(a_1x + b_1y + c_1)^q \exp(-cy). \tag{5.81}$$

This family includes McKay's bivariate gamma distribution [McKay (1934)].

Remarks

In fact, van Uven considered all possible cases, but other solutions do not have both marginals of the same form. The system of bivariate distributions obtained through (5.74) is generally known as the family of *bivariate Pearson distributions.* Any member of this family may be called a Pearson type i distribution, i = I,II,...,VII, if the marginals are type i. For example, the bivariate t-distribution may be called a *bivariate type VII distribution.*

In general, the conditional distribution of a member of the family of Pearson distributions is a univariate Pearson distribution, though it may not have the same form as the marginals [Mardia (1970, p. 10)].

The Fokker–Planck Equation

Another family of bivariate distributions that was constructed from a different equation is due to Wong and Thomas (1962) and Wong (1964). The differential equation concerned is the Fokker–Planck equation in diffusion theory given by

$$\frac{\partial^2}{\partial x^2}[B(x)p] - \frac{\partial}{\partial x}[A(x)p] = \frac{\partial p}{\partial t}, \tag{5.82}$$

where $p = p(x|x_0, t), 0 < t < \infty$, and the variables are x and t rather than x and y. $A(x)$ and $B(x)$ are called the "infinitesimal" mean and variance of the underlying Markov transitional probability density functions; see Chapter 5 of Cox and Miller (1965) for further information and details.

The joint densities $h(x_0, x)$ obtained from the conditional densities p form a family that includes some members of the Pearson system such as the bivariate normal, type I, type II, and Kibble's bivariate gamma. The equilibrium density $f(x) = \lim_{t\to\infty} p(x|x_0, t)$ satisfies the Pearson differential equation (5.82) when $A(x)$ and $B(x)$ are linear and quadratic functions, respectively, and the latter is non-negative.

The Ali–Mikhail–Haq Distribution

Refer to Section 2.3 for this distribution and its derivation from a differential equation [Ali et al. (1978)].

5.16.2 Diagonal Expansion

The diagonal expansion of a bivariate distribution involves representing it as

$$dH(x, y) = dF(x)dG(y) \sum_{i=0}^{\infty} \rho_i \xi_i(x)\eta_i(y). \tag{5.83}$$

ξ_i and η_i are known as the canonical variables and ρ_i as the canonical correlation. When X and Y have finite moments of all orders, sets of orthonormal polynomials $\{P_n\}$ and $\{Q_n\}$ can be constructed with respect to F and G—for example, the Hermite polynomials for normal marginals and shifted Legendre polynomials for uniform (0, 1) marginals.

If

$$\left.\begin{aligned} E[X^n|Y = y] &= \text{a polynomial in } y \text{ of degree} = n \\ E[Y^n|X = x] &= \text{a polynomial in } x \text{ of degree} = n \end{aligned}\right\}, \tag{5.84}$$

then H has a diagonal expression in terms of F and G and their respective orthonormal polynomials.

For given marginals with unbounded supports, it is possible to generate a new bivariate distribution by selecting a new canonical sequence $\{\rho_i\}$ with $\sum \rho_i^2 < \infty$ such as a moment sequence defined on $[0, 1]$ or $[-1, 1]$. See Sections 12.4.4 and 12.4.5 of Hutchinson and Lai (1991) for constructing bivariate distributions with normal and other marginals, respectively. See also Sarmanov (1970) and Lee (1996) for constructing a bivariate exponential distribution using this method.

5.16.3 Bivariate Edgeworth Expansion

Let F be a distribution function with known cumulants κ_i and Φ be the standard normal distribution function. The Edgeworth expansion is a representation of F in terms of Φ and κ_i.

The bivariate Edgeworth series expansion is an extension of the univariate Edgeworth expansion. Briefly, we expand a bivariate density function h in the series of derivatives of the standardized normal density ϕ such that

$$h(x, y) = \phi(x, y; \rho) + \int_{m+n\geq 3} (-1)^{m+n} A_{mn} \frac{D_1^m}{m!} \frac{D_2^n}{n!} \phi(x, y; \rho), \qquad (5.85)$$

where the coefficients A_{mn} may be expressed in terms of the cumulants of X and Y, and $D_1 = \partial/\partial x, D_2 = \partial/\partial y$.

Similarly, the joint distribution function is expanded as

$$H(x, y) = \Phi(x, y; \rho) + \int_{m+n\geq 3} (-1)^{m+n} A_{mn} \frac{D_1^{m-1}}{m!} \frac{D_2^{n-1}}{n!} \phi(x, y; \rho). \qquad (5.86)$$

Thus, h is represented as ϕ proportional to a polynomial in x and y, i.e., $h(x, y) = \phi(x, y) \int_{m,n} a_{mn} x^m y^m$. The distribution obtained by considering terms up to $m + n = 4$ has been given by Pearson (1925). This "fifteen constant" bivariate distribution is also known as the type AA distribution.

Chapter 3 of Mardia (1970) presents a historical account of the bivariate Edgeworth expansion as well as describing how the type AA distribution was fitted to Johannsen's bean data; see also Rodriguez (1983, pp. 235–239). The type AA distribution was also applied by Mitropol'skii (1966, pp. 67–70) to the diameters and heights of pine trees.

5.16.4 An Application to Wind Velocity at the Ocean Surface

For this special application, we feel obliged to follow quite closely the explanations of Frieden (1983, Section 3.15.9) and Cox and Munk (1954).

Cox and Munk photographed from an airplane the sun's glitter pattern on the ocean surface and translated the statistics of the glitter into the statistics of the slope distribution of the ocean surface; that is, of the joint distribution of wave slope in the direction of the wind (X) and transverse to the wind direction (Y). Conceivably, this could be the basis of a method of measuring the wind velocity at the ocean surface.

"If the sea surface were absolutely calm, a single, mirror-like reflection of the sun would be seen at the horizontal point. In the usual case there are thousands of 'dancing' highlights. At each highlight there must be a water facet, possibly quite small, which is so inclined as to reflect an incoming ray from the sun towards the observer. The farther the highlighted facet is from the horizontal specular point, the larger must be this inclination. The width of the glitter patterns is therefore an indication of the maximum slope of the sea surface" [Cox and Munk (1954)]. In fact, these authors measured the variation in brightness within the glitter pattern, rather than computing maximum slopes from the outer boundaries, and thus obtained more detailed information.

In choosing a functional form for $h(x, y)$ in this case, two factors considered are the following:

- The p.d.f. of X should be skewed, as waves tend to lean away from the wind, having gentler slopes on the windward side than on the leeward side.
- There should be no such skew for the p.d.f. of Y because waves transverse to the wind are not directly formed by the wind but rather by leakage of energy from the longitudinal wave motion.

Consequently, the following form of the two-dimensional expansion was fitted to experimental data:

$$\begin{aligned} h(x, y) = f(x)f(y)[1 + \alpha_{12}H_1(x)H_2(y) + \alpha_{30}H_3(y) + \alpha_{04}(y) \\ +\alpha_{22}H_2(x)H_2(y) + \alpha_{40}H_4(x)], \end{aligned} \tag{5.87}$$

where the H_i's are the Hermite polynomials.

5.16.5 Another Application to Statistical Spectroscopy

As a result of analytical and numerical studies showing that the higher bivariate cumulants of the relevant variables are quite small, Kota (1984) concluded that it was meaningful to employ an expansion around a bivariate normal density—especially for a bivariate density of importance in statistical spectroscopy; see also the follow-up work by Kota and Potbhare (1985).

5.17 Concluding Remarks

We have reviewed in this chapter a great many methods of constructing bivariate distributions and have given examples of contexts in which they have been used. Most statisticians, hopefully, would have found something new to them! A particular contribution of this chapter has been the method of organizing the material. It is not, we admit, an elegant and mathematically satisfactory scheme, but it is one that we have found somewhat helpful, and we hope that readers will, too. We first divided methods of construction into popular methods and a miscellaneous group; the first included conditional distributions, compounding, and variables in common, and the second was made up of some inelegant methods, data-guided methods, special methods used in some applied fields, and some potentially useful methods. Finally, each of them had their own subdivisions.

By way of a pointer to the possible future development of the subject, we may remark that, in some areas of statistics, the results that can be obtained are determined by whether one is clever enough to manipulate mathematically rather than any real conceptual depth. For instance, suppose there is a bivariate survival function $\bar{H}_1(x, y)$ and a bivariate p.d.f. $h_2(\theta_1, \theta_2)$. Then, another bivariate survival function can be obtained by compounding as $\iint H_1(\theta_1 x, \theta_2 y) h_2(\theta_1, \theta_2) d\theta_1 d\theta_2$. The results obtainable depend on one's ingenuity in choosing $\bar{H}_1$ and h_2 so that the double integration is tractable. The increasing sophistication and widening availability of packages for computerized algebraic manipulation, such as MACSYMA and REDUCE, gives hope that this limitation may diminish in the years to come; see, for example, Steele (1985), Bryan-Jones (1987), Rayna (1987, pp. 29–31), and Heller (1991) for more on this. Of course, we can ask why we need to have an explicit expression for $\int\int \bar{H} h_2$. One could say that this itself contains all the modeling information and that one should be looking to fit this directly to data.

One can imagine the interfacing of computer algebra packages with those for model fitting, so that for a given $\bar{H}_1$ and h_2, the algebra part solves the double integral and passes the result to the model-fitting part. Because the number-crunching is becoming as fast as it is, the double integral could be evaluated numerically whenever required by the model-fitting package. Although this discussion has been posed in terms of the compounding method for constructing distributions, it applies equally well to other methods of construction as well.

References

1. Abrahams, J., Thomas, J.B.: A note on the characterization of bivariate densities by conditional densities. Communications in Statistics: Theory and Methods **13**, 395–400 (1984)
2. Ahsanullah, M.: Some characterizations of the bivariate normal distribution. Metrika **32**, 215–218 (1985)
3. Ali, M.M., Mikhail, N.N., Haq, M.S.: A class of bivariate distributions including the bivariate logistic. Journal of Multivariate Analysis **8**, 405–412 (1978)
4. Arnold, B.C., Beaver, R.J.: The skew-Cauchy distribution. Statistics and Probability Letters **49**, 285–290 (2000)
5. Arnold, B.C., Castillo, E., Sarabia, J.M.: Conditionally Specified Distributions. Lecture Notes in Statistics, Volume 73. Springer-Verlag, Berlin (1992)
6. Arnold, B.C., Castillo, E., Sarabia, J.M.: Conditional Specification of Statistical Models. Springer-Verlag, New York (1999)
7. Arnold, B.C., Press, S.J.: Bayesian-estimation and prediction for Pareto data. Journal of the American Statistical Association **84**, 1079–1084 (1989).
8. Arnold, B.C., Robertson, C.A.: Autoregressive logistic processes. Journal of Applied Probability **26**, 524–531 (1989)
9. Arnold, B.C., Strauss, D.: Bivariate distributions with conditionals in prescribed exponential families. Journal of the Royal Statistical Society, Series B **53**, 365–375 (1991)
10. Azzalini, A.: The skew-normal distribution and related multivariate families. Scandinavian Journal of Statistics **32**, 159–199 (2005)
11. Azzalini, A.: Skew-normal family of distributions. In: Encyclopedia of Statistical Sciences, Volume 12, S. Kotz, N. Balakrishnan, C.B. Read, and B. Vidakovic (eds.), pp. 7780–7785. John Wiley and Sons, New York (2006)
12. Azzalini, A., Capitanio, A.: Distributions generated by perturbation of symmetry with emphasis on a multivariate skew t distribution. Journal of the Royal Statistical Society, Series B **65**, 367–389 (2003)
13. Azzalini, A., Dal Cappello, T., Kotz, S.: Log-skew-normal and log-skew-t distributions as models for family income data. Journal of Income Distribution **11**, 12–20 (2003)
14. Bhattacharyya, A.: On some sets of sufficient conditions leading to the normal bivariate distribution. Sankhyā **6**, 399–406 (1943)
15. Blake, I.F.: An Introduction to Applied Probability. John Wiley and Sons, New York (1979)
16. Block, H.W., Langberg, N.A., Stoffer, D.S.: Bivariate exponential and geometric autoregressive and autoregressive moving average models. Advances in Applied Probability **20**, 798–821 (1988)
17. Branco, M.D., Dey, P.K.: A general class of multivariate skew-elliptical distributions. Journal of Multivariate Analysis **79**, 99–113 (2001)
18. Bryan-Jones, J.: A tutorial in computer algebra for statisticians. The Professional Statistician **6**, 5–8 (1987)
19. Bryson, M.C., Johnson, M.E.: Constructing and simulating multivariate distributions using Khintchine's theorem. Journal of Statistical Computation and Simulation **16**, 129–137 (1982)
20. Castillo, E., Galambos, J.: Lifetime regression-models based on a functional-equation of physical nature. Journal of Applied Probability **24**, 160–169 (1987)
21. Cohen, L., Zaparovanny, Y.I.: Positive quantum joint distributions. Journal of Mathematical Physics **21**, 794–796 (1980)
22. Conolly, B.W., Choo, Q.H.: The waiting time process for a generalized correlated queue with exponential demand and service. SIAM Journal on Applied Mathematics **37**, 263–275 (1979)

23. Cox, C., Munk, W.: Measurement of the roughness of the sea surface from photographs of the sun's glitter. Journal of the Optical Society of America **44**, 838–850 (1954)
24. Cox, D.R., Miller, H.D.: The Theory of Stochastic Processes. Chapman and Hall, London (1965)
25. Damsleth, E., El-Shaarawi, A.H.: ARMA models with double-exponentially distributed noise. Journal of the Royal Statistical Society, Series B **51**, 61–69 (1989)
26. Deheuvels, P., Tiago de Oliveira, J.: On the nonparametric estimation of the bivariate extreme-value distributions. Statistics and Probability Letters **8**, 315–323 (1989)
27. Devroye, L.: Nonuniform Random Variate Generation. Springer-Verlag, New York (1986)
28. Dewald, L.S., Lewis, P.A.W., McKenzie, E.: A bivariate first-order autoregressive time series model in exponential variables (BEAR (1)). Management Science **35**, 1236–1246 (1989)
29. Downton, F.: Bivariate exponential distributions in reliability theory. Journal of the Royal Statistical Society, Series B **32**, 408–417 (1970)
30. Erdelyi, A. (ed.): Tables of Integral Transforms, Volume 1. McGraw-Hill, New York (1954)
31. Everitt, B.S.: Mixture distributions. In: Encyclopedia of Statistical Sciences, Volume 5, S. Kotz and N.L. Johnson (eds.), pp. 559–569. John Wiley and Sons, New York (1985)
32. Finch, P.D., Groblicki, R.: Bivariate probability densities with given margins. Foundations of Physics **14**, 549–552 (1984)
33. Fraser, D.A.S., Streit, F.: A further note on the bivariate normal distribution. Communications in Statistics: Theory and Methods **10**, 1097–1099 (1980)
34. Fréchet, M.: Sur les tableaux de corrélation dont les marges sont données. Annales de l'Université de Lyon, Série 3 **14**, 53–77 (1951)
35. Frees, E.W., Valdez, E.A.: Understanding relationship using copulas. North-America Actuarial Journal **2**, 1–26 (1998)
36. Frieden, B.R., Probability, Statistical Optics, and Data Testing: A Problem-Solving Approach. Springer-Verlag, Berlin (1983)
37. Gaver, D.P.: Multivariate gamma distributions generated by mixture. Sankhyā, Series A **32**, 123–126 (1970)
38. Gaver, D.P.: Point process problems in reliability. In: Stochastic Point Processes: Statistical Analysis, Theory, and Applications, P.A.W. Lewis (ed.), pp. 774–800. John Wiley and Sons, New York (1972)
39. Gaver, D.P., Lewis, P.A.W.: First-order autoregressive gamma sequences and point processes. Advances in Applied Probability **12**, 727–745 (1980)
40. Gelman, A., Speed, T.P.: Characterizing a joint distribution by conditionals. Journal of the Royal Statistical Society, Series B **35**, 185–188 (1993)
41. Genest, C., Rivest, L-P.: A characterization of Gumbel's family of extreme-value distributions. Statistics and Probability Letters **8**, 207–211 (1989)
42. Gnanadesikan, R.: Methods for Statistical Data Analysis of Multivariate Observations. John Wiley and Sons, New York (1977)
43. Gumbel, E.J.: Bivariate exponential distributions. Journal of the American Statistical Association **55**, 698–707 (1960)
44. Gupta, R.D., Richards, D.St.P.: Multivariate Liouville distributions. Journal of Multivariate Analysis **23**, 233–256 (1987)
45. Hagen, E.W.: Common-mode/common-cause failure: A review. Annals of Nuclear Energy **7**, 509–517 (1980)
46. Halperin, M., Wu, M., Gordon, T.: Genesis and interpretation of differences in distribution of baseline characteristics between cases and noncases in cohort studies. Journal of Chronic Diseases **32**, 483–491 (1979)
47. Heller, B.: MACSYMA for Statisticians. John Wiley and Sons, New York (1991)

48. Hoeffding, W.: Masstabinvariante Korrelationstheorie. Schriften des Mathematischen Instituts und des Instituts für Angewandte Mathematik der Universität Berlin **5**, 179–233 (1940)
49. Hougaard, P.: Analysis of Multivariate Survival Data. Springer-Verlag, New York (2000)
50. Hutchinson, T.P., Lai, C.D.: The Engineering Statistician's Guide to Continuous Bivariate Distributions. Rumsby Scientific Publishing, Adelaide (1991)
51. Joe, H.: Parametric families of multivariate distributions with given marginals. Journal of Multivariate Analysis **46**, 262–282 (1993)
52. Joe, H., Hu, T.Z.: Multivariate distributions from mixtures of max-infinitely divisible distributions. Journal of Multivariate Analysis **57**, 240–265 (1996)
53. Jogdeo, K.: Nonparametric methods for regression. Report S330, Mathematics Centre, Amsterdam (1964)
54. Johnson, M.E.: Multivariate Statistical Simulation. John Wiley and Sons, New York (1987)
55. Johnson, M.E., Tenenbein, A.: A bivariate distribution family with specified marginals. Journal of the American Statistical Association **76**, 198–201 (1981)
56. Johnson, N.L.: Bivariate distributions based on simple translation systems. Biometrika **36**, 297–304 (1949)
57. Johnson, N.L., Kotz, S.: Square tray distributions. Statistics and Probability Letters **42**, 157–165 (1999)
58. Johnson, N.L., Kotz, S., Balakrishnan, N.: Continuous Univariate Distributions, Volume 1, 2nd edition. John Wiley and Sons, New York (1994)
59. Jones, M.C.: Marginal replacement in multivariate densities, with applications to skewing spherically symmetric distributions. Journal of Multivariate Analysis **81**, 85–99 (2002)
60. Kendall, M.G., Stuart, A.: The Advanced Theory of Statistics: Volume 2: Inference and Relationship, 4th edition. Griffin, London (1979)
61. Kimeldorf, G., Sampson, A.: One-parameter families of bivariate distributions with fixed marginals. Communications in Statistics **4**, 293–301 (1975)
62. Kota, V.K.B.: Bivariate distributions in statistical spectroscopy studies: Fixed-J level densities, fixed-J averages and spin cut-off factors. Zeitschrift für Physik, Section A: Atoms and Nuclei **315**, 91–98 (1984)
63. Kota, V.K.B., Potbhare, V.: Bivariate distributions in statistical spectroscopy studies-II: Orthogonal polynomials in two variables and fixed E, J expectation values. Zeitschrift für Physik, Section A: Atoms and Nuclei **322**, 129–136 (1985)
64. Kotz, S., Balakrishnan, N., Johnson, N.L.: Continuous Multivariate Distributions, Volume 1: Models and Applications, 2nd edition. John Wiley and Sons, New York (2000)
65. Kotz, S., Nadarajah, D.: Some extreme type elliptical distributions. Statistics and Probababilty Letters **54**, 171–182 (2001)
66. Kowalczyk, T., Pleszczynska, E.: Monotonic dependence functions of bivariate distributions. Annals of Statistics **5**, 1221–1227 (1977)
67. Kwerel, S.M.: Fréchet bounds. In: Encyclopedia of Statistical Sciences, Volume 3, S. Kotz and N.L. Johnson (eds.), pp. 202–209. John Wiley and Sons, New York (1983)
68. Lai, C.D.: Letter to the editor. Journal of Applied Probability **24**, 288–289 (1987)
69. Lai, C.D.: Construction of bivariate distributions by a generalized trivariate reduction technique. Statistics and Probability Letters **25**, 265–270 (1994)
70. Lai, C.D., Xie, M.: A new family of positive dependence bivariate distributions. Statistics and Probability Letters **46**, 359–364 (2000)
71. Langaris, C.: A correlated queue with infinitely many servers. Journal of Applied Probability **23**, 155–165 (1986)
72. Lawrance, A.J., Lewis, P.A.W.: The exponential autoregressive-moving average EARMA(p,q) process. Journal of the Royal Statistical Society, Series B **42**, 150–161 (1980)

73. Lee, M.L.T.: Properties and applications of the Sarmanov family of bivariate distributions. Communications in Statistics: Theory and Methods **25**, 1207–1222 (1996)
74. Ma, C., Yue, X.: Multivariate p-order Liouville distributions: Parameter estimation and hypothesis testing. Chinese Journal of Applied Probabilty and Statistics **11**, 425–431 (1995)
75. Mardia, K.V.: Families of Bivariate Distributions. Griffin, London (1970)
76. Marshall, A.W., Olkin, I.: A multivariate exponential distribution, Journal of the American Statistical Association **62**, 30–44 (1967a)
77. Marshall, A.W., Olkin, I.: A generalized bivariate exponential distribution. Journal of Applied Probability **4**, 291–302 (1967b)
78. Marshall, A.W., Olkin, I.: Multivariate exponential distributions, Marshall-Olkin. In: Encyclopedia of Statistical Sciences, Volume 6, S. Kotz and N.L. Johnson (eds.), pp. 59–62. John Wiley and Sons, New York (1985a)
79. Marshall, A.W., Olkin, I.: A family of bivariate distributions generated by the bivariate Bernoulli distributions. Journal of the American Statistical Association **80**, 332–338 (1985b)
80. Marshall, A.W., Olkin, I.: Families of multivariate distributions. Journal of the American Statistical Association **83**, 834–841 (1988)
81. Mathai, A.M., Moschopoulos, P.G.: On a multivariate gamma distribution. Journal of Multivariate Analysis **39**, 135–153 (1991)
82. Mathai, A.M., Moschopoulos, P.G.: A form of multivariate gamma distribution. Annals of the Institute of Statistical Mathematics **44**, 97–106 (1992)
83. McKay, A.T.: Sampling from batches. Journal of the Royal Statistical Society, Supplement **1**, 207–216 (1934)
84. McLachlan, G.J., Basford, K.E.: Mixture Models: Inference and Applications to Clustering, Marcel Dekker, New York (1988)
85. Mihram, G.A., Hultquist, A.R.: A bivariate warning-time/failure-time distribution. Journal of the American Statistical Association **62**, 589–599 (1967)
86. Mikusiński, P., Sherwood, H., Taylor, M.D.: Shuffles of Min. Stochastica **13**, 61–74 (1992)
87. Mitchell, C.R., Paulson, A.S.: M-M-1 queues with interdependent arrival and services processes. Naval Research Logistics **26**, 47–56 (1979)
88. Mitropol'skii, A.K.: Correlation Equations for Statistical Computations. Consultants Bureau, New York (1966)
89. Nadarajah, S., Kotz, S.: Some truncated bivariate distributions. Acta Applicandae Mathematicae **95**, 205–222 (2007)
90. Narumi, S.: On the general forms of bivariate frequency distributions which are mathematically possible when regression and variations are subjected to limiting conditions I. Biometrika **15**, 77–88 (1923a)
91. Narumi, S.: On the general forms of bivariate frequency distributions which are mathematically possible when regression and variations are subjected to limiting conditions II. Biometrika **15**, 209–211 (1923b)
92. Nataf, A.: Détermination des distributions de probabilités dont les marges sont données. Comptes Rendus de l'Académie des Sciences **255**, 42–43 (1962)
93. Nelsen, R.B.: An Introduction to Copulas. 2nd edition. Springer-Verlag, New York (2006)
94. Niu, S-C.: On queues with dependent interarrival and service times. Naval Research Logistics Quarterly **24**, 497–501 (1981)
95. Oakes, D.: Bivariate survival models induced by frailties. Journal of the American Statistical Association **84**, 487–493 (1989)
96. Patil, G.P.: On a characterization of multivariate distribution by a set of its conditional distributions. Bulletin of the International Statistical Institute **41**, 768–769 (1965)
97. Pearson, K.: The fifteen constant bivariate frequency surface. Biometrika **17**, 268–313 (1925)

98. Pickands, J.: Multivariate extreme value distributions. Bulletin of the International Statistical Institute **49**, 859–878 (Discussion, 894–902) (1981)
99. Raftery, A.E.: A continuous multivariate exponential distribution. Communications in Statistics: Theory and Methods **13**, 947–965 (1984)
100. Raftery, A.E.: Some properties of a new continuous bivariate exponential distribution. Statistics and Decisions, Supplement Issue No. 2, 53–58 (1985)
101. Rayna, G.: REDUCE: Software for Algebraic Computation. Springer-Verlag, New York (1987)
102. Reiss, R-D.: Approximate Distributions of Order Statistics: With Applications to Nonparametric Statistics. Springer-Verlag, New York (1989)
103. Rodriguez, R.N.: Frequency curves, Systems of. In: Encyclopedia of Statistical Sciences, Volume 3, S. Kotz and N.L. Johnson (eds.), pp. 212–225. John Wiley and Sons, New York (1983)
104. Rüschendorf, L.: Construction of multivariate distributions with given marginals. Annals of the Institute of Statistical Mathematics **37**, 225–233 (1985)
105. Sahu, S.K., Dey, D.K., Branco, M.D.: A new class of multivariate skew distributions with applications to Bayesian regression models. The Canadian Journal of Statistics **31**, 129–150 (2003)
106. Sarmanov, I.O.: Gamma correlation process and its properties. Doklady Akademii Nauk, SSSR **191**, 30–32 (in Russian) (1970)
107. Smith, R.L.: Statistics of extreme values. Bulletin of the International Statistical Institute **51** (Discussion, Book 5, 185–192) (1985)
108. Smith, R.L.: Extreme value theory. In: Handbook of Applicable Mathematics, Volume 7, W. Ledermann (ed.), pp. 437–472. John Wiley and Sons, Chichester (1990)
109. Smith, R.L., Tawn, J.A., Yuen, H.K.: Statistics of multivariate extremes. International Statistical Review **58**, 47–58 (1990)
110. Smith, W.P.: A bivariate normal test for elliptical homerange models: Biological implications and recommendations. Journal of Wildlife Management **47**, 611–619 (1983)
111. Steele, J.M.: MACSYMA as a tool for statisticians. In: American Statistical Association, 1985 Proceedings of the Statistical Computing Section. American Statistical Association, Alexandria, Virginia, pp. 1–4 (1985)
112. Stigler, S.M.: Laplace solution: Bulletin Problems Corner. The Institute of Mathematical Statistics Bulletin **21**, 234 (1992)
113. Stoyanov, J.M.: Counterexamples in Probability, 2nd edition. John Wiley and Sons, New York (1997)
114. Tawn, J.A.: Bivariate extreme value theory: Models and estimation. Biometrika **75**, 397–415 (1988)
115. Tawn, J.A.: Modelling multivariate extreme value distributions. Biometrika **77**, 245–253 (1990)
116. Tiago de Oliveira, J.: Statistical decision for bivariate extremes. In: Extreme Value Theory: Proceedings of a Conference held in Oberwolfach, J. Husler and R-D. Reiss (eds.), pp. 246–261. Springer-Verlag, New York (1989a)
117. Tiago de Oliveira, J.: Intrinsic estimation of the dependence structure for bivariate extremes. Statistics and Probability Letters **8**, 213–218 (1989b)
118. Tiku, M.L., Kambo, N.S.: Estimation and hypothesis testing for a new family of bivariate nonnormal distributions. Communications in Statistics: Theory and Methods **21**, 1683–1705 (1992)
119. Titterington, D.M., Smith, A.F.M., Makov, U.E.: Statistical Analysis of Finite Mixture Distributions. John Wiley and Sons, New York (1985)
120. van Uven, M.J.: Extension of Pearson's probability distributions to two variables-I. Proceedings of the Koninklijke Nederlandse Akademie van Wetenschappen **50**, 1063–1070 (Indagationes Mathematicae **9**, 477–484) (1947a)
121. van Uven, M.J.: Extension of Pearson's probability distributions to two variables-II. Proceedings of the Koninklijke Nederlandse Akademie van Wetenschappen **50**, 1252–1264 (Indagationes Mathematicae **9**, 578–590) (1947b)

122. van Uven, M.J.: Extension of Pearson's probability distributions to two variables-III. Proceedings of the Koninklijke Nederlandse Akademie van Wetenschappen **51**, 41–52 (Indagationes Mathematicae **10**, 12–23) (1948a)
123. van Uven, M.J.: Extension of Pearson's probability distributions to two variables-IV. Proceedings of the Koninklijke Nederlandse Akademie van Wetenschappen **51**, 191–196 (Indagationes Mathematicae **10**, 62–67) (1948b)
124. Wei, G., Fang, H.B., Fang, K.T.: The dependence patterns of random variables – elementary algebra and geometric properties of copulas. Technical Report (MATH-190), Hong Kong Baptist University, Hong Kong (1998)
125. Whitmore, G.A., Lee, M-L.T.: A multivariate survival distribution generated by an inverse Gaussian mixture of exponentials. Technometrics **33**, 39–50 (1991)
126. Whitt, W.: Bivariate distributions with given marginals. Annals of Statistics **4**, 1280–1289 (1976)
127. Wilks, S.S.: Mathematical Statistics, 2nd edition. John Wiley and Sons, New York (1963)
128. Wong, E.: The construction of a class of stationary Markoff processes. In: Stochastic Processes in Mathematical Physics and Engineering (Proceedings of Symposia in Applied Mathematics, Volume 16), R.E. Bellman (ed.), pp. 264–272. American Mathematical Society, Providence, Rhode Island, (1964)
129. Wong, E., Thomas, J.B.: On polynomial expansions of second-order distributions. SIAM Journal on Applied Mathematics **10**, 507–516 (1962)
130. Yeh, H.C., Arnold, B.C., Robertson, C.A.: Pareto processes. Journal of Applied Probability **25**, 291–301 (1988)

Chapter 6
Bivariate Distributions Constructed by the Conditional Approach

6.1 Introduction

6.1.1 Contents

In Section 5.6, we outlined the construction of a bivariate p.d.f. as the product of a marginal p.d.f. and a conditional p.d.f., $h(x,y) = f(x)g(y|x)$. This construction is easily understood, and has been a popular choice in the literature, especially when Y can be thought of as being caused by, or predicted from, X. Arnold et al. (1999, p. 1) contend that it is often easier to visualize conditional densities or features of conditional densities than marginal or joint densities. They cite, for example, that it is not unreasonable to visualize that, in the human population, the distribution of heights for a given weight will be unimodal, with the mode of the conditional distribution varying monotonically with weight. Similarly, we may visualize a unimodal distribution of weights for a given height, this time with the mode varying monotonically with the height. Thus, construction of a bivariate distribution using two conditional distributions may be practically useful.

We begin this chapter by considering distributions such that both sets of conditionals are beta, exponential, gamma, Pareto, normal, Student t or some other distributions in Sections 6.2–6.6. Sections 6.7 and 6.8 deal with situations wherein the conditional distributions and moments are specified. Section 6.9 describes the parameter estimation for conditionally specified models. Sections 6.10 and 6.11 give brief accounts of specific distributions constructed by the conditional method such as McKay's bivariate gamma distribution and its variants, Dubey's distribution, Blumen and Ypelaar's distribution, exponential dispersion models, four densities of Barndorff–Nielsen and Blæsield, and continuous bivariate densities with a discontinuous marginal density function. Section 6.12 discusses a common approach where the marginal and conditional distributions are of the same family. In Section 6.13, we consider bivariate distributions when conditional survival functions are speci-

N. Balakrishnan, C.D. Lai, *Continuous Bivariate Distributions*,
DOI 10.1007/b101765_7, © Springer Science+Business Media, LLC 2009

fied. Finally, several papers dealing with applications of these models are summarized in Section 6.14, and, in particular, the fields of meteorology and hydrology provide several examples.

Arnold et al. (1999) have devoted the bulk of their book to a discussion of the joint distributions obtained from a specification of both conditional densities. The present chapter provides in this direction an overview of five chapters of their important book. For an introduction to the subject of conditionally specified distributions, see Arnold et al. (2001).

6.1.2 Pertinent Univariate Distributions

Definition 6.1. X has an exponential distribution if its density function is

$$f(x) = \theta e^{-\theta x}, \quad x > 0, \ \theta > 0,$$

and we denote it by $X \sim \mathrm{Exp}(\theta)$.

Definition 6.2. X has a gamma distribution if its density function is

$$f(x;\theta_1,\theta_2) = x^{\theta_1 - 1} e^{-\theta_2 x} \frac{\theta_2^{\theta_1}}{\Gamma(\theta_1)} = x^{-1} e^{\theta_1 \log x - \theta_2 x} \frac{\theta_2^{\theta_1}}{\Gamma(\theta_1)}$$

for $x > 0$, and we denote it by $X \sim \Gamma(\theta_1, \theta_2)$.

Definition 6.3. X has a beta distribution if its density function is

$$f(x) = \frac{1}{B(p,q)} x^{p-1}(1-x)^{q-1}, \quad 0 < x < 1, \ p, q > 0.$$

Definition 6.4. X has a beta distribution of the second kind, denoted by $B2(p, q, \sigma)$, if it has a density function of the form

$$f(x, \alpha) = \frac{\sigma^q}{B(p,q)} x^{p-1}(\sigma + x)^{-(p+q)}, \quad x > 0, \ p, q, \sigma > 0.$$

Definition 6.5. X has a Cauchy distribution, denoted by $C(\mu, \sigma)$, if its density function is

$$f(x) = \frac{1}{\pi\sigma\left(1 + \left(\frac{x-\mu}{\sigma}\right)^2\right)}, \quad -\infty < x < \infty, \ \sigma > 0, \ \mu \text{ real}.$$

Definition 6.6. A random variable T_α is said to follow a Student t-distribution if its density function is

$$f(x) = \frac{\Gamma[(\alpha+1)/2]}{(\alpha\pi)^{1/2}\Gamma(\alpha/2)} \left(1 + \frac{x^2}{\alpha}\right)^{-(\alpha+1)/2}, \quad -\infty < x < \infty.$$

Definition 6.7. We say that X has an inverse Gaussian distribution if its density function is

$$f(x) = \sqrt{\frac{\eta_2}{\pi}}\, e^{2\sqrt{\eta_1\eta_2}} e^{-\eta_1 x - \eta_2 x^{-1}}, \quad x \geq 0,$$

and we denote it by $X \sim \mathrm{IG}(\eta_1, \eta_2)$.

Definition 6.8. A random variable has a Pareto type II distribution if its density function is

$$f(x, \alpha) = \frac{\alpha}{\sigma}\left(1 + \frac{x}{\sigma}\right)^{-\alpha-1}, \quad x > 0,\ \alpha, \sigma > 0.$$

This distribution is also known as the Lomax distribution, and it will be denoted by $P(\sigma, \alpha)$.

Definition 6.9. We say that X has a generalized Pareto distribution (or Burr type XII), denoted by $X \sim \mathcal{GP}(\sigma, \delta, \alpha)$, if its survival function is of the form

$$\Pr(X > x) = \left\{1 + \left(\frac{x}{\sigma}\right)^{\delta}\right\}^{-\alpha}, \quad x > 0.$$

6.1.3 Compatibility and Uniqueness

It is well known that if we specify the marginal density of X, $f(x)$, and for each possible value of x, specify the conditional density of Y given $X = x$, i.e., $g(y|x)$, then a unique joint density $h(x, y)$ results.

Suppose now that both the families of conditional distribution of X given Y and conditional distribution of X given Y are specified. This would result in over-determining the joint distribution, and so the problem of consistency has to be resolved. We say that the two conditional distributions are *compatible* if there exists at least one joint distribution of (X, Y) with the given families as its conditional distributions.

Necessary and Sufficient Conditions

A bivariate density $h(x, y)$, with conditional densities $f(x|y)$ and $g(y|x)$, will exist if and only if [see Section 1.6 of Arnold et al. (1999)]

1. $\{(x, y) : f(x|y) > 0\} = \{(x, y) : g(y|x) > 0\}$.
2. There exist $a(x)$ and $b(y)$ such that the ratio $\frac{f(x|y)}{g(y|x)} = a(x)b(y)$, where $a(\cdot)$ and $b(\cdot)$ are non-negative integrable functions.
3. $\int a(x)dx < \infty$.

The three conditions specified above are necessary and sufficient conditions for two conditional distributions to be compatible.

Also, the condition $\int a(x)dx < \infty$ is equivalent to $\int [1/b(y)]dy < \infty$, and only one needs to be checked in practice.

In cases in which compatibility is confirmed, the question of possible uniqueness of the compatible distribution still needs to be addressed. Arnold et al. (1999) showed that the joint density $h(x, y)$ is unique if and only if the Markov chain associated with $a(x, y)$ and $b(x, y)$ is indecomposable. Gelman and Speed (1993) have addressed the issue of uniqueness in a multivariate setting.

6.1.4 Early Work on Conditionally Specified Distributions

One of the earliest contributions to the study of conditionally specified models was the work of Patil (1965). This was followed by Besag (1974), Abrahams and Thomas (1984), and then a major breakthrough by Castillo and Galambos (1987a).

6.1.5 Approximating Distribution Functions Using the Conditional Approach

Parrish and Bargmann (1981) have given a general method for evaluating bivariate d.f.'s that "utilizes a factorization of the joint density function into the product of a marginal density function and an associated density, permitting the expressions of the double integral in a form amenable to the use of specialized Gaussian-type quadrature techniques for numerical evaluation of cumulative probabilities." See also Parrish (1981).

As mentioned earlier, conditionally specified distributions are authoritatively treated in Arnold et al. (1999). Sections 6.2–6.9 summarize some of their work. For ease of referring back to this source, much of their notation has been retained here.

6.2 Normal Conditionals

Bivariate distributions having conditional densities of the normal form and yet not the classical normal distribution have been known in the literature for a long time; see Bhattacharyya (1943), for example.

6.2.1 Conditional Distributions

Suppose

$$X \mid (Y=y) \sim N(\mu_1(y), \sigma_1^2(y)) \text{ and } Y \mid (X=x) \sim N(\mu_2^2(x), \sigma_2(x)), \quad (6.1)$$

where

$$E(X|Y=y) = \mu_1(y) = -\frac{B/2 + Hy - Ey^2/2}{C + 2Jy - Fy^2},$$

$$E(Y|X=x) = \mu_2(x) = -\frac{G + Hx + Jx^2}{D + Ex + Fx^2},$$

and

$$\mathrm{var}(X|Y=y) = \sigma_1^2(y) = \frac{-1}{C + 2Jy - Fy^2},$$

$$\mathrm{var}(Y|X=x) = \sigma_2^2(x) = \frac{1}{D + Ex + Fx^2}.$$

6.2.2 Expression of the Joint Density

The joint density corresponding to the specification in (6.1) is

$$h(x,y) = \frac{1}{\sqrt{2\pi}} \exp\left\{\frac{1}{2}[A + Bx + 2Gy + Cx^2 - Dy^2 + 2Hxy + 2Jx^2y - Exy^2 - Fx^2y^2]\right\}, \quad (6.2)$$

where A is the normalizing constant so that $h(x,y)$ is a bivariate density. Equation (6.2) may be reparametrized as

$$h(x,y) = \exp\left\{(1, x, x^2)\begin{pmatrix} m_{00} & m_{01} & m_{02} \\ m_{10} & m_{11} & m_{12} \\ m_{20} & m_{21} & m_{22} \end{pmatrix}\begin{pmatrix} 1 \\ y \\ y^2 \end{pmatrix}\right\}, \quad (6.3)$$

where

$$\begin{aligned} m_{00} &= A/2,\ m_{01} = G,\ m_{02} = -D/2, \\ m_{10} &= B/2,\ m_{11} = H,\ m_{12} = -E/2, \\ m_{20} &= C/2,\ m_{21} = J,\ \ m_{22} = -F/2. \end{aligned} \tag{6.4}$$

6.2.3 Univariate Properties

The two marginals densities are

$$f(x) = \exp\left\{\frac{1}{2}\left[2(m_{20}x^2 + m_{10}x + m_{00}) - \mu_2^2(x)/\sigma_2^2(x)\right]\right\}\sigma_2(x) \tag{6.5}$$

and

$$g(y) = \exp\left\{\frac{1}{2}\left[2(m_{02}y^2 + m_{01}y + m_{00}) - \mu_1^2(y)/\sigma_1^2(y)\right]\right\}\sigma_1(y). \tag{6.6}$$

6.2.4 Further Properties

The normal conditionals distribution has joint density of the form in (6.3), where the constants, the m_{ij}'s, satisfy one of the two sets of conditions

(i) $m_{22} = m_{12} = m_{21} = 0,\ m_{20} < 0,\ m_{02} < 0,\ m_{11}^2 < 4m_{02}m_{20}$ or
(ii) $m_{22} < 0,\ 4m_{22}m_{02} > m_{12}^2,\ 4m_{20}m_{22} > m_{21}^2$.

Models satisfying (i) are classical bivariate normal with normal marginals, normal conditionals, linear regressions, and constant conditional variances. Models that satisfy (ii) have distinctively non-normal marginal densities, constant or nonlinear regressions, and bounded conditional variances.

6.2.5 Centered Normal Conditionals

Conditional Distributions

Suppose

$$X \mid (Y = y) \sim N(0, \sigma_1^2(y)) \text{ and } Y \mid (X = x) \sim N(0, \sigma_2^2(x)), \tag{6.7}$$

where $\sigma_1^2(y) > 0$ and $\sigma_2^2(x) > 0$ are two unknown functions. In fact, these conditionals are the special case of the normal conditionals in (6.1) with $\mu_1(y) = 0, \mu_2(x) = 0$. Or equivalently, the densities can be identified as that obtainable from (6.3) on setting $m_{01} = m_{10} = m_{21} = m_{12} = m_{11} = 0$.

Expression of the Joint Density

The joint density corresponding to the specification in (6.7) is

$$h(x,y) = k(c)\frac{1}{2\pi\sigma_1\sigma_2}\exp\left\{-\frac{1}{2}\left[\left(\frac{x}{\sigma_1}\right)^2 + \left(\frac{y}{\sigma_2}\right)^2 + c\left(\frac{x}{\sigma_1}\right)^2\left(\frac{y}{\sigma_2}\right)^2\right]\right\}, \tag{6.8}$$

where we have denoted $\sigma_1^2(y) = \frac{\sigma_1^2}{1+c\left(\frac{y}{\sigma_2}\right)^2}$ and $\sigma_2^2(x) = \frac{\sigma_2^2}{1+c\left(\frac{x}{\sigma_1}\right)^2}$.

Univariate Properties

The two marginal densities are

$$f(x) = k(c)\frac{1}{\sigma_1\sqrt{2\pi}}\frac{1}{\sqrt{1+c\left(\frac{x}{\sigma_1}\right)^2}}\exp\left[-\frac{1}{2}\left(\frac{x}{\sigma_1}\right)^2\right] \tag{6.9}$$

and

$$g(y) = k(c)\frac{1}{\sigma_2\sqrt{2\pi}}\frac{1}{\sqrt{1+c\left(\frac{y}{\sigma_2}\right)^2}}\exp\left[-\frac{1}{2}\left(\frac{y}{\sigma_2}\right)^2\right], \tag{6.10}$$

where

$$k(c) = \frac{\sqrt{2c}}{U(1/2, 1, 1/2c)},$$

with $U(a,b,c)$ being Kummer's hypergeometric function.

Remarks

- $\left(\frac{X}{\sigma_1}\right)\sqrt{1+c\left(\frac{Y}{\sigma_2}\right)^2} \sim N(0,1)$ and is independent of Y. Similarly, $\left(\frac{Y}{\sigma_2}\right)\sqrt{1+c\left(\frac{X}{\sigma_1}\right)^2} \sim N(0,1)$ and is independent of X.
- $\operatorname{corr}(X^2, Y^2) = \frac{1-2\delta(c)-4c\delta(c)+4c^2\delta^2(c)}{-1-2\delta(c)+4c^2\delta^2(c)}$, where $\delta(c) = \frac{k'(c)}{k(c)}$.

Applications

Arnold and Strauss (1991) considered 30 bivariate observations of slow-firing target data and fitted the centered normal conditionals model to them by using the maximum likelihood method.

References to Illustrations

Several density surface plots and contour plots of the normal conditionals and the centered normal conditionals models are given in Sections 3.4 and 3.5 of Arnold et al. (1999). Gelman and Meng (1991) produced graphs of three bivariate density functions that are not bivariate normal, including a bimodal joint density.

6.3 Conditionals in Exponential Families

Exponential family. An l_1-parameter family of $\{f_1(x;\boldsymbol{\theta}) : \boldsymbol{\theta} \in \Theta\}$ of the form

$$f_1(x;\boldsymbol{\theta}) = r_1(x)\beta_1(\boldsymbol{\theta}) \exp\left\{\sum_{i=1}^{l_1} \theta_i q_{1i}(x)\right\} \tag{6.11}$$

is called an *exponential family* of distributions. Here, Θ is the natural parameter space and the $q_{1i}(x)$'s are assumed to be linearly independent.

Let us consider another l_2-parameter family of $\{f_2(y;\boldsymbol{\tau}) : \boldsymbol{\tau} \in \Upsilon\}$ of the form

$$f_2(y;\boldsymbol{\tau}) = r_2(y)\beta_2(\boldsymbol{\tau}) \exp\left\{\sum_{j=1}^{l_2} \tau_j q_{2j}(y)\right\}, \tag{6.12}$$

where Υ is the natural parameter space and the $q_{2j}(y)$'s are assumed to be linearly independent.

Suppose we are given two conditional densities $f(x|y)$ and $g(y|x)$ such that $f(x|y)$ belongs to the family (6.11) for some $\boldsymbol{\theta}$ that may depend on y and $g(y|x)$ belongs to the family (6.12) for some $\boldsymbol{\tau}$ that may depend on x. It has been shown [see Arnold et al. (1999)] that the corresponding bivariate density is of the form

$$f(x,y) = r_1(x)r_2(y)\exp\{\mathbf{q}^{(1)}(x)M\mathbf{q}^{(2)}(y)\}, \tag{6.13}$$

where

$$\mathbf{q}^{(1)}(x) = (q_{10}(x), q_{11}(x), \ldots, q_{1l_1}(x)),$$
$$\mathbf{q}^{(2)}(y) = (q_{20}(y), q_{21}(y), \ldots, q_{2l_2}(y))$$

with $q_{10}(x) = q_{20}(y) \equiv 1$, and M is an $(l_1+1) \times (l_2+1)$ matrix of constant parameters. Of course, the density is subject to the usual requirement that $\int\int f(x,y)dx\,dy = 1$.

6.3.1 Dependence in Conditional Exponential Families

Let $\tilde{\mathbf{q}}^{(1)}$ and $\tilde{\mathbf{q}}^{(2)}$ denote $\mathbf{q}^{(1)}$ and $\mathbf{q}^{(2)}$ having their respective first element removed. Delete the first row and first column of M and denote the remaining matrix by $\tilde{M}$. Then, $f(x,y)$ is TP$_2$ if

$$[\tilde{\mathbf{q}}^{(1)}(x_1) - \tilde{\mathbf{q}}^{(1)}(x_2)]'\tilde{M}[\tilde{\mathbf{q}}^{(2)}(y_1) - \tilde{\mathbf{q}}^{(2)}(y_2)] \geq 0 \tag{6.14}$$

for every $x_1 < x_2$ and $y_1 < y_2$. Thus, if the $q_{1i}(x)$'s and the $q_{2j}(y)$'s are all increasing functions, then a sufficient condition for TP$_2$ and hence for non-negative correlation is that $\tilde{M} \geq 0$ (i.e., $m_{ij} \geq 0\ \forall i = 1, 2, \ldots, l_1,\ j = 1, 2, \ldots, l_2$). If $\tilde{M} \leq 0$, then negative correlation is assured. If the q_{1i}'s and q_{2j}'s are not monotone, then it is unlikely that any choice for $\tilde{M}$ will lead to a TP$_2$ density, and in such a setting it is quite possible to encounter both positive and negative correlations.

6.3.2 Exponential Conditionals

In this case, $l_1 = l_2 = 1, r_1(t) = r_2(t) = 1,\ t > 0$, and $q_{11}(t) = q_{21}(t) = -t$.

Conditional Distributions

The conditional densities are exponential, i.e.,

$$X \mid (Y = y) \sim \exp[(1 + cy/\sigma_2)/\sigma_1], \tag{6.15}$$

$$Y \mid (X = x) \sim \exp[(1 + cx/\sigma_1)/\sigma_2]. \tag{6.16}$$

Expression of the Joint Density

The joint density corresponding to the specification in (6.15) and (6.16) is

$$h(x,y) = \exp(m_{00} - m_{10}x - m_{01}y + m_{11}xy), \quad x > 0,\ y > 0. \tag{6.17}$$

A more convenient parametrization of this joint density is

$$h(x,y) = k(c)\exp\left[-\frac{x}{\sigma_1} - \frac{y}{\sigma_2} - \frac{cxy}{\sigma_1\sigma_2}\right], \quad x, y > 0,\ c > 0, \tag{6.18}$$

where the constant $k(c)$ is

$$k(c) = \frac{c\exp(-1/c)}{-\mathrm{Ei}(1/c)}, \tag{6.19}$$

in which Ei(·) is the exponential integral function, defined by $\mathrm{Ei}(u) = -\int_u^\infty v^{-1}e^{-v}dv$. [Beware of the lack of standardization of nomenclature and notation for functions such as this. For computation of this function, see Amos (1980).] The joint p.d.f. of (6.18) was first studied in Arnold and Strauss (1988a).

Univariate Properties

The marginal densities are

$$f(x) = \frac{k(c)}{\sigma_1\left(1+\frac{cx}{\sigma_1}\right)}e^{-x/\sigma_1}, \quad x > 0, \tag{6.20}$$

$$g(y) = \frac{k(c)}{\sigma_2\left(1+\frac{cy}{\sigma_2}\right)}e^{-y/\sigma_2}, \quad y > 0, \tag{6.21}$$

which are not exponential in form but $X(1 + cY/\sigma_2)/\sigma_1 \sim \mathrm{Exp}(1)$ and $Y(1 + cX/\sigma_1)/\sigma_2 \sim \mathrm{Exp}(1)$.

For $\sigma_1 = 1$, (6.20) reduces to

$$f = k(c)\frac{\exp(-x)}{1+cx}, \tag{6.22}$$

where k is as defined in (6.19) and similarly for $g(y)$.

Formula for Cumulative Distribution Function

Assuming $\sigma_1 = \sigma_2$, $\bar{H}$ may be written in a compact, though not elementary, form as

$$\bar{H} = \frac{\mathrm{Ei}(c^{-1} + x + y + cxy)}{\mathrm{Ei}(c^{-1})}, \tag{6.23}$$

and

$$h(x, y) = k(c)e^{-(x+y+cxy)}, \quad x, y > 0, \; c \geq 0.$$

Correlation Coefficients

Pearson's product-moment correlation coefficient is $\frac{c+k(c)-k(c)^2}{k(c)[1+c-k(c)]}$, where k is the same function of c as before. This is zero when $c = 0$, the case of independence, and it becomes increasingly negative with increasing c until it reaches approximately -0.32 at about $c = 6$ and then gets less negative, tending slowly to zero as $c \to \infty$.

Relation to Other Distributions

For a more general family, see Arnold and Strauss (1987), and for conditions on the sign of correlation obtainable with such a generalization, one may refer to Arnold (1987b).

Remarks

- Exponential conditional densities were first studied by Abrahams and Thomas (1984) and then (independently) by Arnold and Strauss (1988a). Consequently, it is easy to write down the regression equation; see Inaba and Shirahata (1986).
- With k as before, the joint moment generating function is

$$M(s,t) = \frac{k(c)}{(1-\sigma_1 s)(1-\sigma_2 t)k(\frac{c}{(1-\sigma_1 s)(1-\sigma_2 t)})} . \tag{6.24}$$

- The bivariate failure rate is increasing in both x and y, being given by (with $\sigma_1 = \sigma_2 = 1$)

$$(1+cx)(1+cy)k\left(\frac{c}{(1+cx)(1+cy)}\right) . \tag{6.25}$$

- Castillo and Galambos (1987b) have considered the case of Weibull conditionals. Their joint distribution can be obtained through the relationship

$$(W_1, W_2) = (X^{c_1}, Y^{c_2}).$$

- The distribution of the product XY was derived by Nadarajah (2006).

Fields of Application

As is often true with the exponential distributions, applications in reliability studies are envisaged. Inaba and Shirahata (1986) fitted this distribution to data on white blood cell counts and survival times of patients who died of acute myelogenous leukemia [Gross and Clark (1975, Table 3.3)], comparing it with the fit obtained from a bivariate normal distribution.

6.3.3 Normal Conditionals

This was dealt with in Section 6.2. Essentially, the normal conditionals belong to two-parameter exponential families with $l_1 = l_2 = 2$ and $r_1(t) = r_2(t) = 1$. Also,

$$\underline{q}^{(1)} = \underline{q}^{(2)}(t) = \begin{pmatrix} 1 \\ t \\ t^2 \end{pmatrix},$$

yielding a bivariate density of the form given in (6.3).

6.3.4 Gamma Conditionals

Gamma conditionals belong to exponential families with $l_1 = l_2 = 2$, $r_1(t) = r_2(t) = \frac{1}{t}$, $t > 0$, and $\mathbf{q}^{(1)}(t) = \mathbf{q}^{(2)}(t) = \begin{pmatrix} 1 \\ -t \\ -\log t \end{pmatrix}$.

Conditional Distributions

Suppose

$$X \mid (Y = y) \sim \Gamma(m_{20} + m_{22} \log y - m_{21} y, m_{10} - m_{11} y + m_{12} \log y)$$

and

$$Y \mid (X = x) \sim \Gamma(m_{02} + m_{22} \log x - m_{12} x, m_{01} - m_{11} x + m_{21} \log x).$$

Expression of the Joint Density

The corresponding joint density function is

$$h(x, y) = \frac{1}{xy} \exp \left\{ (1 \quad -x \quad \log x) \; M \begin{pmatrix} 1 \\ y \\ \log y \end{pmatrix} \right\}, \quad x > 0, \; y > 0. \quad (6.26)$$

Arnold et al. (1999) have listed six possible bivariate densities with requisite conditions such that (6.26) is a proper density function. They have been designated them as Model I, Model II, Model IIIA, Model IIIB, Model IV, and Model V.

Univariate Properties

The corresponding marginal density of X is

$$f(x) = \frac{1}{x}\frac{\Gamma(m_{02} + m_{22}\log x - m_{12}x)e^{m_{00}-m_{10}x+m_{20}\log x}}{(m_{01} - m_{11}x + m_{21}\log x)^{m_{02}+m_{22}\log x - m_{12}x}}, \quad x > 0, \tag{6.27}$$

and an analogous expression holds for $g(y)$.

Other Conditional Properties

The regression curves are generally nonlinear, and they are given by

$$E(X|Y = y) = \frac{m_{20} + m_{22}\log y - m_{21}y}{m_{10} + m_{12}\log y - m_{11}y} \tag{6.28}$$

and

$$E(Y|X = x) = \frac{m_{02} + m_{22}\log x - m_{12}x}{m_{01} + m_{21}\log x - m_{11}x}. \tag{6.29}$$

6.3.5 Model II for Gamma Conditionals

Conditional Distributions

Gamma conditionals Model II can be reparametrized so that

$$X \mid (Y = y) \sim \Gamma(r, (1 + cy/\sigma_2)/\sigma_1) \text{ and } Y \mid (X = x) \sim \Gamma(s, (1 + cx/\sigma_1)/\sigma_2). \tag{6.30}$$

Expression of the Joint Density

The joint density function corresponding to the specification in (6.30) is

$$h(x, y) = \frac{k_{r,s}(c)}{\sigma_1^r\sigma_2^s\Gamma(r)\Gamma(s)}x^{r-1}y^{s-1}\exp\left(-\frac{x}{\sigma_1} - \frac{y}{\sigma_2} - c\frac{xy}{\sigma_1\sigma_2}\right), \quad x, y > 0, \tag{6.31}$$

with $r, s > 0$, $\sigma_1, \sigma_2 > 0$, and $c \geq 0$, with $k_{r,s}(c)$ being the normalizing constant. $r, s > 0$ are shape parameters, σ_1 and σ_2 are scale parameters, and c is a dependence parameter such that $c = 0$ corresponds to the case of independence.

Univariate Properties

The corresponding marginal densities are

$$f(x) = \frac{k_{r,s}(c)}{\sigma_1^r \Gamma(r)}(1+cx/\sigma_1)^{-s} x^{r-1} e^{-x/\sigma_1}, \quad x > 0, \tag{6.32}$$

and

$$g(y) = \frac{k_{r,s}(c)}{\sigma_1^s \Gamma(s)}(1+cy/\sigma_2)^{-r} y^{s-1} e^{-y/\sigma_2}, \quad y > 0, \tag{6.33}$$

with

$$k_{r,s}(c) = \frac{c^r}{U(r, r-s+1, 1/c)},$$

where $U(a,b,z)$ is Kummer's confluent hypergeometric function defined by $U(a,b,z) = \frac{1}{\Gamma(a)} \int_0^\infty e^{-zt} t^{a-1}(1+t)^{b-a-1} dt$.

Correlation

It can be shown that the covariance is given by

$$\mathrm{cov}(X,Y) = \sigma_1\sigma_2[(r+s)c\delta_{r,s}(c) - rs + \delta_{r,s}(c) - c^2\delta_{r,s}^2(c)], \tag{6.34}$$

where $\delta_{r,s}(c) = \frac{\partial}{\partial c} \log k_{r,s}(c)$.

6.3.6 Gamma-Normal Conditionals

Conditional Distributions

Suppose

$$X \mid (Y=y) \sim \Gamma(m_{20} + m_{21}y + m_{22}y^2, m_{10} + m_{11}y + m_{12}y^2), \tag{6.35}$$

$$Y \mid (X=x) \sim N(\mu(x), \sigma^2(x)), \tag{6.36}$$

where

$$\mu(x) = \frac{m_{01} - m_{11}x + m_{21}\log x}{2(-m_{02} + m_{12}x - m_{22}\log x)},$$

$$\sigma^2(x) = \frac{1}{2}(-m_{02} + m_{12}x - m_{22}\log x)^{-1}.$$

Expression of the Joint Density

The joint density function corresponding to the specification in (6.35) and (6.36) is

$$h(x,y) = \frac{1}{x} \exp\left\{ (1 \quad -x \quad \log x)\, M \begin{pmatrix} 1 \\ y \\ y^2 \end{pmatrix} \right\}, \qquad x > 0,\ -\infty < y < \infty. \tag{6.37}$$

Models

Three models are possible, and they are labeled as Model I, Model II, and Model III by Arnold et al. (1999).

6.3.7 Beta Conditionals

Conditional Distributions

Suppose $X \mid (Y = x)$ and $Y \mid (X = x)$ belong to beta exponential families with

$$r_1(x) = \frac{1}{x(1-x)}, \quad r_2(y) = \frac{1}{y(1-y)}, \quad 0 < x,\ y < 1,$$

$$q_{11}(x) = \log x,\ q_{21}(y) = \log y,\ q_{12}(x) = \log(1-x),\ q_{22}(y) = \log(1-y).$$

Expression of the Joint Density

The corresponding joint density function is

$$\begin{aligned} h(x,y) = {} & \frac{1}{x(1-x)y(1-y)} \exp\{m_{11} \log x \log y + m_{12} \log x \log(1-y) \\ & + m_{21} \log(1-x) \log y + m_{22} \log(1-x) \log(1-y) + m_{10} \log x \\ & + m_{20} \log(1-x) + m_{01} \log y + m_{02} \log(1-y) + m_{00}\}, \\ & \text{for } 0 < x,\ y < 1, \end{aligned} \tag{6.38}$$

with parameters subject to several requirements, including m_{ij}, $i = 1, 2$, $j = 1, 2$. In order to guarantee integrability of the marginal distributions, we also require $m_{10} > 0, m_{20}, m_{01} > 0$, $m_{02} > 0$.

Other Conditional Properties

We have

$$E(X|Y=y) = \frac{m_{10} + m_{11}\log y + m_{12}\log(1-y)}{(m_{10}+m_{22}) + (m_{11}+m_{21})\log y + (m_{12}+m_{22})\log(1-y)} \tag{6.39}$$

and a similar expression for $E(Y|X=x)$.

6.3.8 Inverse Gaussian Conditionals

The inverse Gaussian conditionals model corresponds to the following choices for r's and q's in (6.13):

$$r_1(x) = x^{-3/2},\ x>0, \qquad r_2(y) = y^{-3/2},\ y>0,$$

$$q_{11}(x) = -x,\ q_{21} = -y,\ q_{12}(x) = -x^{-1},\ q_{22}(y) = -y^{-1}.$$

Conditional Distributions

We have

$$X \mid (Y=y) \sim \mathrm{IG}(m_{10} - m_{11}y - m_{12}y^{-1}, m_{20} - m_{21}y - m_{22}y^{-1}), \tag{6.40}$$

and consequently,

$$E(X|Y=y) = \sqrt{\frac{m_{20} - m_{21}y - m_{22}y^{-1}}{m_{10} - m_{11}y - m_{12}y^{-1}}}. \tag{6.41}$$

A similar expression for $Y \mid (X=x)$ can be presented.

In order to have proper inverse Gaussian conditionals and guarantee that the resulting marginal densities are integrable, we require that $m_{ij} \le 0, i = 1,2, j = 1,2$. In addition, we require

$$m_{10} > -2\sqrt{m_{11}m_{12}},\ m_{20} > -2\sqrt{m_{21}m_{22}},$$
$$m_{01} > -2\sqrt{m_{11}m_{21}},\ m_{02} > -2\sqrt{m_{12}m_{12}}.$$

Expression of the Joint Density

The corresponding joint density function is

$$h(x,y) = (xy)^{-3/2} \exp\left\{m_{11}xy + m_{12}xy^{-1} + m_{21}x^{-1}y + m_{22}x^{-1}y^{-1} - m_{10}x - m_{20}x^{-1} - m_{01}y - m_{02}y^{-1} + m_{00}\right\}, \quad x, y > 0. \tag{6.42}$$

6.4 Other Conditionally Specified Families

6.4.1 Pareto Conditionals

Conditional Distributions

Suppose

$$X \mid (Y=y) \sim P(\sigma_1(y), \alpha) \text{ and } Y \mid (X=x) \sim P(\sigma_1(y), \alpha), \tag{6.43}$$

where

$$\sigma_1(y) = \frac{\lambda_{00} + \lambda_{01}y}{\lambda_{10} + \lambda_{11}y}, \qquad \sigma_2(x) = \frac{\lambda_{00} + \lambda_{10}x}{\lambda_{01} + \lambda_{11}xy.} \tag{6.44}$$

Expression of the Joint Density

The joint density function corresponding to the specification in (6.43) is

$$h(x,y) = K(\lambda_{00} + \lambda_{10}x + \lambda_{01}y + \lambda_{11}xy)^{-(\alpha+1)}, \quad x, y \geq 0, \tag{6.45}$$

where $\lambda_{ij} \geq 0$, $\alpha > 0$, and the constant $1/K$ is expressible in terms of the Gauss hypergeometric function ${}_2F_1(a, b; c; z)$.

Univariate Properties

The marginals are not Pareto in form in general. Instead, that are

$$f(x) = K(\lambda_{01} + \lambda_{11}x)^{-1}(\lambda_{00} + \lambda_{10}x)^{-\alpha},$$
$$g(y) = K(\lambda_{10} + \lambda_{11}y)^{-1}(\lambda_{00} + \lambda_{01}y)^{-\alpha}. \tag{6.46}$$

Special Case: Mardia's Bivariate Pareto Distribution

Arnold et al. (1999) have considered three cases involving different constraints on α and the λ's. A special case in which $\alpha > 1$, $\lambda_{11} = 0$, and all other λ's are positive gives rise to a bivariate distribution with Pareto marginals and Pareto conditionals, with the joint density function

$$h(x, y) = \frac{(\alpha - 1)\alpha}{\sigma_1 \sigma_2} \left(1 + \frac{x}{\sigma_1} + \frac{y}{\sigma_2}\right)^{-(\alpha+1)} . \tag{6.47}$$

This special case is the bivariate Pareto distribution introduced by Mardia (1962).

Remarks

- Pareto conditional distributions are fully covered in Arnold (1987a).
- X is stochastically increasing (SI) or decreasing with Y depending on the sign of $(\lambda_{10}\lambda_{01} - \lambda_{00}\lambda_{11})$.
- $\text{sign}(\rho) = \text{sign}(\lambda_{10}\lambda_{01} - \lambda_{00}\lambda_{11})$, where ρ is Pearson's product-moment correlation coefficient.

6.4.2 Beta of the Second Kind (Pearson Type VI) Conditionals

Beta of the second kind is also known as the inverted beta or inverted Dirichlet distribution.

Conditional Distributions

Suppose

$$X \mid (Y = y) \sim B2(p, q, \sigma_1(y)) \text{ and } Y \mid (X = x) \sim B2(p, q, \sigma_2(x)), \tag{6.48}$$

where σ_i are as defined in (6.44).

Expression of the Joint Density

The joint density function corresponding to the specification in (6.48) is

$$h(x, y) = K \frac{x^{p-1} y^{p-1}}{(\lambda_{01} + \lambda_{10} x + \lambda_{01} y + \lambda_{11} xy)^{p+q}} , \tag{6.49}$$

where the reciprocal normalizing constant $J = K^{-1}$ is as presented in Table 6.1. It is required that $\lambda_{00}, \lambda_{11} \geq 0$ and $\lambda_{10}, \lambda_{01} > 0$.

Table 6.1 Reciprocals of the normalizing constant for beta of the second kind models

$\lambda_{00} = 0$	$J = \dfrac{B(p,q)B(p-q,q)}{\lambda_{10}^{q}\lambda_{01}^{q}\lambda_{11}^{p-q}}$
$\lambda_{11} = 0$	$J = \dfrac{B(p,q)B(p,q-p)}{\lambda_{10}^{q-p}\lambda_{01}^{q}\lambda_{11}^{p}}$
$\lambda_{00}, \lambda_{11} > 0$	$J = \dfrac{B(p,q)^2}{\lambda_{00}^{q-p}\lambda_{10}^{p}\lambda_{11}^{p}} {}_2F_1\left(p, p; p+q, 1 - \dfrac{1}{\theta}\right)$

Note: Here, ${}_2F_1(a, b; c; z)$ is the Gauss hypergeometric series.

Univariate Properties

The marginal densities are given by

$$f(x) \propto \frac{x^{p-1}}{(\lambda_{01} + \lambda_{11}x)^p(\lambda_{00} + \lambda_{10}x)^q}, \quad g(y) \propto \frac{y^{p-1}}{(\lambda_{01} + \lambda_{11}y)^p(\lambda_{00} + \lambda_{10}y)^q}.$$

Conditional Moments

The conditional moments can be shown to be

$$E(X^k|Y = y) = \frac{B(p+k, q-k)}{B(p,q)} \left(\frac{\lambda_{00} + \lambda_{01}y}{\lambda_{00} + \lambda_{11}y}\right)^k,$$

$$E(Y^k|X = x) = \frac{B(p+k, q-k)}{B(p,q)} \left(\frac{\lambda_{00} + \lambda_{01}x}{\lambda_{00} + \lambda_{11}x}\right)^k,$$

provided $q > k$.

Correlation Coefficient

The correlation coefficient is such that

$$\text{sign}(\rho) = \text{sign}(\lambda_{10}\lambda_{01} - \lambda_{00}\lambda_{11}),$$

just as in the Pareto case.

6.4.3 Generalized Pareto Conditionals

The generalized Pareto distribution is also known as a Burr type XII distribution.

Conditional Distributions

Suppose

$$X \mid (Y = y) \sim \mathcal{GP}(\sigma(y), \delta(y), \alpha(y)) \text{ and } Y \mid (X = x) \sim \mathcal{GP}(\tau(x), \gamma(x), \beta(x)). \tag{6.50}$$

Expression of the Joint Density

Assuming $\delta(y) = \delta, \gamma(x) = \gamma$, two classes of joint densities are obtained corresponding to the specification in (6.50).

Model I:

$$h(x, y) = x^{\delta-1}y^{\gamma-1}[\lambda_1 + \lambda_2 x^\delta + \lambda_3 y^\gamma + \lambda_4 x^\delta y^\gamma]^{\lambda_5}, \ x, y > 0, \tag{6.51}$$

and

Model II:

$$\begin{aligned} h(x, y) = x^{\delta-1}y^{\gamma-1} \exp\{&\theta_1 + \theta_2 \log(\theta_5 + x^\delta) + \theta_3 \log(\theta_6 + y^\gamma) \\ &+\theta_4 \log(\theta_5 + x^\delta) \log(\theta_6 + y^\gamma)\}, \quad x, y > 0. \end{aligned} \tag{6.52}$$

For Model I, we require $\lambda_5 < -1$ and $\lambda_1 \geq 0, \lambda_2 > 0, \lambda_3 > 0, \lambda_4 \geq 0$. For Model II, we require $\theta_5, \theta_6 > 0$, $\theta_2, \theta_3 \leq -1$, and $\theta_4 \leq 0$.

Univariate Properties

For Model I, the marginal densities are

$$f(x) = \frac{1}{\delta(-1 - \lambda_5)} x^{\delta-1}(\lambda_3 + \lambda_4 x^\delta)^{-1}(\lambda_1 + \lambda_2 x^\delta)^{\lambda_5+1}, \quad x > 0, \tag{6.53}$$

$$g(y) = \frac{1}{\gamma(-1 - \lambda_5)} y^{\gamma-1}(\lambda_2 + \lambda_4 y^\gamma)^{-1}(\lambda_1 + \lambda_3 y^\gamma)^{\lambda_5+1}, \quad y > 0, \tag{6.54}$$

where we have let $\alpha(y) = \beta(x) = -1 - \lambda_5$. The marginal densities for Model II can be obtained similarly.

6.4.4 Cauchy Conditionals

Conditional Distributions

Suppose

$$\begin{aligned} Y \mid (X = x) &\sim C(\mu_2(x), \sigma_2(x)), \quad \sigma_2(x) > 0, \\ X \mid (Y = y) &\sim C(\mu_1(y), \sigma_1(y)), \quad \sigma_1(x) > 0. \end{aligned} \tag{6.55}$$

Expression of the Joint Density

Let $M = (m_{ij})$, $i, j = 0, 1, 2$, be a matrix of arbitrary constants. Then, two possible classes are discussed in Arnold et al. (1999).

(i) The class with $m_{22} = 0$ leads to an improper distribution

$$h(x, y) \propto (m_{00} + m_{10}x + m_{01}y + m_{20}x^2 + m_{02}y^2 + m_{11}xy)^{-1}.$$

(ii) The class with $m_{22} > 0$, in general, has densities that are quite complex. However, a special case with $m_{10} = m_{01} = m_{11} = m_{12} = m_{21} = 0$ gives

$$h(x, y) = K \frac{1}{m_{00} + m_{20}x^2 + m_{02}y^2 + m_{22}x^2y^2}, \tag{6.56}$$

where

$$K^{-1} = I = \frac{2\pi}{\sqrt{m_{20}m_{02}}} F\left(\frac{\pi}{2}/\alpha\right),$$

with α satisfying the relation

$$\sin^2 \alpha = \frac{m_{20}^2 m_{02}^2 - m_{00}^2 m_{22}^2}{m_{20}^2 m_{02}^2};$$

here, $F(\frac{\pi}{2}/\alpha)$ is the complete elliptical integral of the first kind, which has been tabulated in Abramowitz and Stegun (1994, pp. 608–611). The conditional scale parameters are

$$\sigma_2(x) = \sqrt{\frac{m_{00} + m_{20}x^2}{m_{02} + m_{22}x^2}} \quad \text{and} \quad \sigma_1(y) = \sqrt{\frac{m_{00} + m_{20}y^2}{m_{02} + m_{22}y^2}}.$$

Univariate Properties

The marginals densities are

$$f(x) \propto \frac{1}{\sqrt{(m_{00} + m_{20}x^2)(m_{02} + m_{22}x^2)}},$$
$$g(y) \propto \frac{1}{\sqrt{(m_{00} + m_{20}y^2)(m_{02} + m_{22}y^2)}}. \tag{6.57}$$

Transformation

If $U = \log X$ and $V = \log Y$, then the joint density of U and V is

$$h_{U,V}(u, v) \propto (\alpha e^{-x-y} + \beta e^{-x+y} + \gamma e^{x-y} + \delta e^{x+y})^{-1} \tag{6.58}$$

for $\alpha, \beta, \gamma\delta > 0$.

6.4.5 Student t-Conditionals

Conditional Distributions

Suppose

$$X \mid (Y = y) \sim \mu_1(y) + \sigma_1(y)T_\alpha \text{ and } Y \mid (X = x) \sim \mu_2(x) + \sigma_2(x)T_\alpha, \tag{6.59}$$

where $\sigma_i > 0$, and T_α denotes a Student t-variable with parameter α.

Expression of the Joint Density

The joint density function corresponding to the specification in (6.59) is

$$h(x, y) \propto [(1\ x\ x^2)M(1\ y\ y^2)']^{-(\alpha+1)/2}. \tag{6.60}$$

The location and scale parameters for the conditional densities are given by

$$\mu_1(y) = -\frac{1}{2} \times \frac{b_1(y)}{c_1(y)},$$
$$\mu_2(x) = -\frac{1}{2} \times \frac{\tilde{b}_1(y)}{\tilde{c}_1(y)},$$

and

$$\sigma_1^2(y) = \frac{4a_1(y)c_1(y) - b_1^2(y)}{4\alpha c_1^2(y)},$$
$$\sigma_2^2(x) = \frac{4\tilde{a}_1(y)\tilde{c}_1(y) - \tilde{b}_1^2(y)}{4\alpha\tilde{c}_1^2(y)}.$$

Univariate Properties

The corresponding marginal densities are

$$f(x) \propto \frac{[\tilde{c}_1(x)]^{(\alpha-1)/2}}{[4\tilde{a}_1(x)\tilde{c}_1(x) - \tilde{b}_1^2(x)]^{\alpha/2}}, \tag{6.61}$$

$$g(x) \propto \frac{[c_1(y)]^{(\alpha-1)/2}}{[4a_1(y)c_1(x) - b_1^2(y)]^{\alpha/2}}. \tag{6.62}$$

6.4.6 Uniform Conditionals

Conditional Distributions

Suppose

$$\begin{aligned} X \mid (Y=y) &\sim U(\phi_1(y),\ \phi_2(y)),\ c<y<d,\ \phi_1(y) \le \phi_2(y), \\ Y \mid (X=x) &\sim U(\psi_1(x),\ \psi_2(x)),\ a<x<b,\ \psi_1(x) \le \psi_2(x), \end{aligned} \tag{6.63}$$

where ϕ and ψ are either both decreasing or both increasing, and that the two domains $N_\phi = \{(x,y) : \phi_1(y) < x < \phi_2(y), c < y < d\}$ and $N_\psi = \{(x,y) : \psi_1(x) < y < \psi_2(x), a < x < b\}$ are coincident, so that the compatibility conditions are satisfied.

Expression of the Joint Density

The joint density function corresponding to the specification in (6.63) is

$$h(x,y) = \begin{cases} k & \text{if } (x,y) \in N_\psi \\ 0 & \text{otherwise} \end{cases}, \tag{6.64}$$

where k^{-1} = area of $N_\psi = \int_a^b [\psi_2(x) - \psi_1(x)]dx = \int_a^b [\phi_2(y) - \phi_1(y)]dy < \infty$.

Univariate Properties

The corresponding marginal densities are

$$\begin{aligned} f(x) &= k[\psi_2(x) - \psi_1(x)], \quad a<x<b, \\ g(y) &= k[\phi_2(y) - \phi_1(y)], \quad c<y<d. \end{aligned} \tag{6.65}$$

6.4.7 Translated Exponential Conditionals

A random variable X has a translated exponential distribution if

$$\Pr(X > x) = e^{-\lambda(x-\alpha)}, \quad x > \alpha,$$

where $\lambda > 0$ and $\alpha \in (-\infty, \infty)$, and is denoted by $X \sim \exp(\alpha, \lambda)$.

Conditional Distributions

Suppose

$$X \mid (Y = y) \sim \exp(\alpha(y), \lambda(y)), \quad y \in S(Y), \tag{6.66}$$

and

$$Y \mid (X = x) \sim \exp(\beta(x), \gamma(x)), \quad x \in S(X), \tag{6.67}$$

where $S(X)$ and $S(Y)$ denote the supports of X and Y, respectively. For compatibility, we must assume that

$$D = \{(x, y) : \alpha(y) < x\} = \{(x, y) : \beta(x) < y\}. \tag{6.68}$$

Expression of the Joint Density

The joint density function corresponding to the specifications in (6.66) and (6.67) is

$$h(x, y) = \exp(d + cx - by - axy), \quad (x, y) \in D, \tag{6.69}$$

where $\gamma(x) = ax + b$, $\lambda(y) = ay - c$, $\beta = \alpha^{-1}$, and d is part of the normalizing constant.

Univariate Properties

The corresponding marginal densities are

$$f(x) = \frac{\exp[cx + d - (ax + b)\beta(x)]}{ax + b}, \quad x \in S(X), \tag{6.70}$$

and

$$g(y) = \frac{\exp[-by + d - (ay - c)\alpha(y)]}{ay - c}, \quad x \in S(X). \tag{6.71}$$

Other Regression Properties

The regression curves are given by

$$E(X|Y=y) = \alpha(y) + (ay-c)^{-1}, \quad y \in S(Y), \tag{6.72}$$

and

$$E(Y|X=x) = \beta(x) + (ax+b)^{-1}, \quad x \in S(X). \tag{6.73}$$

6.4.8 Scaled Beta Conditionals

Conditional Distributions

Suppose

$$\begin{aligned} Y \mid (X=x) &\sim (1-x)B(\alpha_1(x), \beta_1(x)), \quad 0<x<1, \\ X \mid (Y=y) &\sim (1-y)B(\alpha_2(x), \beta_2(x)), \quad 0<y<1. \end{aligned} \tag{6.74}$$

Expression of the Joint Density

The joint density function corresponding to the specification in (6.74) is

$$h(x,y) \propto x^{\theta_1-1} y^{\theta_2-1} (1-x-y)^{\theta_3-1} e^{\eta \log x \log y}, \quad x,y \geq 0,\ x+y \leq 1, \tag{6.75}$$

for $\theta_1, \theta_2, \theta_3 > 0$, $\eta \leq 0$, except that if $\eta < 0$ and $\theta_3 > 1, \theta_1$ and θ_2 can be zero, with the support being that part of the unit square wherein $x+y \leq 1$, and $\theta_1, \theta_2, \theta_3 > 0, \eta \leq 0$.

Univariate Properties

The marginals are not beta in form (unless $\eta = 0$). The expressions of the marginal densities are rather complicated; see James (1975). The beta (Dirichlet) distribution is characterized by being that member of this family with at least one of the marginals as univariate beta.

Remarks

This is the distribution with both sets of conditional densities (of Y given $X=x$ and of X given $Y=y$) beta. It is due to James (1975); see also James (1981, pp. 133–134).

Another Distribution

The distribution above interprets the requirement for the conditional distributions to be beta as follows: beta distributions over the range 0 to $1-x$ (for Y) or $1-y$ (for X), but with the parameters being functions of x or y.

Instead, we might interpret it as a beta distribution with some particular distributions with some particular constant exponent but the range being a function of x or y. Abrahams and Thomas (1984) have shown in this case that the joint density must either be $\frac{\Gamma(\theta_1+\theta_2\theta_3)}{\Gamma(\theta_1)\Gamma(\theta_2)\Gamma(\theta_3)}x^{\theta_1-1}y^{\theta_2-1}(1-x-y)^{\theta_3-1}$ or proportional to $(x+y)^{\theta_1-1}(1-x-y)^{\theta_3}$ (the support of which is that part of the unit square wherein $x+y\le 1$ and having uniform marginals).

6.5 Conditionally Specified Bivariate Skewed Distributions

The development of these models was considered in Arnold et al. (2002).

The basic skewed normal density takes the form

$$f(x,\lambda)=2\phi(x)\Phi(\lambda x),\quad -\infty<x<\infty,$$

where $\phi(x)$ and $\Phi(x)$ denote the standard normal density and the distribution functions and where λ is a parameter that governs the skewness of the density. If X has the density above, we then write $X\sim SN(\lambda)$.

6.5.1 Bivariate Distributions with Skewed Normal Conditionals

Assume $X|(Y=y)\sim SN(\lambda^{(1)}(y))$ and $Y|(X=x)\sim SN(\lambda^{(2)}(x))$, for some functions $(\lambda^{(1)}(y))$ and $(\lambda^{(2)}(x))$. Then there must exist densities $f(x)$ and $g(y)$ such that

$$h(x,y)=2\phi(x)\Phi(\lambda^{(1)}(y)x)g(y)=2\phi(y)\Phi(\lambda^{(2)}(x)y)f(x). \tag{6.76}$$

Arnold et al. (2002) identified two types of solutions that satisfy the functional equation (6.76):

Type I. (Independence). If $\lambda^{(1)}(y)=\lambda^{(1)}$ and $\lambda^{(2)}(x)=\lambda^{(2)}$, then

$$f(x)=2\phi(x)\Phi(\lambda^{(2)}x);\ g(y)=2\phi(y)\Phi(\lambda^{(1)}y)$$

and

$$h(x,y)=4\phi(x)\phi(y)\Phi(\lambda^{(2)}x)\Phi(\lambda^{(1)}y). \tag{6.77}$$

The joint density (6.77) is a proper (integrable) model.

Type II. (Dependent case). If $\lambda^{(1)}(y) = \lambda y$ and $\lambda^{(2)}(x) = \lambda x$, then

$$f(x) = \phi(x); \; g(y) = \phi(y)$$

and

$$h(x, y) = 2\phi(x)\phi(y)\Phi(\lambda xy). \tag{6.78}$$

The joint density (6.78) is also a proper (integrable) model.

Univariate Properties

Both X and Y are normally distributed.

Conditional Properties

The expression $h(x, y)$ has skewed normal conditionals. The corresponding regression functions are nonlinear, with the form

$$E(X|Y = y) = \sqrt{\frac{2}{\pi}} \times \frac{\lambda y}{\sqrt{1 + \lambda^2 y^2}},$$

$$E(Y|X = x) = \sqrt{\frac{2}{\pi}} \times \frac{\lambda x}{\sqrt{1 + \lambda^2 x^2}}.$$

Correlation Coefficient

Pearson's correlation coefficient is given by

$$\rho(X, Y) = \operatorname{sign}(\lambda) \times \frac{U(3/2, 2, 1/2\lambda^2)}{2\lambda^2\sqrt{\pi}},$$

where $U(a, b, z)$ represents the *confluent hypergeometric* function, defined as

$$U(a, b, z) = \frac{1}{\Gamma(a)} \int_0^\infty e^{-zt} t^{a-1} (1 + t)^{b-a-1} dt,$$

in which $b > a > 0$ and $z > 0$. It can be verified that $|\rho(X, Y)| \leq 0.63662$.

6.5.2 Linearly Skewed and Quadratically Skewed Normal Conditionals

Arnold et al. (2002) also considered bivariate distributions having conditional densities of the linearly skewed normal conditionals. More generally, bivariate distributions with quadratically and polynomially skewed normal conditionals were also investigated.

6.6 Improper Bivariate Distributions from Conditionals

Recall that the necessary and sufficient compatibility conditions for two conditionally specified distributions were as follows:

(i) $\{(x, y) : f(x|y) > 0\} = \{(x, y) : g(y|x) > 0\}$.
(ii) $f(x|y)/g(y|x) = a(x)b(y)$.
(iii) $a(x)$ in (ii) must be integrable.

An improper bivariate distribution may nevertheless be useful. Arnold et al. (1999, p. 133) have stated that, "In several potential situations, compatibility fails because Condition (ii) is not satisfied. Such 'improper' models may have utility for predictive purposes and in fact are perfectly legitimate models if we relax the finiteness condition in our definition of probability. Many subjective probabilists are willing to make such an adjustment (they can thus pick an integer at random). Another well-known instance in which the finiteness condition could be relaxed with little qualm is associated with the use of improper priors in Bayesian analysis. In that setting, both sets of conditional densities (the likelihood and the posterior) are integrable non-negative densities but for one marginal (prior), and therefore both marginals are non-negative but nonintegrable. For many researchers, these 'improper' models are perfectly possible. All that is required is that $f(x|y)$ and $f(y|x)$ be non-negative and satisfy (i) and (ii). Integrability is not a consideration. A simple example (mentioned in Chapter 1) will help visualize the situation."

Chapter 6 of Arnold et al. (1999) presents several improper bivariate distributions arising from conditionally specified models including certain uniform conditionals as well as exponential-Weibull conditionals. We refer the interested reader to this source.

6.7 Conditionals in Location-Scale Families with Specified Moments

Arnold et al. (1999) have considered conditionals in unspecified families with specified conditional moments. The discussion is based on the work by Narumi (1923a,b), who sought joint densities whose conditionals satisfy

$$f(x|y) = g_1\left(\frac{x-a(y)}{c(y)}\right)\frac{1}{c(y)}, \tag{6.79}$$

$$g(y|x) = g_2\left(\frac{y-b(x)}{b(x)}\right)\frac{1}{d(x)}, \tag{6.80}$$

where $a(y)$ and $b(x)$ are the regression curves and $c(y)$ and $d(x)$ are scedastic curves of X on Y and Y on X, respectively. Two cases have been presented by Arnold et al. (1999, p. 154) and are discussed below.

Case (i) Linear Regressions and Conditional Standard Deviations

We assume

$$a(y) = a_0 + a_1 y,\ b(x) = b_0 + b_1 x,\ c(y) = 1 + cy,\ d(x) = 1 + dx.$$

Narumi (1923a,b) has shown that the joint density function in this case must be of the form

$$h(x,y) = (\alpha + x)^{p_1}(\beta + y)^{p_2}(\gamma + \delta_1 x + \delta_2 y)^q, \quad x, y > 0. \tag{6.81}$$

Case (ii) Linear Regressions and Quadratic Conditional Variances

We assume

$$\begin{aligned} a(y) &= a_0 + a_1 y,\ b(x) = b_0 + b_1 x, \\ c(y) &= \sqrt{1 + c_1 y + c_2 y^2},\ d(x) = \sqrt{1 + d_1 x + d_2 x^2}. \end{aligned}$$

The joint density function in this case is necessarily of the form

$$h(x,y) = (\alpha + \beta x\gamma y + \delta_1 x^2 + \delta_2 xy + \delta_3 y^2)^{-\gamma}. \tag{6.82}$$

6.8 Given One Family of Conditional Distributions and the Regression Function for the Other

6.8.1 *Assumptions and Specifications*

Suppose we are given a family of conditional densities

$$f(x|y) = a(x,y), \quad x \in S(X),\ y \in S(Y), \tag{6.83}$$

and a regression function

$$E(Y|X=x)=\psi(x), \quad x\in S(X). \tag{6.84}$$

Obviously, questions on compatibility and uniqueness of the joint density arise. Several partial answers to those questions have been provided in the literature. Here, we simply present the following theorem due to Wesolowski (1995).

6.8.2 Wesolowski's Theorem

If (X, Y) is a pair of absolutely continuous random variables with $S(X) = S(Y) = (0,\infty)$ and if, for every $y>0$, $X|(Y=y) \sim P\left(\frac{a+by}{1+cy}, \alpha\right)$, where $\alpha \geq 0$, $a \geq 0$, $b>0$, $c\geq 0$, $\alpha>0$, then the distribution is uniquely determined by $E(Y|X=x)=\psi(x)$, $x>0$.

Example 6.10 (Pareto conditionals). If $E(Y|X=x) = \frac{a+x}{(\alpha-1)(b+cx)}$, then (X, Y) must have a Pareto conditionals distribution. If $c=0$, we then have Mardia's bivariate Pareto distribution.

Example 6.11 (Exponential conditionals). We have

$$f(x|y)=a(x,y)=(y+\delta)e^{-(y+\delta)x}, \quad x>0, \tag{6.85}$$

with

$$E(Y|X=x)=\psi(x), \quad x>0,$$

where $\exp[-\int_0^{\psi}(u)du]$ is a Laplace transform; for example, $\psi(x)=(\gamma+x)^{-1}$.

6.9 Estimation in Conditionally Specified Models

In this section, we aim to summarize estimation methods used for conditionally specified models. Because of some special difficulties, several of the techniques are tailor-made for these models. One of the main obstacles is the presence of the normalizing constant m_{00}, which is chosen to make the density integrate to 1. Unfortunately, m_{00} is often an intractable function of the other parameters. In some cases, an explicit expression is available; for example, in the exponential conditionals density and the Pareto conditionals density.

Chapter 9 of Arnold et al. (1999) has outlined the following methods:

- **Maximum likelihood estimate.** The maximum likelihood estimate $\hat{\boldsymbol{\theta}}$ of $\boldsymbol{\theta}$ satisfies

$$\prod_{i=1}^{n} h(X_i, Y_i; \hat{\boldsymbol{\theta}}) = \max_{\boldsymbol{\theta} \in \Theta} \prod_{i=1}^{n} h(X_i, Y_i; \boldsymbol{\theta}). \tag{6.86}$$

Two examples are presented: (i) Centered normal conditionals distribution and (ii) bivariate Pareto conditionals distribution. The method works better when the resulting joint distributions are themselves exponential families of bivariate densities.

- **Pseudolikelihood estimate.** The method is due to Arnold and Strauss (1988b). The technique involves a pseudolikelihood function that does not involve the normalizing constant. The pseudolikelihood estimate of $\boldsymbol{\theta}$ is to maximize the function

$$\prod_{i=1}^{n} f(X_i|Y_i; \boldsymbol{\theta}) g(Y_i|X_i; \boldsymbol{\theta}) \tag{6.87}$$

over the parameter space Θ.

Arnold and Strauss (1988b) have shown that the resulting estimate is consistent and asymptotically normal with a potentially computable asymptotic variance. In exchange for simplicity in calculation (since the conditionals and hence the pseudolikelihood do not involve the normalizing constant), we pay the price in slightly reduced efficiency. The centered normal conditionals distribution has been used to illustrate this method.

- **Marginal likelihood estimate.** It is the unique value of $\boldsymbol{\theta}$ that maximizes the function

$$\prod_{i=1}^{n} f(X_i; \boldsymbol{\theta}) \prod_{i=1}^{n} g(Y_i; \boldsymbol{\theta}) \tag{6.88}$$

over the parameter space Θ.

Castillo and Galambos (1985) have reported on successful use of this approach for the eight parameters of the normal conditionals model given in (6.2).

- **Moment estimate.** This method is very well known. Assuming $\boldsymbol{\theta} = (\theta_1, \ldots, \theta_k)$, we choose k functions $\phi_1, \ldots, \phi_k$ such that

$$E_{\boldsymbol{\theta}}(\phi_i(\mathbf{X})) = g_i(\boldsymbol{\theta}), \ \mathbf{X} = (X, Y), \quad i = 1, 2, \ldots, k. \tag{6.89}$$

We then set up k equations

$$g_i(\boldsymbol{\theta}) = \frac{1}{n} \sum_{j=1}^{n} \phi_i(\mathbf{X}_j), \ \text{ with } \ \mathbf{X}_j = (X_j, Y_j), \quad i = 1, 2, \ldots, k, \tag{6.90}$$

and solve for $\boldsymbol{\theta}$. To avoid repeated recomputations of the normalizing constant, Arnold and Strauss (1988a) treated this constant as an additional parameter θ_0 and set up an additional moment equation. The following three examples have been given: (i) exponential conditionals distribution,

(ii) centered normal conditionals distribution, and (iii) gamma conditionals distribution Model II.

- **Bayesian estimate and pseudo-Bayes approach.** These two approaches have been described in Section 9.9 of Arnold et al. (1999).

6.10 McKay's Bivariate Gamma Distribution and Its Generalization

We now present two examples of a bivariate distribution where both conditionals and both marginals are specified.

6.10.1 Conditional Properties

$Y-x$ conditional on $(X = x)$ has a gamma distribution with shape parameter q, and X/y conditional on $(Y = y)$ has a beta distribution with parameters p and q.

6.10.2 Expression of the Joint Density

The corresponding joint density function is

$$h(x,y) = \frac{a^{p+q}}{\Gamma(p)\Gamma(q)} x^{p-1}(y-x)^{q-1}e^{-ay}, \quad y > x > 0 \tag{6.91}$$

(i.e., the support is a wedge that is half of the positive quadrant), where $a, p, q > 0$. More details on this distribution can be found in Section 8.17.

6.10.3 Dussauchoy and Berland's Bivariate Gamma Distribution

This reduces to McKay's bivariate gamma distribution when $a_1 = a_2 = \beta = 1$. The support is the wedge $y > \beta x > 0$, and the joint density in this case is

$$\frac{\beta a_2^{t_2}}{\Gamma(l_1)\Gamma(l_1 - l_2)}(\beta x)^{l_1 - 1} \exp(-a_2 x)(y - \beta x)^{l_2 - l_1 - 1} \exp\left[-\frac{a_2}{\beta}(y - \beta x)\right]$$
$$\times\ {}_1F_1\left[l_1, l_2 - l_1; \left(\frac{a_1}{\beta} - a_2\right)(y - \beta x)\right],\ \beta \geq 0,\ 0 < a_2 \leq \frac{a_1}{\beta},\ 0 < l_1 < l_2,$$

where ${}_1F_1$ is the confluent hypergeometric function. More details on this distribution can be found in Section 8.18.

Some Variants of Distribution

We now summarize in Table 8.1 some variations on the theme of Y necessarily being positive and X necessarily being 0 and y.

Table 6.2 Distributions specified by marginal and conditional

Reference	Distribution of Y	Distribution of X, given $Y = y$
McKay (1934)	Gamma	Beta over $(0, y)$
Mihram and Hultquist (1967)	Stacy	Beta over $(0, y)$
Block and Rao (1973)	generalized inverted beta*	Beta over $(0, y)$
Ratnaparkhi (1981)†	Stacy, Pareto, or lognormal	Beta or log-gamma over $(0, y)$

* Density $\propto y^{\alpha-1}(1 + y^c)^{-k}$.
† In Ratnaparkhi's paper, the roles of X and Y were reversed from those here.

6.11 One Conditional and One Marginal Specified

6.11.1 Dubey's Distribution

Dubey (1970) gave some properties of the distribution constructed by supposing (i) that Y has a gamma distribution, and (ii) conditional on $Y = y, X$ has a gamma distribution, with constant shape parameter and mean inversely proportional to y.

6.11.2 Blumen and Ypelaar's Distribution

Expression of the Joint Density

The joint density function is

$$h(x, y) = x^{\alpha} y^{x^{\alpha}-1}, \quad x, y \geq 0. \tag{6.92}$$

Univariate Properties

X is uniformly distributed over the range 0 to 1, but this is not true for Y.

Conditional Properties

Conditional on $X = x$, the cumulative distribution of Y is $y^{x^{\alpha}}$.

Remarks

It seems that the motivation of Blumen and Ypelaar (1980) for constructing this distribution was to obtain one that is (i) tractable for studying the properties of Kendall's tau and (ii) reasonably similar to the bivariate normal (after appropriate transformations of the marginals).

6.11.3 Exponential Dispersion Models

Jørgensen (1987) studied general properties of the class of exponential dispersion models that is the multivariate generalization of the error distribution of generalized linear models. Although this is outside our scope, we note that its Section 5 concerns combining a conditional and a marginal distribution, both being exponential dispersion models, to obtain a higher-dimensional exponential dispersion model.

We may add that in the Discussion of Jørgensen's (1987) paper, Seshadri (1987) has mentioned obtaining a bivariate exponential dispersion model with gamma marginals.

6.11.4 Four Densities of Barndorff-Nielsen and Blæsild

We note that, in the course of studying reproductive exponential models, Barndorff-Nielsen and Blæsild (1983) wrote out four examples of bivariate densities constructed by the conditional approach:

Table 6.3 Four densities of Barndorff-Nielsen and Blæsild

Distribution of X	Distribution of Y given $X = x$	Example no.
Exponential	Inverse Gaussian	1.1
Inverse Gaussian	Normal	4.1
Inverse Gaussian	Inverse Gaussian	4.2
Inverse Gaussian	Gamma	4.3

The inverse Gaussian/inverse Gaussian example is also considered by Barndorff-Nielsen (1983, pp. 306–361), who remarked that a special case of it (with two of the four parameters being zero) can be said to be bivariate stable of index $(\frac{1}{2}, \frac{1}{4})$, as when a sample of size n is taken, the distribution of $(n^{-2} \sum x_i, n^{-4} \sum y_i)$ is the same whatever n is.

6.11.5 Continuous Bivariate Densities with a Discontinuous Marginal Density

The conditional approach was used by Romano and Siegel (1986, Section 2.15) to construct (for the fun of it!) a continuous distribution, and Y (conditional on $X = x$) has a normal distribution with mean $1/x$ and constant variance. The density is 0 for $x \leq 0$ and is proportional to $\exp[-x - \frac{1}{2}(y - x^{-})^2]$ for $x > 0$. Romano and Siegel then showed that $h(x, y)$ is continuous everywhere in the plane, but the marginal density $f(x)$, with its jump at $x = 0$, is not continuous.

Also, Clarke (1975) considered a joint density being proportional to $|x| \exp[-(|x| + x^2y^2/2)]$, which is continuous. The marginal density of X turns out to be $e^{-|x|}/2$ if $x \neq 0$ but is 0 if $x = 0$. This example is also in Székely (1986, pp. 216–217). Clarke also constructed an example in which $h(x, y)$ is continuous everywhere but $f(x)$ is nowhere continuous.

6.11.6 Tiku and Kambo's Bivariate Non-normal Distribution

Expression of the Joint Density

The joint density function is

$$h(x,y) = C\frac{1}{\sqrt{k\sigma_1^2\sigma_2^2(1-\rho^2)}}\exp\left[-\frac{1}{2\sigma_1^2(1-\rho^2)}\left\{x-\mu_1-\frac{\rho\sigma_1}{\sigma_2}(y-\mu_2)\right\}^2\right]$$
$$\times\left\{1+\frac{(y-\mu_2)^2}{k\sigma_2^2}\right\}^{-p}, \tag{6.93}$$

where C is the normalizing constant.

Conditional Properties

X given $Y=y$ is normally distributed. More explicitly, it is the conditional distribution that is associated with the bivariate normal density with correlation coefficient ρ and marginal means μ_1, μ_2 and marginal variances σ_1^2, σ_2^2.

Univariate Properties

Y has a Student t-distribution with density

$$g(y) \propto (k\sigma_2^2)^{1/2}\left\{1+\frac{(y-\mu_2)^2}{k\sigma_2^2}\right\}^{-p},$$

where $k = 2p-3$ and $p \geq 2$.

The marginal distribution of X is unknown, however.

Moments

Let $\mu_{i,j}$ be the cross-product central moment of order $i+j$; all odd order moments are zero, and the first few even order moments are as follows:

$$\mu_{2,0} = \sigma_1^2, \quad \mu_{1,1} = \rho\sigma_1\sigma_2, \quad \mu_{0,2} = \sigma_2^2,$$
$$\mu_{4,0} = 3\sigma_1^4 \left\{1 + \frac{2\rho^4}{2p-5}\right\}, \quad \mu_{3,1} = 3\rho\sigma_1^3\sigma_2 \left\{1 + \frac{2\rho^2}{2p-5}\right\},$$
$$\mu_{0,4} = \frac{3(2p-3)}{2p-5}\sigma_2^4, \quad \mu_{2,2} = \sigma_1^2\sigma_2^2 \left\{1 + 2\rho^2 + \frac{6\rho^2}{2p-5}\right\},$$
$$\mu_{1,3} = 3\rho\sigma_1\sigma_2^3 \left\{1 + \frac{2}{2p-5}\right\}.$$

Derivation

Tiku and Kambo (1992) derived this distribution by replacing one of the two marginal distributions in a bivariate normal by a symmetric distribution (related to the t-distribution), resulting in a symmetric bivariate distribution.

Remarks

For the estimation of parameters of this model, one may refer to Tiku and Kambo (1992).

6.12 Marginal and Conditional Distributions of the Same Variate

For bivariate distributions, it is common to combine marginal and/or conditional densities to describe the joint density $h(x, y)$. It is well known that, given the marginal density $f(x)$ of X and the conditional density $g(y|x)$ of Y given $X = x$, there exists a unique joint density $h(x, y) = f(x)g(y|x)$. We have devoted the major part of this chapter to discussing bivariate distributions when both conditional densities are specified. This section describes a different kind of conditional specification.

A paper that is different is that of Seshadri and Patel (1963), which gave some theoretical results on the extent to which knowledge of the marginal distribution of one variate together with knowledge of the conditional distributions of the same variate serves to determine the bivariate distribution. Can we characterize the joint density if we are given one marginal density, say $f(x)$, and the "wrong" family of conditional densities; i.e., $f(x|y)$, $y \in S(Y)$? The answer to this question is "sometimes." We now explore this problem in a general setting. Suppose we are given two functions, $u(x)$ and $a(x, y)$, and we ask ourselves whether there exists a compatible distribution for (X, Y) such that

$$f(x) = u(x), \quad \forall x \in S(X), \tag{6.94}$$

and, for each $y \in S(Y)$,

$$f(x|y) = a(x,y), \quad \forall x \in S(X). \tag{6.95}$$

We may also ask when there is such a compatible joint distribution that is unique.

It is evident that $u(x)$ and $a(x,y)$ will be compatible if there exists a suitable density for Y, say $w(y)$, such that

$$u(x) = \int_{S(Y)} a(x,y)w(y)dy, \quad \forall x \in S(X). \tag{6.96}$$

Thus, $u(x)$ and $a(x,y)$ are compatible if and only if $u(x)$ can be expressed as a mixture of the given conditional densities $\{a(x,y) : y \in S(Y)\}$. Uniqueness of the compatible distribution $h(x,y) = w(y)a(x,y)$ will be encountered if and only if the family of conditional densities is identifiable.

6.12.1 Example

Arnold et al. (1999) presented an example with

$$a(x,y) = ye^{-xy}, \quad x > 0,$$

and

$$u(x,y) = (1+x)^{-2}, \quad x > 0.$$

It can be verified that these are indeed compatible with the density of Y given by

$$w(y) = e^{-y}, \quad y > 0.$$

Identifiability of the family $\{ye^{-xy},\ x,y > 0\}$ may be verified by using the uniqueness property of Laplace transforms, and consequently there is a unique joint density corresponding to the given $a(x)$ and $u(x,y)$, given by

$$h(x,y) = ye^{(x+1)y}, \quad x,y > 0.$$

6.12.2 Vardi and Lee's Iteration Scheme

Suppose now that $a(x)$ and $u(x,y)$ are given. How can we identify in general the corresponding mixing density $w(y)$? Vardi and Lee (1993) provided an iterative scheme for this purpose.

Let $w_0(y)$ be an arbitrary strictly positive density defined on $S(Y)$. For $n = 0, 1, \ldots$, define

$$w_{n+1}(y) = w_n(y) \int_{S(X)} \frac{a(x,y)u(x)}{\int_{S(Y)} w_n(y')a(x,y')dy'}\, dx. \tag{6.97}$$

Vardi and Lee (1993) showed that the iterative scheme in (6.97) will always converge. If $a(x)$ and $u(x,y)$ are compatible, it will converge to an appropriate mixing scheme $w(y)$.

6.13 Conditional Survival Models

So far, we have discussed only conditionally specified bivariate distributions in terms of conditional density functions in which one of them belongs to a particular parametric family, whereas the other belongs to a possibly different parametric family. In the context of bivariate survival models, it is more natural to condition on component survivals (i.e., on events such as $\{X > x\}$ and $\{Y > y\}$) rather than conditioning on a particular value of X and Y. The question of compatibility will spring to our mind immediately, but this has been answered in Arnold et al. (1999) as follows, Two families of conditional survival functions

$$\begin{aligned} \Pr(X > x|Y > y) &= a(x,y), \quad (x,y) \in S(X) \times S(Y), \\ \Pr(Y > y|X > x) &= b(x,y), \quad (x,y) \in S(X) \times S(Y), \end{aligned} \tag{6.98}$$

are compatible if and only if there exist functions $u(x) \in S(X)$ and $v(y) \in S(Y)$ such that

$$\frac{a(x,y)}{b(x,y)} = \frac{u(x)}{v(y)}, \quad (x,y) \in S(X) \times S(Y), \tag{6.99}$$

where $u(x)$ is a one-dimensional survival function. We now present two examples of distributions characterized by conditional survival.

6.13.1 Exponential Conditional Survival Function

Conditional Properties

Suppose

$$\Pr(X > x|Y > y) = \exp[-\theta(y)x], \quad x, y > 0,$$

and

$$\Pr(Y > y|X > x) = \exp[-\tau(x)y], \quad x, y > 0,$$

where $\theta(y) = \alpha + \gamma y$ and $\tau(x) = \beta + \gamma x$.

Expression of the Joint Survival Function

In this case, we have as the joint survival function

$$\bar{H}(x,y) = \exp(\delta+\alpha x+\beta y+\gamma xy),\ \delta > 0, \alpha,\ \beta > 0,\ \gamma \le 0,\ \alpha\beta \ge -\gamma. \quad (6.100)$$

Reparametrizing in terms of marginal scale parameters and an interaction parameter, we have

$$\bar{H}(x,y) = \exp\left[-\left(\frac{x}{\sigma_1} + \frac{y}{\sigma_2} + \theta\frac{xy}{\sigma_1\sigma_2}\right)\right], \quad x,y > 0, \qquad (6.101)$$

where $\sigma_1, \sigma_2 > 0$ and $0 \le \theta \le 1$. This is indeed Gumbel's type I bivariate exponential distribution, discussed in Section 2.10.

6.13.2 Weibull Conditional Survival Function

Conditional Properties

Suppose

$$\Pr(X > x|Y > y) = \exp\{[-x/\sigma_1(y)]^{\gamma_1}\}, \quad x,y > 0,$$

and

$$\Pr(Y > y|X > x) = \exp\{[-y/\sigma_2(x)]^{\gamma_2}\}, \quad x,y > 0,$$

where $\sigma_1(y)^{\gamma_1} = (\alpha + \gamma y^{\gamma_2})^{-1}$ and $\sigma_2(x)^{\gamma_1} = (\beta + \gamma x^{\gamma_1})^{-1}$.

Expression of the Joint Survival Function

In this case, we have as the joint survival function

$$\bar{H}(x,y) = \exp\left\{-\left[\left(\frac{x}{\sigma_1}\right)^{\gamma_1} + \left(\frac{y}{\sigma_2}\right)^{\gamma_2} + \theta\left(\frac{x}{\sigma_1}\right)^{\gamma_1}\left(\frac{y}{\theta_2}\right)^{\gamma_2}\right]\right\}, \quad x,y > 0, \qquad (6.102)$$

where $\sigma_1, \sigma_2 > 0$ and $0 \le \theta \le 1$. If $\gamma_1 = \gamma_2$, then (6.101) reduces to Gumbel's bivariate exponential distribution in (6.100).

6.13.3 Generalized Pareto Conditional Survival Function

Conditional Properties

Suppose

$$\Pr(X > x|Y > y) = [1 + (x/\sigma_1)^{c_1}]^{-k}, \quad x, y > 0,$$

and

$$\Pr(Y > y|X > x) = [1 + (y/\sigma_2)^{c_2}]^{-k}, \quad x, y > 0.$$

Expressions of the Joint Survival Function

Two solutions are possible for the joint survival function, and they are as follows:

$$\bar{H}(x,y) = \left[1 + \left(\frac{x}{\sigma_1}\right)^{c_1} + \left(\frac{y}{\sigma_2}\right)^{c_2} + \theta\left(\frac{x}{\sigma_1}\right)^{c_1}\left(\frac{y}{\sigma_2}\right)^{c_2}\right]^{-k}, \; x, y > 0, \tag{6.103}$$

for positive constants $c_1, c_2, \sigma_1, \sigma_2, k$ and $\theta \in [0, 2]$, and

$$\begin{aligned}\bar{H}(x,y) &= \exp\left\{-\theta_1 \log\left[1 + \left(\tfrac{x}{\sigma_1}\right)^{c_1}\right] - \theta_2 \log\left[1 + \left(\tfrac{y}{\sigma_2}\right)^{c_2}\right]\right. \\ &\quad \left. -\theta_3 \log\left[1 + \left(\tfrac{x}{\sigma_1}\right)^{c_1}\right] \log\left[1 + \left(\tfrac{y}{\sigma_2}\right)^{c_2}\right]\right\}, \quad x, y > 0, \end{aligned} \tag{6.104}$$

for $\theta_1 > 0, \theta_2 > 0, \theta_3 \geq 0, \sigma_1 > 0, \sigma_2 > 0, c_1 > 0, c_2 > 0$.

The bivariate generalized Pareto distribution in (6.103) was first discussed in Durling (1975).

6.14 Conditional Approach in Modeling

6.14.1 Beta-Stacy Distribution

Mihram and Hultquist (1967) discussed the idea of a warning-time variable, X, for Y = the failure time of a component being tested, where $0 < X < Y$. A bivariate distribution was proposed, with Y having Stacy's generalized gamma distribution and X, conditional on $Y = y$, having a beta distribution over the range 0 to y. The resulting joint density is given by

$$h(x,y) = \frac{|c|}{a^{bc}\Gamma(b)B(p,q)} x^{p-1}(y-x)^{q-1} y^{bc-p-q} \exp[-(y/a)^c] \tag{6.105}$$

if $0 < x < y$ and is 0 otherwise.

Pearson's product-moment correlation coefficient is

$$\sqrt{\frac{p^2\text{var}(Y)}{(p+q)^2\text{var}(X)}}, \tag{6.106}$$

where var(X) is related to the moments of Y by

$$\text{var}(X) = \frac{p(p+1)E(Y^2)}{(p+q)(p+q+1)} - \frac{p^2[E(Y)]^2}{(p+q)^2}, \tag{6.107}$$

and the moments of Y are given by

$$E(Y^r) = a^r\Gamma[(bc+r)/p]/\Gamma(b) \quad \text{for} \quad r/c > -b \tag{6.108}$$

and are undefined otherwise.

The generation of random variates from this distribution is straightforward.

Setting $c = 1$ and $bc = p + q$, we obtain McKay's bivariate gamma distribution.

6.14.2 Sample Skewness and Kurtosis

Shenton and Bowman (1977) considered the joint distribution of the sample skewness and kurtosis statistics. It is well known that, in sampling from a normal population, the distributions of $\sqrt{b_1}$ $(= m_3/m_2^{3/2})$ and b_2 $(= m_4/m_2^2)$ are individually well approximated by Johnson's S_U distribution, but little consideration has been given to the joint distribution (m_j being the jth sample central moment). When Shenton and Bowman conducted extensive simulations of $(\sqrt{b_1}, b_2)$, they found that the distribution of $\sqrt{b_1}$ is unimodal for small b_2 but becomes bimodal for large b_2—provided n is not too large (as $n \to \infty$, so $\sqrt{b_1}$ becomes unimodal, whatever b_2 might be). Their approach to the bivariate distribution was to use S_U for the marginal distribution of $\sqrt{b_1}$ and a conditional gamma density for b_2 given the value of $\sqrt{b_1}$. That is,

$$h(\sqrt{b_1}, b_2) = w(\sqrt{b_1})g(b_2|\sqrt{b_1}), \tag{6.109}$$

where w is the density of S_U, and the gamma density g is written in terms of $b_2 - 1 - b_1$ since the constraint $b_2 \geq 1 + b_1$ applies to the relative values of b_2 and $\sqrt{b_1}$,

$$g(b_2|\sqrt{b_1}) = \frac{k}{\Gamma(\theta)}[k(b_2 - 1 - b_1)]^{\theta-1}\exp[-k(b_2 - 1 - b_1)], \tag{6.110}$$

in which θ is a quadratic in $\sqrt{b_1}$. This work has also been described in Section 7.7 of Bowman and Shenton (1986).

6.14.3 Business Risk Analysis

We summarize here the work of Kottas and Lau (1978). The subject is risk analysis in business, by which is meant determining the stochastic characteristics of secondary variables such as profit Z from (i) the stochastic characteristics of primary variables such as sales volume Q, unit price P, unit variable cost V, and fixed cost F and (ii) a functional relationship such as $Z = Q(P - V) - F$. The starting point of Kottas and Lau is:

- Emphasis has traditionally been on estimating the individual stochastic characteristics of the primary variables, with their interdependencies being neglected.
- Even when some attempt has been made to model the dependencies, this has often been done in an unsatisfactory way; for example, by merely specifying a correlation coefficient.

Kottas and Lau reviewed the shortcomings of the product-moment correlation as a measurement of dependence, the specific one is imposing on a model when making a simple and apparently harmless assumption such as bivariate normality or lognormality and the impracticality of obtaining subjective estimates of higher moments if a more general bivariate distribution is permitted.

The alternative that they suggested is what they call a "functional approach," and it consists of getting the dependencies of $E(Y)$ and $\text{var}(Y)$ on x correctly specified. In principle, this might be extended to higher conditional moments but in practice the shape of the conditional distribution of Y is assumed to be independent of x, only the mean and the spread being allowed to change.

To a statistical audience, the points made by Kottas and Lau may seem uncontroversial and hardly worth saying, but it is a well-written article and it brings home the necessity in model construction to always stay closely in touch with what is practical.

6.14.4 Intercropping

This refers to growing two crops simultaneously on the same area of land and harvesting and processing them separately. Mead et al. (1986) have stated, "Amid all the other justifications of the practice of intercropping, the benefit of 'stability' is a recurring theme. However, the concept of stability is variously and poorly defined, and the attempts to express the stability in the quantitative terms have been statistically unconvincing." Their paper is chiefly about refining the notion of stability, which they do in terms of the relative risks associated with intercropping and monocropping systems.

The data analyzed by Mead et al. (1986) consisted of financial returns obtained by (i) intercropping sorghum with pigeonpea and (ii) monocropping sorghum at 51 site-year combinations in India (7 years, 11 areas, many combinations omitted). A scatterplot of the 51 points reveals:

- a strong correlation between the returns of the two cropping systems;
- a higher average return with intercropping; and
- suggestions in the shape of the scatter that the relationship between the two returns is curvilinear and heteroscedastic.

Mead et al. (1986) suggested it was appropriate to quantify the relative risks of the two cropping systems by plotting the risk of "failure" under each system against each other, as the definition of failure varies. In other words, plot $\Pr(Y < t)$ against $\Pr(X < t)$ for various levels of t. For a dataset of 51 points, that can be done satisfactorily directly from the data points.

However, partly to understand their dataset better and partly to provide an approach that would be more satisfactory for lesser amounts of data (given that it had been validated on larger datasets), Mead et al. (1986) went on to:

- fit a bivariate distribution to their scatter of points; and
- calculate a smooth risk vs. risk curve from that distribution.

The approach chosen was (i) to fit a normal distribution to the sum of the returns ($S = X + Y$) from the two systems and (ii) to assume the conditional distribution of the difference in returns between the two systems had a normal distribution also, with the mean and the logarithm of the variance having a quadratic dependence on S. Mead et al. (1986) have presented a contour plot of the resulting distribution.

This method of analysis was repeated on four other datasets that had resulted from intercropping sorghum with various second crops.

6.14.5 Winds and Waves, Rain and Floods

Height and Period of Waves of the Sea

A good deal of empirical data have been published on the joint distribution of wave height and period.

Haver (1985) approached some data collected off northern Norway from the conditional point of view:

- The distribution of wave height X that was chosen was an unusual one, being lognormal for small X and Weibull for large X.

- Given $X = x$, the "spectral peak period" T was assumed to have a lognormal distribution.

Haver did not assume a functional form for the dependence on X of the parameters of the distribution of T; instead, for each of several ranges of X, the mean and variance of $\log T$ were estimated. In Haver's Figure 10, expressions are given for how these are related to X. However, because the expressions are quite messy, in addition to the marginal distribution of X, an explicit formula for the joint distribution of X and T would be grotesquely cumbersome.

Another study of this type was Burrows and Salih (1987). These authors took X to have a Weibull distribution and the conditional distribution of T to be either Weibull or lognormal; it seems that the Weibull distribution was used in shifted form (i.e., three-parameter form). They fitted these and other distributions to data from 18 sites around the British Isles.

For data from the North Sea, Krogstad (1985) took X to have a Weibull distribution and the conditional distribution of T given X to be normal, with constant mean and variance inversely proportional to X.

Myrhaug and Kjeldsen (1984) analyzed data from the North Sea with regard to the joint distribution of the wave height and several other variables—crest from the steepness and period, also assumed to have Weibull distributions. The conditional distributions of vertical asymmetry factor and total wave steepness, in contrast, were taken to be lognormal.

Wind Speeds

It is of interest in the wind energy industry, as mentioned by Kaminsky and Kirchhoff (1988), to estimate the energy available from the wind energy conversion systems at one height from data collected at a lower height.

It is a common practice to assume the wind energy speed has a Rayleigh distribution. In modeling the joint distribution of wind speeds at two heights, Kaminsky and Kirchhoff therefore required the marginal distributions to have this form, at least roughly. In fact, they considered two alternatives:

- X has a Rayleigh distribution and Y has a Rayleigh distribution with an origin at $Y = x$ and a constant scale factor. So, $\Pr(Y < X) = 0$ for this model.
- X has a Rayleigh distribution and Y has a normal distribution with mean $a + bx$ and a constant variance.

Kaminsky and Kirchhoff (1988) presented an empirical contour plot of the bivariate distribution of wind speeds—at heights of 32 ft and 447 ft at a site in Waterford, Connecticut—along with contour plots of the two distributions that had been fitted to the data. The Rayleigh-normal distribution appeared to be a better fit than the Rayleigh-shifted Rayleigh distribution (but it does have two more parameters). The use of the symbolic algebra software package MACSYMA to obtain expressions for the marginal distributions of Y in the two cases was a further feature of interest in this study.

Wind Speed and Wave Height

Liu (1987) was concerned with the joint distribution of wind speed X and wave height Y on the Great Lakes of North America. Here, $Y|X$ was taken as a gamma distribution; separately, the use of (i) empirically obtained equations connecting the parameters of this to wind speed, together with (ii) a histogram of wind speeds were used for the calculation of the joint distribution.

Storm Surge and Wave Height

In a study by Vrijling (1987) regarding the Dutch dikes, a part was played by the joint distribution of the storm surge level of the sea and the significant wave height, with the assumptions that:

- The storm surge level X has an extreme-value distribution; that is, $F(x) = \exp[-e^{-(x-\alpha)/b}]$.
- Given X, wave height is normally distributed, with mean dependent on X and constant variance.

Floods

Correia (1987) has considered the duration and peak discharge of floods of a river. He supposed that:

- Flood duration is exponentially distributed.
- For a given duration, the peak discharge has a normal distribution whose mean is a linear function of duration and whose variance is a constant.

Streamflow and Rain

Clarke (1979, 1980) used McKay's distribution (Section 6.9) with X = annual streamflow and Y = real precipitation. The justification was that

- With McKay's distribution, X, Y, and $Y - X$ (= evaporation) all have gamma distributions, this being a popular univariate choice in hydrology.
- $Y \geq X$ is reasonable on physical grounds (for watertight basins with little over a year storage).

The motivation for Clarke's work was that X is the variable of chief interest, but there were often only a few years of data available for it, with the records of Y being more extensive.

Rain

- According to Etoh and Murota (1986), a rainstorm can be adequately described by three characteristics: duration X, maximum intensity Y, and total amount Z. Further, it can be assumed that $Z \propto XY/2$. Consequently, two random variables suffice. Etoh and Murota made the following assumptions:
  - X has a gamma distribution.
  - $Y = \eta X^{\alpha}$, where η has a gamma distribution and a is a constant (and $0 \leq a \leq 1$, reflecting a less than proportionate increase of maximum intensity with duration).

  Etoh and Murota had some empirical data from Osaka and some results published by Córdova and Rodríguez-Iturbe (1985) for Denver (Colorado) and Boconó (Venezuela). They found that the shapes of the univariate distributions and the values of the correlation coefficients could be approximately reproduced by judicious selection of the parameters of their model.
- Sogawa et al. (1987) were concerned with (i) the annual rainfall and (ii) the annual maximum daily rainfall, each at four places in Nagano prefecture, Japan. In both cases, they used a quadrivariate conditional maximum-entropy distribution.
- The method adopted by Snyder and Thomas (1987) was not exactly that of conditional distributions, but here is a good place to summarize it. The subject was agriculture-related variates, such as monthly rainfall and monthly average temperature. After univariate transformations, Snyder and Thomas (1987) used "a form-free bivariate distribution based on two-dimensional sliding polynomials," which they found to be "necessary to model the bi-modal and heavy-tailed distributions frequently encountered."

References

1. Abrahams, J., Thomas, J.B.: A note on the characterization of bivariate densities by conditional densities. Communications in Statistics: Theory and Methods **13**, 395–400 (1984)
2. Abramowitz, M., Stegun, I.A.: Handbook of Mathematical Functions. Dover, New York (1994)

3. Amos, D.E.: Algorithm 556: Exponential integrals. Transactions on Mathematical Software **6**, 420–428 (1980)
4. Arnold, B.C.: Bivariate distributions with Pareto conditionals. Statistics and Probability Letters **5**, 263–266 (1987a)
5. Arnold, B.C.: Dependence in conditionally specified distributions. Presented at the Hidden Valley Conference on Dependence (1987b)
6. Arnold, B.C., Castillo, E., Sarabia, J.M.: Conditional Specification of Statistical Models. Springer-Verlag, New York (1999)
7. Arnold, B.C., Castillo, E., Sarabia, J.M.: Conditionally specified distributions: An introduction. Statistical Science **16**, 249–265 (2001)
8. Arnold, B.C., Castillo, E., Sarabia, J.M.: Conditionally specified multivariate skewed distributions. Sankhyā, Series A **64**, 206–226 (2002)
9. Arnold, B.C., Strauss, D.J.: Bivariate distributions with conditionals in prescribed exponential families. Technical Report No. 151, Department of Statistics, University of California, Riverside (1987)
10. Arnold, B.C., Strauss, D.: Bivariate distributions with exponential conditionals. Journal of the American Statistical Association **83**, 522–527 (1988a)
11. Arnold, B.C., Strauss, D.: Pseudolikelihood estimation. Sankhyā, Series B **53**, 233–243 (1988b)
12. Arnold, B.C., Strauss, D.: Bivariate distributions with conditionals in prescribed exponential families. Journal of the Royal Statistical Society, Series B **53**, 365–375 (1991)
13. Barndorff-Nielsen, O.: On a formula for the distribution of the maximum likelihood estimator. Biometrika **70**, 343–365 (1983)
14. Barndorff-Nielsen, O., Blæsild, P.: Reproductive exponential families. Annals of Statistics **11**, 770–782 (1983)
15. Besag, J.E.: Spatial interaction and statistical analysis of lattice systems. Journal of the Royal Statistical Society, Series B **36**, 192–236 (1974)
16. Bhattacharyya, A.: On some sets of sufficient conditions leading to the normal bivariate distribution. Sankhyā **6**, 399–406 (1943)
17. Block, H.W., Rao, B.R.: A beta warning-time distribution and a distended beta distribution. Sankhyā, Series B **35**, 79–84 (1973)
18. Blumen, I., Ypelaar, M.A.: A family of bivariate distributions for rank-based statistics with an application to Kendall's tau. In: Proceedings of the Business and Economic Statistics Section, American Statistical Association, pp. 386–390. American Statistical Association, Alexandria, Virgina (1980)
19. Bowman, K.O., Shenton, L.R.: Moment $(\sqrt{b_1}, b_2)$ techniques. In: Goodness-of-Fit Techniques, R.B. D'Agostino and M.A. Stephens (eds.), pp. 279–329. Marcel Dekker, New York (1986)
20. Burrows, R., Salih, B.A.: Statistical modelling of long-term wave climates. In: Twentieth Coastal Engineering Conference, Proceedings, Volume I, B.L. Edge (ed.), pp. 42–56, American Society of Civil Engineers, New York (1987)
21. Castillo, E., Galambos, J.: Modelling and estimation of bivariate distributions with conditionals based on their marginals. Conference on Weighted Distributions, Pennsylvania State University (1985)
22. Castillo, E., Galambos, J.: Bivariate distributions with normal conditionals. In: Proceedings of the IASTED International Symposium, Cairo, M.H. Hamza (ed.), pp. 59–62. Acta Press, Anaheim, California (1987a)
23. Castillo, E., Galambos, J.: Bivariate distributions with Weibull conditionals. Technical Report, Department of Mathematics, Temple University, Philadelphia (1987b)
24. Clarke, L.E.: On marginal density functions of continuous densities. American Mathematical Monthly **82**, 845–846 (1975)
25. Clarke, R.T.: Extension of annual streamflow record by correlation with precipitation subject to heterogeneous errors. Water Resources Research **15**, 1081–1088 (1979)

26. Clarke, R T.: Bivariate gamma distributions for extending annual streamflow records from precipitation: Some large-sample results. Water Resources Research **16**, 863–870 (1980)
27. Córdova, J.R., Rodrìguez-Iturbe, I.: On the probabilistic structure of storm surface runoff. Water Resources Research **21**, 755–763 (1985)
28. Correia, F.N.: Engineering risk in flood studies using multivariate partial duration series. In: Engineering Reliability and Risk in Water Resources, L. Duckstein and E.J. Plate (eds.), pp. 307–332. Nijhoff, Dordrecht (1987)
29. Dubey, S.D.: Compound gamma, beta, and F distributions. Metrika **16**, 27–31 (1970)
30. Durling, F.C.: The bivariate Burr distribution. In: A Modern Course on Statistical Distributions in Scientific Work. Volume I: Models and Structures, G.P. Patil, S. Kotz, and J.K. Ord (eds.), pp. 329–335. Reidel, Dordrecht (1975)
31. Etoh, T., Murota, A.: Probabilistic model of rainfall of a single storm. Journal of Hydroscience and Hydraulic Engineering **4**, 65–77 (1986)
32. Gelman, A., Meng, X-L.: A note on bivariate distributions that are conditionally normal. The American Statistician **45**, 125–126 (1991)
33. Gelman, A., Speed, T.P.: Characterizing a joint distribution by conditionals. Journal of the Royal Statistical Society, Series B **35**, 185–188 (1993)
34. Gross, A.J., Clark, V.A.: Survival Distributions: Reliability Applications in the Biomedical Sciences, John Wiley and Sons, New York (1975)
35. Haver, S. Wave climate off northern Norway. Applied Ocean Research **7**, 85–92 (1985)
36. Inaba, T., Shirahata, S.: Measures of dependence in normal models and exponential models by information gain. Biometrika **73**, 345–352 (1986)
37. James, I.R.: Multivariate distributions which have beta conditional distributions. Journal of the American Statistical Association **70**, 681–684 (1975)
38. James, I.R.: Distributions associated with neutrality properties for random proportions. In: Statistical Distributions in Scientific Work. Volume 4: Models, Structures, and Characterizations, C. Taillie, G.P. Patil, and B.A. Baldessari (eds.), pp. 125–136. Reidel, Dordrecht (1981)
39. Jørgensen, B.: Exponential dispersion models. Journal of the Royal Statistical Society, Series B **49**, 127–145 (Discussion, 145–162) (1987)
40. Kaminsky, F.C., Kirchhoff, R.H.: Bivariate probability models for the description of average wind speed at two heights. Solar Energy **40**, 49–56 (1988)
41. Kottas, J.F., Lau, H-S.: On handling dependent random variables in risk analysis. Journal of the Operational Research Society **29**, 1209–1217 (1978)
42. Krogstad, H.E.: Height and period distributions of extreme waves. Applied Ocean Research **7**, 158–165 (1985)
43. Liu, P.C.: Estimating long-term wave statistics from long-term wind statistics. In: Twentieth Coastal Engineering Conference, Proceedings, Volume I, B.L. Edge (ed.), pp. 512–521. American Society of Civil Engineers, New York (1987)
44. Mardia, K.V.: Multivariate Pareto distributions. Annals of Mathematical Statistics **33**, 1008–1015 (Correction **34**, 1603) (1962)
45. McKay, A.T.: Sampling from batches. Journal of the Royal Statistical Society, Supplement **1**, 207–216 (1934)
46. Mead, R., Riley, J., Dear, K., Singh, S.P.: Stability comparison of intercropping and monocropping systems. Biometrics **42**, 253–266 (1986)
47. Mihram, G.A., Hultquist, A.R.: A bivariate warning-time/failure-time distribution. Journal of the American Statistical Association **62**, 589–599 (1967)
48. Myrhaug, D., Kjeldsen, S.P.: Parametric modelling of joint probability distributions for steepness and asymmetry in deep water waves. Applied Ocean Research **6**, 207–220 (1984)
49. Nadarajah, S.: Eaxct distributions of XY for some bivariate exponential distributions. Statistics **40**, 307–324 (2006)

50. Narurmi, S.: On the general forms of bivariate frequency distributions which are mathematically possible when regression and variations are subjected to limiting conditions I. Biometrika **15**, 77–88 (1923a)
51. Narumi, S.: On the general forms of bivariate frequency distributions which are mathematically possible when regression and variations are subjected to limiting conditions II. Biometrika **15**, 209–211 (1923b)
52. Parrish, R.S.: Evaluation of bivariate cumulative probabilities using moments to fourth order. Journal of Statistical Computation and Simulation **13**, 181–194 (1981)
53. Parrish, R.S., Bargmann, R.E.: A method for the evaluation of cumulative probabilities of bivariate distributions using the Pearson family. In: Statistical Distributions in Scientific Work, Volume 5: Inferential Problems and Properties, C. Taillie, G.P. Patil, and B.A. Baldessari (eds.), pp. 241–257. Reidel, Dordrecht (1981)
54. Patil, G.P.: On a characterization of multivariate distribution by a set of its conditional distributions. Bulletin of the International Statistical Institute **41**, 768–769 (1965)
55. Ratnaparkhi, M.V.: Some bivariate distributions of (X, Y) where the conditional distribution of Y, given X, is either beta or unit-gamma. In: Statistical Distributions in Scientific Work, Volume 4: Models, Structures, and Characterizations, C. Taillie, G.P. Patil, and B.A. Baldessari (eds.), pp. 389-400. Reidel, Dordrecht (1981)
56. Romano, J.P., Siegel, A.F.: Counterexamples in Probability and Statistics. Wadsworth and Brooks/Cole, Monterey, California (1986)
57. Seshadri, V.: Discussion of "Exponential dispersion models" by B. Jørgensen. Journal of the Royal Statistical Society, Series B **49**, p. 156 (1987)
58. Seshadri, V., Patil, G.P.: A characterization of a bivariate distribution by the marginal and the conditional distributions of the same component. Annals of the Institute of Statistical Mathematics **15**, 215–221 (1963)
59. Shenton, L.R., Bowman, K.O.: A bivariate model for the distribution of $\sqrt{b_1}$ and b_2. Journal of the American Statistical Association **72**, 206–211 (1977)
60. Sogawa, N., Araki, M., Terashima, A.: Study on bivariate MEP distribution and its characteristics (in Japanese). Proceedings of the 28th Japanese Conference on Hydraulics, pp. 397–402 (1987)
61. Snyder, W.M., Thomas, A.W.: Stochastic impacts on farming. IV: A form-free bivariate distribution to generate inputs to agricultural models. Transactions of the American Society of Agricultural Engineers **30**, 946–952 (1987)
62. Székely, G.J.: Paradoxes in Probability Theory and Mathematical Statistics. Reidel, Dordrecht (1986)
63. Tiku, M.L., Kambo, N.S.: Estimation and hypothesis testing for a new family of bivariate non-normal distributions. Communications in Statistics: Theory and Methods **21**, 1683–1705 (1992)
64. Vardi, Y., Lee, P.: From image deblurring to optimal investments: Maximum likelihood solutions for positive linear inverse problems. Journal of the Royal Statistical Society, Series B **55**, 569–612 (1993)
65. Vrijling, J.K.: Probabilistic design of water-retaining structures. In: Engineering Reliability and Risk in Water Resources, L. Duckstein and E.J. Plate (eds.), pp. 115–134. Nijhoff, Dordrecht (1987)
66. Wesolowski, J.: Bivariate distributions via a Pareto conditional distribution and a regression function. Annals of the Institute of Statistical Mathematics **47**, 177–183 (1995)

Chapter 7
Variables-in-Common Method

7.1 Introduction

The terms "trivariate reduction" or "variables in common" are used for schemes for constructing of pairs of r.v.'s that start with three (or more) r.v.'s and perform some operations on them to reduce the number to two.

The idea here is to create a pair of dependent random variables from three or more random variables. In many cases, these initial random variables are independent, but occasionally they may be dependent—an example of the latter is the construction of a bivariate t-distribution from two variates that have a standardized correlated bivariate normal distribution and one that has a chi-distribution. An important aspect of this method is that the functions connecting these random variables to the two dependent random variables are generally elementary ones; random variate generation of the latter can therefore be done as easily as for the former.

Different authors have used the terms in slightly different ways. A broad definition of variables in common (or trivariate reduction) is

$$\left.\begin{aligned} X &= T_1(X_1, X_2, X_3) \\ Y &= T_2(X_1, X_2, X_3) \end{aligned}\right\}, \tag{7.1}$$

where X_1, X_2, X_3 are not necessarily independent or identically distributed. A narrow definition is

$$\left.\begin{aligned} X &= X_1 + X_3 \\ Y &= X_2 + X_3 \end{aligned}\right\}, \tag{7.2}$$

where X_1, X_2, X_3 are i.i.d. Another possible definition is

$$\left.\begin{aligned} X &= T(X_1, X_3) \\ Y &= T(X_2, X_3) \end{aligned}\right\} \tag{7.3}$$

N. Balakrishnan, C.D. Lai, *Continuous Bivariate Distributions*,
DOI 10.1007/b101765_8, © Springer Science+Business Media, LLC 2009

with (i) the X_i being independently distributed and having d.f. $F_0(x_i; \lambda_i)$ and (ii) X and Y having distributions $F_0(x; \lambda_1 + \lambda_2)$ and $F_0(y; \lambda_1 + \lambda_3)$, respectively.

Three well-known distributions obtainable in this way are (a) the bivariate normal, from the additive model in (7.2), with the X_i's having normal distributions; (b) Cherian's bivariate gamma distribution, also from (7.2) but with the X_i's having gamma distributions; and (c) Marshall and Olkin's bivariate exponential distribution from (7.3), with the transformation T being the minimum and the X_i's having exponential distributions.

We first present a general description of this method in Section 7.2. In Section 7.3, we describe the additive model, while the generalized additive model is explained in Section 7.4. Models arising from weighted linear combinations of random variables are discussed in Section 7.5. In Section 7.6, bivariate distributions of random variables having a common denominator are detailed. In Sections 7.7 and 7.8, multiplicative trivariate reduction and Khintchine's mixture forms are discussed. While transformations involving the minimum are explained in Section 7.9, some other forms of the variables-in-common technique are discussed in Section 7.10.

7.2 General Description

Let X_i $(i = 1, 2, 3)$ be three independent random variables with distribution functions $F_i(x_i; \lambda_i)$. The F_i's are often assumed to be the same, but the parameters λ_i may be different. Suppose there exists a function T such that

$$\left.\begin{aligned} X &= T(X_1, X_3) \\ Y &= T(X_2, X_3) \end{aligned}\right\}. \tag{7.4}$$

Then, X and Y are said to have a bivariate distribution generated by a *trivariate reduction technique*. Pearson (1897) generated the bivariate normal distribution in this way and Cherian (1941) the bivariate gamma.

More generally, let us define

$$\left.\begin{aligned} X &= T_1(X_1, \dots, X_n) \\ Y &= T_2(X_1, \dots, X_n) \end{aligned}\right\}, \tag{7.5}$$

where X and Y have one or more X_i's in common and the X_i $(i = 1, 2, \dots, n)$ may not be mutually independent. The structure of T is obviously important, but consider only a simple transformation of the X_i's. Usually, the X_i's will be mutually independent, but occasionally they will be allowed to be dependent.

The following example, taken from Section 6 of Sumita and Kijima (1985), is from the field of production engineering. Suppose a machine is alternately producing items or being maintained. Let a period of useful production (of length X_3) followed by a maintenance period (of length X_1) be referred to as a cycle (of length X_1+X_3). The cost incurred during a cycle consists of running costs during the production period (which are proportional to the length of the production period, X_3) and of such things as parts for maintenance (a random variable, X_2). Then, we have the length of cycle $X = X_1 + X_3$ and the total cost $Y = X_2 + cX_3$. Sumita and Kijima assumed X_1, X_2, and X_3 have exponential distributions.

An example from the field of geotechnical engineering is that in determining the probability of failure of a slope, comparison of the total force resisting sliding with the total force tending to induce sliding is required. These have variables in common, such as the weight of the block of rock, its angle to the horizontal, and forces due to water pressure in a tension crack; see Frangopol and Hong (1987).

An example motivated from plant breeding is the following. Suppose we are interested in the true values of a particular characteristic, but we can only observe $Y = X + \varepsilon$, where ε is an error term. What is the distribution of X within the population selected by the requirement that $Y > y$? For the case of ε being normally distributed, see Curnow (1958).

Another form of variable in common may occur in a reliability context when two components may be subjected to the same set of stresses, which will invariably affect the lifetimes of both components.

7.3 Additive Models

7.3.1 Background

The first model we consider is

$$T(X_1, X_3) = X_1 + X_3. \tag{7.6}$$

The X_i's are usually taken to come from the same family of distributions; it may happen that the family is closed under convolution (i.e., the sum X_1+X_3 also belongs to the same family of distributions).

As mentioned in Section 7.2, Pearson (1897) obtained the bivariate normal using the trivariate reduction technique. In his well-known dice problem, Weldon first constructed a bivariate binomial distribution using (7.2), with the X_i's being independent binomial variables.

Cherian (1941) and David and Fix (1961) obtained a bivariate gamma distribution in the same manner. Let $X = X_1 + X_3, Y = X_2 + X_3$, where the X_i's are independent gamma variables with shape parameters λ_i. Then, the joint density of X and Y is a bivariate gamma density.

Eagleson (1964) used a particular additive model in which the sums and the X_i's belong to the same family of distributions to obtain a class of bivariate distributions whose marginals are members of Meixner classes, defined in Section 7.3.2; see also Lancaster (1975, 1983). Meixner's collection of distributions have often appeared in characterization theorems because of their regression properties. Some of these characterizations and properties have been discussed by Lai (1982). The Meixner collection of distributions has also appeared in Morris (1982, 1983). We now give a brief account of the Meixner classes of bivariate distributions.

7.3.2 Meixner Classes

Suppose that X is a centered (i.e., with zero mean) random variable possessing a moment generating function with distribution function G, on which can be defined an orthogonal polynomial system $\{P_n\}$, where $P_n(x) = x^n +$ terms of lower order, such that $\int P_m P_n dG = \delta_{mn} b_m$. Here, δ_{mn} is the Kronecker delta and b_m is a normalizing constant. Meixner (1934) considered these distributions, for which the generating function for their orthogonal polynomials is of the form

$$K(x,t) = \sum_{n=0}^{\infty} P_n(x) t^n / n! = \exp[xu(t)]/M[u(t)], \tag{7.7}$$

where

$$u(t) = t + \text{possibly terms of higher powers of } t$$

is a real power series in t and $M(\cdot)$ is necessarily the moment generating function.

It has been shown by Meixner (1934) [see also Lancaster (1975)] that there are precisely six statistical distributions for which (7.7) is satisfied, and they are:

- positive binomial,
- normal,
- Poisson,
- gamma (transformed),
- negative binomial, and
- Meixner hypergeometric.

The first five are in common use, while the last distribution has been discussed in diverse literature.

Eagleson (1964) showed that if X_i's belong to the same Meixner class and if they are mutually independent, then X and Y obtained by (7.2) also belong to the same Meixner class, and their joint distribution function satisfies the biorthogonal property

$$dH(x,y) = dF(x)dG(y) \sum_{n=0}^{\infty} \rho_n P_n(x) P_n(y). \tag{7.8}$$

Correlation

It is easy to see that the correlation coefficient of X and Y in the additive model is given by

$$\frac{\mathrm{var}(X_3)}{\sqrt{\mathrm{var}(X_1+X_2)\mathrm{var}(X_2+X_3)}}, \tag{7.9}$$

which is always positive. It follows at once that we cannot obtain bivariate distributions with negative correlations with such a scheme; independent marginals can only be obtained by letting (X_3 = a constant) be included in the family. The values X and Y obtained in this way will have linear regression on each other. This is a consequence of a theorem of Rao (1947), which was restated in Lancaster (1975); see also Eagleson and Lancaster (1967).

7.3.3 Cherian's Bivariate Gamma Distribution

Let $X = X_1+X_3, Y = X_2+X_3$, with X_i's being independent standard gamma random variables having shape parameters α_i. In his derivation, Cherian (1941) assumed that $\alpha_1 = \alpha_2$ and the joint density of X and Y is expressed in terms of an integral. Szántai (1986) provided an explicit expression for the joint density function $h(x, y)$ for arbitrary shape parameters in terms of Laguerre polynomials.

7.3.4 Symmetric Stable Distribution

A class of bivariate symmetric stable distributions can be obtained via the additive model. Let X_i's be three mutually independent symmetric stable random variables with characteristic functions $\exp(-\lambda_i|t|^\alpha)$, $\lambda_i \geq 0$, $0 < \alpha \leq 2$. Consider the transformations $X = X_1 + X_3$ and $Y = X_2 + X_3$; then, the joint characteristic function of (X, Y) is

$$\varphi(s,t) = \exp(-\lambda_1|s|^\alpha - \lambda_3|t+s|^\alpha - \lambda_2|t|^\alpha). \tag{7.10}$$

De Silva and Griffiths (1980) constructed a test of independence for this class of bivariate distributions.

7.3.5 Bivariate Triangular Distribution

Eagleson and Lancaster (1967) constructed a bivariate triangular distribution by letting the X_i's have a uniform distribution on $[0, 1]$.

- The marginal p.d.f.'s are

$$f(x) = \begin{cases} x, & 0 \le x \le 1 \\ 2 - x, & 1 \le x \le 2 \end{cases} \text{ and } g(y) = \begin{cases} y, & 0 \le y \le 1 \\ 2 - y, & 0 \le y \le 1 \end{cases}.$$

- The regression is linear, $E(Y|X = x) = \frac{x+1}{2}$.
- $E(Y^2|X = x) = \frac{1}{12} + \frac{x}{2} + c(x)$, where

$$c(x) = \begin{cases} x^2/2, & 0 \le x \le 1 \\ (x^2 - x + 1)/3, & 1 \le x \le 2 \end{cases};$$

see Eagleson and Lancaster (1967). Since this is not a polynomial of the second degree, the canonical variations associated with the diagonal expansion of the bivariate triangular distribution are not polynomials. This example was constructed as a counterexample to the proposition that linear regression implies the canonical variables are polynomials. Griffiths (1978) showed that these canonical variables, though not polynomials themselves, have a relationship with the Legendre polynomials.

7.3.6 Summing Several I.I.D. Variables

What follows generalizes the model we have discussed so far in that more than two variables are added together, but it is a specialization also, as the variables considered are now i.i.d.

Let X_i $(i = 1, 2, \ldots)$ be a sequence of i.i.d. variables, and let us define

$$\left.\begin{aligned} X &= \textstyle\sum_{i\in A} X_i \\ Y &= \textstyle\sum_{i\in B} X_i \end{aligned}\right\}, \tag{7.11}$$

where A and B are subsets of positive integers. The joint distribution of X and Y has a correlation coefficient given by

$$\rho(X, Y) = \frac{n(A \bigcap B)}{[n(A)n(B)]^{1/2}}, \tag{7.12}$$

where $n(A)$ denotes the number of elements in the set A. Clearly, X and Y are independent if $A \bigcap B = \emptyset$ (i.e., $\rho(X, Y) = 0$), and $\rho(X, Y) = 1$ if $A \equiv B$. For further details, see Lancaster (1982).

Example: Moving Averages

Consider a series of simple moving averages (or moving sums) of order k. Let $A = \{s+1, s+2, \ldots, s+k\}$, $B = \{s+2, s+3, \ldots, s+k+1\}$ for any $s \geq 0$. Then, X and Y are two adjacent moving sums.

7.4 Generalized Additive Models

7.4.1 Trivariate Reduction of Johnson and Tenenbein

Johnson and Tenenbein (1979) considered the trivariate reduction of the form

$$\left.\begin{aligned} X &= X_1 + cX_3 \\ Y &= X_2 + cX_3 \end{aligned}\right\},$$

where X_1, X_2 and X_3 are i.i.d. random variables.

The values τ and ρ_S were calculated for the following choices of X_i's:

Exponential:

$$\tau = \frac{2c^2}{(1+c)(1+2c)}, \qquad \rho_S = \frac{c^2(2c^2+9c+6)}{(1+c)^2(1+2c)(2+c)}.$$

Laplace:

$$\tau = \frac{c^2(32c^5+125c^4+161c^3+90c^2+22c+2)}{2(1+c)^3(1+2c)^4},$$

$$\rho_S = \frac{c^2(16c^7+152c^6+588c^5+1122c^4+1104c^3+555c^2+132c+12}{2(1+c)^4(1+2c)^3(2+c)^2}.$$

Uniform:

$$\tau = \begin{cases} \frac{c^2(c^2-6c+10)}{15} & \text{for } 0 \leq c \leq 1 \\ \frac{15c^2-14c+4}{15c^2} & \text{for } 1 \leq c \end{cases},$$

$$\rho_S = \begin{cases} \frac{c^2(19c^2-126c+210)}{210} & \text{for } 0 \leq c \leq 1 \\ \frac{c^7-14c^6+84c^5-280c^4+770c^3-672c^2+238c-24}{210c^3} & \text{for } 1 \leq c \leq 2 \\ \frac{105c^3-105c+52}{105c^3} & \text{for } 2 \leq c \end{cases}.$$

Note that when $c = 1$, Johnson and Tenenbein's trivariate reduction model reduces to the simple additive model considered in Section 7.3.

A generalized additive model also includes the situation in which

$$X = X_1 + aX_3, \quad Y = X_2 + bX_3.$$

In this case, X and Y are positively quadrant dependent provided X_i's are mutually independent, with a and b having the same sign; see Example 1(ii) of Lehmann (1966).

7.4.2 Mathai and Moschopoulos' Bivariate Gamma

Mathai and Moschopoulos (1991) constructed a bivariate gamma distribution whose components are positively correlated and have three-parameter distributions. Denote the three-parameter (shape, scale, and location) gamma by $V_i \sim \Gamma(\alpha_i, \beta_i, \gamma_i)$, $i = 0, 1, 2$, and let

$$X = \frac{\beta_1}{\beta_0} V_0 + V_1, \qquad Y = \frac{\beta_2}{\beta_0} V_0 + V_2.$$

The X and Y so defined have a bivariate distribution with gamma marginals.

Mathai and Moschopoulos (1992) constructed another form of bivariate gamma distribution. Let V_i, $i = 1, 2$, be as defined above. Form

$$X = V_1, \qquad Y = V_1 + V_2.$$

Then X and Y clearly have a bivariate gamma distribution. The construction above is only part of a multivariate setup motivated by considering the joint distribution of the total waiting times of a renewal process.

7.4.3 Lai's Structure Mixture Model

Lai (1994) proposed a method of constructing bivariate distributions by extending a model proposed by Zheng and Matis (1993). The generalized model may be considered as a modified structure mixture model and has the form

$$X = X_1 + I_1 X_3, \quad Y = X_2 + I_2 X_3, \tag{7.13}$$

where X_i's are independent random variables and I_i $(i = 1, 2)$ are indicator random variables that are independent of X_i, but (I_1, I_2) has a joint probability mass function with joint probabilities p_{ij}, $i, j = 0, 1$.

It is easy to verify that

$$I_1 = \begin{cases} 1 & \text{with probability } \ \pi_1 = p_{10} + p_{11} \\ 0 & \text{with probability } \ 1 - \pi_1 = p_{00} + p_{01} \end{cases}$$

and

$$I_2 = \begin{cases} 1 & \text{with probability } \pi_2 = p_{01} + p_{11} \\ 0 & \text{with probability } 1 - \pi_2 = p_{00} + p_{10}. \end{cases}$$

Denote the mean and variance of X_i by μ_i and σ_i^2, respectively. We then obtain the following properties.

Marginal Properties

We have

$$E(X) = \mu_1 + \pi_1\mu_3 \quad \text{and} \quad E(Y) = \mu_2 + \pi_2\mu_3,$$

and

$$\text{var}(X) = \sigma_1^2 + \pi_1\sigma_3^2 + \pi_1(1-\pi_1)\mu_3^2 \quad \text{and} \quad \text{var}(Y) = \sigma_2^2 + \pi_2\sigma_3^2 + \pi_2(1-\pi_1)\mu_3^2.$$

Correlation Coefficient

Pearson's correlation can be shown to be

$$\rho = \frac{p_{11}(\sigma_3^2 + \mu_3^2) - \pi_1\pi_2\mu_3^2}{\{[\sigma_1^2 + \pi\sigma_3^2 + \pi_1(1-\pi_1)\mu_3^2][\sigma_2^2 + \pi\sigma_3^2 + \pi_2(1-\pi_2)\mu_3^2]\}^{1/2}}. \tag{7.14}$$

The correlation can be negative or positive depending on the values of p_{ij}. Lai (1994) has given lower and upper bounds for ρ.

7.4.4 Latent Variables-in-Common Model

In assessing the health of plants, two raters often show more disagreement about the relatively healthy plants than about the less healthy ones. It is reasonable to assume that each rater may commit an error in judgment.

A common idea in fields such as plant science is that there is a true level of health of any particular plant (H, say), and that the two opinions about this are respectively, $X = H + E_1$ and $Y = H + E_2$, where E_1 and E_2 represent errors, independent of each other and of H. This may be termed a model with latent variables in common.

Hutchinson (2000) proposed a generalization in which the variability of the errors is greater for large valued of H than for small values. Then the new model may be written as

$$X = H + E_1 \cdot \exp(a + bH),$$
$$Y = H + E_2 \cdot \exp(a + bH).$$

Here, H, E_1, E_2 can be taken as mutually independent normal distributions. Clearly, now the E's are multiplied by something that is bigger when H is big than when H is small. When $b = 0$, the bivariate normal model will be obtained.

7.4.5 Bivariate Skew-Normal Distribution

A random variable Z is said to be *skew-normal* with parameter λ if its density function is given by

$$\phi(z;\lambda) = 2\phi(z)\Phi(\lambda z), \quad -\infty < z < \infty, \tag{7.15}$$

where $\phi(z)$ and $\Phi(z)$ denote the $N(0,1)$ density and distribution function, respectively. The parameter λ varying in $(-\infty,\infty)$ regulates the skewness and $\lambda = 0$ corresponds to the standard normal density. Azzalini and Dalla-Valle (1996) have shown that the distribution can be derived in two ways:

(1) Let (X, Y) have a bivariate normal density with standardized marginals with correlation δ. Then, the conditional distribution of Y given $X > 0$ has a skew-normal distribution with parameter λ that is a function of δ.

(2) If Y_0 and Y_1 are independent unit normals and $\delta \in (-\infty,\infty)$, then

$$Z = \delta|Y_0| + (1-\delta^2)^{1/2}Y_1$$

is skew-normal, with λ depending on δ.

Bivariate Skew-Normal

Define

$$\left.\begin{aligned} X &= \delta_1|Y_0| + (1-\delta_1^2)^{\frac{1}{2}}Y_1 \\ Y &= \delta_2|Y_0| + (1-\delta_2^2)^{\frac{1}{2}}Y_2 \end{aligned}\right\}, \tag{7.16}$$

where (Y_1, Y_2) has a standardized bivariate normal distribution and Y_0 has a standard normal distribution independent of (Y_1, Y_2). Then, (X, Y) has a bivariate skew-normal distribution with density

$$h(x,y) = 2\phi(x,y;\omega)\Phi(\alpha_1 x + \alpha_2 y), \tag{7.17}$$

where ω is the correlation coefficient between Y_1 and Y_2 that has the standard bivariate normal distribution, and α_i, $i = 1, 2$ depends on ω and the δ's.

Applications

(1) The bivariate skew-normal model has been fitted to a weight versus height dataset of athletes from the Australian Institute of Sport and reported by Cook and Weisberg (1984); see Azzalini and Dalla-Valle (1996) for details.
(2) Gupta and Brown (2001) have established $P(X < Y)$ in the context of a strength–stress model. The bivariate skew-normal model is fitted to a dataset from Roberts (1988), and then the probability that the IQ score for white employees is less than the IQ score of nonwhite employees is estimated.
(3) For further statistical applications of multivariate skew-normal distributions, one may refer to Azzalini and Capitanio (1999).

7.4.6 Ordered Statistics

Jamalizadeh and Balakrishnan (2008) derived the distributions of order statistics from bivariate skew-normal and bivariate skew-t_ν distributions in terms of generalized skew-normal distributions, and used them to obtain explicit expressions for means, variances and covariance. Here, by generalized skew-normal distribution, we mean the distribution of $X|(U_1 < \theta_1 X, U_2 < \theta_2 X)$ when $X \rightsquigarrow N(0,1)$ independently of $(U_1, U_2)^T \rightsquigarrow \mathrm{BVN}(0,0,1,1,\gamma)$. This distribution, which is a special case of the unified multivariate skew-normal distribution introduced by Arellano-Valle and Azzalini (2006), has also been utilized by Jamalizadeh and Balakrishnan (2009) to obtain a mixture representation for the distributions of order statistics from a trivariate normal distribution. These authors also carried out a similar work for order statistics from the trivariate skew-t_ν distribution by showing that they are indeed mixtures of a generalized skew-t_ν distribution.

Remark

- A bivariate (multivariate) skew-Cauchy distribution is discussed in Arnold and Beaver (2000). The derivation is similar to that for the bivariate skew-normal distribution.
- Two other alternative approaches to derive the multivariate skew-normal distribution have been given, one by Jones (2002) and the other by Branco and Dey (2001), who introduces an extra parameter to regulate skewness to obtain a class of multivariate skew-elliptical distributions.

7.5 Weighted Linear Combination

7.5.1 Derivation

Let

$$\left.\begin{array}{l} X = U_1 \\ Y = cU_1 + (1-c)U_2 \end{array}\right\} \tag{7.18}$$

$(0 \le c \le 1)$, where the U_i's are i.i.d. random variables.

7.5.2 Expression of the Joint Density

If the U_i's have a negative exponential distribution, then

$$h(x,y) = \frac{1}{1-c} e^{-x-y+2cx}, \tag{7.19}$$

the support being part of the positive quadrant.

If the U_i's have a Laplace distribution, then

$$h(x,y) = \frac{1}{4(1-c)} e^{(-|x|-|y-cx|)/(1-c)}, \tag{7.20}$$

the support being the whole plane.

If the U_i's have a uniform distribution, then

$$h(x,y) = \frac{1}{1-c}, \tag{7.21}$$

the support being part of the unit square.

For further details, see Johnson and Tenenbein (1979, 1981).

7.5.3 Correlation Coefficients

For Spearman's rank correlation and Kendall's τ, Johnson and Tenenbein (1979, 1981) presented τ and ρ_S for the following three choices of distributions of X_1 and X_2:

Exponential:

$$\tau = c, \quad \rho_S = c(3-c)/(2-c),$$

and hence

$$\rho_S = \tau(3-2\tau)/(2-\tau).$$

Laplace:

$$\tau = c(3 + 3c - 2c^2)/4, \quad \rho_S = c(9 - 18c^2 + 14c^3 - 3c^4)/[2(2 - c)^2].$$

Uniform:

$$\tau = \begin{cases} \dfrac{4c - 5c^2}{6(1 - c)^2}, & 0 \le c \le 0.5 \\ \dfrac{11c^2 + 16c + 1}{6c^2}, & 0.5 \le c \le 1 \end{cases},$$

$$\rho_S = \begin{cases} \dfrac{c(10 - 13c)}{10(1 - c)^2}, & 0 \le c \le 0.5 \\ \dfrac{3c^2 + 16c^2 - 11c + 2}{10c^3}, & 0.5 \le c \le 1 \end{cases}.$$

7.5.4 Remarks

For a given distribution of the U_i's, these distributions have the "monotone regression dependence" property; i.e., the degree to which they are regression dependent is a monotone function of the parameter indexing the family, c [Bilodeau (1989)].

7.6 Bivariate Distributions Having a Common Denominator

7.6.1 Explanation

In this section, we let X_3, independent of X_1 and X_2, be the common denominator of X and Y, which are defined as

$$X = X_1/X_3, \quad Y = X_2/X_3. \tag{7.22}$$

Many of the well-known bivariate distributions are generated this way, and we will give several examples.

Remark: Ratio variables are sometimes known as index variables in some disciplines.

7.6.2 Applications

Turning away from distribution construction for a moment, a similar pair of equations is often used the in data analysis context, with X_3 being some general measurement of size. For example, in economics, X_1 and X_2 may be measures of the total wealth of a country and X_3 its population, and, in biology, X_1 and X_2 may be the lengths of parts of an animal's body and X_3 its overall length. In any particular application, there might be controversy over whether an empirical positive correlation between the two ratios X_1/X_3 and X_2/X_3 is genuine or results spuriously from dividing by the same factor, X_3. This subject is connected to ideas of "neutrality"; see also Pendleton (1986) and Prather (1988). More recently, Kim (1999) considered the correlation between birth rates and death rates of 97 countries from a dataset reported in the UNESCO 1990 Demographic Year Book. In this case, X_1, X_2, and X_3 denote the number of births, number of deaths, and the size of the population, respectively.

7.6.3 Correlation Between Ratios with a Common Divisor

Pearson (1897) investigated the correlation of ratios of bone measurements and found that although the correlation among the original measures was low, the correlations among ratios with common measures were about 0.5. He concluded that "part [of the correlation between ratio variables that] is solely due to the nature of [the] arithmetic ... is spurious" (p. 491).

The issue of spuriousness of correlations between ratio variables that have a common element has been raised by numerous authors across many disciplines, such as psychology, management, etc. Dunlap et al. (1997) have provided an excellent review on the subject.

Let V_X be the coefficient of variation of a random variable X; i.e., $V_X = \sqrt{\text{var}(X)}/E(X)$. Assuming the X_i's are uncorrelated, Pearson's (1897) approximate formula for the correlation between X and Y is

$$\rho(X,Y) \approx \frac{\rho(X_1,X_2)V_{X_1}V_{X_2} - \rho(X_1,X_3)V_{X_1}V_{X_3} - \rho(X_2,X_3)V_{X_2}V_{X_3} + V_{X_3}^2}{\sqrt{(V_{X_1}^2 + V_{X_3}^2 - 2\rho(X_1,X_3)V_{X_1}V_{X_3})(V_{X_2}^2 + V_{X_3}^2 - 2\rho(X_2,X_3)V_{X_2}V_{X_3})}}\,.$$

Kim (1999) presented the exact formula for the correlation between X and Y when the X_i's are independent as

$$\rho(X,Y) = \frac{V_{1/Z_3}^2 \text{sign}(E(X_1))\text{sign}(E(X_2))}{[V_{X_1}^2(1 + V_{1/X_3}^2)V_{1/X_3}^2] + [V_{X_2}^2(1 + V_{1/X_3}^2)V_{1/X_3}^2]}\,. \tag{7.23}$$

If X_1 and X_2 take positive values, or more generally, when $E(X_1)$ and $E(X_2)$ have the same sign, then the formula above becomes

$$\rho(X,Y) = \frac{V^2_{1/Z_3}}{[V^2_{X_1}(1+V^2_{1/X_3})V^2_{1/X_3}] + [V^2_{X_2}(1+V^2_{1/X_3})V^2_{1/X_3}]} . \tag{7.24}$$

The Case Where All the CV's Are Equal

Consider the case where the coefficients of variation of all variables are equal. Dunlap et al. (1997) have shown that Pearson's approximation formula is simplified to

$$\rho(X,Y) = \frac{1-\rho(X_1,X_3)-\rho(X_2,X_3)+\rho(X_1,X_2)}{2(1-\rho(X_1,X_3))^{1/2}(1-\rho(X_2,X_3))^{1/2}} .$$

Even if the three variables X_1, X_2, X_3 are all independent, the correlation among ratios with a common denominator would not equal 0; instead the equation above simplified to 0.5.

7.6.4 Compounding

The denominator-in-common version of the trivariate reduction method of constructing bivariate distribution sets through $X = X_1/X_3$ and $Y = X_2/X_3$ may readily be seen to be equivalent to compounding of a scale parameter. Suppose we instead write it as $X = X_1/\theta$ and $Y = X_2/\theta$. Then, we have

$$\begin{aligned} H(x,y) &= \Pr(X \le x, Y \le y) \\ &= \Pr(X_1 \le \theta x, X_2 \le \theta y) \\ &= \int \Pr(X_1 \le \theta x)\Pr(X_2 \le \theta y) f(\theta)d\theta \\ &= \int F_{X_1}(\theta x)F_{X_2}(\theta y)f(\theta)d\theta; \end{aligned}$$

see, for example, Lai (1987).

7.6.5 Examples of Two Ratios with a Common Divisor

Example 7.1 (Bivariate Cauchy Distribution). Let X_1 and X_2 be two independent normal variates and X_3 independent of X_1 and X_2 be distributed as χ_1 (i.e., chi-distribution with 1 degree of freedom). Then, the joint distribution of $X = X_1/X_3$ and $Y = X_2/X_3$ is a bivariate Cauchy distribution.

Example 7.2 (Bivariate t-Distribution). Let X_1 and X_2 have a joint standard bivariate normal density and X_3, independent of X_1 and X_2, be distributed as χ_ν. Then, the joint distribution of $X = X_1/(X_3/\sqrt{\nu})$ and $Y = X_2/(X_3/\sqrt{\nu})$ is a bivariate t-distribution with ν degrees of freedom.

Example 7.3 (Bivariate F-Distribution). Let X_1, X_2, and X_3 be independent chi-squared random variates with ν_1, ν_2, and ν_3 degrees of freedom, respectively. Then, $X = \frac{X_1/\nu_1}{X_3/\nu_3}$ and $Y = \frac{X_2/\nu_2}{X_3/\nu_3}$ have a joint bivariate t-density; see Mardia (1970, pp. 92–93). We may generalize the distribution above to the case where X_1 and X_2 have noncentrality parameters λ_1 and λ_2, respectively. The correlation structure for this generalized bivariate F-distribution is considered in detail by Feingold and Korsog (1986).

Example 7.4 (Jensen's Bivariate F-Distribution). Let X_1 and X_2 have a correlated chi-squared distribution of Kibble's type with shape parameter $\alpha = n/2$, and X_3, independent of X_1 and X_2, also be chi-squared, with m degrees of freedom. Then, $X = \frac{X_1/n}{X_3/m}$ and $Y = \frac{X_2/n}{X_3/m}$ have a bivariate F-distribution of Krishnaiah's (1964, 1965) type. More generally, let Q_1 and Q_2 follow Jensen's (1970) bivariate chi-squared distribution with degrees of freedom r and s, respectively, and V, independent of Q_1 and Q_2, be a chi-squared variate with ν degrees of freedom. Then, $X = \frac{Q_1/r}{V/\nu}$ and $Y = \frac{Q_2/s}{V/\nu}$ follow Jensen's bivariate F-distribution.

Example 7.5 (Bivariate Pareto Distribution). Suppose X_1 and X_2 are independent unit exponential variates, and X_3, independent of X_1 and X_2, has a gamma distribution. The joint distribution of X and Y is then bivariate Pareto. More generally, if X_1 and X_2 have unit gamma distributions instead, then a bivariate inverted beta distribution is the resulting distribution.

If we suppose (X_1, X_2) has a Farlie–Gumbel–Morgenstern distribution with unit exponential marginals and that X_3 has an independent gamma distribution with shape parameter c, then the pair $X = X_1/X_3$, $Y = X_2/X_3$ has a bivariate distribution with Pareto marginals; see Johnson (1987, pp. 170–171).

Example 7.6 (Bivariate Inverted Beta Distribution). Suppose X_1, X_2, and X_3 are independent gamma variables with shape parameters α_i ($i = 1, 2, 3$). Then, the pair $X = X_1/X_3$, $Y = X_2/X_3$ has the standard inverted beta distribution; see Tiao and Guttman (1965).

7.6.6 Bivariate t-Distribution with Marginals Having Different Degrees of Freedom

The nature of having the same denominator has been generalized by Jones (2002).

Let X_1, X_2 and W_1, W_2 be mutually independent random variables, each X_i following the standard normal distribution and W_i following the chi-squared distribution with n_i degrees of freedom. For the sake of convenience, we let $\nu_1 = n_1$ and $\nu_2 = n_1 + n_2$, so that $\nu_1 \leq \nu_2$. In the case where $\nu_1 = \nu_2$, we define $W_2 \equiv 0$.

Define a pair of random variables as follows:

$$X = \frac{\sqrt{\nu_1} X_1}{\sqrt{W_1}}, \quad Y = \frac{\sqrt{\nu_2} X_2}{\sqrt{W_1 + W_2}}. \tag{7.25}$$

Details on this distribution will be presented in Section 9.3.

7.6.7 Bivariate Distributions Having a Common Numerator

It is conceivable that one may be interested in the correlations among ratios that have a common numerator [i.e., corr $(X_3/X_1, X_3/X_2)$]. Assuming equal CV's, Dunlap et al. (1997) again simplified the approximation formula of Pearson (1897), giving

$$\rho(X, Y) = \frac{1 - \rho(X_1, X_3) - \rho(X_2, X_3) + \rho(X_1, X_2)}{2(1 - \rho(X_1, X_3))^{1/2}(1 - \rho(X_2, X_3))^{1/2}},$$

which was identical to the correlations among ratio variables with a common denominator. It is easy to see that ratios sharing a numerator will be spuriously correlated as badly as those sharing denominators.

7.7 Multiplicative Trivariate Reduction

In this section, we discuss the case where the transformation is multiplication.

7.7.1 Bryson and Johnson (1982)

Bryson and Johnson (1982) [and Chapter 8 of Johnson (1987)] draw attention to Khintchine's theorem, which states that any random variable X has a single mode at the origin if and only if it can be expressed as a product

$$X = ZU, \tag{7.26}$$

where Z and U are independent continuous variables, U having a uniform distribution on the unit interval; see, for instance, Feller (1971, Section V.9). For a given marginal density of X, f, the density f_Z has to be $-zf'(z)$, where f' is the derivative of f. Bryson and Johnson present a multiplicative version of trivariate reduction,

$$\left.\begin{aligned} X &= ZU_1 \\ Y &= ZU_2 \end{aligned}\right\}, \tag{7.27}$$

where (U_1, U_2) has any bivariate distribution that has uniform marginals. Z is referred to as a "generator" variable. Bryson and Johnson found the correlation between X and Y to be

$$\frac{1}{4}\left\{3 - c_X^{-2} + \rho_{(U)}\left(1 + c_X^{-2}\right)\right\}, \tag{7.28}$$

where c_X is the common coefficient of variation between X and Y, and $\rho_{(U)}$ is the correlation between U_1 and U_2. A consequence of Khintchine's theorem is $c_X^2 \geq \frac{1}{3}$; if U_1 and U_2 have normal or other symmetric distributions, they are uncorrelated, though they are independent only if the U_i's are.

Bryson and Johnson (1982) go on to discuss what they call Khintchine mixtures; see Section 7.8 below.

7.7.2 Gokhale's Model

Gokhale (1973) gave some attention to the scheme of construction

$$\left.\begin{aligned} X &= ZV_1 \\ Y &= ZV_2 \end{aligned}\right\}, \tag{7.29}$$

where V_1, V_2, and Z are independent beta variates whose parameters are either

- Respectively $(a, \theta), (a + m, \theta - m)$, and $(a + \theta, b + m - \theta)$, so that X and Y had beta distributions with parameters $(a, b + m)$ and $(a + m, b)$, respectively.
- Respectively $(a + \Delta, b - \Delta), (a + \Delta, b' - \Delta)$, and (a, Δ), so that X and Y had beta distributions with parameters (a, b) and (a, b'), respectively.

7.7.3 Ulrich's Model

Ulrich (1984) considered

$$\left.\begin{aligned} X &= Z_1 V_1 \\ Y &= Z_2 V_2 \end{aligned}\right\}, \tag{7.30}$$

where the Z_i's are independent, having gamma distributions (with unit scale parameter and shape parameter $\alpha_i + \phi$), and the V_i's, independent of the Z_i's but possibly not mutually independent, have beta distributions with parameters α_i and ϕ. The scheme of dependence that Ulrich paid most attention to is that of his beta mixture. He referred to the resulting distribution of (X, Y) as the "bivariate product gamma."

7.8 Khintchine Mixture

This section may not quite fit well with the rest of this chapter, but it does have a similar flavor.

7.8.1 Derivation

Continuing the discussion of bivariate distributions suggested by Bryson and Johnson (1982) and Johnson (1987, Chapter 8) that we started in Section 7.7.1, let

$$\left.\begin{aligned} X &= Z_1 U_1 \\ Y &= Z_2 U_2 \end{aligned}\right\}, \tag{7.31}$$

where the U_i's are uniformly distributed on $(0, 1)$ and either:

- the U_i's are independent and the Z_i's are either identical (with probability p) or independent (with probability $1 - p$), or
- the Z_i's are independent and the U_i's are either identical (with probability q) or independent (with probability $1 - q$).

As before, the Z_i's are referred to as "generator" variables.

7.8.2 Exponential Marginals

If X and Y are to have exponential marginals, Bryson and Johnson gave these results:

- The case of independent U_i's and identical Z_i's gave a p.d.f. of $-\mathrm{Ei}[\max(x, y)]$, where $\mathrm{Ei}(\cdot)$ is the exponential integral.

- The case of independent Z_i's and identical U_i's gives a p.d.f. of

$$\frac{xy}{(x+y)^3}[2+2(x+y)+(x+y)^2]e^{-(x+y)}.$$

- In the fully independent case, the p.d.f. is $e^{-(x+y)}$.

The correlation is $p/2$ if the first and the third are mixed in proportions $p:1-p$, $q/3$ if the second and the third are mixed in proportions $q:1-q$, and $\frac{p}{2}+\frac{q}{3}$ if all three are mixed in proportions $p:q:1-p-q$.

The following five cases have been illustrated (contour and three-dimensional plots of the p.d.f.'s) by Johnson et al. (1981): independent U_i's, independent Z_i's (i.e., $p=q=0$); independent U_i's, $p=0.6$; independent U_i's, identical Z_i's; independent Z_i's, $q=0.6$; independent Z_i's, identical U_i's. The final one has also been shown in Figure 8.2 of Johnson (1987).

7.8.3 Normal Marginals

This case has also been treated by Bryson and Johnson, but the formulas are more complicated than in the exponential case. The following six cases were illustrated (contour and three-dimensional plots of the p.d.f.'s) by Johnson et al. (1981): $q=0, p=0$; $q=0, p=0.25$; $q=0, p=0.5$; $q=0, p=0.74$; $q=0, p=1$; $q=0.25, p=0.75$. The two cases $p=1$ and $q=1$ are illustrated in Figures 8.3 and 8.4 of Johnson (1987).

Three examples in which the U_i's have the Farlie–Gumbel–Morgenstern distribution are illustrated by Bryson and Johnson (1982) and Johnson (1987, Figures 8.5–8.7). The density is given by

$$h(x,y)=\frac{\alpha xy}{2\max(|x|,|y|)}\phi[\max(|x|,|y|)]+\frac{1-\alpha xy}{2}\{1-\Phi[\max(|x|,|y|)]\}. \tag{7.32}$$

These are illustrated in Johnson et al. (1984, pp. 239–242) and Johnson (1986).

7.8.4 References to Generation of Random Variates

Devroye (1986, pp. 603–604) and Johnson et al. (1984, pp. 239–240) have discussed the random generation from these distributions.

7.9 Transformations Involving the Minimum

Let X_i $(i = 1, 2, 3)$ belong to the same one-parameter family of distribution functions $F(x_i; \lambda_i)$. (We assume that the other parameters, if present, are common to all X_i.) We now wish to find the family that is closed under the transformation $T(X_1, X_2) = \min(X_1, X_3)$; i.e., we want to find distribution functions $F(x; \lambda)$ such that

$$\bar{F}(x; \lambda_1)\bar{F}(x; \lambda_3) = \bar{F}(x; \lambda_1 + \lambda_3), \tag{7.33}$$

where $\bar{F}$, as usual, is $1 - F$. This in turn implies that

$$\bar{F}(x; \lambda) = [\bar{F}(x)]^{\lambda}. \tag{7.34}$$

There are several continuous distributions satisfying the above [see Arnold (1967)]—exponential, Pareto, and Weibull. Marshall and Olkin (1967) constructed their bivariate exponential distribution by taking F to be the exponential distribution and defining $X = \min(X_1, X_3)$ and $Y = \min(X_2, X_3)$, thus giving

$$\bar{H}(x, y) = \exp[-\lambda_1 x - \lambda_2 y - \lambda_3 \max(x, y)]. \tag{7.35}$$

The case of T being the maximum can be discussed similarly.

7.10 Other Forms of the Variables-in-Common Technique

7.10.1 Bivariate Chi-Squared Distribution

Let X_1, X_2, X_3 be independent univariate normal variates, and define

$$\left.\begin{aligned} X &= X_1^2 + X_3^2 \\ Y &= X_2^2 + X_3^2 \end{aligned}\right\}. \tag{7.36}$$

Then, X and Y have a joint bivariate chi-squared distribution (with two degrees of freedom), and the joint moment generating function is

$$E(e^{sX+tY}) = \{[1 - 2(s + t)](1 - 2s)(1 - 2t)\}^{-1/2}. \tag{7.37}$$

The joint density is not of a simple form. This is an example of Cherian's construction of a bivariate gamma distribution discussed earlier.

Note that if X_1^2, X_2^2, X_3^2 are each supposed to have a χ_2^2-distribution (i.e., exponential), the joint density function of X and Y takes the simple form

$$h(x,y) = \left(e^{-\max(x,y)/2} - e^{-(x+y)/2}\right)/4. \qquad (7.38)$$

Note that the marginals are not exponential but χ^4-distributions; see Johnson and Kotz (1972, pp. 260–261).

7.10.2 Bivariate Beta Distribution

This example illustrates that X and Y may have more than one variable in common.

Let X_i $(i = 1, 2, 3)$ be independent and have gamma distributions with shape parameters θ_i. Consider

$$\left.\begin{aligned} X &= X_1/(X_1 + X_2 + X_3) \\ Y &= X_2/(X_1 + X_2 + X_3) \end{aligned}\right\}. \qquad (7.39)$$

Then, X and Y have a bivariate beta distribution. We will obtain the same bivariate beta density if the X_i's in (7.39) are three independent beta variates with parameters $(\theta_i, 1)$, respectively, conditional on $X_1 + X_2 + X_3 \le 1$.

7.10.3 Bivariate Z-Distribution

Consider three independent gamma variates X_1, X_2 and X_3 with shape parameters α, β, and ν, respectively. Form two variables X and Y as follows:

$$\left.\begin{aligned} X &= \log X_3 - \log X_1 \\ Y &= \log X_3 - \log X_2 \end{aligned}\right\}. \qquad (7.40)$$

The joint moment generating function of X and Y can be obtained in a straightforward manner as

$$M(s,t) = \frac{\Gamma(\nu + s + t)\Gamma(\alpha - s)\Gamma(\beta - t)}{\Gamma(\alpha)\Gamma(\nu)}. \qquad (7.41)$$

By inverting the moment generating function in (7.41), we obtain as the joint density function of X and Y

$$h(x,y) = \frac{\Gamma(\nu + \alpha + \beta)}{\Gamma(\alpha)\Gamma(\nu)} \frac{e^{-\alpha x - \beta y}}{(1 + e^{-x} + e^{-y})^{\alpha+\beta+\gamma}}. \qquad (7.42)$$

By writing $X = -\log(X_1/X_3)$ and $Y = -\log(X_2/X_3)$, we see that the distribution of (X,Y) is simply a logarithmic transformation of the bivariate inverted beta distribution discussed earlier; see Hutchinson (1979, 1981) and Lee (1981). As the marginals are Z-distributions, we may call (7.42) a bivariate Z-distribution or a generalized logistic distribution; see Malik and Abraham (1973), Lindley and Sinpurwalla (1986), and Balakrishnan (1992).

Some methods specifically oriented toward the reliability context with exponential distribution have also been discussed by Tosch and Holmes (1980), Lawrance and Lewis (1983), and Raftery (1984, 1985).

References

1. Arellano-Valle, R.B., Azzalini, A.: On the unification of families of skew-normal distributions. Scandinavian Journal of Statistics **33**, 561–574 (2006)
2. Arnold, B.C.: A note on multivariate distributions with specified marginals. Journal of the American Statistical Association **62**, 1460–1461 (1967)
3. Arnold, B.C., Beaver, R.J.: The skew-Cauchy distribution. Statistics and Probability Letters **49**, 285–290 (2000)
4. Azzalini, A., Capitanio, A.: Statistical applications of multivariate skew normal distribution. Journal of the Royal Statistical Society, Series B **61**, 579–602 (1999)
5. Azzalini, A., Dalla-Valle, D.: The multivariate skew-normal distribution. Biometrika **83**, 715–726 (1996)
6. Balakrishnan, N. (ed.): Handbook of the Logistic Distribution. Marcel Dekker, New York (1992)
7. Bilodeau, M.: On the monotone regression dependence for Archimedian bivariate uniform. Communications in Statistics: Theory and Methods **18**, 981–988 (1989)
8. Branco, M.D., Dey, D.K.:, A general class of multivariate skew-elliptical distributions. Journal of Multivariate Analysis **79**, 99–113 (2001)
9. Bryson, M.C., Johnson, M.E.: Constructing and simulating multivariate distributions using Khintchine's theorem. Journal of Statistical Computation and Simulation **16**, 129–137 (1982)
10. Cherian, K.C.: A bivariate correlated gamma-type distribution function. Journal of the Indian Mathematical Society **5**, 133–144 (1941)
11. Cook, R.D., Weisberg, S.: An Introduction to Regression Graphics. John Wiley and Sons, New York (1984)
12. Curnow, R.N.: The consequences of errors of measurement for selection from certain non-normal distributions. Bulletin de l'Institut International de Statistique **37**, 291–308 (1958)
13. David, F.N., Fix, E.: Rank correlation and regression in a nonnormal surface. In: Proceedings of the 4th Berkeley Symposium on Mathematical Statistics and Probability, Volume 1, J. Neyman (ed.), pp. 177–197. University of California Press, Berkeley (1961)
14. de Silva, B.M., Griffiths, R.C.: A test of independence for bivariate symmetric stable distributions. Australian Journal of Statistics **22**, 172–177 (1980)
15. Devroye, L.: Nonuniform Random Variate Generation. Springer-Verlag, New York (1986)
16. Dunlap, W.P., Dietz, J., Cortina, J.M.: The spurious correlation of ratios that have common variable: A Monte Carlo examination of Pearson's formula. The Journal of General Psychology **124**, 182–193 (1997)
17. Eagleson, G.K.: Polynomial expansions of bivariate distributions. Annals of Mathematical Statistics **35**, 1208–1215 (1964)
18. Eagleson, G.K., Lancaster, H.O.: The regression system of sums with random elements in common. Australian Journal of Statistics **9**, 119–125 (1967)

19. Feingold, M., Korsog, P.E.: The correlation and dependence between two F statistics with the same denominator. The American Statistician **40**, 218–220 (1986)
20. Feller, W. An Introduction to Probability Theory, Volume 2, 2nd edition. John Wiley and Sons, New York (1971)
21. Frangopol, D.M., Hong, K.: Probabilistic analysis of rock slopes including correlation effects. In: Reliability and Risk Analysis in Civil Engineering, Volume 2, N.C. Lind (ed.), pp. 733–740. Institute for Risk Research, University of Waterloo, Waterloo, Ontario (1987)
22. Gokhale, D.V.: On bivariate distributions with beta marginals. Metron **31**, 268–277 (1973)
23. Griffiths, R.C.: On a bivariate triangular distribution. Australian Journal of Statistics **20**, 183–185 (1978)
24. Gupta, R.C., Brown, N.: Reliability studies of the skew-normal distribution and its application to strength–stress model. Communications in Statistics: Theory and Methods **30**, 2427–2445 (2001)
25. Hutchinson, T.P.: Four applications of a bivariate Pareto distribution. Biometrical Journal **21**, 553–563 (1979)
26. Hutchinson, T.P.: Compound gamma bivariate distributions. Metrika **28**, 263–271 (1981)
27. Hutchinson, T.P.: Assessing the health of plants: Simulation helps us understand observer disagreements. Environmetrics **11**, 305–314 (2000)
28. Jamalizadeh, A., Balakrishnan, N.: On order statistics from bivariate skew-normal and skew-t_ν distributions. Journal of Statistical Planning and Inference **138**, 4187–4197 (2008)
29. Jamalizadeh, A., Balakrishnan, N.: Order statistics from trivariate normal and t_ν-distributions in terms of generalized skew-normal and skew-t_ν distributions. Journal of Statistical Planning and Inference (to appear)
30. Jensen, D.R.: The joint distribution of quadratic forms and related distributions. Australian Journal of Statistics **12**, 13–22 (1970)
31. Johnson, M.E.: Distribution selection in statistical simulation studies. In: 1986 Winter Simulation Conference Proceedings, J.R. Wilson, J.O. Henriksen, and S.D. Roberts (eds.) pp. 253–259. Institute of Electrical and Electronics Engineers, Piscataway, New Jersey (1986)
32. Johnson, M.E.: Multivariate Statistical Simulation. John Wiley and Sons, New York (1987)
33. Johnson, M.E., Bryson, M.C., Mills, C.F.: Some new multivariate distributions with enhanced comparisons via contour and three-dimensional plots. Report LA-8903-MS, Los Alamos Scientific Laboratory, Los Alamos, New Mexico (1981)
34. Johnson, M.E., Tenenbein, A.: Bivariate distributions with given marginals and fixed measures of dependence. Informal Report LA-7700-MS, Los Alamos Scientific Laboratory, Los Alamos, New Mexico (1979)
35. Johnson, M.E., Tenenbein, A.: A bivariate distribution family with specified marginals. Journal of the American Statistical Association **76**, 198–201 (1981)
36. Johnson, M.E., Wang, C., Ramberg, J.: Generation of continuous multivariate distributions for statistical applications. American Journal of Mathematical and Management Sciences **4**, 225–248 (1984)
37. Johnson, N.L., Kotz, S.: Distributions in Statistics: Continuous Multivariate Distributions, John Wiley and Sons, New York (1972)
38. Jones, M.C.: A dependent bivariate t distribution with marginals on different degrees of freedom. Statistics and Probability Letters **56**, 163–170 (2002)
39. Kim, J.H.: Spurious correlation between ratios with a common divisor. Statistics and Probability Letters **44**, 383–386 (1999)
40. Krishnaiah, P.R.: Multiple comparison tests in multivariate cases. Report ARL 64-124, Aerospace Research Laboratories, Wright-Patterson Air Force Base, Ohio (1964)

41. Krishnaiah, P.R.: On the simultaneous ANOVA and MANOVA tests. Annals of the Institute of Statistical Mathematics **17**, 35–53 (1965)
42. Lai, C.D.: Meixner classes and Meixner hypergeometric distributions. Australian Journal of Statistics **24**, 221–233 (1982)
43. Lai, C.D.: Letter to the editor. Journal of Applied Probability **24**, 288–289 (1987)
44. Lai, C.D.: Construction of bivariate distributions by a generalized trivariate reduction technique. Statistics and Probability Letters **25**, 265–270 (1994)
45. Lancaster, H.O.: Joint distributions in the Meixner classes. Journal of the Royal Statistical Society, Series B **37**, 434–443 (1975)
46. Lancaster, H.O.: Dependence, Measures and indices of. In: Encyclopedia of Statistical Sciences, Volume 2, S. Kotz and N.L. Johnson (eds.), pp. 334–339. John Wiley and Sons, New York (1982)
47. Lancaster, H.O.: Special joint distributions of Meixner variables. Australian Journal of Statistics **25**, 298–309 (1983)
48. Lawrance, A.J., Lewis, P.A.W.: Simple dependent pairs of exponential and uniform random variables. Operations Research **31**, 1179–1197 (1983)
49. Lee, P.A.: The correlated bivariate inverted beta distribution. Biometrical Journal **23**, 693–703 (1981)
50. Lehmann, E.L.: Some concepts of dependence. Annals of Mathematical Statistics **37**, 1137–1153 (1966)
51. Lindley, D.V., Singpurwalla, N.D.: Multivariate distributions for the life lengths of components of a system sharing a common environment. Journal of Applied Probability **23**, 418–431 (1986)
52. Malik, H.J., Abraham, B.: Multivariate logistic distributions. Annals of Statistics **1**, 588–590 (1973)
53. Mardia, K.V.: Families of Bivariate Distributions. Griffin, London (1970)
54. Marshall, A.W., Olkin, I.: A multivariate exponential distribution. Journal of the American Statistical Association **62**, 30–44 (1967)
55. Mathai, A.M., Moschopoulos, P.G.: On a multivariate gamma. Journal of Multivariate Analysis **39**, 135–153 (1991)
56. Mathai, A.M., Moschopoulos, P.G.: A form of multivariate gamma distribution. Annals of the Institute of Statistical Mathematics **44**, 97–106 (1992)
57. Meixner, J.: Orthogonale Polynomsysteme mit einer besonderen Gestalt der erzeugenden Funktion. Journal of the London Mathematical Society **9**, 6–13 (1934)
58. Morris, C.N.: Natural exponential families with quadratic variance functions. Annals of Statistics **10**, 65–80 (1982)
59. Morris, C.N.: Natural exponential families with quadratic variance functions: Statistical theory. Annals of Statistics **11**, 515–529 (1983)
60. Pearson, K.: On a form of spurious correlation which may arise when indices are used in the measurement of organs. Proceedings of the Royal Statistical Society of London, Series A **60**, 489–498 (1897)
61. Pendleton, B.F: Ratio correlation. In: Encyclopedia of Statistical Sciences, Volume 7, S. Kotz and N.L. Johnson (eds.), pp. 636–639. John Wiley and Sons, New York (1986)
62. Prather, J.E.: Spurious correlation. In: Encyclopedia of Statistical Sciences, Volume 8, S. Kotz and N.L. Johnson (eds.), pp. 613–614. John Wiley and Sons, New York (1988)
63. Raftery, A.E.: A continuous multivariate exponential distribution. Communications in Statistics: Theory and Methods **13**, 947–965 (1984)
64. Raftery, A.E.: Some properties of a new continuous bivariate exponential distribution. Statistics and Decisions, Supplement Issue No. 2, 53–58 (1985)
65. Rao, C.R.: Note on a problem of Ragnar Frisch. Econometrica **15**, 245–249 (1947)
66. Roberts, H.V.: Data Analysis for Managers with Minitab. Scientific Press, Redwood City, California (1988)

67. Sumita, U., Kijima, M.: The bivariate Laguerre transform and its applications: Numerical exploration of bivariate processes. Advances in Applied Probability **17**, 683–708 (1985)
68. Szántai, T.: Evaluation of special multivariate gamma distribution. Mathematical Programming Study **27**, 1–16 (1986)
69. Tiao, G.G., Guttman, I.: The inverted Dirichlet distribution with applications. Journal of the American Statistical Association **60**, 793–805 (Correction, **60**, 1251–1252) (1965)
70. Tosch, T.J., Holmes, P.T.: A bivariate failure model. Journal of the American Statistical Association **75**, 415–417 (1980)
71. Ulrich, G.: A class of multivariate distributions with applications in Monte Carlo and simulation. In: American Statistical Association, 1984 Proceedings of the Statistical Computing Section, pp. 185–188. American Statistical Association, Alexandria, Virginia (1984)
72. Zheng, Q., Matis, J.H.: Approximating discrete multivariate distributions from known moments. Communications in Statistics: Theory and Methods **22**, 3553–3567 (1993)

Chapter 8
Bivariate Gamma and Related Distributions

8.1 Introduction

Many of the bivariate gamma distributions considered in this chapter may be derived from the bivariate normal in some fashion, such as by marginal transformation. It is well known that a univariate chi-squared distribution can be obtained from one or more independent and identically distributed normal variables and that a chi-squared random variable is a special case of gamma; hence, it is not surprising that a bivariate gamma model is related to the bivariate normal one.

In this chapter, we present many different forms of bivariate gamma distributions that have been introduced in the literature and list their key properties and interconnections between them. In Section 8.2, we describe the form and features of Kibble's bivariate gamma distribution. In Section 8.3, we present Royen's bivariate gamma distribution and point out its close connection with Kibble's form. The bivariate gamma distribution of Izawa and its properties are described in Section 8.4. Next, the bivariate form of Jensen is discussed in Section 8.5. In Section 8.6, the bivariate gamma distribution of Gunst and Webster and its related models are described. The bivariate gamma model of Smith et al. is detailed next, in Section 8.7. The bivariate gamma distribution obtained from the general Sarmanov family and its properties are discussed in Section 8.8. The bivariate gamma model of Loáiciga and Leipnik is detailed next, in Section 8.9. The forms of bivariate gamma distributions of Cheriyan et al., Prékopa and Szántai, and Schmeiser and Lal are described next, in Sections 8.10, 8.11, and 8.12, respectively. The bivariate gamma distribution obtained from the general Farlie–Gumbel–Morgenstern family and its properties are discussed in Section 8.13. The bivariate gamma models of Moran and Crovelli are presented in Sections 8.14 and 8.15, respectively. Some applications of bivariate gamma distributions in the field of hydrology are mentioned in Section 8.16. Next, the bivariate gamma distributions proposed by McKay et al., Dussauchoy and Berland, Mathai and

N. Balakrishnan, C.D. Lai, *Continuous Bivariate Distributions*,
DOI 10.1007/b101765_9,

Moschopoulos, and Becker and Roux and their properties are described in Sections 8.17, 8.18, 8.19, and 8.20, respectively. Some other forms of bivariate gamma models obtained from the variables-in-common technique are mentioned in Section 8.21. The noncentral version of bivariate chi-squared distribution is discussed in Section 8.22. The bivariate gamma distribution of Gaver and its properties are detailed in Section 8.23. The bivariate gamma distributions of Nadarajah and Gupta and Arnold and Strauss are discussed in Sections 8.24 and 8.25, respectively. Finally, in Section 8.26, the bivariate mixture gamma distribution and its characteristics are presented.

8.2 Kibble's Bivariate Gamma Distribution

8.2.1 Formula of the Joint Density

The joint density function is

$$h(x,y) = f_\alpha(x) f_\alpha(y) \frac{\Gamma(\alpha)}{1-\rho} \exp\left\{ \frac{-\rho(x+y)}{1-\rho} \right\} (xy\rho)^{(\alpha-1)/2} I_{\alpha-1}\left(\frac{2\sqrt{xy\rho}}{1-\rho} \right) \tag{8.1}$$

$(x, y \geq 0,\ 0 \leq \rho < 1)$, where $f_\alpha(t) = \frac{1}{\Gamma(\alpha)}\, e^{-t} t^{\alpha-1}$ and $I_\alpha(\cdot)$ is the modified Bessel function of the first kind and order ν. The probability density function may also be expressed in terms of Laguerre polynomials[1] $L_j^{(\alpha-1)}$ as

$$h(x,y) = f_\alpha(x) f_\alpha(y) \sum_{j=0}^{\infty} L_j^{(\alpha-1)}(x) L_j^{(\alpha-1)}(y) \frac{\Gamma(\alpha)\Gamma(j+1)}{\Gamma(j+\alpha)}. \tag{8.2}$$

An alternative expression of the joint density function, obtained by Krishnaiah (1963) [see also Krishnaiah (1983)], is

$$h(x,y) = \frac{(1-\rho)^{-\alpha}}{\Gamma(\alpha)} \sum_{j=0}^{\infty} a_j (xy)^{\alpha+j-1} \exp\left(-\frac{x+y}{1-\rho} \right), \tag{8.3}$$

where $a_j = \frac{1}{\Gamma(\alpha+j)j!} \frac{\rho^j}{(1-\rho)^{2j}}$.

[1] $L_j^\alpha(x) = \sum_{j=0}^{j} \binom{j+\alpha}{j-k} \frac{(-x)^k}{k!} = \sum_{j=0}^{j} \binom{j+\alpha}{k+\alpha} \frac{(-x)^k}{k!}$. Note that $L_j^{(\alpha)}$ has not been normalized with respect to the marginal gamma density function.

8.2.2 Formula of the Cumulative Distribution Function

Expressed as an infinite series in terms of Laguerre polynomials, the joint distribution function is

$$H(x,y) = F_\alpha(x)F_\alpha(y) + \alpha\sum_{j=0}^{\infty}\frac{\rho^{j+1}}{(j+1)}\frac{\Gamma(\alpha+1)\Gamma(j+1)}{\Gamma(j+\alpha+1)}L_j^\alpha(x)L_j^\alpha(y)f_\alpha(x)f_\alpha(y), \tag{8.4}$$

where $F_\alpha(t) = \int_0^t f_\alpha(u)du$; see Lai and Moore (1984) for details.

Alternatively, the joint distribution function can also be expressed as

$$H(x,y) = \frac{(1-\rho)^\alpha}{\Gamma(\alpha)}\sum_{j=0}^{\infty} c_j F_{\alpha+j-1}\left(\frac{x}{1-\rho}\right)F_{\alpha+j-1}\left(\frac{y}{1-\rho}\right), \tag{8.5}$$

where $c_j = \frac{\Gamma(\alpha+j)\rho^j}{j!}$; see Johnson and Kotz (1972, p. 221).

8.2.3 Univariate Properties

The marginal distributions are both gamma with the same shape parameter α.

8.2.4 Correlation Coefficient

The parameter ρ in (8.1) is indeed Pearson's product-moment correlation coefficient.

8.2.5 Moment Generating Function

The joint moment generating function is

$$M(s,t) = [(1-s)(1-t) - \rho st]^{-\alpha}, \quad 0 < \rho < 0. \tag{8.6}$$

Thus, the moments $\mu'_{r,s}$ can be obtained easily from (8.6).

The joint moment generating function in (8.6) was first given by Wicksell (1933), but the explicit form of the density in (8.1) is due to Kibble (1941). For this reason, some authors refer to this distribution as the *Kibble–Wicksell bivariate gamma distribution.* Vere-Jones (1967) showed that this distribution is infinitely divisible.[2]

8.2.6 Conditional Properties

The regression is linear and is given by

$$E(Y|X = x) = \rho(x - \alpha) + \alpha. \tag{8.7}$$

The conditional variance is also linear and is given by

$$\text{var}(Y|X = x) = (1 - \rho)[2\rho x + \alpha(1 - \rho)]; \tag{8.8}$$

see Mardia (1970, p. 88).

8.2.7 Derivation

In the univariate situation, the derivation of the chi-squared distribution as the sum of squared normal variables is well known. Now, let $(X_1, Y_1), \ldots,$ (X_n, Y_n) be a random sample of size n from a bivariate normal distribution with mean $\mathbf{0}$ and variance–covariance matrix

$$\mathbf{\Sigma} = \begin{pmatrix} \sigma_1^2 & \rho_0\sigma_1\sigma_2 \\ \rho_0\sigma_1\sigma_2 & \sigma_2^2 \end{pmatrix}.$$

Define $X = \frac{1}{2\sigma_1^2}\sum_{i=1}^n X_i^2$ and $Y = \frac{1}{2\sigma_2^2}\sum_{i=1}^n Y_i^2$. Then, after replacing $n/2$ by α in the density function, the distribution of (X, Y) turns out to be Kibble's bivariate gamma with $\rho = \rho_0^2$. For a generalization to higher dimensions, one may refer to Krishnamoorthy and Parthasarathy (1951) and Krishnaiah and Rao (1961).

Clearly, the random variate generation is then easy when 2α is a fairly small integer.

---

[2] A bivariate distribution with characteristic function φ is said to be infinitely divisible if $\varphi^{1/n}$ is also a characteristic function for every positive integer n. In terms of r.v.'s, this means that, for each $n \geq 1$, the random variable with characteristic function φ can be written as $\mathbf{X} = \sum_{j=1}^n \mathbf{X}_{nj}$, where $\mathbf{X}_{nj}$ $(1 \leq j \leq n)$ are independent and identically distributed with characteristic function $\varphi^{1/n}$.

8.2.8 Relations to Other Distributions

- Downton's bivariate exponential distribution is a special case of this distribution; see Chapter 10 for pertinent details.
- According to Khan and Jain (1978), the quantity

$$\frac{u}{u+ax+by}f(x,y;u+ax+by) \tag{8.9}$$

is a p.d.f. of interest in the theory of emptiness of reservoirs, with u being the initial content of the reservoir and $f(x,y;t)$ being the p.d.f. for the amounts ax and by for the flows from two sources into the reservoir during time t. Khan and Jain used (8.9), where f is Kibble's density function. These authors then provided an expression for the p.d.f. and obtained the lower-order moments; see also Jain and Khan (1979, pp. 166–167).

8.2.9 Generalizations

- In Jensen's bivariate gamma distribution, (i) the shape parameters of the marginals are different and they are integers or half-integers, and (ii) the bivariate normal distributions used for derivation have different correlation coefficients. For this and further generalizations, one may refer to Section 8.5.
- Malik and Trudel (1985) expressed (8.2) as

$$h(x,y) = (1-\rho)^{\alpha}\sum_{j=0}^{\infty}\frac{\Gamma(\alpha+j)\rho^j(xy)^{\alpha+j-1}}{\Gamma(\alpha)j![\Gamma(\alpha+j)(1-\rho)^{\alpha+j}]^2}\exp\left(-\frac{x+y}{1-\rho}\right). \tag{8.10}$$

They then generalized the density above in the following form:

$$h(x,y) = (1-\rho)^{(\alpha_1+\alpha_2)/2}\times \sum_{j=0}^{\infty}\frac{\Gamma(\frac{\alpha_1+\alpha_2}{2}+j)\rho^j x^{\alpha_1+j-1}y^{\alpha_2+j-1}}{\Gamma(\frac{\alpha_1+\alpha_2}{2})j!\Gamma(\alpha_1+j)\Gamma(\alpha_2+j)(1-\rho)^{\alpha_1+\alpha_2+2j}}\exp\left(-\frac{x+y}{1-\rho}\right). \tag{8.11}$$

The marginals of this distribution, however, are not gamma unless $\alpha_1=\alpha_2$.

8.2.10 Illustrations

Surfaces and contours of a probability density function of Kibble's form have been provided by Smith et al. (1982). Contours of the probability density function for the cases $\rho = 0.5$, $\alpha = 1$ and $\rho = 0.5$, $\alpha = 2$ have been given by Izawa (1965).

8.2.11 Remarks

- It can be easily proved that [Jensen (1969)]

$$\Pr(c_1 \leq X \leq c_2, c_1 \leq Y \leq c_2) \geq \Pr(c_1 \leq X \leq c_2) \Pr(c_1 \leq Y \leq c_2). \tag{8.12}$$

  Jensen called this positive dependence, but we use this term in a different way in Chapter 3. In particular, we have

$$\Pr(X \leq x, Y \leq y) \geq \Pr(X \leq x) \Pr(Y \leq y)$$

  (i.e., X and Y are positively quadrant dependent); see Section 3.4.
- Izawa (1965) presented formulas for the density and moments of the sum, product, and ratio of X and Y.
- For results on the location of the mode, see Brewer et al. (1987).
- For a brief account of this distribution, in the context of others with gamma marginals, one may refer to Krishnaiah (1985).

8.2.12 Fields of Applications

- **Electric counter system.** Lampard (1968) built this distribution in the conditional manner, $h = f(x)g(y|x)$; his context was a system of two reversible counters (i.e., an input can either increase or decrease the cumulative count), with two Poisson inputs (an increase process and a decrease process). Output events occur when either of the cumulative counts decreases to zero. The sequence of time intervals between outputs forms a Markov chain, and the joint distribution of successive intervals is of Kibble's form of bivariate gamma. Lampard also gave an interpretation of the same process in terms of a queueing system.
- **Hydrology**. Phatarford (1976) used this distribution as a model to describe the summer and winter streamflows.
- **Rainfall**. As the gamma distribution is a popular univariate choice for the description of amount of rainfall, Izawa (1965) used Kibble's bivariate gamma distribution to describe the joint distribution of rainfall at two nearby rain gauges.
- **Wind gusts**. Smith and Adelfang (1981) reported an analysis of wind gust data using Kibble's bivariate gamma distribution. The two variates considered were magnitude and length of the gust.

8.2.13 Tables and Algorithms

For α an integer or half-integer, Gunst and Webster (1973) presented a table of upper 5% critical points, and Krishnaiah (1980) gave an algorithm to compute the probability integral. For arbitrary α, an algorithm to compute the probability integral has been given by Lai and Moore (1984).

8.2.14 Transformations of the Marginals

- The joint distribution of $\sqrt{X}$ and $\sqrt{Y}$ is a bivariate chi-distribution, which is also known as a bivariate Rayleigh distribution. This has been studied by Krishnaiah et al. (1963).
- Izawa (1965) has given some attention to a distribution for which certain transformations of the variates—square root, cube root, or logarithm—have Kibble's bivariate gamma distribution.
- By transforming the marginals to be Pareto in form, Mardia (1962) obtained a model that is termed a type 2 bivariate Pareto distribution.

8.3 Royen's Bivariate Gamma Distribution

Royen (1991) considered this bivariate gamma distribution without realizing its close relationship to Kibble's bivariate gamma distribution.

8.3.1 Formula of the Cumulative Distribution Function

The joint cumulative distribution function is

$$\begin{aligned} H(x,y) &= \frac{(1-\rho^2)^\alpha}{\Gamma(\alpha)} \\ &\times \sum_{n=0}^{\infty} \frac{\Gamma(\alpha+n)\rho^{2n}}{n!} F_{\alpha+n}\left(\frac{x}{2(1-\rho^2)}\right) F_{\alpha+n}\left(\frac{y}{2(1-\rho^2)}\right), \end{aligned} \tag{8.13}$$

where $F_\alpha(\cdot)$ is the cumulative distribution function of the standard gamma with shape parameter α.

8.3.2 Univariate Properties

The marginal distributions are gamma with shape parameter α and scale parameter 1/2.

8.3.3 Derivation

Let $\mathbf{R} = \begin{pmatrix} 1 & \rho \\ \rho & 1 \end{pmatrix}$ be a nonsingular correlation matrix, $\mathbf{Y}_1, \ldots, \mathbf{Y}_d$ be independent standard bivariate normal random variables with correlation matrix $\mathbf{R}$, and $\mathbf{Y}$ be the $(2 \times d)$ matrix with columns $\mathbf{Y}_j$, $j = 1, 2, \ldots, d$. Then, according to Royen (1991), the joint cumulative distribution function of the squared Euclidean norms of the row vectors of $\mathbf{Y}$ is the bivariate gamma distribution in (8.13) with shape parameter $\alpha = d/2$.

8.3.4 Relation to Kibble's Bivariate Gamma Distribution

Comparing (8.13) with (8.5), it is clear that Royen's bivariate gamma is the same as Kibble's distribution except that the marginals of the former have a scale parameter 1/2. Two derivations are also identical apart from the latter having a divisor 2 in the derivation.

8.4 Izawa's Bivariate Gamma Distribution

Izawa (1953) proposed a bivariate gamma model that is constructed from gamma marginals allowing for different scale and shape parameters. As this model was published in Japanese, it did not attract much attention in the literature.

8.4.1 Formula of the Joint Density

Taking both scale parameters to be 1 for the sake of simplicity, the joint density function is

$$h(x,y) = \frac{(xy)^{(\alpha_1-1)/2} x^{(\alpha_1-\alpha_2)} \exp\left(-\frac{x+y}{1-\eta}\right)}{\Gamma(\alpha_1)\Gamma(\alpha_1-\alpha_2)(1-\eta)\eta^{(\alpha_1-1)/2}}$$
$$\times \int_0^1 (1-t)^{(\alpha_1-1)/2} t^{(\alpha_1-\alpha_2-1)} e^{\left(\frac{\eta x t}{1-\eta}\right)} I_{\alpha_1}\left(\frac{2\sqrt{\eta x y(1-t)}}{1-\eta}\right) dt, \tag{8.14}$$

for $\alpha_1 \geq \alpha_2$, $\eta = \rho\sqrt{\alpha_1/\alpha_2}$, $0 \leq \rho < 1$, $0 \leq \eta < 1$, where I_α denotes the Bessel function of the first kind and order α; see Izawa (1953), Nagao (1975), and Yue et al. (2001).

8.4.2 Correlation Coefficient

The Pearson product-moment correlation coefficient is ρ, and η is the association parameter.

8.4.3 Relation to Kibble's Bivariate Gamma Distribution

When $\alpha_1 = \alpha_2 = \alpha$, (8.14) reduces to Kibble's bivariate gamma density function in (8.1).

8.4.4 Fields of Application

Yue et al. (2001) have used this distribution in the field of hydrology.

8.5 Jensen's Bivariate Gamma Distribution

8.5.1 Formula of the Joint Density

In this generalization of Kibble's distribution due to Jensen (1970), the joint density function has as a diagonal expansion in terms of Laguerre polynomials

$$h(x,y) = f_{a/2}(x) f_{b/2}(y) \sum_{k=0}^{\infty} \frac{G_k(\boldsymbol{\delta})(k!)^2\Gamma(a/2)\Gamma(b/2)}{\Gamma(k+a/2)\Gamma(k+b/2)} L_k^{(\frac{a}{2}-1)}(x) L_k^{(\frac{b}{2}-1)}(y), \tag{8.15}$$

where a and b are positive integers such that $a \leq b$, f_α is the standard gamma density as before, and

$$G_k(\boldsymbol{\delta}) = G_k(\delta_1, \delta_2, \ldots, \delta_a) = \sum_{j_1, j_2, \ldots, j_a} c_{1j_1} c_{2j_2} \cdots c_{aj_a}, \tag{8.16}$$

in which the sum is taken over all integer partitions[3] of k in the second subscript of c, and

$$c_{mj_m} = \frac{\delta_m^{j_m} \Gamma(j_m + \frac{1}{2})}{\Gamma(j_m + 1)\Gamma(\frac{1}{2})}.$$

The density function for the equicorrelated case (i.e., all the δ's are equal) with $a = b$ was discussed in Section 8.2; for the case where $a \neq b$, see Krishnamoorthy and Parthasarathy (1951).

8.5.2 Univariate Properties

The marginals are again gamma distributions, but in this case with different shape parameters, $a/2$ and $b/2$.

8.5.3 Correlation Coefficient

Pearson's product-moment correlation is

$$\rho = \frac{\rho_1^2 + \rho_2^2 + \cdots + \rho_a^2}{\sqrt{ab}}, \tag{8.17}$$

where $\rho_j^2 = \delta_j > 0$ and ρ_j is the correlation coefficient of the bivariate normal distribution that is involved in this derivation; see Section 8.5.5 below.

8.5.4 Characteristic Function

The joint characteristic function is

$$\varphi(s,t) = (1 - it)^{-(b-a)/2} \prod_{j=1}^{a} [(1 - is)(1 - it) + st\rho_j^2]^{-1/2}. \tag{8.18}$$

[3] An integer partition of k with a group is a vector $(j_1, j_2, \ldots, j_a)$ such that $\sum_{i=1}^{a} j_i = k$, $0 \leq j_i \leq k$. Each vector is a distinct partition. For example, if $a = k = 2$, then all possible partitions are $(0,2)$, $(2,0)$, and $(1,1)$.

In the equicorrelated case $\rho_1^2 = \rho_2^2 = \cdots = \rho_a^2 = \eta$, (8.18) reduces to

$$\varphi(s,t) = (1-it)^{-(b-a)/2}[(1-is)(1-it) + st\eta]^{-a/2}, \tag{8.19}$$

and the correlation in this case is $\eta\sqrt{a/b}$.

8.5.5 Derivation

This distribution may be derived as follows. Let $\mathbf{Z}$ be a normal random vector with $a+b$ components, having zero means and general positive definite variance–covariance matrix $\mathbf{\Sigma}$, partitioned as $\mathbf{Z}' = (\mathbf{Z}_1', \mathbf{Z}_2')$, $\mathbf{\Sigma} = \begin{pmatrix} \Sigma_{11} & \Sigma_{12} \\ \Sigma_{21} & \Sigma_{22} \end{pmatrix}$, where $\mathbf{Z_1}$ and $\mathbf{Z_2}$ are $(a \times 1)$ and $(b \times 1)$ normal vectors, with $a \le b$, respectively. Here, $\mathbf{\Sigma_{11}}$ and $\mathbf{\Sigma_{22}}$ are identity matrices, and $\mathbf{\Sigma_{12}} = \mathbf{\Sigma_{21}'} = (\mathbf{D}\ \mathbf{0})$, where $\mathbf{D}$ has the ρ's down the diagonal and zeros elsewhere. Then, the quadratic forms $Q_1 = \frac{1}{2}\mathbf{Z_1'}\mathbf{\Sigma_{11}^{-1}}\mathbf{Z_1}$ and $Q_2 = \frac{1}{2}\mathbf{Z_1'}\mathbf{\Sigma_{22}^{-1}}\mathbf{Z_2}$ jointly follow Jensen's bivariate gamma distribution.

8.5.6 Illustrations

For some graphical illustrations of this bivariate gamma distribution, one may refer to Smith et al. (1982) and Tubbs (1983b).

8.5.7 Remarks

Jensen (1970) showed that this bivariate gamma distribution can be expanded diagonally in terms of orthogonal polynomials (in fact, orthonormal polynomials) as

$$h(x,y) = f_{a/2}(x) f_{b/2}(y) \sum_{j=0}^{\infty} M_j \mathcal{L}_j^{(\frac{a}{2}-1)}(x) \mathcal{L}_j^{(\frac{b}{2}-1)}(y), \tag{8.20}$$

where $\mathcal{L}_j^{(\frac{a}{2}-1)}(x)$ and $\mathcal{L}_j^{(\frac{b}{2}-1)}(y)$ are the normalized Laguerre[4] polynomials, and the canonical coefficients are

$$M_j = \frac{j!\sqrt{\Gamma(a/2)\Gamma(b/2)}}{\sqrt{\Gamma(a/2+j)\Gamma(b/2+j)}} G_j(\boldsymbol{\delta}). \tag{8.21}$$

8.5.8 Fields of Application

Smith et al. (1982) and Tubbs (1983b) have used this bivariate gamma distribution to model wind gusts. An advantage of this distribution is that the shape parameters of the marginal gamma distributions can be unequal.

8.5.9 Tables and Algorithms

Tables of upper 5% critical points have been presented by Gunst and Webster (1973). An algorithm for calculating the probability integral of this distribution has been given by Smith et al. (1982).

8.6 Gunst and Webster's Model and Related Distributions

Gunst and Webster (1973) considered Jensen's bivariate gamma distribution in the case where the ρ_i^2's are either zero or η. Let m be the number of nonzero ρ_i^2's.

---

[4] Any orthogonal function or polynomial with respect to a weight function f can be normalized to give $\int \theta_i(x)\theta_j(x)f(x)dx = \delta_{ij}$, where δ_{ij} is 1 if $i = j$ and 0 otherwise. The Laguerre polynomials $L_j^{(\alpha-1)}$ were defined in footnote 1. The normalized Laguerre polynomials are $\mathcal{L}_j^{(\alpha-1)} = L_j^{(\alpha-1)} / \sqrt{\binom{j+\alpha-1}{j}} = \left\{\sqrt{\frac{\Gamma(\alpha)j!}{\Gamma(j+\alpha)}}\right\} L_j^{(\alpha-1)}$. They can then be written as $\mathcal{L}_j^{(\alpha-1)}(x) = \left\{\frac{\Gamma(\alpha)\Gamma(\alpha+j)}{j!}\right\}^{1/2} \sum_{k=0}^{j} (-1)^k \binom{j}{k} \frac{x^k}{\Gamma(\alpha+k)}$. Kotz et al. (2000, p. 436) used $L_j^{(\alpha-1)}$ to denote the normalized Laguerre polynomial, and hence their notation is different from ours.

8.6.1 Case 3 of Gunst and Webster

Set $a = m + n$, and $b = m + p$, with the m nonzero and $\rho_i^2 = \eta$. Then, the joint density function is given by

$$h(x, y) = \frac{(1-\eta)^{-m/2}}{\Gamma(m/2)\Gamma(n/2)\Gamma(p/2)} \times \sum_{j=0}^{\infty}\sum_{k=0}^{\infty}\sum_{l=0}^{\infty} \alpha_{jkl} \frac{\eta^{j+k+l}}{(1-\eta)^{2j+k+l}} x^{N_1+N_2+1} y^{N_1+N_2+1} \times \exp\left(-\frac{x+y}{2(1-\eta)}\right), \tag{8.22}$$

where $N_1 = \frac{m}{2} + j - 1, N_2 = \frac{n}{2} + k - 1, N_3 = \frac{p}{2} + l - 1$, and

$$\alpha_{jkl} = \frac{2^{-(2N_1+N_2+N_3-4)}}{j!k!l!} \times \frac{\Gamma(N_1+1)\Gamma(N_2+1)\Gamma(N_3+1)}{\Gamma(N_1+N_2+2)\Gamma(N_1+N_3+2)}.$$

The correlation coefficient in this case is $\eta/m\sqrt{ab}$. For the case where m, n, and p are not necessarily integers, Krishnaiah and Rao (1961) and Krishnaiah (1983) rewrote the m.g.f. in (8.6) as

$$M(s, t) = (1-s)^{-\alpha}(1-t)^{-\alpha}\{1 - \rho st[(1-s)(1-t)]^{-1}\}^{-\alpha}.$$

Then the first two α's were replaced by α_1 and α_2, with $\alpha_i \geq \alpha > 0$, to give

$$M(s, t) = (1-s)^{-\alpha_1}(1-t)^{-\alpha_2}\{1 - \rho st[(1-s)(1-t)]^{-1}\}^{-\alpha}. \tag{8.23}$$

It is clear from (8.23) that the marginal gamma distributions have shape parameters α_1 and α_2. The m.g.f. above was inverted to obtain the density

$$h(x, y) = f_{\alpha_1}(x) f_{\alpha_2}(y) \sum_{j=0}^{\infty} \rho^j j! \frac{\Gamma(\alpha+j)}{\Gamma(\alpha)} \frac{\Gamma(\alpha_1)}{\Gamma(\alpha_1+j)} \frac{\Gamma(\alpha_2)}{\Gamma(\alpha_2+j)} \times L_j^{(\alpha_1-1)}(x) L_j^{(\alpha_2-1)}(y), \tag{8.24}$$

which is an alternative expression for the joint density function in (8.22). [Note that the Laguerre polynomial $L_j(x, \alpha)$ defined in Krishnaiah (1983) is $j! L_j^{\alpha-1}(x)$.] Sarmanov (1974) also constructed the same bivariate gamma distribution.

8.6.2 Case 2 of Gunst and Webster

In this case, we set $a = m$, and $b = m + p$. This is the equicorrelated case of Jensen's bivariate gamma, i.e., all the δ's are equal. The joint density function is given by

$$h(x,y) = \frac{(1-\eta)^{-m/2}}{\Gamma(m/2)\Gamma(p/2)} \times \sum_{j=0}^{\infty}\sum_{k=0}^{\infty} \alpha_{jk} \frac{\eta^{j+k}}{(1-\eta)^{2j+k}} x^{N_1} y^{N_1+N_2+1} \exp\left(-\frac{x+y}{2(1-\eta)}\right), \tag{8.25}$$

where $N_1 = \frac{m}{2} + j - 1, N_2 = \frac{m}{2} + k - 1$, and $a_{jk} = \frac{2^{-(2N_1+N_2+3)}\Gamma(N_2+1)}{j!k!\Gamma(N_1+N_2+2)}$. The correlation coefficient in this case is $\eta\sqrt{a/b}$.

8.7 Smith, Aldelfang, and Tubbs' Bivariate Gamma Distribution

Smith et al. (1982) extended Case 2 of Gunst and Webster to the case where m and p are not necessarily integers. Replacing $a/2$ and $b/2$ by γ_1 and γ_2, respectively, they showed that the joint density function can be written as

$$h(x,y) = \frac{x^{\gamma_1-1}y^{\gamma_2-1}\exp[(x+y)/(1-\eta)]}{(1-\eta)^{\gamma_1}\Gamma(\gamma_1)\Gamma(\gamma_2-\gamma_1)} \sum_{k=0}^{\infty} a_k I_{\gamma_2+k-1}\left(\frac{2\sqrt{2\eta xy}}{1-\eta}\right), \tag{8.26}$$

where $a_k = \frac{(\nu y)^k\Gamma(\gamma_2-\gamma_1+k)(1-\eta)^{\gamma_2-1}}{k!(\nu xy)^{(\gamma_2+k-1)/2}}$, and η is a dependency parameter satisfying $0 < \eta < 1$ and $\eta = \rho(\gamma_2/\gamma_1)^{1/2}$, in which ρ is the correlation coefficient between X and Y; see Brewer et al. (1987) and Smith et al. (1982) for further details. (The expression for a_k given in those papers seems to be incorrect, however.)

Remarks

- Brewer et al. (1987) gave some results concerning the location of the mode of distributions (8.26).
- See Tubbs (1983a) for the distribution of the ratio X/Y.
- Smith et al. (1982) considered an application of the distribution to gust modeling.

- Yue (2001) studied the applicability of the distribution to flood frequency analysis.
- Nadarajah (2007) questioned the convergence of the series in the expression for the joint p.d.f.

8.8 Sarmanov's Bivariate Gamma Distribution

Sarmanov (1970a,b) introduced asymmetrical bivariate gamma distributions that extend Kibble's bivariate gamma distribution in (8.2).

8.8.1 Formula of the Joint Density

The joint density function is

$$h(x,y) = f_{\alpha_1}(x) f_{\alpha_2}(y) \sum_{j=0}^{\infty} a_j \mathcal{L}_j^{(\alpha_1-1)}(x) \mathcal{L}_j^{(\alpha_2-1)}(y), \tag{8.27}$$

for $x, y \geq 0, \alpha_1 \geq \alpha_2$, where

$$a_j = \lambda^j \left\{ \frac{\Gamma(\alpha_2)\Gamma(\alpha_1 + j)}{\Gamma(\alpha_1 + j)\Gamma(\alpha_2)} \right\}^{1/2}, \quad 0 \leq \lambda < 1.$$

8.8.2 Univariate Properties

The marginals are gamma distributions with shape parameters α_1 and α_2. Note that $\mathcal{L}_j^{(\alpha-1)}(\cdot)$ are the orthonormal Laguerre polynomials with respect to the gamma density f_α.

8.8.3 Correlation Coefficient

Pearson's coefficient of correlation is

$$\operatorname{corr}(X, Y) = \rho = \lambda\sqrt{\alpha_2/\alpha_1} = a_1.$$

8.8.4 Derivation

This distribution can be derived by generalizing the diagonal expansion of Kibble's bivariate gamma density in (8.2) by choosing an appropriate canonical sequence a_i, as discussed in Lancaster (1969).

8.8.5 Interrelationships

Interrelationships between the distributions of Kibble (1941), Jensen (1970), Gunst and Webster (1973), Smith et al. (1982), Krishnaiah (1983), and Malik and Trudel (1985) are as presented below, in which GW stands for Gunst and Webster and MT stands for Malik and Trudel.

Sarmanov (d) $\downarrow$ Kibble

Krishnaiah (a) $\xrightarrow{(b)}$ Smith et al. $\xrightarrow{(c)}$ Kibble $\xleftarrow{(c)}$ MT (e)

Krishnaiah (a) $\downarrow$ GW Case 3(a); Smith et al. $\downarrow$ GW Case 2

GW Case 3(a) $\xrightarrow{(b)}$ GW Case 2 $\xleftarrow{(f)}$ Jensen (g)

Notes: The last two downward arrows indicate that the α_i are restricted to be integers or half-integers.

(a) Parameter α, no greater than α_1 or α_2, is present.
(b) Parameter α is dropped.
(c) α_1 and α_2 are set to be equal.
(d) α_1 and α_2 are not necessarily equal.
(e) The marginals are not gamma distributions.
(f) The correlations are not equal.
(g) $\alpha_1 \leq \alpha_2$, the α_i being integers or half-integers. $\rho_1, \rho_2, \ldots, \rho_{2\alpha_1}$ are nonzero but may be different.

We further note the following:

- Royen's bivariate gamma is essentially the same as Kibble's bivariate gamma distribution, except the marginals are nonstandard gamma with scale parameter 1/2.
- Kibble's bivariate gamma is a special case of Izawa's bivariate gamma model.

8.9 Bivariate Gamma of Loáiciga and Leipnik

Another unsymmetrical bivariate generalization of Kibble's bivariate gamma with different shape and scale parameters was introduced by Loáiciga and Leipnik (2005).

8.9.1 Formula of the Joint Density

The joint density function is

$$h(x,y) = \sum_{n=0}^{\infty}\sum_{k=0}^{n}\sum_{j=0}^{n} A_{nkj} x^{\lambda_1'+k-n} y^{\lambda_2'+j-n} \exp\left(-\frac{x}{b_1}-\frac{y}{b_2}\right) \tag{8.28}$$

for $x > 0$ and $y > 0$, where $\lambda_i = \alpha_i(n+\gamma), \lambda_i' = \lambda_i - 1$, and A_{nkj} are given by

$$A_{nkj} = \frac{(-1)^{n+k+j}\beta^n (n!)^2}{b_1^{k+\lambda_1'+1} b_2^{j+\lambda_2'+1}\Gamma(\lambda_1)\Gamma(\lambda_2)} \binom{-\gamma}{n}\binom{\lambda_1'}{n-k}\binom{\lambda_2'}{n-j}. \tag{8.29}$$

Here $\gamma\alpha_1$ and $\gamma\alpha_2$ are the marginal shape parameters of X and Y, respectively, with $\alpha_1, \alpha_2 \geq 0$; γ is a (collective) positive shape parameter of the joint distribution; and $b_1, b_2 > 0$ are shape parameters.

8.9.2 Univariate Properties

Both X and Y have gamma distributions with shape parameters $\gamma\alpha_j$ and scale parameters b_j, $j = 1, 2$, respectively.

8.9.3 Joint Characteristic Function

$$\varphi(s,t) = [(1-isb_1)^{\alpha_1}(1-itb_2)^{\alpha_2} + \beta st]^{-\gamma}. \tag{8.30}$$

8.9.4 Correlation Coefficient

Pearson's product-moment correlation coefficient is

$$\rho = \frac{\beta}{b_1 b_2 \sqrt{\alpha_1\alpha_2}}.$$

8.9.5 Moments and Joint Moments

$$\mu_i = \alpha_i b_i \gamma, \qquad \sigma_i = \alpha_i b_i^2 \gamma;$$
$$\mu_{3,0} = 2b_1^3 \gamma \alpha_1, \qquad \mu_{0,3} = 2b_2^3 \alpha_2;$$
$$\mu_{2,1} = 2\beta \alpha_1 b_1, \qquad \mu_{1,2} = 2\beta \alpha_2 b_2.$$

Remarks

- In their original derivation, a location parameter ξ_i for each marginal is included so that the characteristic function has the form

$$\varphi(s,t) = e^{(i\xi_1 s + i\xi_2 t)}[(1 - isb_1)^{\alpha_1}(1 - ib_2 t)^{\alpha_2} + \beta st]^{-\gamma}.$$

- Equation (8.30) shows that the distribution is indeed a generalization of Kibble's bivariate gamma with $\alpha_1 = \alpha_2 = 1$, $b_1 = b_2 = b$, and $\gamma = \rho$.
- The distribution X/Y and its moments were derived in Loáiciga and Leipnik (2005). The p.d.f. of the ratio was fitted to correlated bacteria densities in stream water.
- Nadarajah and Kotz (2007a) commented that the sums and products are required in hydrology and then went on to derive the distributions of $X+Y$ and XY when the joint density is given by (8.28).

8.9.6 Application to Water-Quality Data

Loáiciga and Leipnik (2005) have successfully fitted the probability distribution of X/Y to the water-quality data collected from Las Palmas Creek, Santa Barbara, California. The aim of their investigation was to study the ratio of fecal coliforms (FC) to fecal streptococcus (FS). FC and FS are enteric bacteria that live in the intestinal tract of warm-blooded animals and are frequently used as indicators of fecal contamination of water bodies. A total of 38 pairs of 100-ml water aliquots were collected. In each pair, one was analyzed for FC and the other for FS. The authors found that both FC and FS can be adequately modeled by univariate gamma distributions.

8.10 Cheriyan's Bivariate Gamma Distribution

Kotz et al. (2000) have referred to this distribution as *Cheriyan and Ramabhadran's bivariate gamma distribution.*

8.10.1 Formula of the Joint Density

The joint density function is

$$h(x,y) = \frac{e^{-(x+y)}}{\Gamma(\theta_1)\Gamma(\theta_2)\Gamma(\theta_3)} \int_0^{\min(x,y)} (x-z)^{\theta_1}(y-z)^{\theta_2-1} z^{\theta_3-1} e^z dz. \quad (8.31)$$

8.10.2 Univariate Properties

The marginal distributions are gamma with shape parameters $\alpha_1 = \theta_1 + \theta_3$ and $\alpha_2 = \theta_2 + \theta_2$.

8.10.3 Correlation Coefficient

Pearson's product-moment correlation is $\frac{\theta_3}{\sqrt{(\theta_1+\theta_2)(\theta_2+\theta_3)}}$.

Dabrowska (1982) has discussed the behavior of the monotone quadrant dependence function (see Section 3.5.3 for definition and details)—whether the tendency for small values of Y to associate with small values of X is bigger or smaller than the tendency of big values of Y to associate with big values of X, for example.

8.10.4 Moment Generating Function

The joint moment generating function is

$$M(s,t) = (1-s)^{-\theta_1}(1-t)^{-\theta_2}(1-s-t)^{-\theta_3}. \quad (8.32)$$

8.10.5 Conditional Properties

The conditional distribution of Y given X is the sum of two independent random variables, one distributed as $X \times$(standard beta variable, with parameters θ_3 and θ_1) and the other as a standard gamma variable with shape parameter θ_2. The regression is linear and is $E(Y|X=x) = \frac{\theta_3}{\theta_1+\theta_1}x + \theta_2$, and the conditional variance is quadratic and is $\frac{\theta_1\theta_3}{(\theta_1+\theta_3)^2(1+\theta_1+\theta_3)}x^2 + \theta_2$; see Johnson and Kotz (1972, p. 218).

8.10.6 Derivation

This distribution can be derived by the trivariate reduction method. Let $X_i \sim \text{gamma}(\theta_i, 1)$ for $i = 1, 2, 3$, and let the X_i's be mutually independent. Then, $X = X_1 + X_3$ and $Y = X_2 + X_3$ have this joint distribution.

8.10.7 Generation of Random Variates

The trivariate reduction method is very easy to use to generate bivariate random variates from this distribution; see Devroye (1986, pp. 587–588). Consequently, this distribution could be used to generate a bivariate gamma population when the marginals (gamma) and the correlation coefficient are specified; see Schmeiser and Lal (1982).

8.10.8 Remarks

- This distribution originated with Cheriyan, who considered the case in which $\theta_1 = \theta_2$.
- Ramabhadran (1951) also obtained the same distribution and then discussed the multivariate form.
- Independently, Cheriyan (1941) obtained this distribution and derived a number of its properties. In particular, they derived explicit expressions for $h(x, y)$ for five combinations of small values of θ_1, θ_2, and θ_3. For $\theta_1 = \theta_2 = 1$ and θ_3 an integer,

$$
\begin{aligned}
&h(x, y) \\
&= e^{-(x+y)}(-1)^{\theta_3}\left[1 - e^{\omega}\left\{1 - \frac{\omega}{1!} + \frac{\omega^2}{2!} + \cdots + (-1)^{\theta_3 - 1}\frac{\omega^{\theta_3 - 1}}{(\theta_3 - 1)!}\right\}\right],
\end{aligned}
\tag{8.33}
$$

  where $\omega = \min(x, y)$.
- The joint probability density function has a different expression for $x < y$ and for $x > y$; see Moran (1967).
- The joint density can be expanded in terms of Laguerre polynomials as shown by Eagleson (1964) and Mardia (1970).
- Ghirtis (1967) referred to this distribution as the *double-gamma distribution* and studied some properties of estimators of this distribution.
- Jensen (1969) showed that

$$\Pr(a \le X \le b, a \le Y \le b) \ge \Pr(a \le X \le b)\Pr(a \le Y \le b) \tag{8.34}$$

for any $0 \le a < b$. Another way of expressing (8.34) is

$$\Pr(a \le Y \le b | a \le X \le b) \ge \Pr(a \le Y \le b),$$

which means that if it is known that X is between a and b, then it increases the probability that Y is between a and b. Letting either $a = 0$ or $b = \infty$ in (8.34), we conclude that X and Y are PQD. In fact, this result follows directly from Lehmann (1966); see Section 7.4.

- Mielke and Flueck (1976) and Lee et al. (1979) discussed the distribution of X/Y.
- The class of bivariate gamma distributions having diagonal expansions, considered by Griffiths (1969), includes the forms of Cheriyan.

8.11 Prékopa and Szántai's Bivariate Gamma Distribution

Prékopa and Szántai (1978) introduced a multivariate gamma distribution as the distribution of the multivariate vector $\mathbf{Y} = \mathbf{AX}$, where $\mathbf{X}$ has independent standard gamma components and the matrix $\mathbf{A}$ consists of nonzero vectors having components 0 or 1.

Szántai (1986) considered the bivariate case of this multivariate gamma family with the structure

$$X = X_1 + X_3 \qquad \text{and} \qquad Y = X_2 + X_3,$$

where X_1, X_2, and X_3 are independent gamma random variables having shape parameters α_1, α_2, and α_3, respectively.

8.11.1 Formula of the Cumulative Distribution Function

The joint cumulative distribution function is

$$H(x, y) = \int_0^{\min(x,y)} F_{\alpha_1}(x - z) F_{\alpha_2}(y - z) f_{\alpha_3}(z) dz. \tag{8.35}$$

8.11.2 Formula of the Joint Density

The joint density function is

$$h(x,y) = f_{\alpha_1+\alpha_3}(x) f_{\alpha_2+\alpha_3}(y) \sum_{r=0}^{\infty} r! \frac{\Gamma(\alpha_1+r)}{\Gamma(\alpha_1)} \frac{\Gamma(\alpha_1+\alpha_3)}{\Gamma(\alpha_1+\alpha_3+r)} \frac{\Gamma(\alpha_2+\alpha_3)}{\Gamma(\alpha_2+\alpha_3+r)} \times \mathcal{L}_r^{(\alpha_1+\alpha_3-1)}(x) \mathcal{L}_r^{(\alpha_2+\alpha_3-1)}(y), \tag{8.36}$$

where $\mathcal{L}_r^{(\alpha-1)}$ are the orthonormal Laguerre polynomials defined on the gamma density with shape parameter α.

8.11.3 Univariate Properties

The marginal distributions are gamma with shape parameters $\alpha_1 + \alpha_3$ and $\alpha_2 + \alpha_3$, respectively.

8.11.4 Relation to Other Distributions

Clearly, the bivariate distribution of this model is identical to Cheriyan's bivariate gamma distribution. In contrast to Cheriyan's result, Szántai (1986) has given an explicit expression for the joint density function.

8.12 Schmeiser and Lal's Bivariate Gamma Distribution

Schmeiser and Lal (1982) developed an algorithm that enables us to generate bivariate distributions that have

- given gamma marginals with parameters $(\beta_i, \alpha_i), i = 1, 2$ (β_i are scale parameters and α_i are shape parameters),
- any specified correlation coefficient ρ, and
- linear or nonlinear regression curves.

8.12.1 Method of Construction

Let X_1, X_2, and Z be three independent standard gamma variables with shape parameters δ_1, δ_2, and γ, respectively, and let U be an independent uniform random variable on $(0, 1)$. Also, $V = U$ or $V = 1 - U$. Define

$$X = \frac{G_{\lambda_1}^{-1}(U) + Z + X_1}{\beta_1}, \qquad Y = \frac{G_{\lambda_2}^{-1}(V) + Z + X_2}{\beta_2}, \tag{8.37}$$

where $G_\lambda(\cdot)$ is the distribution function of a standard gamma random variable with shape parameter λ and $G_\lambda^{-1}(\cdot)$ is the inverse function of $G_\lambda(\cdot)$.

For $\lambda_i \geq 0, \delta_i \geq 0, \gamma > 0$, the parameters are selected according to

$$\begin{cases} \gamma + \lambda_i + \delta_i = \alpha_i, \qquad i = 1, 2 \\ E\left\{G_{\lambda_1}^{-1}(U)G_{\lambda_2}^{-1}(V) - \lambda_1\lambda_2 + \gamma\right\} = \rho\sqrt{\alpha_1\alpha_2}. \end{cases}$$

8.12.2 Correlation Coefficient

Pearson's product-moment correlation coefficient is given by

$$\rho = \frac{E\left\{G_{\lambda_1}^{-1}(U)G_{\lambda_2}^{-1}(V) - \lambda_1\lambda_2 + \gamma\right\}}{\sqrt{\alpha_1\alpha_2}}.$$

8.12.3 Remarks

- This is another example of constructing a pair of random variables using the variables-in-common method.
- Schmeiser and Lal (1982) also developed an algorithm called GBIV, which determines the parameter values as well as generating the random vector (X, Y).

8.13 Farlie–Gumbel–Morgenstern Bivariate Gamma Distribution

The bivariate gamma distribution of F-G-M type was discussed by D'Este (1981) and Gupta and Wong (1989).

8.13.1 Formula of the Joint Density

The joint density function is

$$h(x, y) = f(x)g(y)[1 + \lambda\{2F(x) - 1\}\{2G(y) - 1\}], \quad |\lambda| \leq 1, \tag{8.38}$$

where $F(x)$ and $G(y)$ are the marginal cumulative distribution functions and $f(x)$ and $g(y)$ are the corresponding density functions.

8.13.2 Univariate Properties

The marginal densities $f(x)$ and $g(y)$ are gamma densities with shape parameters α_1 and α_2, respectively.

8.13.3 Moment Generating Function

The joint moment generating function is

$$M(s,t) = (1-s)^{-\alpha_1}(1-t)^{-\alpha_2}\left[1+\frac{2I(\alpha_1,0;(1-s)^{-1})}{I(\alpha_1,0;1)}\frac{2I(\alpha_2,0;(1-t)^{-1})}{I(\alpha_2,0;1)}\right], \tag{8.39}$$

where $I(a,k;x)=\int_0^x \frac{z^{a-1}}{(z+1)^{2a+k}}dz$; see Gupta and Wong (1989).

8.13.4 Correlation Coefficient

Pearson's product-moment correlation coefficient is

$$\rho + \lambda K(\alpha_1)K(\alpha_2),$$

where

$$K(\alpha) = 1/\left\{2^{2\alpha-1}B(\alpha,\alpha)\sqrt{\alpha}\right\}$$

and $B(\alpha,\beta)$ is the complete beta function.

8.13.5 Conditional Properties

The regression is nonlinear and is given by

$$E(X|Y=y) = \alpha_1 + \frac{\lambda\alpha_1\Gamma(\alpha+1/2)}{(\alpha_1+1)\sqrt{\pi}}\{2G(y)-1\}.$$

A similar expression can be presented for the regression of Y on X.

8.13.6 Remarks

Kotz et al. (2000, p. 441) have presented expressions for the joint moments.

8.14 Moran's Bivariate Gamma Distribution

8.14.1 Derivation

Moran (1969) derived a bivariate gamma distribution by using the following two steps:

(1) Use marginal transformation first to transform the standard bivariate normal with correlation ρ into a copula $C(u, v)$.
(2) Use inverse transform $X = F^{-1}(U), Y = G^{-1}(V)$ to find the joint distribution function of X and Y. In fact, the cumulative distribution function is given by $H(x, y) = C(F(x), G(y))$. Here, F is the marginal gamma distribution function with shape parameter α_1 and scale parameter λ_1 and G is the other marginal gamma distribution with shape parameter α_2 and scale parameter λ_2.

8.14.2 Formula of the Joint Density

The joint density function is

$$h(x, y) = \frac{1}{\sqrt{(1-\rho^2)}} f(x) g(y) \exp\left\{-\frac{(\rho x')^2 - 2\rho x' y' + (\rho y')^2}{2(1-\rho^2)}\right\}, \quad x, y \geq 0, \tag{8.40}$$

where $x' = \Phi^{-1}(F(x))$ and $y' = \Phi^{-1}(G(y))$, with Φ being the distribution function of the standard normal.

8.14.3 Computation of Bivariate Distribution Function

Yue (1999) presented a procedure to compute the bivariate distribution function. Effectively, this is done through generation of marginal gammas using Jonk's gamma generator that is written in MATLAB code.

8.14.4 Remarks

- Moran's model is a special case of the bivariate meta-Gaussian model proposed by Kelly and Krzysztofowicz (1997).
- This is an example of obtaining a bivariate distribution using copulas.

8.14.5 Fields of Application

Yue et al. (2001) presented a review of several bivariate gamma models including those of Moran, Izawa, Smith et al., and F-G-M models, and illustrated their applications in hydrology.

8.15 Crovelli's Bivariate Gamma Distribution

Crovelli (1973) proposed a bivariate gamma distribution having the joint density

$$h(x,y) = \begin{cases} \beta_1\beta_2 e^{-\beta_2 y}(1 - e^{-\beta_1 x}) & \text{for } 0 \le \beta_1 x \le \beta_2 y \\ \beta_1\beta_2 e^{-\beta_1 x}(1 - e^{-\beta_2 y}) & \text{for } 0 \le \beta_2 y \le \beta_1 x \end{cases} .$$

8.15.1 Fields of Application

Crovelli (1973) used this bivariate distribution to model the joint distribution of storm depths and durations.

8.16 Suitability of Bivariate Gammas for Hydrological Applications

A bivariate gamma distribution whose marginals have different scale and shape parameters may be useful to model multivariate hydrological events such as floods and storms. Yue et al. (2001) considered four models (Izawa, Moran, Smith et al., and F-G-M) and discussed their advantages and limitations. Using both real and generated flood data, they found that Izawa, Moran, and Smith et al. models with five parameters (two shape, two scale, and one correlation parameter) are suitable to describe two positively correlated flood characteristics (such as flood peak and flood volume or flood volume and flood duration), whereas the Moran and F-G-M models are able to describe both positively and negatively correlated random variables. However, the applicability of the latter model is somewhat limited because of the limited range of correlation it can attain; also see Long and Krzysztofowicz (1992).

8.17 McKay's Bivariate Gamma Distribution

8.17.1 Formula of the Joint Density

The joint density function is

$$h(x, y) = \frac{a^{p+q}}{\Gamma(p)\Gamma(q)} x^{p-1}(y - x)^{q-1}e^{-ay}, \quad y > x > 0 \tag{8.41}$$

(i.e., the support is a wedge that is half of the positive quadrant), where $a, p, q > 0$.

8.17.2 Formula of the Cumulative Distribution Function

The p.d.f. in (8.41) may be expressed in terms of the transcendental function known as Fox's H function. Hence, as done by Kellogg and Barnes (1989, Section 4.6), the joint distribution function can also be expressed in terms of Fox's function.

8.17.3 Univariate Properties

The marginal distributions of X and Y are gamma, with shape parameters p and $p + q$, respectively, but they have a common scale parameter a.

8.17.4 Conditional Properties

$Y - x$, conditional on $(X = x)$, has a gamma distribution with shape parameter q. X/y, conditional on $(Y = y)$, has a beta distribution with parameters p and q.

Correlation Coefficient

Pearson's product-moment correlation coefficient is $\sqrt{p/(p + q)}$.

8.17.5 Methods of Derivation

- McKay (1934) derived this distribution as follows: Let $(X_1, X_2, \ldots, X_N)$ be a random sample from a normal population. Suppose s_N^2 is the sample variance and s_n^2 is the variance in a subsample of size n. Then, s_N^2 and s_n^2 jointly have McKay's bivariate gamma distribution.
- As a member of Pearson's system of bivariate distributions, it may be derived by a differential equation; see Section 5.15 for details.
- It was derived by the conditional approach as a special case of beta-Stacy distribution by Mihram and Hultquist (1967).

Illustrations

Plots of the probability density surface for three cases—$a = 2.0$, $p = q = 0.5$; $a = p = q = 0.5$; $a = 1.0$, $p = 0.2$, $q = 0.8$—have been provided by Kellogg and Barnes (1989).

8.17.6 Remarks

- This is also known as the bivariate Pearson type III distribution, although in van Uven's designation, it is type IVa.
- One of the examples that Parrish and Bargmann (1981) gave to illustrate their method of evaluating d.f.'s was this distribution.
- The exact distributions of the sums, products, and ratios for McKay's bivariate gamma distributions were obtained by Gupta and Nadarajah (2006).

8.18 Dussauchoy and Berland's Bivariate Gamma Distribution

This is an extension of McKay's bivariate gamma distribution.

8.18.1 Formula of the Joint Density

The support is the wedge $y > \beta x > 0$, and within this wedge, the joint density is

$$h(x,y) = \frac{\beta a_2^{t_2}}{\Gamma(l_1)\Gamma(l_1 - l_2)} (\beta x)^{l_1 - 1} \exp(-a_2 x)(y - \beta x)^{l_2 - l_1 - 1}$$
$$\times \exp\left[-\frac{a_2}{\beta}(y - \beta x)\right] {}_1F_1\left[l_1, l_2 - l_1; \left(\frac{a_1}{\beta} - a_2\right)(y - \beta x)\right],$$
$$\beta \geq 0;\ 0 < a_2 \leq \frac{a_1}{\beta};\ 0 < l_1 < l_2,$$

where ${}_1F_1$ is the confluent hypergeometric function.

This distribution reduces to McKay's bivariate gamma distribution when $a_1 = a_2 = \beta = 1$.

Remarks

- The marginal distributions of X and Y are gamma with shape parameters l_1 and l_2, respectively.
- Pearson's product-moment correlation coefficient is $\frac{\beta a_2}{a_1}\sqrt{l_1/l_2}$.
- The plots of the probability density surface (seven cases) were given by Berland and Dussauchoy (1973).
- The density has been written above in a form that makes clear the independence of X and $Y - \beta x$.
- For more details, see Dussauchoy and Berland (1972), Berland and Dussauchoy (1973), and Dussauchoy and Berland (1975) for the multivariate case.
- Berland and Dussauchoy (1973) applied this distribution to the joint distribution of the charge transported by a microdischarge (of electricity between two electrodes) and the interval of time between two of them.

Some Variants of this Distribution

We now summarize some variations in Table 8.1 on the theme of Y necessarily being positive, and X necessarily being between 0 and y.

Table 8.1 Distributions specified by marginal and conditional

Reference	Distribution of Y	Distribution of X, given $Y = y$
McKay (1934)	Gamma	Beta over $(0, y)$
Mihram and Hultquist (1967)	Stacy	Beta over $(0, y)$
Block and Rao (1973)	Generalized inverted beta*	Beta over $(0, y)$
Ratnaparkhi (1981)†	Stacy, Pareto, or lognormal	Beta or log-gamma over $(0, y)$

* Density $\propto y^{\alpha-1}(1 + y^c)^{-k}$.

† In Ratnaparkhi's paper, the roles of X and Y were reversed from those here.

8.19 Mathai and Moschopoulos' Bivariate Gamma Distributions

We discuss bivariate versions of two multivariate gamma distributions proposed by Mathai and Moschopoulos (1991, 1992). To simplify our presentation, we assume that the location parameter of the gamma variable is zero. Also, our scale parameter beta here is defined differently from that of Mathai and Moschopoulos.

8.19.1 Model 1

Method of Construction

Mathai and Moschopoulos (1991) constructed a bivariate gamma distribution, whose components are positively correlated, as follows.

Let V_i be a gamma variable with shape parameter α_i and scale parameter β_i, having as its density $\frac{1}{\Gamma(\alpha_i)}\beta_i^{\alpha_i}e^{-\beta_i v_i}$, $i = 0, 1, 2$. Define

$$X = \frac{\beta_0}{\beta_1}V_0 + V_1, \qquad Y = \frac{\beta_0}{\beta_2}V_0 + V_2.$$

Then, X and Y have a bivariate distribution with gamma marginals.

Joint Moment Generating Function

The joint moment generating function is

$$M(s,t) = (1 - \beta_1^{-1}s)^{-\alpha_1}(1 - \beta_2^{-1}t)^{-\alpha_2}(1 - \beta_1^{-1}s - \beta_2^{-1}t)^{-\alpha_0}. \qquad (8.42)$$

Univariate Properties

X is distributed as gamma with shape parameter $\alpha_0+\alpha_1$ and scale parameter β_1, while Y is distributed as gamma with shape parameter $\alpha_0 + \alpha_2$ and scale parameter β_2.

Correlation Coefficients

Pearson's product-moment correlation coefficient is

$$\operatorname{corr}(X,Y)=\rho=\frac{\alpha_0}{\sqrt{(\alpha_0+\alpha_1)(\alpha_0+\alpha_2)}}.$$

Conditional Properties

The regression is linear and is given by

$$E(X|Y=y)=E(X)+\frac{\alpha_0\beta_2}{\beta_2(\alpha_0+\alpha_2)}(y-E(Y)).$$

A similar expression can be presented for the regression of Y on X.

Relations to Other Distributions

This is a slight extension of Kibble's bivariate gamma distribution. If $\beta_i=1$, it reduces to Kibble's case, and if $\beta_i=1/2$, it becomes Royen's bivariate gamma distribution.

8.19.2 Model 2

Method of Construction

Mathai and Moschopoulos (1992) constructed another form of multivariate gamma distribution. The special case of the bivariate version is as follows. Let V_i, $i=1,2$, be defined as above but with the same scale parameter. Form

$$X=V_1, \qquad Y=V_1+V_2;$$

then, X and Y clearly have a bivariate gamma distribution. The above construction above is only part of a multivariate setup motivated by the consideration of the joint distribution of the total waiting times of a renewal process.

Formula of the Joint Density

The joint density function is

$$h(x,y)=\frac{\beta^{(\alpha_1+\alpha_2)}}{\Gamma(\alpha_1)\Gamma(\alpha_2)}x^{\alpha_1-1}(y-x)^{\alpha_2-1}e^{-\beta y}. \tag{8.43}$$

Marginal Properties

The marginal distributions of X and Y are gamma, with shape parameters α_1 and $\alpha_1 + \alpha_2$, respectively, and with a common scale parameter β.

Relation to Other Distributions

The bivariate case of this multivariate gamma is simply McKay's bivariate gamma distribution.

8.20 Becker and Roux's Bivariate Gamma Distribution

8.20.1 Formula of the Joint Density

The joint density function is

$$h(x,y) = \begin{cases} \frac{\beta'\alpha^a}{\Gamma(a)\Gamma(b)} x^{a-1}[\beta'(y-x)+\beta x]^{b-1}\exp[-\beta' y-(\alpha+\beta-\beta')x], \; 0<x<y \\ \frac{\alpha'\beta^b}{\Gamma(a)\Gamma(b)} y^{b-1}[\alpha'(x-y)+\alpha y]^{a-1}\exp[-\alpha' x-(\alpha+\beta-\alpha')y], \; 0<y<x \end{cases}. \tag{8.44}$$

8.20.2 Derivation

Let us restate Freund's model as follows. Suppose that shocks that knock out components A and B, respectively, are governed by Poisson processes. Let us further assume the following:

- For component A, the Poisson process has rate α when component B is functioning and rate α' after component B has failed.
- For component B, the Poisson process has rate β when component A is functioning and rate β' after component A has failed. Becker and Roux (1981) generalized Freund's distribution by supposing that the components did not fail after a single shock but that it took a and b shocks, respectively, to destroy them. (The numbers a and b are deterministic, and not random.) The resulting joint density is the one given in (8.44).

8.20.3 Remarks

- The original model proposed by Becker and Roux (1981) was slightly reparametrized by Steel and le Roux (1987) to a form that is more amenable for practical applications.
- When $a = b = 1$, the model abovereduces to Freund's (1961) bivariate exponential distribution; see Chapter 10 for pertinent details.

8.21 Bivariate Chi-Squared Distribution

8.21.1 Formula of the Cumulative Distribution Function

The joint cumulative distribution function is

$$H(x, y) = \sum_{j=0}^{\infty} c_j \Pr[\chi^2_{n-1+2j} \leq (1-\rho)^{-1}x] \times \Pr[\chi^2_{n-1+2j} \leq (1-\rho)^{-1}y], \quad (8.45)$$

for $x, y \geq 0$, $0 \leq \rho \leq 1$, where

$$c_j = \frac{\Gamma\left(\frac{1}{2}(n-1)+j\right)(1-\rho)^{\frac{1}{2}(n-1)}\rho^j}{\Gamma\left(\frac{1}{2}(n-1)\right) j!}.$$

Note that c_i, $i = 0, 1, \ldots$, are terms in the expression of the negative binomial

$$\left(\frac{1}{1-\rho}, \frac{\rho}{1-\rho}\right)^{-(n-1)/2},$$

so that $\sum_{j=0}^{\infty} c_j = 1$. Thus, the joint distribution of X and Y can be regarded as a mixture of joint distributions, with weights c_j, in which X and Y are independent χ^2_{n-1+2j} distributions.

8.21.2 Univariate Properties

Both marginals have chi-squared distributions with $n-1$ degrees of freedom.

8.21.3 Correlation Coefficient

Pearson's product-moment correlation coefficient is $\text{corr}(X, Y) = \rho = \rho_0^2$.

8.21.4 Conditional Properties

X, conditional on $(Y = y)$, is distributed as $(1-\rho)\times$(noncentral χ^2 with $(n-1)$ degrees of freedom and noncentrality parameter $\rho y(1-\rho)^{-1}$). Therefore, the regression is linear and is given by

$$E(X|Y = y) = (n-1)(1-\rho) + \rho y. \tag{8.46}$$

Also, the conditional variance is linear and is given by

$$\text{var}(X|Y = y) = 2(n-1)(1-\rho)^2 + 4\rho(1-\rho)y. \tag{8.47}$$

A similar expression can be presented for Y, conditioned on $(X = x)$.

8.21.5 Derivation

Let (X_i, Y_i), $i = 1, 2, \ldots, n$, be n independent random vectors, each having a standard bivariate normal distribution with correlation coefficient ρ_0. Further, let $X = \sum_{i=1}^n (X_i - \bar{X})^2$ and $Y = \sum_{i=1}^n (Y_i - \bar{Y})^2$, where $\bar{X}$ and $\bar{Y}$ are the sample means of X_i and Y_i, respectively. Then, X and Y have a joint cumulative distribution function as given in (8.45); see, for example, Vere-Jones (1967) and Moran and Vere-Jones (1969).

8.21.6 Remarks

- The bivariate distribution is also called the *generalized Rayleigh distribution*; see, for example, Miller (1964).
- The joint distribution $\sqrt{X}$ and $\sqrt{Y}$ is a bivariate chi-distribution studied by Krishnaiah et al. (1963).
- A more general bivariate gamma can be obtained by replacing $(n-1)$ in (8.48) by ν, which should be positive but need not be an integer.
- X/Y is distributed as a mixture, with the same proportions as c_j, of $F_{n-1+2j,n-1+2j}$ distributions.

8.22 Bivariate Noncentral Chi-Squared Distribution

Let (X_i, Y_i), $i = 1, 2, \ldots, n$, be n independent random vectors having bivariate normal distributions with means (μ_i, μ_i), identical variances σ^2, and correlation ρ_0. Further, let $X = \sum_{i=1}^n X_i^2/\sigma^2$ and $Y = \sum_{i=1}^n Y_i^2/\sigma^2$. Krishnan (1976) then showed that their joint distribution has density function

$$h(x,y) = \frac{k}{4} \exp\left[-\frac{x+y}{2(1-\rho)}\right] \sum_{i=0}^{\infty} d_i I_{f_i}\left\{\frac{\sqrt{(\rho x y)}}{1-\rho}\right\} \times I_{f_i}\left\{\frac{\sqrt{\lambda x}}{1+\sqrt{\rho}}\right\} I_{f_i}\left\{\frac{\sqrt{\lambda y}}{1+\sqrt{\rho}}\right\}, \tag{8.48}$$

where $\lambda = \sum_{i=1}^n \mu_i^2$ is the noncentrality parameter, $\rho = \rho_0^2$, I_{f_i} is the modified Bessel function of the first kind and order $f_i = \frac{1}{n} + i - 1$, and k and d_i are given by

$$k = \exp\left(-\frac{\lambda}{1+\sqrt{\rho}}\right) \left[\frac{2(1+\sqrt{\rho})^2}{\lambda\sqrt{\rho}}\right]^{\frac{n}{2}-1} / (1-\rho),$$

$$d_i = \binom{n+i-3}{n-3} \left(\frac{n}{2} + i - 1\right) \Gamma\left(\frac{n}{2} - 1\right).$$

Krishnan (1976) also showed that the joint moment generating function is

$$M(s,t) = [1 - 2(s+t) + 4st(1-\rho)]^{-n/2} \exp\left\{\frac{\lambda[s+t-4st(1-\sqrt{\rho})]}{1-2(s+t)+4st(1-\rho)}\right\}. \tag{8.49}$$

When $\lambda = 0$, we obtain Kibble's bivariate gamma distribution.

8.23 Gaver's Bivariate Gamma Distribution

We present here the bivariate version of Gaver's (1970) multivariate gamma distribution.

8.23.1 Moment Generating Function

The joint moment generating function is

$$M(s,t) = \left(1 - \frac{\alpha+1}{\alpha}s - \frac{\alpha+1}{\alpha}t + \frac{\alpha+1}{\alpha}st\right)^{-k}, \quad k, \alpha > 0. \tag{8.50}$$

8.23.2 Derivation

Let X and Y have the same gamma distribution with the shape parameter $\theta + k$ (θ is an integer, and $k > 0$ need not be an integer). For a given value of θ, X and Y are independent. Assuming that θ has a negative binomial distribution with probability generating function $(\frac{\alpha}{1+\alpha-z})^k$, the joint moment generating function of X and Y is obtained as given in (8.50).

8.23.3 Correlation Coefficients

Pearson's product-moment correlation coefficient is $\text{corr}(X, Y) = \rho = \frac{1}{1+\alpha}$.

8.24 Bivariate Gamma of Nadarajah and Gupta

Nadarajah and Gupta (2006) introduced two new gamma distributions based on a characterizing property involving products of gamma and beta random variables. Both joint density functions involve the Whittaker function defined by

$$W_{\lambda,\mu} = \frac{a^{\mu+1/2}\exp(-a/2)}{\Gamma(\mu-\lambda+1/2)} \int_0^\infty t^{\mu-\lambda-1/2}(1+t)^{\mu+\lambda-1/2}\exp(-at)dt.$$

8.24.1 Model 1

Formula of the Joint Density

The joint density function is

$$\begin{aligned} h(x,y) &= C\Gamma(b)(xy)^{c-1}\left(\frac{x}{\mu_1}+\frac{y}{\mu_2}\right)^{\frac{a-1}{2}-c} \exp\left\{-\frac{1}{2}\left(\frac{x}{\mu_1}+\frac{y}{\mu_2}\right)\right\} \\ &\times W_{c-b+\frac{1-a}{2},c-\frac{a}{2}}\left(\frac{x}{\mu_1}+\frac{y}{\mu_2}\right), \qquad x>0,\ y>0, \end{aligned}$$

where C is a constant given by $C^{-1} = (\mu_1\mu_2)^c\Gamma(\text{c})\Gamma(\text{a})\Gamma(\text{b})$.

When $b = 1$, then the joint p.d.f. reduces to a simpler form:

$$h(x,y) = C(xy)^{c-1}\left(\frac{x}{\mu_1}+\frac{y}{\mu_2}\right)\Gamma\left(2c-a, \frac{x}{\mu_1}+\frac{y}{\mu_2}\right).$$

Method of Derivation

Assume that W is beta distributed with shape parameters a and b. Assume further that U and V are gamma distributed with common shape parameter c and scale parameters $1/\mu_1$ and $1/\mu_2$, respectively, with $c = a + b$. Then $X = UW$, $Y = VW$ have the joint density function given above.

Correlation Coefficient

Pearson's product-moment correlation coefficient is

$$\mathrm{corr(X, Y)} = \rho = \frac{\sqrt{ab}}{a+b+1}.$$

Other Properties

Product moments and conditional distributions are also given in Nadarajah and Gupta (2006).

8.24.2 Model 2

Formula for the Joint Density

The joint density function is

$$\begin{aligned} h(x,y) &= C\Gamma(b_1)\Gamma(b_2)\mu^{\frac{b_1+b_2-c+1}{2}} x^{\frac{a_1+b_2-3}{2}} y^{a_2-1} \exp\left(-\frac{x}{2\mu}\right) \\ &\times \textstyle\sum_{j=0}^{\infty} \frac{(-1)^j (\mu x)^{-j/2} y^j}{j!\Gamma(b_2-j)} W_{\frac{b_2-b_1-c-j-1}{2}, \frac{b_1+b_2-c-j}{2}} \left(\frac{x}{\mu}\right), \end{aligned}$$

for $x \geq y > 0$, where C is a constant given by $C^{-1} = \mu^c \Gamma(\mathrm{c}) B(a_1, b_1) B(a_2, b_2)$.

The corresponding expression for $0 < x \leq y$ can be obtained from the last equation for the joint density by symmetry; i.e., interchange x with y, a_1 with a_2, and b_1 with b_2.

If both $b_1 = 1$ and $b_2 = 1$, then the joint density above reduces to

$$h(x,y) = C\mu^{2-c} x^{a_1-1} y^{a_2-1} \Gamma\left(2-c, \frac{x}{\mu}\right),$$

where $\Gamma(a, x)$ is the incomplete gamma function.

Method of Derivation

Assume that U and V are beta distributed with shape parameters (a_1, b_1) and (a_2, b_2), respectively, where $a_1 + b_1 = a_2 + b_2 = c$. Assume further that W is gamma distributed with shape parameter c and scale parameter $1/\mu$. Then $X = UW$, $Y = VW$ have the joint density function given above.

Correlation Coefficient

Pearson's product-moment correlation coefficient is

$$\mathrm{corr(X, Y)} = \rho = \frac{\sqrt{a_1 a_2}}{c}.$$

Other Properties

Product moments and conditional distributions are also given in Nadarajah and Gupta (2006).

8.25 Arnold and Strauss' Bivariate Gamma Distribution

This is a slight generalization of Arnold and Strauss' (1988) bivariate distribution with exponential conditionals.

Formula of the Joint Density

The joint density function is

$$h(x, y) = K x^{\alpha-1} y^{\beta-1} \exp\{-(ax + by + cxy)\} \tag{8.51}$$

for $x > 0$, $y > 0$, $\alpha > 0$, $\beta > 0$, $a > 0$, $b > 0$, and $c > 0$, where K is the normalizing constant such that

$$\frac{1}{K} = b^{\alpha-\beta} c^{-\alpha} \Gamma(\alpha)\Gamma(\alpha)\Psi\left(\alpha, \alpha - \beta + 1, \frac{ab}{c}\right).$$

Here Ψ is the Kummer function defined by

$$\Psi(a, b, z) = \frac{1}{\Gamma} \int_0^\infty t^{\alpha-1}(1+t)^{b-a-1} \exp(-zt) dt.$$

8.25.1 Remarks

- The distribution above was considered by Nadarajah (2005, 2006).
- The distributions of XY and $X/(X+Y)$ were considered by Nadarajah (2005).
- The Fisher information matrix and tools for numerical computation of the derivation were also derived by Nadarajah (2006).

8.26 Bivariate Gamma Mixture Distribution

8.26.1 Model Specification

Let X have a gamma density

$$f(x|\nu,\gamma) = \frac{1}{\Gamma(\nu)}\gamma^{\nu}x^{\nu-1}e^{-\gamma x}, \quad x > 0,$$

with shape parameter $\nu > 0$ and random scale parameter γ taking two distinct values, γ_1 and γ_2. Similarly, Y has a gamma density

$$g(y|\alpha,\beta) = \frac{1}{\Gamma(\alpha)}\beta^{\alpha}y^{\alpha-1}e^{-\beta y}, \quad y > 0,$$

with shape parameter $\alpha > 0$, and β is a random scale parameter taking two distinct values β_1 and β_2.

For given (γ,β), we assume that X and Y are independent but γ and β are correlated, having a joint probability mass function $\Pr(\gamma=\gamma_j,\beta=\beta_j) = p_{\gamma_i\beta_j}$, $i,j=1,2$.

8.26.2 Formula of the Joint Density

The joint density function is [see Jones et al. (2000), where the scale parameter is defined differently]

$$\begin{aligned} h(x,y) = x^{\nu-1}y^{\alpha-1}\Big[& a\gamma_1^{\nu}\beta_1^{\alpha}e^{-(\gamma_1 x+\beta_1 y)} + b\gamma_1^{\nu}\beta_2^{\alpha}e^{-(\gamma_x+\beta_2 y)} \\ &+ c\gamma_2^{\nu}\beta_1^{\alpha}e^{-(\gamma_2 x+\beta_1 y)} + (1-a-b-c)\gamma_2^{\nu}\beta_2^{\alpha}e^{-(\gamma_2 x+\beta_2 y)}\Big], \end{aligned} \tag{8.52}$$

where $a = p_{\gamma_1\beta_1}$, $b = p_{\gamma_1\beta_2}$, $c = p_{\gamma_2\beta_1}$, and $d = p_{\gamma_2\beta_2} = 1-a-b-c$.

8.26.3 Formula of the Cumulative Distribution Function

The joint cumulative distribution function is

$$H(x,y) = \frac{1}{\Gamma(\nu)\Gamma(\alpha)} \{a\Gamma_{\gamma_1 x}(\nu)\Gamma_{\beta_1 y}(\alpha) + b\Gamma_{\gamma_1 x}(\nu)\Gamma_{\beta_2 y}(\alpha) + c\Gamma_{\gamma_2 x}(\nu)\Gamma_{\beta_1 x}(\alpha) + (1-a-b-c)\Gamma_{\gamma_2 x}(\nu)\Gamma_{\beta_2 x}(\alpha)\}, \tag{8.53}$$

where $\Gamma(\nu) = \int_0^t x^{\nu-1}e^{-x}dx$ is the incomplete gamma function.

8.26.4 Univariate Properties

The marginal densities are

$$f(x) = \pi_1 f_1(x) + (1-\pi_1)f_2(x), \quad \pi_1 = a+b,$$

$$g(y) = \pi_2 g_1(y) + (1-\pi_2)g_2(y), \quad \pi_2 = a+c,$$

where $f_i(x) = f(x|\nu,\gamma_i)$, $g_i(y) = g(y|\alpha,\beta_i)$. Consequently, we have

$$E(X) = \nu[\pi_1/\gamma_1 + (1-\pi_1)/\gamma_2], \quad E(Y) = \nu[\pi_2/\beta_1 + (1-\pi_2)/\beta_2].$$

8.26.5 Moments and Moment Generating Function

The joint moment generating function is

$$M(s,t) = a(1-s/\gamma_1 s)^{-\nu}(1-t/\beta_1)^{-\alpha} + b(1-s/\gamma_1 s)^{-\nu}(1-t/\beta_2)^{-\alpha} + c(1-s/\gamma_2 s)^{-\nu}(1-t/\beta_1)^{-\alpha} + d(1-s/\gamma_2 s)^{-\nu}(1-t/\beta_2)^{-\alpha}, \tag{8.54}$$

where $d = (1-a-b-c)$. The product moments (about zero) are given by

$$\mu_{ij} = \frac{\Gamma(j+\nu)\Gamma(i+\alpha)}{\Gamma(\nu)\Gamma(\alpha)} \left\{a\gamma_1^{-i}\beta_1^{-j} + b\gamma_2^{-i}\beta_1^{-j} + c\gamma_2^{-i}/\beta_1^{-j} + d\gamma_2^{-i}\beta_2^{-j}\right\},$$

where $d = (1-a-b-c)$.

8.26.6 Correlation Coefficient

Pearson's product-moment correlation coefficient is

$$\rho = \nu\alpha \cdot \mathrm{corr}(\gamma, \beta)\sqrt{\frac{\mathrm{var}(\gamma)\mathrm{var}(\beta)}{\mathrm{var}(X)\mathrm{var}(Y)}}\ ; \tag{8.55}$$

ρ is bounded above by

$$\rho_{\max} = \left\{1 + \frac{(\gamma_1+\gamma_2)^2}{\nu(\gamma_1-\gamma_2)^2}\right\}^{-1/2}\left\{1 + \frac{(\beta_1+\beta_2)^2}{\nu(\beta_1-\beta_2)^2}\right\}^{-1/2}, \tag{8.56}$$

which is attainable if and only if $\gamma_1/\gamma_2 = \beta_1/\beta_2$ at $a = \gamma_1/(\gamma_1+\gamma_2), b = c = 0$.

The minimum of ρ occurs at approximately $b = c = 0.5$ if ν, α, and γ_1/γ_2, β_1/β_2 are similar.

8.26.7 Fields of Application

Tocher (1928) presented a number of large bivariate datasets concerning the milk yields of dairy cows. The bivariate gamma mixture model of Jones et al. (2000) has been used to model these data very well.

8.26.8 Mixtures of Bivariate Gammas of Iwasaki and Tsubaki

Using an integrating method to satisfy the integrability condition of the quasi-score function, Iwasaki and Tsubaki (2005) derived a bivariate distribution that can be expressed as a mixture of a discrete distribution whose probability mass is concentrated at the origin and independent gamma density functions.

8.27 Bivariate Bessel Distributions

There are two kinds of univariate Bessel distributions. Let U_1 and U_2 be two independent chi-squared random variables with common degrees of freedom ν; see for example, Johnson et al. (1994, pp. 50–51)

1. The first kind of Bessel distribution corresponds to $a_1U_1 + a_2U_2$ for $a_1 > 0$, $a_2 > 0$.

2. The second kind of Bessel distribution corresponds to $a_1U_1 - a_2U_2$ for $a_1 > 0$, $a_2 > 0$.

Let U, V, W be three independent chi-squared random variables with common degrees of freedom ν. Nadarajah and Kotz (2007b) have constructed four bivariate Bessel functions as follows:

(1) For $\alpha_1 > \beta_1 > 0$ and $\alpha_2 > \beta_2 > 0$, define

$$X = \alpha_1 U + \beta_1 V, \qquad Y = \alpha_2 U + \beta_2 V.$$

(2) For $\alpha_1 > \beta_1 > 0$ and $\alpha_2 > \beta_2 > 0$, define

$$X = \alpha_1 U + \beta_1 W, \qquad Y = \alpha_2 V + \beta_2 W.$$

(3) For $\alpha_1 > 0, \beta_1 > 0, \alpha_2 > 0$, and $\beta_2 > 0$, define

$$X = \alpha_1 U - \beta_1 V, \qquad Y = \alpha_2 U - \beta_2 V.$$

(4) For $\alpha_1 > 0, \beta_1 > 0, \alpha_2 > 0$, and $\beta_2 > 0$, define

$$X = \alpha_1 U - \beta_1 W, \qquad Y = \alpha_2 V - \beta_2 W.$$

The marginals of (1) and (2) belong to the Bessel distribution of the first kind, whereas the marginals of (3) and (4) are of the Bessel distribution of the second kind.

Explicit expressions as well as the contour plots for the four joint distributions are given in their equations (7), (10), (12), and (13), respectively. The product moments of these distributions were also derived.

References

1. Arnold, B.C., Strauss, D.: Pseudolikelihood estimation. Sankhyā, Series B **53**, 233–243 (1988)
2. Becker, P.J., Roux, J.J.J.: A bivariate extension of the gamma distribution. South African Statistical Journal **15**, 1–12 (1981)
3. Berland, R., Dussauchoy, A.: Aspects statistiques des régimes de microdécharges electriques entre électrodes métalliques placées dans un vide industriel. Vacuum **23**, 415–421 (1973)
4. Block, H.W., Rao, B.R.: A beta warning-time distribution and a distended beta distribution. Sankhyā, Series B **35**, 79–84 (1973)
5. Brewer, D.W., Tubbs, J.D., Smith, O.E.: A differential equations approach to the modal location for a family of bivariate gamma distributions. Journal of Multivariate Analysis **21**, 53–66 (1987). [Also appears as Chapter III of J.D. Tubbs, D.W. Brewer, and O.E. Smith (1983), Some Properties of a Five-Parameter Bivariate Probability Distribution, NASA Technical Memorandum 82550, Marshall Space Flight Center, Huntsville, Alabama.]
6. Cheriyan, K.C.: A bivariate correlated gamma-type distribution function. Journal of the Indian Mathematical Society **5**, 133-144 (1941)

7. Crovelli, R.A.: A bivariate precipitation model. In: 3rd Conference on Probability and Statistics in Atmospheric Science, pp. 130–134. American Meteorological Society, Boulder, Colorado (1973)
8. Dabrowska, D.: Parametric and nonparametric models with special schemes of stochastic dependence. In: Nonparametric Statistical Inference, Volume I, B.V. Gnedenko, M.L. Puri and I. Vincze (eds.), Colloquia Mathematica Societatis János Bolyai Volume 32, pp. 171–182. North-Holland, Amsterdam (1982)
9. David, F.N., Fix, E.: Rank correlation and regression in a nonnormal surface. In: Proceedings of the 4th Berkeley Symposium on Mathematical Statistics and Probability, Volume 1, J. Neyman (ed.), pp. 177–197. University of California Press, Berkeley(1961)
10. D'Este, G.M.: A Morgenstern-type bivariate gamma distribution. Biometrika **68**, 339–340 (1981)
11. Devroye, L.: Nonuniform Random Variate Generation. Springer-Verlag, New York (1986)
12. Dussauchoy, A., Berland, R.: Lois gamma à deux dimensions. Comptes Rendus de l'Academie des Sciences, Paris, Série A **274**, 1946–1949 (1972)
13. Dussauchoy, A., Berland, R.: A multivariate gamma-type distribution whose marginal laws are gamma, and which has a property similar to a characteristic property of the normal case. In: A Modern Course on Distributions in Scientific Work, Vol. I: Models and Structures, G.P. Patil, S. Kotz and J.K. Ord (eds.), pp. 319–328, Reidel, Dordrecht (1975)
14. Eagleson, G.K.: Polynomial expansions of bivariate distributions. Annals of Mathematical Statistics **35**, 1208–1215 (1964)
15. Freund, J.E.: A bivariate extension of the exponential distribution. Journal of the American Statistical Association **56**, 971–977 (1961)
16. Gaver, D.P.: Multivariate gamma distributions generated by mixture. Sankhyā, Series A **32**, 123–126 (1970)
17. Ghirtis, G.C.: Some problems of statistical inference relating to the double-gamma distribution. Trabajos de Estadística **18**, 67–87 (1967)
18. Griffiths, R.C.: The canonical correlation coefficients of bivariate gamma distributions. Annals of Mathematical Statistics **40**, 1401–1408 (1969)
19. Gunst, R.F., Webster, J.T.: Density functions of the bivariate chi-square distribution. Journal of Statistical Computation and Simulation **2**, 275–288 (1973)
20. Gupta, A.K., Nadarajah, S.: Sums, products and ratios for McKay's bivariate gamma distribution. Mathematical and Computer Modelling **43**, 185–193 (2006)
21. Gupta, A.K., Wong, C.F.: On a Morgenstern-type bivariate gamma distribution. Metrika **31**, 327–332 (1989)
22. Iwasaki, M., Tsubaki, H.: A new bivariate distribution in natural exponential family. Metrika **61**, 323–336 (2005)
23. Izawa, T.: The bivariate gamma distribution. Climate and Statistics **4**, 9–15 (in Japanese) (1953)
24. Izawa, T.: Two or multidimensional gamma-type distribution and its application to rainfall data. Papers in Meteorology and Geophysics **15**, 167–200 (1965)
25. Jain, G.C., Khan, M.S.H.: On an exponential family. Statistics **10**, 153–168 (1979)
26. Jensen, D.R.: An inequality for a class of bivariate chi-square distributions. Journal of the American Statistical Association **64**, 333–336 (1969)
27. Jensen, D.R.: The joint distribution of quadratic forms and related distributions. Australian Journal of Statistics **12**, 13–22 (1970)
28. Johnson, N.L., Kotz, S.: Distributions in Statistics: Continuous Multivariate Distributions. John Wiley and Sons, New York (1972)
29. Johnson, N.L., Kotz, S., Balakrishnan, N.: Continuous Univariate Distributions, Volume 1, 2nd edition. John Wiley and Sons, New York (1994)
30. Jones, G., Lai, C.D., Rayner, J.C.W.:A bivariate gamma mixture distribution. Communications in Statistics: Theory and Methods **29**, 2775–2790 (2000)

31. Kellogg, S.D., Barnes, J.W.: The bivariate H-function distribution. Mathematics and Computers in Simulation **31**, 91–111 (1989)
32. Kelly, K.S., Krzysztofowicz, R.: A bivariate meta-Gaussian density for use in hydrology. Stochastic Hydrology and Hydraulics **11**, 17–31 (1997)
33. Khan, M.S.H., Jain, G.C.: A class of distributions in the first emptiness of a semi-infinite reservoir. Biometrical Journal **20**, 243–252 (1978)
34. Kibble, W.F.: A two-variate gamma type distribution. Sankhyā **5**, 137–150 (1941)
35. Kotz, S., Balakrishnan, N., Johnson, N.L.: Continuous Multivariate Distributions, Volume 1: Models and Applications, 2nd edition. John Wiley and Sons, New York (2000)
36. Krishnaiah, P.R.: Multivariate gamma distributions. In: Encyclopedia of Statistical Sciences, Volume 6, S. Kotz and N.L. Johnson (eds.), pp. 63–66. John Wiley and Sons, New York (1985)
37. Krishnaiah, P.R.: Multivariate gamma distributions and their applications in reliability. Technical Report No. 83-09, Center for Multivariate Analysis, University of Pittsburgh (1983)
38. Krishnaiah, P.R.: Computations of some multivariate distributions. In: Handbook of Statistics, Volume 1, P.R. Krishnaiah (ed.), pp. 745–971. North-Holland, Amsterdam (1980)
39. Krishnaiah, P.R.: Simultaneous tests and the efficiency of balanced incomplete block designs. Report No. ARL 63-174, Aerospace Research Laboratories, Wright-Patterson Air Force Base, Ohio (1963)
40. Krishnaiah, P.R., Hagis, P., Steinberg, L.: A note on the bivariate chi-distribution. SIAM Review **5**, 140–144 (1963)
41. Krishnaiah, P.R., Rao, M.M.: Remarks on a multivariate gamma distribution. American Mathematical Monthly **68**, 342–346 (1961)
42. Krishnamoorthy, A.S., Parthasarathy, M.: A multivariate gamma-type distribution. Annals of Mathematical Statistics **22**, 549–557 (Correction **31**, 229) (1951)
43. Krishnan, M.: The noncentral bivariate chi-squared distribution and extensions. Communications in Statistics: Theory and Methods **5**, 647–660 (1976)
44. Lai, C.D., Moore, T.: Probability integrals of a bivariate gamma distribution. Journal of Statistical Computation and Simulation **19**, 205–213 (1984)
45. Lampard, D.G.: A stochastic process whose successive intervals between events form a first-order Markov chain-I. Journal of Applied Probability **5**, 648–668 (1968)
46. Lancaster, H.O.: The Chi-Squared Distribution. John Wiley and Sons, New York (1969)
47. Lee, R-Y., Holland, B.S., Flueck, J.A.: Distribution of a ratio of correlated gamma random variables. SIAM Journal of Applied Mathematics **36**, 304–320 (1979)
48. Lehmann, E.L.: Some concepts of dependence. Annals of Mathematical Statistics **37**, 1137–1153 (1966)
49. Loáciga, H.A., Leipnik, R.B.: Correlated gamma variables in the analysis of microbial densities in water. Advances in Water Resources **28**, 329–335 (2005)
50. Long, D., Krzysztofowicz, R.: Farlie–Gumbel–Morgenstern bivariate densities: Are they applicable in hydrology? Stochastic Hydrology and Hydraulics **6**, 47–54 (1992)
51. Malik, H.J., Trudel, R.: Distributions of the product and the quotient from bivariates t, F and Pareto distribution. Communications in Statistics: Theory and Methods **14**, 2951–2962 (1985)
52. Mardia, K.V.: Families of Bivariate Distributions. Griffin, London (1970)
53. Mardia, K.V.: Multivariate Pareto distributions. Annals of Mathematical Statistics **33**, 1008–1015 (Correction **34**, 1603) (1962)
54. Mathai, A.M., Moschopoulos, P.G.:. On a multivariate gamma. Journal of Multivariate Analysis **39**, 135–153 (1991)
55. Mathai, A.M., Moschopoulos, P.G.: A form of multivariate gamma distribution. Annals of the Institute of Statistical Mathematics **44**, 97–106 (1992)

56. McKay, A.T.: Sampling from batches. Journal of the Royal Statistical Society, Supplement **1**, 207–216 (1934)
57. Mielke, P.W., Flueck, J.A.: Distributions of ratios for some selected bivariate probability functions. In: American Statistical Association, Proceedings of the Social Statistics Section, pp. 608–613. American Statistical Association, Alexandria, Virginia (1976)
58. Mihram, G.A., Hultquist, A.R.: A bivariate warning-time/failure-time distribution. Journal of the American Statistical Association **62**, 589–599 (1967)
59. Miller, K.S.: Multidimensional Gaussian Distribution. John Wiley and Sons, New York (1964)
60. Moran, P.A.P.: Statistical inference with bivariate gamma distributions. Biometrika **56**, 627–634 (1969)
61. Moran, P.A.P., Vere-Jones, D.: The infinite divisibility of multivariate gamma distributions. Sankhyā, Series A **40**, 393–398 (1969)
62. Moran, P.A.P.:Testing for correlation between non-negative variates. Biometrika **54**, 385–394 (1967)
63. Nadarajah, S.: Products and ratios for a bivariate gamma distribution. Applied Mathematics and Computation **171**, 581–595 (2005)
64. Nadarajah, S.: FIM for Arnold and Strauss's bivariate gamma distribution. Computational Statistics and Data Analysis **51**, 1584–1590 (2006)
65. Nadarajah, S.: Comment on "Sheng Y. 2001. A bivariate gamma distribution for use in multivariate flood frequency analysis, Hydrological Processes 15(6):1033–1045." Hydrological Processess **21**, 2957 (2007)
66. Nadarajah, S., Gupta, A.K.: Some bivariate gamma distributions. Applied Mathematics Letters **19**, 767–774 (2006)
67. Nadarajah, S., Kotz, S.: A note on the correlated gamma distribution of Loaiciga and Leipnik. Advances in Water Resources **30**, 1053–1059 (2007a)
68. Nadarajah, S., Kotz, S.: Some bivariate Bessel distributions. Applied Mathematics and Computation **187**, 332–339 (2007b)
69. Nagao, M.: Multivariate probability distributions in statistical hydrology: Bivariate statistics as a center. In: Summer Training Conference on Hydraulics, Hydraulic Committee of Japanese Society of Civil Engineering, Hydraulics Series 75-A-4:1–19 (in Japanese) (1975)
70. Parrish, R.S., Bargmann, R.E.: A method for the evaluation of cumulative probabilities of bivariate distributions using the Pearson family. In: Statistical Distributions in Scientific Work. Volume 5: Inferential Problems and Properties, C. Taillie, G.P. Patil, and B.A. Baldessari (eds.), pp. 241–257. Reidel, Dordrecht (1981)
71. Phatarford, R.M.: Some aspects of stochastic reservoir theory. Journal of Hydrology **30**, 199–217 (1976)
72. Prékopa, A., Szántai, T.: A new multivariate gamma distribution and its fitting to empirical streamflow data. Water Resources Research **14**, 19–24 (1978)
73. Ramabhadran, V.R.: A multivariate gamma-type distribution. Sankhyā **11**, 45–46 (1951)
74. Ratnaparkhi, M.V.: Some bivariate distributions of (X, Y) where the conditional distribution of Y, given X, is either beta or unit-gamma. In: Statistical Distributions in Scientific Work, Volume 4: Models, Structures, and Characterizations, C. Taillie, G.P. Patil, and B.A. Baldessari (eds.), pp. 389–400. Reidel, Dordrecht (1981)
75. Royen, T.: Expansions for the multivariate chi-square distribution. Journal of Multivariate Analysis **38**, 213–232 (1991)
76. Sarmanov, I.O.: Gamma correlation process and its properties. Doklady Akademii Nauk, SSSR **191**, 30–32 (in Russian) (1970a)
77. Sarmanov, I.O.: An approximate calculation of correlation coefficients between functions of dependent random variables. Matematicheskie Zametki **7**, 617–625 (in Russian). English translation in Mathematical Notes, Academy of Sciences of USSR **7**, 373–377 (1970b)

78. Sarmanov, I.O.: New forms of correlation relationships between positive quantities applied in hydrology. In: Mathematical Models in Hydrology Symposium, IAHS Publication No. 100, pp. 104–109. International Association of Hydrological Sciences (1974)
79. Schmeiser, B.W., Lal, R.: Bivariate gamma random vectors. Operations Research **30**, 355–374 (1982)
80. Smith, O.E., Adelfang, S.I.: Gust model based on the bivariate gamma probability distribution. Journal of Spacecraft and Rockets **18**, 545–549 (1981)
81. Smith, O.E., Adelfang, S.I., Tubbs, J.D.: A bivariate gamma probability distribution with application to gust modeling. NASA Technical Memorandum 82483, Marshall Space Flight Center, Huntsville, Alabama (1982)
82. Steel, S.J., le Roux, N.J.: A reparameterisation of a bivariate gamma extension. Communications in Statistics: Theory and Methods **16**, 293–305 (1987)
83. Szántai, T.: Evaluation of special multivariate gamma distribution function. Mathematical Programming Study **27**, 1–16 (1986)
84. Tocher, J.F.: An investigation of the milk yield of dairy cows: Being a statistical analysis of the data of the Scottish Milk Records Association for the years 1908, 1909, 1911, 1912, 1920, and 1923. Biometrika **20**, 106–244 (1928)
85. Tubbs, J.D.: A method for determining if unequal shape parameters are necessary in a bivariate gamma distribution. Chapter II of J.D. Tubbs, D.W. Brewer, and O.E. Smith, Some Properties of a Five-Parameter Bivariate Probability Distribution, NASA Technical Memorandum 82550, Marshall Space Flight Center, Hunstville, Alabama (1983a)
86. Tubbs, J.D.: Analysis of wind gust data. Chapter IV of J.D. Tubbs, D.W. Brewer, and O.E. Smith, Some Properties of a Five-Parameter Bivariate Probability Distribution. NASA Technical Memorandum 82550, Marshall Space Flight Center, Huntsville, labama (1983b)
87. Vere-Jones, D.: The infinite divisibility of a bivariate gamma distribution. Sankhyā, Series A **29**, 421–422 (1967)
88. Wicksell, S.D.: On correlation functions of type III. Biometrika **25**, 121–133 (1933)
89. Yue, S.: Applying bivariate normal distribution to flood frequency analysis. Water International **24**, 248–254 (1999)
90. Yue, S.: A bivariate gamma distribution for use in multivariate frequency analysis. Hydrological Processess **15**, 1033–1045 (2001)
91. Yue, S., Ouarda, T.B.M.J., Bobée, B.: A review of bivariate gamma distributions for hydrological application. Journal of Hydrology **246**, 1–18 (2001)

Chapter 9
Simple Forms of the Bivariate Density Function

9.1 Introduction

When one considers a bivariate distribution, it is perhaps common to think of a joint density function rather than a joint distribution function, and it is also conceivable that such a density may be simple in expression, while the corresponding distribution function may involve special functions, can be expressed only as an infinite series, and sometimes may even be more complicated. Such distributions form the subject matter of this chapter. Although the standard form of these densities is simple, their generalizations are often not so simple. To include these generalizations would undoubtedly place the title of this chapter under question, but the alternative of leaving them out would be remiss. Therefore, for the sake of completeness, generalized forms of these simple densities will also be included in this discussion.

In Section 9.2, we describe the classical bivariate t-distribution and its properties. The noncentral version of the bivariate t-distribution is discussed next, in Section 9.3. In Section 9.4, the bivariate t-distribution having as its marginals t-distributions having different degrees of freedom is presented and some of its properties are detailed. The bivariate skew t-distributions of Jones and Branco and Dey are discussed in Sections 9.5 and 9.6, respectively. Next, the bivariate t-/skew t-distribution and its properties are discussed in Section 9.7. A family of bivariate heavy-tailed distributions is presented in Section 9.8. In Sections 9.9–9.12, the bivariate Cauchy, F, Pearson type II, and finite range distributions, respectively, are all described in detail. In Sections 9.13 and 9.14, the classical bivariate beta and Jones' form of bivariate beta distributions are presented along with their properties. The bivariate inverted beta distribution and its properties are detailed in Section 9.15. The bivariate Liouville, logistic, and Burr distributions and their characteristics and properties are presented in Sections 9.16–9.18, respectively. Rhodes' distribution is the topic of discussion of Section 9.19. Finally, the bivariate distribution with support above the diagonal proposed recently by Jones and Larsen (2004)

N. Balakrishnan, C.D. Lai, *Continuous Bivariate Distributions*,
DOI 10.1007/b101765_10,

is described in Section 9.20, where its properties and applications are also pointed out.

Many of the distributions in this chapter belong to Pearson's system and thus can be derived by the differential equation method described in Section 5.16.1. It is a common practice to refer to Pearson distributions by the form of their marginals—thus, for example, a bivariate type II has type II marginal distributions. But van Uven's designation is also used. The following table clarifies the nomenclature we have used.

Common name	van Uven's designation	Pearson marginals
t	III$a\alpha$	VII
F (inverted beta)	II$a\beta$	VI
	III$a\beta$	II
beta (Dirichlet)	II$a\alpha$	I and I, or I and II
McKay's bivariate gamma	IVa	III
	II$a\gamma$	VI
	IIb	V and VI
normal	VI	normal

Elderton and Johnson (1969, p. 138), Johnson and Kotz (1972, Table 1 in Chapter 34), and Rodriguez (1983) have presented versions of the table above in which expressions for the densities, supports, and restrictions on the parameters are also included.

9.2 Bivariate t-Distribution

9.2.1 Formula of the Joint Density

The joint density function is

$$h(x,y) = \frac{1}{2\pi\sqrt{1-\rho^2}}\left[1 + \frac{1}{\nu(1-\rho^2)}(x^2 - 2\rho xy + y^2)\right]^{-(\nu+2)/2} \tag{9.1}$$

for $\nu > 0,\ -1 < \rho < 1,\ x, y > 0$.

9.2.2 Univariate Properties

Both marginal distributions are t-distributions with the same degrees of freedom ν.

9.2.3 Correlation Coefficients

For $\nu > 2$, Pearson's product-moment correlation coefficient is ρ. For $0 < \nu \le 2$, ρ represents the gradient of the major axis of elliptical contours.

This distribution is an example of zero correlation not necessarily implying independence; see also Sections 9.2.6 and 9.2.9.

9.2.4 Moments

From the basic construction of this distribution described below in Section 9.2.6, the product moments are easily found to be

$$\mu'_{r,s} = E(X^r Y^s) = \nu^{(r+s)/2} E(X_1^r X_2^s) E(S^{-(r+s)}), \tag{9.2}$$

where $E(X_1^r X_2^s)$ is simply the (r,s)th product moment of the standard bivariate normal distribution with correlation coefficient ρ and

$$E(S^{(r+s)}) = 2^{-(r+s)/2} \Gamma\left(\frac{\nu - r - s}{2}\right) / \Gamma(\nu/2). \tag{9.3}$$

If X and Y are independent (i.e., $\rho = 0$), then $\mu'_{r,s}$ is zero unless both r and s are even, in which case it is given by

$$\nu^{(r+s)/2} \frac{[1 \cdot 3 \cdot 5 \cdots (2r-1)][1 \cdot 3 \cdot 5 \cdots (2s-1)]}{(\nu-2)(\nu-4)\cdots(\nu-r-s)}\ ; \tag{9.4}$$

see Johnson and Kotz (1972, pp. 135–136) for details.

The characteristic function of this distribution is given by Sutradhar (1986).

9.2.5 Conditional Properties

When $X = x$, the linear transformation of Y, viz. $U = \left[\frac{\nu(\nu+1)}{\nu+x^2}\right]^{1/2} \frac{Y - \rho x}{\sqrt{1-\rho^2}}$, has a t-distribution with $\nu + 1$ degrees of freedom. The regression is linear and is given by $E(Y|X = x) = \rho x$, and the conditional variance is quadratic and is given by $\frac{\nu}{\nu-1}(1-\rho^2)(1 + x^2/\nu)$; see Mardia (1970, p. 92).

9.2.6 Derivation

This distribution is derived from the trivariate reduction method as follows. Let (X_1, X_2) have the standardized bivariate normal distribution, with correlation coefficient ρ, and S, independent of X_1 and X_2, be distributed as χ_ν (i.e., the square root of a χ^2_ν-variate). Then $X = X_1\sqrt{\nu}/S$ and $Y = X_2\sqrt{\nu}/S$ follow the bivariate t-distribution in (9.1).

9.2.7 Illustrations

Devlin et al. (1976) have presented contour plots of the density in (9.1), while Johnson (1987, pp. 119–122, 124) has presented illustrations of the density surface.

9.2.8 Generation of Random Variates

The generation of random variates from this bivariate t-distribution has been discussed by many authors, including Johnson et al. (1984, p. 235), Văduva (1985), and Johnson (1987, pp. 120–121).

9.2.9 Remarks

- This is also known as the Pearson type VII distribution, though the density of the latter usually appears in the form
$$h(x,y) = \frac{-m\sqrt{(1-\rho^2)}}{\pi k^m}(k + x^2 - 2xy\rho + y^2)^{m-1} \qquad (9.5)$$
for $m < 0$; $-1 < \rho < 1; k > 0$.
- For the special case where $\rho = 0$ and $\nu = 1$, the bivariate Cauchy distribution is obtained; see Section 9.9 for more details.
- For $\rho = 0$, X^2 and Y^2 have a bivariate F-distribution; see Section 9.10 for more details.
- As $\nu \to \infty$, this distribution tends to a bivariate normal distribution.
- The contours of the probability density are ellipses. One may refer to Chapter 13 for more details on elliptical distributions.
- The variable $(X^2 - 2\rho XY + Y^2)/[2(1-\rho^2)]$ has an F-distribution with $(2, \nu)$ degrees of freedom; see Johnson et al. (1984).

- For this distribution, zero correlation does not imply independence of X and Y. This is so because though X_1 and X_2 having a bivariate normal distribution with correlation ρ become independent when $\rho = 0$, the denominator variable S is in common. In fact, apart from the bivariate normal, all the elliptically contoured bivariate distributions discussed in Chapter 13 have this property.
- The distributions of XY and X/Y have been discussed by Malik and Trudel (1985) and Wilcox (1985).
- For probability inequalities connected with bivariate and multivariate t-distributions, one may refer to Tong (1980, Section 3.1).

9.2.10 Fields of Application

- While this distribution is not often used to fit data, tables of its percentage points are required in the applications of multiple comparison procedures, ranking and selection procedures, and estimation of rank parameters. For a more detailed discussion, one may refer to Dunnett and Sobel (1954), Gupta (1963), Johnson and Kotz (1972, p. 145), and Chen (1979).
- Pearson (1924) fitted the distribution to two sets of data on the number of cards of a given suit that two players of whist hold in their hands.
- Econometricians make extensive use of systems of linear simultaneous equations and then commonly assume the stochastic terms, the disturbances, to have a multivariate normal distribution. Concerned with the possibility that the actual distribution has thicker tails than the normal, and hence that too much weight is given to outliers by conventional methods of estimation, Prucha and Kelejian (1984) proposed alternative methods based on multivariate t and other thick-tailed distributions.

9.2.11 Tables and Algorithms

Johnson and Kotz (1972, pp. 137–140) have listed many references to tables. Some recent tables include those of Chen (1979), Gupta et al. (1985), Wilcox (1986), and Bechhofer and Dunnett (1987).

For numerical computation of multivariate t probabilities over convex regions, see Somerville (1998). A generalization of Plackett's formula was derived by Genz (2004) for efficient numerical computations of the bivariate and trivariate t probabilities.

Genz and Bretz (2002) gave a comparison of methods for the computation of multivariate t probabilities.

9.2.12 Spherically Symmetric Bivariate t-Distribution

If $\rho = 0$, then (9.1) simply becomes

$$h(x,y) = \frac{1}{2\pi}\nu^{(\nu+2)/2}\left\{\nu + (x^2+y^2)\right\}^{-(\nu+2)/2}, \tag{9.6}$$

which is a spherically symmetric bivariate distribution. By replacing ν inside the bracket in (9.6) by a^2 and adjusting the normalizing constant, we obtain

$$h(x,y) = \frac{1}{2\pi}a^{\nu}\nu\left(a^2+x^2+y^2\right)^{-(\nu+2)/2}. \tag{9.7}$$

This is the form of bivariate t-distribution that is considered by Wesolowski and Ahsanullah (1995). For a review of spherically symmetric distributions, one may refer to Fang (1997).

9.2.13 Generalizations

- Poly (or multiple) t-distributions are those densities that correspond to the product of two or more terms like the right-hand side of (9.1); see Press (1972).

9.3 Bivariate Noncentral t-Distributions

Johnson and Kotz (1972, Chapter 37) considered the derivation of a more general form of bivariate t-distribution of the form

$$\left.\begin{array}{l} X = (X_1+\delta_1)\sqrt{\eta_1}/S_1 \\ Y = (X_2+\delta_2)\sqrt{\eta_2}/S_2 \end{array}\right\}, \tag{9.8}$$

where the δ's are noncentrality parameters, the X's have a joint normal distribution with a common variance σ^2, and the S_i/σ's have a joint chi-distribution. The cumulative distribution has been derived by Krishnan (1972). Ramig and Nelson (1980) have presented tables of the integral when $S_1 = S_2$.

The correlation coefficient ρ in this case is between -1 and 1.

9.3.1 Bivariate Noncentral t-Distribution with $\rho = 1$

Consider

$$X = \frac{Z + \delta_1}{\sqrt{Y/\nu}}, \qquad Y = \frac{Z + \delta_2}{\sqrt{Y/\nu}}, \tag{9.9}$$

where Z is a standard normal variable. The correlation coefficient is 1, which is not surprising since the two numerators $Z + \delta_1$ and $Z + \delta_2$ are mutually completely dependent. The joint distribution of X and Y in (9.9) seems to have been first discussed by Owen (1965). Some applications and properties, including tables, have been presented by Chou (1992).

9.4 Bivariate t-Distribution Having Marginals with Different Degrees of Freedom

The nature of using the same denominator to derive the bivariate t-distribution has been generalized by Jones (2002a). Specifically, let X_1, X_2 and W_1, W_2 be mutually independent random variables, the X_i's following the standard normal distribution and W_i's following the chi-squared distribution with n_i degrees of freedom. For the sake of convenience, let $\nu_1 = n_1$ and $\nu_2 = n_1 + n_2$ so that $\nu_1 \leq \nu_2$. In the case $\nu_1 = \nu_2$, we simply define $W_2 \equiv 0$.

Define a pair of random variables X and Y as

$$X = \sqrt{\nu_1} X_1 / \sqrt{W_1}, \qquad Y = \sqrt{\nu_2} X_2 / \sqrt{W_1 + W_2}. \tag{9.10}$$

Formula of the Joint Density

The joint density function is

$$h(x,y) = C_{12} \frac{{}_2F_1(\frac{1}{2}\nu_2 + 1, \frac{1}{2}n_2; \frac{1}{2}(\nu_2 + 1); (x^2/\nu_1)/\{1 + x^2/\nu_1 + y^2/\nu_2\})}{\{1 + x^2/\nu_1 + y^2/\nu_2\}^{\nu_2/2+1}}, \tag{9.11}$$

where

$$C_{12} = \frac{1}{\pi} \frac{\Gamma\frac{1}{2}(\nu_1 + 1)\Gamma\frac{1}{2}(\nu_2 + 1)}{\sqrt{\nu_1\nu_2}\Gamma\frac{1}{2}(\nu_1)\Gamma\frac{1}{2}(\nu_2 + 1)}$$

and ${}_2F_1(a, b; c; z)$ is the Gauss hypergeometric distribution.

Univariate Properties

The marginal distributions of X and Y are t-distributions with ν_1 and ν_2 degrees of freedom, respectively.

Joint Product Moments

The general (r, s)th product moment is given by

$$E(X^r Y^s) = \frac{\nu_1^{r/2}\nu_2^{s/2}\Gamma(\frac{1}{2}(r+1))\Gamma(\frac{1}{2}(s+1))\Gamma(\frac{1}{2}(\nu_1 - r))\Gamma(\frac{1}{2}(\nu_2 - r - s))}{\pi\Gamma(\frac{1}{2}\nu_1)\Gamma(\frac{1}{2}(\nu_2 - r))} \quad (9.12)$$

if r and s are both even and is zero otherwise.

Correlation Coefficient

Like the spherically symmetric bivariate t-distribution in (9.7) above (with the correlation coefficient between X_1 and X_2 being zero), X and Y are uncorrelated and yet not independent in this case as well.

Conditional Properties

Denote $u_1 = 1 + x^2/\nu_1$. Then, the conditional density of Y, given $X = x$, is

$$g(y|x) = C_{2|1}\frac{{}_2F_1(\frac{1}{2}\nu_2 + 1, \frac{1}{2}n_2; \frac{1}{2}(\nu_2 + 1); (u_1 - 1)/(u_1 + y^2/\nu_2))}{(u_1 + y^2/\nu_2)^{\nu_2/2+1}}, \quad (9.13)$$

where

$$C_{2|1} = \frac{u_1^{(\nu_1+1)}\Gamma(\frac{1}{2}\nu_2 + 1)}{\sqrt{\pi\nu_2}\Gamma(\frac{1}{2}(\nu_2 + 1))}.$$

In a similar way, with $u_2 = 1 + y^2/\nu_2$, the conditional density of X, given $Y = y$, is

$$f(x|y) = C_{1|2}\frac{{}_2F_1(\frac{1}{2}\nu_2 + 1, \frac{1}{2}n_2; \frac{1}{2}(\nu_2 + 1); 1 - u_2/(u_2 + x^2/\nu_1))}{(u_2 + x^2/\nu_1)^{\nu_2/2+1}}, \quad (9.14)$$

where

$$C_{1|2} = \frac{u_2^{(\nu_2+1)}\Gamma(\frac{1}{2}\nu_1 + 1)\Gamma(\frac{1}{2}\nu_2)\Gamma(\frac{1}{2}\nu_2 + 1)}{\sqrt{\pi\nu_1}\Gamma(\frac{1}{2}(\nu_1))\Gamma^2(\frac{1}{2}(\nu_2 + 1))}.$$

Illustrations

Jones (2002a) has presented a contour plot of the density when $\nu_1 = 2$ and $\nu_2 = 5$.

9.5 Jones' Bivariate Skew t-Distribution

The bivariate skew t-distribution constructed by Jones (2001) is described here. This, incidentally, differs from another bivariate distribution that is also known as a bivariate skew t-distribution. The derivation of the latter is in the same spirit as that of the bivariate skew-normal distribution described in Section 7.4. In order to make a distinction, we shall call the latter the *bivariate skew t-distribution.* It has been discussed by Branco and Dey (2001), Azzalini and Capitanio (2003), and Kim and Mallick (2003), and it will be the subject of the next section.

9.5.1 Univariate Skew t-Distribution

A skew t-distribution, defined by Jones (2001) and studied further by Jones and Faddy (2003), has as its density function

$$f(t) = \frac{1}{2^{c-1}B(a,b)c^{1/2}}\left\{1+\frac{t}{(c+t^2)^{1/2}}\right\}^{a+1/2}\left\{1-\frac{t}{(c+t^2)^{1/2}}\right\}^{b+1/2} \tag{9.15}$$

for $a, b > 0$ and $c = a + b$. When $a = b$, f in (9.15) reduces to a standard t-density with $2a$ degrees of freedom.

9.5.2 Formula of the Joint Density

The joint density function is

$$\begin{aligned} h(x,y) = K_v &\left\{\frac{2\left(x+\sqrt{w_1+x^2}\right)^{2\nu_1}}{w_1^{\nu_1}\sqrt{w_1+x^2}}\right\}\left\{\frac{2\left(y+\sqrt{w_2+y^2}\right)^{2\nu_2}}{w_2^{\nu_2}\sqrt{w_2+y^2}}\right\} \\ &\times\left\{1+\frac{\left(x+\sqrt{w_1+x^2}\right)^2}{w_1}+\frac{\left(y+\sqrt{w_2+y^2}\right)^2}{w_2}\right\}^{-n}, \end{aligned} \tag{9.16}$$

where $w_i = \nu_0 + \nu_i$, $i = 1, 2$, $n = \nu_0 + \nu_1 + \nu_2$, and

$$K_v = \Gamma(n)/\{\Gamma(\nu_0)\Gamma(\nu_1)\Gamma(\nu_2)\},$$

the multinomial coefficient. When $\nu_0 = \nu_1 = \nu_2 = \nu/2$, say, then the density in (9.16) becomes a bivariate symmetric t-density distribution having the usual t marginals, given by

$$h(x,y) = 4\Gamma(3\nu/2)\Gamma(\nu/2)^{-3}\nu^{\nu/2}\left\{\frac{\left(x+\sqrt{\nu+x^2}\right)^{\nu}}{\sqrt{\nu+x^2}}\right\}\left\{\frac{\left(y+\sqrt{\nu+y^2}\right)^{\nu}}{\sqrt{\nu+y^2}}\right\}$$
$$\times\left\{\nu+\left(x+\sqrt{\nu+x^2}\right)^2+\left(y+\sqrt{\nu+y^2}\right)^2\right\}^{-3\nu/2}. \qquad (9.17)$$

Remarks

In Jones (2001), (9.16) is called the bivariate t-/skew t-distribution. However, he called their marginals a "skew t" variable. In order to be consistent with the acronym for the marginals, we have named (9.16) as Jones' bivariate skew t-distribution.

9.5.3 Correlation and Local Dependence for the Symmetric Case

Pearson's correlation coefficient is given by

$$\rho = \frac{(2\nu-3)}{8}\left(\frac{\Gamma((\nu-1)/2)}{\Gamma(\nu/2)}\right), \quad \nu > 2. \qquad (9.18)$$

It is conjectured that this correlation is a monotonically increasing function of $\nu > 0$.

The local dependence function defined by $\gamma(x,y) = \partial^2 \log h(x,y)/\partial x\, \partial y$ is given by

$$\gamma(x,y) = \frac{6\nu\left(x+\sqrt{\nu+x^2}\right)^2\left(y+\sqrt{\nu+y^2}\right)^2}{\sqrt{(\nu+x^2)(\nu+y^2)}\left\{\nu+(x+\sqrt{\nu+x^2})^2+\left(y+\sqrt{\nu+y^2}\right)^2\right\}^2}. \qquad (9.19)$$

Note that $\gamma(x,y) > 0$.

9.5.4 Derivation

Let W_i, $i = 0,1,2$, be mutually independent χ^2 random variables with $2\nu_i$ degrees of freedom as specified above. Define $X = \frac{\sqrt{\omega_1}}{2}\left(\sqrt{\frac{W_1}{W_0}} - \sqrt{\frac{W_0}{W_1}}\right)$; similarly, $X = \frac{\sqrt{\omega_2}}{2}\left(\sqrt{\frac{W_2}{W_0}} - \sqrt{\frac{W_0}{W_2}}\right)$, where $\omega_i = \nu_0 + \nu_i$, $i = 1,2$. Then, X and Y have a joint density as given in (9.16).

9.6 Bivariate Skew t-Distribution

The bivariate skew t-distribution presented here differs from Jones' bivariate skew t-distribution discussed in the preceding section. The distribution presented in this section is derived by adding an extra parameter to the bivariate t-distribution to regulate skewness. As mentioned in the last section, we call this distribution as the bivariate skew t-distribution.

9.6.1 Formula of the Joint Density

The joint density function is

$$h(x,y) = h_T(x,y;\nu)T_1\left(\alpha_1 x + \alpha_2 y\left(\frac{\nu+2}{Q+\nu}\right)^{1/2};\nu+2\right), \qquad (9.20)$$

where $Q = (x^2 - 2\rho xy + y^2)/(1-\rho^2)$, $h_T(x,y;\nu)$ is the bivariate t-density in (9.1), and $T_1(x;\nu+2)$ is the cumulative distribution function of the Student t-distribution with $\nu+2$ degrees of freedom.

9.6.2 Moment Properties

Azzalini and Capitanio (2003) discussed the likelihood inference and presented moments of this distribution up to the fourth order as well. Kim and Mallick (2003) derived the moment properties when the bivariate skew t has a nonzero mean vector $\boldsymbol{\mu} \neq \mathbf{0}$.

9.6.3 Derivation

Let $\boldsymbol{Z}$ denote the standard bivariate skew-normal variable having probability density function $2\phi(\boldsymbol{z};\Omega)\Phi(\boldsymbol{\alpha}'\boldsymbol{z})$, where ϕ is the standard bivariate normal density with correlation matrix Ω, Φ is the distribution function of the standard normal, and $\boldsymbol{\alpha} = (\alpha_1,\alpha_2)'$.

Let $\boldsymbol{X} = (X,Y)'$ and $V \sim \chi^2_\nu$. Then, $\boldsymbol{X} = V^{-1/2}\boldsymbol{Z}$ has its density function as given in (9.20); see, for example, Kim and Mallick (2003).

9.6.4 Possible Application due to Flexibility

It has been stated by several authors that, by introducing a skewness parameter to a symmetric distribution, the new bivariate distribution would bring additional flexibility for modeling skewed data. This will be useful for regression and calibration problems when the corresponding error distribution exhibits skewness.

9.6.5 Ordered Statistics

Jamalizadeh and Balakrishnan (2008b) derived the distributions of order statistics from bivariate skew t_ν-distribution in terms of generalized skew-normal distributions, and used them to obtain explicit expressions for means, variances and covariance. Here, by generalized skew-normal distribution, we mean the distribution of $X|(U_1 < \theta_1 X, U_2 < \theta_2 X)$ when $X \rightsquigarrow N(0,1)$ independently of $(U_1, U_2)^T \rightsquigarrow \text{BVN}(0,0,1,1,\gamma)$. This distribution, which is a special case of the unified multivariate skew-normal distribution introduced by Arellano-Valle and Azzalini (2006), has also been utilized by Jamalizadeh and Balakrishnan (2009) to obtain a mixture representation for the distributions of order statistics from the trivariate skew t_ν-distribution in terms of generalized skew t_ν-distributions.

9.7 Bivariate t-/Skew t-Distribution

This model was proposed by Jones (2002b) based on a marginal replacement scheme. The idea is to replace one of the marginals of the spherically symmetric bivariate t-distribution of (9.1) by the univariate skew t-distribution as specified by (9.15).

9.7.1 Formula of the Joint Density

The joint density function is

$$h(x,y) = \frac{\Gamma((\nu+2)/2)}{\Gamma((\nu+1)/2)B(a,c)(a+c)^{1/2}2^{a+c-1}(\pi\nu)^{1/2}} \times \frac{(1+\nu^{-1}x^2)^{(\nu+1)/2}\left(1+\frac{x}{(a+c+x^2)^{1/2}}\right)^{a+1/2}\left(1-\frac{x}{(a+c+x^2)^{1/2}}\right)^{c+1/2}}{(1+\nu^{-1}(x^2+y^2))^{(\nu+2)/2}}; \tag{9.21}$$

here, a, c, and ν are all positive. Equation (9.21) becomes the spherically symmetric bivariate t-density in (9.6) when $a = c = \nu/2$.

9.7.2 Univariate Properties

The marginal distribution of X is the skew t-distribution presented in (9.15) with parameters a and c. The marginal distribution of Y is symmetric and can be well approximated by a t-distribution with the same variance.

9.7.3 Conditional Properties

The conditional distribution of Y, given $X = x$, matches that of the bivariate t-distribution and is a univariate t-distribution with $\nu + 1$ degrees of freedom scaled by a factor $\{(\nu+1)^{-1}(x^2+\nu)\}^{1/2}$.

9.7.4 Other Properties

- X and Y are uncorrelated.
- The local dependence function is the same as that of the bivariate t-distribution in (9.6).
- $\text{corr}(|X|, |Y|) > 0$.

9.7.5 Derivation

This distribution can be derived by the marginal replacement scheme, i.e., multiply (9.6) by (9.15) and divide by the density of the standard univariate t-distribution.

9.8 Bivariate Heavy-Tailed Distributions

9.8.1 Formula of the Joint Density

The joint density function is

$$h(x, y) = (1 + x^2)^{-c_1/2}(1 + y^2)^{-c_2/2}(1 + x^2 + y^2)^{-c/2} \tag{9.22}$$

for $x, y \geq 0$, $c_1, c_2, c > 0$.

9.8.2 Univariate Properties

Let $s_1 = c + c_1, s_2 = c + c_2, s_3 = c + c_1 + c_2$, and further

$$\psi_c(x) = (1 + x^2)^{-c/2} \text{ and } \psi_c^*(x) = (1 + x^2)^{-c/2} \log(2 + x^2).$$

1. If $s_1 < s_3 - 1$, then $f(x) = \psi_{s_1}(x)$.
2. If $s_1 = s_3 - 1$, then $f(x) = \psi_{s_1}^*(x)$.
3. If $s_1 > s_3 - 1$, then $f(x) = \psi_{s_3-1}(x)$.

9.8.3 Remarks

- The first two terms on the right-hand side of (9.22) correspond to independent univariate t-densities, while the last term corresponds to a bivariate t-density.
- Le and O'Hagan (1998) have discussed various other properties of this family of distributions, and, in particular, they have expounded the difference between this distribution and the class of v-spherical distributions of Fernandez, Osiewalski, and Steel (1995), which also possesses a heavy tail.

9.8.4 Fields of Application

This distribution provides resolutions for conflicting information in a Bayesian setting; see O'Hagan and Le (1994).

9.9 Bivariate Cauchy Distribution

This distribution, a special case of the bivariate t-distribution when $\rho = 0$ and $\nu = 1$, is of limited interest, as it has no correlation parameter.

9.9.1 Formula of the Joint Density

The joint density function is

$$h(x,y) = \frac{1}{2\pi}(1 + x^2 + y^2)^{-3/2}, \quad x, y \in \mathbf{R}. \tag{9.23}$$

Of course, location and scale factors can readily be introduced into (9.23) if required.

9.9.2 Formula of the Cumulative Distribution Function

The joint cumulative distribution function is

$$H(x,y) = \frac{1}{4} + \frac{1}{2\pi}\left(\tan^{-1} x + \tan^{-1} y + \tan^{-1} \frac{xy}{\sqrt{(1 + x^2 + y^2}}\right). \tag{9.24}$$

9.9.3 Univariate Properties

Both marginals are Cauchy, and therefore their means and standard deviations do not exist; consequently, some other measures of location and spread need to be used in this case.

9.9.4 Conditional Properties

The conditional density of Y, given $X = x$, is

$$g(y|x) = \frac{1}{2}(1 + x^2)/(1 + x^2 + y^2)^{3/2}.$$

Hence, $Y/\sqrt{\frac{1}{2}(1+x^2)}$, conditional on $X = x$, has a t-distribution with two degrees of freedom. The distribution of any linear combination of X and Y is Cauchy as well; see Ferguson (1962).

9.9.5 Illustrations

Contours of the density have been presented by Devlin et al. (1976). Plots of the density as well as the contours after transformation to uniform marginals have been provided by Barnett (1980). Johnson et al. (1984) have presented the contours after transformation to normal marginals.

9.9.6 Remarks

- For bivariate distributions with Cauchy conditionals, see Section 6.4 and also Chapter 5 of Arnold et al. (1999).
- Sun and Shi (2000) have considered the tail dependence in the bivariate Cauchy distribution.

9.9.7 Generation of Random Variates

For generation of random variates from this distribution, one may refer to Devroye (1986, p. 555) and Johnson et al. (1984).

9.9.8 Generalization

Jamalizadeh and Balakrishnan (2008a) proposed a *generalized bivariate Cauchy distribution* as the distribution of $(W_1, W_2)^T = \left(\frac{U_2}{|U_1|}, \frac{U_3}{|U_1|}\right)$, where $(U_1, U_2, U_3)^T$ has a standard trivariate normal distribution with correlation matrix $\mathbf{R}$. They then showed the joint distribution function of $(W_1, W_2)^T$ to be, for $(t_1, t_2) \in \mathbb{R}^2$,

$$F(t_1, t_2; \mathbf{R}) = \frac{1}{4\pi}\left\{\cos^{-1}\left(-\frac{\rho_{23} - \rho_{12}t_1 - \rho_{13}t_2 + t_1t_2}{\sqrt{1 - 2\rho_{12}t_1 + t_1^2}\sqrt{1 - 2\rho_{13}t_2 + t_2^2}}\right)\right.$$
$$+ \tan^{-1}\left(\frac{t_1 - \rho_{12}}{\sqrt{1 - \rho_{12}^2}}\right) + \tan^{-1}\left(\frac{t_2 - \rho_{13}}{\sqrt{1 - \rho_{13}^2}}\right)$$
$$+ \cos^{-1}\left(-\frac{\rho_{23} + \rho_{12}t_1 + \rho_{13}t_2 + t_1t_2}{\sqrt{1 + 2\rho_{12}t_1 + t_1^2}\sqrt{1 + 2\rho_{13}t_2 + t_2^2}}\right)$$
$$\left. + \tan^{-1}\left(\frac{t_1 + \rho_{12}}{\sqrt{1 - \rho_{12}^2}}\right) + \tan^{-1}\left(\frac{t_2 + \rho_{13}}{\sqrt{1 - \rho_{13}^2}}\right)\right\}.$$

In the special case when $\rho_{12} = \rho_{13} = 0$ and $\rho_{23} = \rho$, this distribution reduces to the standard bivariate Cauchy distribution discussed, for example, in Fang, Kotz, and Ng (1990); in this case, the above joint distribution function simplifies to

$$\frac{1}{2\pi}\left\{\cos^{-1}\left(-\frac{\rho + t_1t_2}{\sqrt{1 + t_1^2}\sqrt{1 + t_2^2}}\right) + \tan^{-1}(t_1) + \tan^{-1}(t_2)\right\}, \quad (t_1, t_2) \in \mathbb{R}^2.$$

9.9.9 Bivariate Skew-Cauchy Distribution

Consider three independent standard Cauchy random variables W_1, W_2, and U. Let $\boldsymbol{W} = (W_1, W_2)$. Arnold and Beaver (2000) constructed a basic bivariate skew-Cauchy distribution by considering the conditional distribution of $\boldsymbol{W}$ given $\lambda_0 + \lambda_1'\boldsymbol{W} > U$.

The basic bivariate skew-Cauchy distribution has a joint density of the form

$$h(x, y) = \psi(x)\psi(y)\Psi(\lambda_0 + \lambda_{11}x + \lambda_{12}y)/\Psi\left(\frac{\lambda_0}{1 + \lambda_{11} + \lambda_{12}}\right),$$

where $\psi(u) = \frac{1}{\pi(1+u^2)}$, $\Psi(u) = \frac{1}{2} + \frac{1}{\pi}\tan^{-1} u, u\,\text{real}$, and $\lambda_1' = (\lambda_{11}, \lambda_{12})$.

9.10 Bivariate F-Distribution

The distribution has been widely studied. It is also known as the bivariate inverted beta or the bivariate inverted Dirichlet distribution [Kotz et al. (2000, pp. 492–497)].

9.10.1 Formula of the Joint Density

The joint density function is

$$h(x,y) = Kx^{(\nu_1-2)}y^{(\nu_2-2)/2}\left(1+\frac{\nu_1 x+\nu_2 y}{\nu_0}\right)^{-(\nu_0+\nu_1+\nu_2)/2}, \quad x,y \geq 0, \tag{9.25}$$

where the ν's are positive and referred to as the "degrees of freedom," and the constant K is given by

$$\Gamma\left(\frac{\nu_0+\nu_1+\nu_2}{2}\right)\nu_0^{-(\nu_0+\nu_1+\nu_2)/2}\frac{\nu_0^{\nu_0/2}\nu_1^{nu_1/2}\nu_2^{\nu_2/2}}{\Gamma(\nu_0/2)\Gamma(\nu_1/2)\Gamma(\nu_2/2)}.$$

9.10.2 Formula of the Cumulative Distribution Function

There is no elementary form for $H(x,y)$, but it may be written in terms of F_2, Appell's hypergeometric functions of two variables; see Amos and Bulgren (1972) and Hutchinson (1979, 1981).

9.10.3 Univariate Properties

The marginal distributions of X and Y are F-distributions with (ν_1,ν_0) and (ν_2,ν_0) degrees of freedom, respectively.

9.10.4 Correlation Coefficients

Pearson's product-moment correlation is $\sqrt{\frac{\nu_1\nu_2}{(\nu_0+\nu_1-2)(\nu_0+\nu_2-2)}}$ for $\nu_0 > 4$.

9.10.5 Product Moments

The (r,s)th product moment is given by

$$E(X^rY^s) = \frac{\Gamma(\frac{1}{2}\nu_0 - r - s)(\frac{1}{2}\nu_1+r)(\frac{1}{2}\nu_2+s)}{\Gamma(\nu_0/2)\Gamma(\nu_1/2)\Gamma(\nu_2/2)(\nu_1/\nu_0)^r(\nu_2/\nu_0)^s}$$

if $r + s < \nu_0/2$ and is undefined otherwise; see Nayak (1987).

9.10.6 Conditional Properties

The expression $(\nu_0 + \nu_1)Y/(\nu_0 + \nu_1 x)$, conditional on $X = x$, has an F-distribution with degrees of freedom $(\nu_2, \nu_0 + \nu_1)$. The regression is linear and is given by $E(Y|X = x) = (\nu_0 + \nu_1 x)/(\nu_0 + \nu_1 - 2)$ for $\nu_0 > 0$; see Mardia (1970, p. 93) and Nayak (1987).

9.10.7 Methods of Derivation

This distribution may be obtained by transforming the bivariate t-distribution in (9.6). More precisely, if (X, Y) is the bivariate t-variate with ν_0 degrees of freedom and $\rho = 0$, then (X^2, Y^2) has a bivariate F-distribution with degrees of freedom ν_0, 1, and 1. However, this method does not lead to a bivariate F-distribution with other values of ν_1 and ν_2.

Alternatively, we may consider a trivariate reduction technique with $X = \frac{X_1/\nu_1}{X_0/\nu_0}$ and $Y = \frac{X_2/\nu_2}{X_0/\nu_0}$, where X_0, X_1, and X_2 are independent chi-squared variables with degrees of freedom ν_0, ν_1, and ν_2, respectively. Then, X and Y have a bivariate F-distribution with degrees of freedom ν_0, ν_1, and ν_2. The distribution may also be obtained by the method of compounding (equivalent to the method of trivariate reduction in some situations). For further details, see Adegboye and Gupta (1981).

9.10.8 Relationships to Other Distributions

- It is related to the bivariate t-distribution as indicated earlier.
- The bivariate inverted beta distribution (see Section 9.15) is essentially the bivariate F-distribution, written in a slightly different form.
- It is a special case of the bivariate Lomax distribution.
- For the distributions of XY and X/Y, one may refer to Malik and Trudel (1985).
- A noncentral generalization has been given by Feingold and Korsog (1986). This is obtained by letting $X = \frac{X_1/\nu_1}{X_0/\nu_0}$, $Y = \frac{X_2/\nu_2}{X_0/\nu_0}$, where X_0, X_1, and X_2 have noncentral chi-squared distributions.
- Another generalization is Krishnaiah's (1964, 1965) bivariate F-distributions, obtained by the trivariate reduction method just mentioned, but with X_1 and X_2 now being correlated (central) chi-squared variates; viz. their joint distribution is Kibble's bivariate gamma (see Section 8.2).

- A generalization of Krishnaiah's bivariate F-distribution is Jensen's bivariate F, which is obtained through two quadratic forms from a multivariate distribution and a chi-squared distribution; see Section 8.5 for more details.
- The distribution of $V = \min(X, Y)$ was studied in detail by Hamdy et al. (1988).

9.10.9 Fields of Application

The distribution is rarely used to fit data. However, tables of its percentage points are required in the analysis of variance and experimental design in general; see Johnson and Kotz (1972, pp. 240–241). This distribution is closely related to the bivariate beta distribution, and the application of the latter to compositional data is sometimes expressed in such a way that bivariate F is the one that gets applied; see Ratnaparkhi (1983). However, the distribution of $V = \min(X, Y)$ arises in many statistical problems including analysis of variance, selecting and ordering populations, and in some two-stage estimation procedures [Hamdy et al. (1988)].

9.10.10 Tables and Algorithms

Amos and Bulgren (1972) recognized that the cumulative distribution can be expressed in terms of Appell's F_2 function. Tiao and Guttman (1965) expressed the integral in terms of a finite sum of incomplete beta functions.

Hewitt and Bulgren (1971) have shown that if ν_1 and ν_2 are equal, then for any a and b such that $0 \leq a \leq b < \infty$,

$$\Pr(a < X \leq b, a < Y \leq b) \geq \Pr(a < X \leq b) \Pr(a < Y \leq b), \tag{9.26}$$

meaning that X and Y are positively quadrant dependent. Numerical studies carried out they show that the right-hand side of (9.26) is quite a good approximation to the left-hand side. Accuracy increases as ν_0 increases, but decreases as ν_1 and ν_2 increase. Hamdy et al. (1988) have presented an algorithm to compute the lower and upper percentage points of $\min(X, Y)$; see also the references therein.

9.11 Bivariate Pearson Type II Distribution

9.11.1 Formula of the Joint Density

The joint density function is

$$h(x,y) = \frac{(\nu+1)}{\pi\sqrt{1-\rho^2}}\left[1 - \frac{x^2 - 2\rho xy + y^2}{1-\rho^2}\right]^{\nu}, \tag{9.27}$$

where $\nu > 1$, $-1 < \rho < 1$, and (x, y) is in the ellipse $x^2 - 2\rho xy + y^2 = 1 - \rho^2$, which itself lies within the unit square.

9.11.2 Univariate Properties

The marginals are of Pearson type II with density $f(x) = (1-x^2)^{\nu+\frac{1}{2}}/B(\frac{1}{2}, \nu+\frac{3}{2})$, $-1 < x < 1$ and a similar expression for $g(y)$. The distribution is also known as the *symmetric beta distribution.* A simple linear transformation $Z = (X+1)/2$ reduces a Pearson type II distribution to a standard beta distribution.

9.11.3 Correlation Coefficient

The variable ρ in (9.27) is indeed Pearson's product-moment correlation.

9.11.4 Conditional Properties

The conditional distribution of one variable, given the other, is also of Pearson type II.

9.11.5 Relationships to Other Distributions

Let $U = (aX - bY)^2$ and $V = (aY - bX)^2$, where $a = \frac{\sqrt{1+\rho}+\sqrt{1-\rho}}{2\sqrt{1-\rho^2}}$ and $b = \frac{\sqrt{1+\rho}-\sqrt{1-\rho}}{2\sqrt{1-\rho^2}}$. Then, U and V have a bivariate beta distribution with joint density $\dfrac{n+1}{\pi}\dfrac{(1-u-v)^n}{\sqrt{uv}}$.

9.11.6 Illustrations

Johnson (1986; 1987, pp. 111–117, 123) has presented plots of the density.

9.11.7 Generation of Random Variates

Johnson (1987, pp. 115–116, 123) and Johnson et al. (1984, p. 235) have discussed generation of random variates from this distribution.

9.11.8 Remarks

- This is type IIIaβ in van Uven's classification.
- Along with the bivariate normal and t-distributions, this distribution is a well-known member of the class of elliptically contoured distributions.
- The quantity $(X^2 - 2\rho XY + Y^2)/(1-\rho^2)$ has a beta(1, $n+1$) distribution; see, for example, Johnson et al. (1984).
- The cumulative distribution has a diagonal expansion in terms of orthogonal (Gegenbauer) polynomials; see McFadden (1966).
- The expression for Rényi and Shannon entropies for a bivariate Pearson type II distribution was given in Nadarajah and Zografosb (2005).

9.11.9 Tables and Algorithms

An algorithm for computing the bivariate probability integral can be developed using the results of Parrish and Bargmann (1981). Joshi and Lalitha (1985) have developed a recurrence formula for the evaluation of $\bar{H}$.

9.11.10 Jones' Bivariate Beta/Skew Beta Distribution

Consider a special case of the bivariate Pearson type II distribution for which $\rho = 0$. Then, letting $b = \nu + \frac{3}{2}$, (9.27) becomes

$$h(x,y) = \frac{\Gamma(b+1/2)}{\Gamma(b-1/2)\pi}\left(1 - x^2 - y^2\right)^{b-3/2}, \quad b > 1/2, \tag{9.28}$$

which is a spherically symmetric distribution.

Each marginal is a univariate symmetric beta (Pearson type II) with density function

$$\frac{1}{B(b,1/2)}(1-x^2)^{b-1}, \quad 1<x<1. \tag{9.29}$$

Jones (2002b) obtained an asymmetric beta density by multiplying (9.28) by $(1+x)^{a-b}(1-x)^{c-b}$ and renormalizing suitably to give

$$\frac{1}{B(a,c)2^{a+c-1}}(1+x)^{a-1}(1-x)^{c-1}, \quad 1<x<1. \tag{9.30}$$

Jones (2001) constructed a new bivariate distribution by the marginal replacement scheme, specifically by replacing the marginal density of X in (9.28) by (9.30), resulting in a bivariate beta/skew beta distribution (X has a skew beta distribution) with joint density function

$$h(x,y)=\frac{\Gamma(b)(1+x)^{a-b}(1-x)^{c-b}}{B(a,c)\Gamma(b-1/2)2^{a+c-1}\sqrt{\pi}}(1-x^2-y^2)^{b-3/2} \tag{9.31}$$

for $0<x^2+y^2<1$, $a>0$, $b>1/2$, $c>0$. By construction, X has a skew beta density given in (9.29), and the marginal distribution of Y is a symmetric beta. The conditional distribution of Y, given $X=x$, is also a rescaled symmetric beta over the interval $(-\sqrt{1-x^2},\sqrt{1-x^2})$.

9.12 Bivariate Finite Range Distribution

The bivariate finite range distribution has been discussed by Roy (1989, 1990) and Roy and Gupta (1996).

9.12.1 Formula of the Survival Function

The joint survival function is

$$\bar{H}(x,y)=(1-\theta_1 x-\theta_2 y-\theta_3 xy)^p, \tag{9.32}$$

where $\theta_1>0, \theta_2>0$, $p-1\ge\theta_3/(\theta_1\theta_2)\ge -1$, $0\le x\le\theta_1^{-1}$, $0\le y\le(1-\theta_1 x)/(\theta_2+\theta_3 x)$.

9.12.2 Characterizations

The joint survival function in (9.32) can be characterized either through a constant bivariate coefficient of variation $C_i(x,y) = \{V_i(x,y)\}^{1/2}/M_i(x,y)$, where $V_1(x,y) = \text{var}(X-x|X>x, Y>y)$, $V_2(x,y) = \text{var}(Y-y|X>x, Y>y)$, $M_1(x,y) = E(X-x|X>x, Y>y)$ and $M_2(x,y) = E(Y-y|X>x, y>y)$ or by a constant product of mean residual lives and hazard rates.

Case 1. $1/\sqrt{3} \le C_1(x,y) = C_2(x,y) = k < 1$ if and only if (X,Y) has a bivariate finite range distribution in (9.32) with $p = 2k^2/(1-k^2)$. Also, $0 < k < 1\sqrt{3}$ if and only if X and Y are mutually independent with $\theta_3 = -\theta_1\theta_2$.

Case 2. Let $r_1(x,y) = -\frac{\partial}{\partial x}\log \bar{H}(x,y)$ and $r_2(x,y) = -\frac{\partial}{\partial y}\log \bar{H}(x,y)$. Then, $0 < 1 - r_i(x,y)M_i(x,y) = k \le 1/2$ $(i = 1,2)$ if and only if (X,Y) has a bivariate finite range distribution in (9.32). Also, $\frac{1}{2} \le k < 1$ if and only if X and Y have independent finite range distributions.

9.12.3 Remarks

- The distribution in (9.32) has been referred to as a *bivariate rescaled Dirichlet distribution* by Ma (1996).
- The bivariate finite range distribution, bivariate Lomax, and Gumbel's bivariate exponential are three distributions that are characterized either through a constant bivariate coefficient of variation or by a constant product of mean residual lives and hazard rates.

9.13 Bivariate Beta Distribution

9.13.1 Formula of the Joint Density

The joint density function is

$$h(x,y) = \frac{\Gamma(\theta_1+\theta_2+\theta_3)}{\Gamma(\theta_1)\Gamma(\theta_2)\Gamma(\theta_3)} x^{\theta_1-1} y^{\theta_2-1} (1-x-y)^{\theta_3-1} \qquad (9.33)$$

for $x, y \ge 0$, $x + y \le 1$. This distribution is often known as the *bivariate Dirichlet distribution*; see Chapter 49 of Kotz et al. (2000).

9.13.2 Univariate Properties

The marginal distributions of X ad Y are beta$(\theta_1, \theta_2 + \theta_3)$ and beta$(\theta_2, \theta_2, \theta_1 + \theta_3)$, respectively.

9.13.3 Correlation Coefficient

Pearson's product-moment correlation coefficient is $-\sqrt{\frac{\theta_1\theta_2}{(\theta_1+\theta_3)(\theta_2+\theta_3)}}$. Thus, as might be expected from its support and its application to joint distributions of proportions, this distribution is unusual in being oriented toward negative correlation—to get positive correlation, we would have to change X to $-X$ or Y to $-Y$.

9.13.4 Product Moments

The product moments are given by

$$\mu'_{r,s} = \frac{\Gamma(\theta_1 + r)\Gamma(\theta_2 + s)\Gamma(\theta_1 + \theta_2 + \theta_3)}{\Gamma(\theta_1 + \theta_2 + \theta_3 + r + s)\Gamma(\theta_1)\Gamma(\theta_2)}; \tag{9.34}$$

see Wilks (1963, p. 179).

9.13.5 Conditional Properties

The expression $Y/(1 - x)$, conditional on $X = x$, has a beta(θ_2, θ_3) distribution.

9.13.6 Methods of Derivation

This distribution may be defined by the trivariate reduction method as follows. If $X_i \sim$ Gamma$(\theta_i, 1)$, then $X_1/(X_1+X_2+X_3)$ and $X_2/(X_1+X_2+X_3)$, conditional on $X_1 + X_2 + X_3 \leq 1$, have a bivariate beta distribution; see Loukas (1984).

9.13.7 Relationships to Other Distributions

- This distribution is related to the bivariate Pearson type I distribution; it is often referred to as a bivariate Dirichlet distribution.
- The relation between this distribution and the bivariate Pearson type II was mentioned earlier in Section 9.11.4.
- The conditional distributions are beta; see James (1975).

9.13.8 Illustrations

Hoyer and Mayer (1976) and Kellogg and Barnes (1989) have illustrated the density and contours.

9.13.9 Generation of Random Variates

Because of the method of derivation described above in Section 9.13.5, generation of variates is straightforward as mentioned by Devroye (1986, pp. 593–596), see also Macomber and Myers (1978) and Văduva (1985).

9.13.10 Remarks

- The variates X and Y are "neutral" in the following sense: X and $Y/(1-X)$ are independent, and the distribution being symmetric in x and y, so are Y and $X/(1-Y)$.
- $H(x, y)$ has a diagonal expansion in terms of orthogonal (shifted Jacobi) polynomials; see Lee (1971).
- If (i) $h(x, y)$ takes the product form $a_1(x)a_2(y)a_3(1-x-y)$, (ii) at least one of the a_i is a power function, and (iii) the regressions $E(Y|X)$ and $E(X|Y)$ are both linear, then $h(x, y)$ is the bivariate beta distribution; see Rao and Sinha (1988).
- $X+Y$ has a beta distribution with parameters $\theta_1+\theta_2$ and θ_3. Also, $X+Y$ is independent of X/Y, which has an inverted beta distribution.
- Kotz et al. (2000) have given a comprehensive treatment of multivariate Dirichlet distributions in Chapter 49 of their book.
- It is a member of the bivariate Liouville family of distributions to be discussed in Section 9.16 below.
- Provost and Cheong (2000) considered the distribution of a linear combination $\lambda_1 X + \lambda_2 Y$.

9.13.11 Fields of Application

- The distribution mainly arises in the context of a trivariate reduction of three quantities that must sum to 1 (for example, the probabilities of events or the proportions of substances in a mixture) that are mutually exclusive and collectively exhaustive. When considering just two of these quantities, a bivariate beta distribution may be a natural model to adopt.
- Mosimann (1962) and others have studied spurious correlations or correlations among proportions in relation to various types of pollen and grain and types of vegetation in general. See also the work of Narayana (1992) for an illuminating numerical example that was mentioned above.
- Sobel and Uppuluri (1974) utilized a Dirichlet distribution for the distribution of sparse and crowded cells closely related to occupancy models.
- Chatfield (1975) presented a particular example for the general context just mentioned. The subject is the joint distribution of brand shares; that is, the proportion of brands $1, 2, \ldots, n$ of some consumer product that are bought by customers. (The bivariate distribution on (9.33) will arise for $n = 3$.) Chatfield mentioned that the following two conditions are approximately correct in most product fields:

  - A consumer's rates of buying different brands are independent.
  - A consumer's brand shares are independent of his/her total rate of buying.

  The joint distribution of brand shares must then follow the multivariate beta distribution because of the following characterization theorem. Suppose Y_i are independent positive r.v.'s and that $T = \sum_i Y_i$ and $X_i = Y_i/T$; then, each X_i is independent of T, and the joint distribution of X's is multivariate beta. See Goodhardt et al. (1984) for a more comprehensive account of work in this field.
- Wrigley and Dunn (1984) showed that the Dirichlet model provides a good fit to a consumer-panel survey dataset from a study on urban consumer purchasing behavior.
- Hoyer and Mayer (1976) used this distribution in modeling the proportions of the electorate who vote for candidates in a two-candidate election (these two proportions adding to less than 1 because of abstentions). They say that this distribution "is sufficiently versatile to model many natural phenomena, yet it demonstrates a degree of simplicity such that a candidate who is reasonably adept at estimating probabilities could easily use our model to make a fairly accurate estimate of the actual joint distribution of proportions of his and his opponent's vote for a fixed set of political strategies."
- A-Grivas and Asaoka (1982) used a bivariate beta distribution to describe the joint distribution of two soil strength parameters.
- Modeling activity times in a PERT (Program Evaluation and Review Technique) network. A PERT network involves a collection of activities and

each activity is often modeled as a random variable following a beta distribution; see Monhor (1987).

- In Bayesian statistics, the beta distribution is a popular choice for a prior because it is a conjugate with respect to the binomial distribution; i.e., the posterior distribution is also beta. Similarly, the multivariate beta and multinomial distributions go together in the same manner. An example of such an analysis is by Apostolakis and Moieni (1987). These authors considered a system of three identical components subject to shocks that knock out $0, 1, 2$, or 3 of them in a style of Marshall and Olkin's model. Apostolakis and Moieni supposed that the state of knowledge regarding the vector of probabilities (p_0, p_1, p_2, p_3) could be described by a multivariate beta distribution.
- Lange (1995) applied the Dirichlet distribution to forensic match probabilities. The Dirichlet distribution is also relevant to the related problem of allele frequency estimation.

9.13.12 Tables and Algorithms

For algorithms evaluating the cumulative distribution function, one may refer to Parrish and Bargmann (1981), who used this distribution as an illustration of their general technique for evaluation of bivariate cumulative bivariate probabilities. Yassaee (1979) also evaluated the probability integral of the bivariate beta distribution by using a program that is used for evaluating the probability integral of the inverted beta distribution given earlier by Yassaee (1976).

9.13.13 Generalizations

- Connor and Mosimann (1969) and Lochner (1975) considered the generalized density of the form

$$h(x,y) = [\mathrm{B}(\alpha_1,\beta_1)\mathrm{B}(\alpha_2,\beta_2)]^{-1}x^{\alpha_1-1}y^{\alpha_2-1}(1-x)^{\alpha_1-(\alpha_2+\beta_2)}(1-x-y)^{\beta_2-1} \tag{9.35}$$

for $x, y \geq 0, x + y \leq 1$. When $\alpha_2 = \beta_1 - \beta_2$, it reduces to the standard bivariate beta density in (9.33). Since the generalized bivariate beta distribution has a more general covariance structure than the bivariate beta distribution, the former turns out to be more practical and useful. Wong (1998) has studied this distribution further.
- The bivariate Tukey lambda distribution, briefly considered by Johnson and Kotz (1973), is the joint distribution of the variables

$$\left.\begin{array}{l} X = [U^{\lambda} - (1-U)^{\lambda}]/\lambda \\ Y = [V^{\mu} - (1-V)^{\mu}]/\mu \end{array}\right\}, \tag{9.36}$$

where (U, V) has a bivariate beta distribution. The resulting distribution is a mess ["mathematically not very elegant" according to Johnson and Kotz, and "almost intractable" according to James (1975)].

- If $(U < V)$ again has a bivariate beta distribution, a distribution of (X, Y) is defined implicitly by $U = \sqrt{(XY)}, V = \sqrt{(1-X)(1-Y)}$; this is briefly mentioned by Mardia (1970, p. 88).
- Ulrich (1984) proposed a "bivariate beta mixture" distribution, which he used for a robustness study. Within each rectangle that the unit square is divided into, the p.d.f. is proportional to the product of a beta distribution of Y; the constants of proportionality are different for the different rectangles.
- Attributing an idea by Salvage, Dickey (1983) gave some attention to the distribution of the variables obtained by (first) scaling and (second) renormalizing to sum to unity,

$$\left.\begin{array}{l} X = aU/(aU + bV) \\ Y = bV/(aU + bV) \end{array}\right\}, \tag{9.37}$$

  with (U, V) having a bivariate beta distribution.
- For another generalization, one may refer to Nagarsenker (1970).
- Lewy (1996) also extended the bivariate beta to what he called a *delta-Dirichlet distribution*. The development of delta-Dirichlet distributions originated in sampling problems relating to the estimation of the species composition of the biomass within the Danish industrial fishery and with evaluation of the accuracy of estimates.

9.14 Jones' Bivariate Beta Distribution

This distribution was first proposed by Jones (2001) and independently by Olkin and Liu (2003).

9.14.1 Formula of the Joint Density

The joint density function is

$$h(x, y) = \frac{\Gamma(a+b+c)}{\Gamma(a)\Gamma(b)\Gamma(c)} \frac{x^{a-1}y^{b-1}(1-x)^{b+c-1}(1-y)^{a+c-1}}{(1-xy)^{a+b+c}}. \tag{9.38}$$

9.14.2 Univariate Properties

The marginal distributions are standard beta distributions with parameters (a, c) and (b, c), respectively.

9.14.3 Product Moments

Olkin and Liu (2003) showed that

$$E(X^k Y^l) = {}_3F_2(a + k, b + l, s; s + k, s + l; 1), \tag{9.39}$$

where ${}_3F_2$ is the generalized hypergeometric distribution function defined by ${}_3F_2(a, b, c; d, e; z) = \sum_k \frac{(a)_k (b)_k c_k}{(d)_k (e)_k} \frac{z^k}{k!}$.

9.14.4 Correlation and Local Dependence

Letting $k = l = 1$ in (9.39), we have

$$E(XY) = \frac{ab}{s} \frac{\Gamma(a + c)\Gamma(b + c)}{\Gamma(a + b + c)} {}_3F_2(a + 1, b + 1, s; s + 1, s + 1; 1)$$

and $E(X)E(Y) = \frac{ab}{(a+c)(b+c)}$, from which the correlation can be found, although numerical computations are required. Table 1 of Olkin and Liu (2003) provides correlation coefficient values for various choices of a, b, and c.

Note. $E(XY)$ was also derived in Jones (2001).

$$\gamma(x, y) = \frac{a + b + c}{(1 - xy)^2} \ .$$

9.14.5 Other Dependence Properties

Olkin and Liu (2003) showed that h is TP_2 (also known as LRD; see Section 3.4.6 for a definition). Thus, X and Y are PQD.

9.14.6 Illustrations

Density surfaces have been given by Olkin and Liu (2003) for several choices of a, b, and c. Two contour plots of the density have been given by Jones (2001).

9.15 Bivariate Inverted Beta Distribution

9.15.1 Formula of the Joint Density

The joint density function is

$$h(x,y) = \frac{\Gamma(\alpha_1+\alpha_2+\alpha_3)}{\Gamma(\alpha_1)\Gamma(\alpha_2)\Gamma(\alpha_3)} \frac{x^{\alpha_1-1}y^{\alpha_2-1}}{(1+x+y)^{\alpha_1+\alpha_2+\alpha_3}}, \quad x,y \geq 0. \tag{9.40}$$

It is also commonly known as the bivariate inverted Dirichlet distribution.

9.15.2 Formula of the Cumulative Distribution Function

The joint cumulative distribution function is

$$\begin{aligned} H(x,y) &= \frac{\Gamma(\alpha_1+\alpha_2+\alpha_3)x^{\alpha_1}y^{\alpha_2}}{\Gamma(\alpha_1+1)\Gamma(\alpha_2+2)\Gamma(\alpha_3)} \\ &\quad \times F_2(\alpha_1+\alpha_2+\alpha_3;\alpha_1,\alpha_2;\alpha_1+1,\alpha_2+1;-x,-y) \qquad (9.41) \\ &= \frac{\Gamma(\alpha_1+\alpha_2+\alpha_3)}{\Gamma(\alpha_1+1)\Gamma(\alpha_2+2)\Gamma(\alpha_3)} \frac{x^{\alpha_1}y^{\alpha_2}}{(1+x+y)^{\alpha_1+\alpha_2+\alpha_3}} \\ &\quad \times F_2\left(\alpha_1+\alpha_2+\alpha_3;1,1;\alpha_1+1,\alpha_2+1;\frac{x}{1+x+y},\frac{y}{1+x+y}\right), \qquad (9.42) \end{aligned}$$

where F_2 is Appell's hypergeometric function of two variables.

9.15.3 Derivation

Suppose X_1, X_2 and X_3 are independent gamma variables with shape parameters α_i $(i = 1, 2, 3)$. Then the pair $X = X_1/X_3$, $Y = X_2/X_3$ has the standard inverted beta distribution; see Tiao and Guttman (1965). This is evidently an example of the construction of a bivariate distribution by the trivariate reduction method.

9.15.4 Tables and Algorithms

Yassaee (1976) presented a computer program for calculating the probability integral of the inverted beta distribution.

For computation of $H(x, y)$, see Ong (1995).

9.15.5 Application

The inverted beta distribution is used in the calculation of confidence regions for variance ratios of random models for balanced data; see Sahai and Anderson (1973).

9.15.6 Generalization

Nagarsenker (1970) discussed the generalized density

$$h(x, y) \propto \frac{x^{\alpha_1 - 1} y^{\alpha_2 - 1}}{(1 + x + y)^{(\alpha_1/\beta_1) + (\alpha_2/\beta_2) + \alpha_3}}.$$

9.15.7 Remarks

- Comparing (9.25) and (9.40), we see this is effectively the bivariate F-distribution discussed in Section 8.11. Another account is due to Ratnaparkhi (1983).
- It is also a special case of a bivariate Lomax distribution.
- It is also a member of the bivariate Liouville family of distributions.

9.16 Bivariate Liouville Distribution

Liouville distributions seem to be one of those classes of distributions that have attracted much attention in recent years. Marshall and Olkin's (1979) book was perhaps the first place where Liouville distributions were discussed, briefly. Shortly thereafter, Sivazlian (1981) presented results on marginal distributions and transformation properties of Liouville distributions. Anderson and Fang (1982, 1987) discussed Liouville distributions arising from quadratic forms. The first comprehensive discussion of these distributions was provided by Fang et al. (1990). A series of papers by Gupta and Richards (1987, 1991, 1992, 1995, 1997, 2001a,b), along with Gupta et al. (1996), provides a rich source of information on Liouville distributions and their properties, matrix extensions, some other generalizations, and their applications to statistical reliability theory.

The family of bivariate Liouville distributions are often regarded as companions of the Dirichlet (beta) family because they were derived by Liouville through an application of a well-known extension of the Dirichlet integral. The family includes the well-known bivariate beta and bivariate inverted beta distributions. Gupta and Richards (2001b) provided a history of the development of the Dirichlet and Liouville distributions.

9.16.1 Definitions

Two definitions can be provided as follows. X and Y have a bivariate Liouville distribution if their joint density is proportional to [Gupta and Richards (1987)]

$$\psi(x+y)x^{a_1-1}y^{a_2-1}, \quad x>0,\ y>0,\ 0<x+y<b. \tag{9.43}$$

Thus

$$h(x,y)=\frac{C\Gamma(a)}{\Gamma(a_1)\Gamma(a_2)}x^{a_1-1}y^{a_2-1}\psi(x+y) \tag{9.44}$$

where $a=a_1+a_2$, $C^{-1}=\int_0^b t^{a-1}\psi(t)dt$, and ψ is a suitable non-negative function defined on $(0,b)$.

An alternative definition, as given in Fang et al. (1990), is as follows. Let $\boldsymbol{X}=(X,Y)'$ and $\boldsymbol{Y}=(Y_1,Y_2)'$. Then, $\boldsymbol{X}=(X,Y)'$ has a bivariate Liouville distribution if it has a stochastic representation $\boldsymbol{X}\stackrel{d}{=}R\boldsymbol{Y}$, where $R=X+Y$ has a univariate Liouville distribution and $\boldsymbol{Y}=(Y_1,Y_2)'$ is independent of R and has a beta density function

$$\frac{\Gamma(a_1+a_2)}{\Gamma(a_1)\Gamma(a_2)}x^{a_1-1}(1-x)^{a_2-1},\ 0\le x\le 1.$$

Using another expression, we can present

$$X\stackrel{d}{=}RY_1=(X+Y)Y_1;\quad Y\stackrel{d}{=}RY_2=(X+Y)Y_2,\quad Y_1+Y_2=1. \tag{9.45}$$

The density function of the bivariate Liouville distribution may also be written as

$$\frac{C\Gamma(a)}{\Gamma(a_1)\Gamma(a_2)}\frac{x^{a_1-1}y^{a_2-1}}{(x+y)^{a-1}}\phi(x+y),\ a_1+a_2=a, \tag{9.46}$$

defined over the simplex $\{(x,y): x\geq 0, y\geq 0, 0\leq x+y\leq b\}$ if and only if ϕ is defined over $(0,b)$.

The density generator ψ is related to the function ϕ as

$$\psi(t)=\frac{\Gamma(a)}{t^{a-1}}\phi(t),\quad a=a_1+a_2. \tag{9.47}$$

The generator in (9.47) satisfies the condition

$$\int_0^\infty \frac{t^{a-1}}{\Gamma(a)}\psi(t)dt=\int_0^\infty \phi(t)dt<\infty. \tag{9.48}$$

Ratnaparkhi (1985) called this distribution the bivariate Liouville–Dirichlet and presented the examples summarized below:

$\psi(t)$	b	Resulting bivariate distributions
$(1-t)^{a_3-1}$	1	Beta
$(1+t)^{-a-a_3}$	∞	Inverted beta (F)
$t^{a-1}e^{-t}$	∞	Gamma, $h(x,y)\propto(x+y)^{\alpha_3}x^{\alpha_1}y^{a_2}e^{-(x+y)}$
$(-\log t)^{a_3-1}$	1	"Unit-gamma-type" $h(x,y)\propto x^{\alpha_1-1}y^{a_2-1}[-\log(x+y)]^{a_3-1}$

The joint density in the third example corresponds to the distribution of correlated gamma variables; see, for example, Marshall and Olkin (1979).

9.16.2 Moments and Correlation Coefficient

The moments and covariance structure of the bivariate Liouville distribution can be derived easily; see Gupta and Richards (2001a). Because Y_1 and Y_2 are both beta and Y_i and R are independent, we readily find

$$E(X)=E(RY_1)=\frac{a_1}{a}E(R),\ E(Y)=E(RY_2)=\frac{a_2}{a}E(R) \tag{9.49}$$

and

$$\mathrm{var}(X)=\frac{a_1}{a^2(a+1)}\left\{a(a_1+1)\mathrm{var}(R)+a_2(E(R))^2\right\}. \tag{9.50}$$

A similar expression can be presented for var(Y). Furthermore,

$$\mathrm{cov}(X,Y)=\frac{a_1a_2}{a^2(a+1)}\left\{a\ \mathrm{var}(R)-(E(R))^2\right\}. \tag{9.51}$$

Denote the coefficient of variation of R by $\text{cv}(R) = \frac{\sqrt{\text{var}(R)}}{E(R)}$. Then, the covariance is negative if $\text{cv}(R) < 1/\sqrt{a}$. Gupta and Richards (2001a) have presented a sufficient condition for this inequality to hold.

- If (X, Y) has a bivariate beta distribution (a member of the bivariate Liouville family), then the above-mentioned sufficient condition holds and so we have X and Y negatively correlated, which is a well-known result.
- Let $\psi(t) = t^\alpha(1-t)^\beta$, $0 < t < 1$, where α and β are chosen so that $\text{cv}(R) = \frac{1}{\sqrt{a}}$. In this case, X and Y are uncorrelated but not independent.
- If $\psi(t) = e^{-t}t^\alpha$, $t > 0$ so that X and Y have a correlated bivariate gamma distribution of Marshall and Olkin (1979), then $\text{cv}(R) = \frac{1}{\sqrt{a}}$ implies $\alpha = 0$, which is equivalent to X and Y being independent.

9.16.3 Remarks

The bivariate Liouville distribution arises in a variety of statistical and probability contexts, some of which are listed below:

- Bivariate majorization—Marshall and Olkin (1979) and Diaconis and Perlman (1990).
- Total positivity and correlation inequalities—Aitchison (1986) and Gupta and Richards (1987, 1991).
- Statistical reliability theory—Gupta and Richards (1991).
- Stochastic partial orderings—Gupta and Richards (1992).
- For other properties, such as stochastic representations, transformation properties, complete neutrality, marginal and conditional distributions, regressions, and characterization, one may refer to Gupta and Richards (1987).

Fang et al. (1990) showed that if $\boldsymbol{X}$ has a bivariate Liouville distribution, then the condition that X and Y are independent is equivalent to X and Y being distributed as gamma with a common scale parameter.

Kotz et al. (2000) have provided an excellent summary on the multivariate Liouville distributions.

9.16.4 Generation of Random Variates

For generation of random variates, one may refer to Devroye (1986, pp. 596–599).

9.16.5 Generalizations

Gupta et al. (1996) introduced a sign-symmetric Liouville distribution, but the joint density function does not have a simple form.

9.16.6 Bivariate pth-Order Liouville Distribution

Ma and Yue (1995) introduced a bivariate pth-order Liouville distribution having a joint density function of the form

$$c\theta^{-a}x^{a_1-1}y^{a_2-1}\psi\left(\frac{(x^p+y^p)^{1/p}}{\theta}\right), \qquad x,y,p,\theta>0, \tag{9.52}$$

where $a = a_1 + a_2$, $0 \le x + y < b \le \infty$, and $\psi(\cdot)$ is a non-negative measurable function on $(0,\infty)$ such that $0 < \int_0^\infty \psi(t)t^{a-1}dt < \infty$.

For $p = 1$, it is the usual bivariate Liouville distribution. The bivariate Lomax distribution of Nayak (1987) with density

$$\frac{c}{\theta^a}x^{a_1-1}y^{a_2-1}\left(1+\frac{1}{\theta}(x+y)\right)^{-(a+l)},$$

where $\psi(t) = (1+t)^{(a+l)}$, $l > 0$, is a special case. Ma and Yue (1995) demonstrated how the parameter θ can be estimated by using their methods.

In the case where $\alpha_1 = \alpha_2 = p$, (9.52) is the bivariate l_p-norm symmetric distribution introduced by Fang and Fang (1988, 1989) and Yue and Ma (1995). Roy and Mukherjee (1988) discussed the case $p = 2$ as an extension of a class of generalized mixtures of exponential distributions.

9.16.7 Remarks

- X and Y can be viewed as a univariate dependent sample of random lifetimes of a coherent system or proportional hazards model when the joint density is given by (9.45).
- Ma et al. (1996), in addition to discussing the basic properties and the dependence structure of a multivariate pth-order Liouville distribution, also discussed the multivariate order statistics induced by ordering the l_p-norm.
- Ma and Yue (1995) also discussed the estimation of the parameter θ.

9.17 Bivariate Logistic Distributions

The work that is commonly cited on this subject is that of Gumbel (1961). He proposed three bivariate logistic distributions:

$$H(x,y) = \frac{1}{1+e^{-x}+e^{-y}}, \quad x,y \in \mathbf{R}, \tag{9.53}$$

$$H(x,y) = \exp\left[-\left\{\log(1+e^{-x})^{1/\alpha} + \log(1+e^{-y})^{1/\alpha}\right\}\right]^{\alpha}, \quad x,y \in \mathbf{R}, \tag{9.54}$$

and

$$H(x,y) = (1+e^{-x})^{-1}(1+e^{-y})^{-1}\left\{1+\alpha e^{-x-y}(1+e^{-x})^{-1}(1+e^{-y})^{-1}\right\} \tag{9.55}$$

for $x, y \in \mathbf{R}$ and $-1 < \alpha < 1$.

9.17.1 Standard Bivariate Logistic Distribution

The distribution in (9.53) is known as the standard bivariate logistic distribution.

Formula of the Joint Density

The joint density function is

$$h(x,y) = \frac{2e^{-x}e^{-y}}{(1+e^{-x}+e^{-y})^3}, \quad x,y \in \mathbf{R}. \tag{9.56}$$

Conditional Properties

The conditional density of X, given $Y = y$, can be shown to be

$$f(x|y) = \frac{2e^{-x}(1+e^{-y})^2}{(1+e^{-x}+e^{-y})^3},$$

and a similar expression can be presented for $g(y|x)$. The regression of X on Y is

$$E(X|Y=y) = 1 - \log(1+e^{-y}).$$

Correlation Coefficient

Pearson's product-moment correlation coefficient is

$$\operatorname{corr}(X,Y)=\rho=\frac{1}{2},$$

which reveals the restrictive nature of this bivariate logistic distribution.

Moment Generating Function

The joint moment generating function is given by

$$M(s,t)=\Gamma(1+s+t)\Gamma(1-s)\Gamma(1-t).$$

Derivation

Let U, V, and W be independent and identically distributed extreme value random variables with density function $e^{-x}e^{e^{-x}}, -\infty<x<\infty$. Then, the joint density function of $X=V-U$ and $Y=W-U$ is the standard bivariate logistic distribution. This, incidentally, is another example of the construction of a bivariate distribution by the variable-in-common scheme.

Relationships to Other Distributions

The copula density that corresponds to the standard bivariate logistic distribution is

$$c(u,v)=\frac{2uv}{(u+v-uv)^3}; \tag{9.57}$$

see, for example, Nelsen (1999, p. 24). Now, let us consider Mardia's bivariate Pareto distribution with the joint density (after reparametrization)

$$h(x,y)=\frac{(\alpha-1)\alpha}{\sigma_1\sigma_2}\left(1+\frac{x}{\sigma_1}+\frac{y}{\sigma_2}\right)^{-(\alpha+1)}.$$

For $\alpha=1$, the copula density that corresponds to the distribution above is given by

$$c(u,v)=\frac{2(1-u)(1-v)}{\{(1-u)+(1-v)-(1-u)(1-v)\}^3}.$$

Rotating this surface about $(\frac{1}{2},\frac{1}{2})$ by π radians, we obtain the copula in (9.57).

9.17.2 Archimedean Copula

The bivariate logistic distribution that corresponds to (9.54) is an Archimedean copula (see Section 1.5 for a definition) with generator $\varphi(u) = (-\log u)^{1/\alpha}$. This copula was termed the Gumbel–Hougaard copula earlier in Section 2.6.

9.17.3 F-G-M Distribution with Logistic Marginals

The distribution in (9.55) is the well-known Farlie–Gumbel–Morgenstern distribution with logistic marginals. The bivariate F-G-M distribution was discussed in detail in Section 2.2.

9.17.4 Generalizations

- Satterthwaite and Hutchinson (1978) extended the standard bivariate logistic to the form
$$H(x,y) = (1 + e^{-x} + e^{-y})^{-c}, \quad x, y \in \mathbf{R},\ c > 0. \tag{9.58}$$
This is only a marginal transformation of the bivariate Pareto distribution.
- Arnold (1990, 1992) constructed a generalization of a bivariate logistic model through geometric minimization of the form
$$\bar{H}(x,y) = (1 + e^{x} + e^{y} + \theta e^{x+y})^{-1}, \quad 0 \le \theta \le 2. \tag{9.59}$$
If geometric maximization is considered instead, we obtain
$$H(x,y) = (1 + e^{-x} + e^{-y} + \theta e^{-x-y})^{-1}, \quad 0 \le \theta \le 2. \tag{9.60}$$
The distribution in (9.60) reduces to (9.54) when $\theta = 0$.

We note that the bivariate model in (9.59) was first derived by Ali et al. (1978), and its corresponding copula was given in Section 2.3.

9.17.5 Remarks

The multivariate extension of (9.53) was discussed by Malik and Abraham (1973). Kotz et al. (2000) therefore refers to this distribution as the Gumbel–

Malik–Abraham distribution. Section 10 of Chapter 51 of Kotz et al. (2000) also discusses several generalizations of multivariate beta distributions.

9.18 Bivariate Burr Distribution

Bivariate Burr distributions with Burr type III or type XII marginals have received some attention in the literature. Two main methods have been used for their construction:

- The Farlie–Gumbel–Morgenstern method.
- Compounding, either as a straightforward generalization of the construction of the bivariate Pareto distribution (abbreviated as P in the following table), or the bivariate method which Hutchinson (1979, 1981) showed underlies the Durling–Burr distribution (abbreviated as D).

The following table lists some sources where more details can be found; see also Sections 2.8 and 2.9. A brief account of these distributions has been given by Rodriguez (1983, pp. 241–244).

Marginals	Construction	References
XII	Compounding (P)	Takahasi (1965), Crowder (1985)
XII	Compounding (D)	Durling (1975), Bagchi and Samanta (1985)
XII	F-G-M	Bagchi and Samanta (1985)
III	Compounding (P)	Rodriguez (1980), Rodriguez and Taniguchi (1980)
III	Compounding (D)	Rodriguez (1980)
III	Compounding*	Rodriguez (1980)
III	F-G-M	Rodriguez (1980)
III	F-G-M, extended	Rodriguez (1980)

* $\int_0^\infty \min[1, (x/\lambda)^c]dF(\lambda)$, where $F(\lambda = (1-k)(1-\lambda^{-c})^{-k} + k(1+\lambda^{-c})^{-k-1}$. Rodriguez (1980, p. 39) makes only passing mention of these.

9.19 Rhodes' Distribution

9.19.1 Support

The region of support of this distribution is all x, y such that $1 - \frac{x}{a} + \frac{y}{b} > 0$ and $1 + \frac{x}{a'} - \frac{y}{b'} > 0$.

9.19.2 Formula of the Joint Density

The joint density function is

$$h(x, y) \propto \left(1 - \frac{x}{a} + \frac{y}{b}\right)^{p} \left(1 + \frac{x}{a'} - \frac{y}{b'}\right)^{p'} e^{-tx-my}. \tag{9.61}$$

9.19.3 Derivation

Starting with two independent variables having not necessarily identical gamma distributions, let X be a linear combination of them and Y be some other linear combination of them. The result then is that (X, Y) has Rhodes' distribution.

9.19.4 Remarks

For the properties of this distribution, see Mardia (1970, pp. 40, 94–95). Rhodes (1923) fitted this distribution to barometric heights observed at Southampton and Laudale; see Pearson and Lee (1897).

9.20 Bivariate Distributions with Support Above the Diagonal

Jones and Larsen (2004) proposed and studied a general family of bivariate distributions that is based on, but greatly extends, the joint distribution of order statistics from independent and identically distributed univariate variables.

9.20.1 Formula of the Joint Density

The joint density function is

$$h(x, y) = \frac{\Gamma(a+b+c)}{\Gamma(a)\Gamma(b)\Gamma(c)} k(x)k(y)K^{a-1}(x)(K(y)-K(x))^{b-1}(1-K(y))^{c-1} \tag{9.62}$$

on $x < y$, where $a, b, c > 0$. Here, K is the distribution function from which the random sample is drawn and $k = K'$ is the corresponding density function. Furthermore, it is assumed that K is a symmetric univariate distribution.

9.20.2 Formula of the Cumulative Distribution Function

The joint distribution function $H(x, y)$ can be expressed in terms of an incomplete two-dimensional beta function.

9.20.3 Univariate Properties

The marginal density functions are

$$f(x) = \frac{\Gamma(a+b+c)}{\Gamma(a)\Gamma(b+c)} k(x) K^{a-1}(x)(1 - K(x))^{b+c-1}$$

and

$$g(y) = \frac{\Gamma(a+b+c)}{\Gamma(a+b)\Gamma(c)} k(y) K^{a+b-1}(y)(1 - K(y))^{c-1}.$$

9.20.4 Other Properties

- If K has a uniform distribution on $[0, 1]$, then the joint density in (9.62) has a link to the bivariate beta distribution; see Jones and Larsen (2004).
- The local dependence function is

$$\gamma(x, y) = \frac{(b-1)k(x)k(y)}{(K(y) - K(x))^2}, \quad x < y.$$

  It follows that γ is positive or negative depending on whether $b > 1$ or $b < 1$.
- $E(Y|X = x)$ is nondecreasing in x for all $b > 0$ and so $\mathrm{cov}(X, Y) \geq 0$ for all $b > 0$; Jones and Larsen (2004) have provided a proof.

9.20.5 Rotated Bivariate Distribution

Consider a rotated version of (9.62) obtained through rotating the two axes anticlockwise by $45°$; i.e., we wish to find the joint distribution of $W = X + Y$ and $Z = Y - X > 0$.

Formula of the Joint Density

Let $h_{W,Z}$ denote the joint density function of W and Z. Jones and Larsen (2004) have shown that

$$h_{W,Z}(w,z) = \frac{\Gamma(a+b+c)}{2\Gamma(a)\Gamma(b)\Gamma(c)} k\left(\frac{w-z}{2}\right) k\left(\frac{w+z}{2}\right) K^{a-1}\left(\frac{w-z}{2}\right) \times\left(K\left(\frac{w+z}{2}\right) - K\left(\frac{w-z}{2}\right)\right)^{b-1} \left(1 - K\left(\frac{w+z}{2}\right)\right)^{c-1} \tag{9.63}$$

for $-\infty < w < \infty$, $z > 0$.

The marginal distributions of the rotated bivariate distribution in (9.63) appear to be intractable analytically. Note that, however, $E(W) = E(Y) + E(X)$, $E(Z) = E(Y) - E(X)$, and $\operatorname{cov}(X,Y) = \operatorname{var}(Y) - \operatorname{var}(X)$.

Special Case where $a = c$

Since $h_{W,Z}(-w,z;a,b,c) = h_{W,Z}(w,z;c,b,a)$, there is symmetry in the w direction if $a = c$. For this special case, $\operatorname{var}(Y) = \operatorname{var}(X)$, which implies that $\operatorname{cov}(X,Y) = 0$, but X and Y are not independent.

9.20.6 Some Special Cases

We now consider some special cases of the density in (9.62).

(i) Bivariate Skew t-Distribution

Suppose K has Student's t-distribution with two degrees of freedom,

$$k(x) = \frac{1}{(2+x^2)^{3/2}}, \quad K(x) = \frac{1}{2}\left(1 + \frac{x}{\sqrt{2+x^2}}\right).$$

Then, (9.62) reduces to the bivariate skew t-distribution discussed in Section 9.5.

(ii) Bivariate log *F* Distribution

If $k(x)$ is the density of the logistic distribution, then a bivariate $\log F$ distribution is obtained. The univariate $\log F$ distribution may be found in Brown et al. (2002), for example.

9.20.7 Applications

The bivariate log F distribution proves to be a good fit to the temperature data of Jolliffe and Hope (1996). Jones and Larsen (2004) have also listed several potential applications of this family of distributions.

References

1. Adegboye, O.S., Gupta, A.K.: On a multivariate F-distribution and its application. In: American Statistical Association, 1981 Proceedings of the Social Statistics' Section, pp. 314–317. American Statistical Association, Alexandria, Virgina (1981)
2. Aitchison, J.: The Statistical Analysis of Compositional Data. Chapman and Hall, London (1986)
3. A-Grivas, D., Asaoka, A.: Slope safety prediction under static and seismic loads. Journal of the Geotechnical Engineering Division, Proceedings of the American Society of Civil Engineers **108**, 713–729 (1982)
4. Ali, M.M., Mikhail, N.N., Haq, M.S.: A class of bivariate distributions including the bivariate logistic. Journal of Multivariate Analysis **8**, 405–412 (1978)
5. Amos, D.E., Bulgren, W.G.: Computation of a multivariate F distribution. Mathematics of Computation **26**, 255–264 (1972)
6. Anderson, T.W., Fang, K.T.: Distributions of quadratic forms and Cochran's theorem for elliptically contoured distributions and their applications. Technical Report No. 53, Department of Statistics, Stanford University, Stanford, California (1982)
7. Anderson, T.W., Fang, K.T.: Cochran's theorem for elliptically contoured distributions. Sankhyā, Series A **49**, 305–315 (1987)
8. Apostolakis, G., Moieni, P.: The foundations of models of dependence in probabilistic safety assessment. Reliability Engineering **18**, 177–195 (1987)
9. Arellano-Valle, R.B., Azzalini, A.: On the unification of families of skew-normal distributions. Scandinavian Journal of Statistics **33**, 561–574 (2006)
10. Arnold, B.C.: A flexible family of multivariate Pareto distributions. Journal of Statistical Planning and Inference **24**, 249–258 (1990)
11. Arnold, B.C.: Multivariate logistic distributions. In: Handbook of the Logistic Distribution, N. Balakrishnan (ed.), pp. 237–261. Marcel Dekker, New York (1992)
12. Arnold, B.C., Beaver, R.J.: The skew-Cauchy distribution. Statistics and Probability Letters **49**, 285–290 (2000)
13. Arnold, B.C., Castillo, E., Sarabia, J.M.: Conditional Specification of Statistical Models. Springer-Verlag, New York (1999)
14. Azzalini, A., Capitanio, A.: Distributions generated by perturbation of symmetry with emphasis on multivariate skew t distribution. Journal of the Royal Statistical Society, Series B **65**, 367–390 (2003)

15. Bagchi, S.B., Samanta, K.C.: A study of some properties of bivariate Burr distributions. Bulletin of the Calcutta Mathematical Society **77**, 370–383 (1985)
16. Barnett, V.: Some bivariate uniform distributions. Communications in Statistics—Theory and Methods **9**, 453–461 (Correction **10**, 1457) (1980)
17. Bechhofer, R.E., Dunnett, C.W.: Four Sets of Tables: Percentage Points of Multivariate Student t Distributions. (Selected Tables in Mathematical Statistics, Volume 11.) American Mathematical Society, Providence, Rhode Island (1987)
18. Branco, M.D., Dey, D.K.: A general class of multivariate skew elliptical distributions. Journal of Multivariate Analysis **79**, 99–113 (2001)
19. Brown, B.M., Spears, F.M., Levy, L.B.: The log F: A distribution for all seasons. Computational Statistics **17**, 47–58 (2002)
20. Chatfield, C.: A marketing application of a characterization theorem. In: A Modern Course on Distributions in Scientific Work, Volume 2—Model Building and Model Selection, G.P. Patil, S. Kotz, and J.K. Ord (eds.), pp. 175–185. Reidel, Dordrecht (1975)
21. Chen, H.J.: Percentage points of multivariate t distribution with zero correlations and their application. Biometrical Journal **21**, 347–359 (1979)
22. Chou, Y.M.: A bivariate noncentral t-distribution with applications. Communications in Statistics: Theory and Methods **21**, 3427–3462 (1992)
23. Connor, R.J., Mosimann, J.E.: Concepts of independence for proportions with a generalization of the Dirichlet distribution. Journal of the American Statistical Association **64**, 194–206 (1969)
24. Crowder, M.: A distributional model for repeated failure time measurements. Journal of the Royal Statistical Society, Series B **47**, 447–452 (1985)
25. Devlin, S.J., Gnanadesikan, R., Kettenring, J.R.: Some multivariate applications of elliptical distributions. In: Essays in Probability and Statistics, S. Ikeda, T. Hayakawa, H. Hudimoto, M. Okamoto, M. Siotani. and S. Yamamoto (eds.), pp. 365–393. Shinko Tsucho, Tokyo (1976)
26. Devroye, L.: Non-uniform Random Variate Generation. Springer-Verlag, New York (1986)
27. Diaconis, P., Perlman, M.D.: Bounds for tail probabilities of weighted sums of independent gamma random variables. In: Topics in Statistical Dependence, H.W. Block et al. (eds.), pp. 147–166. Institute of Mathematical Statistics, Hayward, California (1990)
28. Dickey, J.M.: Multiple hypergeometric functions: Probabilistic interpretations and statistical uses. Journal of the American Statistical Association **78**, 628–637 (1983)
29. Dunnett, C.W., Sobel, M.: A bivariate generalization of Student's t-distribution, with tables for certain special cases. Biometrika **41**, 153–169 (1954)
30. Durling, F.C.: The bivariate Burr distribution. In: A Modern Course on Statistical Distributions in Scientific Work. Volume I—Models and Structures, G.P. Patil, S. Kotz, and J. K. Ord (eds.), pp. 329–335. Reidel, Dordrecht (1975)
31. Elderton, W.P., Johnson, N.L.: Systems of Frequency Curves. Cambridge University Press, Cambridge (1969)
32. Fang, K.T.: Elliptically contoured distributions. In: Encyclopedia of Statistical Sciences, Update Volume 1, S. Kotz, C.B. Read, and D.L. Banks (eds.) pp. 212–218. John Wiley and Sons, New York (1997)
33. Fang, K.T., Fang, B.Q.: Some families of multivariate symmetric distributions related to exponential distributions. Journal of Multivariate Analysis **24**, 109–122 (1988)
34. Fang, K.T., Fang, B.Q.: A characterization of multivariate l_1-norm symmetric distribution. Statistics and Probability Letters **7**, 297–299 (1989)
35. Fang, K.T., Kotz, S., Ng, K.W.: Symmetric Multivariate and Related Distributions. Chapman and Hall, London (1990)
36. Feingold, M., Korsog, P.E.: The correlation and dependence between two F statistics with the same denominator. The American Statistician **40**, 218–220 (1986)

37. Ferguson, T.S.: A representation of the symmetric bivariate Cauchy distribution. Annals of Mathematical Statistics **33**, 1256–1266 (1962)
38. Fernandez, C., Osiewalski, J., Steel, M.F.J.: Modeling and inference with v-spherical distributions. Journal of the American Statistical Association **90**, 1331–1340 (1995)
39. Genz, A.: Numerical computation of rectangular bivariate and trivariate normal and t probabilities. Statistics and Computating **14**, 251–260 (2004)
40. Genz, A., Bretz, F.: Comparison of methods for the computation of multivariate t probabilities. Journal of Computational and Graphical Statistics **11**, 950–971 (2002)
41. Goodhardt, G.J., Ehrenberg, A.S.C., Chatfield, C.: The Dirichlet: A comprehensive model of buying behaviour. Journal of the Royal Statistical Society, Series A **147**, 621–643 (Discussion 643–655) (1984)
42. Gumbel, E.J.: Bivariate logistic distributions. Journal of the American Statistical Association **56**, 335–349 (1961)
43. Gupta, R.D., Misiewicz, J.K., Richards, D.St.P.: Infinite sequences with sign-symmetric Liouville distributions. Probability and Mathematical Statistics **16**, 29–44 (1996)
44. Gupta, R.D., Richards, D.St.P.: Multivariate Liouville distributions. Journal of Multivariate Analysis **23**, 233–256 (1987)
45. Gupta, R.D., Richards, D.St.P.: Multivariate Liouville distributions, II. Probability and Mathematical Statistics **12**, 291–309 (1991)
46. Gupta, R.D., Richards, D.St.P.: Multivariate Liouville distributions, III. Journal of Multivariate Analysis **43**, 29–57 (1992)
47. Gupta, R.D., Richards, D.St.P.: Multivariate Liouville distributions, IV. Journal of Multivariate Analysis **54**, 1–17 (1995)
48. Gupta, R.D., Richards, D.St.P.: Multivariate Liouville distributions, V. In: Advances in the Theory and Practice of Statistics: A Volume in Honor of Samuel Kotz, N.L. Johnson and N. Balakrishan (eds.), pp. 377–396. John Wiley and Sons, New York (1997)
49. Gupta, R.D., Richards, D.St.P.: The covariance structure of the multivariate Liouville distributions. Contempory Mathematics **287**, 125–138 (2001a)
50. Gupta, R.D., Richards, D.St.P.: The history of the Dirichlet and Liouville distributions. International Statistical Reviews **69**, 433–446 (2001b)
51. Gupta, S.S.: Probability integrals of multivariate normal and multivariate t. Annals of Mathematical Statistics **34**, 792–828 (1963)
52. Gupta, S.S., Panchapakesan, S., Sohn, J.K.: On the distribution of the Studentized maximum of equally correlated normal random variables. Communications in Statistics—Simulation and Computation **14**, 103–135 (1985)
53. Hamdy, H.I., Son, M.S., Al-Mahmeed, M.: On the distribution of bivariate F-variable. Computational Statistics and Data Analysis **6**, 157–164 (1988)
54. Hewett, J., Bulgren, W.G.: Inequalities for some multivariate f-distributions with applications. Technometrics **13**, 397–402 (1971)
55. Hoyer, R.W., Mayer, L.S.: The equivalence of various objective functions in a stochastic model of electoral competition. Technical Report No. 114, Series 2, Department of Statistics, Princeton University, Princeton, New Jersey (1976)
56. Hutchinson, T.P.: Four applications of a bivariate Pareto distribution. Biometrical Journal **21**, 553–563 (1979)
57. Hutchinson, T.P.: Compound gamma bivariate distributions. Metrika **28**, 263–271 (1981)
58. Jamalizadeh, A., Balakrishnan, N.: On a generalization of bivariate Cauchy distribution. Communications in Statistics: Theory and Methods **37**, 469–474 (2008a)
59. Jamalizadeh, A., Balakrishnan, N.: On order statistics from bivariate skew-normal and skew-t_ν distributions. Journal of Statistical Planning and Inference **138**, 4187–4197 (2008b)

60. Jamalizadeh, A., Balakrishnan, N.: Order statistics from trivariate normal and t_ν-distributions in terms of generalized skew-normal and skew-t_ν distributions. Journal of Statistical Planning and Inference (to appear)
61. James, I.R.: Multivariate distributions which have beta conditional distributions. Journal of the American Statistical Association **70**, 681–684 (1975)
62. Johnson, M.E.: Multivariate Statistical Simulation. John Wiley and Sons, New York (1987)
63. Johnson, M.E.: Distribution selection in statistical simulation studies. In: 1986 Winter Simulation Conference Proceedings, J.R. Wilson, J.O. Henriksen, and S.D. Roberts (eds.), pp. 253–259. Institute of Electrical and Electronics Engineers, Piscataway, New Jersey (1986)
64. Johnson, M.E., Wang, C., Ramberg, J.: Generation of continuous multivariate distributions for statistical applications. American Journal of Mathematical and Management Sciences **4**, 225–248 (1984)
65. Johnson, N.L., Kotz, S.: Distributions in Statistics: Continuous Multivariate Distributions. John Wiley and Sons, New York (1972)
66. Johnson, N.L., Kotz, S.: Extended and multivariate Tukey lambda distributions. Biometrika **60**, 655–661 (1973)
67. Jolliffe, I.T., Hope, P.B.: Bounded bivariate distributions with nearly normal marginals. The American Statistician **50**, 17–20 (1996)
68. Jones, M.C.: Multivariate t and the beta distributions associated with the multivariate F distribution. Metrika **54**, 215–231 (2001)
69. Jones, M.C.: A dependent bivariate t distribution with marginals on different degrees of freedom. Statistics and Probability Letters **56**, 163–170 (2002a)
70. Jones, M.C.: Marginal replacement in multivariate densities, with applications to skewing spherically symmetric distributions. Journal of Multivariate Analysis **81**, 85–99 (2002b)
71. Jones, M.C., Faddy, M.J.: A skew extension of the t-distribution, with applications. Journal of the Royal Statistical Society, Series B **65**, 159–174 (2003)
72. Jones, M.C., Larsen, P.V.: Multivariate distributions with support above the diagonal. Biometrika **91**, 975–986 (2004)
73. Joshi, P.C., Lalitha, S.: A recurrence formula for the evaluation of a bivariate probability. Journal of the Indian Statistical Association **23**, 159–169 (1985)
74. Kellogg, S.D., Barnes, J.W.: The bivariate H-function distribution. Mathematics and Computers in Simulation **31**, 91–111 (1989)
75. Kim, H.M., Mallick, B.K.: Moments of random vectors with skew t distribution and their quadratic forms. Statistics and Probability Letters **63**, 417–423 (2003)
76. Kotz, S., Balakrishnan, N., Johnson, N.L.: Continuous Multivariate Distributions, Volume 1: Models and Applications, 2nd edition. John Wiley and Sons, New York (2000)
77. Krishnaiah, P.R.: Multiple comparison tests in multivariate cases. Report ARL 64-124. Aerospace Research Laboratories, Wright-Patterson Air Force Base, Ohio (1964)
78. Krishnaiah, P.R.: On the simultaneous ANOVA and MANOVA tests. Annals of the Institute of Statistical Mathematics **17**, 35–53 (1965)
79. Krishnan, M.: Series representations of a bivariate singly noncentral t-distribution. Journal of the American Statistical Association **67**, 228–231 (1972)
80. Lange, K.: Application of Dirichlet distribution to forensic match probabilities. Genetica **96**, 107–117 (1995)
81. Le, H., O'Hagan, A.: A class of bivariate heavy-tailed distribution. Sankhyā, Series B, Special Issue on Bayesian Analysis **60**, 82–100 (1998)
82. Lee, P.A.: A diagonal expansion for the 2-variate Dirichlet probability density function. SIAM Journal on Applied Mathematics **21**, 155–163 (1971)
83. Lewy, P.: A generalized Dirichlet distribution accounting for singularities of variables. Biometrics **52**, 1394–1409 (1996)

84. Lochner, R.H.: A generalized Dirichlet distribution in Bayesian life testing. Journal of the Royal Statistical Society, Series B **37**, 103–113 (1975)
85. Loukas, S.: Simple methods for computer generation of bivariate beta random variables. Journal of Statistical Computation and Simulation **20**, 145–152 (1984)
86. Ma, C.: Multivariate survival functions characterized by constant product of mean remaining lives and hazard rates. Metrika **44**, 71–83 (1996)
87. Ma, C., Yue, X.: Multivariate p-oder Liouville distributions: Parameter estimation and hypothesis testing. Chinese Journal of Applied Probability and Statistics **11**, 425–431 (1995)
88. Ma, C., Yue, X., Balakrishnan, N.: Multivariate p-order Liouville distributions: Definition, properties, and multivariate order statistics induced by ordering l_p-norm. Techical Report, McMaster University, Hamilton, Ontario, Canada (1996)
89. Macomber, J.H., Myers, B.L.: The bivariate beta distribution: Comparison of Monte Carlo generators and evaluation of parameter estimates. In: 1978 Winter Simulation Conference, Volume 1, H.J. Highland, N.R. Nielsen, and L.G. Hull (eds.), pp. 142–152. Institute of Electrical and Electronics Engineers, New York (1978)
90. Malik, H.J., Abraham, B.: Multivariate logistic distributions. Annals of Statistics **1**, 588–590 (1973)
91. Malik, H.J., Trudel, R.: Distributions of the product and the quotient from bivariates t, F and Pareto distribution. Communications in Statistics: Theory and Methods **14**, 2951–2962 (1985)
92. Mardia, K.V.: Families of Bivariate Distributions. Griffin, London (1970)
93. Marshall, A.W., Olkin, I.: Inequalities: Theory of Majorization and Its Applications. Academic Press, New York (1979)
94. McFadden, J.A.: A diagonal expansion in Gegenbauer polynomials for a class of second-order probability densities. SIAM Journal on Applied Mathematics **14**, 1433–1436 (1966)
95. Monhor, D.: An approach to PERT: Application of Dirichlet distribution. Optimization **18**, 113–118 (1987)
96. Mosimann, J.E.: On the compound multinomial distribution, the multivariate β-distribution and correlations among proportions. Biometrika **49**, 65–82 (1962)
97. Nadarajah, S., Zografos, K.: Expressions for Rényi and Shannon entropies for bivariate distributions. Information Sciences **170**, 73–189 (2005)
98. Nagarsenker, B.N.: A generalisation of beta densities and their multivariate analogues. Metron **28**, 156–168 (1970)
99. Narayana, A.: A note on parameter estimation in the multivariate beta distribution. Computer Mathematics and Applications **24**, 11–17 (1992)
100. Nayak, T.K.: Multivariate Lomax distribution: Properties and usefulness in reliability theory. Journal of Applied Probability **24**, 170–177 (1987)
101. Nelsen, R.B.: An Introduction to Copulas. Springer-Verlag, New York (1999)
102. O'Hagan, A., Le, H.: Conflicting information and a class of bivariate heavy-tailed distributions. In: Aspects of Uncertainty: A Tribute to D.V. Lindley. P.R. Freeman and A.F.M. Smith (eds.), pp. 311–327, John Wiley and Sons, Chichester, England (1994)
103. Olkin, I., Liu, R.: A bivariate beta distribution. Statistics and Probability Letters **62**, 407–412 (2003)
104. Ong, S.H.: Computation of bivariate gamma and inverted beta distribution functions. Journal of Statistical Computation and Simulation **51**, 153–163 (1995)
105. Owen, D.B.: A special case of a bivariate non-central t-distribution. Biometrika **52**, 437–446 (1965)
106. Parrish, R.S., Bargmann, R.E.: A method for the evaluation of cumulative probabilities of bivariate distributions using the Pearson family. In: Statistical Distributions in Scientific Work, Volume 5—Inferential Problems and Properties, C. Taillie, G.P. Patil, and B.A. Baldessari (eds.), pp. 241–257. Reidel, Dordrecht (1981)

107. Pearson, K.: On a certain double hypergeometric series and its representation by continuous frequency surfaces. Biometrika **16**, 172–188 (1924)
108. Pearson, K., Lee, A.: On the distribution of frequency (variation and correlation) of the barometric height at divers stations. Philosophical Transactions of the Royal Society of London, Series A **190**, 423–469 (1897)
109. Press, S.J.: Applied Multivariate Analysis. Holt, Reinhart and Winston, New York (1972)
110. Provost, S.B., Cheong, Y.H.: On the distribution of linear combinations of a Dirichlet random vector. Canadian Journal of Statistics **28**, 417–425 (2000)
111. Prucha, I.R., Kelejian, H.H.: The structure of simultaneous equation estimators: A generalization towards nonnormal disturbances. Econometrica **52**, 721–736 (1984)
112. Ramig, P.F., Nelson, P.R.: The probability integral for a bivariate generalization of the non-central t. Communications in Statistics—Simulation and Computation **9**, 621–631 (1980)
113. Rao, B.V., Sinha, B.K.: A characterization of Dirichlet distributions. Journal of Multivariate Analysis **25**, 25–30 (1988)
114. Ratnaparkhi, M.V.: Inverted Dirichlet distribution. In: Encyclopedia of Statistical Sciences, Volume 4, S. Kotz and N.L. Johnson (eds.), pp. 256–259. John Wiley and Sons, New York (1983)
115. Ratnaparkhi, M.V.: Liouville-Dirichlet distribution. In: Encyclopedia of Statistical Sciences, Volume 5, S. Kotz and N.L. Johnson (eds.), pp. 86–87. John Wiley and Sons, New York (1985)
116. Rhodes, E.C.: On a certain skew correlation surface. Biometrika **14**, 355–377 (1923)
117. Rodriguez, R.N.: Multivariate Burr III distributions, Part I: Theoretical properties. Research Publication GMR-3232, General Motors Research Laboratories, Warren, Michigan (1980)
118. Rodriguez, R.N.: Frequency surfaces, systems of. In: Encyclopedia of Statistical Sciences, Volume 3, S. Kotz and N.L. Johnson (eds.), pp. 232–247. John Wiley and Sons, New York (1983)
119. Rodriguez, R.N., Taniguchi, B.Y.: A new statistical model for predicting customer octane satisfaction using trained rater observations. (With Discussion) Paper No. 801356, Society of Automotive Engineers (1980)
120. Roy, D.: A characterization of Gumbel's bivariate exponential and Lindley and Singpurwalla's bivariate Lomax distributions. Journal of Applied Probability **26**, 886–891 (1989)
121. Roy, D.: Correction to "A characterization of Gumbel's bivariate exponential and Lindley and Singpurwalla's bivariate Lomax distributions." Journal of Applied Probability **27**, 736 (1990)
122. Roy, D., Gupta, R.P.: Bivariate extension of Lomax and finite range distributions through characterization approach. Journal of Multivariate Analysis **59**, 22–33 (1996)
123. Roy, D., Mukherjee, S.C.: Generalized mixtures of exponential distributions. Journal of Applied Probability **27**, 510–518 (1988)
124. Sahai, H., Anderson, R.L.: Confidence regions for variance ratios of random models for balanced data. Journal of the American Statistical Association **68**, 951–952 (1973)
125. Satterthwaite, S.P., Hutchinson, T.P.: A generalisation of Gumbel's bivariate logistic distribution. Metrika **25**, 163–170 (1978)
126. Sivazlian, B.D.: On a multivariate extension of the gamma and beta distributions. SIAM Journal of Applied Mathematics **41**, 205–209 (1981)
127. Sobel, M., Uppuluri, V.R.R.: Sparse and crowded cells and Dirichlet distribution. Annals of Statistics **2**, 977–987 (1974)
128. Somerville, P.N.: Numerical computation of multivariate normal and multivariate-t probabilities over convex regions. Journal of Computational and Graphical Statistics **7**, 529–544 (1998)
129. Sun, B.-K., Shi, D.-J.: Tail dependence in bivariate Cauchy distribution (in Chinese). Journal of Tianjin University **33**, 432–434 (2000)

130. Sutradhar, B.C.: On the characteristic function of multivariate Student t-distribution. Canadian Journal of Statistics **14**, 329–337 (1986)
131. Takahasi, K. Note on the multivariate Burr's distribution. Annals of the Institute of Statistical Mathematics **17**, 257–260 (1965)
132. Tiao, G.G., Guttman, I.: The inverted Dirichlet distribution with applications. Journal of the American Statistical Association **60**, 793–805 (Correction **60**, 1251–1252) (1965)
133. Tong, Y.L.: Probability Inequalities in Multivariate Distributions. Academic Press, New York (1980)
134. Ulrich, G.: A class of multivariate distributions with applications in Monte Carlo and simulation. In: American Statistical Association, 1984 Proceedings of the Statistical Computing Section, pp. 185–188. American Statistical Association, Alexandria, Virginia (1984)
135. Vǎduva, I.: Computer generation of random vectors based on transformation of uniformly distributed vectors, In: Proceedings of the Seventh Conference on Probability Theory, pp. 589–598. Editura Academiei, Bucarest and VNU Science Press, Utrecht (1985)
136. Wesolowski, J., Ahsanullah, M.: Conditional specification of multivariate Pareto and Student distribution. Communications in Statistics: Theory and Methods **24**, 1023–1031 (1995)
137. Wilcox, R.R.: Percentage points of the product of two correlated t variates. Communications in Statistics: Simulation and Computation **14**, 143–157 (1985)
138. Wilcox, R.R.: Percentage points of the bivariate t distribution for non-positive correlations. Metron **44**, 115–119 (1986)
139. Wilks, S.S.: Mathematical Statistics, 2nd edition, John Wiley and Sons, New York (1963)
140. Wong, T.-T.: Generalized Dirichlet distribution in Bayesian analysis. Applied Mathematics and Computation **97**, 165–181 (1998)
141. Wrigley, N., Dunn, R.: Stochastic panel-data models of urban shopping behavior: 2. Multistore purchasing patterns and Dirichlet model. Environment and Planning A **16**, 759–778 (1984)
142. Yassaee, H.: Probability integral of inverted Dirichlet distribution and its applications. Compstat **76**, 64–71 (1976)
143. Yassaee, H.: On probability integral of Dirichlet distributions and their applications. Preprint, Arya-Mehr University of Technology, Tehran, Iran (1979)
144. Yue, X., Ma, C.: Multivariate l_p-norm symmetric distributions. Statistics and Probability Letters **26**, 281–283 (1995)

Chapter 10
Bivariate Exponential and Related Distributions

10.1 Introduction

The vast majority of the bivariate exponential distributions arise in the reliability context one way or another. When we talk of reliability, we have in mind the failure of an item or death of a living organism. We especially think of time elapsing between the equipment being put into service and its failure. In the bivariate or multivariate context, we are concerned with dependencies between two failure times, such as those of two components of an electrical, mechanical, or biological system.

Just as the univariate exponential distribution is important in describing the lifetime of a single component [see, e.g., Balakrishnan and Basu (1995)], bivariate distributions with exponential marginals are also used quite extensively in describing the lifetimes of two components together. Bivariate exponential distributions often arise from shocks that knock out or cause cumulative damage to components that will knock out the components eventually. The numbers of shocks N_1 and N_2 that are required to knock out components 1 and 2, respectively, usually have a bivariate geometric distribution. Marshall and Olkin's and Downton's bivariate exponential distributions are prime examples of models that can be derived in this manner. A notable exception is Freund's bivariate exponential, which cannot be obtained from such a bivariate geometric compounding scheme. Bivariate exponential mixtures may also arise in a reliability context with two components sharing a common environment.

Distributions with exponential marginals may, of course, be obtained by starting with any bivariate distribution of a familiar form and then transforming the X and Y axes appropriately. In particular, this may be done with any of the copulas presented earlier in Chapter 2—in the expression of C, we simply need to replace x by $1 - e^{-x}$ and y by $1 - e^{-y}$.

Surveys of bivariate exponential distributions and their applications to reliability have been given by Basu (1988) and Balakrishnan and Basu (1995).

N. Balakrishnan, C.D. Lai, *Continuous Bivariate Distributions*,

DOI 10.1007/b101765_11,

Chapter 47 of Kotz et al. (2000) presents an excellent treatment on bivariate and multivariate exponential distributions.

In Section 10.2, we first present the three forms of bivariate exponential distributions introduced by Gumbel. Freund's bivariate exponential distribution and its properties are discussed in Section 10.3. In Section 10.4, the extension of Freund's distribution due to Hashino and Sugi is described. The well-known Marshall and Olkin bivariate exponential distribution and related issues are discussed in Section 10.5. As Marshall and Olkin's distribution contains a singular part, Block and Basu proposed an absolutely continuous bivariate exponential distribution. This model is presented in Section 10.6. In Section 10.7, Sarkar's bivariate exponential distribution is described. Next, in Section 10.8, a comparison of different properties of the models of Marshall and Olkin, Block and Basu, Sarkar, and Freund is made, and some basic differences and commonalities are pointed out. In Sections 10.9 and 10.10, the generalized forms (which include both Freund and Marshall-Olkin distributions) proposed by Friday and Patil and Tosch and Holmes, respectively, are presented. The system of exponential mixture distributions due to Lawrance and Lewis and its characteristic properties are discussed in Section 10.12. The bivariate exponential distributions obtained from Raftery's scheme are mentioned in Section 10.13. In Section 10.14, the bivariate exponential distributions derived by Iyer et al. by using auxiliary random variables forming linear structures are presented, and their differing correlation structures are highlighted. Another well-known bivariate exponential distribution, known as the Moran–Downton model in the literature, and its related developments are detailed in Section 10.15. The bivariate exponential distributions of Sarmanov, Cowan, Singpurwalla and Youngren, and Arnold and Strauss are presented in Sections 10.16–10.19, respectively. Several different forms of mixtures of bivariate exponential distributions have been considered in the statistical as well as applied fields, and Section 10.20 presents these forms. Section 10.21 describes details on bivariate exponential distributions connected with geometric compounding schemes. Different concepts of the lack of memory property associated with different forms of bivariate exponential distributions are described next in Section 10.22. Section 10.23 briefly discusses the effect of parallel redundancy in systems with dependent exponential components. In Section 10.24, the role of bivariate exponential distributions as a stress-strength model is explained. Finally, the bivariate Weibull distributions and their properties are presented in Section 10.25.

10.2 Gumbel's Bivariate Exponential Distributions

Gumbel (1960) introduced three types of bivariate exponential distributions, and these are described in this section.

10.2.1 Gumbel's Type I Bivariate Exponential Distribution

The joint cumulative distribution function is

$$H(x,y) = 1 - e^{-x} - e^{-y} + e^{-(x+y+\theta xy)}, \quad x, y \geq 0, 0 \leq \theta \leq 1. \tag{10.1}$$

This distribution was discussed earlier in Section 2.10.

10.2.2 Characterizations

Along with the bivariate Lomax distribution and bivariate finite range distribution, Gumbel's type I bivariate exponential distribution can be characterized through

- constant product of bivariate mean remaining (residual) lives and hazard rates [see Roy (1989), Ma (1996), Roy and Gupta (1996)] and
- constant coefficient of variation of bivariate residual lives; see Roy and Gupta (1996).

10.2.3 Estimation Method

By introducing scale parameters to the marginal distributions, the survival function corresponding to (10.1) (after relabeling θ by α) becomes

$$\bar{H}(x,y) = \exp\left\{-\frac{x}{\theta_1} - \frac{y}{\theta_2} - \frac{\alpha xy}{\theta_1\theta_2}\right\}, \ x, y > 0, \ \theta_1, \theta_2 > 0, \ 0 < \alpha < 1. \tag{10.2}$$

Castillo et al. (1997) have discussed methods for estimating the parameters in (10.2).

10.2.4 Other Properties

- The correlation coefficient is given in Section 2.10.
- The copula $C(u,v)$ is given by (2.47).
- The product moments were derived by Nadarajah and Mitov (2003).
- The Fisher information matrix was derived by Nadarajah (2006a).
- It is easy to show that X and Y are NQD (negative quadrant dependent); see Lai and Xie (2006, p. 324).

- Kotz et al. (2003b) derived the distributions of $T_1 = \min(X, Y)$ and $T_2 = \max(X, Y)$. In particular, it was shown that

$$E(T_1) = e^{1/\theta}\sqrt{\frac{\pi}{\theta}}\left[1 - \Phi\left(\sqrt{2/\theta}\right)\right]$$

  and that

$$E(T_2) = 2 - e^{1/\theta}\sqrt{\frac{\pi}{\theta}}\left[1 - \Phi\left(\sqrt{2/\theta}\right)\right].$$

  Further, it was shown that $E(T_2)$ is almost linearly increasing in ρ.
- Franco and Vivo (2006) discussed log-concavity of the extremes. (The distribution that has a log-concave density has an increasing likelihood ratio.)

10.2.5 Gumbel's Type II Bivariate Exponential Distribution

The F-G-M bivariate distributions were discussed in detail earlier in Section 2.2. Gumbel's type II bivariate exponential distribution is simply an F-G-M model with exponential marginals. The density function is given by

$$h(x, y) = e^{-x-y}\left\{1 + \alpha(2e^{-x} - 1)(2e^{-y} - 1)\right\}, \quad |\alpha| < 1. \tag{10.3}$$

Bilodeau and Kariya (1994) observed that the density functions of both type I and type II are of the form

$$h(x, y) = \lambda_1\lambda_2 g(\lambda_1 x, \lambda_2 y; \theta)e^{-\lambda_x - \lambda_2}.$$

Fisher Information

Nagaraja and Abo-Eleneen (2002) derived expressions for the elements of the Fisher information matrix for the three elements of the Gumbel type II bivariate exponential distribution. They observed that the improvement in the efficiency of the maximum likelihood estimate of the mean of X due to availability of the covariate values as well as the knowledge of the nuisance parameters is limited for this distribution.

Other Properties

- The copula is given by (2.1).

- The distributions of the maximum and minimum statistics are well known and can be easily derived; see, for example, Lai and Xie (2006, p. 310). Clearly, they can be expressed as mixtures of two or more exponential distributions.
- Franco and Vivo (2006) discussed log-concavity of the extreme statistics $\min(X,Y)$ and $\max(X,Y)$.

10.2.6 Gumbel's Type III Bivariate Exponential Distribution

The joint cumulative distribution function is

$$H(x,y) = 1 - e^{-x} - e^{-y} + \exp\left\{-(x^m + y^m)^{1/m}\right\}, \quad x,y \geq 0,\ m \geq 1. \quad (10.4)$$

The survival function is

$$\bar{H}(x,y) = \exp\left\{-(x^m + y^m)^{1/m}\right\}.$$

The corresponding joint density function is

$$\begin{aligned} h(x,y) &= (x^m + y^m)^{-2+(1/m)} x^{m-1} y^{m-1} \left\{(x^m + y^m)^{1/m} + m - 1\right\} \\ &\quad \times \exp\left\{-(x^m + y^m)^{1/m}\right\}, \quad x,y \geq 0,\ m > 1. \end{aligned} \quad (10.5)$$

If $m = 1$, X and Y are mutually independent. Lu and Bhattacharyya (1991 a,b) have studied this bivariate distribution in detail and in particular provided several inferential procedures for this model.

Some Other Properties

- Baggs and Nagaraja (1996) have derived the distributions of the maximum and minimum statistics; in particular, the minimum is exponentially distributed, but the maximum statistic T_2 is a generalized mixture of three or fewer exponentials.
- Franco and Vivo (2006) discussed the log-concavity property of T_2.
- The copula that corresponds to this distribution is known as the Gumbel–Hougaard copula as given in (2.30).
- The Gumbel–Hougaard copula is max-stable and hence an extreme-value copula. It is the only Archimedean extreme-value copula [Nelsen (2006, p. 143)].

10.3 Freund's Bivariate Distribution

This distribution is often given the acronym BEE (bivariate exponential extension) because it is not a bivariate exponential distribution in the traditional sense, as the marginals are not exponentials. We note that the Friday and Patil distribution in Section 10.9 is also known as BEE.

10.3.1 Formula of the Joint Density

The joint density function is

$$h(x,y) = \begin{cases} \alpha\beta' \exp[-(\alpha+\beta-\beta')x - \beta' y] & \text{for } x \le y \\ \alpha'\beta \exp[-(\alpha+\beta-\alpha')y - \alpha' x] & \text{for } x \ge y \end{cases}, \tag{10.6}$$

where $x, y \ge 0$ and the parameters are all positive.

10.3.2 Formula of the Cumulative Distribution Function

The joint cumulative distribution function corresponding to (10.6) is

$$H(x,y) = \begin{cases} \frac{\alpha}{\alpha+\beta-\beta'} \exp[-(\alpha+\beta-\beta')x - \beta' y] + \frac{\beta-\beta'}{\alpha+\beta-\beta'} \exp[-(\alpha+\beta)y] & \text{for } x \le y \\ \frac{\beta}{\alpha+\beta-\alpha'} \exp[-(\alpha+\beta-\alpha')y - \alpha' x] + \frac{\alpha-\alpha'}{\alpha+\beta-\alpha'} \exp[-(\alpha+\beta)x] & \text{for } x \ge y \end{cases}, \tag{10.7}$$

where $x, y \ge 0$.

10.3.3 Univariate Properties

The marginal distributions are not exponential, but they are mixtures of exponentials. Hence, (10.6) is often known as Freund's bivariate exponential extension, or a bivariate exponential mixture distribution, as it is called by Kotz et al. (2000, p. 356). The expression for the marginal density $f(x)$ is

$$f(x) = \frac{(\alpha-\alpha')(\alpha+\beta)}{\alpha+\beta-\alpha'} e^{-(\alpha+\beta)y} + \frac{\alpha'\beta}{\alpha+\beta-\alpha'} e^{-\alpha' x}, \tag{10.8}$$

provided $\alpha + \beta \neq \alpha'$, and naturally a similar expression for $g(y)$ holds with β and β' changed to α and α', respectively. The special case of $\alpha + \beta = \alpha'$ gives $f(x) = (\alpha'\beta x + \alpha)e^{-\alpha' x}$.

The mean and variance of this distribution are $\frac{\alpha'+\beta}{\alpha'(\alpha+\beta)}$ and $\frac{\alpha'^2+2\alpha\beta+\beta^2}{\alpha'^2(\alpha+\beta)^2}$, respectively.

10.3.4 Correlation Coefficient

Pearson's correlation coefficient is given by

$$\frac{\alpha'\beta' - \alpha\beta}{\sqrt{(\alpha'^2 + 2\alpha\beta + \beta^2)(\beta'^2 + 2\alpha\beta + \alpha^2)}}, \tag{10.9}$$

which is restricted to the range $-\frac{1}{3}$ to 1.

10.3.5 Conditional Properties

The conditional densities can be derived, but they are quite cumbersome. We refer our readers to Kotz et al. (2000, p. 357) for more details.

10.3.6 Joint Moment Generating Function

The joint m.g.f. is

$$M(s,t) = (\alpha + \beta - s - t)^{-1}\left[\frac{\alpha'\beta}{\alpha' - s} + \frac{\alpha\beta'}{\beta' - t}\right]. \tag{10.10}$$

10.3.7 Derivation

This distribution was originally derived by Freund (1961) from a reliability consideration as follows. Suppose a system has two components A and B whose lifetimes X and Y have exponential densities $\alpha e^{-\alpha x}$ and $\beta e^{-\beta y}$, respectively. Further, suppose that the only dependence between X and Y arises from failure of either component changing the parameter of the life distribution of the other component; more specifically, when A fails, the parameter for Y becomes β', and when B fails, the parameter for X becomes α'. Then, the joint density of X and Y is as presented in (10.6).

We may restate Freund's model in terms of a shock model. Suppose that the shocks that knock out components A and B, respectively, are governed by two Poisson processes:

- For component A, the Poisson process has a rate α when component B is functioning and rate α' after component B fails.
- For component B, the Poisson process has a rate β when component A is functioning and rate β' after component A fails.

Freund's model may realistically represent systems in which the failure of one component puts an additional burden on the remaining one (e.g., kidneys) or, alternatively, the failure of one may relieve somewhat the burden on the other (e.g., competing species). A special case of Freund's bivariate distribution was also derived by Block and Basu (1974); see Section 10.6 for pertinent details.

10.3.8 Illustrations

Conditional density plots have been presented by Johnson and Kotz (1972, p. 265).

10.3.9 Other Properties

- For distributions of the minimum and maximum statistics, see Baggs and Nagaraja (1996).
- The exact distribution of the product XY is given in Nadarajah (2006b).
- For sums, products, and ratios for Freund's bivariate exponential distribution, see Gupta and Nadarajah (2006).
- For an expression of the Rényi and Shannon entropy for Freund's bivariate exponential distribution, see Nadarajah and Zografos (2005).

10.3.10 Remarks

- For a test of symmetry and independence, one may refer to O'Neill (1985).
- There is some interest in the reliability literature in the probability of system failure when two components are in parallel and repair or replacement of a failed component takes a finite time. In this situation, the probability that the working component fails before the failed one is repaired is of importance. Biswas and Nair (1984) have considered this situation when Freund's distribution is applicable; see also Adachi and Kodama (1980) and Goel et al. (1984).

- For parallel systems, Klein and Moeschberger (1986) made some calculations of the errors resulting from erroneously assuming component lifetimes have independent exponential distributions when in fact they jointly have Freund's distribution.
- The study of Klein and Basu (1985) referred to in Section 10.5.9 below also included bias reduction techniques for the estimation of $\bar{H}$ when (10.7) holds.
- Besides the variants and generalizations of this distribution that are described in Sections 10.3.12–10.3.16 and Section 10.4 below, we note a complicated generalization given by Holla and Bhattacharya (1965) that involves replacement of failed components.

10.3.11 Fields of Application

This distribution is useful as a reliability model. It was applied to analyze the data of Barlow and Proschan (1977) concerning failures of Caterpillar tractors; see also O'Neill (1985). For an application in distribution substation locations, see Khodr et al. (2003).

10.3.12 Transformation of the Marginals

The power-transformed version of Fruend's distribution has been considered by Spurrier and Weier (1981), concentrating on the performance of maximum likelihood estimates (which are not in closed form).

Hashino and Sugi's (1984) extension of this distribution was used with power-transformed observations by Hashino (1985); see Section 10.4 for more details.

10.3.13 Compounding

Roux and Becker (1981) obtained a compound distribution, which they called a bivariate Bessel distribution, by assuming that $\alpha'' = 1/\alpha'$ is exponentially distributed with density $\exp(-\alpha''/\gamma)/\gamma$, and similarly, $\beta'' = 1/\beta'$ has density $\exp(-\beta''/\delta)/\delta$. The resulting density is given by

$$h(x,y) = \begin{cases} 2\beta\gamma^{-1}\exp[-(\alpha+\beta)y]K_0\left(2\frac{\sqrt{(x-y)}}{\gamma}\right) & \text{for } 0 < y < x \\ 2\alpha\delta^{-1}\exp[-(\alpha+\beta)x]K_0\left(2\frac{\sqrt{(y-x)}}{\gamma}\right) & \text{for } 0 < x < y \end{cases}, \tag{10.11}$$

where K_0 is the modified Bessel function of the third kind of order zero.

10.3.14 Bhattacharya and Holla's Generalizations

In Model I of Bhattacharya and Holla (1963), it is supposed that when one component fails, the distribution of the other's lifetime becomes Weibull, not exponential. The density is then proportional to $(y-x)^{q-1}\exp[-\delta(y-x)^q - (\alpha+\beta)x]$ for $0 < x < y$, with an analogous expression for $0 < y < x$. In Model II, the distribution of the other component's lifetime becomes gamma after the failure of one component. The density is in this case proportional to $(y-x)^{q-1}\exp[\delta(y-x) - (\alpha+\beta)x]$ for $0 < x < y$, with an analogous expression for $0 < y < x$.

10.3.15 Proschan and Sullo's Extension of Freund's Model

Proschan and Sullo (1974) considered an extension in which one assumes the existence of a common cause of failure (i.e., a shock from a third source that destroys both components). This additional assumption is similar to that of Marshall and Olkin's model to be discussed in Section 10.5 below. It is easy to see that Proschan and Sullo's extension (often denoted by PSE) subsumes both Freund's bivariate exponential and Marshall and Olkin's BVE model.

$$h(x,y) = \begin{cases} \alpha\upsilon\exp[-(\theta-\upsilon)x - \upsilon y] & \text{for } x < y, \\ \eta\beta\exp[-(\theta-\eta)y - \eta x] & \text{for } x > y, \\ \gamma\exp(-\theta x), & \text{for } x = y. \end{cases}$$

Here, $\theta = \alpha+\beta+\gamma$, $\eta = \alpha'+\gamma$, and $\upsilon = \beta'+\gamma$. When $\gamma = 0$, it gives Freund's model. For $\alpha = \alpha'$ and $\beta = \beta'$ it gives the BVE model.

The resulting model retains the lack of memory property (10.21) that is enjoyed by Marshall and Olkin's model. Some inference results were derived for this extension by Hanagal (1992).

10.3.16 Becker and Roux's Generalization

Becker and Roux (1981) generalized Freund's model by supposing that the components did not fail after a single shock but that it took a and b shocks, respectively, to destroy them. (These numbers a and b are deterministic, not random.)

The resulting density function is

$$h(x,y) = \begin{cases} \frac{\beta'\alpha^a}{\Gamma(a)\Gamma(b)} x^{a-1}[\beta'(y-x)+\beta x]^{b-1}\exp[-\beta' y-(\alpha+\beta-\beta')x], \; 0<x<y, \\ \frac{\alpha'\beta^b}{\Gamma(a)\Gamma(b)} y^{b-1}[\alpha'(x-y)+\alpha y]^{a-1}\exp[-\alpha' x-(\alpha+\beta-\alpha')y], \; 0<y<x; \end{cases}$$

see also Steel and Roux (1987).

10.4 Hashino and Sugi's Distribution

10.4.1 Formula of the Joint Density

For $x, y \geq 0$, the joint density is given by

$$h(x,y) = \begin{cases} \alpha\beta' \exp[-\beta' y-(\alpha+\beta-\beta')x] & \text{for } 0\le x\le y, \text{with } x\le\gamma, \\ \alpha'\beta \exp[-\alpha' x-(\alpha+\beta-\alpha')y] & \text{for } 0\le y\le x, \text{with } y\le\gamma, \\ ab' \exp[-b'(y-\delta)-(a+b-b')(x-\delta)] & \text{for } \gamma\le x\le y, \\ a'b \exp[-a'(x-\delta)-(a+b-a')(y-\delta)] & \text{for } \gamma\le y\le x, \end{cases} \tag{10.12}$$

where all the parameters are positive. In fact, there are just six free parameters because of continuity conditions at $X=\gamma$ and $Y=\gamma$.

10.4.2 Remarks

An English account of this extension of Freund's distribution is given by Hashino (1985), who has attributed this model to Hashino and Sugi (1984). Hashino has presented expressions of the marginal density of Y, the marginal cumulative distribution of Y, the conditional cumulative distribution function of X given Y, and the joint cumulative distribution function H.

The distribution was not motivated by a reliability application; rather, it was intended to provide a tractable bivariate distribution that is somewhat analogous to the univariate piecewise-exponential distribution.

10.4.3 An Application

The Osaka district in Japan suffers from typhoons. When these occur, the river, in its tidal reaches, rises for two reasons: the rain that drains into it, and the storm surge that comes in from the sea. The study by Hashino was of the peak rainfall intensity and the maximum storm surge for 117 typhoons occurring over an 80-year period. In fitting the density in (10.12), X and Y were transformed to X^m/σ_{xm} and Y^m/σ_{ym}, respectively, with σ's being standard deviations of the transformed variables. Hashino found large differences (a factor of more than 2) between return periods $1/H(x, y)$ calculated using the fitted distribution and $1/[F(x)G(y)]$ calculated by assuming independence.

Two minor points: (i) It appears that the typhoons included in the study were restricted to those for which the storm surge exceeded a certain level; Hashino did not discuss whether this truncation of the sample had any effect on the conclusions. (ii) The correlation coefficient, given by Hashino (viz., -0.02 is calculated for the distribution by applying it only to large values of X and Y [i.e., the last expression in (10.12 and not for the distribution as a whole].

10.5 Marshall and Olkin's Bivariate Exponential Distribution

It is one of the most widely studied bivariate exponential distributions. The acronym BVE is often used in the literature to designate this distribution. It is comprehensively studied in Section 2.4 of Chapter 47 in Kotz et al. (2000).

10.5.1 Formula of the Cumulative Distribution Function

The upper right volume under the probability density surface is given by [see Marshall and Olkin (1967a)]

$$\bar{H}(x,y) = \exp[-\lambda_1 x - \lambda_2 y - \lambda_{12} \max(x,y)], \quad x, y \geq 0, \tag{10.13}$$

where all λ's are positive.

10.5.2 Formula of the Joint Density Function

This takes slightly different forms depending on whether x or y is bigger:

$$h(x,y) = \begin{cases} \lambda_2(\lambda_1 + \lambda_{12}) \exp[-(\lambda_1 + \lambda_{12})x - \lambda_2 y] & \text{for } x > y, \\ \lambda_1(\lambda_2 + \lambda_{12}) \exp[-\lambda_1 x + (\lambda_2 + \lambda_{12})y] & \text{for } y > x, \\ \text{Singularity along the diagonal} & \text{for } x = y. \end{cases} \tag{10.14}$$

The amount of probability for the singular part is $\lambda_{12}/(\lambda_1 + \lambda_2 + \lambda_{12})$.

The singularity[1] in this case is due to the possibility of X exactly equaling Y. In the reliability context, this corresponds to the simultaneous failure of the two components.

10.5.3 Univariate Properties

Both marginal distributions are exponential.

10.5.4 Conditional Distribution

The conditional density of Y given $X = x$ is

$$h(y|x) = \begin{cases} \frac{\lambda_1(\lambda_2+\lambda_{12})}{\lambda_1+\lambda_{12}} e^{-\lambda_2 y - \lambda_{12}(y-x)} \text{ for } y > x, \\ \\ \lambda_2 e^{-\lambda_2 y} \lambda_1 \text{ for } y < x. \end{cases}$$

10.5.5 Correlation Coefficients

Pearson's product-moment correlation coefficient is $\lambda_{12}/(\lambda_1 + \lambda_2 + \lambda_{12})$. The rank correlation coefficients were given in Chapter 2.

[1] For a bivariate distribution, a singularity is a point with positive probability or a line such that every segment has positive probability. (We are not concerned here with more complicated forms of singularity.)

10.5.6 Derivations

Fatal Shocks

Suppose there is a two-component system subject to shocks that may knock out the first component, the second component, or both of them. If these shocks result from independent Poisson processes with parameters λ_1, λ_2, and λ_{12}, respectively, Marshall and Olkin's distribution results. Equivalently, $X = \min(Z_1, Z_3)$ and $Y = \min(Z_2, Z_3)$, where the Z's are independent exponential variates. Thus, this is an example of the trivariate reduction method.

Nonfatal Shocks

It could be that the shocks sometimes knock out a component and sometimes not.[2] Consider events in the Poisson process with rate θ that cause failure to the ith component (but not the other) with probability p_i ($i = 1, 2$) and cause failure to both components with probability p_{12}, where $1 - p_1 - p_2 - p_{12} > 0$. If $\lambda_i = p_i\theta$ and $\lambda_{12} = p_{12}\theta$, then the times to failure X and Y of components 1 and 2 have their joint survival function as in (10.13); see Marshall and Olkin (1985) for a representation like this.

10.5.7 Fisher Information

Nagaraja and Abo-Eleneen (2002) obtained the Fisher information for the three parameters of this model. They observed that the improvement in the efficiency of the maximum likelihood estimator of the mean of X due to the availability of the covariate as well as the knowledge of the nuisance parameter is quite substantial.

10.5.8 Estimation of Parameters

- Arnold (1968) proposed consistent estimators of λ_1, λ_2, and λ_{12}.
- For the maximum likelihood estimation of parameters, one may refer to Bemis et al. (1972), Proschan and Sullo (1974, 1976), and Bhattacharyya and Johnson (1971, 1973). Proschan and Sullo (1976) also proposed estimators based on the first iteration of the maximum of the log-likelihood

[2] The term *nonfatal shock* model is perhaps unfortunate, as it may suggest that the shocks are injurious, whereas in fact it is usually assumed that they are either fatal or do not have an effect at all.

function. Awad et al. (1981) proposed "partial maximum likelihood estimators." Chen et al. (1998) investigated the asymptotic properties of the maximum likelihood estimators based on mixed censored data.
- Hanagal and Kale (1991a) constructed consistent moment-type estimators. Hanagal and Kale (1991b) also discussed tests for the hypothesis $\lambda_{12} = 0$.
- For other references on estimation, see pp. 363–367 of Kotz et al. (2000).

10.5.9 Characterizations

Block (1977b) proved that X and Y with exponential marginals have Marshall and Olkin's bivariate exponential distribution if and only if one of the following two equivalent conditions holds:

- $\min(X, Y)$ has an exponential distribution,
- $X - Y$ and $\min(X, Y)$ are independent.

Some other characterizations have been established by Samanta (1975), Obretenov (1985), Azlarov and Volodin (1986, Chapter 9), Roy and Mukherjee (1989), and Wu (1997).

10.5.10 Other Properties

- The joint moment generating function is

$$M(s,t) = \frac{(\lambda + s + t)(\lambda_1 + \lambda_{12})(\lambda_2 + \lambda_{12}) + \lambda_{12}st}{(\lambda_1 + \lambda_{12} - s)(\lambda_2 + \lambda_{12} - t)}.$$

- $\min(X, Y)$ is exponential and $\max(X, Y)$ has a survival function given by

$$e^{-(\lambda_1+\lambda_{12})x} + e^{-(\lambda_2+\lambda_{12})x} - e^{-(\lambda_1+\lambda_2+\lambda_{12})x}, \quad x > 0;$$

  see Downton (1970) and Nagaraja and Baggs (1996).
- The aging properties of minimum and maximum statistics were discussed by Franco and Vivo (2002), who showed that $\max(X, Y)$ is a generalized mixture of three exponential components. The distribution is neither ILR (increasing likelihood ratio) nor DLR (decreasing likelihood ratio). Because the minimum statistic is exponentially distributed, it is therefore both ILR and DLR.
- The exact distribution of the product XY is given in Nadarajah (2006b).
- An expression for Rényi and Shannon entropy for this distribution was obtained by Nadarajah and Zografos (2005).

- The distribution is not infinitely divisible except in the degenerate case when $\lambda_1 = 0$ (or $\lambda_2 = 0$) or when $\lambda_{12} = 0$ (in the latter case, X and Y are independent).
- For dependence concepts for Marshall and Olkin's bivariate distribution, see Section 3.4 for details.
- Beg and Balasubramanian (1996) have studied the concomitants of order statistics arising from this bivariate distribution.
- By letting $\theta_i = 1/\lambda_i, i = 1, 2$, Boland (1998) has shown that $c_1X + c_2Y$ is "stochastically arrangement increasing" in $\boldsymbol{c} = (c_1, c_2)'$ and $\boldsymbol{\theta} = (\theta_1, \theta_2)'$.
- It has the lack of memory property given below in (10.21).

10.5.11 Remarks

- This distribution was first derived by Marshall and Olkin (1967a). It is sometimes denoted simply by BVE.
- $\bar{H}(x, y)$ can be expressed as

$$\bar{H}(x, y) = \frac{\lambda_1 + \lambda_2}{\lambda} \bar{H}_a(x, y) + \frac{\lambda_{12}}{\lambda} \bar{H}_s(x, y), \tag{10.15}$$

where $\lambda = \lambda_1 + \lambda_2 + \lambda_{12}$ and H_s and H_a are the singular and absolutely continuous parts[3] of $\bar{H}$ given by

$$\bar{H}_s(x, y) = \exp[-\lambda \max(x, y)], \tag{10.16}$$

$$\begin{aligned} \bar{H}_a(x, y) = & \frac{\lambda}{\lambda_1 + \lambda_2} \exp[-\lambda_1 x - \lambda y - \lambda \max(x, y)] \\ & - \frac{\lambda_{12}}{\lambda_1 + \lambda_2} \exp[-\lambda \max(x, y)]. \end{aligned} \tag{10.17}$$

- For tests of independence, see Kumar and Subramanyam (2005) and the references therein.
- Lu (1997) proposed a new plan for life-testing two-component parallel systems under Marshall and Olkin's bivariate exponential distribution.
- Earlier, Ebrahimi (1987) also discussed accelerated life tests based on Marshall and Olkin's model.
- In the "competing risks" context [for an explanation of this, see Chapter 9 of Cox and Oakes (1984)], this distribution is fully identified, provided it is known which observations correspond to failure from both causes together as well as which correspond to failure from each cause alone. This is because the distribution arises from three kinds of shocks acting

[3] That is, referring respectively to $Y = X$ (with positive probability) and $Y \neq X$ (where the p.d.f. is finite). More formally, a bivariate distribution H is absolutely continuous if the joint density exists almost everywhere.

independently, and that leading to failure of Type 1 and Type 2 together can simply be treated as a failure of Type 3; see David and Moeschberger (1978, Section 4.4).

- In collecting data where this distribution is to be applied, it may happen that the nature of the second failure is indeterminate; i.e., it is not known whether the second shock would or would not have knocked out both components had both still been functioning. This leads to difficulties in estimating the λ's; see Shamseldin and Press (1984).
- Klein and Moecshberger (1988) made some calculations of errors resulting from wrongly assuming that component lifetimes have independent exponential distributions when in fact they jointly have Marshall and Olkin's distribution. They carried out the calculations for both series and parallel systems.
- According to Klein and Basu (1985), if interest centers on estimating $\bar{H}$, the matter is not as simple as merely substituting good estimates of the model parameters into (10.13), as the resulting estimate may be biased to an unacceptable degree. So, Klein and Basu discussed some methods of bias reduction.
- This distribution and the associated shock model quickly received attention in the reliability literature; see Harris (1968). Some developments since then include the following. A brief report on a two-component system with Marshall and Olkin's distributions for both life and repair times is due to Ramanarayanan and Subramanian (1981). Osaki (1980), Sugasaw and Kaji (1981), and Goel et al. (1985) have presented some results for a two-component system in which failures follow this model, but other distributions (such as those of inspection, repair, and interinspection times) are arbitrary. Ebrahimi (1987) has given some results for the case where the two-component system is tested at a number of different stress levels, s_j, and failures follow the Marshall–Olkin distribution, with each λ being proportional to s_j^2. Osaki et al. (1989) have presented some results for availability measures of systems in which two units are in series, failure of unit 1 shuts off unit 2 but not vice versa, with the lifetimes following the Marshall–Olkin distribution, the units have arbitrary repair-time distributions, and two alternative assumptions are made about the repair discipline.
- Another account of this distribution is given by Marshall and Olkin (1985).
- A parametric family of bivariate distributions for describing the lifelengths of a system of two dependent components operating under a common environment when component conditional lifetime distribution follows Marshall and Olkin's bivariate exponential and the environment follows an inverse Gaussian distribution was derived by Al-Mutairi (1997).

10.5.12 Fields of Application

Among many applications of Marshall and Olkin's distribution, we note especially the fields of nuclear reactor safety, competing risks, and reliability.

Certainly, the idea of simultaneous failure of two components is far from being merely of academic interest. Hagen (1980) has presented a review in the context of nuclear power and has pointed out that introducing redundancy into a system reduces random component failure to insignificance, leading to the common-mode/common-cause type being predominant among system failures.

Rai and Van Ryzin (1984) applied this distribution as a tolerance distribution in a quantal response context to the occurrence of bladder and liver tumors in mice exposed to one of several alternative dosages of a carcinogen. Actually, the distribution was (i) used in the form with Weibull marginals and (ii) mixed with a finite probability of tumors occurring even at zero dose.

Kotz et al. (2000) have provided a list of references for each of the three primary applications mentioned above.

10.5.13 Transformation to Uniform Marginals

Cuadras and Augé (1981) proposed the following joint distribution, whose support is the unit square:

$$H(x,y) = \begin{cases} x^{1-c}y & \text{for } x \geq y, \\ xy^{1-c} & \text{for } x < y. \end{cases} \tag{10.18}$$

The corresponding joint density is

$$h(x,y) = \begin{cases} (1-c)x^{-c} & \text{for } x > y, \\ (1-c)y^{-c} & \text{for } x < y, \\ \text{singularity along the diagonal } x = y. \end{cases} \tag{10.19}$$

Cuadras and Augé did not refer to Marshall and Olkin, and so it is likely that they were not aware that their distribution was a transformation of one that is already known. Conway (1981) gave an illustration of the Marshall and Olkin distribution after transformation to uniform marginals, and that becomes an illustration of the Cuadras and Augé distribution.

10.5.14 Transformation to Weibull Marginals

As with other distributions having exponential marginals, this one is sometimes generalized by changing them to Weibull; see, for example, Marshall and Olkin (1967a), Moeschberger (1974), and Lee (1979).

10.5.15 Transformation to Extreme-Value Marginals

This distribution is sometimes met in the form with extreme value marginals.

10.5.16 Transformation of Marginals: Approach of Muliere and Scarsini

First, consider the univariate case. Muliere and Scarsini (1987) presented a general version of the lack of memory property as follows:

$$\bar{F}(s * t) = \bar{F}(s)\bar{F}(t), \tag{10.20}$$

where $*$ is any binary operation that is associative (i.e., such that $(x*y)*z = x*(y*z)$). Examples include the following:

- The operation $*$ being addition leads to the usual lack of memory characterization of exponential distribution: If $\bar{F}(s+t) = \bar{F}(s)\bar{F}(t)$, then $\bar{F}(x) = e^{-\lambda t}$.
- If $x * y = (x^\alpha + y^\alpha)^{1/\alpha}$, then the Weibull distribution $\bar{F}(x) = \exp(-\lambda x^\alpha)$ results.
- If $x * y = xy$, then the Pareto distribution $\bar{F}(x) = x^{-\lambda}$ results.

In the bivariate case, consider first the following version of the bivariate lack of memory property:

$$\bar{H}(s_1 + t, s_2 + t) = \bar{H}(s_1, s_2)\bar{H}(t, t). \tag{10.21}$$

For more on this, see Section 10.22, but if we assume the marginals are exponential, the solution is the Marshall and Olkin distribution. Now consider

$$\bar{H}(s_1 * t, s_2 * t) = \bar{H}(s_1, s_2)\bar{H}(t, t) \tag{10.22}$$

together with (10.20) for each marginal. The solution is then

$$\bar{H}(s,t) = \exp\left\{-\lambda_1 a(s) - \lambda_2 a(t) - \lambda_{12} a[\max(s,t)]\right\}, \tag{10.23}$$

with $a(\cdot)$ being a (strictly increasing) function corresponding to the operation $*$; i.e., $a(x * y) = a(x) + a(y)$. Examples include the following:

- The operation $*$ being addition leads to the Marshall and Olkin distribution.
- If $x * y = (x^\alpha + y^\alpha)^{1/\alpha}$, the Weibull version of the Marshall and Olkin distribution results, i.e., $\bar{H}(x, y) = \exp[-\lambda_1 x^\alpha - \lambda_2 y^\alpha - \lambda_{12} \max(x^\alpha, y^\alpha)]$.
- If $x * y = xy$, then the result is $\bar{H}(x, y) = x^{-\lambda_1} y^{-\lambda_2} [\max(x, y)]^{-\lambda_{12}}$, the Pareto version of Marshall and Olkin's distribution. For related developments, one may refer to Sections 6.2.1 and 6.2.3 of Arnold (1983).

10.5.17 Generalization

Johnson and Kotz (1972, p. 267) have credited Saw (1969) for the proposal of replacing $\max(x, y)$ in (10.13) by an increasing function of $\max(x, y)$. One choice leads to

$$\bar{H}(x, y) = [1 + \max(x, y)]^{\lambda_{12}} \exp[\lambda_1 x - \lambda_2 y - \lambda_{12} \max(x, y)]. \quad (10.24)$$

Marshall and Olkin (1967b) considered some generalizations of (10.13), including

$$\bar{H}(x, y) = \exp\{-\lambda_1 x - \lambda_2 y - \lambda_{12} \max[x, y + \min(x, \delta)]\}, \quad \delta \geq 0. \quad (10.25)$$

Ohi and Nishida (1979), following an idea of Itoi et al. (1976), considered the case where component i ($i = 1, 2$) needs k_i shocks before it fails. The bivariate life distribution that results is called a bivariate Erlang distribution (BVEr). Ohi and Nishida then showed that:

- X and Y are positively regression dependent (see Section 3.4.4 for this concept).
- BVEr is bivariate NBU but not bivariate IFR. Here bivariate NBU is defined as a joint distribution that satisfies the inequality $\bar{H}(x+t, y+t) \leq \bar{H}(x, y)\bar{H}(t, t)$ for all $x, y, t \geq 0$.

Hyakutake (1990) suggested incorporating location parameters ξ_1 and ξ_2 in the BVE. The joint survival function is

$$\bar{H}(x, y) = e^{-\lambda_1(x-\xi_1)-\lambda_2(y-\xi_2)-\lambda_{12}\max(x-\xi_1, y-\xi_2)}, \quad x > \xi_1, \; y > \xi_2.$$

Ryu (1993) extended Marshall and Olkin's model such that the new joint distribution is absolutely continuous and need not be memoryless. The new marginal distribution has an increasing failure rate, and the joint distribution exhibits an aging pattern.

10.6 ACBVE of Block and Basu

10.6.1 Formula of the Joint Density

The joint density is

$$h(x,y)=\begin{cases}\frac{\lambda_1\lambda(\lambda_2+\lambda_{12})}{\lambda_1+\lambda_2}\exp[-\lambda_1 x-(\lambda_2+\lambda_{12})y] & \text{if } x<y,\\ \frac{\lambda_2\lambda(\lambda_1+\lambda_{12})}{\lambda_1+\lambda_2}\exp[-(\lambda_1+\lambda_{12})x-\lambda_2 y] & \text{if } x>y,\end{cases} \tag{10.26}$$

where $x, y \geq 0$, the λ's are positive, and $\lambda = \lambda_1 + \lambda_2 + \lambda_{12}$.

10.6.2 Formula of the Cumulative Distribution Function

The upper right volume under the probability density surface is given by

$$\begin{aligned}\bar{H}(x,y) &= \frac{\lambda}{\lambda_1+\lambda_2}\exp[-\lambda_1 x-\lambda_2 y-\lambda_{12}\max(x,y)]\\ &-\frac{\lambda_{12}}{\lambda_1+\lambda_2}\exp[-\lambda\max(x,y)].\end{aligned} \tag{10.27}$$

10.6.3 Univariate Properties

The marginals are not exponential but rather a negative mixture of two exponentials given by

$$\bar{F}(x)=\frac{\lambda}{\lambda_1+\lambda_2}\exp[-(\lambda_1+\lambda_{12}x)]-\frac{\lambda_{12}}{\lambda_1+\lambda_2}\exp(-\lambda x), \tag{10.28}$$

and a similar expression holds for $\bar{G}(y)$ as well.

10.6.4 Correlation Coefficient

Pearson's product-moment correlation coefficient is

$$\frac{(\lambda_1+\lambda_2)^2(\lambda_1+\lambda_{12})(\lambda_2+\lambda_{12})-\lambda^2\lambda_1\lambda_2}{\sqrt{[(\lambda_1+\lambda_2)^2(\lambda_1+\lambda_{12})^2+\lambda_2(\lambda_2+2\lambda_1)\lambda^2][(\lambda_1+\lambda_2)^2(\lambda_2+\lambda_{12})^2+\lambda_1(\lambda_1+2\lambda_2)\lambda^2]}}\,. \tag{10.29}$$

We feel that the expression presented by Block and Basu (1974) may be in error.

10.6.5 Moment Generating Function

The m.g.f. may be obtained from (10.10) (by using substitutions given in Section 10.6.6) to be

$$M(s,t) = \frac{1}{\lambda_1+\lambda_2}\frac{\lambda}{\lambda-(s+t)}\left[\frac{\lambda_1(\lambda_2+\lambda_{12})}{\lambda_2+\lambda_{12}-t}+\frac{(\lambda_1+\lambda_{12})\lambda_2}{\lambda_1+\lambda_{12}-s}\right]. \tag{10.30}$$

10.6.6 Derivation

This distribution was derived by Block and Basu (1974) by omitting the singular part of Marshall and Olkin's distribution; see also Block (1975). Alternatively, it can be derived by Freund's method, with

$$\left.\begin{array}{l} \alpha\ = \lambda_1+\lambda_{12}[\lambda_1/(\lambda_1+\lambda_2)] \\ \alpha' = \lambda_1+\lambda_{12} \\ \beta\ = \lambda_2+\lambda_{12}[\lambda_2/(\lambda_1+\lambda_2)] \\ \beta' = \lambda_2+\lambda_{12} \end{array}\right\}. \tag{10.31}$$

10.6.7 Remarks

- $\min(X,Y)$ is an exponential variate.
- $X-Y$ and $\min(X,Y)$ are independent variables.
- The lack of memory property holds.
- For inferential methods, see Hanagal and Kale (1991a), Hanagal (1993), Achcar and Santander (1993), and Achcar and Leandro (1998).
- Achcar (1995) has discussed accelerated life tests based on bivariate exponential distributions.
- The exact distributions of sum $R = X+Y$, the product $P = XY$, and the ratio $W = X/(X+Y)$, and the corresponding moment properties are derived by Nadarajah and Kotz (2007) when X and Y follow Block and Basu's bivariate exponential distribution.
- From the expression for $\bar{H}(x,y)$, it is easy to show that the distribution is PQD.

10.6.8 Applications

Gross and Lam (1981) considered this distribution to be suitable in cases such as the following:

- lengths of tumor remission when a patient receives different treatments on two occasions,
- lengths of time required for analgesics to take effect when patients with headaches receive different ones on two occasions.

Gross and Lam were then concerned primarily with developing hypothesis tests for equality of marginal means. They also made the following suggestion for determining whether Block and Basu's distribution is appropriate or not:

- Test whether $\min(X,Y)$ has an exponential distribution.
- Test whether $X-Y$ and $\min(X,Y)$ are uncorrelated.
- Test whether $X-Y$ has the distribution given by their Eq. (4.1).

These three properties, except with independence replacing zero correlation in the second of them, together characterize the Block and Basu distribution.

Block and Basu's bivariate exponential distribution was applied by Nadarajah and Kotz (2007) to drought data.

10.7 Sarkar's Distribution

10.7.1 Formula of the Joint Density

For (x,y) in the positive quadrant, the joint density function $h(x,y)$ is given by

$$\begin{cases} \frac{\lambda_1\lambda}{(\lambda_1+\lambda_2)^2}\exp[-\lambda_1 x-(\lambda_2+\lambda_{12})y] \\ \times[(\lambda_1+\lambda_2)(\lambda_2+\lambda_{12})-\lambda_2\lambda\exp(-\lambda_1 y)][A(\lambda_1 x)]^{\gamma}[A(\lambda_2 y)]^{-(1+\gamma)} \text{ if } x\le y, \\ \\ \frac{\lambda_2\lambda}{(\lambda_1+\lambda_2)^2}\exp[-(\lambda_1+\lambda_{12})x-\lambda_2 y] \\ \times[(\lambda_1+\lambda_2)(\lambda_2+\lambda_{12})-\lambda_1\lambda\exp(-\lambda_2 y)][A(\lambda_1 x)]^{-(1+\gamma)}[A(\lambda_2 y)]^{\gamma} \text{ if} x\ge y, \end{cases} \tag{10.32}$$

where the λ's are positive, $\lambda=\lambda_1+\lambda_2+\lambda_{12}$, $\gamma=\lambda_{12}/(\lambda_1+\lambda_2)$, and $A(z)=1-\exp(-z)$.

10.7.2 Formula of the Cumulative Distribution Function

The joint survival function is given by

$$\bar{H}(x,y) = \begin{cases} \exp[-(\lambda_2+\lambda_{12})y]\{1-[A(\lambda_1 x)]^{1+\gamma}\}[A(\lambda_2 y)]^{-\gamma} & \text{if } x \le y, \\ \exp[-(\lambda_1+\lambda_{12})y]\{1-[A(\lambda_1 x)]^{-\gamma}\}[A(\lambda_2 x)]^{1+\gamma} & \text{if } x \ge y, \end{cases} \tag{10.33}$$

$H(x,y)$ is absolutely continuous in this case.

10.7.3 Univariate Properties

Both the marginal distributions are exponential.

10.7.4 Correlation Coefficient

An expression for Pearson's correlation coefficient has been given by Sarkar (1987) but is rather complicated.

10.7.5 Derivation

This distribution, sometimes denoted by ACBVE$_2$, was derived by Sarkar through the following conditions of characterization:

- The bivariate distribution is absolutely continuous.
- X and Y are exponential variates with parameters $\lambda_1+\lambda_{12}$ and $\lambda_2+\lambda_{12}$, respectively.
- $\min(X,Y)$ is exponential with parameter $\lambda=\lambda_1+\lambda_2+\lambda_{12}$.
- $\min(X,Y)$ is independent of $g(X,Y)$ for some g of the form $l(x)-l(y)$, where l is an increasing function.

10.7.6 Relation to Marshall and Olkin's Distribution

This distribution is obtained from Marshall and Olkin's distribution by requiring absolute continuity of the distribution function and by replacing the condition of independence of $\min(X,Y)$ and $X-Y$ by the modified condition above. Also, it does not possess the lack of memory property now.

10.8 Comparison of Four Distributions

At this point, we compare the properties of the Marshall and Olkin, Block and Basu, Sarkar, and Freund distributions in the following table.

	Marshall and Olkin	Block and Basu	Sarkar	Freund
Exponential marginals	√		√	
Absolutely continuous		√	√	√
Bivariate lack of memory, (10.21)	√	√		√
$\min(X, Y)$ is exponential	√	√	√	√
$\min(X, Y)$ is independent of $X - Y$	√	√	modified	√

10.9 Friday and Patil's Generalization

Friday and Patil (1977) proposed a distribution that subsumes both Freund's and Marshall and Olkin's distributions with joint survival function

$$\bar{H}(x, y) = \gamma \bar{H}_A(x, y) + (1 - \gamma)\bar{H}_B(x, y), \tag{10.34}$$

where $\bar{H}_A$ is the survival function corresponding to Freund's distribution (10.7), and $\bar{H}_B$ is the singular distribution $\exp[-(\alpha + \beta)\max(x, y)]$. More explicitly, we have

$$\bar{H}(x, y) = \begin{cases} \theta_1 \exp[-(\alpha + \beta - \beta')x - \beta' y] + (1 - \theta_1)\exp[-(\alpha + \beta)y] & \text{for } x \leq y, \\ \theta_2 \exp[-\alpha' x - (\alpha + \beta - \alpha')y] + (1 - \theta_2)\exp[-(\alpha + \beta)x] & \text{for } x \geq y, \end{cases} \tag{10.35}$$

where $\theta_1 = \gamma\alpha(\alpha + \beta - \beta')^{-1}$, $\theta_2 = \gamma\beta(\alpha + \beta - \alpha')^{-1}$, and $0 \leq \gamma \leq 1$. This distribution is another one that has the lack of memory property in (10.21). It is sometimes denoted by BEE.

Friday and Patil also showed that only two independent standard exponential variates are needed to generate a pair (X, Y) with their distribution as in (10.35), and thus the same is true for Freund's and Marshall and Olkin's distributions. They then examined the computational efficiency of their scheme. Some further results have been given by Itoi et al. (1976).

The model of Platz (1984) is another one that includes both Marshall and Olkin and Freund models and in addition one-out-of-three and two-out-of-three systems with identical components.

Remarks

- Of course, the Friday and Patil bivariate exponential distribution also includes Block and Basu's ACBVE.
- The distributions of the maximum and minimum statistics are given in Baggs and Nagaraja (1996). The maximum is either a generalized mixture of three or fewer exponentials or a mixture of gamma and exponentials. Franco and Vivo (2002) considered their IFR and DFR properties.
- Franco and Vivo (2007) gave a comprehensive study on the aging properties of the extreme statistics $\min(X, Y)$ and $\max(X, Y)$.
- Sun and Basu (1993) have shown that among the bivariate exponential distributions with constant total failure rates and constant $\Pr[X > Y \mid \min(X, Y) = t]$, the Friday and Patil distribution is the largest family.
- The proposed infinitesimal generator representation of Wang (2007) can be used to characterize the bivariate exponential distributions of Freund, Marshall and Olkin, Block and Basu, and Friday and Patil.

10.10 Tosch and Holmes' Distribution

The model of Tosch and Holmes (1980) generalizes both the Marshall and Olkin and Freund models. It permits simultaneous failure of both components, and the residual lifetime of one component is not independent of the status (working or failed) of the other component. Stated formally,

$$\left.\begin{array}{l} X \ \min(U_1, U_2) + U_3 I_{\{U_1 > U_2\}} \\ Y \ \min(U_1, U_2) + U_4 I_{\{U_1 \leq U_2\}} \end{array}\right\}, \tag{10.36}$$

where the U's are non-negative mutually independent r.v.'s and $I_{\{.\}}$ is the indicator variable, i.e., it is 1 if the condition within the brackets is true and zero if it is false. In other words, if component 1 is the first to fail, then its lifetime X is U_1 and the second component's extra lifetime is U_4; conversely, if component 2 is the first to fail, its lifetime is U_2 and the first component's extra lifetime is U_3. The cumulative distribution of (10.36) cannot be easily obtained in general. However, it can be found when U_1 and U_2 are exponential variables with scale parameters α and β, respectively, and U_3 and U_4 are exponential variables apart from discontinuity at the origin (i.e., $\Pr(U_3 \leq t) = 1 - q + q[1 - \exp(\alpha' t)]$ and $\Pr(U_4 \leq t) = 1 - q + q[1 - \exp(-\beta' t)]$, with $0 \leq q \leq 1$).

10.11 A Bivariate Exponential Model of Wang

Wang (2007) used a counting process approach for characterizing a system of two dependent component failure rates. The components are subjected to a series of Poisson shocks. The distribution in question was derived by specifying the entries of the infinitesimal generator of a continuous time generator (Q matrix).

For a two-component system, state 0 denotes no failure and states $1_1, 1_2$, and 2 denote the failure of components 1, 2, and both components, respectively. The corresponding failure rates are λ_1, λ_2, and λ_{12}, respectively. The failure rate of the surviving component changes from λ_i to λ_i' after the other component fails. The infinitesimal generator of the model is

$$Q = \begin{matrix} 0 \\ 1_1 \\ 1_2 \\ 2 \end{matrix} \begin{pmatrix} -(\lambda_1+\lambda_2+\lambda_{12}) & \lambda_1 & \lambda_2 & \lambda_{12} \\ 0 & -(\lambda_2'+\lambda_{12}) & 0 & (\lambda_2'+\lambda_{12}) \\ 0 & 0 & -(\lambda_1'+\lambda_{12}) & (\lambda_1'+\lambda_{12}) \\ 0 & 0 & 0 & 0 \end{pmatrix}. \tag{10.37}$$

10.11.1 Formula of the Joint Density

Let $Q = \frac{\lambda_1+\lambda_2}{\lambda} Q_a + \frac{\lambda_{12}}{\lambda} Q_s$. Then

$$h(x,y) = \begin{cases} \frac{\lambda_1\lambda(\lambda_2'+\lambda_{12})}{\lambda_1+\lambda_2} \exp[-(\lambda_1+\lambda_2-\lambda_2')x - (\lambda_2'+\lambda_{12})y], \; 0 < x < y, \\ \frac{\lambda_2\lambda(\lambda_1'+\lambda_{12})}{\lambda_1+\lambda_2} \exp[-(\lambda_1'+\lambda_{12})x - (\lambda_1+\lambda_2-\lambda_1')y], \; 0 < y < x, \end{cases} \tag{10.38}$$

is the joint density function that corresponds to Q_a and $g(t) = \lambda \exp(-\lambda t)$ corresponds to the Q_s matrix.

10.11.2 Univariate Properties

Marginals are not exponentially distributed.

10.11.3 Remarks

Wang (2007) has shown:

- If $\lambda_1' = \lambda_1$ and $\lambda_2' = \lambda_2$, then Q_a corresponds to the infinitesimal generator of the Block and Basu distribution and Q is the infinitesimal generator of the Marshall and Olkin distribution.
- If $\lambda_{12} = 0$, then Q_a corresponds to the infinitesimal generator of the Freund distribution.
- Let $0 \le \gamma \le 1$ and set $\lambda_{12} = 0$. Then $\gamma Q_a + (1-\gamma) Q_s$ is the infinitesimal generator for the Friday and Patil distribution.

10.12 Lawrance and Lewis' System of Exponential Mixture Distributions

Lawrance and Lewis' (1983) models are easy to simulate, can represent a broad range of correlation structures, and are analytically tractable.

10.12.1 General Form

To begin with, we note that if E_1 and E_2 are i.i.d. standard exponential variates and (independently of E_1 and E_2) if I is 0 or 1 with probabilities β and $1-\beta$, respectively, then $\beta E_1 + I E_2$ is also a standard exponential variate.

The general form of this model [see Lawrance and Lewis (1983)] is

$$\left.\begin{aligned} X &= \beta_1 V_1 E_1 + I_1 E_2 \\ Y &= I_2 E_1 + \beta_2 V_2 E_2 \end{aligned}\right\}, \tag{10.39}$$

where E_1 and E_2 are independent and exponentially distributed, V_1 and V_2 are each either 0 or 1 (not necessarily independent of each other) with $\Pr(V_i = 1) = \alpha_i$, and I_1 and I_2 are each either 0 or 1 (not necessarily independent of each other) with $\Pr(I_i = 1) = (1-\beta_i)/[1-(1-\alpha_i)\beta_i]$.

Lawrance and Lewis termed the model in (10.39) the EP+ model. They had focused on three special cases, denoted by EP1, EP3, and EP5.

10.12.2 Model EP1

This takes $\alpha_1 = \alpha_2 = 1, \beta_1 = \beta_2 (= \beta)$, and $I_1 = I_2 (= I)$. Thus,

$$\left.\begin{aligned} X &= \beta E_1 + I E_2 \\ Y &= I E_1 + \beta E_2 \end{aligned}\right\} \tag{10.40}$$

with $\Pr(I = 1) = 1 - \beta$.

The joint density in this case is

$$h(x,y) = I_{\{\beta y < x < y/\beta\}} \frac{1}{1+\beta} \exp\left(-\frac{x+y}{1+\beta}\right) + \frac{1}{\beta} \exp\left(-\frac{x+y}{\beta}\right), \quad 0 < \beta \leq 1 \tag{10.41}$$

where $I_{\{\cdot\}}$ is the indicator function, as earlier. Lawrance and Lewis illustrated this density for $\beta = 0.5$. The product-moment correlation is $3\beta(1-\beta)$, and so it is at $\beta = 0.5$ that it reaches its maximum. The grade correlation (Spearman's rho) is given by $3\beta(1-\beta)(8+7\beta+\beta^2)/[(1+\beta)^2(2+\beta)^2]$.

10.12.3 Model EP3

This takes $\alpha_1 = \alpha_2 = 1$, with (I_1, I_2) having maximum possible dependency. The last statement means that the possible combination of values occurs with the following probabilities:

	$I_2 = 0$	$I_2 = 1$
$I_1 = 0$	$\min(\beta_1, \beta_2)$	$\max(\beta_1 - \beta_2, 0)$
$I_1 = 1$	$\max(\beta_2 - \beta_1, 0)$	$\min(1-\beta_1, 1-\beta_2)$

Then,

$$\left.\begin{aligned} X &= \beta_1 E_1 + I_1 E_2 \\ Y &= I_2 E_1 + \beta_2 E_2 \end{aligned}\right\} \tag{10.42}$$

with probabilities of the various combinations of values of I_1 and I_2 being as above.

Lawrance and Lewis presented expressions for both the product-moment correlation and the grade (Spearman's ρ) correlation.

10.12.4 Model EP5

This takes $\alpha_1 = \alpha_2\ (= \alpha), \beta_1 = \beta_2\ (= \beta), V_1 = V_2\ (= V)$, and $I_1 = I_2\ (= I)$. Thus,

$$\left.\begin{aligned} X &= \beta V E_1 + I E_2 \\ Y &= I E_1 + \beta V E_2 \end{aligned}\right\} \tag{10.43}$$

with $\Pr(V = 1) = \alpha$ and $\Pr(I = 1) = (1-\beta)/[1-(1-\alpha)\beta]$.

The product-moment correlation in this case is $3\alpha\beta(1-\alpha\beta)$.

10.12.5 Models with Negative Correlation

Lawrance and Lewis also discussed a number of analogous models that have negative correlations, still with exponential marginals.

10.12.6 Models with Uniform Marginals

Lawrance and Lewis also discussed thedistributions above after they were transformed to have uniform marginals. They presented an illustration of the EP1 model (with $\beta = 0.32$) after such a transformation.

10.12.7 The Distribution of Sums, Products, and Ratios

Nadarajah and Ali (2006) derived the exact distribution of $R = X + Y$, $P = XY$, and $W = X/(X+Y)$ when X and Y follow Lawrance and Lewis' bivariate exponential distribution.

10.12.8 Mixture Models

Models that can exhibit either positive or negative dependence can be obtained easily by mixing one of those having positive correlation with one of those having negative correlation.

10.12.9 Models with Line Singularities

Models that are like Marshall and Olkin's distribution in that there is a nonzero probability that $Y = X$ may be readily constructed.

Let (X_1, X_2) be a pair of variates with standard exponential marginals, such as those described above. Let E be an independent standard exponential variate. Let (I_1, I_2) be an indicator pair, possibly completely or partially dependent, with marginal probabilities $\Pr(I_i = 1) = 1 - \beta_i$. Three methods of obtaining a distribution having a line singularity are as follows:

$$\left.\begin{aligned} X &= I_1 X_1 + (1 - I_1)E \\ Y &= I_2 X_2 + (1 - I_2)E \end{aligned}\right\}, \tag{10.44}$$

$$\left.\begin{aligned} X &= I_1X_1 + \beta_1 E \\ Y &= I_2X_2 + \beta_2 E \end{aligned}\right\}, \tag{10.45}$$

$$\left.\begin{aligned} X &= \min(X_1, E) \\ Y &= \min(X_2, E) \end{aligned}\right\}. \tag{10.46}$$

10.13 Raftery's Scheme

In its general form, Raftery's (1984, 1985) scheme of obtaining a bivariate distribution with exponential marginals is given by

$$\left.\begin{aligned} X &= (1 - p_{10} - p_{11})U + I_1 W \\ Y &= (1 - p_{01} - p_{11})V + I_2 W \end{aligned}\right\}, \tag{10.47}$$

where U, V, W are independent and exponentially distributed, and in addition, they are independent of I_i. I_1 and I_2 are each either 0 or 1, with probabilities as set out below:

	$I_2 = 0$	$I_2 = 1$
$I_1 = 0$	p_{00}	p_{01}
$I_1 = 1$	p_{10}	p_{11}

Raftery showed the correlation to be $2p_{11} - (p_{01} + p_{11})(p_{10} + p_{11})$. There is also an extension of the model to permit negative correlation. Raftery then paid special attention to the following cases.

10.13.1 First Special Case

This sets $p_{01} = p_{10} = 0$, so that

$$\left.\begin{aligned} X &= (1 - p_{11})U + IW \\ Y &= (1 - p_{11})V + IW \end{aligned}\right\}. \tag{10.48}$$

10.13.2 Second Special Case

This sets $p_{01} = 0, p_{10} = 1 - p_{11}$, so that

$$\left.\begin{aligned} X &= W \\ Y &= (1 - p_{11})V + I_2 W \end{aligned}\right\}, \tag{10.49}$$

and the distribution in this case is a mixture of independence and weighted linear combination.

10.13.3 Formula of the Joint Density

The joint density function can be obtained in explicit form but is quite messy; see Raftery (1984).

10.13.4 Formula of the Cumulative Distribution Function

The joint survival function that corresponds to (10.48) with $\delta = p_{11}$ is

$$\bar{H}(x,y) = \begin{cases} e^{-x} + \frac{1-\delta}{1+\delta} e^{-x/(1-\delta)} \{e^{y\delta/(1-\delta)} - e^{-y/(1-\delta)}\} & \text{for } x \geq y, \\ e^{-y} - \frac{1-\delta}{1+\delta} e^{-y/(1-\delta)} \{e^{x\delta/(1-\delta)} - e^{-x/(1-\delta)}\} & \text{for } x \leq y. \end{cases} \tag{10.50}$$

10.13.5 Derivation

The distribution aries from a shock model in the following manner. Consider a system that has two components, S_1 and S_2, each of which can be functioning normally, unsatisfactory, or have failed. The system is subject to three kinds of shock, governed by independent Poisson processes. These kinds of shocks cause normal components to become unsatisfactory, an unsatisfactory S_1 to fail, and an unsatisfactory S_2 to fail, respectively.

10.13.6 Illustrations

Contours of the joint density have been represented by Raftery (1984).

10.13.7 Remarks

- Generation of random variates is easy by following the method of construction given in (10.47).
- The distribution can attain the Fréchet bounds.
- O'Cinneide and Raftery (1989) have shown that this distribution is an example of a bivariate phase-type distribution; see Assaf et al. (1984).
- The distributions of the extreme statistics $\min(X, Y)$ and $\max(X, Y)$ were given in Baggs and Nagaraja (1996), and their aging properties were discussed in Franco and Vivo (2002). See also Baggs and Nagaraja (1996) for their elementary aging properties.
- Bhattacharyya (1997) adopted Raftery's bivariate exponential construction to propose an absolutely continuous bivariate model for modeling survival data with random censoring when the censoring pattern and the failure pattern are dependent and follow exponential distributions with different means.

10.13.8 Applications

Raftery (1984) applied a Weibull version of this model to fit two datasets:

- Two hundred forty-nine pairs of successive failure times of a computer [Cox and Lewis (1966, p. 16)]; it was found that $p_{01} = p_{10}$ is suitable in this case.
- Proportions of a population who were without a car and who were foreign-born for the 88 unincorporated places with a population greater than 25000 found in the 1960 U.S. Census [Tukey (1977, p. 323)]; it was found that $p_{01} = 0$ is suitable in this case.

10.14 Linear Structures of Iyer et al.

Iyer et al. (2002) derived bivariate exponential distributions using auxiliary random variables that form linear structures. Two types of bivariate exponential models were developed. One gives positive correlations, and the other yields negative correlations. The bivariate models they developed are based on the work of Gaver and Lewis (1980).

10.14.1 Positive Cross Correlation

X and Y are linearly related in the form

$$Y = aX + Z, \quad a > 0, \tag{10.51}$$

where a is a constant and X and Z are independent. In fact, $Z = IE$, where I is the indicator variable (Bernoulli variable) with $\Pr(I = 1) = (1 - a\rho_1)$, $\rho_1 > 0$, and E is exponential with parameter λ_y; so, I and E are independent, so Z is a discontinuous exponential at the origin with distribution function $a\rho_1 + (1 - a\rho_1)(1 - e^{\lambda_y z})$.

Univariate Properties

X and Y have exponential distributions with parameters λ_x and λ_y, respectively.

10.14.1.1 Correlation Coefficient

Pearson's product-moment correlation is simply $\rho = a\rho_1$.

10.14.2 Negative Cross Correlation

The aim is to obtain a negative cross correlation between X and Y. The model considered is then

$$\left.\begin{array}{l} X = aP + V \\ Y = bQ + W \end{array}\right\} \tag{10.52}$$

for $a, b \geq 0$. Here, P and Q are independent of each other and so are Q and W. The following three models were focused on.

Model 1

P and Q are antithetic exponential variables given by

$$P = -\frac{1}{\lambda_p} \log U, \qquad Q = -\frac{1}{\lambda_q} \log U,$$

where U is uniform on $(0, 1)$. V is the product of a Bernoulli variable with mean $(1 - a\frac{\lambda_x}{\lambda_p})$ and an exponential with parameter λ_x. Similarly, W is the product of a Bernoulli variable with mean $(1 - a\frac{\lambda_y}{\lambda_q})$ and an exponential with parameter λ_y. Then, it turns out that

$$\mathrm{corr}(X,Y) = \rho = \frac{ab\lambda_x\lambda_y}{\lambda_p\lambda_q}\left(1 - \frac{\pi^2}{6}\right), \quad 0 \le \frac{a\lambda_x}{\lambda_p}, \frac{a\lambda_y}{\lambda_a} \le 1.$$

It follows that $\left(1 - \frac{\pi^2}{6}\right) \le \rho \le 0$. As a special case, instead of assuming V and W to be independent, we could have $W = V$ so that $\mathrm{cov}(X,Y) = ab\,\mathrm{cov}(P,Q) + \sigma_v^2$, where σ_v^2 is the variance of V.

Model 2

V and W are antithetic such that

$$V = \begin{cases} 0 & \text{if } U \le c \\ -\frac{1}{\lambda_v}\log\left(\frac{1-U}{1-c}\right) & \text{if } U > c \end{cases}, \qquad W = \begin{cases} 0 & \text{if } U \le d \\ -\frac{1}{\lambda_w}\log\left(\frac{U}{d}\right) & \text{if } U > d \end{cases},$$

for $0 \le c, d \le 1$. X and Y are exponential variables with parameters $\lambda_x = \lambda_v$ and $\lambda_y = \lambda_w$, and P and Q are independent exponential variables with λ_p and λ_q such that $c = \frac{a\lambda_x}{\lambda_p}$ and $1 - d = \frac{b\lambda_y}{\lambda_q}$. Then, it turns out that

$$\rho = \begin{cases} \int_c^d \log\frac{1-u}{1-c}\log\frac{u}{d}du - (1-c)d & \text{if } c < d \\ -(1-c)d & \text{if } d \le c \end{cases}.$$

The magnitude of negative correlation from Model 2 can exceed $(\pi^2/6) - 1$.

Model 3

In this model, we can make both P and Q and V and W antithetic, with P and Q being independent of V and W.

10.14.3 Fields of Application

This bivariate exponential model is useful in introducing dependence between the interarrival and service times in a queueing model and in a failure process involving multicomponent systems.

10.15 Moran–Downton Bivariate Exponential Distribution

This bivariate exponential distribution was first introduced by Moran (1967) and then popularized by Downton (1970). In fact, it is a special case of Kibble's bivariate gamma distribution discussed in Section 8.2. Many authors simply call it Downton's bivariate exponential distribution.

10.15.1 Formula of the Joint Density

The joint density function is

$$h(x,y) = \frac{1}{1-\rho}\exp[-(x+y)/(1-\rho)]I_0\left(\frac{2\sqrt{xy\rho}}{1-\rho}\right), \quad x,y \geq 0, \qquad (10.53)$$

where I_0 is the modified Bessel function of the first kind of order zero.

10.15.2 Formula of the Cumulative Distribution Function

Expressed as an infinite series, the joint cumulative distribution function is

$$H(x,y) = (1-e^{-x})(1-e^{-y}) + \sum_{j=0}^{\infty}\frac{\rho^{j+1}}{(j+1)^2}L_j^{(1)}(x)L_j^{(1)}(y)xye^{-(x+y)} \qquad (10.54)$$

for $x,y \geq 0$, where the $L_j^{(1)}$ are Laguerre polynomials defined earlier in Section 8.2.1.

10.15.3 Univariate Properties

Both marginal distributions are exponential.

10.15.4 Correlation Coefficients

The value ρ in (10.53) is in fact Pearson's product-moment correlation. As to the estimation of ρ, Al-Saadi and Young (1980) obtained the maximum likelihood estimator, the method of moments estimator, the sample correlation estimator, and the two bias-reduction estimators; see also Nagao and Kadoya (1971).

Balakrishnan and Ng (2001a) proposed two modified bias-reduction estimators, $\tilde{\rho}_5$ and $\tilde{\rho}_6$, and their jackknifed versions, $\tilde{\rho}_{5,J}$ and $\tilde{\rho}_{6,J}$, respectively. They carried out an extensive simulation study and found that both jackknife estimators reduce the bias substantially. Although $\tilde{\rho}_{6,J}$ seems to be the best estimator in terms of bias, it has a larger mean squared error. Overall, $\tilde{\rho}_6$ seems to be the best estimator, as it possesses a small bias as well as a smaller mean squared error than that of $\tilde{\rho}_{6,J}$. For the bivariate as well as multivariate forms of the Moran–Downton exponential distribution, Balakrishnan and Ng (2001b) studied the properties of estimators proposed by Al-Saadi and Young (1980) and Balakrishnan and Ng (2001a). They also used these estimators to propose pooled estimators in the multi-dimentional case and compared their performance with maximum likelihood estimators by means of Monte Carlo simulations.

10.15.5 Conditional Properties

The regression $E(Y|X = x)$ and the conditional variance are both linear in x; see Nagao and Kadoya (1971).

10.15.6 Moment Generating Function

The joint moment generating function is

$$M(s, t) = [(1 - s)(1 - t) - \rho st]^{-1}. \tag{10.55}$$

10.15.7 Regression

The regression is linear and is given by

$$E(Y|X = x) = 1 + \rho(x - 1). \tag{10.56}$$

10.15.8 Derivation

In the context of reliability studies, Downton (1970) used a successive damage model to derive this distribution as follows. Consider a system of two components, each being subjected to shocks, the interval between successive ones having an exponential distribution. Suppose the number of shocks N_1 and N_2 required to fail follows a bivariate geometric distribution with joint probability generating function

$$P(z_1, z_2) = \frac{z_1 z_2}{1 + \alpha + \beta + \gamma - \alpha z_1 - \beta z_2 - \gamma z_1 z_2}. \tag{10.57}$$

Write

$$(X, Y) = \left(\sum_{i=1}^{N_1} X_i, \sum_{i=1}^{N_2} Y_i \right), \tag{10.58}$$

where X_i and Y_i are the intershock intervals, mutually independent exponential variates. Then, the component lifetimes (X, Y) have a joint density as in (10.53).

Gaver (1972) gave a slightly different motivation for this distribution. He supposed that two types of shocks are occurring on an item of equipment, fatal and nonfatal. Repairs are only made after a fatal shock has occurred; repairs of all the nonfatal defects are also made then. If it is assumed that the two types of shocks both follow Poisson processes and the time for repair is the sum of the random number of exponential variates, then the time to failure and time to repair have Downton's bivariate exponential distribution. Expressed concisely, the positive correlation arises because the longer the time to fail, the longer the cumulated nonfatal damage.[4]

10.15.9 Fisher Information

We now introduce marginal parameters μ_1 and μ_2 to X and Y, respectively. Let $l = \log h(x, y)$ denote the log-likelihood function and

$$Q = E\left[\frac{\sqrt{\rho \mu_1 \mu_2 x y}}{1-\rho} I_1 \left\{ \frac{2\sqrt{\rho \mu_1 \mu_2 x y}}{1-\rho} \right\} I_0^{-1} \left\{ \frac{2\sqrt{\rho \mu_1 \mu_2 x y}}{1-\rho} \right\} \right]^2 .$$

[4] Gaver (1972) described another model that contrasts with this one because it leads to negative correlation between exponential variates. Instead of failures occurring because of shocks from outside, they occur due to built-in defects. Let it be supposed that, when a failure occurs, a detailed inspection of the equipment is made, and all the defects are discovered and repaired. A short time to failure is likely to have arisen because there were many built-in defects and is thus likely to be associated with a lengthy time of repair. Conversely, a long time to failure probably comes because there is only one defect, or very few, in which case the repair time will be short.

Shi and Lai (1998) have derived explicit formulas for the Fisher information matrix, and these are as follows:

$$\begin{aligned} E\left(\frac{\partial l}{\partial \mu_i}\right)^2 &= \frac{1}{\mu_i^2}\left\{\frac{2-4\rho}{(1-\rho)^2} - 1 + Q\right\},\ i=1,2,\\ E\left(\frac{\partial l}{\partial \rho}\right)^2 &= \frac{1}{(1-\rho)^2}\left\{-\frac{2+6\rho}{(1-\rho)^2} - 1 + \left(1+\frac{1}{\rho}\right)^2 Q\right\},\\ E\left(\frac{\partial l}{\partial \mu_i}\frac{\partial l}{\partial \rho}\right) &= \frac{1}{\mu_i(1-\rho)}\left\{\frac{1-5\rho}{(1-\rho)^2} - 1 + \left(1+\frac{1}{\rho}\right)Q\right\},\ i=1,2,\\ E\left(\frac{\partial l}{\partial \mu_1}\frac{\partial l}{\partial \mu_2}\right) &= \frac{1}{\mu_1\mu_2}\left\{\frac{1-3\rho}{(1-\rho)^2} - 1 + Q\right\}. \end{aligned}$$

10.15.10 Estimation of Parameters

We have discussed statistical inference on the correlation coefficient ρ in Section 10.15.4.

Suppose now that the scale parameters of X and Y are λ_1 and λ_2, respectively. Iliopoulos (2003) then considered the estimation of $\lambda = \lambda_2/\lambda_1$, which is the ratio of the means of the two marginal distributions. For Bayesian estimation of the ratio, see Iliopoulos and Karlis (2003).

10.15.11 Illustrations

An example of the surface of probability density has been given by Nagao and Kadoya (1971).

10.15.12 Random Variate Generation

Let

$$\left.\begin{aligned} X_1^* &= X(1-\rho)\\ X_2^* &= Y(1-\rho) \end{aligned}\right\} \tag{10.59}$$

for $0 \le \rho \le 1$, where X and Y are standard exponential variates. The joint characteristic function $\varphi(s,t)$ of X_1^* and X_2^* satisfies the relation

$$\varphi(s,t) = \psi(s)\psi(t)[(1-\rho) + \rho\psi(s,t)], \tag{10.60}$$

where $\psi(t)$ is the c.f. of the marginals, given by $[1 - it(1-\rho)]^{-1}$. To avoid the intermediate generation of bivariate normal variates, Paulson (1973) proposed the following method of random variate generation. Suppose $\varphi_n(s,t)$, $n = 1, 2, \ldots$, is a sequence of characteristic functions that satisfies the recurrence relation $\varphi_n(s,t) = \psi(s)\psi(t)[(1-\rho) + \rho\varphi(s,t)]$. This corresponds to the vector-valued r.v. $\mathbf{Y_n} = \mathbf{U_n} + \mathbf{V_n}\mathbf{Y_{n-1}}$, where $\{\mathbf{U_n}\}$ is a sequence of independent bivariate r.v.'s whose components are independent standard exponential variates, $\{\mathbf{V_n}\}$ is a sequence of matrix-valued r.v.'s that take on the value $\begin{pmatrix} 0 & 0 \\ 0 & 0 \end{pmatrix}$ and $\begin{pmatrix} 1 & 0 \\ 0 & 1 \end{pmatrix}$ with probabilities $(1-\rho)$ and ρ, respectively, $\mathbf{V_n}$ and $\mathbf{U_n}$ being mutually independent and $\mathbf{Y_0} = \mathbf{0}$ a vector of zero.

It is clear that $\varphi_n(s,t)$ converges to $\varphi(s,t)$, and hence $\{\mathbf{Y_n}\}$ converges in distribution to $(X_1^*, X_2^*)'$. Hence, (X_1^*, X_2^*) can be generated as accurately as desired by choosing an appropriate value of n; the value $n = 10$ seems to be quite satisfactory for ρ between 0 and 0.6. Finally, we set $(X,Y) = ((1-\rho)^{-1}X_1^*, (1-\rho)^{-1}X_2^*)$.

10.15.13 Remarks

- This distribution is a special case of the bivariate exponential distributions of Hawkes (1972), Paulson (1973), and Arnold (1975a) and Kibble's bivariate gamma as pointed out at the onset of this section.
- A detailed study was made by Nagao and Kadoya (1971).
- $X + Y$ is expressible as the sum of two independent exponential variates with parameters $(1+\sqrt{\rho})^{-1}$ and $(1-\sqrt{\rho})^{-1}$, respectively; see Lai (1985).
- A formula related to the use of this distribution in the "competing risk" context can be found in David and Moeschberger (1978, Section 4.2).
- $H(x,y) - F(x)G(y)$ increases as ρ increases; see Lai and Xie (2006, p. 323).
- The exact distribution of the product XY was obtained by Nadarajah (2006b).
- Sums, products, and ratios for Downton's bivariate exponential distribution were derived by Nadarajah (2006c).
- The distributions of the extreme statistics $\min(X,Y)$ and $\max(X,Y)$ were studied in Downton (1970).
- $\mu_{(2)} = \mathrm{E}\,(\max(X,Y))$ of the Moran–Downton distribution were compared with F-G-M and Marshall and Olkin's bivariate exponential distributions; see Kotz et al. (2003b).
- X and Y are SI (stochastically increasing) and thus are PQD; see Example 3.6 in Chapter 3.
- Hunter (2007) examined the effect of dependencies in the arrival process on the steady-state queue length process in single-server queueing models with exponential service time distribution. Four different models for the arrival process, each with marginally distributed exponential interarrivals to the queueing system, are considered. Two of these models are based on the upper and lower bounding joint distribution functions given by the Fréchet bounds for bivariate distributions with specified marginals, the third is based on Downton's bivariate exponential distribution, and the fourth is based on the usual $M/M/1$ model.

- Brusset and Temme (2007) obtained an analytically closed form of a quadratic objective function arising from a stochastic decision process under Moran and Downton's bivariate exponential distribution. The authors claimed that such objective functions often arise in operations research, supply chain management, or any other setting involving two random variables.
- The expression for the cumulative distribution function (10.54) shows that the joint distribution can be expanded diagonally in terms of Laguerre orthonormal polynomials.

10.15.14 Fields of Application

- **Queueing systems.** Consider a single-server queueing system such that the interarrival time X and the service time Y have exponential distributions, as is a common assumption in this context. If it is desired to introduce positive correlation (arising from cooperative service) into the model, Downton's distribution is a suitable choice; see Conolly and Choo (1979). Langaris (1986) applied it to a queueing system with infinitely many servers.
- **Markov dependent process.** Let X_i denote the time interval between the ith and $(i+1)$th events. Assuming that $\{X_i\}$ is a Markov chain, then $N(t)$ = number of events that occur in $(0, t]$ is a Wold point process. Lai (1978) used Downton's bivariate exponential model to describe the joint distribution of the lengths of successive time intervals.
- **Hydrology.** Nagao and Kadoya (1971) claimed that this distribution can be used for such pairs of hydrological quantities as a streamflow at two points on a river or rainfall at two locations; see also Yue et al. (2001).
- **Intensity and duration of a storm of rainfall.** Córdova and Rodríguez-Iturbe (1985) claimed that the exponential distributions for these variables have been shown to be sufficiently realistic. They argued that independence should not be assumed.

  - Firstly, it is empirically not true (correlations of 0.3 and 0.33 being found in datasets from Boconó, Venezuela, and Boston, Massachusetts).
  - Secondly, some important quantities are highly dependent on correlation: quantities such as the mean and variance of storm depth (product of intensity and duration) and the probability of nonzero storm surface runoff (if the soil is sufficiently dry and the storm is sufficiently small, there is no surface runoff).

- **Height and period of water waves.**

- The Rayleigh distribution, a special case of the Weibull, has a cumulative distribution function $F(x) = 1 - \exp(-x^2)$ and p.d.f. $f(x) = 2x\exp(-x^2)$ for $x > 0$. It is a common choice to describe both the periods squared and heights of waves of the sea—especially the latter, partly because there is theoretical support in the case of a narrow band spectrum. Consequently, Battjes (1971) suggested that a bivariate distribution with Rayleigh marginals may be used for the joint distribution of these variables and put forward the model in (10.53), appropriately transformed.
- Kimura (1981) suggested the Weibull-marginals version of this distribution for the height and period of water waves. Kimura performed experiments in which random waves were generated in a wave tank and their heights and periods measured and cross-tabulated. Although Kimura claimed that this distribution "shows good applicability for the principal part of the joint distribution," he also admitted that it failed at the edge—the main method of comparing theory with data in this work was by means of the conditional distribution of height given certain values of period, and the case presented showed reasonable agreement at $T = 0.8, 1.0, 1.2$, but poor agreement at $T = 0.6, 1.4$ (where T is the period expressed in units of its root-mean-square).
- Burrows and Salih (1987) included the Weibull-marginals version of this distribution among those they fitted to data from around the British Isles.

• **Height of water waves.** Kimura and Seyama (1985) used the Rayleigh-marginals version of this distribution to model the joint distribution of successive wave heights. Their concern was the overtopping of a sea wall that may occur when a group of high waves attacks it.

10.15.15 Tables or Algorithms

The algorithm for the probability integral of this distribution has been provided by Lai and Moore (1984). Tables for the conditional distribution function have been given by Nagao and Kadoya (1971).

10.15.16 Weibull Marginals

Kimura (1981) has given some properties of this distribution when the marginals are transformed to be a Weibull distribution. An expression for the general mixed moment has been given as well, from which the correlation coefficient can be readily obtained.

10.15.17 A Bivariate Laplace Distribution

The difference of i.i.d. exponential variates has a Laplace distribution with p.d.f. $f(x) = e^{-|x|/2}$ (with scale parameter omitted). This property has been used by Ulrich and Chen (1987) to obtain a distribution with Laplace marginals by setting

$$\left.\begin{array}{l} U = X_1 - X_2 \\ V = Y_1 - Y_2 \end{array}\right\}, \tag{10.61}$$

where the (X_i, Y_i) comes from distribution (10.53).

The joint m.g.f. can easily be shown to be

$$[(1-s)(1-t) - \rho st]^{-1}[(1+s)(1+t) - \rho st]^{-1},$$

but the p.d.f. that Ulrich and Chen obtained by inverting this is quite messy, involving a double infinite series.

Since it is easy to generate Downton variates (as $X = W_1^2 + W_2^2, Y = X_1^2 + Z_2^2$, where (W_i, Z_i) has a bivariate normal distribution), it is easy to generate variates from the Ulrich and Chen distribution by using (10.61).

10.16 Sarmanov's Bivariate Exponential Distribution

A general family of bivariate distributions with arbitrary marginals was introduced by Sarmanov (1966); the special case with exponential marginals was further studied in Lee (1996).

10.16.1 Formula of the Joint Density

The joint density function is

$$h(x, y) = f(x)g(y)\left\{1 + \omega\phi_1(x)\phi_2(y)\right\}, \tag{10.62}$$

where $\int_{-\infty}^{\infty} \phi_1(x)f(x)dx = 0$, $\int_{-\infty}^{\infty} \phi_2(y)g(y)dy = 0$, and ω satisfies the condition that $1 + \omega\phi_1(x)\phi_2(y) \geq 0$ for all x and y.

Lee (1996) gives the expression for the joint p.d.f. when the marginals are exponential,

$$f(x,y) = \lambda_1\lambda_2 e^{-(\lambda_1 x+\lambda_2 y)}\left\{1+\omega\left(e^{-x}-\tfrac{\lambda_1}{1+\lambda_1}\right)\left(e^{-y}-\tfrac{\lambda_2}{1+\lambda_2}\right)\right\}, \quad (10.63)$$

where $\frac{-(1+\lambda_1)(1+\lambda_2)}{\max(\lambda_1\lambda_2,1)} \leq \omega \leq \frac{(1+\lambda_1)(1+\lambda_2)}{\max(\lambda_1,\lambda_2)}$; $\phi_1(x) = e^{-x} - \frac{\lambda_1}{1+\lambda_1}$ and $\phi_2(y) = e^{-y} - \frac{\lambda_2}{1+\lambda_2}$.

Note: Similar to Moran and Downton's bivariate exponential, Sarmanov's bivariate distribution also has a diagonal expansion in terms of the orthogonal polynomials associated with their margianls.

10.16.2 Other Properties

Lee (1996) discussed four main properties of the Sarmanov family, two of which are of particular interest to us.

(a) The conditional distribution of Y given $X = x$ is

$$\Pr(Y \leq y|X = x) = G(y) + \omega\phi_1(x)\int_{-\infty}^{y} G(t)\phi_2(t)\,dt.$$

(b) The regression of Y on X is

$$E\,(Y|X = x) = \mu_Y + \omega\nu_Y\phi_1(x)$$

where $\nu_X = \int_{-\infty}^{\infty} t\phi_1(t)f(t)\,dt$, $\nu_Y = \int_{-\infty}^{\infty} t\phi_2(t)g(t)\,dt$.

(c) Further, it was shown that h is TP_2 if $\omega\phi_1'(x)\phi_2'(y) \geq 0$ for all x and y and RR_2 if $\omega\phi_1'(x)\phi_2'(y) \leq 0$ for all x and y. Here ϕ_1' and ϕ_2' are derivatives of ϕ_1 and ϕ_2, respectively.

For exponential marginals, we have

$$\begin{aligned} F(x,y) &= \\ (1-e^{-\lambda_1 x})(1-e^{-\lambda_2 y}) &+ \frac{\omega\lambda_1\lambda_2}{(1+\lambda_1)(1+\lambda_2)}\left(e^{-\lambda_1 x}-e^{-(\lambda_1+1)x}\right)\left(e^{-\lambda_2 y}-e^{-(\lambda_2+1)y}\right) \\ &\geq F_X(x)F_Y(y), \end{aligned}$$

whence X and Y are shown to be PQD if $0 \leq \omega \leq \frac{(1+\lambda_1)(1+\lambda_2)}{\max(\lambda_1,\lambda_2)}$.

10.17 Cowan's Bivariate Exponential Distribution

10.17.1 Formula of the Cumulative Distribution Function

The joint survival function is given by

$$\bar{H}(x,y) = \exp\left[-\frac{1}{2}\left(x+y+\sqrt{(x+y)^2-4\eta xy-4xy}\right)\right], \quad x,y \geq 0, \tag{10.64}$$

for $0 \leq \eta \leq 1$. Obviously, scale parameters can be introduced into this model if desired.

10.17.2 Formula of the Joint Density

The joint density function is

$$h(x,y) = \frac{1-\eta}{2S^3}\left\{4\eta xy + S[S(x+y)+x^2+y^2+2\eta xy]\right\}\exp[-(x+y+S)/2], \tag{10.65}$$

where $S^2 = (x+y)^2 - 4\eta xy$.

10.17.3 Univariate Properties

Both the marginal distributions are exponential.

10.17.4 Correlation Coefficients

Pearson's product-moment correlation coefficient is

$$-1 + \frac{2}{\eta}\left[1 + \frac{1-\eta}{\eta}\log(1-\eta)\right].$$

Spearman's correlation is

$$\frac{3}{8+\eta}\left[4-\eta-\frac{8(1-\eta)}{\xi}\log\frac{(\eta-\xi)(3\eta+\xi)}{(\eta+\xi)(3-\eta-\xi)}\right],$$

where $\xi = \sqrt{\eta(8+\eta)}$.

10.17.5 Conditional Properties

The conditional mean and standard deviation of Y, given $X = x$, are not of simple form, but graphs of these functions have been given by Cowan

(1987). A graph of $E(Y|X = x)$ when the marginals have been transformed to uniforms has also been presented by Cowan.

10.17.6 Derivation

The following derivation was presented by Cowan (1987). He derived it as the joint distribution of distances, in two directions separated by an angle α, to the nearest lines of a Poisson process in the plane. The association parameter η is $(1 + \cos \alpha)/2$.

10.17.7 Illustrations

Cowan (1987) presented contours of the p.d.f. and the cumulative distribution function for the case $\alpha = \pi/6$.

10.17.8 Remarks

An expression for the joint characteristic function of X and Y has been given by Cowan (1987). The minimum of X and Y is exponentially distributed.

10.17.9 Transformation of the Marginals

The cumulative distribution, when the marginals are transformed to uniform, is

$$H(u, v) = \sqrt{uv} \exp \left(-\frac{1}{2} \sqrt{(\log uv)^2 - 4\eta \log u \log v} \right). \tag{10.66}$$

10.18 Singpurwalla and Youngren's Bivariate Exponential Distribution

Singpurwalla and Youngren (1993) introduced the following form of bivariate exponential distribution.

10.18.1 Formula of the Cumulative Distribution Function

The joint survival function is given by

$$\bar{H}(x,y) = \sqrt{\frac{1 - m\min(x,y) + m\max(x,y)}{1 + m(x+y)}}\exp\{-m\max(x,y)\} \quad (10.67)$$

for $x, y \geq 0$, where m is a common parameter.

10.18.2 Formula of the Joint Density

The joint density function is

$$h(x,y)$$
$$= m^2 e^{-mx}\frac{(1+mx)\{(1-mx)^2 - m^2y^2\} + \{1+m(x-y)\}^2 - my(1+mx)}{\{1+m(x-y)\}^{3/2}\{1+m(x+y)\}^{5/2}}$$

on the sets of points $x > y$; on the set of points $y > x$, x is replaced by y and vice versa in the expression above. The joint density is undefined on the line $x = y$, which is similar to Marshall and Olkin's bivariate exponential distribution.

10.18.3 Univariate Properties

Both marginal distributions are exponential.

10.18.4 Derivation

It arises naturally in a shot-noise process environment.

10.18.5 Remarks

For further discussion on this bivariate distribution, one may refer to Kotz and Singpurwalla (1999).

10.19 Arnold and Strauss' Bivariate Exponential Distribution

The joint distribution was derived by Arnold and Strauss (1988). See also Arnold et al. (1999, p. 80) and Section 6.3.2 for other details.

10.19.1 Formula of the Joint Density

The joint density function is

$$h(x,y) = C(\beta_3)\beta_1\beta_2 e^{-\beta_1 x-\beta_2 y-\beta_1\beta_2 xy},\ x,y>0, \beta_i>0\,(i=1,2), \beta_3 \geq 0,$$

where $C(\beta_3) = \int_0^\infty \frac{e^{-u}}{1+\beta_3 u}du$. Alternatively, the density may be expressed as

$$h(x,y) = K\exp\{mxy - ax - by\},$$

where, for convergence, we must have $a, b > 0$ and $m \leq 0$, and K is a normalizing constant.

10.19.2 Formula of the Cumulative Distribution Function

The survival function is

$$\bar{H}(x,y) = \frac{C(\beta_3)e^{-\beta_1 x-\beta_2 y-\beta_1\beta_2 xy}}{(1+\beta_1\beta_3 x)(1+\beta_2\beta_3 y)C\left(\frac{\beta_3}{(1+\beta_1\beta_3 x)(1+\beta_2\beta_3 y)}\right)}.$$

10.19.3 Univariate Properties

Both marginals are not exponentials. See (6.20) and (6.21) for details.

10.19.4 Conditional Distribution

Both conditional distributions are exponentials.

10.19.5 Correlation Coefficient

In this case, we have $\rho \leq 0$; i.e., X and Y are negatively correlated.

10.19.6 Derivation

The derivation was based on the requirement that $X|Y = y$ and $Y|X = x$ are both exponential. Arnold and Strauss' model was motivated by the view that a researcher often has a better insight into the forms of conditional distributions rather than the joint distribution.

10.19.7 Other Properties

- extreme statistics were derived by Navarro et al. (2004). They were also given in Lai and Xie (2006, p. 313).
- The distribution of the product XY was derived in Nadarajah (2006b).
- The exact form of the Rényi and Shannon entropy of the distribution was given by Nadarajah and Zografos (2005).

10.20 Mixtures of Bivariate Exponential Distributions

Some bivariate distributions in this chapter (for example, Freund's bivariate exponential distribution) do not have exponential marginals. Often, their marginals are mixtures of exponential distributions. In this section, we consider various bivariate exponential distributions being mixed by another distribution.

10.20.1 Lindley and Singpurwalla's Bivariate Exponential Mixture

Lindley and Singpurwalla (1986) constructed a bivariate exponential mixture in the reliability context. Consider a system of two components (in series, or in parallel) that operates in an environment whose characteristics may affect its reliability. Suppose that, in environment i, the components' lifetimes are exponentially distributed with mean lifetime $1/\lambda_i$. Assume that λ has a gamma distribution over the population of environments. Then, the joint density of the lifetimes X and Y of the two components is of bivariate Pareto form (see Section 2.8.2),

$$h(x,y) = \frac{(a+1)(a+2)b^{a+1}}{(b+x+y)^{a+3}}, \tag{10.68}$$

where a and b are the parameters of the gamma distribution of λ. [If the components are in series, we will be especially interested in $\bar{H}(t,t)$, whereas if they are in parallel, $1 - H(t,t)$ will be of primary concern.]

This subject is taken further at other points in this book. Of course, Section 7.6 put this idea into the variables-in-common form. Generalizations of (10.68) can be given by taking the compounded (mixed) distributions to be gamma instead of exponential.

10.20.2 Sankaran and Nair's Mixture

Sankaran and Nair (1993) derived a bivariate exponential mixture distribution via two dependent exponential components operated in a random environment characteristic η. For a fixed η, X and Y have a type I bivariate Gumbel distribution with joint survival function

$$\bar{H}(x,y|\eta) = \exp[-\eta(\alpha_1 x + \alpha_2 y + \theta xy)]. \tag{10.69}$$

If η has a gamma distribution with scale parameter m and shape parameter p, then the resulting mixture distribution is given by

$$\bar{H}(x,y) = (1 + a_1 x + a_2 y + bxy)^{-p}, \quad x, y \geq 0, \tag{10.70}$$

where $a_i = \alpha_i/m, i = 1, 2$ and $b = \theta/m$. Equation (10.70) is simply the bivariate Lomax distribution discussed in Section 2.8.

10.20.3 Al-Mutairi's Inverse Gaussian Mixture of Bivariate Exponential Distribution

Al-Mutairi (1997) derived a parametric family of bivariate distributions for describing lifelengths of a system of two dependent components operating in a common environment where the conditional lifetime distribution follows Marshall and Olkin's bivariate exponential, and the common environment follows an inverse Gaussian distribution. Marshall and Olkin's bivariate exponential and Whitmore and Lee's (1991) bivariate distributions are then shown to be members of this family.

Al-Mutairi (1997) has given an excellent review and summary of bivariate exponential mixtures derived from the environment factor being a mixing distribution.

10.20.4 Hayakawa's Mixtures

Using a finite population of exchangeable two-component systems based on the indifference principle, Hayakawa (1994) proposed a class of bivariate exponential distributions that includes the Freund, Marshall and Olkin, and Block and Basu models as special cases. For an infinite population, Hayakawa's bivariate distributions can be written as

$$\bar{H}(x,y) = \int \bar{H}(x,y|\phi)dG(\phi),$$

where $\bar{H}(x,y|\phi)$ can be decomposed into an absolutely continuous part H_a and a singular part H_s, and G is the distribution function of the parameter ϕ.

This class of distributions includes mixtures of Freund's, Marshall and Olkin's, and Friday and Patil's distributions.

10.21 Bivariate Exponentials and Geometric Compounding Schemes

10.21.1 Background

Many bivariate exponential distributions may arise in one of the following two ways: first as a consequence of a random shock model due to Arnold (1975b) and second from a characteristic function equation due to Paulson (1973) and Paulson and Uppuluri (1972a,b). Block (1977a) used a bivariate geometric compounding mechanism to unify the approaches of previous authors.[5] Before describing it, we shall briefly describe probability generating functions and the bivariate geometric distributions.

10.21.2 Probability Generating Function

Let N be a non-negative integer-valued random variable. Then, the probability generating function (p.g.f.) of N is defined as $P(s) = E(s^N)$. Similarly,

[5] Marshall and Olkin (1967a), Downton (1970), Hawkes (1972), Paulson (1973), and Arnold (1975a,b). Another relevant work is that of Ohi and Nishida (1978).

the p.g.f. of (N_1, N_2) is defined as $P(s_1, s_2) = E(s_1^{N_1} s_2^{N_2})$. It is easy to show that if $\psi(t)$ is the c.f. of the i.i.d. r.v.'s X_i, then $P[\psi(t)]$ is the c.f. of the compound r.v. $\sum_{i=1}^{N} X_i$, which is the sum of a random number of X_i's.

10.21.3 Bivariate Geometric Distribution

A random variable N has a geometric distribution if

$$\Pr(N = n) = p^{n-1}(1-p)$$

for all positive integers n and some probability $p \in (0, 1)$.

If X's are i.i.d. r.v.'s with an exponential distribution, then $\sum_{i=1}^{N} X_i$ also has an exponential distribution if N has a geometric distribution.[6] Here, we say that a random variable (N_1, N_2) has a bivariate geometric distribution if the marginals are geometric distributions. (We are not concerned with the specific bivariate structure.)

10.21.4 Bivariate Geometric Distribution Arising from a Shock Model

Suppose we have two components receiving shocks in discrete cycles. (No assumption is made at this stage concerning the time interval between successive shocks.) In each cycle, there is a shock to both components in such a way that with probability p_{11} both components survive, with probability p_{10} the first survives and the second fails, with probability p_{01} the first fails and the second survives, and with probability p_{00} both fail. By conditioning on the outcome of the first cycle [Hawkes (1972) and Arnold (1975b)], we find that the number of shocks (N_1, N_2) to failure of components 1 and 2 satisfies the following functional equation in the p.g.f.:

$$g(s_1, s_2) = s_1 s_2 [p_{00} + p_{01} g(1, s_2) + p_{10} g(s_1, 1) + p_{11} g(s_1, s_2)]. \quad (10.71)$$

The survival function $\bar{H}(n_1, n_2) = \Pr(N_1 > n_1, N_2 > n_2)$ associated with (10.71) is given by

$$\bar{H}(n_1, n_2) = \begin{cases} p_{11}^{n_1} (p_{01} + p_{11})^{n_2 - n_1} & \text{if } n_1 \le n_2 \\ p_{11}^{n_2} (p_{10} + p_{11})^{n_1 - n_2} & \text{if } n_2 \le n_1 \end{cases}, \quad (10.72)$$

---

[6] The proof of this is simple. The p.g.f. of N is $\frac{(1-p)s}{1-ps}$ and the c.f. of X is $\psi(t) = (1 - i\mu t)^{-1}$, where $\mu = E(X)$. It follows from above that the c.f. of $\sum_{i=1}^{N} X_i$ is simply $\varphi(t) = P[\varphi(t)] = (1 - i\frac{\mu t}{1-p})^{-1}$, which is the c.f. of the exponential distribution with mean $\mu/(1-p)$.

where $p_{00} + p_{01} + p_{10} + p_{11} = 1$, $p_{10} + p_{11} < 1$, and $p_{01} + p_{11} < 1$. The p.g.f. of (N_1, N_2) is given by

$$g(s_1, s_2) = \frac{p_{00} s_1 s_2}{1 - p_{11} s_1 s_2} + \frac{p_{01}(p_{00} + p_{10}) s_2}{1 - (p_{01} + p_{11}) s_2} + \frac{p_{10}(p_{00} + p_{01}) s_1}{1 - (p_{10} + p_{11}) s_1}. \tag{10.73}$$

Equation (10.73) was also derived by Esary and Marshall (1973).

If $s_1 s_2$ in (10.73) is replaced by the characteristic function of any bivariate distribution with exponential marginals, $\psi(t_1, t_2)$,[7] we obtain the c.f. $\varphi(t_1, t_2) = E[\exp(it_1 X + it_2 Y)]$, which satisfies the functional equation

$$\varphi(t_1, t_2) = \psi(t_1, t_2)[p_{00} + p_{01}\psi(0, t_2) p_{10}\psi(t_1, 0) + p_{11}\psi(t_1, t_2)]. \tag{10.74}$$

By using the idea that the characteristic function of a random sum is the composition of ψ and P, we see that (10.74) corresponds to the compounding of the distribution with c.f. ψ with respect to the bivariate geometric distribution given in (10.72). In other words, (10.74) is the characteristic function equation of the bivariate random variable

$$(X, Y) = \left(\sum_{i=1}^{N_1} X_{i1}, \sum_{i=1}^{N_2} X_{i2} \right), \tag{10.75}$$

where (N_1, N_2) has the bivariate geometric distribution given in (10.72) and is independent of (X_{i1}, X_{i2}) $(i = 1, 2 \ldots)$, which are independent and identically distributed having exponential c.f. $\psi(t_1, t_2)$. (X, Y) in (10.75) has a bivariate exponential distribution since univariate geometric sums of exponential variables are exponential. X can now be interpreted in the following way (with a similar interpretation for Y). Component 1 is subjected to shocks that arrive according to a Poisson process. The probability that a shock will cause failure is $p_{01} + p_{00}$. Then, X represents the lifetime of the first component. For the bivariate exponential distribution of Marshall and Olkin, $X_{i1} = X_{i2}$ for all i.

10.21.5 Bivariate Exponential Distribution Compounding Scheme

The aim here is to provide a common framework (which we call the compounding scheme) for constructing various well-known bivariate exponential distributions. Two ingredients are used: (i) an input bivariate exponential with c.f. ψ and (ii) a set of non-negative parameters p_{ij} $(i, j = 0, 1)$ such that $p_{00} + p_{01} + p_{10} + p_{11} = 1, p_{01} + p_{11} < 1$, and $p_{10} + p_{11} < 1$. This leads to for-

[7] Thus, we can call this c.f. ψ a "generator" or an "input" function. In order to obtain a desired output, φ, we need to choose an appropriate ψ together with appropriate p_{00}, p_{01}, p_{01}, and p_{11}.

mulation (10.74). Alternatively, in the formulation of (10.75), we may regard the bivariate exponential distribution we want to construct as the compound distribution and the bivariate geometric distribution as the compounding distribution. Table 10.1 [adapted from Block (1977a)] summarizes the ways in which various bivariate distributions satisfy the equation and hence fit into the compounding scheme. This gives the name of these distributions as the "input" bivariate exponential distributions. The distributions are arranged in order from simplest to most complex. (The first entry indicates a pair of independent exponential variates that may be obtained from a pair of mutually completely dependent and identical exponential variates.)

Table 10.1 Bivariate exponential distributions arising from a compounding scheme. (The p_{ij}'s in the first row, ψ and φ, are as we have defined, and the notation is as in the original publications)

Distribution	$\psi(t_1, t_2)$	p_{00}	p_{01}	p_{10}
Independent marginals	$[1 - i\theta(t_1, t_2)]^{-1}$	0	θ/θ_1	θ/θ_2
Marshall and Olkin (1967a)	$[1 - i\theta(t_1, t_2)]^{-1}$	$\theta\lambda_{12}$	$\theta\lambda_1$	$\theta\lambda_2$
Downton (1970)	$[1 - \frac{it_1}{\mu_1(1+\gamma)}]^{-1}$ $\times[1 - \frac{it_2}{\mu_1(1+\gamma)}]^{-1}$	$(1+\gamma)^{-1}$	0	0
Hawkes (1972)	$(1 - iP_1t_1/\mu_1)^{-1}$ $\times(1 - iP_2t_1/\mu_2)^{-1}$ $(P_1 = p_{11} + p_{10},$ $P_2 = p_{11} + p_{01})$	p_{11}	p_{10}	p_{01}
Paulson (1973)	$(1 - i\theta_1t_1)^{-1}(1 - i\theta_2t_2)^{-1}$	a	c	b
Arnold (1975b)	$\psi(t_1, t_2) \in E_{n-1}^{(2)}$	p_{11}	p_{10}	p_{01}

$E_n^{(2)} = \left\{\psi(t_1, t_2) \in E_{n-1}^{(2)}\right\}$ with $E_0^{(2)} = \{\psi(t_1, t_2) = [1 + i\theta(t_1 + t_2)]^{-1}\}$

Some observations can be made from Table 10.1 about the bivariate exponential distributions. First, it is clear that Downton's distribution is a special case of the Hawkes and Paulson distributions. The latter two distributions can be seen to be the same, but they were derived differently. Arnold's class contains all of the exponential distributions given in the table. The first and second are in $E_1^{(2)}$; then, since the first is in $E_1^{(2)}$, it follows that the third, fourth, and fifth are in $E_2^{(2)}$.

10.21.6 *Wu's Characterization of Marshall and Olkin's Distribution via a Bivariate Random Summation Scheme*

Wu (1997) characterized Marshall and Olkin's bivariate exponential distribution using the same bivariate geometric distribution (10.72), which has a joint probability function

$$\Pr(N_1 = m, N_2 = n) = \begin{cases} p_{11}^{n-1}(p_{10}+p_{11})^{m-n-1}p_{10}(p_{01}+p_{00}) & \text{if } m > n \\ p_{11}^{m-1}p_{00} & \text{if } m = n \\ p_{11}^{m-1}(p_{01}+p_{11})^{n-m-1}p_{01}(p_{10}+p_{00}) & \text{if } m < n \end{cases}. \tag{10.76}$$

Let $\{X_{1i}\}$ and $\{X_{2i}\}$ be two sequences of random variables such that $E(X_{i1}) = \frac{1}{\lambda_1+\lambda_{12}}$ and $E(X_{2i}) = \frac{1}{\lambda_2+\lambda_{12}}$. Let (N_1, N_2) have a general bivariate geometric distribution in (10.76) with $p_{01} = \lambda_1\theta, p_{10} = \lambda_2\theta$ $(\theta > 0)$ and $p_{00} + p_{01} + p_{10} + p_{11} = 1$, $p_{10} + p_{11} < 1, p_{01} + p_{11} < 1$. Then, the distribution of

$$\left(p_{00}+p_{01}\right)\sum_{i=1}^{N_1} X_{1i}, (p_{00}+p_{10})\sum_{i=1}^{N_2} X_{2i}\right)$$

converges weakly, as $\theta \to 0$, to Marshall and Olkin's bivariate exponential distribution.

10.22 Lack of Memory Properties of Bivariate Exponential Distributions

The univariate exponential distribution is characterized by the functional equation

$$\bar{F}(s+\delta) = \bar{F}(s)\bar{F}(y), s, \delta > 0. \tag{10.77}$$

This is referred to as the *lack of memory* (LOM) property.

Equation (10.77) can be rewritten as

$$\Pr(X > s + \delta | X > \delta) = \Pr(X > s), \tag{10.78}$$

so that the probability of surviving an additional time s for a component of age δ is the same as for a new component. A bivariate analogue of (10.78) can be written as

$$\Pr(X > s_1 + \delta, Y > s_2 + \delta | X > \delta, Y > \delta) = \Pr(X > s_1, Y > s_2) \tag{10.79}$$

which asserts that the joint survival probability of a pair of components, each of age δ, is the same as that of a pair of new components. We may write (10.79) as

$$\bar{H}(s_1+\delta, s_2+\delta) = \bar{H}(s_1, s_2)\bar{H}(\delta, \delta), \;\; s_1, s_2, \delta > 0. \tag{10.80}$$

This is also termed the LOM property, which was briefly discussed in Section 10.5.16. The bivariate exponential of Marshall and Olkin is the only distribution with exponential marginals that satisfies (10.80); see, for example, Barlow and Proschan (1981, pp. 130–131). This functional equation has many possible solutions if the requirement of exponential marginals is not imposed. The class of possible solutions of the equation is characterized by Ghurye and Marshall (1984). Apart from Marshall and Olkin's BVE, other known solutions of (10.80) are the following:

- Freund's bivariate distribution (described in Section 10.3),
- ACBVE of Block and Basu (see Section 10.6),
- PSE of Proschan and Sullo (described in Section 10.3.15), and
- BEE of Friday and Patil (1977), which includes Freund's and Marshall and Olkin's BVE and ACBVE as special cases; see Section 10.9.

The formula for the BEE is given in (10.35), and the survival function of the PSE (Proschan–Sullo extension) is given by

$$\bar{H}(x,y) = \begin{cases} (\lambda_1+\lambda_2-\lambda_2')^{-1}\left\{\lambda_1 e^{-(\lambda_1+\lambda_2-\lambda_2')x-(\lambda_0+\lambda_2')y} + (\lambda_2-\lambda_2')e^{-\lambda y}\right\} \\ \hfill \text{for } x \le y \\ (\lambda_1+\lambda_2-\lambda_1')^{-1}\left\{\lambda_2 e^{-(\lambda_0+\lambda_1')x-(\lambda_1+\lambda_2-\lambda_1')y} + (\lambda_1-\lambda_1')e^{-\lambda x}\right\} \\ \hfill \text{for } x \ge y, \end{cases}$$

where $\lambda_0, \lambda_1, \lambda_2, \lambda_1', \lambda_2'$ are all positive, $\lambda = \lambda_0+\lambda_1+\lambda_2$, and $x, y \ge 0$. Its density is given in Section 10.3.15.

The LOM property in (10.80) is characterized by Block and Basu (1974) and Block (1977b). The characterization theorem can be stated as follows. Let (X, Y) be a non-negative bivariate random vector with absolutely continuous marginal distribution functions F and G, and let $U = \min(X, Y)$ and $V = X - Y$. Then, the LOM property holds if and only if there is a $\theta > 0$ such that

- U and V are independent and
- U is an exponential variate with mean θ^{-1}.

Further, if the LOM property holds, the distribution of V is given by

$$\Pr(V \le v) = \begin{cases} F(v) + \theta^{-1}f(v) & \text{if } v \ge 0 \\ 1 - G(-v) - \theta^{-1}g(-v) & \text{if } v < 0 \end{cases}, \tag{10.81}$$

where f and g are the density functions corresponding to F and G, respectively. Another point is that if $\bar{H}(x,y)$ has the LOM property, so does the survival function $\frac{1}{2}[\bar{H}(x,y)+\bar{H}(y,x)]$.

10.22.1 Extended Bivariate Lack of Memory Distributions

Ghurye (1987) provided an extended version of the LOM property by imposing

$$\bar{H}(x,y) = \bar{A}(\min(x,y))\bar{K}(x-y), \quad x,y \geq 0,$$

where

$$\bar{K}(\omega) = \begin{cases} \bar{G}(\omega) & \text{for } \omega > 0 \\ \bar{H}(|\omega|) & \text{for } \omega < 0 \end{cases}$$

and $\bar{A}, \bar{G}, \bar{H}$ are survival functions of $\min(X,Y), X$, and Y, respectively.

Yet another extension of the LOM property was obtained by Ghurye (1987) by generalizing (10.80) to

$$\bar{H}(x+t, y+t) = \bar{H}(x,y)\bar{H}(t,t)\bar{B}(t;x,y),$$

where $\bar{B}$ is an age factor.

Another extension of the LOM property is due to Raja Rao et al. (1993), and they called it the "setting the clock back to zero property." The type I bivariate Gumbel exponential distribution possesses this particular property. Incidentally, this bivariate exponential distribution is characterized by another form of bivariate lack of memory as well.

10.23 Effect of Parallel Redundancy with Dependent Exponential Components

Suppose X and Y are two lifetimes of a parallel system of two components. Kotz et al. (2003a) considered the effectiveness of redundancy when two components are dependent. They have shown that the degree of correlation affects the increase in the mean time for parallel redundancy when the two component lifetimes are positively quadrant dependent. Let $T = \max(X,Y)$ and $E(T)$ represent the mean time to failure of the parallel system.

Suppose now that X and Y are both exponentially distributed with unit mean and have a joint bivariate distribution specified by

(1) F-G-M bivariate exponential distribution,
(2) Marshall and Olkin's bivariate exponential distribution, and
(3) Downton's bivariate exponential distribution.

It has been shown that for $0 \leq \alpha \leq 1$, X and Y of the F-G-M distributions are PQD whereas the two-component lifetimes of the two other distributions are always PQD. Table 10.2 summarizes the comparisons among the three distributions under consideration.

Table 10.2 Mean lifetime $E(T)$ and range of correlation for three bivariate exponential distributions

Bivariate distribution	Mean lifetime	Range of ρ
F-G-M	$1.5 - \rho/3$	$0 \leq \rho < 1/4$
Marshall and Olkin	$1.5 - \rho/2$	$0 \leq \rho \leq 1$
Downton	$1 + \frac{\sqrt{(1-\rho}}{2}$	$0 \leq \rho < 1$

It can be easily shown that

$$1.5 - \frac{\rho}{2} \leq 1 + \frac{1}{2}(1-\rho)^{1/2}, \quad 0 \leq \rho < 1,$$

and

$$1.5 - \frac{\rho}{3} \leq 1 + \frac{1}{2}(1-\rho)^{1/2}, \quad 0 \leq \rho < 3/4.$$

It follows at once that Downton's model yields a higher mean time to failure than either Marshall and Olkin's model or the F-G-M model.

10.23.1 Mean Lifetime under Gumbel's Type I Bivariate Exponential Distribution

The joint survival function is

$$\bar{H}(x,y) = e^{-x-y-\theta xy}, \quad x, y \geq 0, \ 0 \leq \theta \leq 1.$$

Clearly, X and Y are NQD (negatively quadrant dependent). Kotz et al. (2003b) showed that

$$E(T) = 2 - e^{1/\theta}\sqrt{\frac{\pi}{\theta}}\left[1 - \Phi\left(\sqrt{\frac{2}{\theta}}\right)\right], \tag{10.82}$$

where $\Phi(\cdot)$ is the standard normal distribution function. Table 10.3 below provides some numerical values of $E(T)$ for some selected values of ρ.

Table 10.3 Some numerical values of $E(T)$ for Gumbel's type I model

ρ	0	−0.1	−0.25	−0.4
$E(T)$	1.5	1.527	1.570	1.615

10.24 Stress–Strength Model and Bivariate Exponential Distributions

10.24.1 Basic Idea

Let X be the strength of a component subject to a stress Y. The component fails if at any moment the applied stress (or load) is greater than its strength. The stress is a function of the environment to which the component is subjected, whereas strength depends on material properties, manufacturing procedures, and so on. The reliability R that the strength of a component exceeds the stress is

$$R = \Pr(X > Y) \tag{10.83}$$

This model was considered by Birnbaum (1956) and has since found an increasing number of applications in many different areas, especially in the structural and aircraft industries. Johnson (1988) has given a review on this subject. A similar formulation occurs in hydrology. Let X be the input of a pollutant into a river of flow Y, and assume that the flora and fauna of the river are sensitive to the concentration of the pollutant. Then, $\Pr(X > cY)$ is the relevant quantity; see, for example, Plate and Duckstein (1987, pp. 56–58).

In many situations, the distribution of Y (or both X and Y) is completely known, except possibly for a few unknown parameters, and it is desired to obtain parameter estimates. Church and Harris (1970), Downton (1973), Owen et al. (1964), Govindarajulu (1968), and Reiser and Guttman (1986, 1987) have all considered the problem of stress and strength under the assumption that X and Y have independent normal distributions. Because in many physical situations, especially in the reliability context, exponential and related distributions provide more realistic models, it is desirable to obtain estimators of R for these cases. Some results for the exponential case are given by Tong (1974), Kelley et al. (1976), and Basu (1981).

Most of the authors have assumed that X and Y are independent. However, it is more realistic to assume some form of dependence between X and Y since they may be influenced by a common environmental factor. We shall now evaluate R for two models in which X and Y are correlated.

For theory and applications of the stress–strength model, see the monograph by Kotz et al. (2003b).

10.24.2 Marshall and Olkin's Model

Suppose X and Y have the joint bivariate exponential distribution of Marshall and Olkin given by (10.13). In the notation of Section 10.5.6, $X > Y$ if and only if $Z_2 < \min(Z_1, Z_3)$. Hence,

$$R = \Pr\{Z_2 < \min(Z_1, Z_2)\} = \lambda_2/\lambda,$$

where $\lambda = \lambda_1 + \lambda_2 + \lambda_{12}$; see Basu (1981). Similarly, $\Pr(X < Y) = \lambda_1/\lambda$. Also, $\Pr(X \geq Y) = R + \Pr(X = Y) = (\lambda_2 + \lambda_{12})/\lambda$. Awad et al. (1981) have given estimators for $\Pr(X < Y)$, $\Pr(X > Y)$, and $\Pr(X = Y)$.

10.24.3 Downton's Model

Suppose X and Y have a joint density given by (10.53). Lai (1985) showed that

$$R = (1 - \rho^2) \sum_{i=0}^{\infty} \frac{B_\alpha(i+1, i+1)}{B(i+1, i+1)}, \tag{10.84}$$

where $\alpha = \mu_2/(\mu_1 + \mu_2)$, μ_1 and μ_2 are parameters of X and Y, respectively, and B_x is the incomplete beta function.

Note that for $\rho = 0$ (i.e., when X and Y are independent), $R = \mu_2/(\mu_1 + \mu_2)$, as expected [Tong (1974)].

10.24.4 Two Dependent Components Subjected to a Common Stress

Consider a parallel system of two components having strengths X and Y that are subjected to a common stress Z that is independent of the strength of the components. Then the reliability of the system R is given by $R = \Pr(Z < \max(X, Y))$. Hanagal (1996) estimated R when (X, Y) have different bivariate exponential models proposed by Marshall and Olkin (1967a), Block and Basu (1974), Freund (1961), and Proschan and Sullo (1974). The distribution of Z is assumed to be either exponential or gamma. The asymptotic normal (AN) distributions of these estimates were obtained. Hanagal (1996) also gave a numerical study for obtaining the MLE of R in all four bivariate models when the common stress (Z) is exponentially distributed.

10.24.5 A Component Subjected to Two Stresses

Hanagal (1999) considered the reliability of a component subjected to two different stresses that are independent of the strength of a component. The distribution of stresses follows a bivariate exponential distribution. If Z is the strength of a component subjected to two stresses (X, Y), then the reliability of the component is given by $R = \Pr\{(X+Y) < Z\}$. Hanagal estimated R when (X, Y) follows different bivariate exponential models proposed by Marshall and Olkin (1967a), Block and Basu (1974), Freund (1961), and Proschan and Sullo (1974). The distribution of Z is assumed to be exponential. The asymptotic normality of these estimates of R was obtained.

10.25 Bivariate Weibull Distributions

Because the univariate Weibull distribution is obtained from the univariate exponential by a simple transformation of the variable, bivariate distributions with Weibull marginals can readily be obtained by starting with any of the bivariate distributions having exponential marginals and then transforming X and Y appropriately.

There are many types of bivariate Weibull distributions, and they can be categorized into five classes, as follows. In each case, X and Y are individually taken to have Weibull distributions.

- **Class C1.** X and Y are independent.
- **Class C2.** $X = \min(X_1, X_2)$, $Y = \min(X_2, X_3)$, where the X_i's are independent, but not necessarily identically distributed, Weibull variates.
- **Class C3.** $\min(aX, bY)$ has a Weibull distribution for every $a > 0$ and $b > 0$.
- **Class C4.** $\min(X, Y)$ has a Weibull distribution.
- **Class C5.** The class of all bivariate distributions with Weibull marginals.

Lee (1979) described the classes above and showed that the inclusions C1 $\subset$ C2 $\subset$ C3 $\subset$ C4 $\subset$ C5 are strict. Another comprehensive treatment can be found in Block and Savits (1980). For a brief overview, see Jensen (1985).

Applications can easily be imagined in any of the fields where bivariate distributions with exponential marginals are used, especially those such as reliability, where the univariate Weibull is a popular generalization of the univariate exponential. The Weibull distribution, and others derived by the extreme-value approach, can be plausibly applied to the strength of materials. Warren (1979) suggested using a bivariate distribution with Weibull marginals for the joint distribution of modulus of elasticity and modulus of rupture for lumber.

Hougaard (1986, 1989) presented a bivariate (in a multivariate setting) distribution with joint survival function

$$\bar{H}(x,y) = \exp\left\{-\left(\theta_1 x^p + \theta_2 y^p\right)^k\right\}, \ p \geq 0, k \geq 0, x, y \geq 0. \tag{10.85}$$

For the Gumbel form of bivariate Weibull distribution, Begum and Khan (1977) have discussed the marginal and joint distributions of concomitants of order statistics and their single moments.

We note that it is easy to generate a bivariate Weibull distribution by a marginal transformation, a popular method for constructing a bivariate model with specified marginals [Lai (2004)].

10.25.1 Marshall and Olkin (1967)

This is obtained from the power law transformation of the well-known bivariate exponential distribution (BVE) studied in Marshall and Olkin (1967a). The joint survivor function with Weibull marginals is given as

$$\bar{H}(x,y) = \exp\{-[\lambda_1 x^{\alpha_1} + \lambda_2 y^{\alpha_2} + \lambda_{12} \max(x^{\alpha_1}, y^{\alpha_2})]\}, \tag{10.86}$$

where $\lambda_i > 0$, $\alpha_i \geq 0$, $\lambda_{12} \geq 0$; $i = 1, 2$. This bivariate Weibull reduces to the bivariate exponential distribution when $\alpha_1 = \alpha_2 = 1$.

Lu (1992) considered Bayes estimation for the model above for censored data.

10.25.2 Lee (1979)

A related model due to Lee (1979) involves the transformation $X = X_1/c_1$, $Y = X_2/c_2$ assuming (X_1, X_2) has a joint survival function given in (10.86) and $\alpha_1 = \alpha_2 = \alpha$. The new model (X, Y) has a joint survival function given by

$$\bar{H}(x,y) = \exp\left\{-\left[\lambda_1 c_1^\alpha x^\alpha + \lambda_2 c_2^\alpha y^\alpha + \lambda_{12} \max(c_1^\alpha x^\alpha, c_2^\alpha y^\alpha)\right]\right\}, \tag{10.87}$$

where $c_i > 0, \lambda_i > 0, \lambda_{12} \geq 0$.

Yet another related model due to Lu (1989) has the survival function

$$\bar{H}(x,y) = \exp\left\{-\lambda_1 x^{\alpha_1} - \lambda_2 y^{\alpha_2} - \lambda_0 \max(x,y)^{\alpha_0}\right\}, \tag{10.88}$$

where $\lambda' s > 0, \alpha_i \geq 0$; $i = 0, 1, 2$. This can be seen as a slight modification (or generalization) of Marshall and Olkin's bivariate exponential distribution due to the exponent in the third term having a new parameter.

10.25.3 Lu and Bhattacharyya (1990): I

A general model proposed by Lu and Bhattacharyya (1990) has the form

$$\bar{H}(x,y) = \exp\left\{-(x/\beta_1)^{\alpha_1} - (y/\beta_2)^{\alpha_2} - \delta w(x,y)\right\}, \tag{10.89}$$

where $\alpha_i > 0$, $\beta_i \geq 0$, $\delta \geq 0$; $i = 1, 2$.

Different forms for the function of $w(t_1, t_2)$ yield a family of models. One form for $w(x, y)$ is the following:

$$w(x,y) = \left[(x/\beta_1)^{\alpha_1/m} + (y/\beta_2)^{\alpha_2/m}\right]^m, \quad m > 0. \tag{10.90}$$

This yields the following survival function for the model:

$$\bar{H}(x,y) = \exp\left\{-(x/\beta_1)^{\alpha_1} - (y/\beta_2)^{\alpha_2} - \delta[(x/\beta_1)^{\alpha_1/m} + (y/\beta_2)^{\alpha_2/m}]^m\right\}. \tag{10.91}$$

10.25.4 Farlie–Gumbel–Morgenstern System

The Farlie–Gumbel–Morgenstern system of distributions [Hutchinson and Lai (1990, Section 5.2) and Kotz et al. (2000, Section 44.13)] is given by

$$\bar{H}(x,y) = \bar{F}(x)\bar{G}(y)\left\{1 + \gamma\left[1 - \bar{F}(x)\right]\left[1 - \bar{G}(y)\right]\right\}, \quad -1 < \gamma < 1. \tag{10.92}$$

With $\bar{F}(x) = \exp\{-x^{\alpha_1}\}$, $\bar{G}(y) = \exp\{-y^{\alpha_2}\}, \alpha_i > 0$; this yields a bivariate Weibull model with the marginals being a standard Weibull in the sense of Johnson et al. (1994).

10.25.5 Lu and Bhattacharyya (1990): II

A different type of bivariate Weibull distribution due to Lu and Bhattacharyya (1990) is given by

$$\bar{H}(x,y) = \left[1 + \left[\{[\exp(x/\beta_1)^{\alpha_1}] - 1\}^{1/\gamma} + \{\exp[(y/\beta_2)^{\alpha_2}] - 1\}^{1/\gamma}\right]^{\gamma}\right]^{-1}. \tag{10.93}$$

This model has a random hazard interpretation, but for no value of γ, the model yields independence between the two variables.

10.25.6 Lee (1979): II

Lee (1979) proposed the bivariate Weibull distribution

$$\bar{H}(x,y) = \exp\{-(\lambda_1 x^{\alpha_1} + \lambda_2 y^{\alpha_2})^{\gamma}\}, \tag{10.94}$$

where $\alpha_i > 0$, $0 < \gamma \le 1$, $\lambda_i > 0$, $x, y \ge 0$.

The model was used by Hougaard (1986) to analyze tumor data.

A slight reparametrization of the model by letting $\lambda_i = \left(\frac{1}{\theta_i}\right)^{\beta_i/\delta}$, $\alpha_i = \beta_i/\delta, i = 1, 2$, and $\gamma = \delta$ in (10.94) gives

$$\bar{H}(x,y) = \exp\left\{-\left[\left(\frac{x}{\theta_1}\right)^{\beta_1/\delta} + \left(\frac{y}{\theta_2}\right)^{\beta_2/\delta}\right]^{\delta}\right\}. \tag{10.95}$$

The model has been applied to an analysis of field data under a two-dimensional warranty in which age and usage are used simultaneously to determine the eligibility of a warranty claim [Jung and Bai (2007)].

The product moments were derived by Nadarajah and Mitov (2003). Surprisingly, their expressions are rather simple.

10.25.7 Comments

- Johnson et al. (1999) proposed to use the bivariate Weibull model (10.95) as a candidate to model the strength properties of lumber.
- Johnson and Lu (2007) used a "proof load design" to estimate the parameters of the preceding model.
- The bivariate Weibull observations (X, Y) from the distribution (10.95) can be obtained through

$$X = U^{\delta/\beta_1} V^{1/\beta_1} \theta_1, \quad Y = (1-U)^{\delta/\beta_2} V^{1/\beta_2} \theta_2,$$

  where U and V are independent uniform variates.

10.25.8 Applications

Applications can easily be imagined in any of the fields where bivariate distributions with exponential marginals are used, especially those such as reliability where the univariate Weibull is a popular generalization of the univariate exponential. The Weibull distribution, and others derived by the extreme-value approach, can be plausibly applied to the strength of materials. Warren (1979) suggested using a bivariate distribution with Weibull marginals for the joint distribution of modulus of elasticity and modulus of rupture for lumber.

10.25.9 Gamma Frailty Bivariate Weibull Models

Bjarnason and Hougaard (2000) considered two gamma frailty bivariate Weibull models. A frailty model is a random effects model for survival data. The key assumption is that the dependence between two individual lifetime variables X and Y is caused by the frailty Z representing unobserved common risk factors and that conditional on Z, X, and Y are independent. Because the frailty is not observed, it is assumed to follow some distribution, typically a gamma distribution.

In their paper, Bjarnason and Hougaard assumed that Z has a gamma distribution with both scale and shape parameters given by δ and that, conditional on Z, X and Y are Weibull with scale parameters $Z\lambda_1$, $Z\lambda_2$, respectively but with a common shape parameters γ. It is easy to show that the joint (unconditional) survival function of X and Y is given by

$$\bar{H}(x,y) = \{1 + (\lambda_1 x^\gamma + \lambda_2 y^\gamma)/\delta\}^{-\delta}. \qquad (10.96)$$

Equation (10.96) is clearly a bivariate Burr distribution. For $\gamma = 1$, it reduces to a bivariate Pareto distribution. The authors then proceeded to derive the Fisher information for the distribution above.

A second gamma frailty model was derived by Bjarnason and Hougaard (2000) that gives rise to a bivariate Weibull model that has a Clayton copula:

$$\bar{H}(x,y) = \left(e^{\lambda_1 x^\gamma/\delta} + e^{\lambda_2 y^\gamma/\delta} - 1\right)^{-\delta}. \qquad (10.97)$$

The Fisher information was also found for this bivariate Weibull model by the authors.

Both of these two models were used on the catheter infection data of McGilchrist and Aisbett (1991).

10.25.10 Bivariate Mixture of Weibull Distributions

Patra and Dey (1999) constructed a class of bivariate (in the multivariate setting) in which each component has a mixture of Weibull distributions.

10.25.11 Bivariate Generalized Exponential Distribution

A univariate distribution with survival function $S(x) = (1 - e^{-x})^\theta$, $x \geq 0$, $\theta > 0$, is called a generalized exponential distribution, and is denoted by GED(θ); see Gupta and Kundu (1999). Sarhan and Balakrishnan (2007) then constructed a bivariate generalized exponential distribution with joint survival function of the form

$$S(x_1, x_2) = e^{-\theta_0 z} \left\{1 - (1 - e^{-x_1})^{\theta_1}\right\} \left\{1 - (1 - e^{-x_2})^{\theta_2}\right\},$$
$$x_1, x_2 > 0, \;\; \theta_0, \theta_1, \theta_2 > 0,$$

where $z = \max(x_1, x_2)$, and then discussed many of its properties such as marginals, conditionals and moments. They also discussed mixtures of these distributions.

References

1. Achcar, J.A.: Inferences for accelerated tests considering a bivariate exponential distribution. Statistics **26**, 269–283 (1995)
2. Achcar, J.A., Leandro, R.A.: Use of Markov chain Monte Carlo methods in Bayesian analysis of the Block and Basu bivariate exponential distribution. Annals of the Institute of Statistical Mathematics **50**, 403–416 (1998)
3. Achcar, J.A., Santander, L.A.M.: Use of approximate Bayesian methods for the Block and Basu bivariate exponential distribution. Journal of the Italian Statistical Society **3**, 233–250 (1993)
4. Adachi, K., Kodama, M.: Availability analysis of two-unit warm standby system with inspection time. Microelectronics and Reliability **20**, 449–455 (1980)
5. Al-Mutairi, D.K.: Properties of an inverse Gaussian mixture of bivariate exponential distribution and its generalization. Statistics and Probability Letters **33**, 359–365 (1997)
6. Al-Saadi, S.D., Young, D.H.: Estimators for the correlation coefficient in a bivariate exponential distribution. Journal of Statistical Computation and Simulation **11**, 13–20 (1980)
7. Arnold, B.C.: Parameter estimation for a multivariate exponential distribution. Journal of the American Statistical Association **63**, 848–852 (1968)
8. Arnold, B.C.: A characterization of the exponential distribution by geometric compounding. Sankhyā, Series A **37**, 164–173 (1975a)
9. Arnold, B.C.: Multivariate exponential distributions based on hierarchical successive damage. Journal of Applied Probability **12**, 142–147 (1975b)
10. Arnold, B.C.: Pareto Distributions. International Co-operative Publishing House, Fairland, Maryland (1983)
11. Arnold, B.C., Castillo, E., Sarabia, J.M.: Conditional Specification of Statistical Models. Springer-Verlag, New York (1999)
12. Arnold, B.C., Strauss, D.: Pseudolikelihood estimation. Sankhyā, Series B **53**, 233–243 (1988)
13. Assaf, D., Langberg, N.A., Savits, T.H., Shaked, M.: Multivariate phase-type distributions. Operations Research **32**, 688–702 (1984)

14. Awad, A.M., Azzam, M.M., Hamdan, M.A.: Some inference results on $\Pr(X < Y)$ in the bivariate exponential model. Communications in Statistics: Theory and Methods **10**, 2515–2525 (1981)
15. Azlarov, T.A., Volodin, N.A.: Characterization Problems Associated with the Exponential Distribution. Springer-Verlag, New York (Original Soviet edition was dated 1982) (1986)
16. Baggs, G.E., Nagaraja, H.N.: Reliability properties of order statistics from bivariate exponential distributions. Communications in Statistics: Stochastic Models **12**, 611–631 (1996)
17. Balakrishnan, N., Basu, A.P. (eds.): The Exponential Distribution: Theory, Methods and Applications. Taylor and Francis, Philadelphia (1995)
18. Balakrishnan, N., Ng, H.K.T.: Improved estimation of the correlation coefficient in a bivariate exponential distribution. Journal of Statistical Computation and Simulation **68**, 173–184 (2001a)
19. Balakrishnan, N., Ng, H.K.T.: On estimation of the correlation coefficient in Moran–Downton multivariate exponential distribution. Journal of Statistical Computation and Simulation **71**, 41–58 (2001b)
20. Barlow, R.E., Proschan, F.: Techniques for analyzing multivariate failure data. In: The Theory and Applications of Reliability, Volume 1, C.P. Tsokos and I.N. Shimi (eds.), pp. 373–396. Academic Press, New York (1977)
21. Barlow, R.E., Proschan, F.: Statistical Theory of Reliability and Life Testing. To Begin With, Silver Spring, Maryland (1981)
22. Basu, A.P.: The estimation of $P(X < Y)$ for distributions useful in life testing. Naval Research Logistics Quarterly **28**, 383–392 (1981)
23. Basu, A.P.: Multivariate exponential distributions and their applications in reliability. In: Handbook of Statistics, Volume 7, Quality Control and Reliability, P.R. Krishnaiah, and C.R. Rao (eds.), pp. 467–476. North-Holland, Amsterdam (1988)
24. Battjes, J.A.: Run-up distributions of waves breaking on slopes. Journal of the Waterways, Harbors and Coastal Engineering Division, Transactions of the American Society of Civil Engineers **97**, 91–114 (1971)
25. Becker, P.J., Roux, J.J.J.: A bivariate extension of the gamma distribution. South African Statistical Journal **15**, 1–12 0(1981)
26. Beg, M.I., Balasubramanian, K.: Concomitants of order statistics in the bivariate exponential distributions of Marshall and Olkin. Calcutta Statistical Association Bulletin **46**, 109–115 (1996)
27. Begum, A.A., Khan, A.H.: Concomitants of order statistics from Gumbel's bivariate Weibull distributions. Calcutta Statistical Association Bulletin **47**, 132–138 (1977)
28. Bemis, B.M., Bain, L.J., Higgins, J.J.: Estimation and hypothesis testing for the parameters of a bivariate exponential distribution. Journal of the American Statistical Association **67**, 927–929 (1972)
29. Bhattacharya, S.K., Holla, M.S.: Bivariate life-testing models for two component systems. Annals of the Institute of Statistical Mathematics **15**, 37–43 (1963)
30. Bhattacharyya, A.: Modelling exponential survival data with dependent censoring. Sankhyā, Series A **59**, 242–267 (1997)
31. Bhattacharyya, G.K., Johnson, R.A.: Maximum likelihood estimation and hypotheisis testing in the bivariate exponential model of Marshall and Olkin. Technical Report No. 276, Department of Statistics, University of Wisconsin, Madison (1971)
32. Bhattacharyya, G.K., Johnson, R.A.: On a test of independence in a bivariate exponential distribution. Journal of the American Statistical Association **68**, 704–706 (1973)
33. Bilodeau, M., Kariya, T.: LBI tests of independence in bivariate exponential distributions. Annals of the Institute of Statistical Mathematics **46**, 127–136 (1994)
34. Birnbaum, Z.W.: On a use of the Mann–Whitney statistic. In: Proceedings of the 3rd Berkeley Symposium on Mathematical Statistics and Probability, Volume 1, J. Neyman (ed.), pp. 13–17. University of California Press, Berkeley (1956)

35. Biswas, S., Nair, G.: A generalization of Freund's model for a repairable paired component based on a bivariate Geiger Muller (G.M.) counter. Microelectronics and Reliability **24**, 671–675 (1984)
36. Bjarnason, H., Hougaard, P.: Fisher information for two gamma frailty bivariate Weibull models. Lifetime Data Analysis **6**, 59–71 (2000)
37. Block, H.W.: Continuous multivariate exponential extensions. In: Reliability and Fault Tree Analysis. Theoretical and Applied Aspects of System Reliability and Safety Assessment, R.E. Barlow, J.R. Fussell, and N.D. Singpurwalla (eds.), pp. 285–306. Society for Industrial and Applied Mathematics, Philadelphia (1975)
38. Block, H.W.: A family of bivariate life distributions. In: The Theory and Applications of Reliability, Volume 1, C.P. Tsokos, and I.N. Shimi (eds.), pp. 349–371. Academic Press, New York (1977a)
39. Block, H.W.: A characterization of a bivariate exponential distribution. Annals of Statistics **5**, 808–812 (1977b)
40. Block, H.W., Basu, A.P.: A continuous bivariate exponential extension. Journal of the American Statistical Association **69**, 1031–1037 (1974)
41. Block, H.W., Savits, T.H.: Multivariate increasing failure rate distributions. Annals of Probability **8**, 793–801 (1980)
42. Boland, P.J.: An arrangement increasing property of the Marshall–Olkin bivariate exponential. Statistics and Probability Letters **37**, 167–170 (1998)
43. Brusset, X., Temme, N.M.: Optimizing an objective function under a bivariate probability model. European Journal of Operational Research **179**, 444–458 (2007)
44. Burrows, R., Salih, B.A.: Statistical modelling of long term wave climates. In: Twentieth Coastal Engineering Conference, Proceedings, Volume I, B.L. Edge (ed.), pp. 42–56. American Society of Civil Engineers, New York (1987)
45. Chen, D., Lu, J.C., Hughes-Oliver, J.M., Li, C.S.: Asymptotic properties of maximum likelihood estimates for bivariate exponential distribution and mixed censored data. Metrika **48**, 109–125 (1998)
46. Church, J.D., Harris, B.: The estimation of reliability from stress–strength relationship. Technometrics **12**, 49–54 (1970)
47. Conolly, B.W., Choo, Q.H.: The waiting time process for a generalized correlated queue with exponential demand and service. SIAM Journal on Applied Mathematics **37**, 263–275 (1979)
48. Conway, D.: Bivariate distribution contours. In: Proceedings of the Business and Economic Statistics Section, pp. 475–480. (1981)
49. Cordóva, J.R., Rodriguez-Iturbe, I.: On the probabilistic structure of storm surface runoff. Water Resources Research **21**, 755–763 (1985)
50. Cowan, R.: A bivariate exponential distribution arising in random geometry. Annals of the Institute of Statistical Mathematics **39**, 103–111 (1987)
51. Cox, D.R., Lewis, P.A.W.: The Statistical Analysis of Series of Events. Chapman and Hall, London (1966)
52. Cox, D.R., Oakes, D.: Analysis of Survival Data. Chapman and Hall, London (1984)
53. Cuadras, C.M., Augé, J.: A continuous general multivariate distribution and its properties. Communications in Statistics: Theory and Methods **10**, 339–353 (1981)
54. David, H.A., Moeschberger, M.L.: The Theory of Competing Risks. Griffin, London (1978)
55. Downton, F.: Bivariate exponential distributions in reliability theory. Journal of the Royal Statistical Society, Series B **32**, 408–417 (1970)
56. Downton, F.: The estimation of $Pr(Y < X)$ in the normal case. Technometrics **15**, 551–558 (1973)
57. Ebrahimi, N.: Analysis of bivariate accelerated life test data from bivariate exponential of Marshall and Olkin. American Journal of Mathematical and Management Sciences **6**, 175–190 (1987)

58. Esary, J.D., Marshall, A.W.: Multivariate geometric distributions generated by a cumulative damage process. Report NP555EY73041A, Naval Postgraduate School, Monterey, California (1973)
59. Franco, M., Vivo, J.M.: Reliability properties of series and parallel systems from bivariate exponential models. Communications in Statistics: Theory and Methods **39**, 43–52 (2002)
60. Franco, M., Vivo, J.M.: Log-concavity of the extremes from Gumbel bivariate exponential distributions. Statistics **40**, 415–433 (2006)
61. Franco, M., Vivo, J.M.: Generalized mixtures of gamma and exponentials and reliability properties of the maximum from Friday and Patil bivariate model. Communications in Statistics: Theory and Methods **36**, 2011–2025 (2007)
62. Freund, J.E.: A bivariate extension of the exponential distribution. Journal of the American Statistical Association **56**, 971–977 (1961)
63. Friday, D.S., Patil, G.P.: A bivariate exponential model with applications to reliability and computer generation of random variables. In: The Theory and Applications of Reliability, Volume 1, C.P. Tsokos and I.N. Shimi (eds.), pp. 527–549. Academic Press, New York (1977)
64. Gaver, D.P.: Point process problems in reliability. In: Stochastic Point Processes: Statistical Analysis, Theory, and Applications, P.A.W. Lewis (ed.), pp. 774–800. John Wiley & Sons, New York (1972)
65. Gaver, D.P., Lewis, P.A.W.: First-order autoregressive gamma sequences and point processes. Advances in Applied Probability **12**, 727–745 (1980)
66. Ghurye, S.G.: Some multivariate lifetime distributions. Advances in Applied Probability **19**, 138–155 (1987)
67. Ghurye, S.G., Marshall, A.W.: Shock processes with aftershocks and multivariate lack of memory. Journal of Applied Probability **21**, 786–801 (1984)
68. Goel, L.R., Gupta, R., Singh, S.K.: A two-unit parallel redundant system with three modes and bivariate exponential lifetimes. Microelectronics and Reliability **24**, 25–28 (1984)
69. Goel, L.R., Gupta, R., Singh, S.K. Availability analysis of a two-unit (dissimilar) parallel system with inspection and bivariate exponential life times. Microelectronics and Reliability **25**, 77–80 (1985)
70. Govindarajulu, Z.: Distribution-free confidence bounds for $P(X < Y)$. Annals of the Institute of Statistical Mathematics **20**, 229–238 (1968)
71. Gross, A.J., Lam, C.F.: Paired observations from a survival distribution. Biometrics **37**, 505–511 (1981)
72. Gumbel, E.J.: Bivariate exponential distributions. Journal of the American Statistical Association **55**, 698–707 (1960)
73. Gupta, A.K., Nadarajah, S.: Sums, products, and ratios for Freund's bivariate exponential distribution. Applied Mathematics and Computation **173**, 1334–1349 (2006)
74. Gupta, R.D., Kundu, D.: Generalized exponential distribution. Australian and New Zealand Journal of Statistics **41**, 173–188 (1999)
75. Hagen, E.W.: Common-mode/common-cause failure: A review. Annals of Nuclear Energy **7**, 509–517 (1980)
76. Hanagal, D.D.: Some inference results in modified Freund's bivariate exponential distribution. Biometrical Journal **34**, 745–756 (1992)
77. Hanagal, D.D.: Some inference results in an absolutely continuous multivariate exponential model of Block and Basu. Statistics and Probability Letters **16**, 177–180 (1993)
78. Hanagal, D.D.: Estimation of system reliability from stress–strength relationship. Communications in Statistics: Theory and Methods **25**, 1783–1797 (1996)
79. Hanagal, D.D.: Estimation of reliability of a component subjected to bivariate exponential stress. Statistical Papers **40**, 211–220 (1999)

80. Hanagal, D.D., Kale, B.K.: Large sample tests of independence for an absolutely continuous bivariate exponential model. Communications in Statistics: Theory and Methods **20**, 1301–1313 (1991a)
81. Hanagal, D.D., Kale, B.K.: Large sample tests of λ_3 in the bivariate exponential distribution. Statistics and Probability Letters **12**, 311–313 (1991b)
82. Hashino, M.: Formulation of the joint return period of two hydrologic variates associated with a Poisson process. Journal of Hydroscience and Hydraulic Engineering **3**, 73–84 (1985)
83. Hashino, M., Sugi, Y.: Study on combination of bivariate exponential distributions and its application (in Japanese). Scientific Papers of the Faculty of Engineering, University of Tokushima **29**, 49–57 (1984)
84. Harris, R.: Reliability applications of a bivariate exponential distribution. Operations Research **16**, 18–27 (1968)
85. Hawkes, A.G.: A bivariate exponential distribution with applications to reliability. Journal of the Royal Statistical Society, Series B **34**, 129–131 (1972)
86. Hayakawa, Y.: The construction of new bivariate exponential distributions from a Bayesian perspective. Journal of the American Statistical Association **89**, 1044–1049 (1994)
87. Holla, M.S., Bhattacharya, S.K.: A bivariate gamma distribution in life testing. Defence Science Journal (India) **15**, 65–74 (1965)
88. Hougaard, P.: A class of multivariate failure time distributions. Biometrika **73**, 671–678 (Correction **75**, 395) (1986)
89. Hougaard, P.: Fitting multivaraite failure time distribution. IEEE Transactions on Reliability **38**, 444–448 (1989)
90. Hunter, J.: Markovian queues with correlated arrival processes. Asia-Pacific Journal of Operational Research **24**, 593–611 (2007)
91. Hutchinson, T.P., Lai, C.D.: Continuous Bivariate Distributions, Emphasizing Applications. Rumsby Scientific Publishing, Adelaide, Australia (1990)
92. Hyakutake, H.: Statistical inferences on location parameters of bivariate exponential distributions. Hiroshima Mathematical Journal **20**, 527–547 (1990)
93. Iliopoulos, G.: Estimation of parametric functions in Downton's bivariate exponential distribution. Journal of Statistical Planning and Inference **117**, 169–184 (2003)
94. Iliopoulos, G., Karlis, D.: Simulation from the Bessel distribution with applications. Journal of Statistical Computation and Simulation **73**, 491–506 (2003)
95. Itoi, T., Murakami, T., Kodama, M., Nishida, T.: Reliability analysis of a parallel redundant system with two dissimilar correlated units. Technology Reports of Osaka University **26**, 403–409 (1976)
96. Iyer, S.K., Manjunath, D., Manivasakan, R.: Bivariate exponential distributions using linear structures. Sankhyā, Series A **64**, 156–166 (2002)
97. Jensen, D.R.: Multivariate Weibull distributions. In: Encyclopedia of Statistical Sciences, Volume 6, S. Kotz and N.L. Johnson (eds.), pp. 132–133. John Wiley and Sons, New York (1985)
98. Johnson, N.L., Kotz, S.: Distributions in Statistics: Continuous Multivariate Distributions. John Wiley and Sons, New York (1972)
99. Johnson, N.L., Kotz, S., Balakrishnan, N.: Continuous Univariate Distributions, Vol. 1, 2nd edition, John Wiley and Sons, New York (1994)
100. Johnson, R.A.: Stress–strength models for reliability. In: Handbook of Statistics, Volume 7, Quality Control and Reliability, P.R. Krishnaiah and C.R. Rao (eds.), pp. 27–54. North-Holland, Amsterdam (1988)
101. Johnson, R.A., Evans, J.W., Green, D.W.: Some bivariate distributions for modeling the strength properties of lumber, Research Paper FPL–RP-575, Forest Product Laboratory, The United States Department of Agriculture (1999)
102. Johnson, R.A., Lu, W.: Proof load designs for estimation of dependence in bivariate Weibull model. Statistics and Probability Letters **77**, 1061–1069 (2007)

103. Jung, M., Bai, D.S.: Analysis of field data under two-dimensional warranty. Reliability Engineering and System Safety **92**, 135–143 (2007)
104. Kelley, G.D., Kelley, J.A., Schucany, W.R.: Efficient estimation of $P(Y < X)$ in the exponential case. Technometrics **18**, 359–360 (1976)
105. Khodr, H.M., Melián, J.A., Quiroz, A.J., Picado, D.C., Yusta, J.M., Urdanetaet, A.J.: A probabilistic methodology for distribution substation location. IEEE Transactions on Power Systems **18**, 388–393 (2003)
106. Kimura, A.: Joint distribution of the wave heights and periods of random sea waves. Coastal Engineering in Japan **24**, 77–92 (1981)
107. Kimura, A., Seyama, A.: Statistical properties of short-term overtopping. In: Nineteenth Coastal Engineering Conference, Proceedings, Volume I, B.L. Edge (ed.), pp. 532–546. American Society of Civil Engineers, New York (1985)
108. Klein, J.P., Basu, A.P.: Estimating reliability for bivariate exponential distributions. Sankhyā, Series B **47**, 346–353 (1985)
109. Klein, J.P., Moeschberger, M.L.: The independence assumption for a series or parallel system when component lifetimes are exponential. IEEE Transactions on Reliability **R-35**, 330–334 (1986)
110. Klein, J.P., Moeschberger, M.L.: Bounds on net survival probabilities for dependent competing risks. Biometrics **44**, 529–538 (1988)
111. Kotz, S., Balakrishnan, N., Johnson, N.L.: Continuous Multivariate Distributions, Volume 1, 2nd edition. John Wiley and Sons, New York (2000)
112. Kotz, S., Lai, C.D., Xie, M.: On the effect of redundancy for systems with dependent components. IIE Transactions **35**, 1103–1110 (2003a)
113. Kotz, S., Lumelskii, Y., Pensky, M.: The Stress-Strength Model and Its Generalizations: Theory and Applications. World Scientific Publishing, River Edge, New Jersey (2003b)
114. Kotz, S., Singpurwalla, N.D.: On a bivariate distribution with exponential marginals. Scandinavian Journal of Statistics **26**, 451-464 (1999)
115. Kumar, A., Subramanyam, A.: Tests of independence in a bivariate exponential distribution. Metrika **61**, 47–62 (2005)
116. Lai, C.D.: An example of Wold's point processes with Markov-dependent intervals. Journal of Applied Probability **15**, 748–758 (1978)
117. Lai, C.D.: On the reliability of a standby system composed of two dependent exponential components. Communications in Statistics: Theory and Methods **14**, 851–860 (1985)
118. Lai, C.D.: Constructions of continuous bivariate distributions. Journal of the Indian Society for Probability and Statistics **8**, 21–43 (2004)
119. Lai, C.D., Moore, T.: Probability integrals of a bivariate gamma distribution. Journal of Statistical Computation and Simulation **19**, 205–213 (1984)
120. Lai, C.D., Xie, M.: Stochastic Ageing and Dependence for Reliability. Springer-Verlag, New York (2006)
121. Langaris, C.: A correlated queue with infinitely many servers. Journal of Applied Probability **23**, 155–165 (1986)
122. Lawrance, A.J., Lewis, P.A.W.: Simple dependent pairs of exponential and uniform random variables. Operations Research **31**, 1179–1197 (1983)
123. Lee, L.: Multivariate distributions having Weibull properties. Journal of Multivariate Analysis **9**, 267–277 (1979)
124. Lee, M.L.T.: Properties and applications of the Sarmanov family of bivariate distributions. Communications in Statistics: Theory and Methods **25**, 1207–1222 (1996)
125. Lindley, D.V., Singpurwalla, N.D.: Multivariate distributions for the life lengths of components of a system sharing a common environment. Journal of Applied Probability **23**, 418–431 (1986)
126. Lu, J.C.: Weibull extensions of the Freund and Marshall–Olkin bivariate exponential models. IEEE Transactions on Reliability **38**, 615–619 (1989)

127. Lu, J.C.: Bayes parameter estimation for the bivariate Weibull model of Marshall and Olkin for censored data. IEEE Transactions on Reliability **41**, 608–615 (1992)
128. Lu, J.C.: A new plan for life-testing two-component parallel systems. Statistics and Probability Letters **34**, 19–32 (1997)
129. Lu, J.C., Bhattacharyya, C.K.: Some new constructions of bivariate Weibull models. Annals of the Institute of Statistical Mathematics **42**, 543–559 (1990)
130. Lu, J., Bhattacharyya, G.K.: Inference procedures for bivariate exponential distribution model of Gumbel. Statistics and Probability Letters **12**, 37–50 (1991a)
131. Lu, J., Bhattacharyya, G.K.: Inference procedures for bivariate exponential distribution model of Gumbel based on life test of component and system. Journal of Statistical Planning and Inference **27**, 283–296 (1991b)
132. Ma, C.: Multivariate survival functions characterized by constant product of mean remaining lives and hazard rates. Metrika **44**, 71–83 (1996)
133. Marshall, A.W., Olkin, I.: A multivariate exponential distribution. Journal of the American Statistical Association **62**, 30–44 (1967a)
134. Marshall, A.W., Olkin, I.: A generalized bivariate exponential distribution. Journal of Applied Probability **4**, 291–302 (1967b)
135. Marshall, A.W., Olkin, I.: Multivariate exponential distributions, Marshall–Olkin. In: Encyclopedia of Statistical Sciences, Volume 6, S. Kotz and N.L. Johnson (eds.), pp. 59–62. John Wiley and Sons, New York (1985)
136. McGilchrist, C.A., Aisbett, C.W.: Regression with frailty in survival analysis. Biometrics **47**, 461–466 (1991)
137. Moeschberger, M.L.: Life tests under dependent competing causes of failure. Technometrics **16**, 39–47 (1974)
138. Moran, P.A.P.: Testing for correlation between non-negative variates. Biometrika **54**, 385–394 (1967)
139. Muliere, P., Scarsini, M.: Characterization of a Marshall–Olkin type class of distributions. Annals of the Institute of Statistical Mathematics **39**, 429–441 (1987)
140. Nadarajah, S.: Information matrix for the bivariate Gumbel distribution. Applied Mathematics and Computation **172**, 394–405 (2006a)
141. Nadarajah, S.: Exact distributions of XY for some bivariate exponential distributions. Statistics **40**, 307–324 (2006b)
142. Nadarajah, S.: Sums, products, and ratios for Downton's bivariate exponential distribution. Stochastic Environmental Research and Risk Assessment **20**, 164–170 (2006c)
143. Nadarajah, S., Ali, M.M.: The distribution of sums, products and ratios for Lawrance and Lewis's bivariate exponential random variables. Computational Statistics and Data Analysis **50**, 3449–3463 (2006)
144. Nadarajah, S., Kotz, S.: Block and Basu's bivariate exponential distribution with application to drought data. Probability in the Engineering and Informational Sciences **21**, 143–145 (2007)
145. Nadarajah, S., Mitov, K.: Product moments of multivariate random vectors. Communications in Statistics: Theory and Methods **32**, 47–60 (2003)
146. Nadarajah, S., Zografos, K.: Expressions for Rényi and Shannon entropies for bivariate distributions. Information Sciences **170**, 173–189 (2005)
147. Nagao, M., Kadoya, M.: Two-variate exponential distribution and its numerical table for engineering application. Bulletin of the Disaster Prevention Research Institute, Kyoto University **20**, 183–215 (1971)
148. Nagaraja, H.N., Abo-Eleneen, Z.A.: Fisher information in the Farlie–Gumbel–Morgenstern type bivariate exponential distribution. In: Uncertainty and Optimality, K.B. Misra (ed.), pp. 319–330. World Scientific Publishing, River Edge, New Jersey (2002)
149. Navarro, J., Ruiz, J.M., Sandoval, C.J.: Distributions of k-out-of-n systems with dependent components. In: International Conference on Distribution Theory, Order Statistics, and Inference in Honor of Barry C. Arnold, June 16–18, 2004 (2004)

150. Nelsen, R.B.: An Introduction to Copulas, 2nd edition. Springer-Verlag, New York (2006)
151. Obretenov, A.: Characterization of the multivariate Marshall–Olkin exponential distribution. Probability and Mathematical Statistics **6**, 51–56 (1985)
152. O'Cinneide, C.A., Raftery, A.E.: A continuous multivariate exponential distribution that is multivariate phase type. Statistics and Probability Letters **7**, 323–325 (1989)
153. Ohi, F., Nishida, T.: Bivariate shock models and its application to the system reliability analysis. Mathematica Japonica **23**, 109–122 (1978)
154. Ohi, F., Nishida, T.: Bivariate Erlang distribution functions. Journal of the Japan Statistical Society **9**, 103–108 (1979)
155. O'Neill, T.J.: Testing for symmetry and independence in a bivariate exponential distribution. Statistics and Probability Letters **3**, 269–274 (1985)
156. Osaki, S.: A two-unit parallel redundant system with bivariate exponential lifetimes. Microelectronics and Reliability **20**, 521–523 (1980)
157. Osaki, S., Yamada, S., Hishitani, J.: Availability theory for two-unit nonindependent series systems subject to shut-off rules. Reliability Engineering and System Safety **25**, 33–42 (1989)
158. Owen, D.B., Crasewell, R.J., Hanson, D.L.: Nonparametric upper confidence bounds for $\Pr\{Y < X\}$ when X and Y are normal. Journal of the American Statistical Association **59**, 906–924 (1964)
159. Patra, K., Dey, D.K.: A multivariate mixture of Weibull distributions in reliability modelling. Statistics and Probability Letters **45**, 225–235 (1999)
160. Paulson, A.S.: A characterization of the exponential distribution and a bivariate exponential distribution. Sankhyā, Series A **35**, 69–78 (1973)
161. Paulson, A.S., Uppuluri, V.R.R.: A characterization of the geometric distribution and a bivariate geometric distribution. Sankhyā, Series A **34**, 297–300 (1972a)
162. Paulson, A.S., Uppuluri, V.R.R.: Limits laws of a sequence determined by a random difference equation governing a one-component system. Mathematical Biosciences **13**, 325–333 (1972b)
163. Plate, E.J., Duckstein, L.: Reliability in hydraulic design. In: Engineering Reliability and Risk in Water Resources, L. Duckstein and E.J. Plate (eds.), pp. 27–60. Nijhoff, Dordrecht (1987)
164. Platz, O.: A Markov model for common-cause failures. Reliability Engineering **9**, 25–31 (1984)
165. Proschan, F., Sullo, P.: Estimating the parameters of a bivariate exponential distribution in several sampling situations. In: Reliability and Biometry: Statistical Analysis of Life Lengths, F. Proschan and R.J. Serfling (eds.), pp. 423–440. Society for Industrial and Applied Mathematics, Philadelphia (1974)
166. Proschan, F., Sullo, P.: Estimating the parameters of a multivariate exponential distribution. Journal of the American Statistical Association **71**, 465–472 (1976)
167. Raftery, A.E.: A continuous multivariate exponential distribution. Communications in Statistics: Theory and Methods **13**, 947–965 (1984)
168. Raftery, A.E.: Some properties of a new continuous bivariate exponential distribution. Statistics and Decisions, Supplement Issue No. 2, 53–58 (1985)
169. Rai, K., Van Ryzin, J.: Multihit models for bivariate quantal responses. Statistics and Decisions **2**, 111–129 (1984)
170. Raja Rao, B., Damaraju, C.V., Alhumound, J.M.: Setting the clock back to zero property of a class of bivariate distributions. Communications in Statistics: Theory and Methods **22**, 2067–2080 (1993)
171. Ramanarayanan, R., Subramanian, A.: A 2-unit cold standby system with Marshall–Olkin bivariate exponential life and repair times. IEEE Transactions on Reliability **30**, 489–490 (1981)
172. Reiser, B., Guttman, I.: Statistical inference for $\Pr(Y < X)$: The normal case. Technometrics **28**, 253–257 (1986)

173. Reiser, B., Guttman, I.: A comparison of three point estimators for $\Pr(Y < X)$ in the normal case. Computational Statistics and Data Analysis **5**, 59–66 (1987)
174. Roux, J.J.J., Becker, P.J.: Compound distributions relevant to life testing. In: Statistical Distributions in Scientific Work, Volume 4: Models, Structures, and Characterizations, C. Taillie, G.P. Patil, and B.A. Baldessari (eds.), pp. 111–124. Reidel, Dordrecht (1981)
175. Roy, D.: A characterization of Gumbel's bivariate exponential and Lindley and Singpurwalla's bivariate Lomax distributions. Journal of Applied Probability **26**, 886–891 (1989)
176. Roy, D., Gupta, R.P.: Bivariate extension of Lomax and finite range distributions through characterization approach. Journal of Multivariate Analysis **59**, 22–33 (1996)
177. Roy, D., Mukherjee, S.P.: Characterizations of some bivariate life-distributions. Journal of Multivariate Analysis **28**, 1–8 (1989)
178. Ryu, K.W.: An extension of Marshall and Olkin's bivariate exponential distribution. Journal of the American Statistical Association **88**, 1458–1465 (1993)
179. Samanta, M.: A characterization of the bivariate exponential distribution. Journal of the Indian Society of Agricultural Statistics **27**, 67–70 (1975)
180. Sarhan, A.M., Balakrishnan, N.: A new class of bivariate distributions and its mixture. Journal of Multivariate Analysis **98**, 1508–1527 (2007)
181. Sarmanov, O.V.: Generalized normal correlation and two-dimensional Frechét classes. Doklady (Soviet Mathematics) **168**, 596–599 (1966)
182. Sankaran, P.G., Nair, U.N.: A bivariate Pareto model and its applications to reliability. Naval Research Logistics **40**, 1013–1020 (1993)
183. Sarkar, S.K.: A continuous bivariate exponential distribution. Journal of the American Statistical Association **82**, 667–675 (1987)
184. Saw, J.G.: A bivariate exponential density and a test that two identical components in parallel behave independently. Technical Report No. 22, Department of Industrial and Systems Engineering, University of Florida, Gainesville (1969)
185. Shamseldin, A.A., Press, S.J.: Bayesian parameter and reliability estimation for a bivariate exponential distribution, parallel sampling. Journal of Econometrics **24**, 363–378 (1984)
186. Singpurwalla, N.D., Youngren, M.A.: Multivariate distributions induced by dynamic environments. Scandinavian Journal of Statistics **20**, 251–261 (1993)
187. Spurrier, J.D., Weier, D.R.: Bivariate survival model derived from a Weibull distribution. IEEE Transactions on Reliability **30**, 194–197 (1981)
188. Steel, S.J., Le Roux, N.J.: A reparameterisation of a bivariate gamma extension. Communications in Statistics: Theory and Methods **16**, 293–305 (1987)
189. Sugasaw, Y., Kaji, I.: Light maintenance for a two unit parallel redundant system with bivariate exponential lifetimes. Microelectronics and Reliability **21**, 661–670 (1981)
190. Sun, K., Basu, A.P.: Characterizations of a family of bivariate distributions. In: Advances of Reliability, A.P. Basu (ed.), pp. 395–409. Elsevier Science, Amsterdam (1993)
191. Tong, H.: A note on the estimation of $\Pr(Y < X)$ in the exponential case. Technometrics **16**, 625 (Correction **17**, 395) (1974)
192. Tosch, T.J., Holmes, P.T.: A bivariate failure model. Journal of the American Statistical Association **75**, 415–417 (1980)
193. Tukey, J.W.: Exploratory Data Analysis. Addison-Wesley, Reading, Massachusetts (1977)
194. Ulrich, G., Chen, C-C.: A bivariate double exponential distribution and its generalizations. In: Proceedings of the Statistical Computing Section, pp. 127–129 (1987)
195. Wang, R.T.: A reliability model for the multivariate exponential distributions. Journal of Multivariate Analysis **98**, 1033–1042 (2007)
196. Warren, W.G.: Some recent developments relating to statistical distributions in forestry and forest products research. In: Statistical Distributions in Ecological Work, J.K. Ord, G.P. Patil, and C. Taillie (eds.), pp. 247–250. International Co-operative Publishing House, Fairland, Maryland (1979)

197. Whitmore, G.A., Lee, M-L.T.: A multivariate survival distribution generated by an inverse Gaussian mixture of exponentials. Technometrics **33**, 39–50 (1991)
198. Wu, C.: New characterization of Marshall–Olkin-type distributions via bivariate random summation scheme. Statistics and Probability Letters **34**, 171–178 (1997)
199. Yue, S., Ouarda, T.B.M.J., Bobée, B.: A review of bivariate gamma distributions for hydrological application. Journal of Hydrology **246**, 1–18 (2001)

Chapter 11
Bivariate Normal Distribution

11.1 Introduction

In introductory statistics courses, one has to know why the (univariate) normal distribution is important—especially that the random variables that occur in many situations are approximately normally distributed and that it arises in theoretical work as an approximation to the distribution of many statistics, such as averages of independent random variables. More or less, the same reasons apply to the bivariate normal distribution. "But the prime stimulus has undoubtedly arisen from the strange tractability of the normal model: a facility of manipulation which is absent when we consider almost any other multivariate data-generating mechanism."—Barnett (1979). We may also note the following views expressed by different authors:

- "In multivariate analysis, the only distribution leading to tractable inference is the multivariate normal"—Mardia (1985).
- "The only type of bivariate distribution with which most of us feel familiar (other than the joint distribution of a pair of independent random variables) is the bivariate normal distribution"—Anscombe (1981, p. 305).
- "But who has ever seen a multivariate normal sample?" asks Barnett (1979) rhetorically, and then goes on to present, without any conscious bias in their selection, three bivariate datasets from the published literature that all turn out to be grossly non-normal.
- "The only sure defense against a successful disproof of the assumption of multivariate normality is to abstain from collecting, or presenting, too much data!"—wording adapted from Burnaby (1966, p. 109).

The origins of the bivariate normal are found in the first half of the nineteenth century in the work of Laplace, Plana, Gauss, and Bravais. Seal (1967) and Lancaster (1972) have given accounts of these developments. The latter pointed out that the early authors derived the bivariate normal as the joint distribution of the linear forms of independently distributed normal variables but did not define a coefficient of correlation; the distribution was used as a

N. Balakrishnan, C.D. Lai, *Continuous Bivariate Distributions*,
DOI 10.1007/b101765_12,

basis for a theory of measurement error. Francis Galton (b.1822, d.1911), in analyzing the measurements of the heights of parents and their adult children, studied the structure of a bivariate normal density function. He observed that the marginal distributions of the data were normal and the contours of equal frequency were ellipses. He was the first to recognize the need for a measure of correlation in bivariate data. Since his time, the growth in the use of the bivariate normal has been enormous, so as to produce the comments already quoted.

In Section 11.2, we present some basic formulas and properties of the bivariate normal distribution. In Section 11.3, different methods of deriving bivariate normal distributions are mentioned. Some well-known characterizations of the bivariate normal distributions are listed in Section 11.4. Distributions, moments, and other properties of order statistics arising from a bivariate normal distribution are discussed in Section 11.5. While some available illustrations of the bivariate normal are described in Section 11.6, relationships to some other distributions are mentioned in Section 11.7. Next, the estimation of parameters of the bivariate normal distribution is discussed in Section 11.8. In Section 11.9, some other interesting properties of the bivariate normal distribution are briefly mentioned. Some specialized fields in which the bivariate normal model is applied in interesting ways are listed in Section 11.10, while common applications of the bivariate normal distribution are mentioned in Section 11.11. In Section 11.12, different computational methods and algorithms that are available for computating of the bivariate normal distribution function are discussed. Many different test procedures and graphical methods are available for assessing the validity of the bivariate normal distribution, and these are detailed in Section 11.13. Distributions with normal conditionals and bivariate skew-normal distributions are described in Sections 11.14 and 11.15, respectively. Some univariate transformations on a bivariate normal random vector and the resulting distributions are discussed in Section 11.16. In Section 11.17, the truncated bivariate normal distribution and its properties are presented. The bivariate normal mixture distributions and related issues are described in Section 11.18. In Section 11.19, some bivariate non-normal distributions with normal marginals are presented. Finally, in Section 11.21, the bivariate inverse Gaussian distribution and its properties are discussed.

For further information, interested readers may refer to Johnson and Kotz (1972, Chapter 36), Kotz et al. (2000, Chapter 46), Kendall and Stuart (1977, Chapter 15; 1979, Chapters 18 and 26), and Patel and Read (1982, Chapter 10).

11.2 Basic Formulas and Properties

11.2.1 Notation

In this chapter, we use ϕ and Φ to denote the p.d.f. and cumulative distribution function of the standardized univariate normal distribution, and similarly, ψ and Ψ denote the p.d.f. and c.d.f. of the standardized bivariate normal distribution. The upper right volume under the probability density surface is denoted by $L(x, y; \rho)$.

We also denote the means and variances by $E(X) = \mu_1, E(Y) = \mu_2$, $\text{var}(X) = \sigma_1^2$, and $\text{var}(Y) = \sigma_2^2$.

11.2.2 Support

This is the region of values of X and Y over which the p.d.f. is nonzero. For brevity, we refer to the three most common regions of support as: *the unit square* (meaning $0 \le x, y \le 1$), *the positive quadrant* (meaning $x, y \ge 0$), and *the whole plane* (meaning any real values of x and y). For the bivariate normal, the support is the whole plane.

11.2.3 Formula of the Joint Density

The joint density function is

$$\psi(x, y; \rho) = \frac{1}{2\pi\sqrt{1-\rho^2}} \exp\left[-\frac{1}{2(1-\rho^2)}(x^2 - 2\rho xy + y^2)\right]. \tag{11.1}$$

Location and scale parameters can be introduced, as usual, by replacing x and y by $(x-\mu_1)/\sigma_1$ and $(y-\mu_2)/\sigma_2$, respectively. The contours of this joint density are elliptical.

The general (nonstandardized) form of the density thus obtained is given by

$$\begin{aligned}\psi(x, y; \rho) = \frac{1}{2\pi\sigma_1\sigma_2\sqrt{1-\rho^2}} \exp\Bigg[&-\frac{1}{2(1-\rho^2)}\left(\frac{x-\mu_1}{\sigma_1}\right)^2 \\ &-2\rho\left(\frac{x-\mu_1}{\sigma_2}\right)\left(\frac{y-\mu_2}{\sigma_2}\right) + \left(\frac{y-\mu_2}{\sigma_2}\right)^2\Bigg].\end{aligned} \tag{11.2}$$

Equation (11.1) may be expressed as

$$\psi(x,y;\rho) = \phi(x)\phi\left(\frac{y-\rho x}{\sqrt{1-\rho^2}}\right) = \phi(y)\phi\left(\frac{x-\rho y}{\sqrt{1-\rho^2}}\right) \tag{11.3}$$

since the conditional distribution of Y, given $X = x$, is normal with mean ρx and variance $1-\rho^2$ and that of X, given $Y = y$, is normal with mean ρy and variance $1-\rho^2$.

11.2.4 Formula of the Cumulative Distribution Function

An expression in terms of elementary functions does not exist for the joint cumulative distribution function; see Section 11.12 for different methods of computation.

Mukherjea et al. (1986) wrote the joint cumulative distribution function corresponding to (11.2) but with zero means in the form

$$\Psi(x,y;\rho) = \frac{ab\sqrt{1-\rho^2}}{2\pi}\int_{-\infty}^{x}\int_{-\infty}^{y} e^{-\frac{1}{2}(a^2u^2-2ab\rho uv+b^2v^2)}du\,dv, \tag{11.4}$$

where $\sigma_1\sqrt{1-\rho^2} = 1/a$, and $\sigma_2\sqrt{1-\rho^2} = 1/b$, and presented the following properties for the partial derivatives for $H(x,y)$ in (11.4):

$$\frac{\partial^2\Psi(x,y;\rho)}{\partial x\partial y} = \frac{ab\sqrt{1-\rho^2}}{2\pi}e^{-\frac{1}{2}a^2x^2-2ab\rho xy+b^2y^2}, \tag{11.5}$$

$$\begin{aligned}\frac{\partial\Psi(x,y;\rho)}{\partial x} &= \frac{a\sqrt{1-\rho^2}}{2\pi}e^{-\frac{1}{2}a^2(1-\rho^2)x^2}\Phi(by-a\rho x)\\ &= \frac{a\sqrt{1-\rho^2}}{2\pi}e^{-\frac{1}{2}a^2(1-\rho^2)x^2}\{1-\Phi(a\rho x-by)\},\end{aligned} \tag{11.6}$$

and

$$\begin{aligned}\frac{\partial\Psi(x,y;\rho)}{\partial x} &= \frac{b\sqrt{1-\rho^2}}{2\pi}e^{-\frac{1}{2}b^2(1-\rho^2)y^2}\Phi(ax-b\rho y)\\ &= \frac{b\sqrt{1-\rho^2}}{2\pi}e^{-\frac{1}{2}b^2(1-\rho^2)x^2}\{1-\Phi(b\rho y-ax)\}.\end{aligned} \tag{11.7}$$

Sungur (1990) has noted the property that

$$\frac{d\Psi(x,y;\rho)}{d\rho} = \psi(x,y;\rho). \tag{11.8}$$

11.2.5 Univariate Properties

Both marginal distributions are normal.

11.2.6 Correlation Coefficients

The parameter ρ in (11.1) is Pearson's product-moment correlation coefficient. Further, Kendall's τ and Spearman's ρ_S may be expressed in terms of ρ as

$$\tau = \frac{2}{\pi}\sin^{-1}\rho, \quad \rho_S = \frac{6}{\pi}\sin^{-1}\frac{\rho}{2}.$$

11.2.7 Conditional Properties

Both conditional distributions are normal; the regression is linear and the conditional variance is constant, and they are given by

$$E(Y|X = x) = \mu_2 + \rho\sigma_2(x - \mu_1)/\sigma_1, \tag{11.9}$$
$$\text{var}(Y|X = x) = \sigma_2^2(1 - \rho^2), \tag{11.10}$$

in which location and scale parameters have been included.

11.2.8 Moments and Absolute Moments

Assuming the distribution is in the standardized form as in (11.1), we have as the joint moment generating function

$$M(s,t) = E(e^{sX+tY}) = \exp\left[\frac{1}{2}(s^2 + 2st\rho + t^2)\right]. \tag{11.11}$$

A recurrence relation for the central product moments is given by

$$\mu_{m,n} = (m+n-1)\rho\mu_{m-1,n-1} + (m-1)(n-1)(1-\rho^2)\mu_{m-2,n-2}, \tag{11.12}$$

where $\mu_{m,n}$ is the central product moment, $E[(X - \mu_1)^m(Y - \mu_2)^n]$. It is convenient to write $\mu_{m,n}$ in different forms according to whether m and n are both even, both odd, or one is even and the other is odd as follows:

$$\mu_{2m,2n} = \frac{(2m)!(2n)!}{2^{m+n}} \sum_{j=0}^{\min(m,n)} \frac{(2\rho)^{2j}}{(m-j)!(n-j)!(2j)!}, \tag{11.13}$$

$$\mu_{2m+1,2n+1} = \frac{(2m+1)!(2n+1)!}{2^{m+n}} \sum_{j=0}^{\min(m,n)} \frac{(2\rho)^{2j}}{(m-j)!(n-j)!(2j+1)!}, \tag{11.14}$$

$$\mu_{2m,2n+1} = 0. \tag{11.15}$$

Pearson and Young (1918) tabulated the values of $\mu_{m,n}$ for m and n up to 10.

Let ν_{mn} denote the joint absolute moment, $E(|X^m Y^n|)$, and set $\tau = \sqrt{1-\rho^2}$. Then, we have:

$$\nu_{11} = 2(\tau + \rho \sin^{-1}\rho)/\pi, \qquad \nu_{22} = 1 + 2\rho^2,$$
$$\nu_{12} = (1+\rho^2)\sqrt{2/\pi}, \qquad \nu_{23} = 2(1+3\rho^2)\sqrt{2/\pi},$$
$$\nu_{13} = 2[(\tau(2+\rho^2) + 3\rho \sin^{-1}\rho)]/\pi, \qquad \nu_{24} = 3(1+4\rho^2)\sqrt{2/\pi},$$
$$\nu_{14} = [(3+6\rho^2) - \rho^4]\sqrt{2/\pi}, \qquad \nu_{33} = 2[(4+11\rho^2)\tau + 3\rho(3+2\rho^2) \times \sin^{-1}\rho]/\pi.$$

For further formulas up to $m + n \leq 12$, see Nabeya (1951). Generally,

$$\nu_{mn} = \pi^{-1} 2^{(m+n)/2} (1-\rho^2)^{m+n+1} \sum_{k=0}^{\infty} \Gamma\left(\frac{m+1}{2} + k\right) \Gamma\left(\frac{n+1}{2} + k\right) \frac{(2\rho)^{2k}}{(2k)!}, \tag{11.16}$$

which can alternatively be expressed in terms of Gauss' hypergeometric function ${}_2F_1$.

More extensive collections of formulas can be found, for example, in Johnson and Kotz (1972, pp. 91–93), Patel and Read (1982, Section 10.4), and Kendall and Stuart (1977, paragraphs 3.27–3.29).

11.3 Methods of Derivation

The bivariate normal distribution can be derived in many ways, and we present here five of those.

11.3.1 Differential Equation Method

The bivariate normal density may be obtained by solving a pair of partial differential equations, $\frac{\partial \log h}{\partial x} = \frac{L_1}{Q}$ and $\frac{\partial \log h}{\partial y} = \frac{L_2}{Q}$, where h is the joint p.d.f.

of X and Y, L_1 and L_2 are linear functions of both x and y, and Q is a quadratic function of x and y.

11.3.2 Compounding Method

If X and Y have independent univariate normal distributions, each with mean μ and standard deviation 1, then the joint distribution of X and Y is circular normal centered at (μ, μ). Now, if μ itself has a normal distribution with mean 0 and standard deviation σ, then (X, Y) is bivariate normal with mean at $(0, 0)$, variances of $1 + \sigma^2$, and correlation coefficient as $\sigma/\sqrt{1+\sigma^2}$.

11.3.3 Trivariate Reduction Method

Let X_i $(i = 1, 2, 3)$ be three independent univariate normal random variables. Then, $X = X_1 + X_3$ and $Y = X_2 + X_3$ have a bivariate normal distribution, and thus the bivariate normal distribution is a classic example of the trivariate reduction method.

11.3.4 Bivariate Central Limit Theorem

Let $(X_1, X_2), \ldots, (X_n, Y_n)$ be i.i.d. random vectors with finite second moments and correlation coefficient ρ, the same as in the parent distribution. Then, the bivariate normal distribution results as the joint limiting distribution of the sample means.

11.3.5 Transformations of Diffuse Probability Distributions

Puente and Klebanoff (1994) constructed bivariate Gaussian distributions as transformations of diffuse probability distributions via space-filling fractal interpolating functions; see also Puente (1997).

11.4 Characterizations

The bivariate normal distribution has been characterized in a number of different ways, and we list here some of them:

- All cumulants and cross-cumulants of order higher than 2 are zero.
- For any constants a and b both of which are not zero, $aX + bY$ has a normal distribution [Johnson and Kotz (1972, p. 59)].
- Suppose X and Y have a bivariate exponential-type distribution.[1] Then, (X, Y) is bivariate normal if and only if (i) $E(X|Y = y)$ and $E(Y|X = x)$ are both linear and (ii) $X+Y$ is normal [Johnson and Kotz (1972, p. 86)].
- Brucker (1979) showed that (X, Y) has a bivariate normal distribution if and only if the conditional distribution of each component, given the value of the other component, is normal, with linear regression and constant variance. Fraser and Streit (1980) gave a modification of Brucker's conditions.
- If $X - aY$ and Y are independent and $Y - bX$ and X are independent, for all a, b such that $ab \neq 0$ or 1, then (X, Y) has a normal distribution [Rao (1975, pp. 1–13)].
- That the sample mean vector $(\bar{X}, \bar{Y})$ of a random sample from a bivariate population and the elements (S_1^2, S_2^2, R) that determine the sample variance–covariance matrix, where

$$R = \frac{\sum(X_i - \bar{X})(Y_i - \bar{Y})}{\sqrt{\sum(X_i - \bar{X})^2 \sum(Y_i - \bar{Y})^2}},$$

  are independent characterizes the sampling population as bivariate normal [Kendall and Stuart (1977, p. 413)].
- Let X, Y, U_1, and U_2 be random variables and a and b be constants such that (i) $Z_1 = X + aY + U_1$ and (Y, U_1, U_2) are independent and (ii) $Z_2 = bX+Y+U_2$ and (X, U_1, U_2) are independent. Then, (Z_1, Z_2) has a bivariate normal distribution if $a \neq 0$, $b \neq 0$; further, (Z_1, Z_2) and (U_1, U_2) are independent [Khatri and Rao (1976, pp. 83–84)].
- Holland and Wang (1987) have shown that for a bivariate density function $h(x, y)$ defined on R^2, if

  (a) $\frac{\partial^2 \log h(x,y)}{\partial x \partial y} = \lambda$ (constant),
  (b) $\int_{-\infty}^{\infty} h(x, y)dy = \frac{1}{\sqrt{2\pi}} e^{-x^2/2}$, and
  (c) $\int_{-\infty}^{\infty} h(x, y)dx = \frac{1}{\sqrt{2\pi}} e^{-y^2/2}$,

  then $h(x, y)$ is the standard bivariate normal density function with correlation coefficient $\rho = \frac{\sqrt{1+4\lambda^2}-1}{2\lambda}$.

[1] For this class of distribution, the density is of the form $h = a(x, y) \exp[x\theta_1 + y\theta_2 - q(\theta_1, \theta_2)]$, where θ_1 and θ_2 are two parameters and $a(x, y) \geq 0$ is a function of x and y [Bildikar and Patil (1968)].

- Hamedani (1992) presented 18 different characterizations of the bivariate normal, many of which do not possess straightforward generalizations to the multivariate case.
- Ahsanullah and Wesolowski (1992) discussed a characterization of the bivariate normal distribution by normality of one conditional distribution and some properties of conditional moments of the other variables. If $E(|X|) < \infty, Y|(X = x) \sim N(\alpha x + \beta, \sigma^2)$, and $E(X|Y = y) = \gamma y + \delta$ for some real numbers α, β, γ, and δ with $\alpha \neq 0, \gamma \neq 0$, and $\sigma > 0$, then (X, Y) is distributed as bivariate normal. Ahsanullah and Wesolowski (1992) also presented a slight extension of this result.
- Ahsanullah et al. (1996) presented a bivariate non-normal vector (X, Y) with normal marginal distributions, correlation coefficient ρ, and $\mathrm{corr}(X^2, Y^2) = \rho^2$. Note that if (X, Y) is distributed as normal with correlation coefficient ρ, then X^2 and Y^2 will have correlation ρ^2, but this fourth-moment relation is too weak to characterize a bivariate normal distribution with zero mean vector. However, with additional conditions on X and Y such as finiteness of the second and fourth moments, the conditional distribution of Y given that $(X = x)$ is normal with linear mean, and $E(Y^2|X = x) = b + cx^2$ for constants b and c, $\mathrm{corr}(X^2, Y^2) = \rho^2$ is sufficient to characterize bivariate normality.
- Let X and Y be independent random variables, and let $U = \alpha X + \beta Y$ and $V = \gamma X + \delta Y$, where α, β, γ, and δ are some real numbers such that $\alpha\delta - \beta\gamma \neq 0$. Then, Kagan and Wesolowski (1996) showed that X and Y are normal random variables if the conditional distribution of U given V is normal (with probability 1).
- Castillo and Galambos (1989) [see also Arnold et al. (1999)] established the following interesting conditional characterization of the bivariate normal distribution.

  X and Y have a bivariate normal distribution if and only if all conditional distributions, both of X given Y and Y given X, are normal and any one of the following properties holds:

  (i) $\sigma_2^2(x) = \mathrm{var}(Y|X = x)$ or $\sigma_1^2 = \mathrm{var}(X|Y = y)$ is constant;
  (ii) $\lim_{y\to\infty} y^2\sigma_1^2(y) = \infty$ or $\lim_{x\to\infty} x^2\sigma_1^2(x) = \infty$;
  (iii) $\lim_{y\to\infty} \sigma_1(y) \neq 0$ or $\lim_{x\to\infty} \sigma_2(x) \neq 0$; or
  (iv) $E(Y|X = x)$ or $E(X|Y = y)$ is linear and nonconstant.

- More advanced forms of characterizations can be found in Johnson and Kotz (1972, pp. 59–62) and Mathai and Pederzoli (1977, Chapter 10).

11.5 Order Statistics

Suppose X and Y have a bivariate normal distribution specified by (11.2) with $\rho^2 \neq 1$. Let $Z_{(1)} = \min(X, Y)$ and $Z_{(2)} = \max(X, Y)$. Cain (1994) showed that the distribution function of $Z_{(1)}$ is

$$F_{Z_{(1)}}(x) = \Phi\left(\frac{x-\mu_1}{\sigma_1}\right) + \int_{(x-\mu_1)/\sigma_1}^{\infty} \Phi\left(\frac{x-\mu_2-\rho\sigma_2 u}{\sigma_2\sqrt{1-\rho^2}}\right)\phi(u)du. \quad (11.17)$$

From (11.17), the probability density function of $Z_{(1)}$ can be expressed as

$$f_{Z_{(1)}} = f_1(x) + f_2(x), \quad (11.18)$$

where

$$f_1(x) = \frac{1}{\sigma_1}\Phi\left\{\frac{-\left(\frac{x-\mu_2}{\sigma_2}\right)+\rho\left(\frac{x-\mu_1}{\sigma_1}\right)}{\sqrt{1-\rho^2}}\right\}\phi\left(\frac{x-\mu_1}{\sigma_1}\right)$$

and

$$f_2(x) = \frac{1}{\sigma_2}\Phi\left\{\frac{-\left(\frac{x-\mu_1}{\sigma_1}\right)+\rho\left(\frac{x-\mu_2}{\sigma_2}\right)}{\sqrt{1-\rho^2}}\right\}\phi\left(\frac{x-\mu_2}{\sigma_2}\right);$$

note that $\int_{-\infty}^{\infty} f_2(x)dx = \Pr(X > Y)$.

From (11.18), the moment generating function of $Z_{(1)}$ is

$$M_{Z_{(1)}} = M_1(t) + M_2(t), \quad (11.19)$$

where

$$M_1(t) = e^{t\mu_1+\frac{1}{2}t^2\sigma_1^2}\Phi\left\{\frac{\mu_2-\mu_1-t(\sigma_1^2-\rho\sigma_1\sigma_2)}{\sqrt{\sigma_2^2-2\rho\sigma_1\sigma_2+\sigma_1^2}}\right\}$$

and

$$M_2(t) = e^{t\mu_2+\frac{1}{2}t^2\sigma_2^2}\Phi\left\{\frac{\mu_1-\mu_2-t(\sigma_2^2-\rho\sigma_1\sigma_2)}{\sqrt{\sigma_2^2-2\rho\sigma_1\sigma_2+\sigma_1^2}}\right\}.$$

It now follows that

$$E(Z_{(1)}) = \mu_1\Phi\left\{\frac{\mu_2-\mu_1}{\sqrt{\sigma_2^2-2\rho\sigma_1\sigma_2+\sigma_1^2}}\right\} + \mu_2\Phi\left\{\frac{\mu_1-\mu_2}{\sqrt{\sigma_2^2-2\rho\sigma_1\sigma_2+\sigma_1^2}}\right\}$$
$$-\sqrt{\sigma_2^2-2\rho\sigma_1\sigma_2+\sigma_1^2}\phi\left\{\frac{\mu_1-\mu_2}{\sqrt{\sigma_2^2-2\rho\sigma_1\sigma_2+\sigma_1^2}}\right\}; \quad (11.20)$$

similarly,

$$E(Z_{(1)}^2) = (\mu_1^2 + \sigma_1^2)\Phi\left\{\frac{\mu_2 - \mu_1}{\sqrt{\sigma_2^2 - 2\rho\sigma_1\sigma_2 + \sigma_1^2}}\right\}$$
$$+(\mu_2^2 + \sigma_2^2)\Phi\left\{\frac{\mu_1 - \mu_2}{\sqrt{\sigma_2^2 - 2\rho\sigma_1\sigma_2 + \sigma_1^2}}\right\}$$
$$-(\mu_1 + \mu_2)\sqrt{\sigma_2^2 - 2\rho\sigma_1\sigma_2 + \sigma_1^2\phi}\left\{\frac{\mu_1 - \mu_2}{\sqrt{\sigma_2^2 - 2\rho\sigma_1\sigma_2 + \sigma_1^2}}\right\}. \tag{11.21}$$

Cain and Pan (1995) extended Cain's (1994) results by establishing a recurrence relation for $\mu_r' = E(Z_{(1)}^r)$.

Gupta and Gupta (2001) also derived the distributions of the extreme statistics $\min(X, Y)$ and $\max(X, Y)$; in particular, they showed that both extreme statistics have the IFR property.

11.5.1 Linear Combination of the Minimum and the Maximum

Let $W = a_1 Z_{(1)} + a_2 Z_{(2)}$, where a_1 and a_2 are constants. Define $b_i = 1/a_i$ for $i = 1, 2$. Nagaraja (1982) showed that, for $a_i \neq 0$, the density of W can be expressed as

$$f_W = \begin{cases} f_1(w) & \text{if } b_1 + b_2 > 0 \\ f_1(-w) & \text{if } b_1 + b_2 < 0 \end{cases}, \tag{11.22}$$

where

$$f_1(w) = \tfrac{2}{\sqrt{\zeta}}\phi\left(\tfrac{w}{\sqrt{\zeta}}\right)\Phi(\eta w)$$

with

$$\eta = \frac{b_1 b_2 (b_1 - b_2)}{b_1 + b_2}\sqrt{\frac{1-\rho}{(1+\rho)\delta}} \quad \text{and} \quad \zeta = a_1^2 + 2\rho a_1 a_2 + a_2^2.$$

11.5.2 Concomitants of Order Statistics

Let $(X_1, Y_1), \ldots, (X_n, Y_n)$ be a random sample from a bivariate normal distribution in (11.2). If the sample is ordered by the X-value, then the Y-value associated with the ith order statistic $X_{(i)}$ is called the concomitant of the ith order statistic and is denoted by $Y_{[i]}$; see David (1981).

It is well known that X_i and Y_i are linked by the regression model

$$Y_i = \mu_2 + \rho\frac{\sigma_2}{\sigma_1}(X_i - \mu_1) + \varepsilon_i, \tag{11.23}$$

where $|\rho| < 1$ and X_i and ε_i are independent. It follows that $E(\varepsilon_i) = 0$ and $\text{var}(\varepsilon_i) = \sigma_2^2(1 - \rho^2)$.

It follows from (11.23) that

$$Y_{[r]} = \mu_2 + \rho\frac{\sigma_2}{\sigma_1}(X_{(r)} - \mu_1) + \varepsilon_{[r]}, \qquad r = 1, 2, \ldots, n, \tag{11.24}$$

where $\varepsilon_{[r]}$ denotes the specific ε_i associated with $X_{(r)}$. It then follows from Watterson (1959) and Sondhauss (1994) that

$$\begin{aligned} E(Y_{[r]}) - \mu_2 &= \rho(E(Y_{(r)}) - \mu_1), \\ \text{var}(Y_{[r]}) - \sigma_2^2 &= \rho^2\{\text{var}Y_{(r)} - \sigma_2^2\}, \end{aligned}$$

and

$$\text{cov}(Y_{[r]}, Y_{[s]}) = \rho^2\text{cov}(Y_{(r)}, Y_{(s)}), \text{ for } r \neq s.$$

For asymptotic results on these concomitant order statistics, see Nagaraja and David (1994). An extensive review on this topic has been given by David and Nagaraja (1998).

Linder and Nagaraja (2003) considered the situation where a bivariate normal random sample of size n is subjected to type II censoring on one of the variates so that only a set of p order statistics and their concomitants are observed. They then obtained close approximations to the distributions of sample variances of the observed order statistics and their concomitants through gamma distributions.

Rather than ordering a bivariate data through one component and looking at the other component as a concomitant to order statistic, Balakrishnan (1993) considered the case when the ordering of n pairs of observations are instead based on ordering through a linear combination of the two components. He has discussed various properties of order statistics induced by the ordering of such a linear combination in the case of a bivariate normal distribution, and has also generalized these results to the multivariate normal case.

Lien and Balakrishnan (2003) developed a conditional correlation analysis for order statistics from a bivariate normal distribution, and applied their results to evaluate the presence of inventory effects in futures markets.

It is of interest to mention here that by starting with any univariate density function $f(x)$ and an associated orthogonal function $g(x)$, Balasubramanian and Balakrishnan (1995) described a method of construction of bivariate and multivariate distributions that have many desirable properties such as closure under marginal and conditional, and also interestingly closure under concomitants of order statistics of any component.

11.6 Illustrations

Practically every introductory statistics textbook has some sort of illustration of the bivariate normal distribution. The following are particularly noteworthy:

- Contours of density: Johnson and Kotz (1972, pp. 88–90), Johnson (1987, pp. 51–53), and Kotz et al. (2000, p. 256).
- Plot of density surface: Rodriguez (1982), Johnson and Kotz (1972, pp. 89–90), and Kotz et al. (2000, pp. 257–258).
- Contour plot of the uniform representation of the bivariate normal: Barnett (1980).
- Contours and the three-dimensional plots after the marginals have been transformed to be exponential: Johnson et al. (1981).
- Zelen and Severo (1960) have plotted graphs of $L(h, 0; \rho)$ for various ranges and ρ. Equidistributional contours ($L(x, y; \rho) = \alpha$) for a standard bivariate normal distribution with $\alpha = 0.25$ are presented by Kotz et al. (2000, p. 272).

11.7 Relationships to Other Distributions

- Let $H(x, y)$ be a ϕ-bounded [see Lancaster (1969) for a definition] bivariate distribution, with standard marginals. Then the density function can be written as a mixture (finite or infinite) of bivariate normal densities as

$$h(x, y) = \int_{-1}^{1} \psi(x, y; \rho) d\mu(\rho), \tag{11.25}$$

  where $\psi(x, y; \rho)$ is the standardized bivariate normal density and $\mu(\cdot)$ is a distribution function over $[-1, 1]$.
- Kibble's bivariate gamma distribution may be obtained from the bivariate normal distribution; see Section 8.2 for pertinent details.
- Suppose (X_1, X_2) has the standardized bivariate normal distribution and X_0, independent of X_1 and X_2, has a chi-squared distribution with ν degrees of freedom. Then, $X = X_1/\sqrt{X_0/\nu}$ and $Y = X_2/\sqrt{X_0/\nu}$ jointly have a bivariate t-distribution with ν degrees of freedom; see Section 9.2 for details on this distribution.

11.8 Parameter Estimation

If all five parameters $\mu_1, \mu_2, \sigma_1, \sigma_2$, and ρ in (11.2) are unknown, then the maximum likelihood estimators are

$$\hat{\mu}_1 = \bar{X},\ \hat{\mu}_2 = \bar{Y},\ \hat{\sigma}_1 = S_1,\ \hat{\sigma}_2 = S_2,\ \hat{\rho} = R, \tag{11.26}$$

respectively, where S_i^2 are the sample variances (with n as the divisor) and R is the sample correlation coefficient.

If the values of some of the parameters are known, different estimators of the remaining parameters are obtained, and the cases that are of interest are:

(i) One mean, say μ_1, is known.
(ii) μ_1 and μ_2 are known.
(iii) μ_1 and σ_1 are known.
(iv) μ_1, σ_1, and ρ are known.
(v) μ_1, σ_1, μ_2, σ_2 are known.
(vi) $\sigma_1 = \sigma_2$ (but common value unknown).
(vii) $\mu_1 = \mu_2$, $\sigma_1 = \sigma_2$ (common values unknown).
(viii) $\sigma_1^2\sigma_2^2(1-\rho^2) = \theta^2$ (known).
(ix) $\mu_1 = \mu_2$ (common value unknown).
(x) Information is missing.

For a detailed account of all these developments, one may refer to Chapter 46 of Kotz et al. (2000).

More recently, the following estimation problems have been discussed in the literature:

(i) $\sigma = \sigma_1^2/\sigma_2^2$ with unknown marginal means. Iliopoulos (2001) derived a uniformly minimum variance unbiased estimator of the ratio σ as

$$\delta_U = \frac{n-3+2T}{n-1}S, \tag{11.27}$$

where $S = A_{11}/A_{22} = R^2 = A_{12}^2/(A_{11}A_{22})$ with $A_{11} = \sum_{i=1}^n (X_i - \bar{X})^2$, $A_{22} = (Y_i - \bar{Y})^2$, and $A_{12} = \sum_{i=1}^n (X_i - \bar{X})(Y_i - \bar{Y})$.

(ii) $\mu_1 = \mu_2 = \mu$, unknown, and σ_1 and σ_2 are possibly unequal. Yu et al. (2002) considered in this setting the problem of estimating the common mean μ based on paired data as well as on one of the marginals. Two double sampling schemes with the second-stage sampling being either a simple random sampling (SRS) or a ranked set sampling (RSS) were considered. Yu then proposed two common mean estimators and found that, under normality, the proposed RSS common mean estimator is always superior to the proposed SRS common mean estimator and other existing estimators such as the RSS regression estimator proposed earlier by Yu and Lam (1997).

(iii) Al-Saleh and Al-Ananbeh (2007) considered estimation of the means of the bivariate normal using moving extreme ranked set sampling with a concomitant variable.

11.8.1 Estimate and Inference of ρ

The maximum likelihood estimate of ρ based on simple random samples from a bivariate normal population is simply the well-known sample correlation coefficient r given by (4.3),

$$\hat{\rho} = r = \frac{\sum_{i=1}^{n}(x_i - \bar{x})(y_i - \bar{y})}{\sqrt{\sum_{i=1}^{n}(x_i - \bar{x})^2 \sum_{i=1}^{n}(y_i - \bar{y})^2}}.$$

- There are many examples where one or both of X and Y may be difficult to observe or measure directly. Zheng and Modarres (2006) described several situations where the traditional sample correlation coefficient cannot be used practically for estimating ρ. They proposed a robust estimate of the correlation coefficient for a bivariate normal distribution using ranked set sampling. They showed that this estimate is at least as efficient as the corresponding estimate based on the simple random sampling and highly efficient compared with the maximum likelihood estimate using balanced ranked set sampling. Moreover, the estimate is robust to common ranking errors.
- Evandt et al. (2004) proposed to use a little-known robust estimator $\hat{\rho} = \sin((\pi/2)\hat{\tau})$ that was shown to be at least as good as Spearman's rho ρ_S when the possibility of outliers must be taken into consideration.
- Sun and Wong (2007) proposed a likelihood-based high-order asymptotic method to obtain a confidence interval for ρ.
- Tsou (2005) provided a suitable simple adjustment to the bivariate normal likelihood function inferences for the inference of the correlation coefficient. The resulting inference procedure is asymptotically valid for practically all continuous bivariate distributions so long as they have a finite fourth moment.
- Testing independence of X and Y under a bivariate normal model is equivalent to testing $\rho = 0$. A popular test statistic under this hypothesis is $r\sqrt{n-2}/\sqrt{1-r^2}$, which has a t-distribution with $n-2$ degrees of freedom. Another popular test statistic for a given ρ, derived by way of a variance stabilizing transformation, is

$$\sqrt{n-3}\left\{\log\frac{1+r}{1-r} - \log\frac{1+\rho}{1-\rho}\right\}\Big/2,$$

which has approximately a standard normal distribution; see Bickel and Doksum (1977).

- Three run-based and two rank-based nonparametric tests were proposed by Kim and Balakrishnan (2005) for testing independence between lifetimes and covariates from censored bivariate normal samples.

11.8.2 Estimation Under Censoring

For the bivariate normal distribution, when the available samples are either type II right censored or progressively type II right censored on one variable and concomitants being available on the other variable, various inferential methods such as maximum likelihood estimation of the parameters, EM-algorithm for the numerical determination of the MLEs, confidence intervals and tests of hypothesis have been discussed by Balakrishnan and Kim (2004, 2005a,b,c).

11.9 Other Interesting Properties

- X and $Y - E(Y|X)$ are independent.
- $\rho = 0$ if and only if X and Y are independent. Here ρ describes the strength of the linear relationship between X and Y.
- Let (X_1, Y_1) and (X_2, Y_2) be two standard bivariate normal random vectors, with correlation coefficients ρ_1 and ρ_2, respectively. If $\rho_1 \geq \rho_2$, then $\Pr(X_1 > x, Y_1 > y) \geq \Pr(X_2 > x, Y_2 > y)$, which is known as Slepian's inequality; see Gupta (1963a, p. 805). Alternatively, $\Pr(X_1 < x, Y_1 < y) \geq \Pr(X_2 < x, Y_2 < y)$. In this case, we say (X_1, Y_1) has larger quadrant dependence than (X_2, Y_2). In other words, the bivariate normal distribution with fixed marginals is ordered by quadrant dependence (see Section 3.9) through the correlation coefficient ρ.
- By letting $\rho_2 = 0$ in the inequality above, we have $\Pr(X \leq x, Y \leq y) \geq \Pr(X \leq x, Y \leq y)$ if $\rho \geq 0$ for all x, y; of course, there is a similar inequality if $\rho \leq 0$. Moreover, $\Pr(|X| \leq x, |Y| \leq y) \geq \Pr(|X| \leq x)\Pr(|Y| \leq y)$, where $x \geq 0, y \geq 0$. Tong (1980, pp. 8–15) considered these inequalities for multivariate, rather than bivariate, distributions, so that the form of the variance–covariance matrix played an important part in his discussion.
- $X + Y$ has a univariate normal distribution; more generally, so does $aX + bY$.
- The magnitude of the vector sum—i.e., $\sqrt{X^2 + Y^2}$—has a Rayleigh distribution if X and Y are i.i.d. normal variates with zero means. For the case where X and Y have the general bivariate normal distribution, Chou and Corotis (1983) have presented some results.
- $\psi(x, y; \rho)$ can be expanded diagonally in terms of ρ and the Hermite polynomials in the form

$$\psi(x, y; \rho) = \phi(x)\phi(y) \sum_{j=0}^{\infty} \rho^j H_j(x) H_j(y); \tag{11.28}$$

see Cramér (1946, p. 133). Here, H_j's are normalized so that the integral $\int \phi(x)H_i(x)H_j(x)dx$ is 1 if $i = j$ and is 0 if $x \neq j$; also, $H_0 = 1$. Since

$$-\frac{d}{dx}\left[H_{j-1}(x)\phi(x)\right] = \sqrt{j}H_j(x)\phi(x) \tag{11.29}$$

[see Kendall and Stuart (1979, p. 326)], it follows that $\Psi(x, y)$ also has a diagonal expansion in terms of ρ and the Hermite polynomials in the form

$$\Psi(x, y; \rho) = \phi(x)\phi(y) \sum_{j=1}^{\infty} \frac{\rho^j}{j} H_{j-1}(x) H_{j-1}(y) + \Phi(x)\Phi(y). \tag{11.30}$$

- Considering only the Hermite polynomial of order 1, the diagonal expansion in (11.28) may be approximated by

  $$\psi(x, y; \rho) \simeq \phi(x)\phi(y)(1 + \rho xy).$$

  This is the same first-order approximation to the standard bivariate normal density proposed by Sungur (1990).
- $T = (1 - \rho^2)^{-1}(X^2 - 2\rho XY + Y^2)$ has an exponential distribution with mean 2. Hence, the integral of (11.1) over the interior of the ellipse $x^2 - 2\rho xy + y^2 = k$ is $\Pr[T \leq k(1 - \rho^2)] = 1 - \exp[-\frac{1}{2}k(1 - \rho^2)]$ [see Johnson and Kotz (1972, p. 16)].
- Any bivariate distribution obtained from the bivariate normal by separate transformations of X and Y has a correlation that in absolute value cannot exceed $|\rho|$ [Kendall and Stuart (1979, p. 600)]; see Section 11.16.5 for more details.
- If Ψ_1, Ψ_2, Ψ_3 and Ψ_4 are four bivariate normal distribution functions such that $\Psi_1\Psi_2 \equiv \Psi_3\Psi_4$, then Ψ_3, Ψ_4 are the same as Ψ_1, Ψ_2 [see Anderson and Ghurye (1978)].
- If (X, Y) has a standardized bivariate normal distribution, then the ratio X/Y has a Cauchy distribution with p.d.f. $\frac{\sqrt{1-\rho^2}}{\pi(1-2\rho u+u^2)}$. The bivariate normal is not the only distribution for which this is true; see Section 9.14 of Springer (1979). Hinkley (1969) approximated the cumulative distribution function of X/Y when (X, Y) is not standardized. For subsequent developments, one may refer to Springer (1979, Section 4.8.3), Aroian (1986), and references therein. For the distribution of the product XY, see Craig (1936), Haldane (1942), Aroian (1947, 1978), and Springer (1979, Section 4.8.3).
- For multidimensional central limit theorems, see Heyde (1985). (The subject is apparently not so interesting when dealing with ordinary numbers

because so much of unidimensional theory carries over without any difficulty.)

11.10 Notes on Some More Specialized Fields

- For formulas relating to the application of this distribution in the "competing risk" context, one may refer to David and Moeschberger (1978, Chapter 4).
- The quantization of a two-dimensional random variable is of interest to electrical engineers. In this connection, we quote from Bucklew and Gallagher (1978): "Consider a two-dimensional random variable **X** whose bivariate density is circularly symmetric and we desire to represent this quantity by a finite set of values. One possible representation of **X** leads to a Cartesian coordinate system expression wherein we individually quantize the two rectangular components of the random variable. Another common representation leads to a polar coordinate representation where we quantize the magnitude and phase angle of X. We obtain a simple criterion by which to determine whether polar format or rectangular format gives a smaller mean square quantization error."
- The following description of planar random movement is adapted from van Zyl (1987). A particle, starting from the origin, jumps a random length U on the plane with all directions for the jumps being equally likely. Thus, in an obvious notation, after n such jumps, the particle is at coordinates $(\sum_{i=1}^{n} u_i \cos\theta_i, \sum_{i=1}^{n} u_i \sin\theta_i)$, where θ is uniformly distributed over the range 0 to 2π. Van Zyl then derived the characteristic function of the distribution of the position, discussed the normal approximation, and presented approximate results for the distance of the particle from the origin.

11.11 Applications

There are numerous applications for the bivariate and multivariate normal distributions. Chapter 19 of Hutchinson and Lai (1991) gives brief accounts of over 30 subject areas where the bivariate normal distribution has been used. See also Chapters 20–23 of the same monograph. A quick Google search will yield hundreds of applications over many disciplines such as agriculture, biology, engineering, economics and finance, the environment, genetics, medicine, psychology, quality control, reliability and survival analysis, sociology, physical sciences, and technology.

11.12 Computation of Bivariate Normal Integrals

11.12.1 The Short Answer

The short answer is as follows:

- Tables of bivariate integral: the National Bureau of Standards (1959) and Japanese Standard Association (1972) have tables of the function L.
- Computer program: Donnelly's (1973) program is widely available and is written in FORTRAN. Perhaps Baughman's (1988) program supersedes Donnelly's and is a more current algorithm. Section 11.12.3 presents a comparison of various algorithms.

The remainder of this section expands on this and is divided into several subsections on algorithms, and then tables, computer programs, and references to reviews of the subject are also presented. We consider only L and related quantities and not, for instance, integrals over an offset circle, for which one may refer to Groenewoud et al. (1967) and Patel and Read (1982, Section 10.3).

In passing, we note that the computation of the *uni*variate normal d.f. is not straightforward when one is interested in the tails of the distribution as, for example, in safety contexts; see Rosenblueth (1985). However, MINITAB has a built-in procedure to compute the normal probability integrals pretty accurately.

11.12.2 Algorithms—Rectangles

In the early development of the subject, interest centered on the upper right volume under a density surface, what would be referred to in the reliability context as the "survival function." It is conventional to use the symbol L for this:

$$L(h,k;\rho) = \Pr(X > h, Y > k) = \int_h^\infty \int_k^\infty \psi(x,y;\rho)dy\,dx. \tag{11.31}$$

A special case is $L(0,0;\rho) = \frac{1}{4} + \frac{1}{2\pi}\sin^{-1}\rho$. We also have

$$L(h,k;\rho) = L(h,0;\rho_{hk}) + L(k,0;\rho_{hk}) - \frac{1}{2}(1-\delta_{hk}),$$

where δ_{hk} is 0 or 1 depending on whether h and k have the same sign or not, and $\rho_{hk} = \frac{(\rho h - k)a_h}{\sqrt{h^2 - 2\rho hk + k^2}}$, with a_h being 1 or -1 depending on whether h is positive or negative.

Relationships involving the distribution function Ψ are as follows:

$$\Psi(h,k;\rho) = L(-h,-k;\rho) = L(h,k;\rho) + \Phi(h) + \Phi(k) - 1,$$
$$\Psi(-h,k;\rho) = L(h,-k;\rho) = 1 - \Phi(h) - L(h,k;\rho).$$

Pearson (1901) presented a method for evaluating L as a power series in ρ involving tetrachoric functions. His method provides a good approximation for small $|\rho|$. Computation of L can also be through the functions V and T which will be discussed shortly.

Drezner (1978) presented a simple algorithm for Ψ based on numerical integration of the density function using the Gauss quadrature method.

Divgi (1979) calculated Ψ by transforming X and Y into two independent standard normal variates and then approximating $1 - \Phi(x)$ by $x\phi(x)\sum_{k=0}^{n} d_{nk}x^k$, where d_{nk} are the coefficients given by the author.

Bouver and Bargmann (1979) approximated the bivariate integral by

$$(b-a)\sum_{i=1}^{n} \frac{w_i}{2}\frac{e^{-x^2/2}}{\sqrt{2\pi}}\Phi\left(\frac{k-x\rho}{\sqrt{1-\rho^2}}\right), \tag{11.32}$$

where $x = a + h(1+c_i)/2$ and $a = -10$ with step size $h = (b-a)/30$. Here, w_i and c_i are the weights and abscissas of the Gauss–Legendre numerical integration rule. Bjerager and Skov (1982) have given another approximate formula for Ψ.

A simple way of obtaining a close approximation is to write $\Psi(h,k;\rho) = \Pr(X < h)\Pr(Y < k|X < h)$. Now work out the mean and standard deviation of $Y|X < h$, and to find $\Pr(Y < k|X < h)$, assume it has a normal distribution. The repeated application of this is a familiar strategy to approximate the n-dimensional cumulative normal distribution function (for $n \geq 3$) since there is a paucity of competing methods. Mee and Owen (1983) investigated its usefulness in the bivariate case. In particular, they gave guidance as to how the calculation should be performed—as $\Psi(h,k;\rho)$, $\Psi(k,h;\rho)$, $\Phi(k) - \Psi(-h,k;-\rho)$, or $\Phi(h) - \Psi(-k.h;\rho)$, depending on the values of h and k.

Foulley and Gianola (1984) approximated L terms of ten positive roots of Hermite polynomials of order 20.

Wang (1987) proposed a method for computing the bivariate normal probability integral over any rectangular region, with reasonable accuracy and without the need for any numerical integration. By observing that the cross-product ratio for infinitesimal rectangular regions of the bivariate normal is constant, the method involves an iterative proportional fitting algorithm for the row and column marginal totals in a two-way table that is constructed from discretized normal probabilities. A table of integrals when $\rho = 1/2$ has been given by Wang.

Rom and Sarkar (1990) proposed a modification of Wang's contingency approach. They developed a new algorithm utilizing quadrature and the association model to approximate the diagonal probabilities. The off-diagonal probabilities are then approximated using this model. They claim that their approach has several advantages over Wang's method.

Drezner and Wesolowski (1990) proposed an algorithm that is efficient for the whole range of correlation coefficients. The method uses Gaussian quadrature based on only five points and results in a maximum error of 2×10^{-7}. Albers and Kallenberg (1994) also discussed simple approximations to $L(h, k; \rho)$ for large values of ρ.

Lin (1995) proposed the simple approximation for $L(h, 0; \rho)$

$$L(h, 0; \rho) \simeq \frac{1}{\sqrt{8a}} e^{b^2/(4a)} \left\{1 - \Phi\left(\sqrt{2a}\left(h + \frac{b}{2a}\right)\right)\right\}, \tag{11.33}$$

where $a = 0.5 + 0.416\rho^2/(1-\rho^2)$ and $b = -0.717\rho/\sqrt{1-\rho^2}$. Lin (1995) also suggested an even simpler approximation, given by

$$L(h, 0; \rho) \simeq \frac{1}{\sqrt{8a}} e^{b^2/(4a)} \frac{0.5e^{-a^2(h+\frac{b}{2a})^2}}{1 + 0.91\{\sqrt{2a}(h + \frac{b}{2a})\}^{1.12}}. \tag{11.34}$$

Lin has shown that the accuracy of these approximations is quite sufficient for many practical situations.

Results in terms of the bivariate (and multivariate) Mills' ratio—that is, the ratio of tail volume L to bounding ordinate ψ—have been given by Savage (1962) and Ruben (1964). Savage derived upper and lower bounds for this ratio, and Ruben presented an asymptotic expansion.

Derivative Fitting Procedure

Zhang (1994) presented a derivative fitting procedure for computing the c.d.f. $\Psi(h, k, \rho) = \int_{-\infty}^{h}\int_{-\infty}^{k} \psi(x, y, \rho)dx\,dy$, which can be written as [Gupta (1963a)]

$$\Psi(h, k, \rho) = \int_0^\rho \psi(h, k, z)dz + \Phi(h)\Phi(k). \tag{11.35}$$

Expand (11.35) approximately as a polynomial of ρ as

$$\Psi^*(h, k, \rho) = a_0 + \sum_{i=1}^{m} a_i \rho^i, \tag{11.36}$$

where a_i are functions of h and k and are independent of ρ.

Set $\frac{\partial \Psi^*}{\partial \rho} = \frac{\partial \Psi}{\partial \rho}$. Since $\frac{\partial^2 \psi}{\partial x \partial y} = \frac{\partial \psi}{\partial \rho}$ [Gupta (1963a)], it follows that

$$\frac{\partial \Psi}{\partial \rho} = \int_{-\infty}^{h}\int_{-\infty}^{k} \frac{\partial^2 \phi}{\partial x \partial y} dx\,dy = \psi(h, k, \rho).$$

It now follows from (11.35) that

$$\sum_{i=1}^{m} i a_i \rho^{i-1} = \psi(h, k, \rho). \tag{11.37}$$

To determine m a_i's, m fitting points ρ_i are taken from the interval $[0, 1)$. Substituting those ρ's into (11.37) yields

$$\begin{bmatrix} 1 & 2\rho_1 & 3\rho_1^2 & \cdots & m\rho_1^{m-1} \\ 1 & 2\rho_2 & 3\rho_2^2 & \cdots & m\rho_2^{m-1} \\ \vdots & \vdots & \vdots & & \vdots \\ 1 & 2\rho_m & 3\rho_m^2 & \cdots & m\rho_m^{m-1} \end{bmatrix} \begin{bmatrix} a_1 \\ a_2 \\ \vdots \\ a_m \end{bmatrix} = \begin{bmatrix} b_1 \\ b_2 \\ \vdots \\ b_m \end{bmatrix}, \tag{11.38}$$

where $b_i = \psi(h, k, \rho_i)$. Rewrite (11.38) in a common matrix notation as $C\boldsymbol{a} = \boldsymbol{b}$ so that $\boldsymbol{a} = C^{-1}\boldsymbol{b}$. The term a_0 is given by $\Psi(h, k, 0) = \Phi(h)\Phi(k)$.

It is now clear that (11.35) can now be written as

$$\Psi^*(h, k, \rho) = \Phi(h)\Phi(k) + R'C^{-1}\boldsymbol{b}. \tag{11.39}$$

The author reported that the approximation to the cumulative bivariate normal by $\Psi^*(h, k, \rho)$ is quite accurate (up to six significant digits) for $-.75 \leq \rho \leq 0.75$ and very poor when ρ is close to 1 or -1. He also provided a technique to improve the numerical accuracy when $|\rho| > 0.75$.

Bounds on $L(h, k; \rho)$

Willink (2004) has presented inequalities for the upper bivariate normal tail probability $L(h, k; \rho)$ for use in bounding the probability integral $\Psi(h, k; \rho)$. The author considered them relatively simple and more widely applicable than the existing bounds with similar performance, and they have superior performance if $|\rho|$ is small or $\Psi(h, k; \rho)$ is very large. The upper bound is tight when $\Psi(h, k; \rho)$ is large and has a simple form when $h = k$.

For $h > 0$, $\rho \geq 0$, the two lower bounds are, respectively,

$$L(h, k; \rho) \geq \Phi(-h) - \frac{\sqrt{(1-\rho^2)}}{\rho}\phi(h) \times \left[G\left(\frac{k - \rho h}{\sqrt{1-\rho^2}}\right) - G\left(\frac{k - \rho h}{\sqrt{1-\rho^2}} - a\right)\right] \tag{11.40}$$

for $a = \frac{\rho}{\sqrt{(1-\rho^2)}} \cdot \frac{\Phi(-h)}{\phi(h)}$ and $G(x) = \int_{-\infty}^{x} \Phi(y)\,dy$, and

$$L(h, k; \rho) \geq \Phi(-h) - \frac{\phi(h)}{h}\Phi\left(\frac{k-\rho h}{\sqrt{1-\rho^2}}\right) + \frac{\phi(k)}{h}\exp\left[\frac{(k-h/\rho)^2}{2}\right]\Phi\left(\frac{k-h/\rho}{\sqrt{1-\rho^2}}\right). \quad (11.41)$$

The upper bound concerned is given by

$$L(h, k; \rho) \leq \Phi(-h) - \left[\Phi\left(\frac{\rho h - k}{\sqrt{1-\rho^2}}\right) + \rho \exp\left[\frac{h^2-k^2}{2}\right]\Phi\left(\frac{\rho k - h}{\sqrt{1-\rho^2}}\right)\right]. \quad (11.42)$$

Because it is expected that (11.40) will perform better than (11.42), Willink (2004) proposed to form a hybrid lower bound that is the maximum of the two individual bounds. Thus, the bounds are now given by

$$\max\{\text{RHS}(11.40), \text{RHS}(11.41)\} \leq L(h, k; \rho) \leq \text{RHS}(11.42), \ \ h > 0, \ \ \rho \geq 0. \quad (11.43)$$

In particular, the bounds are simple when $h = k$. By letting $\theta = \sqrt{\frac{1-\rho}{1+\rho}}$, we can show that (11.43) now becomes

$$\Phi(-h)\Phi(-\theta h) \leq L(h, h; \rho) \leq \Phi(-h)\Phi(-\theta h) \leq L(h, h; \rho)(1+\rho), \quad h > 0, \rho \geq 0. \quad (11.44)$$

11.12.3 Algorithms: Owen's T Function

For computational purposes, it is easier to work with Owen's T function than with V (to be discussed in Section 11.12.4), where

$$\begin{aligned} T(h, \lambda) &= \frac{\pi}{2}\int_0^\lambda (1+x^2)^{-1}\exp[-h^2(1+x^2)/2]dx \\ &= \frac{1}{2\pi}\tan^{-1}\lambda - V(h, \lambda h). \end{aligned} \quad (11.45)$$

Here $\frac{1}{2\pi}\tan^{-1}\lambda$ is the integral of the circular normal density $\psi(x, y; 0)$ over the sector in the positive quadrant that is bounded by the lines $y = 0$ and $y = \lambda x$.

We note the following relations that show that the computation of T for $h \geq 0$ and $0 \leq \lambda \leq 1$ is sufficient to obtain T for any other values of the arguments:

$$T(\lambda h, \lambda^{-1}) = \frac{1}{2}[\Phi(h) + \Phi(\lambda h)] - \Phi(h)\Phi(\lambda h) - T(h, \lambda),$$
$$T(h, \lambda) = T(-h, \lambda) = -T(h, -\lambda),$$
$$T(h, 0) = 0.$$

Further, Ψ can be expressed in terms of Φ and T, as follows:

$$\Psi(h, k; \rho) = \frac{1}{2}[\Phi(h) + \Phi(k)] - T\left(h, \frac{k - \rho h}{h\sqrt{1 - \rho^2}}\right) - T\left(h, \frac{h - \rho k}{k\sqrt{1 - \rho^2}}\right) - b,$$

where $b = 0$ if $hk > 0$ or $hk = 0$ with $h + k \geq 0$ and $b = \frac{1}{2}$ otherwise.

Owen (1980) presented a collection of integral formulas involving the normal distribution. Most of these concern the univariate function, but there are several relating to Ψ and/or T.

Much effort has been devoted to searching for an accurate approximation for T. Owen (1956) showed that

$$T(h, \lambda) = \frac{1}{2\pi}\left(\tan^{-1}\lambda - \sum_{j=0}^{\infty} c_j \lambda^{2j+1}\right), \tag{11.46}$$

with

$$c_j = \frac{(-1)^j}{2j + 1}\left[1 - e^{-h^2/2}\sum_{i=0}^{j}\frac{(\frac{1}{2}h^2)^i}{i!}\right].$$

For small values of h and λ, convergence is rapid, and this formula is useful for computing T. Amos (1969) has given instructive comparisons of computer times to calculate Ψ using various formulas; he recommended that (11.46) is generally preferable.

Borth (1973) agreed with the use of (11.46) as an approximation to T with a desired accuracy of 10^{-7} if $h \leq 1.6$ or $\lambda \leq 0.3$. For faster convergence with higher values of h or λ, he gave the following modification. Approximating $(1 + x^2)^{-1}$ over the range -1 to 1 by a polynomial of degree $2m$, $\sum_{k=0}^{m} a_{2k} I_{2k} x^{2k}$. Then, T may be approximated by

$$\frac{1}{2\pi}\exp\left(\frac{-h^2}{2}\right)\sum_{k=0}^{m} a_{2k} I_{2k}(w)\left(\frac{h}{\sqrt{2}}\right)^{-(2k+1)}, \tag{11.47}$$

where $w = h\lambda/\sqrt{2}$ and I is obtained by the iterative relation

$$I_{2k}(w) = \frac{1}{2}[(2k - 1)I_{2k-2}(w) - w^{2k-1}\exp(-w^2)]$$

with $I_0(w) = \sqrt{\pi}[\Phi(w\sqrt{2}) - \frac{1}{2}]$. Borth recommended this modification if $h > 1.6$ and $\lambda > 0.3$, noting that if $h > 5.2$, then $T < 10^{-7}$, and that the required accuracy is attained if $m = 6$. This use of Owen's algorithm with Borth's modification does combine speed of computation with accuracy.

Sowden and Ashford (1969) suggested a composite method of computing L, incorporating Owen's algorithm (11.46).

For small h and λ, Daley (1974) claimed that Simpson's numerical integration rule to evaluate (11.45) yields a better result than the power series expansion in (11.46).

Young and Minder (1974) also gave an algorithm for calculating T over all values of h and λ. (There have been corrections and improvements to this by various authors.)

Approximation for $T(h, \lambda)$ When h Is Small

Young and Minder's algorithm has been modified and extended by several authors, including Hill (1978), Thomas (1979), and Chou (1985). Boys (1989) found that Chou's modified version of Young and Minder (1974) does not provide accurate results when h is small and λ is large. He therefore provided an approximation based on the first few terms in an asymptotic expansion of $T(h, \cdot)$ for small h and defining $a = h\lambda$ to give

$$\begin{aligned} T(h, a/h) \simeq \frac{1}{4} - \frac{1}{2\pi}\left[\frac{\exp(-a^2/2)}{a} + \sqrt{2\pi}\left\{\Phi(a) - \frac{1}{2}\right\}\right] h \\ + \frac{1}{12\pi}\left[\frac{a^2+2}{a^3}\exp\left(\frac{-a^2}{2}\right) + \sqrt{2\pi}\left\{\Phi(a) - \frac{1}{2}\right\}\right] h^3. \end{aligned} \quad (11.48)$$

Comparison of Algorithms for Bivariate Normal Probability Integrals

As there are several numerical algorithms available to compute the bivariate normal integrals, a practitioner is often faced with a decision to select an optimal procedure in terms of speed and accuracy. Unfortunately, high accuracy comes at the cost of computational time. Terza and Welland (1991) carried out a comparison of eight approximation algorithms with regard to the accuracy and speed trade-off. The eight procedures used in the comparison are: Owen (1956), Young and Minder (1974), Daley (1974), Drezner (1978), Divgi (1979), Bouver and Bargmann (1979), Parrish and Bargmann (1981), and Welland and Terza (1987). We have discussed all except the last two. We also note that Owen's algorithm is implemented by the IMSL subroutine DBNRDF. Terza and Welland (1991) produced 12 tables and drew the following conclusions from their numerical results:

> The method developed by Divgi (1979) emerges as the clear method of choice, achieving 14-digit accuracy ten and a half times faster than its nearest competitor. Furthermore, in the time required by Divgi's approximation to reach this level of precision, none of the other methods can support more than 3-digit accuracy.

Wang and Kennedy (1990) disagreed somewhat with the findings of Terza and Welland (1991) and stated in their paper, "Although it appears that the accuracy comparisons were successfully made in this study, the possibility exists that variation in levels of accuracy of the basis algorithm over different regions might have caused erroneous conclusions to be made when comparing the algorithms for achieved accuracy. What is needed in studies of this type is a base algorithm which provides a computed value along with a useful bound for the error in the value. In other words, a self-validating computational method and associated algorithm is needed to provide numbers for use in comparing accuracy of competing algorithms." Wang and Kennedy (1990) then carried out a comparison of several algorithms over a rectangle based on self-validated results from interval analysis. They concluded that even the most accurate of the algorithms currently in use for the bivariate normal is substantially less accurate and no more accurate than a Taylor series approximation for computing probabilities over rectangles.

11.12.4 Algorithms: Triangles

$V(h,k)$ is defined as

$$V(h,k) = \frac{1}{2\pi}\int_0^h \int_0^{kx/h} \exp\{-(x^2+y^2)/2\}\, dydx, \qquad h,k \ge 0. \qquad (11.49)$$

This is the integral of the standard circular normal density over the triangle with vertices $(0,0)$, $(h,0)$, and (h,k). Clearly, $V(0,k) = 0 = V(h,0)$. Then, the following relation holds between L and V:

$$L(h,k;\rho) = 1 - \frac{1}{2}\{\Phi(h)+\Phi(k)\} - \frac{1}{2\pi}\cos^{-1}\rho + V\left(h, \frac{k-\rho h}{\sqrt{1-\rho^2}}\right)$$
$$+V\left(k, \frac{k-\rho k}{\sqrt{1-\rho^2}}\right),$$

and other relations involving the function V are as follows:

$$\begin{aligned} V(h,k) &= -V(-h,k) = V(-h,-k) = -V(h,-k), \\ V(h,k) + V(k,h) &= \left[\Phi(h) - \frac{1}{2}\right][\Phi(h) - 1], \\ V(h,\infty) &= \frac{1}{2}\left[\Phi(h) - \frac{1}{2}\right], \\ V(\infty,k) &= 0. \end{aligned}$$

Nicholson (1943) presented tables of V to six decimal places.

11.12.5 Algorithms: Wedge-Shaped Domain

Grauslund and Lind (1986) considered the integral of the standard circular normal density over a wedge-shaped domain given by

$$I(h,k) = \int\int_D \phi(x)\phi(y) dx\, dy, \tag{11.50}$$

where the integral is taken over the region D defined by $y \geq k, x \geq hy/k$.

The function I in (11.50) can be reduced to a single integral in the form

$$I(h,k) = k\phi(k) \int_h^\infty \frac{\phi(x)}{k^2 + x^2} dx. \tag{11.51}$$

Then the following identities hold:

$$I(h,k) = \Phi(-k) - I(-h,k) = \frac{1}{2} - I(h,-k) = \Phi(-h)\Phi(-k) - I(k,h).$$

The integral may be evaluated by numerical methods. Alternatively, Grauslund and Lind obtained a simple approximation that they claimed to be suitable for many technical applications. Introduce the function

$$I_1(h,k) = \frac{k}{2h}\left[\Phi\left(-\sqrt{(h^2+k^2)/2}\right)\right]^2,$$

which is, when $h > k$, a first approximation and a lower bound of I. In order to attain greater accuracy, write I in the form

$$I(h,k) = c(h,k)I_1(h,k),$$

where c is a correction factor function. Grauslund and Lind presented a table of c in terms of h and k and in addition two approximations:

- $c = 1.053$ if $2 \leq h \leq 8$ and $0 \leq k \leq 8$.
- $c = a_1 + a_2 k + a_3 h^2 + a_4 hk + a_5 h^3 + a_6 h^2 k + a_7 hk^2$, the values of the a's having been given by the authors.

Gideon and Gurland (1978) considered essentially the same function, though in different notation. They approximated it by $d(r, \theta)\frac{1}{2}[1 - \Phi(r)]$, where $r = \sqrt{h^2 + k^2}$, $\theta = \tan^{-1}\frac{k}{h}$, and $d(r, \theta)$ is approximated by $b_0\theta + (b_1 + b_2 r)r\theta + (b_3 + b_4 r)r\theta^3 + (b_5 + b_6 r)r\theta^5$, the values of the b's having been given by the authors (being different for different ranges of r).

11.12.6 Algorithms: Arbitrary Polygons

Cadwell (1951) was possibly the first one to consider the bivariate normal integral over an arbitrary polygon. His procedure was to transform X and Y into independent standard normal variates, by a rotation of axes followed by a change of scale, and then make use of the V function discussed earlier. His work was developed more formally by the National Bureau of Standards (1959); see also Johnson and Kotz (1972, pp. 99–100). We note that a linear transformation of the type mentioned above will leave unaltered the property of the polygon whether it is convex, simple, or self-intersecting.

11.12.7 Tables

We now summarize the major sets of tables relevant to the bivariate normal integral. For more details, one may refer to National Bureau of Standards (1959) and Greenwood and Hartley (1962, pp. 119–122).

Pearson (1901)	$L(h, k; \rho)$
Nicholson (1943)	$V(h, j)$
National Bureau of Standards (1959)	$L(h, k; \rho), V(h, \lambda h)$, and $V(\lambda h, h)$
Japanese Standards Association (1972)	$L(h, k; \rho)$ and $V(h, \lambda h)$
Owen(1956, 1962)	$T(h, \lambda)$
Smirnov and Bol'shev (1962)	$T(h, \lambda)$ and $T(h, 1)$

11.12.8 Computer Programs

The following list provides various programs that are available in the literature, with the last column indicating the type of code.

Donnelly (1973)	$L(h, k; \rho)$	FORTRAN
Young and Minder (1974) (corrections and remarks by other authors)	$T(h, \lambda)$	FORTRAN
Divgi (1979)	$L(h, k; \rho)$	
Bouver and Bargmann (1979)	$\Psi(h, k)$	FORTRAN
DiDonato and Hageman (1982)	Integral over polygon	
Baughman (1988)	$L(h, k; \rho)$	FORTRAN
Boys (1989)	$T(h, \lambda)$	FORTRAN
Drezner and Wesolowski (1990)	$L(h, k; \rho)$	FORTRAN
Goedhart and Jansen (1992)	$T(h, \lambda)$	FORTRAN

It should also be mentioned that the IMSL package includes a routine for evaluating the bivariate normal integral, and so do NWA, STATPAK, which is microcomputer-oriented [and for which we rely upon Siegel and O'Brien (1985) for the information], and STATLIB [Brelsford and Relies (1981, p. 370)]. STATLIB's method is somewhat unsophisticated, being based on Simpson's rule integration for the univariate cumulative normal function.

Computation of Bivariate Normal Integral Using R

R has been a very popular statistical package in recent years. The bivariate normal integrals can be computed by the function mvt in the R package mvtnorm. For implementation details, download the document Using mvtnorm from

`http://cran.r-project.org/web/packages/mvtnorm/index.html`.

11.12.9 Literature Reviews

Extensive literature reviews of the subject may be found in National Bureau of Standards (1959), Gupta (1963a,b), Johnson and Kotz (1972, Chapter 36), Martynov (1981), Patel and Read (1982, Chapter 10), and Kotz et al. (2000).

11.13 Testing for Bivariate Normality

Some of the discussion in this section is of general issues of discerning shape in empirical bivariate data but is included here since the bivariate normal is so often the benchmark in such situations.

11.13.1 How Might Bivariate Normality Fail?

As was pointed out earlier, the bivariate normal has been used extensively in empirical research. The question arises as to how we can know if the two random variables have a joint distribution that is bivariate normal. Among reviews assessing normality with bivariate (or higher-dimensional) data, we call attention to Kowalski (1970), Andrews et al. (1971), Gnanadesikan (1977), Mardia (1980), Small (1985), Csörgö (1986), and Looney (1995). There are many possible ways of departing from the bivariate normal; as a first step, we may classify them as:

- failure of marginals to be normal,
- normal marginals but failure to be bivariate normal, or
- failure to be normal after univariate transformations have made the marginals normal.

Broadly speaking, there are two methods of checking bivariate normality: graphical procedures, and formal tests of significance. The structure of the main part of this section is as follows:

- Graphical checks.
- Formal tests—univariate normality.
- Formal tests—bivariate normality.
- Tests of bivariate normality after transformations.
- Some comments and suggestions.

However, we shall first make some remarks about outliers.

11.13.2 Outliers

A failure to be bivariate normal may apparently be due to "outliers"—one observation or a few that seem to be separated from the others.

- The concept of an outlier in the bivariate or multivariate case is by no means as straightforward as it is for a univariate sample, and one may refer to Barnett and Lewis (1984, Chapter 9) for a further discussion on this. Other accounts are those of Hawkins (1980, Chapter 8) and Barnett (1983b).
- One important idea is to represent a multivariate observation $\mathbf{x}$ by some distance measure $(\mathbf{x}-\mathbf{x_0})'\boldsymbol{\Omega}^{-1}(\mathbf{x}-\mathbf{x_0})$, where $\mathbf{x_0}$ is a measure of location and $\boldsymbol{\Omega}$ is a measure of scatter; here, $\mathbf{x_0}$ might be the population mean (if known) or the sample mean, and $\boldsymbol{\Omega}$ might be the population or sample variance–covariance matrix. The bivariate observations can then be ordered according to this measure.

- For a transformation approach to handling outliers, see Barnett (1983a) and Barnett and Lewis (1984, Section 9.3.4). The idea in this case involves obtaining standard normal variates U_1 and U_2 by means of the equation $F(x) = \Phi(u_1), G(y|X = x) = \Phi(u_2)$.
- For outlier detection when the "linear structured model" applies—i.e., there are unobserved variates Z_1 and Z_2 connected by $Z_2 = \alpha + \beta Z_1$, with observed variates $X = Z_1 + \epsilon_1$ and $Y = Z_2 + \epsilon_2$, the interested reader may refer to Barnett (1985).
- For the "influence" approach to outlier detection, see Chernick (1983). The idea is to identify which observations have the biggest effect on some statistics of interest, such as the mean or the correlation. Chernick has reported an application to monthly consumption/generation data for power plants.
- Bacon-Shone and Fung (1987) have proposed a graphical method that they claim is good at detecting multiple outliers.
- For multivariate "trimming" (i.e., removal of extreme values), one may refer to Ruppert (1988).
- Building on univariate ideas of Green (1976), Mathar (1985) has discussed the classification of bivariate and multivariate distributions as "outlier-prone": A multivariate distribution is absolutely outlier-resistant if, with increasing sample size, the difference between the largest and the second-largest distances from the origin converges to zero in probability; it is relatively outlier resistant if the corresponding ratio converges to one in probability. However, with these definitions, the outlier behavior of a multivariate distribution is determined by its marginals, the dependence structure does not affect it. Consequently, it can hardly be said to be of multidimensional relevance if the marginals contain all the information.
- As an example of an applied work, we draw attention to Clark et al. (1987), whose variates were diastolic blood pressure; a feature of their investigation was how the identification of observations as outliers or not changed as various covariates (such as the nature of activity when blood pressure was measured) were taken into consideration.

11.13.3 Graphical Checks

Univariate Plotting

To check univariate normality, arrange the observations in order of size, calculate suitable plotting positions, and plot on the special graph paper that is available for this purpose (or convert the plotting positions to equivalent normal deviates and use ordinary graph paper). A straight line in such a plot indicates normality. An excellent account of this kind of technique is by D'Agostino (1986a); see also Harter (1984), Sievers (1986), and (for censored data) Michael and Schucany (1986). The plotting position $i/(n+1)$ is often used for the ith observation in an ordered sample size n, but there does not seem to be any consensus with regard to the best choice in the applied [see Cunnane (1978)] as well as the statistical literature.

Motivated by the phenomenon that the log returns of many financial problems are normally distributed, Hazelton (2003) proposed a normal log-density plot to assess normality. The idea is to plot the kernel density estimate and compare it with the log of a normal density.

Jones (2004) proposed an alternative by Hazelton (2003) based on his early work [Jones and Daly (1995)] by plotting

$$\log\{\phi[\Phi^{-1}((i-1/2)/n)]\} \quad \text{against} \quad x_{(i)},\ i=1,\ldots,n.$$

The last plot is simpler and is known as the normal log density probability plot.

Scatterplots

A well-known method that can be used to check bivariate normality is to draw a scatter diagram. If the sample observations do come from a bivariate normal distribution, the points will lie roughly within an elliptical region with a heavier concentration near the middle and with a gradually decreasing concentration away from the middle. A scatterplot may indicate non-normality or reveal outliers that, if included in the analysis, may give a spurious indication of non-normality or reveal outliers, or perhaps conceal a real departure from normality. For a listing of programs written in APL that create a scatterplot and superimpose contours of the bivariate normal p.d.f., see Bouver and Bargmann (1981). For the "sharpening" of a scatterplot to reveal its structure more clearly, see Section 11.18.8. For a review of "convex hulls" and other methods of "peeling" bivariate data (with rectangles or ellipses), one may refer to Green (1981, 1985).

For many ideas about elaborating scatterplots to bring out their meaning more clearly, see Chambers et al. (1983, especially Chapter 4).

In many applications, however, a scatterplot will be inconclusive and a formal test of goodness-of-fit may be required. Note also that a necessary condition for a bivariate normal is that the conditional means be linear and the conditional variances constant. Therefore, a plot of these statistics can be helpful in assessing bivariate normality as well.

F-Probability Plot

Ahn (1992) introduced an F-probability plot for checking multivariate normality. The plot is based on the squared jackknife distances which have an exact finite sampling distribution when a sample is taken from a multivariate normal distribution, and to provide test statistics, the F-probability plot correlation coefficient and the F-probability plot intercept. The former can be used to measure the linearity of the F-probability plot and the latter to detect extreme observations, and thus they can be used as numerical measures to assess multivariate normality.

Radii and Angles

Another approach to assessing bivariate normality, which is based on radii and angles, has been discussed by Gnanadesikan (1977, Chapter 5). The rationale for this method is as follows. Let $(X_1, X_2)'$ denote the bivariate normal vector with variance–covariance matrix $\mathbf{\Sigma}$. First, transform the original variates X_1 and X_2 to independent [2] normal variates X and Y using

$$\begin{pmatrix} X \\ Y \end{pmatrix} = \mathbf{\Sigma}^{-1/2} \begin{pmatrix} X_1 - \mu_1 \\ X_2 - \mu_2 \end{pmatrix}. \tag{11.52}$$

Second, transform (X, Y) to polar coordinates (R, Θ). Then, under the hypothesis of bivariate normality, R^2 has a χ^2_2-distribution (i.e., exponential with mean 2), and Θ has a uniform distribution over the range 0 to 2π. These consequences may be tested graphically—by plotting sample quantiles of R^2 against quantiles of the exponential distribution and similarly by plotting sample quantiles of $\mathit{\Theta}$ against quantiles of the uniform distribution. For illustration, see Gnanadesikan (1977, Exhibits 28i,j, 29d,e). If bivariate normality holds, the two plots should be approximately linear. However, if $\mu' = (\mu_1, \mu_2)$ and $\mathbf{\Sigma}$ are estimated, the distributional properties of R and Θ are only approximate. For $n \geq 25$, the approximation is usually good. The radii-and-angles approach, though informal, is an informative graphical aid. However, as in the case of scatterplots, the test may be inconclusive, particularly with small samples.

Project Pursuit

Alhough it is aimed at the multivariate situation rather than only the bivariate case, the method of projection pursuit [Friedman and Tukey (1974), Friedman and Stuetzle (1982), and Tukey and Tukey (1981)] should be mentioned. The strategy is to "pursue the projection"—i.e., find the vector—that most clearly reveals the non-normality of the data. At each step, an augmenting function is estimated as the ratio of the data to the model when projected

[2] There are infinitely many ways to transform a bivariate normal vector into two independent normal variates by decomposing $\mathbf{\Sigma}$ into products of two matrices. In addition, $\mathbf{\Sigma} = \mathbf{\Sigma}^{1/2}\mathbf{\Sigma}^{1/2}$ as represented by (11.52), and the Choleski decomposition $\mathbf{\Sigma} = \mathbf{LL}'$, where $\mathbf{L}$ is a lower triangular matrix, are popular.

onto a certain vector. The final model is the product of the initial model (such as the multivariate normal) and a series of augmenting functions.

In somewhat the same style is an idea described by Gnanadesikan (1977, pp. 142–143), in which univariate Box–Cox transformations—see Section 11.13.6 below—are repeatedly applied in different directions.

The Kernel Method

Silverman (1986, Chapter 4) has argued persuasively that both two-dimensional histograms and scatterplots are poor aids to grasping the structure of bivariate data and has therefore proposed the "kernel" method as an improvement. He has described this method of estimating a density as being the sum of bumps centered at the observations.

- Choose a kernel function $K(\cdot)$ and a window with width w.
- The density at a point $\mathbf{x}$ is then estimated to be

$$\hat{h}(\mathbf{x}) = \frac{1}{2w^2} \sum_{i=1}^{n} K\left(\frac{\mathbf{x} - \mathbf{x}_i}{w}\right), \tag{11.53}$$

  where $\mathbf{x_1}, \mathbf{x_2}, \ldots, \mathbf{x}_n$ are the n data points in the sample.
- $K(\cdot)$ is usually a radially symmetric unimodal p.d.f. such as the bivariate normal.
- The data should be rescaled to avoid extreme differences of spread in the various coordinate directions (or else the single smoothing parameter w should be replaced by a vector).
- Appropriate computer graphs are then used to display $\hat{h}(\mathbf{x})$ as a surface or contour plot.

Although one may fear that different $K(\cdot)$ and w may give different results, Silverman has provided an extensive discussion about these choices. He has given an example [also in Silverman (1981)] of 100 data points drawn from a bivariate normal mixture, smoothed using window widths 1.2, 2.2, and 2.8; the first appears undersmoothed, the last oversmoothed. And let us remember that the classes from which a histogram is constructed are arbitrary, too, and that the scales chosen for a scatterplot affect the subjective impression it gives. Other accounts of this are by Everitt and Hand (1981, Section 5.3) and Chambers et al. (1983, Section 4.10).

Tanabe et al. (1988) have presented FORTRAN subroutines for computing bivariate (and univariate) density estimates, using a Bayesian nonparametric method.

Haar Distribution

In the article by Wachter (1983), we notice the following paragraph: "On the whole, Haar measures have gained prominence in statistics with the realization that many consequences of multivariate normal assumptions do not depend on normality itself but only on rotational symmetry. For graphically based data analysis, symmetry assumptions are often preferable to parametric distributional assumptions such as normality. Thus, curiously enough, data analytic emphasis in multivariate statistics has promoted ties with highly mathematical theory of Haar distributions."

11.13.4 Formal Tests: Univariate Normality

Why Test?

As mentioned earlier, graphical checks of bivariate normality may be inconclusive, and hence a formal test of significance will perhaps give a more objective guide to the suitability of the bivariate normal distribution. Once again, as mentioned earlier, there are many different ways in which an empirical distribution may deviate from bivariate normality. This suggests that equally many techniques may be needed to spot such deviations.[3] According to Small (1985), "There is no single best method, and choice should be guided by what departure might be expected a priori or would have the most serious consequences."

We deal here with tests on the marginals to see if they are normal (remember that marginal normality is a necessary, though not sufficient, condition for bivariate normality). Fuller accounts are due to D'Agostino (1982, 1986b), Koziol (1986), and (for censored data) Michael and Schucany (1986).

Chi-Squared Test

Group the observations into a number of ranges of the variate. Determine the expected numbers that would fall into these groups under the normal distribution, and then calculate the statistic $\sum(O - E)^2/E$. The advantage of this is its ease and elementary nature. A minor technical disadvantage is that one never knows precisely how to carry out the grouping—having too few groups loses much information, whereas having too many groups means there are few observations per group, and hence the chi-squared approximation to the test statistic may be dubious. But the major disadvantage is the loss of information concerning the ordering of the groups and the consequent loss

[3] But there is a danger here—if lots of different tests are conducted, a "significant' result is quite likely to come about purely by chance.

of power. As an example, suppose the pattern of residuals was $++---+$ $++---++$. If X^2 just failed to indicate statistical significance, we would nevertheless suspect that the empirical distribution differed from the normal in kurtosis.

For more on this, see Moore (1986).

Moment Tests

The skewness ($\sqrt{b_1}$) and kurtosis (b_2) statistics are defined by $\sqrt{b_1} = m_3/m_2^{3/2}$ and $b_2 = m_4/m_2^2$, where m_i is the ith sample moment about the mean.

The values $\sqrt{b_1}/\sqrt{6/n}$ and $(b_2 - 3)/\sqrt{24/n}$ are both asymptotically normal. Consequently, D'Agostino and Pearson (1973) suggested adding together the squares of the standardized variates corresponding to these sample statistics and treating the results as χ_2^2-variate (i.e., exponential with mean 2), this test being sensitive to departures from normality in both skewness and kurtosis, though naturally not as powerful with regard to either as a specific test for that feature would be. And, of course, it is not so sensitive to departures from normality other than those that are reflected by the skewness and kurtosis statistics.

Tests based on $\sqrt{b_1}$ and b_2 have been reviewed by Bowman and Shenton (1986) and D'Agostino (1986b).

For a further discussion on $\sqrt{b_1}$ and its interpretation, see Rayner et al. (1995).

Z-Test of Lin and Mudholkar

The mean and the variance of a random sample are independently distributed if and only if the parent population is normal. This characterization was used as a basis for Lin and Mudholkar (1980) to develop a test, termed the Z test, for the composite hypothesis of normality against asymmetric alternatives.

Tests Based on the Empirical Distribution Function

The best-known among these tests is the Kolmogorov–Smirnov test.

Using the notation F_n for the empirical d.f. based on a sample size of n and F for the hypothesized distribution, the test involves calculating

$$D_n = \sup_{-\infty<x<\infty} |F_n(x) - F(x)|. \tag{11.54}$$

The Cramér–von Mises test is based on

$$W_n^2 = \int_{\infty}^{\infty} |F_n(x) - F(x)|^2 A[F(x)]dF(x), \tag{11.55}$$

where A is a non-negative weight function. There are other alternatives as well.

There are many difficulties with such tests. One is their frequent absence from elementary textbooks and computer packages. Another is perhaps that no one understands their properties when the parameters of the distribution have been estimated from the sample and that the sample size is small to moderate. Another is the labor involved. However, the chapter by Stephens (1986a) seems very thorough and helpful. See also Paulson et al. (1987) for other tests for multivariate normality based on empirical distribution functions.

Probability Plots

Tests have been proposed [for example, by Shapiro and Wilk (1965)] that are based on how far a probability plot of the type described in the beginning of this subsection is from a straight line (i.e., on its correlation coefficient). For a review of this approach, see Stephens (1986b).

CPIT Plots

CPIT stands for conditional probability integral transformation. One may refer to Quesenberry (1986a,b) for this method.

Jarque and Bera Test

A popular normality test that is based on the sample moments was proposed by Jarque and Bera (1980, 1987) and Bera and Jarque (1981).

The test statistic is given by

$$JB = n\left(\frac{\alpha_3^2}{6} + \frac{(\alpha_4 - 3)^2}{24}\right),$$

where

$$\alpha_3 = \frac{n^{-1}\sum_{i=1}^{n}(x_i - \bar{x})^3}{s^3}$$

and

$$\alpha_4 = \frac{n^{-1}\sum_{i=1}^{n}(x_i - \bar{x})^4}{s^4},$$

with $s^2 = \sum_{i=1}^{n}(x_i - \bar{x})^2/2$. Using the notation given under the moment tests, α_3 is simply $\sqrt{b_1}$ and α_2 is simply b_2.

Thadewald and Büning (2007) have investigated the power of several normality tests, including those of Shapiro and Wilk, Kolmogorov and Smirnov, and Cramér and von Mises. They concluded that the Jarque and Bera test is superior in power to its competitors for symmetric distributions with medium to long tails and for slightly long skewed distributions with long tails. However, the test is poor for distributions with short tails, especially if the shape is bimodal.

Zhang's Omnibus Test

Zhang (1999) proposed a test statistic Q for testing normality based on the ratio of two unbiased estimators of the standard deviation, q_1 and q_2. Mingoti and Neves (2003) discussed some properties of q_1 and q_2 and showed that the variance of q_1 increases as the true population variance increases. Huang and Wei (2007) have shown that q_1 is normally distributed so that the normality percentage points for Q are no longer appropriate. Using simulations, Huang and Wei (2007) recalculated the percentage points for Q.

11.13.5 Formal Tests: Bivariate Normality

Chi-Squared Test

It is easy, in principle, to compare observed and expected numbers in discretized bivariate distributions by means of Pearson's X^2 or a similar statistic, but the disadvantages mentioned in the subsection relating to the univariate case still remain.

Tests Based on the Empirical Distribution Function

Such tests have been proposed in the literature, but they do not seem to have received wide acceptance yet. One of them involves a statistic of Cramér–von Mises type, described by Pettitt (1979), who also discusses (on pp. 707–708) the kind of departure from normality that the test is and is not sensitive to. A related approach is via Rosenblatt's (1952) multivariate probability integral transformation or conditional probability integral transformation; see Quesenberry (1986a) and the references therein.

Tests Based on the Empirical Characteristic Function

Tests of this type have been put forward by Csörgö (1984, 1986) and Baringhaus and Henze (1988). The former also mentions he has an analogous test for Marshall and Olkin's (1967) distributions; see also Csörgö (1989). For a more recent treatment on this topic, see Naito (1996).

Malkovich and Afifi's (1973) Tests

These authors generalized the univariate skewness and kurtosis statistics, and the W statistics proposed by Shapiro and Wilk (1965), to the bivariate case using Roy's union–intersection principle [for which see Arnold (1988)]. They made use of the property of the bivariate normal distribution that any linear combination of X and Y is univariate normal. Formally, they defined bivariate skewness as $\max_{\mathbf{c}}[\beta_1(\mathbf{c})]$, where

$$\beta_1(\mathbf{c}) = \frac{\{E[c_1(X-\mu_1)+c_2(Y-\mu_2)]^3\}^2}{[\mathrm{var}(c_1X+c_2Y)]^2},$$

and bivariate kurtosis as $\max_{\mathbf{c}}\{[\beta_2(\mathbf{c})-3]^2\}$, where

$$\beta_2(\mathbf{c}) = \frac{E[c_1(X-\mu_1)+c_2(Y-\mu_2)]^4}{[\mathrm{var}(c_1X+c_2Y)]^2}$$

for some vector $\mathbf{c}' = (c_1, c_2)$. Using Roy's principle, one retains the null hypothesis of bivariate normality if $\max_{\mathbf{c}}[b_1(\mathbf{c})] \le k_{b_1}$ and $\max_{\mathbf{c}}\{[b_2(\mathbf{c}) - k]^2\} \le k_{b_2}$, where $b_1(\mathbf{c})$ and $b_2(\mathbf{c})$ are the sample counterparts of $\beta_1(\mathbf{c})$ and $\beta_2(\mathbf{c})$ here, with k being constants, such that $k \to 3$ as the sample size becomes infinitely large.

According to Bera and John (1983), these tests are conceptually simple but computationally burdensome.

Malkovich and Afifi (1973) introduced a measure of skewness based on an i.i.d. sample $\mathbf{x}_1, \ldots, \mathbf{x}_n$ of points in $\mathbb{R}^d$ as follows. For $\mathbf{u} \in \Omega_d$, where Ω_d is the unit d-dimensional sphere $\{\mathbf{x} \in \mathbb{R}^d : \|\mathbf{x}\| = 1\}$, let $b_1(\mathbf{u})$ denote the measure of skewness in the sample in the $\mathbf{u}$-direction given by

$$b_{1,n}(\mathbf{u}) = \frac{n\left\{\sum_{i=1}^n (\mathbf{u}^T(\mathbf{x}_i - \bar{\mathbf{x}}))^3\right\}^2}{\left\{\sum_{i=1}^n (\mathbf{u}^T(\mathbf{x}_i - \bar{\mathbf{x}}))^2\right\}^3},$$

where $\bar{\mathbf{x}} = \frac{1}{n}\sum_{i=1}^n \mathbf{x}_i$ is the sample mean. Their measure of skewness is then

$$b_{1,n}^* = \sup_{\mathbf{u}\in\Omega_d} b_{1,n}(\mathbf{u})$$

which is equivalent to

$$\sup_{\mathbf{u}\in\Omega_d}\left(\frac{1}{\sqrt{n}}\sum_{i=1}^{n}(\mathbf{u}^T\mathbf{z}_i)^3\right)^2 = \sup_{\mathbf{u}\in\Omega_d}\left(\mathbf{c}_{1,n}(\mathbf{u})\right)^2,$$

where $\mathbf{z}_1 = \mathbf{S}^{-1/2}(\mathbf{x}_i - \bar{\mathbf{x}})$, with $\mathbf{S}$ denoting the sample covariance matrix. Motivated by this, Balakrishnan et al. (2007) considered a signed measure of skewness statistic as

$$\mathbf{T_n} = \int_{\Omega_d} \mathbf{u}c_{1,n}(\mathbf{u})d\lambda(\mathbf{u}),$$

where λ is a rotationally invariant probability measure on Ω_d, and proposed a chi-square statistic $Q_n = \mathbf{T}_\mathbf{n}^\mathrm{T}\mathbf{D}^{-1}\mathbf{T_n}$, where $\mathbf{D}$ is an estimate of the covariance matrix of $\mathbf{T_n}$, for testing for symmetry of the population distribution. They also evaluated the power performance of this test empirically.

Cox and Small's Test

These authors based their method on the extent of nonlinearity of the regression line. Specifically, it involves the coefficients of quadratic terms when Y is regressed on X and X^2 and X is regressed on Y and Y^2. A statistic that is asymptotically χ_2^2 (i.e., exponential with mean 2) can be calculated. A disadvantage of this procedure is that the bivariate normal is not the only distribution having normal marginals and linear regressions, there are many others. (Any mixture of two bivariate normal distributions having the same means and standard deviations provides an example.)

Hawkins' (1981) Procedure

In this paper, a procedure was proposed that can be used to test for normality and homoscedasticity simultaneously. Considerable use of this has been made in the book by McLachlan and Basford (1988).

Invariant Tests

Loosely speaking, a test procedure that is unaltered under arbitrary affine transformation of the underlying data is considered to be an invariant test. Thus, tests for bivariate normality based on departures from the empirical distribution of the D_i^2 from their postulated chi-squared cumulative distributions are invariant. Here, $D_i^2 = (X_i - \bar{X})'S^{-1}(Y_i - \bar{Y})$, where S is the sample covariance matrix. Koziol (1982) pointed out that in addition to the test described, a Cramér–von Mises test and normality tests yb Malkovich and Afifi (1973) and Hawkins (1981) are all members of a family of invariant tests. For more on invariant and consistent tests for multivariate normality, see Henze and Zirkler (1990). Recently, Henze (2002) presented a critical review of multivariate normality.

Bera and John's (1983) Tests

These authors considered the bivariate Pearson family of distributions (which includes the bivariate normal). They then used Rao's (1948) score principle to develop four tests for bivariate normality. Each of the test statistics was shown to have an asymptotic chi-distribution. They also compared the powers of their tests with those of Mardia's (1970b) tests (to be discussed shortly).

Bivariate Skewness and Kurtosis

Many test statistics involve sample product moments and are asymptotically distributed as chi-squared.

Let $(x_1, y_1), (x_2, y_2), \ldots, (x_n, y_n)$ be n pairs of independent observations from a bivariate population. Define

$$\begin{pmatrix} z_{1i} \\ z_{2i} \end{pmatrix} = \begin{pmatrix} \hat{\sigma}_1^2 & r\hat{\sigma}_1\hat{\sigma}_2 \\ r\hat{\sigma}_1\hat{\sigma}_2 & \hat{\sigma}_2^2 \end{pmatrix}^{-1/2} \begin{pmatrix} x_i - \bar{x} \\ y_i - \bar{y} \end{pmatrix} \tag{11.56}$$

for $i = 1, 2, \ldots, n$, where $\bar{x}$ and $\bar{y}$ are the sample means, $\hat{\sigma}^2$'s are the maximum likelihood estimates of the variances (i.e., they are the sample variances with n in the divisor), and r is the sample correlation coefficient. Next, let us denote

$$m_{ij} = \frac{1}{n} \sum_{k=1}^{n} z_{1k}^i z_{2k}^j. \tag{11.57}$$

We may then define test statistics in terms of sample product moments. Mardia's (1970b) tests are based on sample measures of bivariate skewness and kurtosis, defined as

$$b_{1,2} = m_{30}^2 + m_{03}^2 + 3(m_{21}^2 + m_{12}^2) \tag{11.58}$$

and

$$b_{2,2} = \frac{1}{n} \sum_{j=1}^{n} (z_{1j}^2 + z_{2j}^2)^2 = m_{40} + m_{04} + 2m_{22}, \tag{11.59}$$

respectively.

Univariate skewness and kurtosis are functions of the third and fourth central moments, respectively. Here, $b_{1,2}$ is a function of product moments of order 3, and $b_{2,2}$ is a function of product moments of order 4, thus indicating their appropriateness as bivariate skewness and kurtosis measures, respectively. But note that, in each case, the first two terms are functions of univariate statistics—so the formulas do not give measures free of these. Tests derived from these measures are large-sample tests [Mardia (1974)]; indeed, Reyment (1971) confirmed that Mardia's procedures do not stabilize until a large number of observations have been included in the sample. Mardia (1974) presented tables of critical values of $b_{1,2}$ and $b_{2,2}$ in (11.58) and (11.59) for several choices of n and levels of significance.

Mardia has formulated several other tests; among them are S_W^2 and C_W^2, which combines $b_{1,2}$ and $b_{2,2}$ [see Mardia and Foster (1983, pp. 212–213) and Mardia (1985)].

Monte Carlo comparisons of the behavior of measures introduced by Mardia, Malkovich and Afifi, and others, in circumstances where the distribution is a mixing of two bivariate normals, were made by Isogai (1983a,b), who then attempted to clarify the meaning of these statistical diagnostic tools for measuring non-normality. Other comparisons of several proposals include those by Ulrich (1984) and Booker et al. (1984).

Schwager (1985) discussed notions of multivariate skewness and kurtosis proposed by Mardia, Malkvoich and Afifi, Isogai, and others.

According to the assessment of Looney (1986), $b_{1,2}$ and $b_{2,2}$ are the most thoroughly developed tests for bivariate normality, including a published algorithm by Mardia and Zemroch (1975). But they appear to be less powerful than Bera and John's (1983) tests.

For more recent developments, see Móri et al. (1993), and Henze (1994, 1997a). Henze (1997b) has considered a weighted sum of Mardia's measure of multivariate skewness and a sample version of a skewness measure introduced by Móri et al. (1993).

Use Tests for Univariate Normality to Assess Multivariate Normality

There are several techniques for assessing multivariate normality based on well-known tests for univariate normality. For example, Mudholkar et al. (1992) developed a multivariate adaption of the Lin and Mudholkar (1980) z-test of univariate normality, The p-variate adaption of the Shapiro–Wilk test of normality has been considered by Mudholkar et al. (1995). Looney (1995) has also described several such tests.

Best and Rayner's Comparisons

Koziol (1986, 1987) discussed certain statistics in the style of Neyman's "smoothed" tests. Best and Rayner (1988) presented Koziol's formulas adapted for the bivariate case. The first of these is simply $nb_{1,2}/6$ and is thus equivalent to Mardia's test based on $b_{1,2}$. The second is defined by

$$\hat{U}_4^2 = n\left[\frac{(m_{22}-1)^2}{4} + \frac{m_{31}^2 + m_{13}^2}{6} + \frac{(m_{04}-3)^2 + (m_{40}-3)^2}{24}\right]. \tag{11.60}$$

Best and Rayner (1988) compared the (approximate) powers of the test statistics $\hat{U}_3^2, \hat{U}_4^2, (U_3^2 + U_4^4)$, $b_{2,2}$, and S_W^2 with the level of significance set at 5%, under the following alternatives:

- X and Y being independent lognormal variables,
- X and Y being independent uniform variables,
- X and Y being independent t_4 variables (i.e., having the t-distribution with 4 degrees of freedom), and
- (X, Y) having various bivariate normal mixture distributions.

Their conclusion was that no single statistic dominates, although $(\hat{U}_3^2 + \hat{U}_4^2)$ usually does better than S_W^2, a statistic recommended by Mardia.

Asymptotically χ_2^2?

For a number of proposed tests, it happens that the test statistic is asymptotically distributed as χ_2^2. For the two tests they investigated, Mason and Young (1985) found that this approximation can be conservative, and lead to inappropriate rejection of normality, when the population parameters in the formulas are replaced by their sample estimates.

Comparison of Tests for Bivariate Normality with Unknown Parameters by Transformation to a Univariate Statistic

Versluis (1996) has compared 15 tests for bivariate normality with unknown parameters. The bivariate dataset will first be transformed into a univariate statistic. For their test #1 to test #12, the dataset is transformed into the set of variables $\{z_i\}$ as

$$z_i = \frac{1}{1-R^2}\left\{\left(\frac{x_i - \hat{\mu}_1}{\hat{\sigma}_1}\right)^2 - \frac{2R(x_i - \hat{\mu}_1)(y_i - \mu_2)}{\hat{\sigma}_1\hat{\sigma}_2} + \left(\frac{y_i - \hat{\mu}_2}{\sigma_2}\right)^2\right\}, \tag{11.61}$$

where $\hat{\mu}_i, \hat{\sigma}_i$, and R are as defined in (11.26) (see Section 11.8).

The first 12 statistics are the Kolmogorov–Smirnov test, Cramér–von Mises test, Kuiper test, Watson test, Anderson–Darling test, Rényi test (L1), Rényi test (L2), Rényi test (U1), Rényi test (U2), Brain–Shapiro test, and Shapiro–Wilk–Stephens test with the test statistic given by

$$T_{SWS} = \frac{4(n-1)^2}{n(n+1)\left(\sum_i^n z_i^2\right) - 4n(n-1)^2}. \tag{11.62}$$

The last three tests are the bivariate Shapiro–Malkovich test, Shapiro–Malkovich skewness test (b_1), and Shapiro–Malkovich kurtosis test (b_2). The last two are simply those presented above under Malkovich and Afifi (1973). Based on this comparative study, Versluis (1996) found that the Shapiro–Wilk–Stephens test in (11.62) performs very well for all the alternative distributions considered.

Computational Aspect of Normality Tests: FORTRAN Subroutines and SAS Procedures

As recommended by many authors, a reasonable first step in assessing multivariate normality is to test each variable for univariate normality. Of the many procedures available for assessing univariate normality, two of the most commonly used are (1) an examination of skewness and kurtosis and (2) the Shapiro–Wilk [Shapiro and Wilk (1965)] W test. Looney (1995) suggested that a next logical step after testing each of the variables for univariate normality is to apply some computationally simple tests for the multivariate case that are based on the two univariate tests just mentioned. Looney went on to argue that, given the availability of the reliable software for performing these tests [for example, by Royston (1982) and D'Agostino et al. (1990)], computational algorithms for the multivariate normality tests can be developed with a minimal effort.

Looney (1995) described the specific FORTRAN subroutines and SAS procedures and functions that were used for each of the following four normality tests that are based on tests for univariate normality. (The resulting SAS macros and FORTRAN programs are available at no charge from this author):

- Royston's (1983) H test: A multivariate extension of the Shapiro–Wilk test;
- Small's (1980) Q_1 and Q_2: Multivariate extensions of univariate skewness ($\sqrt{b_1}$) and kurtosis (b_2);
- Srivastava's (1984) measures of multivariate skewness and kurtosis;
- Srivastava and Hui's (1987) Shapiro–Wilk tests.

11.13.6 Tests of Bivariate Normality After Transformation

A popular approach[4] to understanding the bivariate distribution with non-normal marginals is to (i) transform the marginals to normality, (ii) check that the bivariate distribution is roughly bivariate normal in appearance, and then (iii) proceed with the analysis under the assumption of bivariate normality. Separate statements can then be made about the univariate transformations that were necessary and the conclusions drawn from the bivariate transformed observations. This procedure, we feel, has much to recommend it; a technical disadvantage is that the properties of tests for bivariate normality are even less understood when applied to raw observations, we believe. Occasionally, a more fundamental objection arises when we have an explicit model for how the distribution is constructed, albeit with some uncertainties, as when we are assuming a trivariate reduction model but do not know the forms of component distributions; in such a case, the bivariate distribution is intimately tied to its marginals, and a procedure that separates the treatment of the individual variables from the treatment of their association may be thoroughly undesirable.

If we follow this strategy, we have to decide how to transform the marginals to normality:

- Do we enforce exact normality by calculating $\Phi^{-1}[F(x)]$ for each observation? If so, then a question will arise as to how to interpret this in the context of a sample, i.e., whether the best estimate of F is $i/(n+1)$ when x is the ith smallest observation in the sample of size n or something else.
- Alternatively, do we insist on some easily comprehended transformation, such as the logarithm, or a power function? If so, how much effort should we put into searching for the best transformation? Should we just try one or two of the best-known ones? Should we consider a whole parametric family, allowing the data to determine the parameter that gives the best fit?

As to easily comprehended transformations, we note the following:

- Probably the most popular single choice is the logarithm.
- Johnson's (1949) system of bivariate distributions consists of the following transformations applied to the marginals of the bivariate normal: logit, $\sinh^{-1}$, log, and none. If choosing from this set, we can then work with the original observations and one of Johnson's distributions, if we prefer doing that to working with transformed observations and the bivariate normal. Further details of this system are presented in Section 11.16.2.

---

[4] See, especially, Kowalski (1970). The main portions of this paper were tests for univariate normality; tests for bivariate normality; the coordinate transformation to normality and its estimation; summary of results of tests for bivariate normality; and application to correlation theory.

- The most popular parametric family is that of Box and Cox (1964); see Box and Tiao (1973, Chapter 10) and, for computer programs, Howarth and Earle (1979), Liem (1980), and the references contained therein. Andrews et al. (1971) extended this to bivariate distributions. The family is that of power transformations, with the logarithm as a special case. The following discussion presents more details.

The Box–Cox transformation of (x, y) to (z_1, z_2) is as follows:

$$z_1 = \begin{cases} (x^{\lambda_1} - 1)/\lambda_1 & \text{for } \lambda_1 \neq 0 \\ \log x & \text{for } \lambda_1 = 0 \end{cases}, \tag{11.63}$$

$$z_2 = \begin{cases} (x^{\lambda_2} - 1)/\lambda_2 & \text{for } \lambda_2 \neq 0 \\ \log x & \text{for } \lambda_2 = 0 \end{cases}. \tag{11.64}$$

One might choose the λ's in (11.63) and (11.64) so as to make the z's as (univariate) normal as possible—the method of Box and Cox would be applied to each variable separately. But perhaps bivariate normality is not optimized thereby. A likelihood approach to achieving joint normality is as follows. First, express the bivariate density function of (X, Y) in terms of the bivariate normal density with mean $\boldsymbol{\mu}$ and variance–covariance matrix $\boldsymbol{\Sigma}$. Next, find the log-likelihood function of $\boldsymbol{\mu}$, $\boldsymbol{\Sigma}$, and $\boldsymbol{\lambda}$, where $\boldsymbol{\lambda}' = (\lambda_1, \lambda_2)$. By keeping λ_1 and λ_2 fixed temporarily, we can find the maximum likelihood estimates $\hat{\boldsymbol{\mu}}$ and $\hat{\boldsymbol{\Sigma}}$, and the maximized log-likelihood function is then

$$L^*(\lambda_1, \lambda_2) = -\frac{n}{2} \log |\hat{\boldsymbol{\Sigma}}| + (\lambda_1 - 1) \sum_{i=1}^{n} \log x_i + (\lambda_2 - 1) \sum_{i=1}^{n} \log y_i. \tag{11.65}$$

Next, the maximum likelihood estimates $\hat{\lambda}_1$ and $\hat{\lambda}_2$ may be obtained numerically by maximizing (11.65) with respect to λ_1 and λ_2. Andrews et al. (1971) showed that $2[L^*(\hat{\lambda}_1, \hat{\lambda}_2) - L^*(1, 1)]$ has asymptotically a χ^2_2-distribution (i.e., exponential distribution with mean 2). Rejection of the null hypothesis $\boldsymbol{\lambda}' = (1, 1)$ implies non-normality of the original data.

11.13.7 Some Comments and Suggestions

This subject, in our opinion, is in a rather unsatisfactory state. The various pieces of knowledge do not seem to be well integrated and there seem to be gaps between them. There seems to be rather little experience with tests that have been put forward. Hence, there is a lack of knowledge about their properties—behavior with samples of small to moderate size and with observations that are merely crudely grouped are two areas we have in mind. Moreover, even if we knew how to measure non-normality, would we also know how much non-normality is present?

One thing that helps sometimes is a careful thought. Stimulated by contact with data and by knowledge of various types of departures from bivariate normality that have been discussed theoretically, one can sometimes understand what is revealed by the data.

If data and theory are not sufficient in the abstract, then the context may serve to focus ideas. In particular:

- Does the mechanism generating the data direct attention to a particular class of distributions, such as those constructed by compounding or from a univariate distribution of X and a set of conditional distributions of Y given X?
- To what use is the result going to be put? For instance, is the correlation coefficient important? Is the survival function $\Pr(X > x, Y > y)$ important?

Many univariate distributions form a hierarchy—for example, the exponential is a special case of the gamma, which is a special case of Stacy's generalized gamma, which in turn can have a shift parameter included, and so on. This provides a natural environment for testing the goodness-of-fit: fit a three- or four-parameter distribution, and test whether the parameter values are consistent with some special case that corresponds to one- or two-parameter distributions. Such a procedure is much less common with bivariate distributions because such hierarchies are not so well known. Nonetheless, it is a desirable one, if applicable.

If more specific procedures do not come to mind, we suggest the one mentioned above. In summary:

- Consider the marginals. Ask what shape they have. Answer this by various forms of probability plotting, calculation of moments, and comparison of the goodness-of-fit of members of a hierarchy of distributions.
- If marginals appear to be non-normal, transform them to normality.
- Does bivariate normality hold? Answer this by using various graphical procedures and calculations (of moments, for example).

We note that if a scatter diagram prepared after transformation to marginal normality still fails to be bivariate normal, then a comparison with Figure 1 of Johnson et al. (1984, p. 242) may provide some insight into the underlying bivariate distributions.

Other choices of the standard form for the marginals may be equally or more suitable than the normal. The uniform is the obvious competitor—in this case, Mardia (1970a, p. 81) has suggested focusing attention on the regression and scedastic curves.

11.14 Distributions with Normal Conditionals

Arnold et al. (1999) have devoted a chapter of their book (Chapter 3) on bivariate distributions having conditional densities of the normal form. In general, these distributions do not have normal marginals. A special case that has a simple joint density function is known as the "bivariate normal" with centered normals. This distribution has been studied by Sarabia (1995), and its properties were discussed in Section 6.2.5. The bivariate distributions with normal conditionals were discussed in Section 6.2.

11.15 Bivariate Skew-Normal Distribution

There are at least two versions of bivariate skew-normal distributions.

11.15.1 Bivariate Skew-Normal Distribution of Azzalini and Dalla Valle

The density function of the bivariate skew-normal distribution, as given by Azzalini and Dalla Valle (1996), is

$$h(x, y) = 2\psi(x, y; \omega)\Psi(\lambda_1 x + \lambda_2 y), \tag{11.66}$$

where

$$\lambda_1 = \frac{\delta_1 - \delta_2\omega}{\sqrt{(1-\omega^2)(1-\omega^2-\delta_1^2-\delta_2^2+2\delta_1\delta_2\omega)}} \quad \text{and} \quad \lambda_2 = \frac{\delta_2 - \delta_1\omega}{\sqrt{(1-\omega^2)(1-\omega^2-\delta_1^2-\delta_2^2+2\delta_1\delta_2\omega)}}$$

and ψ and Ψ are the bivariate normal density and univariate normal distribution, respectively. For a more detailed discussion, see Section 7.4.5.

The joint distribution of LBM (lean body mass) and BMI (body mass index) of a sample of 202 Australian athletes was fitted in Azzalini and Dalla Valle (1996) by a bivariate skew-normal distribution.

11.15.2 Bivariate Skew-Normal Distribution of Sahu et al.

Sahu et al. (2003) developed a new class of bivariate (multivariate) skew-normal distributions using transformation and conditioning. Azzalini and Dalla Valle (1996) obtained their skew-normal distribution by conditioning on one suitable random variable being greater than zero, whereas Sahu et al. (2003) condition on as many random variables as the dimension of the normal variables. In the one-dimensional case, both families are identical.

Formula of the Joint Density

Let $\boldsymbol{z}=(x,y)$.

$$h(x,y) = 4|\Sigma + D^2|^{-1/2}\psi\left\{(\Sigma + D^2)^{-1/2}(\boldsymbol{z} - \boldsymbol{\mu})\right\}\Pr(\boldsymbol{V} > \boldsymbol{0})$$

where $\Sigma = \begin{pmatrix} \sigma_1^2 & \sigma_1\sigma_2\rho \\ \sigma_1\sigma_2\rho & \sigma_2^2 \end{pmatrix}$ is the covariance matrix for the bivariate normal and D is the diagonal matrix with elements δ_1 and δ_2, which can be both positive or both negative. Here $\boldsymbol{V}$ is distributed as a bivariate normal with mean matrix $D(\Sigma+D^2)^{-1}(\boldsymbol{z}-\boldsymbol{\mu})$ and covariance matrix $I - D(\Sigma+D^2)^{-1}D$.

If $\Sigma = \mathrm{diag}(\sigma_1^2, \sigma_1^2)$, the X and Y are independent, having density

$$\begin{aligned} h(x,y) &= 2(\sigma_1^2 + \delta_1^2)^{-1/2}\phi\left(\tfrac{x-\mu_1}{\sqrt{\sigma_1^2+\delta_1^2}}\right)\Phi\left(\tfrac{\delta_1}{\sigma_1}\tfrac{x-\mu_1}{\sqrt{\sigma_1^2+\delta_1^2}}\right) \\ &\quad \times 2(\sigma_2^2 + \delta_2^2)^{-1/2}\phi\left(\tfrac{y-\mu_2}{\sqrt{\sigma_2^2+\delta_2^2}}\right)\Phi\left(\tfrac{\delta_2}{\sigma_2}\tfrac{y-\mu_2}{\sqrt{\sigma_2^2+\delta_2^2}}\right). \end{aligned}$$

Moment Generating Function

The moment generating function is given by

$$M(s,t) = 4\Psi(D\boldsymbol{t})\exp\left\{\boldsymbol{t}'\boldsymbol{\mu} + \boldsymbol{t}(\Sigma + D^2)\boldsymbol{t}/2\right\},$$

where $\boldsymbol{t} = (s,t)$,

$$\begin{aligned} E(X) &= \mu_1 + (2/\pi)^{1/2}\delta_1, \quad E(Y) = \mu_2 + (2/\pi)^{1/2}\delta_2, \\ \mathrm{var}(X) &= \sigma_1^2 + (1 - 2/\pi)\delta_1, \quad \mathrm{var}(Y) = \sigma_1^2 + (1 - 2/\pi)\delta_2, \\ \mathrm{corr}(X,Y) &= \frac{\rho\sigma_1\sigma_2}{\sqrt{\sigma_1^2 + (1 - 2/\pi)\delta_1}\sqrt{\sigma_1^2 + (1 - 2/\pi)\delta_2}}. \end{aligned}$$

Remark. Sahu et al. (2003) pointed out that because the matrix D is assumed to be diagonal, the introduction of skewness does not affect the correlation structure. It changes the values of the correlations, but the structure remains the same.

Applications

Ghosh et al. (2007) have considered the bivariate random effect model using this skew-normal distributions with applications to HIV-RNA that are in the blood as well as seminal plasma for HIV-AIDS patients.

11.15.3 Fundamental Bivariate Skew-Normal Distributions

A new class of multivariate skew-normal distributions, fundamental skew-normal distributions, and their canonical version, was developed by Arellano-Valle and Genton (2005). It contains the product of independent univariate skew-normal distributions as a special case. The joint distribution does not have an explicit form.

11.15.4 Review of Bivariate Skew-Normal Distributions

Azzalini (2005) provides a comprehensive review of the skew-normal distribution and related skew-elliptical families. The article also provides applications to many practical problems. An introductory overview of the subject is given by Azzalini (2006).

11.16 Univariate Transformations

11.16.1 The Bivariate Lognormal Distribution

If $\log X$ and $\log Y$ have a bivariate normal distribution with means μ_1 and μ_2, variances σ_1^2 and σ_2^2, and correlation ρ, then

$$E(Y|X=x) = x^{\rho\sigma_2/\sigma_1} \exp\Big[-\tfrac{1}{2}(1-\rho^2)\sigma_2^2 + \mu_2 - \rho\sigma_2\mu_1/\sigma_1\Big], \tag{11.67}$$

$$\operatorname{var}(Y|X=x) = \omega'(\omega'-1)x^{2\rho\sigma_2/\sigma_1} \exp[2(\mu_2 - \rho\sigma_2\mu_1/\sigma_1)], \tag{11.68}$$

where $\omega' = \exp[(1-\rho^2)\sigma_2^2]$; see Johnson and Kotz (1972, p. 19). The joint moments are given by

$$\mu'_{ij} = E(X^iY^j) = \exp[i\mu_1 + j\mu_2 + \tfrac{1}{2}(i^2\sigma_2^2 + 2ij\rho\sigma_1\sigma_2 + j^2\sigma_2)]. \tag{11.69}$$

In particular, we have as the covariance and correlation

$$\operatorname{cov}(X,Y) = [\exp(\rho\sigma_1\sigma_2) - 1]\exp[\mu_1 + \mu_2 + (\sigma_1^2 + \sigma_2^2)/2] \tag{11.70}$$

and

$$\operatorname{corr}(X,Y) = \frac{\exp(\rho\sigma_1\sigma_2) - 1}{\sqrt{[\exp(\sigma_1^2 - 1][\exp(\sigma_2^2) - 1]}}. \tag{11.71}$$

For the meaningfulness or otherwise of the correlation coefficient in this case, see Section 11.16.5. Thomopoulos and Longinow (1984) have listed the basic properties of the bivariate lognormal distribution. The application they envisaged for it is in structural reliability analyses in which load and resistance are correlated.

Applications of Bivariate Lognormal Distributions

- Sizes and shapes of animals often can be modeled by the bivariate lognormal distribution. For a review, see Section 19.3.1 of Hutchinson and Lai (1990).
- Basford and McLachlan (1985) used a mixture of bivariate lognormal distributions in analyzing AHF activity and AHF-like antigen in normal women and hemophilia A carriers.
- Schneider and Holst (1983) and Holst and Schneider (1985) have used the bivariate lognormal distribution to describe the diameter D and length L of airborne man-made mineral fibers; see also Cheng (1986).
- Hiemstra and Creese (1970) wanted to simulate chronological sequences of precipitation data. In doing this, they assumed bivariate normal distributions of several pairs of variables, including duration and amount of precipitation.
- Cloud-seeding experiments commonly use a target and a control area. An analysis of an experiment in Colorado was reported by Mielke et al. (1977), who took X to be the precipitation in the target area and Y to be the precipitation in the control area, and assumed these variates to follow a bivariate lognormal distribution.
- Kmietowicz (1984) applied the bivariate lognormal distribution to a cross-tabulation of household size and income in rural Iraq and found it gave a satisfactory fit.
- Burmaster (1998) used bivariate lognormal distributions for the joint distribution of water ingestion and body weight for three groups of women (controls, pregnant, and lactating,all 15–49 years of age) in the United States.
- Yue (2002) used the bivariate lognormal normal distribution as a model for the joint distribution of storm peak (maximum rainfall intensity) and storm amount (volume). The model was found to be appropriate for describing multiple episodic events at the Motoyama meteorological station

in Japan. The data consisted of 96-year daily rainfall data from 1896 to 1993 (except the years 1939 and 1940). See also an earlier paper by Yue (2000) where the bivariate lognormal model was used for fitting correlated food peaks and volumes and correlated food volumes and durations.

- Lien and Balakrishnan (2006) considered the random vector (X, Y) to have a bivariate lognormal distribution with parameters $\mu_X, \mu_Y, \sigma_X, \sigma_Y, \rho$; that is, the random vector $(\ln X, \ln Y)$ has a bivariate normal distribution with means (μ_x, μ_Y), variances (σ_X^2, σ_Y^2), and correlation coefficient ρ. Then, under a multiplicative constraint of the form $X^a Y^b \leq K$, they derived explicit expressions for single and product moments and showed that the coefficients of variation always decrease regardless of the multiplicative constraint imposed. They also evaluated the effects of such a constraint on the variances and covariance, and presented conditions under which the correlation coefficient increases, and finally applied these results to futures hedging analysis and some other financial applications.

11.16.2 Johnson's System

Derivation

Bivariate distributions with specified marginals may be obtained from the bivariate normal by stretching and compressing the X and Y axes as required. Johnson (1949) constructed what has become a well-known system of distributions as follows. The bivariate distributions are denoted by S_{IJ}, in which one variable has an S_I-distribution and the other has an S_J-distribution, where I and J can be B, U, L, or N (standing for bounded, unbounded, lognormal, and normal). Thus, the variables

$$Z_1 = \gamma_1 + \delta_1 a_I \left(\frac{X - \xi_1}{\lambda_1} \right), \tag{11.72}$$

$$Z_2 = \gamma_2 + \delta_2 a_J \left(\frac{Y - \xi_2}{\lambda_2} \right), \tag{11.73}$$

where $a_B(y) = \log[y/(1-y)]$, $a_U(y) = \sinh^{-1} y$, $a_L(y) = \log y$, and $a_N(y) = y$, are standardized (unit) normal variables with correlation coefficient ρ.

Chapter 5 of Johnson's (1987) book discussed this system in detail, and so we can be brief here. The great advantage is that the simplicity of derivation makes variate generation for simulation studies equally simple; see also Rodriguez (1983, pp. 239–240).

For the case S_{NL}, see Yuan (1933), Crofts (1969), Crofts and Owen (1972), Suzuki (1983), and Suzuki et al. (1984).

Formula of the Joint Density

The joint distribution of X and Y, defined through 11.72 and 11.73, has nine parameters: γ_1, γ_2, δ_1, δ_2, $\xi_1, \xi_2, \lambda_1, \lambda_2$, and ρ. The standard form of the distribution is obtained by taking $\xi_1 = \xi_2 = 0, \lambda_1 = \lambda_2 = 1$. The joint density is

$$h(x, y) = \delta_1 \delta_2 \psi[\gamma_1 + \delta_1 f_I(x), \gamma_2 + \delta_2 f_J(y); \rho]. \tag{11.74}$$

Univariate Properties

X and Y have S_I and S_J distributions, respectively.

Conditional Properties

The conditional distribution of Y, given $X = x$, is of the same system (S_J) as Y but with γ_2, δ_2 replaced by $\frac{1}{\sqrt{1-\rho^2}}\{\gamma_2 - \rho[\gamma_1 + \delta_1 f_I(x)]\}$, $\frac{1}{\sqrt{1-\rho^2}}\delta_2$, respectively; see Johnson and Kotz (1972, pp. 15–17).

The regressions and conditional variances are given by Mardia (1970a, pp. 25–26), and a table of median regressions has been given by Johnson and Kotz (1972, p. 17) and Rodriguez (1983).

References to Illustrations

Johnson (1987, p. 63) has remarked that, "The only difficulty in employing the system in simulation work is to specify appropriate parameter combinations to meet the needs of particular applications." To assist with this, he has given numerous contour plots for the S_{LL}, S_{UU}, and S_{BB} cases. (In number, 24, 40, and 84 contour plots, plus five density surface plots of S_{BB} distributions). These may be equally useful as an aid to distribution selection when wondering whether any are suitable to fit an empirical dataset or not.

DeBrota et al. (1988) have described software for fitting the univariate Johnson system to data and to assist in making a subjective visual choice of an appropriate member of the system in the absence of data. They mention at the end that they are developing corresponding multivariate software.

Applications of Johnson's System

- The dimension of trees such as height, diameter, and volume; see Warren (1979), Schreuder and Hafley (1977), and Hafley and Buford (1985).
- Policy analysis; see Wilson (1983).

11.16.3 The Uniform Representation

An explicit expression for this is, for example, in Barnett (1980) and Johnson et al. (1981). Barnett has illustrated contours of the p.d.f. for the case $\rho = 0.8$.

11.16.4 The g and h Transformations

We mentioned earlier Tukey's g and h family of univariate distributions. Johnson (1987, pp. 205–206) has suggested applying this transformation to the marginals of a bivariate normal distribution.

11.16.5 Effect of Transformations on Correlation

It is well known that if we start with a bivariate normal distribution and apply any nonlinear transformation to the marginals, Pearson's product-moment correlation coefficient is smaller (in absolute magnitude) in the resulting distribution than in the original normal distribution. Rank correlation coefficients are, of course, unaltered, provided the transformations are monotone.

An extensive quantitative study of the effect of the marginal transformations on the correlation coefficient was reported by Der Kiureghian and Liu (1986). Their dependent variable was the ratio ρ/ρ_t, where ρ is the correlation coefficient in the normal distribution and ρ_t is the correlation coefficient in the distribution after transformation of the marginals. They gave a series of empirically derived formulas for calculating this ratio based on ρ_t and δ_t, where δ_t is the coefficient of variation in the transformed distribution. (Note that they were envisaging ρ_t being known and ρ being wanted and not vice versa.) The marginal distributions they considered were the following: uniform, shifted exponential, shifted Rayleigh, type I largest value, type I smallest value, lognormal, gamma, type II largest value, and type III smallest values.

- The simplest formulas were $\frac{\rho}{\rho_t}=$ constant for the one with the marginal being normal and the other being one of the first five in the list above.
- At the other extreme, the most complicated formula was $\frac{\rho}{\rho_t}$, as a 19-term polynomial in ρ, δ_{t_1}, and δ_{t_2}, for the one with both marginals being type II largest value.

Bhatt and Dave (1964) gave an expression for the correlation between the variates that result when two standard normal variates with correlation ρ are subjected to arbitrary polynomial transformations. The expression is in terms of ρ and the coefficients when the transformations are written in terms of Hermite polynomials. In the special case where both transformations are quadratic (i.e., $a_0 + a_1x + a_2x^2$ and $b_0 + b_1y + b_2y^2$), the correlation becomes

$$\frac{a_1b_1\rho + 2a_2b_2\rho^2}{\sqrt{(a_1^2 + 2a_2^2)(b_1^2 + 2b_2^2)}}. \tag{11.75}$$

For the case where both transformations were cubic, see Vale and Maurelli (1983).

As to the lognormal distribution, on setting $\rho = -1$ and $\rho = 1$ in (11.71), we find

$$\frac{\exp(-\sigma_1\sigma_2) - 1}{\sqrt{[\exp(\sigma_1^2 - 1][\exp(\sigma_2^2) - 1]}} \le \text{corr}(X, Y) \le \frac{\exp(\sigma_1\sigma_2) - 1}{\sqrt{[\exp(\sigma_1^2 - 1][\exp(\sigma_2^2) - 1]}}. \tag{11.76}$$

Adapting from Romano and Siegel (1986, Section 4.22), "This has some striking implications. If, for example, we restrict ourselves to the family of distributions with $\sigma_1 = 1$ and $\sigma_2 = 4$ but we allow any values for the means and the correlation between $\log X$ and $\log Y$, then the correlation between X and Y is constrained to lie in the interval from -0.000251 to 0.01372! Such a result raises a serious question in practice about how to interpret the correlation between lognormal random variables. Clearly, small correlations may be very misleading because a correlation of 0.01372 indicates, in fact, X and Y are perfectly functionally (but nonlinearly) related."

The general shape of the univariate gamma distribution makes it a competitor of the lognormal for fitting to data. Moran (1967) has discussed the range of correlations possible in a bivariate distribution with gamma marginals having specified shape parameters.

Lai et al. (1999) have carried out a robustness study of the sample correlation of the bivariate lognormal case. Their simulation (confirmed by numerical analysis) indicates that the bias in estimating the population correlation coefficient of the lognormal can be very large, particularly if $\rho \neq 0$.

We have already seen in (11.71) what the correlation in a normal distribution becomes when the variates are exponentiated. Bhatt and Dave (1965) have given some results for the correlation between $\sum_{i=0}^{n} a_i \exp(\alpha_i X)$ and $\sum_{i=0}^{n} b_i \exp(\beta_i Y)$, with $\alpha_0 = \beta_0 = 0$, and (X, Y) having a standard bivariate normal distribution with correlation ρ. Special cases mentioned include:

- $a_0 + a \cosh \alpha x + b \cosh \beta y$, for which the correlation is

$$\sinh^2(\rho\alpha\beta/2)[\sinh(\beta^2/2)],$$

  and
- $a_0 + a \sinh \alpha x$, $b_0 + b \sinh \beta y$, which is Johnson's S_U distribution; in this case, the correlation is found to be $\sinh(\rho\alpha\beta)/\sqrt{\sinh(\alpha^2)\sinh(\beta^2)}$. For the general case, the distribution of (X, Y) not being standardized, see Eq. (5.2) of Johnson (1987).

Lindqvist (1976) argued that the correlation coefficients—in particular, when they are used as inputs to a factor analysis—should not be based on variates that are skewed. He has given a computer program that chooses and applies a transformation of the form log(X+ constant) if the skewness in the raw data is unacceptably large. Factor analyses of 13 constituents of 566 rock specimens were performed on such transformed data and on data for which all variates had simply been log-transformed, and results of the former were found preferable.

On the other hand, McDonald (1960) argued that the change in r when making transformations even to bivariate data that are grossly non-normal, such as ones encountered in hydrology, is usually of little practical importance. McDonald's evidence was from precipitation data from Arizona. (However, of the 14 correlations investigated, in half of them taking log transformation changed r by at least 0.05. As one usually wants to be confident about the first decimal place, perhaps it would be wise to go to the trouble of choosing the right transformation, despite what McDonald says.[5])

11.17 Truncated Bivariate Normal Distributions

11.17.1 Properties

The most common form of truncation of a standardized bivariate normal distribution is single truncation, from above or below, with respect to one of the variables. We shall consider the case where $X > h$. Thus, the support is $h < X < \infty$, $-\infty < Y < \infty$, and the p.d.f. is evidently $\psi(x, y; \rho)/\Phi(-h)$.

The marginal density of the truncated variable is obviously $\phi(x)/\Phi(-h)$. The marginal density of Y is $\frac{\phi(y)}{\Phi(-h)}\Phi\left(\frac{-h+\rho y}{\sqrt{1-\rho^2}}\right)$; see Chou and Owen (1984, p. 2538).

Let E_T and var_T denote the mean and variance after truncation. Also, let $q(h)$ be the hazard rate (failure rate) $\phi(h)/\Phi(-h)$, i.e., the inverse of Mills' ratio. Then, we have

$$E_T = q(h), \tag{11.77}$$

$$E_T(Y) = \rho q(h), \tag{11.78}$$

$$\text{var}(X) = 1 - q(h)[q(h) - h], \tag{11.79}$$

$$\text{var}(Y) = 1 - \rho^2 q(h)[q(h) - h]; \tag{11.80}$$

see Rao et al. (1968, pp. 434–435). Pearson's product-moment correlation is

5 If one is solely interested in a measure of correlation, not in the marginal distributions, one might calculate a rank correlation from the data and then convert it to an equivalent ρ by $\tau = \frac{2}{\pi}\sin^{-1}\rho$ or $\rho_S = \frac{6}{\pi}\sin^{-1}\frac{\rho}{2}$.

$$\rho_T = \rho\sqrt{\mathrm{var}_T(X)/\mathrm{var}_T(Y)} = \rho\left(\rho^2 + \frac{1-\rho^2}{\mathrm{var}_T(X)}\right)^{-1/2}. \tag{11.81}$$

Since $\mathrm{var}_T(X) \le \mathrm{var}_T(Y)$, it follows that $|\rho_T| \le |\rho|$.

The conditional distribution of Y, given $X = x$, is normal with mean ρx and standard deviation $1/\sqrt{1-\rho^2}$, i.e., a single truncation does not affect the regression. However, the regression of X on Y is given by

$$E_T(X|Y=y) = \rho y + \sqrt{1-\rho^2}\,q\left(\frac{h-\rho y}{\sqrt{1-\rho^2}}\right) \tag{11.82}$$

[Johnson and Kotz (1972, p. 113)].

The moment generating function is

$$M(s,t) = \frac{\Phi(s+\rho t-h)}{\Phi(-h)}\exp[(s^2+2\rho st+t^2)/2]. \tag{11.83}$$

For further results, including truncations on both variables, see Johnson and Kotz (1972, Section 36.7) and the references cited therein. Regier and Hamdan (1971) and Gajjar and Subrahmaniam (1978) have given a number of results, both algebraic and numerical in nature, for the case of single truncation in both variables. Kovner and Patil (1973) obtained expressions for the moments up to order 4 when both variables are doubly truncated. For some formulas relating to the truncated bivariate lognormal distribution, see Lien (1985) and Shah and Parikh (1964).

Nath (1972) derived the moments of a linearly truncated bivariate normal distribution such that the support is of the form $w_1X + w_2Y \ge a$.

Brunden (1978) discussed the probability contours and a goodness-of-fit test for the singly truncated bivariate normal distribution.

11.17.2 Application to Selection Procedures

The context envisaged here is that of the quality of performance of a manufactured item or perhaps of an employee. The items that are put into service, or the employees who are hired, are those that score above some threshold level on a screening test. Some measure of performance in service becomes available at some later date; there is substantial, but less than perfect, correlation between scores of two tests. (Academics will immediately think of students' performance at high school and performance at college, for example.)

If it is assumed that the joint distribution of performances in the unselected population is bivariate normal, then the relevant distribution for items in service is that for which properties are given above. Equation (11.81), for example, tells us what the correlation will be.[6]

This area is extensively discussed in National Bureau of Standards (1959, Section 2.6). More recent work includes the following:

- *Problem*: Knowing the marginal properties and correlation, determine k from known values of h and p, where $\Pr(Y > k|X > h) = p$. Chou and Owen (1984) obtained an approximation using the method of Cornish-Fisher expansion. As this involves the bivariate cumulants κ_{ij}, Chou and Owen gave a method for calculating these: $\kappa_{ij} = \rho^j \kappa_{k+j\,0}$, where κ_{l0} is given in their table for $l = 1$ to 8 and $h = -3.0(0.2)3.0$; see also Odeh and Owen (1980).
- *Problem*: Knowing the marginal properties and correlation, determine h from known values of k, ζ, l, and m such that we are assured (with degree of confidence ζ) that the number of units satisfying $Y \leq k$ is at least l in the group of m units satisfying $X \leq h$ (there being as many units available for screening as are necessary). This may be thought of as the problem facing a supplier who wants to reject as few of his items as possible, subject to being reasonably confident that the proportion of substandard items is low. Owen et al. (1981) obtained the required results; see also Madsen (1982).
- *Problem*: What if there is an upper limit of acceptability for Y as well as a lower one? See Li and Owen (1979).
- *Problem*: What if the mean and standard deviation of X are not known in advance, but have to be estimated from a preliminary sample? See Owen and Haas (1978) and Odeh and Owen (1980) for relevant discussions.
- Davis and Jalkanen (1988) gave a practical example of reduced correlation in a truncated sample. The subject was amounts of gold and silver in samples from drill holes from a gold field. For the whole sample, the correlation between these quantities was 0.61, but for the 28% of samples that contained the most gold—and thus of most interest—the correlation was only 0.26.

Another account, which details further developments in these directions, is due to Owen (1988).

---

[6] But if the distribution is not bivariate normal, then it is possible for the correlation to be increased rather than decreased. Suppose $Y = a + bX + \varepsilon$. Then, the correlation between X and Y is $b^2/(b^2 + \sigma_\varepsilon/\sigma_X)$. Consequently, if truncation of X increases the variance of X rather than decreasing it, the correlation is increased. This happens for distributions of non-negative r.v.'s that have nonzero density at the origin and a coefficient of variation greater than 1; this class includes decreasing hazard rate distributions. These points were made by Mullooly (1988).

11.17.3 Truncation Scheme of Arnold et al. (1993)

Arnold et al. (1993) considered a truncated bivariate normal model in which both tails of Y are truncated so that the joint density of (X, Y) is now given by

$$h(x,y) = \begin{cases} \dfrac{\psi(x,y;\rho)}{\Phi\left(\frac{b-\mu_2}{\sigma_2}\right) - \Phi\left(\frac{a-\mu_2}{\sigma_2}\right)}, & -\infty < x < \infty, a < y < b \\ 0 \qquad\qquad , & \text{otherwise.} \end{cases} \tag{11.84}$$

Denoting $\beta = \frac{b-\mu_2}{\sigma_2}$ and $\alpha = \frac{a-\mu_2}{\sigma_2}$, they obtained the marginal distribution of X as

$$f(x) = \frac{1}{\sigma_1} k\left(\frac{x-\mu_1}{\sigma_1}\right), \tag{11.85}$$

where

$$k(z) = \frac{\phi(y)\left\{\Phi\left(\frac{\beta-\rho y}{\sqrt{1-\rho^2}}\right) - \Phi\left(\frac{\beta-\rho y}{\sqrt{1-\rho^2}}\right)\right\}}{\Phi(\beta) - \Phi(\alpha)}. \tag{11.86}$$

Clearly, $k(z)$ is the density function of $Z = (X - \mu_1)/\sigma_1$. Note that the expression in (11.86) coincides with that of Chou and Owen (1984) for the case where $\beta = \infty$.

For the case where $\alpha = 0$ and $\beta = \infty$, the density in (11.86) becomes

$$k(z) = 2\phi(z)\Phi\left(\frac{\rho y}{\sqrt{1-\rho^2}}\right) = 2\phi(y)\Phi(\lambda y), \tag{11.87}$$

which is Azzalini's (1985) skew-normal distribution.

11.17.4 A Random Right-Truncation Model of Gürler

Gürler (1996) considered a random truncation of a bivariate normal model in the context of survival analysis.

In a random right-truncation model, one observes the i.i.d. samples of (Y, T) only if $(Y \leq T)$, where Y is the variable of interest and T is an independent variable that prevents the complete observation of Y. Gürler (1996) proposed an estimator for the bivariate survival function of (X, Y) and a nonparametric estimator for the so-called bivariate reverse-hazard vector. An application of the suggested estimators is presented for transfusion-related AIDS (TR-AIDS) data on the incubation time.

11.18 Bivariate Normal Mixtures

11.18.1 Construction

Suppose we mix together two bivariate normal distributions. The density is then

$$h(x,y) = p\psi_1(x,y;\rho_1) + (1-p)\psi_2(x,y;\rho_2), \ 0 \le p \le 1, \tag{11.88}$$

and we shall denote the means μ_{X_i} and μ_{Y_i}, the standard deviations σ_{X_i} and σ_{Y_i}, and the correlation coefficients ρ_i, $i = 1, 2$. Johnson (1987, pp. 56–61) has provided 60 contour plots, with different mixing proportions, means, standard deviations, and correlations, to indicate the range of appearance of bivariate mixtures. The covariance is

$$p\rho_1\sigma_{X_1}\sigma_{Y_1} + (1-p)\rho_2\sigma_{X_2}\sigma_{Y_2} + p(1-p)(\mu_{X_1} - \mu_{X_2})(\mu_{Y_1} - \mu_{Y_2}) \tag{11.89}$$

[Johnson (1987, p. 57)].

When p is close to 0 or 1, a bivariate normal mixture can be considered as a single bivariate normal that has been "contaminated" by a mixture with a small proportion of another one.

The subject of cluster analysis may be viewed as an attempt to fit a mixture of normal distributions to a dataset. Usually, there are many more variables than two, so we shall not discuss this other than to mention the paper by Wolfe (1970), which explicitly treats the subject in this manner.

Tarter and Silvers (1975) described an interactive computer graphical method for decomposing mixtures consisting of two or more bivariate normal components; see also Titterington et al. (1985, pp. 142–145). Though largely on the univariate case, the book by Everitt and Hand (1981) deals with the multivariate case to some extent. It is especially useful in regard to methods of parameter estimation. Much of the same remarks could be made about the book by Titterington et al. (1985). The Appendix to McLachlan and Basford (1988) contains FORTRAN programs for fitting mixtures of multivariate normal distributions; there is much more material of interest in this book, especially on testing for multivariate normality and identification of outliers.

11.18.2 References to Illustrations

Johnson (1987, pp. 56–61) has given 60 contour plots: in all of them, $\mu_{X_1} = \mu_{Y_1} = 0$, $\sigma_{X_1} = \sigma_{Y_1} = \sigma_{X_1} = \sigma_{Y_1} = 1$; all combinations ($2 \times 5 \times 6$) are shown of (i) $p = 0.5$ or 0.9, (ii) $\rho_1 = -0.9, -0.5, 0.0, 0.5$, or 0.9, and (iii) $(\rho_2, \mu_{X_2}, \mu_{Y_2}) = (0, 0.5, 0.5)$, $(0, 1.0, 1.0)$,$(0, 1.5, 1.5)$,$(0, 2.0, 2.0)$,$(0.5, 1.0, 1.0)$ or $(0.9, 1.0, 1.0)$.

There is also an illustration of a three-component density in Everitt (1985).

11.18.3 Generalization and Compounding

One may generalize the form in (11.88) to a finite mixture of the form $h(x, y) = \sum_{i=1}^{n} p_i \psi_i(x, y)$, in which $\sum_i p_i = 1$. A mixture of infinitely many bivariate normal, with the same mean leads to a class of elliptical compound bivariate normal distributions.

11.18.4 Properties of a Special Case

We shall now assume that the two component distributions are equal in their vectors of means and standard deviations. Without loss of any generality, we shall take the means to be 0 and the standard deviations to be 1.

- The correlation is $p\rho_1 + (1 - p)\rho_2$.
- The conditional distribution of Y, given $X = x$, is a mixture of $N(\rho_1 x, 1 - \rho_1^2)$ and $N(\rho_2 x, 1 - \rho_2^2)$ in the proportions $p : 1 - p$.
- The regression of Y on X is linear, i.e., $E(Y|X = x) = [p\rho_1 + (1 - p)\rho_2]x$.
- The conditional variance is a quadratic function of x given by

$$\text{var}(Y|X = x) = 1 - [p\rho_1^2 + (1 - p)\rho_2^2] + \{p\rho_1^2 + (1 - p)\rho_2^2\}x^2.$$

- For a more detailed discussion on these conditional properties, see Kowalski (1973).
- The special case where $p = 0.5$ and $\rho_1 = -\rho_2$ has been considered by several authors, including Lancaster (1959) and Sarmanov (1966). It is an example of a bivariate distribution with normal marginals whose variates are uncorrelated yet dependent; its canonical correlation coefficient has the property $c_n = \rho^n$ when n is even and 0 when n is odd (here $\rho = \rho_1 = -\rho_2$). Another example of a bivariate distribution with dependent normal marginals having zero correlation has been given by Ruymgaart (1973).

11.18.5 Estimation of Parameters

Let r be the sample correlation coefficient. There are a number of papers on the distribution of r in random samples from mixtures of two bivariate normal distributions. Johnson et al. (1995, p. 561–567) contains a good discussion as well as several references on the subject. Simulations showed that r is biased toward zero as an estimator of ρ. In general, the expression for the distribution of r is complicated.

Linday and Basak (1993) demonstrated that one can quickly (in computer time) and efficiently estimate the parameters of this distribution using the method of moments.

11.18.6 Estimation of Correlation Coefficient for Bivariate Normal Mixtures

Let r be the sample correlation coefficient. There are a number of papers on the distribution of r in random samples from mixtures of two bivariate normal distributions. Johnson et al. (1995, pp. 561–567) contains a good discussion as well as several references on the subject. Simulations showed that r is biased toward zero as an estimator of ρ. In general, the expression for the distribution of r is complicated.

Consider the bivariate mixture model in (11.88) with $\mu_{X_i} = \mu_{Y_i} = 0$, $i = 1, 2$, $\sigma_{X_1} = \sigma_{Y_1} = 1$ and $\sigma_{X_2} = \sigma_{Y_2} = k$. This model is known as the 'gross error model' in the robustness studies. Let $\rho_1 = \rho$ and $\rho_2 = \rho'$. It is well known [see Devlin et al. (1975)] that the sample correlation coefficient is strongly biased for ρ, being very sensitive to the presence of outliers in the data, and hence it is necessary to use a robust estimator in this case.

Shevlyakov and Vilchevski (2002) proposed a minimax variance estimator of ρ given by

$$r_{\text{tr}} = \left(\sum_{i=n_1+1}^{n-n_2} u_{(i)}^2 - \sum_{i=n_1+1}^{n-n_2} v_{(i)}^2 \right) \Bigg/ \left(\sum_{i=n_1+1}^{n-n_2} u_{(i)}^2 + \sum_{i=n_1+1}^{n-n_2} v_{(i)}^2 \right), \tag{11.90}$$

where $u_{(i)}$ and $v_{(i)}$ are the ith order statistics of the robust principal variables $u = (x+y)/\sqrt{2}$ and $v = (x-y)/\sqrt{2}$, respectively. The authors call the estimator above the "trimmed correlation coefficient."

Equation (11.90) yields the following limiting cases: (i) the sample correlation r with $n_1 = 0, n_2 = 0$, and with the classical estimators (the sample means for location and standard deviation for scale) in its inner structure; and (ii) the median correlation coefficient r_{med} with $n_1 = n_2 = (n-1)/2$.

Li et al. (2006) considered robust estimation of the correlation coefficient for ε-contaminated bivariate normal distributions.

Recently, Nagar and Castañeda (2002) derived the non-null distribution of r by first considering the jth moment of $1 - r^2$. Then, by using the inverse Mellin transformation, the density of $1 - r^2$ will be obtained, from which the density of r will be derived.

Estimation of Correlation Coefficient Based on Selected Data

In psychometrics, one often encounters data that may not be considered random but selected according to some explanatory variable. Hägglund and Larsson (2006) considered maximum likelihood estimates when data arise from a bivariate normal distributions that is truncated in an extreme way. Two methods were tried on both simulated and real data.

11.18.7 Tests of Homogeneity in Normal Mixture Models

Consider a mixture of two bivariate normal populations with identical variance–covariance matrices $\mathbf{\Sigma}$ but possibly having different mean vectors $\boldsymbol{\mu_1}=(\mu_{X_1}, \mu_{Y_1})'$ and $\boldsymbol{\mu_2}=(\mu_{X_2}, \mu_{Y_2})'$. Assuming the mixing proportion p to be known, Goffinet et al. (1992) studied the behavior of the likelihood ratio test statistic for testing the null hypothesis $\boldsymbol{\mu_1} = \boldsymbol{\mu_2}$. This is equivalent to testing whether one is sampling from a mixture of two distributions or from a single distribution.

There is much interest in testing homogeneity versus mixture. Some of the key references are Lindsay (1995), Chen and Chen (2001), Chen et al. (2001), and Qin and Smith (2006).

Like Goffinet et al. (1992), Qin and Smith (2006) also considered the likelihood ratio test assuming that the variance–covariance matrix is known, with the mixing proportion p being bounded away from 0 or 1 (i.e., $0 < p < 1$).

Chuang and Mendell (1997) also studied the likelihood ratio test statistic when $\boldsymbol{\mu_1} \neq \boldsymbol{\mu_2}$ under the alternative hypothesis but $\mu_{X_1} = \mu_{Y_1}$ and $\mu_{X_2} = \mu_{Y_2}$.

11.18.8 Sharpening a Scatterplot

"Sharpening" a scatterplot aims to reveal its structure more clearly by increasing the impact of points that are typical at the expense of atypical points. Green (1988) has presented an example of 150 points generated from a mixture of two normal distributions in which the two clusters show up more clearly when the points in regions of low estimated probability density are plotted with smaller symbols. A similar example is shown by Tukey and Tukey (1981); see also Chambers et al. (1983, especially Section 4.10).

11.18.9 Digression Analysis

Digression analysis places emphasis on the data rather than the distribution, and it regresses Y on X rather than treating them symmetrically. Instead of fitting one regression line to empirical points (x_i, y_i), it fits two lines, with each point being supposed to be associated with the line that is nearest to it. This is a natural, though not entirely appropriate, thing to do if it is assumed that the points are from two populations mixed together. Thus, what is done (when the regressions are straight lines) is to minimize

$$\sum \min\{[y - (\alpha_1 + \alpha_2 x)]^2, [y - (\alpha_3 + \alpha_4 x)]^2\} \tag{11.91}$$

with respect to the parameters α's. For more on this, one may refer to Mustonen (1982).

"Switching regression" is another phrase used for much the same thing [see, for example, Quandt and Ramsey (1978)]. Regression methods do not necessarily rely on any bivariate distribution $H(x, y)$, of course, and rather on the conditional distribution of T given X.

11.18.10 Applications

- An important application of bivariate normal mixtures to problems in genetics is described in Qin and Smith (2006). He et al. (2006) showed that bivariate mixtures can be useful alternatives to univariate methods to detect differential gene expression in exploratory data analysis. See also McLachlan et al. (2005) for similar applications.
- Zerehdaran et al. (2006) used bivariate mixture models to study the relationships between body weight (BW) and ascites indicator traits in broilers.
- Alexander and Scourse (2004) used the bivariate normal mixture to model the log prices of two assets.
- For other applications, see Lindsay (1995).
- Bivariate normal mixtures were used by McLaren et al. (2008) to analyze joint population distributions of transferrin saturation (TS) and serum ferritin concentration (SF) measured in hemochromatosis and iron overload screening (HEIRS).

11.18.11 Bivariate Normal Mixing with Bivariate Lognormal

Schweizer et al. (2007) used a simple mixture of the bivariate normal and the bivariate lognormal to model the depth and velocity of a stream reach. The resulting joint distribution provided a good fit to the survey data from 92 stream reaches in New Zealand. The study has an important application for instream habitat assessment.

11.19 Nonbivariate Normal Distributions with Normal Marginals

Various distributions have been constructed to show that normal marginals are necessary but not sufficient for the joint distribution to be *the* bivariate normal (in the sense of (11.1)). Actually, what we need to do is no more than starting with any bivariate distribution with continuous marginals and then transforming the marginals to be normal. Other distributions illustrate that linearity of the regression is not a sufficient condition, either. In this section, we collect some classroom examples that are intended to dispel possible misconceptions.

11.19.1 Simple Examples with Normal Marginals

A list of such examples was presented by Kowalski (1973). Here, we shall illustrate how some such examples can be obtained by manipulating (mixing and shifting probability masses) the bivariate normal distribution.

Example 1 [Lancaster (1959)]: The joint density is

$$h(x, y) = [\psi(x, y; \rho) + \psi(x, y; -\rho)]/2. \tag{11.92}$$

A special case of this is given by Van Yzeren (1972) when $\rho = \frac{1}{2}$. The following corresponds to $\rho = 1$. Let X be a normal variate, and let Y be X or $-X$ with equal probability [Broffitt (1986)]. More generally, $\alpha\psi(x, y; \rho_1) + (1 - \alpha)\psi(x, y; \rho_2)$ also has normal marginal distributions.

Example 2 [Anderson (1958, p. 37)]: Let (X, Y) have a standardized uncorrelated bivariate normal distribution. Draw identical circles in each of the four quadrants of the plane, each having the same position with respect to the origin. With the circles being numbered clockwise around the origin, transfer the probability mass in circle 1 to circle 2, and transfer the probability mass in circle 3 to circle 4. The resulting distribution still possesses normal marginals.

Example 3 [Flury (1986)]: Let X and Y be independent normal variates. Divide the plane into octants bordered by $x = 0, y = 0$, $x + y = 0$, and $x - y = 0$. Shade alternate octants. Transfer all the probability masses in the blank areas to the next shaded area. The resulting distribution has normal marginals. This can be easily extended to X and Y being correlated—the probability masses in the blank areas are now transferred by reflecting them about their boundary line $x = 0$ or $y = 0$; see also Romano and Siegel (1986, Section 2.10).

11.19.2 Normal Marginals with Linear Regressions

Linearity of regressions, even in association with marginal normality, is not a sufficient condition for bivariate normality. It is easy to show that $\alpha\psi + (1 - \alpha)\psi_2$ (from Example 1) has linear regression [Kowalski (1973)], for example.

Example 4 [Ruymgaart (1973)]: Consider the joint density

$$h(x, y) = \psi(x, y; 0) + \lambda u(x)u(y), \tag{11.93}$$

where $u(t) = \sin|t|$ for $-2\pi \leq t \leq 2\pi$ and is 0 otherwise, and λ is chosen to prevent h from assuming negative values. In this example, the regressions are linear (especially flat). Furthermore, X and Y are uncorrelated. Nevertheless, $h(x, y)$ in (11.93) is still not the bivariate normal density in (11.1).

11.19.3 Linear Combinations of Normal Marginals

It is well known (see Section 11.4.4) that (X, Y) has a bivariate normal distribution if and only if all linear combinations of X and Y are univariate normal. Melnick and Tenenbein (1982) gave an example in which n linear combinations of two normal marginals are normal, for a large n, yet the distribution is not bivariate normal.

11.19.4 Uncorrelated Nonbivariate Normal Distributions with Normal Marginals

The variates of Examples 1 and 4 are uncorrelated. Another example is as follows.

Example 5 [Melnick and Tenenbein (1982)]: Let X have a standard normal distribution and Y be defined as

$$Y = \begin{cases} X & \text{if} |X| \le 1.54 \\ -X & \text{if} |X| > 1.54 \end{cases} . \tag{11.94}$$

Then, X and Y are uncorrelated. Here, 1.54 (correct to three significant digits) is the solution to the integral equation $\Phi(c) = 0.75 + c\phi(c)$.

11.20 Bivariate Edgeworth Series Distribution

General bivariate non-normal distributions, allowing varying degrees of skewness and kurtosis on the two components, can be produced through bivariate Edgeworth series distribution; see Gayen (1951). The joint density of this bivariate Edgeworth series distribution is

$$f_{\text{ES}}(x,y) = \left\{ 1 + \sum_{\substack{j=0 \\ j+k=3,4,6}}^{3} \sum_{k=0}^{3} \frac{(-1)^{j+k} A_{j,k}}{j!k!} D_x^j D_y^k \right\} f(x,y), \ -\infty < x, y < \infty,$$

where $f(x,y)$ is the standard bivariate normal density function with correlation ρ, D_x and D_y are partial derivative operators, and $A_{j,k}$'s are parameters which are functions of the population cumulants. A method of simulating data from this bivariate distribution, with prescribed marginals, has been discussed by Kocherlakota, Kocherlakota, and Balakrishnan (1986). This bivariate non-normal distribution has been used, for example, in examining the robustness properties of the SPRT (Sequential Probability Ratio Test) for correlation coefficient by Kocherlakota, Kocherlakota, and Balakrishnan (1985).

11.21 Bivariate Inverse Gaussian Distribution

This is related to the bivariate normal distribution through some form of inverse transformation; see the derivation in Section 11.21.5 below.

11.21.1 Formula of the Joint Density

The joint density is

$$h = \frac{1}{4\pi}\sqrt{\frac{\lambda_1\lambda_2}{x^3y^3(1-\nu^2)}}\left\{\exp\left[\frac{-1}{2(1-\rho^2)}\left(\frac{\lambda_1}{\mu_1^2}\frac{(x-\mu_1)^2}{x} - \frac{2\nu}{\mu_1\mu_2}\sqrt{\frac{\lambda_1\lambda_2}{xy}}(x-\mu_1)(y-\mu_2)\right.\right.\right.$$
$$\left.\left. + \frac{\lambda_2}{\mu_2^2}\frac{(y-\mu_2)^2}{y}\right)\right] + \exp\left[\frac{-1}{2(1-\rho^2)}\left(\frac{\lambda_1}{\mu_1^2}\frac{(x-\mu_1)^2}{x} + \frac{2\nu}{\mu_1\mu_2}\sqrt{\frac{\lambda_1\lambda_2}{xy}}(x-\mu_1)(y-\mu_2)\right.\right.$$
$$\left.\left.\left. + \frac{\lambda_2}{\mu_2^2}\frac{(y-\mu_2)^2}{y}\right)\right]\right\}, \qquad x, y \geq 0, \ -1 < \nu < 1, \tag{11.95}$$

where the μ's and λ's are positive parameters.

11.21.2 Univariate Properties

The marginals are inverse Gaussian (Wald) with means μ_i and variances μ_i^3/λ_i.

11.21.3 Correlation Coefficients

This distribution is unusual in that Pearson's product-moment correlation is always zero. But, X and Y are independent if and only if the parameter ν is zero. ν is the correlation coefficient of the underlying standard bivariate normal distribution.

11.21.4 Conditional Properties

The regression is constant; i.e., $E(Y|X = x) = \mu_2$. The conditional variance is not constant, and takes large values at extreme values of x, and it is given by

$$\text{var}(Y|X = x) = \frac{\mu_2^3}{\lambda_2}\left\{(1-\nu^2) + \frac{\nu^2\lambda_1(x-\mu_1)^2}{\mu_1^2 x}\right\}. \tag{11.96}$$

11.21.5 Derivations

Let (Z_1, Z_2) have a standard bivariate normal distribution with correlation ν. Define $U = Z_1^2$ and $V = Z_2^2$. (Clearly, U and V both have chi-squared distributions with 1 degree of freedom.) Then, (U, V) has Kibble's bivariate gamma distribution (see Section 8.2) with $\alpha = \frac{1}{2}$. Consider the two-to-one transformations $U = (X-\mu_1)^2/(\mu_1^2 X)$ and $V = (Y-\mu_2)^2/(\mu_2^2 Y)$; then, upon solving for X and Y, their joint density turns out to be the one in (11.95).

11.21.6 References to Illustrations

Surfaces and contour plots of the density have been presented by Kocherlakota (1986).

11.21.7 Remarks

- For further results, including joint moments and infinite series representations of the distribution function, see Kocherlakota (1986).
- Wasan and Buckholtz (1973) derived a partial differential equation that, when solved under suitable boundary conditions, leads to a density of a bivariate inverse Gaussian process; they gave two examples, one with independent variables and one with dependent variables. For the latter, the joint density is

$$h(x,y) = \frac{s(t-s)}{2\pi\sqrt{(x-y)^2y^3}} \exp\left\{-\frac{[x-y-(t-s)]^2}{2(x-y)} - \frac{(y-s)^2}{2y}\right\} \quad (11.97)$$

  for $x > y > 0$, with $t > s > 0$; see also Wasan (1968). The density in (11.97) is the joint density of (X, Y), where $X = Z + Y$ and Z and Y are independent inverse Gaussian variates.
- Another bivariate inverse Gaussian distribution has been described by Al-Hussaini and Abd-el-Hakim (1981). For this, the support is naturally the positive quadrant again, and the p.d.f. is

$$h(x,y) = f(x)g(y)[1 + \rho\Lambda(x,y)], \quad (11.98)$$

  where the parameter ρ equals the product-moment correlation coefficient, f and g are univariate inverse Gaussian densities with parameters (λ_1, μ_1) and (λ_2, μ_2), respectively, and

$$\Lambda(x,y) = 8\sqrt{\frac{\lambda_1\lambda_2}{\mu_1^3\mu_2^3}}(x-\mu_1)(y-\mu_2)\exp\left\{-\left[\frac{\lambda_1(x-\mu_1)^2}{2\mu_1^2x} + \frac{\lambda_2(y-\mu_2)^2}{2\mu_2^2y}\right]\right\}. \quad (11.99)$$

  Kocherlakota (1986) has argued that (11.95) is a more natural extension of the univariate distribution than the form in (11.98).
- Banerjee (1977) has mentioned a bivariate inverse Gaussian distribution for which the regression takes the form $E(Y|X = x) = \alpha - \beta/x$.
- Iyengar and Patwardhan (1988) have reviewed the inverse Gaussian distribution, and one section of their paper is the bivariate case; they regarded the proposal of Al-Hussaini and Abd-el-Hakim as rather artificial and do not mention that of Kocherlakota; see Iyengar (1985).

- Barndorff-Nielsen et al. (1991) constructed several types of multivariate inverse Gaussian distributions. Univariate marginals are of the same type.
- The multivariate inverse Gaussian distribution proposed by Minami (2003) was derived through a multivariate inverse relationship with multivariate Gaussian distributions and characterized as the distribution of the location at a certain stopping time of a multivariate Brownian motion. In Minami (2003), it was shown that the multivariate inverse Gaussian distribution is also a limiting distribution of multivariate Lagrange distributions, which are a family of waiting-time distributions under certain conditions.

References

1. Ahn, S.K.: F-probability plot and its application to multivariate normality. Communications in Statistics: Theory and Methods **21**, 997–1023 (1992)
2. Ahsanullah, M., Bansal, N., Hamedani, G.G., Zhang, H.: A note on bivariate normal distribution. Report, Rider University, Lawrenceville, New Jersey (1996)
3. Ahsanullah, M., Wesolowski, J.: Bivariate normality via Gaussian conditional structure. Report, Rider College, Lawrenceville, New Jersey (1992)
4. Albers, W., Kallenberg, W.C.M.: A simple approximation to the bivariate normal distribution with large correlation coefficient. Journal of Multivariate Analysis **49**, 87–96 (1994)
5. Alexander, C., Scourse, A.: Bivariate normal mixture spread option valuation. Quantative Finance **4**, 637–648 (2004)
6. Al-Hussaini, E.K., Abd-el-Hakim, N.S.: Bivariate inverse Gaussian distribution. Annals of the Institute of Statistical Mathematics **33**, 57–66 (1981)
7. Al-Saleh, M.F., Al-Ananbeh, A.M.: Estimation of the means of the bivariate normal using moving extreme ranked set sampling with concomitant variable. Statistical Papers **48**, 179–195 (2007)
8. Amos, D.E.: On computation of bivariate normal distribution. Mathematics of Computation **23**, 655–659 (1969)
9. Anderson, T.W.: An Introduction to Multivariate Analysis. John Wiley and Sons, New York (1958)
10. Anderson, T.W., Ghurye, S.G.: Unique factorization of products of bivariate normal cumulative distribution functions. Annals of the Institute of Statistical Mathematics **30**, 63–70 (1978)
11. Andrews, D.F., Gnanadesikan, R., Warner, J.L.: Transformations of multivariate data. Biometrics **27**, 825–840 (1971)
12. Anscombe, F.J.: Computing in Statistical Science through APL. Springer-Verlag, New York (1981)
13. Arellano-Valle, R.B., Genton, M.G.: On fundamental skew distributions. Journal of Multivariate Analysis **96**, 93–116 (2005)
14. Arnold, S.F.: Union–intersection principle. In: Encyclopedia of Statistical Sciences Volume 9, S. Kotz and N.L. Johnson (eds.), pp. 417–420. John Wiley and Sons, New York (1988)
15. Arnold, B.C., Beaver, R.J., Groenveld, R.A., Meeker, W.Q.: The nontruncated marginal of a truncated bivariate normal distribution. Psychometrika **58**, 471–488 (1993)
16. Arnold, B.C., Castillo, E., Sarabia, J.M.: Conditional Specification of Statistical Models. Springer-Verlag, New York (1999)

17. Aroian, L.A.: The probability function of the product of two normally distributed variables. Annals of Mathematical Statistics **18**, 265–271 (1947)
18. Aroian, L.A.: Mathematical forms of the distribution of the product of two normal variables. Communications in Statistics: Theory and Methods **7**, 165–172 (1978)
19. Azzalini, A.: A class of distributions which includes normal ones. Scandinavian Journal of Statistics **12**, 171–178 (1985)
20. Azzalini, A.: The skew-normal distribution and related multivariate families. Scandinavian Journal of Statistics **32**, 159–199 (2005)
21. Azzalini, A.: Skew-normal family of distributions. In: Encyclopedia of Statistical Sciences, Volume 12, S. Kotz, N. Balakrishnan, C.B. Read, and B. Vidakovic (eds.), pp. 7780–7785. John Wiley and Sons, New York (2006)
22. Azzalini, A., Dalla Valle, A.: The multivariate skew-normal distribution. Biometrika **83**, 715–726 (1996)
23. Bacon-Shone, J., Fung, W.K.: A new graphical method for detecting single and multiple outliers in univariate and multivariate data. Applied Statistics **36**, 153–162 (1987)
24. Balakrishnan, N.: Multivariate normal distribution and multivariate order statistics induced by ordering linear combinations. Statistics and Probability Letters **17**, 343–350 (1993)
25. Balakrishnan, N., Brito, M.R., Quiroz, A.J.: A vectorial notion of skewness and its use in testing for multivariate symmetry. Communications in Statistics: Theory and Methods **36**, 1757–1767 (2007)
26. Balakrishnan, N., Kim, J.-A.: EM algorithm for Type-II right censored bivariate normal data. In: Parametric and Semiparametric Models with Applications to Reliability, Survival Analysis, and Quality of Life, M.S. Nikulin, N. Balakrishnan, M. Mesbah, and N. Limnios (eds.), pp. 177–210. Birkhäuser, Boston (2004)
27. Balakrishnan, N., Kim, J.-A.: Point and interval estimation for bivariate normal distribution based on progressively Type-II censored data. Communications in Statistics: Theory and Methods **34**, 1297–1347 (2005a)
28. Balakrishnan, N., Kim, J.-A.: EM algorithm and optimal censoring schemes for progressively Type-II censored bivariate normal data. In: Advances in Ranking and Selection, Multiple Comparisons and Reliability, N. Balakrishnan, N. Kannan, and H.N. Nagaraja (eds.), pp. 21-45. Birkhäuser, Boston (2005b)
29. Balakrishnan, N., Kim, J.-A.: Nonparametric tests for independence between lifetimes and covariates from censored bivariate normal samples. Communications in Statistics: Simulation and Computation **34**, 685–710 (2005c)
30. Balasubramanian, K., Balakrishnan, N.: On a class of multivariate distributions closed under concomitance of order statistics. Statistics and Probability Letters **23**, 239–242 (1995)
31. Banerjee, A.K.: A bivariate inverse Gaussian distribution (Preliminary report). (Abstract only.) Institute of Mathematical Statistics Bulletin **6**, 138–139 (1977)
32. Baringhaus, L., Henze, N.: A consistent test for multivariate normality based on the empirical characteristic function. Metrika **35**, 339–348 (1988)
33. Barndorff-Nielsen, O.E., Blæsild, P., Seshadri, V.: Multivariate distributions with generalized inverse Gaussian marginals and associate Poisson mixture. Canadian Journal of Statistics **20**, 109–120 (1991)
34. Barnett, V.: Some outlier tests for multivariate samples. South African Statistical Journal **13**, 29–52 (1979)
35. Barnett, V.: Some bivariate uniform distributions. Communications in Statistics: Theory and Methods **9**, 453–461 (Correction **10**, 1457) (1980)
36. Barnett, V.: Reduced distance measures and transformation in processing multivariate outliers. Australian Journal of Statistics **25**, 64–75 (1983a)
37. Barnett, V.: Principles and methods for handling outliers in data sets. In: Statistical Methods and the Improvement of Data Quality, T. Wright (ed.), pp. 131–166. Academic Press, New York (1983b)

38. Barnett, V.: Detection and testing of different types of outlier in linear structural relationships. Australian Journal of Statistics **27**, 151–162 (1985)
39. Barnett, V., Lewis, T.: Outliers in Statistical Data, 2nd edition. John Wiley and Sons, Chichester (1984)
40. Basford, K.E., McLachlan, G.J.: Likelihood estimation with normal mixture models. Applied Statistics **34**, 282–289 (1985)
41. Baughman, A.L.: A FORTRAN function for the bivariate normal integral. Computer Methods and Programs in Biomedicine **27**, 169–174 (1988)
42. Bera, A., Jarque, C.: Efficient tests for normality, homoscedasticity and serial independence of regression residuals: Monte Carlo evidence. Economic Letters **7**, 313–318 (1981)
43. Bera, A., John, S.: Tests for multivariate normality with Pearson alternatives. Communications in Statistics: Theory and Methods **12**, 103–117 (1983)
44. Best, D.J., Rayner, J.C.W.: A test for bivariate normality. Statistics and Probability Letters **6**, 407–412 (1988)
45. Bhatt, N.M., Dave, P.H.: A note on the correlation between polynomial transformations of normal variates. Journal of the Indian Statistical Association **2**, 177–181 (1964)
46. Bhatt, N.M., Dave, P.H.: Change in normal correlation due to exponential transformations of standard normal variates. Journal of the Indian Statistical Association **3**, 46–54 (1965)
47. Bickel, P.J., Doksum, K.A.: Mathematical Statistics: Basic Ideas and Selected Topics. Holden-Day, Oakland (1977)
48. Bildikar, S., Patil, G.P.: Multivariate exponential-type distributions. Annals of Mathematical Statistics **39**, 1316–1326 (1968)
49. Bjerager, P., Skov, K.: A simple formula approximating the normal distribution function. In: Euromech 155: Reliability Theory of Structural Engineering Systems, pp. 217–231. Danish Engineering Academy, Lyngby (1982)
50. Booker, J.M., Johnson, M.E., Beckman, R.J.: Investigation of an empirical probability measure based test for multivariate normality. In: American Statistical Association, 1984 Proceedings of the Statistical Computing Section, pp. 208–213. American Statistical Association, Alexandria, Virginia (1984)
51. Borth, D.M.: A modification of Owen's method for computing the bivariate normal integral. Applied Statistics **22**, 82–85 (1973)
52. Bouver, H., Bargmann, R.E.: Comparison of computational algorithms for the evaluation of the univariate and bivariate normal distribution. In: Proceedings of Computer Science and Statistics: 12th Annual Symposium on the Interface, J. F. Gentleman (ed.), pp. 344–348. (1979)
53. Bouver, H., Bargmann, R.E.: Evaluation and graphical application of probability contours for the bivariate normal distribution. In: American Statistical Association, 1981 Proceedings of the Statistical Computing Section, pp. 272–277. (1981)
54. Bowman, K.O., Shenton, L.R.: Moment $(\sqrt{b_1}, b_2)$ techniques. In: Goodness-of-Fit Techniques, R.B. D'Agostino and M.A. Stephens (eds.), pp. 279–329. Marcel Dekker, New York (1986)
55. Box, G.E.P., Cox, D.R.: An analysis of transformations. Journal of the Royal Statistical Society, Series B **26**, 211–243 (Discussion, 244–252) (1964)
56. Box, G.E.P., Tiao, G.C.: Bayesian Inference in Statistical Analysis. Addison-Wesley, Reading, Massachusetts (1973)
57. Boys, R.: Algorithm AS R80: A remark on Algorithm AS 76: An integral useful in calculating noncentral t and bivariate normal probabilities. Applied Statistics **38**, 580–582 (1989)
58. Brelsford, W.M., Relies, D.A.: STATLIB: A Statistical Computing Library. Prentice-Hall, Englewood Cliffs, New Jersey (1981)
59. Broffitt, J.D.: Zero correlation, independence, and normality. The American Statistician **40**, 276–277 (1986)

60. Brucker, J.: A note on the bivariate normal distribution. Communications in Statistics: Theory and Methods **8**, 175–177 (1979)
61. Brunden, M.N.: The probability contours and a goodness-of-fit test for the singly truncated bivariate normal distribution. Communications in Statistics: Theory and Methods **7**, 557–572 (1978)
62. Bucklew, J.A., Gallagher, N.C.: Quantization of bivariate circularly symmetric densities. In: Proceedings of the Sixteenth Annual Allerton Conference on Communication, Control, and Computing, pp. 982–990. Department of Electrical Engineering and Coordinated Science Laboratory, University of Illinois, Urbana-Champaign (1978)
63. Burmaster, D.E.: Lognormal distributions for total water intake and tap water intake by pregnant and lactating women in the United States. Risk Analysis **18**, 215–219 (1998)
64. Burnaby, T.P.: Growth-invariant discriminant functions and generalised distances. Biometrics **22**, 96–110 (1966)
65. Cadwell, J.H.: The bivariate normal integral. Biometrika **38**, 475–479 (1951)
66. Cain, M.: The moment generating function of the minimum of bivariate normal random variables. The American Statistician **48**, 124–125 (1994)
67. Cain, M., Pan, E.: Moments of the minimum bivariate normal random variables. The Mathematical Scientist **20**, 119–122 (1995)
68. Castillo, E., Galambos, J.: Conditional distribution and bivariate normal distribution. Metrika **36**, 209–214 (1989)
69. Chambers, J.M., Cleveland, W.S., Kleiner, B., Tukey, P.A.: Graphical Methods for Data Analysis. Wadsworth, Belmont, California (1983)
70. Chen, H., Chen, J.: Large sample distribution of the likelihood ratio test for normal mixtures. Statistics and Probability Letters **52**, 125–133 (2001)
71. Chen, H., Chen, J., Kalbfleisch, J.D. A modified likelihood ratio test for homogeneity in finite mixture models. Journal of the Royal Statistical Society, Series B **63**, 19–29 (2001)
72. Cheng, Y-S.: Bivariate lognormal distribution for characterizing asbestos fiber aerosols. Aerosol Science and Technology **5**, 359–368 (1986)
73. Chernick, M.R.: Influence functions, outlier detection, and data editing. In: Statistical Methods and the Improvement of Data Quality, T. Wright (ed.), pp. 167–176. Academic Press, New York (1983)
74. Chou, K.C., Corotis, R.B.: Generalized wind speed probability distribution. Journal of Engineering Mechanics **109**, 14–29 (1983)
75. Chou, Y-M.: Remark AS R55: A remark on Algorithm AS 76: An integral useful in calculating noncentral t and bivariate normal probabilities. Applied Statistics **34**, 100–101 (1985)
76. Chou, Y-M., Owen, D.B.: An approximation to percentiles of a variable of the bivariate normal distribution when the other variable is truncated, with applications. Communications in Statistics: Theory and Methods **13**, 2535–2547 (1984)
77. Chuang, R-J., Mendell, N.R.: The approximate null distribution of the likelihood ratio test for a mixture of two bivariate normal distributions with equal covariance. Communications in Statistics: Simulation and Computation **26**, 631–648 (1997)
78. Clark, L.A., Denby, L., Pregibon, D., Harshfield, G.A., Pickering, T.G., Blank, S., Laragh, J.H.: A data-based method for bivariate outlier detection: Application to automatic blood pressure recording devices. Psychophysiology **24**, 119–125 (1987)
79. Craig, C.C.: On the frequency function of xy. Annals of Mathematical Statistics **7**, 1–15 (1936)
80. Cramér, H.: Mathematical Methods of Statistics. Princeton University Press, Princeton, New Jersey (1946)
81. Crofts, A.E.: An investigation of normal–lognormal distributions. Technical Report No. 32, Department of Statistics, Southern Methodist University, Dallas (1969)
82. Crofts, A.E., Owen, D.B.: Large sample maximum likelihood estimation in a normal-lognormal distribution. South African Statistical Journal **6**, 1–10 (1972)

83. Csörgö, S.: Testing by the empirical characteristic function: A survey. In: Asymptotic Statistics 2. Proceedings of the Third Prague Symposium on Asymptotic Statistics, P. Mandl and M. Hušková (eds.), pp. 45–56. Elsevier, Amsterdam (1984)
84. Csörgö, S.: Testing for normality in arbitrary dimension. Annals of Statistics **14**, 708–723 (1986)
85. Csörgö, S.: Consistency of some tests for multivariate normality. Metrika **36**, 107–116 (1989)
86. Cunnane, C.: Unbiased plotting positions: A review. Journal of Hydrology **37**, 205–222 (1978)
87. D'Agostino, R.B.: Departures from normality, tests for. In: Encyclopedia of Statistical Sciences, Volume 2, S. Kotz and N.L. Johnson (eds.), pp. 315–324. John Wiley and Sons, New York (1982)
88. D'Agostino, R.B.: Graphical analysis. In: Goodness-of-Fit Techniques, R.B. D'Agostino and M.A. Stephens (eds.) pp. 7–62. Marcel Dekker, New York (1986a)
89. D'Agostino, R.B.: Tests for the normal distribution. In: Goodness-of-Fit Techniques, R.B. D'Agostino and M.A. Stephens (eds.) pp. 367–419. Marcel Dekker, New York (1986b)
90. D'Agostino, R.B., Belanger, A., D'Agostino, R.B., Jr.: A suggestion for using powerful and informative tests of normality. The American Statistician **44**, 316–321 (1990)
91. D'Agostino, R., Pearson, E.S.: Tests for departure from normality, empirical results for the distributions of b_2 and $\sqrt{b_1}$. Biometrika **60**, 613–622 (1973)
92. Daley, D.J.: Computation of bi- and trivariate normal integrals. Applied Statistics **23**, 435–438 (1974)
93. David, H.A.: Order Statistics, 2nd edition. John Wiley and Sons, New York (1981)
94. David, H.A., Moeschberger, M.L.: The Theory of Competing Risks. Griffin, London (1978)
95. Davis, B.M., Jalkanen, G.J.: Nonparametric estimation of multivariate joint and conditional spatial distributions. Mathematical Geology **20**, 367–381 (1988)
96. David, H.A., Nagaraja, H.N.: Concomitants of order statistics. In: Handbook of Statistics, Volume 16: Order Statistics: Theory and Methods. N. Balakrishnan and C.R. Rao (eds.), pp. 487–513. North-Holland, Amsterdam (1998)
97. DeBrota, D.J., Dittus, R.S., Roberts, S.D., Wilson, J.R., Swain, J.J., Venkatraman, S.: Input modeling with the Johnson system of distributions. In: 1988 Winter Simulation Conference Proceedings, M.A. Abrams, P.L. Haigh, and J.C. Comfort (eds.), pp. 165–179. Institute of Electrical and Electronics Engineers, Piscataway, New Jersey (1988)
98. Devlin, S.J., Gnanadesikan, R., Kettenring, J.R.: Robust estimation and outlier detection with correlation coefficients. Biometrika **62**, 531–545 (1975)
99. Der Kiureghian, A., Liu, P-L.: Structural reliability under incomplete probability information. Journal of Engineering Mechanics **112**, 85–104 (1986)
100. DiDonato, A.R., Hageman, R.K.: A method for computing the integral of the bivariate normal distribution over an arbitrary polygon. SIAM Journal on Scientific and Statistical Computing **3**, 434–446 (1982)
101. Divgi, D.R.: Calculation of univariate and bivariate normal probability functions. Annals of Statistics **7**, 903–910 (1979)
102. Donnelly, T.G.: Algorithm 462: Bivariate normal distribution. Communications of the Association for Computing Machinery **16**, 638 (1973)
103. Drezner, Z.: Computation of the bivariate normal integral. Mathematics of Computation **32**, 277–279 (1978)
104. Drezner, Z., Wesolowski, J.: On the computation of the bivariate normal integral. Journal of Statistical Computation and Simulation **35**, 101–107 (1990)
105. Evandt, O., Coleman, S., Ramalhoto, M.F., van Lottum, C.: A little-known robust estimator of the correlation coefficient and its use in a robust graphical test for bivariate normality with applications in aluminium industry. Quality and Reliability Engineering International **20**, 433–456 (2004)

106. Everitt, B.S.: Mixture distributions. In: Encyclopedia of Statistical Sciences, Volume 5, S. Kotz and N.L. Johnson (eds.), pp. 559–569. John Wiley and Sons, New York (1985)
107. Everitt, B.S., Hand, D.J.: Finite Mixture Distributions. Chapman and Hall, London (1981)
108. Flury, B.K.: On sums of random variables and independence. The American Statistician **40**, 214–215 (1986)
109. Foulley, J.L., Gianola, D.: Estimation of genetic merit from bivariate "all or none" responses. Génétique, Sélection. Evolution **16**, 285–306 (1984)
110. Fraser, D.A.S., Streit, F.: A further note on the bivariate normal distribution. Communications in Statistics: Theory and Methods **10**, 1097–1099 (1980)
111. Friedman, J.H., Stuetzle, W.: Projection pursuit methods for data analysis. In: Modern Data Analysis, R.L. Launer and A.F. Siegel (eds.), pp. 123–147. Academic Press, New York (1982)
112. Friedman, J.H., Tukey, J.W.: A projection pursuit algorithm for exploratory data analysis. IEEE Transactions on Computing **23**, 881–890 (1974)
113. Gajjar, A.V., Subrahmaniam, K.: On the sample correlation coefficient in the truncated bivariate normal population. Communications in Statistics: Simulation and Computation **7**, 455–477 (1978)
114. Gayen, A.K.: The frequency distribution of the product-moment correlation coefficient in random samples of any size drawn from non-normal universes. Biometrika **38**, 219–247 (1951)
115. Ghosh, P., Branco, M.D., Chakraborty, H.: Bivariate random effect using skew-normal distribution with application to HIV-RNA. Statistics in Medicine **26**, 1225–1267 (2007)
116. Gideon, R.A., Gurland, J.: A polynomial type approximation for bivariate normal variates. SIAM Journal on Applied Mathematics **34**, 681–684 (1978)
117. Gnanadesikan, R.: Methods for Statistical Data Analysis of Multivariate Observations. John Wiley and Sons, New York (1977)
118. Goedhart, P.W., Jansen, M.J.W.: A remark on Algorithm AS 76: An integral useful in calculating noncentral t and bivariate normal probabilities. Applied Statistics **42**, 496–497 (1992)
119. Goffinet, B., Loisel, P., Laurent, B.: Testing in normal mixture models when the proportions are known. Biometrika **79**, 842–846 (1992)
120. Grauslund, H., Lind, N.C.: A normal probability integral and some applications. Structural Safety **4**, 31–40 (1986)
121. Green, P.J.: Peeling bivariate data. In: Interpreting Multivariate Data, V. Barnett (ed.), pp. 3–19. John Wiley & Sons, Chichester (1981)
122. Green, P.J.: Peeling data. In: Encyclopedia of Statistical Sciences, Volume 6, S. Kotz and N.L. Johnson (eds.), pp. 660–664. John Wiley and Son, New York (1985)
123. Green, P.J.: Sharpening data. In: Encyclopedia of Statistical Sciences, Volume 8, S. Kotz and N.L. Johnson (eds.), pp. 431–433, John Wiley and Sons, New York (1988)
124. Green, R.F.: Outlier-prone and outlier-resistant distributions. Journal of the American Statistical Association **71**, 502–505 (1976)
125. Groenewoud, C., Hoaglin, D.C., Vitalis, J.A.: Bivariate Normal Offset Circle Probability Tables with Offset Ellipse Transformations. Cornell Aeronautical Laboratory, Buffalo, New York (1967)
126. Gupta, P.L., Gupta, R.C.: Failure rate of the minimum and maximum of a multivariate normal distribution. Metrika **53**, 39–49 (2001)
127. Gupta, S.S.: Probability integrals of multivariate normal and multivariate t. Annals of Mathematical Statistics **34**, 792–828 (1963a)
128. Gupta, S.S.: Bibliography on the multivariate normal integrals and related topics. Annals of Mathematical Statistics **34**, 829–838 (1963b)
129. Gürler, G.: Bivariate estimation with right-truncated data. Journal of the American Statistical Association **91**, 1152–1165 (1996)

130. Hafley, W.L., Buford, M.A.: A bivariate model for growth and yield prediction. Forest Science **31**, 237–247 (1985)
131. Hägglund, G., Larsson, R.: Estimation of the correlation coefficient based on selected data. Journal of Educational and Behavioral Statistics **31**, 377–411 (2006)
132. Haldane, J.B.S.: Moments of the distribution of powers and products of normal variates. Biometrika **32**, 226–242 (1942)
133. Hamedani, G.G.: Bivariate and multivariate normal characterizations: A brief survey. Communications in Statistics: Theory and Methods **21**, 2665–2688 (1992)
134. Harter, H.L.: Another look at plotting positions. Communications in Statistics: Theory and Methods **13**, 1613–1633 (1984)
135. Hawkins, D.M.: Identification of Outliers. Chapman and Hall, London (1980)
136. Hawkins, D.M.: A new test for multivariate normality and homoscedasticity. Technometrics **23**, 105–110 (1981)
137. Hazelton, M.L.: A graphical tool for assessing normality. The American Statistical Association **57**), 285–288 (2003)
138. He, Y., Pan, W., Lin, J.Z.: Cluster analysis using multivariate normal mixture models to detect differential gene expression with microarray data. Computational Statistics and Data Analysis **51**, 641–658 (2006)
139. Henze, N.: On Mardia's kurtosis test for multivariate normality. Communications in Statistics: Theory and Methods **23**, 1031–1045 (1994)
140. Henze, N.: Limit laws for multivariate skewness in the sense of Móri, Rohatgi and Székely. Statistics and Probability Letters **33**, 299–307 (1997a)
141. Henze, N.: Extreme smoothing and testing for multivariate normality. Statistics and Probability Letters **35**, 203–213 (1997b)
142. Henze, N.: Invariant tests for multivariate normality: A critical review. Statistics Papers **43**, 476–506 (2002)
143. Henze, N., Zirkler, B.: A class of invariant and consistent tests for multivariate normality. Communications in Statistics: Theory and Methods **19**, 3595–3617 (1990)
144. Heyde, C.C.: Multidimensional central limit theorems. In: Encyclopedia of Statistical Sciences, Volume 5, S. Kotz and N.L. Johnson (eds.), pp. 643–646. John Wiley and Sons, New York (1985)
145. Hiemstra, L.A.V., Creese, R.C.: Synthetic generation of seasonal precipitation. Journal of Hydrology **11**, 30–46 (1970)
146. Hill, I.D.: Remark AS R26: A remark on Algorithm AS 76: An integral useful in calculating noncentral t and bivariate normal probabilities. Applied Statistics **27**, 239 (1978)
147. Hinkley, D.V.: On the ratio of two correlated normal random variables. Biometrika **56**, 635–639 (Correction **57**, 683) (1969)
148. Holland, P.W., Wang, Y.J.: Dependence function for continuous bivariate densities. Communications in Statistics: Theory and Methods **16**, 863–876 (1987)
149. Holst, E., Schneider, T.: Fibre size characterization and size analysis using general and bivariate log-normal distributions. Journal of Aerosol Science **16**, 407–413 (1985)
150. Howarth, R.J., Earle, S.A.M.: Application of a generalized power transformation to geochemical data. Mathematical Geology **11**, 45–62 (1979)
151. Huang, Y-T., Wei, P-F.: A remark on the Zhang Omnibus test for normality. Journal of Applied Statistics **34**, 177–184 (2007)
152. Hutchinson, T.P., Lai, C.D.: Continuous Bivariate Distributions, Emphasising Applications. Rumsby Scientific Publishing, Adelaide (1991)
153. Iliopoulos, G.: Decision theoretic estimation of the ratio of variances in a bivariate normal distribution. Annals of the Institute of Statistical Mathematics **53**, 436–446 (2001)
154. Isogai, T.: Monte Carlo study on some measures for evaluating multinormality. Reports of Statistical Application Research, Union of Japanese Scientists and Engineers **30**, 1–10 (1983a)

155. Isogai, T.: On measures of multivariate skewness and kurtosis. Mathematica Japonica **28**, 251–261 (1983b)
156. Iyengar, S.: Hitting lines with two-dimensional Brownian motion. SIAM Journal on Applied Mathematics **45**, 983–989 (1985)
157. Iyengar, S., Patwardhan, G.: Recent developments in the inverse Gaussian distribution. In: Handbook of Statistics, Volume 7, Quality Control and Reliability, P. R. Krishnaiah and C. R. Rao (eds.), pp. 479–490. North-Holland, Amsterdam (1988)
158. Japanese Standards Association: Statistical Tables and Formulas with Computer Applications. Japanese Standards Association, Tokyo (1972)
159. Jarque, C., Bera, A.: Efficient tests for normality, homoscedasticity and serial independence of regression residuals. Economics Letters **6**, 255–259 (1980)
160. Jarque, C., Bera, A.: A test for normality of observations and regression residuals. International Statistical Reviews **55**, 163–172 (1987)
161. Johnson, M.E.: Multivariate Statistical Simulation. John Wiley and Sons, New York (1987)
162. Johnson, M.E., Bryson, M.C., Mills, C.F.: Some new multivariate distributions with enhanced comparisons via contour and three-dimensional plots. Report LA-8903-MS, Los Alamos Scientific Laboratory, Lost Almamos, New Mexico (1981)
163. Johnson, M.E., Wang, C., Ramberg, J.: Generation of continuous multivariate distributions for statistical applications. American Journal of Mathematical and Management Sciences **4**, 225–248 (1984)
164. Johnson, N.L.: Bivariate distributions based on simple translation systems. Biometrika **36**, 297–304 (1949)
165. Johnson, N.L., Kotz, S.: Distributions in Statistics: Continuous Multivariate Distributions. John Wiley and Sons, New York (1972)
166. Johnson, N.L., Kotz, S., Balakrishnan, N.: Continuous Univariate Distributions, Volume 2, 2nd edition. John Wiley and Sons, New York (1995)
167. Jones, M.C.: "A graphical tool for assessing normality" by M.L. Hazelton, The American Statistician **57**, 285–288 (2003). Comment by Jones. The American Statistician **58**, 176–177 (2004)
168. Jones, M.C., Daly, F.: Density probability plots. Communications in Statistics: Simulation and Computation **24**, 911–927 (1995)
169. Kagan, A., Wesolowski, J.: Normality via conditional normality of linear forms. Statistics and Probability Letters **29**, 229–232 (1996)
170. Kendall, M.G., Stuart, A.: The Advanced Theory of Statistics, Vol. 1: Distribution Theory, 4th edition. Griffin, London (1977)
171. Kendall, M.G., Stuart, A.: The Advanced Theory of Statistics, Vol. 2: Inference and Relationship, 4th edition. Griffin, London (1979)
172. Khatri, C.G., Rao, C.R.: Characterizations of multivariate normality, I: Through independence of some statistics. Journal of Multivariate Analysis **6**, 81–94 (1976)
173. Kim, J-A., Balakrishnan, N.: Nonparametric tests for independence between lifetimes and covariates from censored bivariate normal samples. Communications in Statistics: Simulation and Computation **34**, 685–710 (2005)
174. Kmietowicz, Z.W.: The bivariate lognormal model for the distribution of household size and income. The Manchester School of Economic and Social Studies **52**, 196–210 (1984)
175. Kocherlakota, K., Kocherlakota, S., Balakrishnan, N.: Random number generation from a bivariate Edgeworth series distribution. Computational Statistics Quarterly **2**, 97–105 (1986)
176. Kocherlakota, S.: The bivariate inverse Gaussian distribution: An introduction. Communications in Statistics: Theory and Methods **15**, 1081–1112 (1986)
177. Kocherlakota, S., Kocherlakota, K., Balakrishnan, N.: Effects of nonnormality on the SPRT for the correlation coefficient: Bivariate Edgeworth series distribution. Journal of Statistical Computation and Simulation **23**, 41–51 (1985)

178. Kovner, J.L., Patil, S.A.: On the moments of the doubly truncated bivariate normal population with application to ratio estimate. Journal of the Indian Society of Agricultural Statistics **25**, 131–140 (1973)
179. Kotz, S., Balakrishnan, N., Johnson, N.L.: Continuous Multivariate Distributions, Volume 1, 2nd edition. John Wiley and Sons, New York (2000)
180. Kowalski, C.J.: The performance of some rough tests for bivariate normality before and after coordinate transformations to normality. Technometrics **12**, 517–544 (1970)
181. Kowalski, C.J.: Non-normal bivariate distributions with normal marginals. The American Statistician **27**, 103–106 (1973)
182. Koziol, J.A.: A class of invariant procedures for assessing multivariate normality. Biometrika **69**, 423–427 (1982)
183. Koziol, J.A.: Assessing multivariate normality: A compendium. Communications in Statistics: Theory and Methods **15**, 2763–2783 (1986)
184. Koziol, J.A.: An alternative formulation of Neyman's smooth goodness-of-fit tests under composite alternative. Metrika **34**, 17–24 (1987)
185. Lai, C.D., Rayner, J.C.W., Hutchinson, T.P.: Robustness of the sample correlation- The bivariate lognormal case. Journal of Applied Mathematics and Decision Sciences **3**, 7–19 (1999)
186. Lancaster, H.O.: Zero correlation and independence. Australian Journal of Statistics **21**, 53–56 (1959)
187. Lancaster, H.O.: The Chi-Squared Distribution, John Wiley and Sons, New York (1969)
188. Lancaster, H.O.: Development of notion of statistical dependence. Mathematical Chronicle **2**, 1–16 (1972)
189. Li, L., Owen, D.B.: Two-sided screening procedures in the bivariate case. Technometrics **21**, 79–85 (1979)
190. Li, Zh.V., Shevlyakov, G.L., Shin, V.I.: Robust estimation of a correlation coefficient for epsilon-contaminated bivariate normal distributions. Automation and Remote Control **67**, 1940–1957 (2006)
191. Liem, T.C.: A computer program for Box–Cox transformations in regression models with heteroscedastic and autoregressive residuals. The American Statistician **34**, 121 (1980)
192. Lien, D.H.D.: Moments of truncated bivariate log-normal distributions. Economics Letters **19**, 243–247 (1985)
193. Lien, D., Balakrishnan, N.: Conditional analysis of order statistics from a bivariate normal distribution with an application to evaluating inventory effects in future market. Statistics and Probability Letters **63**, 249–257 (2003)
194. Lien, D., Balakrishnan, N.: Moments and properties of multiplicatively constrained bivariate lognormal distribution with applications to futures hedging. Journal of Statistical Planning and Inference **136**, 1349–1359 (2006)
195. Lin, C.C., Mudholkar, G.S.: A simple test of normality against asymmetric alternatives. Biometrika **67**, 455–461 (1980)
196. Lin, J-T.: A simple approximation for bivariate normal integral. Probability in the Engineering and Information Sciences **9**, 317–321 (1995)
197. Linder, R.S., Nagaraja, H.N.: Impact of censoring on sample variances in a bivariate normal model. Journal of Statistical Planning and Inference **114**, 145–160 (2003)
198. Lindsay, B.G.: Mixture Models: Theory, Geometry and Applications. IMS, Hayward, California (1995)
199. Lindsay, B.G., Basak, P.: Multivariate normal mixtures: A fast consistent method of moments. Journal of the American Statistical Association **88**, 468-476 (1993)
200. Lindqvist, L.: SELLO, a FORTRAN IV program for the transformation of skewed distributions to normality. Computers and Geosciences **1**, 129–145 (1976)
201. Looney, S.W.: A review of techniques for assessing multivariate normality. In: American Statistical Association, 1986 Proceedings of the Statistical Computing Section, pp. 280–285. American Statistical Association, Alexandria, Virginia (1986)

202. Looney, S.W.: How to use tests for univariate normality to assess multivariate normality. Journal of Statistical Theory and Practice **49**, 64–70 (1995)
203. Madsen, R.W.: A selection procedure using a screening variate. Technometrics **24**, 301–306 (1982)
204. Malkovich, J.F., Afifi, A.A.: On tests for multivariate normality. Journal of the American Statistical Association **68**, 176–179 (1973)
205. Mardia, K.V.: Families of Bivariate Distributions. Griffin, London (1970a)
206. Mardia, K.V.: Measures of multivariate skewness and kurtosis with applications. Biometrika **57**, 519–530 (1970b)
207. Mardia, K.V.: Applications of some measures of multivariate skewness and kurtosis for testing normality and robustness studies. Sankhyā, Series B **36**, 115–128 (1974)
208. Mardia, K.V.: Tests of univariate and multivariate normality. In: Handbook of Statistics, Volume 1, Analysis of Variance, P.R. Krishnaiah (ed.), pp. 279–320. North-Holland, Amsterdam (1980)
209. Mardia, K.V.: Mardia's test of multinormality. In: Encyclopedia of Statistical Sciences, Volume 5, S. Kotz and N.L. Johnson (eds.), pp. 217–221. John Wiley and Sons, New York (1985)
210. Mardia, K.V., Foster, K.: Omnibus tests for multinormality based on skewness and kurtosis. Communications in Statistics: Theory and Methods **12**, 207–221 (1983)
211. Mardia, K.V., Zemroch, P.J.: Algorithm AS 84: Measures, of multivariate skewness and kurtosis. Applied Statistics **24**, 262–265 (1975)
212. Marshall, A.W., Olkin, I.: A multivariate exponential distribution. Journal of the American Statistical Association **62**, 30–44 (1967)
213. Martynov, G.V.: Evaluation of the normal distribution function. Journal of Soviet Mathematics **17**, 1857–1875 (1981)
214. Mason, R.L., Young, J.C.: Re-examining two tests for bivariate normality. Communications in Statistics: Theory and Methods **14**, 1531–1546 (1985)
215. Mathai, A.M., Pederzoli, G.: Characterizations of the Normal Probability Law. John Wiley & Sons, New York (1977)
216. Mathar, R.: Outlier-prone and outlier-resistant multidimensional distributions. Statistics **16**, 451–456 (1985)
217. McDonald, J.E.: Remarks on correlation methods in geophysics. Tellus **12**, 176–183 (1960)
218. McLachlan, G.J., Basford, K.E.: Mixture Models: Inference and Applications to Clustering. Marcel Dekker, New York (1988)
219. McLachlan, G.J., Bean, R.W., Ben-Tovim Jones, L., Zhu, X.: Using mixture models to detect differentially expressed genes. Australian Journal of Experimental Agriculture **45**, 859–866 (2005)
220. McLaren C.E., Gordeuk, V.R., Chen, W.P., Barton, J.C., Action, R.T., Speechley, M., Castro, O., Adams, P.C., Sniveley, B.M., Harris, E.L., Reboussin, D.M., McLachlan, G.J., Bean, R.: Bivariate mixture modeling of transferrin saturation and serum ferritin concentration in Asians, African Americans, Hispanics, and whites in the hemochromatosis and iron overload screening (HEIRS). Translational Research **151**, 97–109 (2008)
221. Mee, R.W., Owen, D.B.: A simple approximation for bivariate normal probabilities. Journal of Quality Technology **15**, 72–75 (1983)
222. Mielke, P.W., Williams, J.S., Wu, S-C.: Covariance analysis technique based on bivariate log-normal distribution with weather modification applications. Journal of Applied Meteorology **16**, 183–187 (1977)
223. Melnick, E.L. Tenenbein, A.: Misspecifications of the normal distribution. The American Statistician **36**, 372–373 (1982)
224. Michael, J.R., Schucany, W.R.: Analysis of data from censored samples. In: Goodness-of-Fit Techniques, R.B. D'Agostino and M.A. Stephens (eds.), pp. 461–496, Marcel Dekker, New York (1986)

225. Minami, M.: A multivariate extension of inverse Gaussian distribution derived from inverse relationship. Communications in Statistics: Theory and Methods **32**, 2285–2304 (2003)
226. Mingoti, S.A., Neves, Q.F.: A note on the Zhang omnibus test for normality based on the Q statistic. Journal of Applied Statistics **30**, 335–341 (2003)
227. Moore, D.S.: Tests of chi-squared type. In: Goodness-of-Fit Techniques, R.B. D'Agostino and M.A. Stephens (eds.), pp. 63–95. Marcel Dekker, New York (1986)
228. Moran, P.A.P.: Testing for correlation between non-negative variates. Biometrika **54**, 385–394 (1967)
229. Móri, T.F., Rohatgi, V.K., Szrkely, G.J.: On multivariate skewness and kurtosis. Theory of Probability and Its Applications **38**, 547–551 (1993)
230. Mudholkar, G.S., McDermott, M., Srivastava, D.K.: A test of p-variate normality. Biometrika **79**, 850–854 (1992)
231. Mudholkar, G.S., Srivastava, D.K., Lin, C.T.: Some p-variate adaptions of the Shapiro–Wilk test of normality. Communications in Statistics: Theory and Methods **24**, 953–985 (1995)
232. Mukherjea, A., Nakassis, A., Miyashita, J.: Identification of parameters by the distribution of the maximum random variable: The Anderson-Ghurye theorems. Journal of Multivariate Analysis **18**, 178–186 (1986)
233. Mullooly, J.P.: The variance of left-truncated continuous nonnegative distributions. The American Statistician **42**, 208–210 (1988)
234. Mustonen, S.: Digression analysis. In: Encyclopedia of Statistical Sciences, Volume 2, S. Kotz and N.L. Johnson (eds.), pp. 373–374. John Wiley and Sons, New York (1982)
235. Nabeya, S.: Absolute moments in 2-dimensional normal distribution. Annals of the Institute of Statistical Mathematics **3**, 2–6 (1951)
236. Nagar, D.K., Castañeda, M.E.: Distribution of correlation coefficient under mixture normal model. Metrika **55**, 183–190 (2002)
237. Nagaraja, H.N.: A note on linear function of ordered correlated normal random variables. Biometrika **69**, 284–285 (1982)
238. Nagaraja, H.N., David, H.A.: Distribution of the maximum of concomitants of selected order statistics. Annals of Statistics **22**, 478–494 (1994)
239. Naito, K.: On weighting the studentized empirical characteristic function for testing normality. Communications in Statistics: Simulation and Computation **25**, 201–213 (1996)
240. Nath, G.B.: Moments of a linearly truncated bivariate normal distribution. Australian Journal of Statistics **14**, 97–102 (1972)
241. National Bureau of Standards Tables of the Bivariate Normal Distribution Function and Related Functions, Applied Mathematics Series, No. 50. U.S. Government Printing Office, Washington, D.C. (1959)
242. Nicholson, C.: The probability integral for two variables. Biometrika **33**, 59–72 (1943)
243. Odeh, R.E., Owen, D.B.: Tables for Normal Tolerance Limits, Sampling Plans, and Screening. Marcel Dekker, New York (1980)
244. Owen, D.B.: Tables for computing bivariate normal probabilities. Annals of Mathematical Statistics **27**, 1075–1090 (1956)
245. Owen, D.B.: Handbook of Statistical Tables. Addison-Wesley, Reading, Massachusetts (1962)
246. Owen, D.B.: A table of normal integrals. Communications in Statistics: Simulation and Computation **9**, 389–419 (Additions and Corrections **10**, 537–541) (1980)
247. Owen, D.B.: Screening by correlated variates. In: Encyclopedia of Statistical Sciences, Volume 8, S. Kotz and N.L. Johnson (eds.), pp. 309–312. John Wiley and Sons, New York (1988)
248. Owen, D.B., Haas, R.W.: Tables of the normal conditioned on t-distribution. In: Contributions to Survey Sampling and Applied Statistics. Papers in Honor of H.O. Hartley, H.A. David (ed.), pp. 295–318. Academic Press, New York (1978)

249. Owen, D.B., Li, L., Chou, Y.M.: Prediction intervals for screening using a measured correlated variable. Technometrics **23**, 165–170 (1981)
250. Parrish, R.S., Bargmann, R.E.: A method for the evaluation of cumulative probabilities of bivariate distributions using the Pearson family. In: Statistical Distributions in Scientific Work, Volume 5: Inferential Problems and Properties, C. Taillie, G.P. Patil, and B.A. Baldessari (eds.), pp. 241–257. Reidel, Dordrecht (1981)
251. Patel, J.K., Read, C.B.: Handbook of the Normal Distribution. Marcel Dekker, New York (1982)
252. Paulson, A.S., Roohan, P., Sullo, P.: Some empirical distribution function tests for multivariate normality. Journal of Statistical Computation and Simulation **28**, 15–30 (1987)
253. Pearson, K.: Mathematical contributions to the theory of evolution-VII: On the correlation of characters not quantitatively measurable. Philosophical Transactions of the Royal Society of London, Series A **195**, 1–47 (1901)
254. Pearson, K., Young, A.W.: On the product moments of various orders of the normal correlation surface of two variates. Biometrika **12**, 86–92 (1918)
255. Pettitt, A.N.: Testing for bivariate normality using the empirical distribution function. Communications in Statistics: Theory and Methods **8**, 699–712 (1979)
256. Puente, C.E.: The remarkable kaleidoscopic decompositions of the bivariate Gaussian distribution. Fractals **5**, 47–61 (1997)
257. Puente, C.E., Klebanoff, A.D.: Gaussians everywhere. Fractals **2**, 65–79 (1994)
258. Qin, Y.S., Smith, B.: The likelihood ratio test for homogeneity in bivariate normal mixtures. Journal of Multivariate Analysis **97**, 474–491 (2006)
259. Quandt, R.E., Ramsey, J.B.: Estimating mixtures of normal distributions and switching regressions. Journal of the American Statistical Association **73**, 730–738 (Discussion, 738–752) (1978)
260. Quesenberry, C.P.: Probability integral transformations. In: Encyclopedia of Statistical Sciences, Volume 7, S. Kotz and N.L. Johnson (eds.), pp. 225–231. John Wiley and Sons, New York (1986a)
261. Quesenberry, C.P.: Some transformation methods in goodness-of-fit. In: Goodness-of-Fit Techniques, R.B. D'Agostino, and M.A. Stephens (eds.), pp. 235–277. Marcel Dekker, New York (1986b)
262. Rao, B.R., Garg, M.L., Li, C.C.: Correlation between the sample variances in a singly truncated bivariate normal distribution. Biometrika **55**, 433–436 (1968)
263. Rao, C.R.: Large sample tests of statistical hypotheses concerning several parameters with applications to problems of estimation, Proceedings of the Cambridge Philosophical Society **44**, 50–57 (1948)
264. Rao, C.R.: Some problems in the characterization of the multivariate normal distribution. In: A Modern Course on Distributions in Scientific Work, Volume 3: Characterizations and Applications, G.P. Patil, S. Kotz, and J.K. Ord (eds.), pp. 1–13. Reidel, Dordrecht (1975)
265. Rayner, J.C.W., Best, D.J., Mathews, K.L.: Interpreting the skewness coefficient. Communications in Statistics: Theory and Methods **24**, 593–600 (1995)
266. Regier, M.H., Hamdan, M.A.: Correlation in a bivariate normal distribution with truncation in both variables. Australian Journal of Statistics **13**, 77–82 (1971)
267. Reyment, R.A.: Multivariate normality in morphometric analysis. Mathematical Geology **3**, 357–368 (1971)
268. Rodriguez, R.N.: Correlation. In: Encyclopedia of Statistical Sciences, Volume 2, S. Kotz and N.L. Johnson (eds.), pp. 193–204. John Wiley and Sons, New York (1982)
269. Rodriguez, R.N.: Frequency surfaces, systems of. In: Encyclopedia of Statistical Sciences, Volume 3, S. Kotz and N.L. Johnson (eds.), pp. 232–247. John Wiley and Sons, New York (1983)
270. Rom, D., Sarkar, S.K.: Approximating probability integrals of multivariate normal using association models. Journal of Statistical Computation and Simulation **35**, 109–119 (1990)

271. Romano, J.P., Siegel, A.F.: Counterexamples in Probability and Statistics. Wadsworth and Brooks/Cole, Monterey, California (1986)
272. Rosenblatt, M.: Remarks on a multivariate transformation. Annals of Mathematical Statistics **23**, 470–472 (1952)
273. Rosenblueth, E.: On computing normal reliabilities. Structural Safety **2**, 165–167 (Corrections **3**, 67) (1985)
274. Royston, J.P.: Algorithm AS 181. The W test for normality. Applied Statistics **35**, 232–234 (1982)
275. Royston, J.P.: Some techniques for assessing multivariate normality based on the Shapiro-Wilk W. Applied Statistics **32**, 121–133 (1983)
276. Ruben, H.: An asymptotic expansion for the multivariate normal distribution and Mill's ratio. Journal of Research, National Bureau of Standards **68**, 3–11 (1964)
277. Ruppert, D.: Trimming and Winsorization. In: Encyclopedia of Statistical Sciences, Volume 9, S. Kotz and N.L. Johnson (eds.), pp. 348–353. John Wiley and Sons, New York (1988)
278. Ruymgaart, F.H.: Non-normal bivariate densities with normal marginals and linear regression functions. Statistica Neerlandica **27**, 11–17 (1973)
279. Sahu, S.K., Dey, D.K., Branco, M.D.: A new class of multivariate skew distributions with applications to Bayesian regression models. Canadian Journal of Statistics **31**, 129–150 (2003)
280. Sarabia, J-M.: The centered normal conditional distributions. Communications in Statistics: Theory and Methods **24**, 2889–2900 (1995)
281. Sarmanov, O.V.: Generalized normal correlation and two dimensional Fréchet classes. Soviet Mathematics **7**, 596–599 (Original article was in Russian) (1966)
282. Savage, I.R.: Mills' ratio for multivariate normal distributions. Journal of Research of the National Bureau of Standards-B: Mathematics and Mathematical Physics **66**, 93–96 (1962)
283. Schneider, T., Holst, E.: Man-made mineral fibre size distributions utilizing unbiased and fibre length biased counting methods and the bivariate log-normal distribution. Journal of Aerosol Science **14**, 139–146 (The authors have noted misprints in their equations (20) and (Appendix 1)) (1983)
284. Schreuder, H.T., Hafley, W.L.: A useful bivariate distribution for describing stand structure of tree heights and diameters. Biometrics **33**, 471–478 (1977)
285. Schwager, S.J.: Multivariate skewness and kurtosis. In: Encyclopedia of Statistical Sciences, Volume 6, S. Kotz and N.L. Johnson (eds.), pp. 122–125. John Wiley and Sons, New York (1985)
286. Schweizer, S., Borsuk, M.E., Jowett, I., Reichert, P.: Predicting joint frequency of depth and velocity for instream habitat assessment. River Research and Applications **23**, 287–302 (2007)
287. Seal, H.L.: Studies in the history of probability and statistics. XV. The historical development of the Gauss linear model. Biometrika **54**, 1–24 (1967)
288. Shah, S.M., Parikh, N.T.: Moments of singly and doubly truncated standard bivariate normal distribution. Vidya (Gujarat University) **7**, 82–91 (1964)
289. Shapiro, S.S., Wilk, M.B.: An analysis of variance test for normality (complete samples). Biometrika **52**, 591–611 (1965)
290. Shevlyakov, G.L., Vilchevski, N.O.: Minimax variance estimation of a correlation coefficient for ε-contaminated bivariate normal distributions. Statistics and Probability Letters **57**, 91–100 (2002)
291. Siegel, A.F., O'Brien, F.: Unbiased Monte Carlo integration methods with exactness for low order polynomials. SIAM Journal of Scientific and Statistical Computation **6**, 169–181 (1985)
292. Sievers, G.L.: Probability plotting. In: Encyclopedia of Statistical Sciences, Volume 7, S. Kotz and N.L. Johnson (eds.), pp. 232–237. John Wiley and Sons, New York (1986)

293. Silverman, B.W.: Density estimation for univariate and bivariate data. In: Interpreting Multivariate Data, V. Barnett (ed.), pp. 37–53. John Wiley and Sons, Chichester (1981)
294. Silverman, B.W.: Density Estimation for Statistics and Data Analysis. Chapman and Hall, London (1986)
295. Small, N.J.H.: Marginal skewness and kurtosis for testing multivariate normality. Applied Statistics **29**, 85–87 (1980)
296. Small, N.J.H.: Multivariate normality, testing for. In: Encyclopedia of Statistical Sciences, Volume 6, S. Kotz and N.L. Johnson (eds.), pp. 95–100. John Wiley and Sons, New York (1985)
297. Smirnov, N.V., Bol'shev, L.N.: Tables for Evaluating a Function of a Bivariate Normal Distribution (in Russian). Is'datel'stov Akademii Nauk SSSR, Moscow (1962)
298. Sondhauss, U.: Asymptotische Eigenschaften intermediärer Ordnungs-statistiken und ihrer Konkomitanten, Diplomarbeit. Department of Statistics, Dortmund University, Germany (1994)
299. Sowden, R.R., Ashford, J.R.: Computation of bivariate normal integral. Applied Statistics **18**, 169–180 (1969)
300. Springer, M.D.: The Algebra of Random Variables. John Wiley and Son, New York (1979)
301. Srivastava, M.S.: A measure of skewness and kurtosis and a graphical method for assessing multivariate normality. Statistics and Probability Letters **2**, 263–267 (1984)
302. Srivastava, M.S., Hui, T.K.: On assessing multivariate normality based on Shapiro-Wilk statistic. Statistics and Probability Letters **5**, 15–18 (1987)
303. Stephens, M.A.: Tests based on EDF statistics. In: Goodness-of-Fit Techniques, R.B. D'Agostino and M.A. Stephens (eds.), pp. 97–193. Marcel Dekker, New York (1986a)
304. Stephens, M.A.: Tests based on regression and correlation. In: Goodness-of-Fit Techniques, R.B. D'Agostino and M.A. Stephens (eds.), pp. 195–233. Marcel Dekker, New York (1986b)
305. Sun, Y., Wong, A.C.M.: Interval estimation for the normal correlation coefficient. Statistics and Probability Letters **77**, 1652–1661 (2007)
306. Sungur, E.A.: Dependence information in parametrized copulas. Communications in Statistics: Simulation and Computation **19**, 1339–1360 (1990)
307. Suzuki, M.: Estimation in a bivariate semi-lognormal distribution. Behaviormetrika **13**, 59–68 (1983)
308. Suzuki, M., Iwase, K., Shimizu, K.: Uniformly minimum variance unbiased estimation in a semi-lognormal distribution. Journal of the Japan Statistical Society **14**, 63–68 (1984)
309. Tanabe, K., Sagae, M., Ueda, S.: BNDE, Fortran Subroutines for Computing Bayesian Nonparametric Univariate and Bivariate Density Estimator. Computer Science Monograph No. 24, Institute of Statistical Mathematics, Tokyo (1988)
310. Tarter, M., Silvers, A.: Implementation and applications of a bivariate Gaussian mixture decomposition. Journal of the American Statistical Association, **70**, 47–55 (1975)
311. Terza, J., Welland, U.: A comparison of bivariate normal algorithms. Journal of Statistical Computation and Simulation **39**, 115–127 (1991)
312. Thadewald, T., Büning, H.: Jarque–Bera test and its competitors for testing normality: A power comparison. Journal of Applied Statistics **34**, 87–105 (2007)
313. Thomas, G.E.: Remark AS R30: A remark on Algorithm AS 76: An integral in calculating noncentral t and bivariate normal probabilities. Applied Statistics **28**, 113 (1979)
314. Thomopoulos, N.T., Longinow, A.: Bivariate lognormal probability distribution. Journal of Structural Engineering **110**, 3045–3049 (1984)
315. Titterington, D.M., Smith, A.F.M., Makov, U.E.: Statistical Analysis of Finite Mixture Distributions. John Wiley and Sons, New York (1985)
316. Tong, Y.L.: Probability Inequalities in Multivariate Distributions. Academic Press, New York (1980)

317. Tsou, T.-S.: Robust inference for the correlation coefficient: A parametric method. Communications in Statistics: Theory and Methods **34**, 147–162 (2005)
318. Tukey, P.A., Tukey, J.W.: Data-driven view selection: Agglomeration and sharpening. In: Interpreting Multivariate Data, V. Barnett (ed.), pp. 215–243. John Wiley and Sons, Chichester (1981)
319. Ulrich, G.: A class of multivariate distributions with applications in Monte Carlo and simulation. In: American Statistical Association, 1984 Proceedings of the Statistical Computing Section, pp. 185–188. American Statistical Associate, Alexandria, Virginia (1984)
320. Vale, C.D., Maurelli, V.A.: Simulating multivariate nonnormal distributions. Psychometrika **48**, 465–471 (1983)
321. van Yzeren, J.: A bivariate distribution with instructive properties as to normality, correlation and dependence. Statistica Neerlandica **26**, 55–56 (1972)
322. van Zyl, J.M.: Planar random movement and the bivariate normal density. South African Statistical Journal **21**, 1–12 (1987)
323. Versluis, C.: Comparison of tests for bivariate normality with unknown parameters by transformation to a univariate statistic. Communications in Statistics: Theory and Methods **25**, 647–665 (1996)
324. Wachter, K.W.: Haar distributions. In: Encyclopedia of Statistical Sciences, Volume 3, S. Kotz and N.L. Johnson (eds.), pp. 559–562. John Wiley and Sons, New York (1983)
325. Wang, M., Kennedy, W.J.: Comparison of algorithms for bivariate normal probability over a rectangle based on self-validated results from interval analysis. Journal of Statistical Computation and Simulation **37**, 13–25 (1990)
326. Wang, Y.: The probability integrals of bivariate normal distributions: A contingency table approach. Biometrika **74**, 185–190 (1987)
327. Warren, W.G.: Some recent developments relating to statistical distributions in forestry and forest products research. In: Statistical Distributions in Ecological Work, J.K. Ord, G.P. Patil, and C. Taillie (eds.), pp. 247–250. International Co-operative Publishing House, Fairland, Maryland (1979)
328. Wasan, M.T.: On an inverse Gaussian process. Skandinavisk Aktuarietidskrift **1968**, 69–96 (1968)
329. Wasan, M.T., Buckholtz, P.: Differential representation of a bivariate inverse Gaussian process. Journal of Multivariate Analysis **3**, 243–247 (1973)
330. Watterson, G.A.: Linear estimation in censored samples from multivariate normal populations. Annals of Mathematical Statistics **30**, 814–824 (1959)
331. Welland, U., Terza, J.V.: Bivariate normal approximation: Gauss–Legendre quadrature applied to Sheppard's formula. Working Paper, Department of Economics, Pennsylvania State University, State College, Pennsylvania (1987)
332. Willink, R.: Bounds on bivariate normal distribution. Communications in Statistics: Theory and Methods **33**, 2281–2297 (2004)
333. Wilson, J.R.: Fitting Johnson curves to univariate and multivariate data. In: 1983 Winter Simulation Conference Proceedings, Volume 1, S. Roberts, J. Banks, and B. Schmeiser (eds.), pp. 114–115. Institute of Electrical and Electronics Engineers, Piscataway, New Jersey (1983)
334. Wolfe, J.E.: Pattern clustering by multivariate mixture analysis. Multivariate Behavioral Research **5**, 329–350 (1970)
335. Young, J.C., Minder, C.E.: Algorithm AS 76: An integral useful in calculating noncentral t and bivariate normal probabilities. Applied Statistics **23**, 455–457 (Corrections **27**, 379; **28**, 113; **28**, 336; **34**, 100–101 and **35**, 310–312) (1974). [Reprinted with revisions in P. Griffiths and I.D. Hill (eds.), Applied Statistics Algorithms, pp. 145–148. Ellis Horwood, Chichester (1985).]
336. Yu, P.L.H.: Lam, K.: Regression estimator in ranked set sampling. Biometrics **53**, 1070–1080 (1997)

337. Yu, P.L.H., Sun, Y., Sinha, B.K.: Estimation of the common mean of a bivariate normal population. Annals of the Institute of Statitical Mathematics **54**, 861–878 (2002)
338. Yuan, P-T.: On the logarithmic frequency distribution and the semi-logarithmic correlation surface. Annals of Mathematical Statistics **4**, 30–74 (1933)
339. Yue, S.: The bivariate lognormal distribution to model a multivariate flood episode. Hydrological Process **14**, 2575–2588 (2000)
340. Yue, S.: The bivariate lognormal distribution for describing joint statistical properties of a multivariate storm event. Environmetrics **13**, 811–819 (2002)
341. Zelen, M., Severo, N.C.: Graphs for bivariate normal probabilities. Annals of Mathematical Statistics **31**, 619–624 (1960)
342. Zerehdaran, S., van Grevehof, E.A., van der Waaij, E.H., Bovenhuis, H.: A bivariate mixture model analysis of body weight and ascites traits in broilers. Poultry Science **85**, 32–38 (2006)
343. Zhang, P.: Omnibus test for normality using Q statistic. Journal of Applied Statistics **26**, 519–528 (1999)
344. Zhang, Y.C.: Derivative fitting procedure for computing bivariate normal distributions and some applications. Structural Safety **14**, 173–183 (1994)
345. Zheng, G., Modarres, R.: A robust estimate for the correlation coefficient for bivariate normal distribution using ranked set sampling. Journal of Statistical Planning and Inference **136**, 298–309 (2006)

Chapter 12
Bivariate Extreme-Value Distributions

12.1 Preliminaries

The univariate extreme-value distributions consist of types 1 (Gumbel), 2 (Fréchet), and 3. The three types can be transformed to each other. The type 3 distribution of $(-X)$ is the usual Weibull distribution.

In the bivariate context, marginals are of secondary interest compared with the dependence structure. Tiago de Oliveira (1962/63, 1975a,b, 1980, 1984), Gumbel and Goldstein (1964), Gumbel (1965), Gumbel and Mustafi (1967), and Galambos (1987, Chapter 5, especially Section 5.4) assumed Gumbel marginals, whereas de Haan and Resnick (1977) and Kotz and Nadarajah (2000, Chapter 3) chose Fréchet marginals $F(x) = \exp(-x^{-1})$. All three types can be easily transformed to exponential variates, and in most cases we will follow Pickands (1981), Deheuvels (1983, 1985), Smith (1994), and Tawn (1988a) in choosing exponential marginals.

There are several excellent treatises on bivariate and multivariate extreme value distributions; see, for example, Galambos (1987), Smith (1990, 1994), Kotz and Nadarajah (2000), Coles (2001), and Beirlant et al. (2004).

In Section 12.2, we first introduce the bivariate extreme-value distribution. Next, in Section 12.4, we discuss the classical bivariate extreme-value distribution with Gumbel marginals and its properties. Then, in Sections 12.5–12.7, we discuss the bivariate extreme-value distributions with exponential, Fréchet, and Weibull marginal distributions, respectively. In Section 12.8, we describe the methods of derivation, estimation methods are detailed in Section 12.9, and some references to illustrations are presented in Section 12.10. Section 12.11 describes algorithms for the simulation of random variates from the bivariate extreme-value distribution. Some applications are indicated in Section 12.12 and finally conditionally specified bivariate Gumbel distributions are mentioned in Section 12.13.

N. Balakrishnan, C.D. Lai, *Continuous Bivariate Distributions*,
DOI 10.1007/b101765_13, © Springer Science+Business Media, LLC 2009

12.2 Introduction to Bivariate Extreme-Value Distribution

12.2.1 Definition

Let (X_i, Y_i), $i = 1, 2, \ldots, n$, be n pairs of independent bivariate random variables with $X_{\max} = \max(X_1, \ldots, X_n)$ and $Y_{\max} = \max(Y_1, \ldots, Y_n)$. It is possible to find linear transformations $X_{(n)} = a_n X_{\max} + b_n$ ($a_n > 0$) and $Y_{(n)} = c_n Y_{\max} + d_n$ ($c_n > 0$) such that $X_{(n)}$ (and $Y_{(n)}$) is one of the three types of extreme-value distributions as $n \to \infty$. Then, the limiting joint distribution of $X_{(n)}$ and $Y_{(n)}$ is a bivariate extreme-value distribution.

A general definition of a bivariate extreme-value distribution can be presented through a copula [Pickands (1981)]. Let (X, Y) have a joint bivariate extreme-value distribution with marginals $F(x)$ and $G(y)$; then, the associated copula is given by

$$\begin{aligned} C(u, v) &= \Pr\{F(X) \le u, G(Y) \le v\} \\ &= \exp[\log(uv) A\{\log(u)/\log(uv)\}] \end{aligned} \tag{12.1}$$

for all $0 \le u, v \le 1$ in terms of a convex function A defined on $[0, 1]$ in such a way that $\max(t, 1-t) \le A(t) \le 1$ for all $0 \le t \le 1$. A is known as the *dependence function*, and we will discuss its properties in Section 12.5.2.

12.2.2 General Properties

- Let (X_i, Y_i), $i = 1, 2, \ldots, n$ be a random sample from a bivariate population with a joint distribution whose copula is C. Let $X_{(n)} = \max\{X_i\}$ and $Y_{(n)} = \max\{Y_i\}$. Then the copula that corresponds to $X_{(n)}$ and $Y_{(n)}$ is

  $$C_{(n)}(u, v) = C^n(u^{\frac{1}{n}}, v^{\frac{1}{n}}).$$

  A copula C_* is an extreme-value copula if there exists a copula C such that

  $$C_*(u, v) = \lim_{n \to \infty} C^n(u^{\frac{1}{n}} v^{\frac{1}{n}});$$

  see Nelsen (2006, p. 97).
- Shi (2003) has considered a transformation of variables from the copula above with $S = -\log(UV) A\left(\frac{\log U}{\log(UV)}\right)$, $T = \frac{\log U}{\log(UV)}$. It has been shown that S and T are "essentially" independent; this leads to some stochastic representation for the bivariate extreme-value distribution.

- In many bivariate distributions (such as the bivariate normal), $X_{\max}$ and $Y_{\max}$ may be asymptotically independent (as the sample size tends to infinity) even if X and Y are correlated. This is so if $\bar{H}(xy)/\{1 - H(x,y)\} \to 0$ as $x, y \to \infty$. This result is due to Geffroy (1958/59).
- Let (X, Y) have a bivariate extreme-value distribution. Then, X and Y are PQD.
- Let $H_1(x,y)$ and $H_2(x,y)$ be two bivariate extreme-value distributions, so their weighted geometric mean is

$$[H_1(x,y)]^{\beta}[H_2(x,y)]^{1-\beta}, \quad 0 \le \beta \le 1,$$

see Gumbel and Goldstein (1964).

12.3 Bivariate Extreme-Value Distributions in General Forms

Gumbel (1958, 1965) has described two general forms for bivariate extreme-value distributions in terms of the marginals (univariate extreme-value distributions):

1. Type A

$$H(x,y) = F(x)G(y)\exp\left\{-\theta\left[\frac{1}{\log F(x)} + \frac{1}{\log G(y)}\right]^{-1}\right\}, \quad 1 \le \theta < 1.$$

The corresponding copula is

$$C(u,v) = uv\exp\left(\frac{-\theta}{\log uv}(\log u \log v)\right).$$

2. Type B

$$H(x,y) = \exp\left\{-\left[(-\log F(x))^m + (-\log G(y))^m\right]^{1/m}\right\}, \quad m \ge 1.$$

The copula that corresponds to the type B extreme-value distribution is

$$C(u,v) = \exp\left(-\left[(-\log u)^m + (-\log v)^m\right]^{1/m}\right).$$

It is an extreme-value copula since $C(u^k, v^k) = C^k(u,v)$; in fact, it is the only Archimedean copula that is also an extreme-value copula, as remarked in Example 1.8. It is called the Gumbel–Hougaard copula in Section 2.6.

The type A bivariate extreme-value distribution is known by some as the (Gumbel) mixed model [see, for example, Yue et al. (2000)], whereas the type B bivariate extreme-value distribution is known as the logistic model.

Restricting to the case where both marginals are Gumbel, Yue and Wang (2004) compared these two models by Monte Carlo experiments. Their results indicate that within the range of $0 \le \rho \le 2/3$, both models provide the same joint probabilities and joint return periods, and both may be useful for representing statistical properties of X and Y. When $\rho > 2/3$, only the logistic (type B) model can be applied to the joint distribution of X and Y.

12.4 Classical Bivariate Extreme-Value Distributions with Gumbel Marginals

Three special types are considered in this section—type A, type B, and type C—all having Gumbel marginals. The distributions with exponential marginals will be discussed in Section 12.5.

A bivariate extreme-value distribution with Gumbel marginals has the general form

$$H(x,y) = \exp\left\{-\int_0^1 \min[f_1(s)e^{-x}, f_2(s)e^{-y}]ds\right\}, \tag{12.2}$$

where $f_1(t)$ and $f_2(t)$ are non-negative Lebesgue integrable functions such that $\int_0^1 f_i(t)dt = 1, i = 1, 2$; see, for example, Resnick (1987, p. 272).

12.4.1 Type A Distributions

These distributions are also known as the mixed model.

Formula of the Cumulative Distribution Function

The joint distribution function is

$$H(x,y) = \exp[-e^{-x} - e^{-y} + \theta(e^x + e^y)^{-1}], \quad \theta \le 1, \tag{12.3}$$

which is an increasing function of θ.

Formula of the Joint Density

The joint density function is

$$\begin{aligned} h(x,y) = {} & e^{-(x+y)}[1 - \theta(e^{2x} + e^{2y})(e^x + e^y)^{-2} + 2\theta e^{2(x+y)}(e^x + e^y)^{-3} \\ & + \theta^2 e^{2(x+y)}(e^x + e^y)^{-4}] \exp[-e^{-x} - e^{-y} + \theta(e^x + e^y)^{-1}]. \end{aligned} \tag{12.4}$$

Univariate Properties

The marginal distribution function of X is $F(x) = \exp[-e^{-x}]$, $-\infty < x < \infty$, and a similar expression for $G(y)$. That is, the marginals are both type I extreme-value distributions. Note that the type I extreme-value distribution is also known as the Gumbel distribution. In fact, it is the distribution most commonly referred to in discussions of univariate extreme-value distributions.

Medians and Modes

The median of the common distribution of X and Y is

$$\mu = -\log(\log 2) = 0.36651, \tag{12.5}$$

so that $F(\mu)G(\mu) = \frac{1}{4}$ and

$$H(\mu, \mu) = \exp\left(-2e^{-\mu} + \frac{1}{2}\theta e^{-\mu}\right) = \left(\frac{1}{4}\right)^{1-\theta/4}. \tag{12.6}$$

Also,

$$H(0,0) = (e^{-2})^{1-\theta/4}. \tag{12.7}$$

The value $\tilde{\mu}$, such that $H(\tilde{\mu}, \tilde{\mu}) = \frac{1}{4}$, satisfies the equation

$$\left(2 - \frac{1}{2}\theta\right) e^{-\tilde{\mu}} = 2\log 2, \tag{12.8}$$

and so

$$\tilde{\mu} = \log\left(1 - \frac{1}{4}\theta\right) - \log(\log 2) = \log\left(1 - \frac{1}{4}\theta\right) + 0.3665. \tag{12.9}$$

Since $0 \le \theta \le 1$, $0.3665 - \log(\frac{4}{3}) = 0.0787 \le \tilde{\mu} \le 0.3665$.

The mode of the common distribution of X and Y is at zero. The mode of the joint distribution is at

$$x = y = \log\left[\frac{(2-\theta)(4-\theta)}{2\theta}\left\{\sqrt{\frac{1}{2} + \frac{2}{(2-\theta)^2}} - 1\right\}\right]. \tag{12.10}$$

The numerical values are tabulated, for example, in Table 53.1 of Kotz et al. (2000, p. 627).

Correlation Coefficients

The expression for the product-moment correlation is quite complex. However, Spearman's rho (the grade correlation) is simpler, and is given by

$$\begin{aligned}\rho_S &= 3\left(2-\frac{1}{4}\theta\right)^{-1}\\ &\quad\times\left[1+2\left(2\theta-\frac{1}{4}\theta^2\right)^{-1}\tan^{-1}\left\{\left(2\theta-\frac{1}{4}\theta^2\right)^{1/2}\left(2-\frac{1}{2}\theta\right)^{-1}\right\}\right]-3.\end{aligned} \tag{12.11}$$

There appears to be a misprint in the formula given by Gumbel and Mustafi (1967, p. 583). However, their Table 3 appears to be correct. Some values of Spearman's rho for a few values of θ can also be found in the same table.

12.4.2 Type B Distributions

Type B bivariate extreme-value distributions are also known as logistic models.

Formula of the Cumulative Distribution Function

The joint cumulative distribution function is

$$H(x,y)=\exp\left[-(e^{-mx}+e^{-my})^{1/m}\right],\quad m\ge 1. \tag{12.12}$$

Since $\lim_{m\to\infty}(e^{-mx}+e^{-my})^{1/m}=\max(e^{-x},e^{-y})$, we obtain

$$\begin{aligned}\lim_{m\to\infty}H(x,y)&=\min\left[\exp(-e^{-x}),\exp(-e^{-y})\right]\\ &=\min\big(F(x),G(y)\big).\end{aligned} \tag{12.13}$$

It is clear that, for $m=1$, X and Y are independent.

Formula of the Joint Density

The joint density function is

$$\begin{aligned}h(x,y) &= e^{-m(x+y)}(e^{-mx} + e^{-my})^{-2+1/m}] \\ &\times\{m - 1 + (e^{-mx} + e^{-my})^{1/m}\} \\ &\times \exp[-(e^{-mx} + e^{-my})^{1/m}], \end{aligned} \tag{12.14}$$

for $m \geq 1$.

Univariate Properties

The marginal distributions are both type I extreme-value distributions.

Medians and Modes

With the univariate median μ defined as $F(\mu) = G(\mu) = \frac{1}{2}$, we find, for type B distributions,

$$H(\mu, \mu) = \left(\frac{1}{4}\right)^{1/m} \tag{12.15}$$

and

$$H(0, 0) = (e^{-2})^{1/m} \tag{12.16}$$

[compare these with (12.6) and (12.7)].

The values of $\tilde{\mu}$ such that $H(\tilde{\mu}, \tilde{\mu}) = \frac{1}{4}$ satisfies the equation

$$\exp[-2^{1/m}e^{-\tilde{\mu}}] = \frac{1}{4},$$

and so

$$\tilde{\mu} = -\log(\log 2) - \frac{m-1}{m}\log 2. \tag{12.17}$$

Since $m \geq 1$, $0.3665 - \log 2 = -0.3266 \leq \tilde{\mu} \leq 0.3665$.

The mode of the joint distribution is at

$$x = y = (1 + m^{-1})\log 2 - \log\left[\sqrt{(m-1)^2 + 4} - m + 3\right]. \tag{12.18}$$

Some numerical values have been presented in Table 53.1 of Kotz et al. (2000).

Correlation Coefficients

The Pearson product-moment correlation coefficient is $\rho = 1 - m^{-2}$.

Other Properties

The expression $X - Y$ has a logistic distribution with

$$\Pr(X - Y \leq t) = (1 + e^{-mt})^{-1}. \tag{12.19}$$

Fisher Information Matrix

Shi (1995b) has derived the Fisher information matrix for the multivariate version of the logistic model.

Type B Bivariate Extreme-Value Distribution with Mixed Gumbel Marginals

Escalante-Sandoval (1998) considered a type B bivariate extreme value distribution (12.12) but with the marginals being the mixtures of Gumbel distributions. The joint distribution was found to be useful for performing flood frequency analysis.

12.4.3 Type C Distributions

For these distributions (also known as the biextremal model), the joint distribution function is

$$H(x, y) = \exp\left[-\max\{e^{-x} + (1 - \phi)e^{-y}, e^{-y}\}\right], \quad 0 < \phi < 1. \tag{12.20}$$

The distribution in (12.20) can be generated as the joint distribution of X and

$$Y = \max(X + \log\phi, Z + \log(1 - \phi)),$$

where X and Z are mutually independent variables with each having a Gumbel distribution.

The distribution has a singular component along the line $y = x + \log\phi$ since

$$\Pr[Y = X + \log\phi] = \Pr[Z - X \leq \log\{\phi/(1 - \phi)\}] = \phi. \tag{12.21}$$

Correlation Coefficients

The correlation coefficient is given by

$$\operatorname{corr}(X,Y) = \rho = -6\pi^2 \int_0^\phi (1-t)^{-1} \log t \, dt,$$

and the Spearman correlation is $\rho_S = 3\phi/(2+\phi)$.

Medians and Modes

$$H(\mu,\mu) = \frac{1}{4}(2^\phi) \tag{12.22}$$

and

$$H(0,0) = (e^{-2})^{1-\phi/2}. \tag{12.23}$$

The value $\tilde{\mu}$, such that $H(\tilde{\mu},\tilde{\mu}) = \frac{1}{4}$, is given by

$$\tilde{\mu} = -\log\left(\frac{\log 2}{1-\frac{1}{2}\phi}\right). \tag{12.24}$$

12.4.4 Representations of Bivariate Extreme-Value Distributions with Gumbel Marginals

Tiago de Oliveira (1961) showed that a bivariate distribution with standard type I extreme-value marginals can be defined by a cumulative distribution function of the form

$$H(x,y) = \exp\left\{-(e^{-x}+e^{-y})k(y-x)\right\}, \tag{12.25}$$

where $k(\cdot)$ satisfies the conditions

$$\begin{aligned}
&\lim_{t\to\pm\infty} k(t) = 1,\\
&\frac{d}{dt}\{(1+e^{-t})k(t)\} \le 0,\\
&\frac{d}{dt}\{(1+e^{t})k(t)\} \ge 0,\\
&(1+e^{-t})k''(t) + (1-e^{-t})k'(t) \ge 0.
\end{aligned}$$

Type A is obtained by taking

$$k(t) = 1 - \frac{1}{4}\theta \operatorname{sech}^2\frac{1}{2}t. \tag{12.26}$$

Type B is obtained by taking

$$k(t) = (e^{mt} + 1)^{1/m}(e^t + 1)^{-1}. \qquad (12.27)$$

Type C is obtained by taking

$$k(t) = (e^t + 1)^{-1}\{1 - \phi + \max(e^t, \phi)\}, \quad 0 < \phi < 1. \qquad (12.28)$$

12.5 Bivariate Extreme-Value Distributions with Exponential Marginals

Pickands (1981) [see also Tawn (1988a)] showed a bivariate extreme-value distribution with unit exponential marginals can be expressed via a dependence function.

12.5.1 Pickands' Dependence Function

Here,

$$\bar{H}(x, y) = \exp\left[-(x + y)A\left(\frac{y}{x + y}\right)\right], \quad x, y > 0, \qquad (12.29)$$

where

$$A(w) = \int_0^1 \max[(1 - w)q, w(1 - q)]\frac{dB}{dq}dq, \qquad (12.30)$$

in which B is a positive function on $[0, 1]$. In order to have unit exponential marginals, we need

$$1 = \int_0^1 q\frac{dB}{dq}dq = \int_0^1 (1 - q)\frac{dB}{dq}. \qquad (12.31)$$

[To deduce this, we successively set $x = 0$ and $y = 0$ in (12.29). We then find that $A(0)$ and $A(1)$ must both be 1 and put these values into (12.31).] It follows from (12.31) that $\frac{1}{2}B$ is the distribution function of a random variable with mean $\frac{1}{2}$. We call A the dependence function of (X, Y), in accordance with the usage of Pickands (1981) and Tawn (1988a). [Do not confuse the with any other meaning of the term; for example, that of Oakes and Manatunga (1992).]

For accounts of the connections between various dependence functions, see Deheuvels (1984) and Weissman (1985).

12.5.2 Properties of Dependence Function A

1. $A(0) = A(1) = 1$.
2. $\max(w, 1-w) \leq A(w) \leq 1,\ 0 \leq w \leq 1$.
3. $A(w) = 1$ implies that X and Y are independent. $A(w) = \max(w, 1-w)$ implies that X and Y are equal, i.e., $\Pr(X = Y) = 1$.
4. A is convex; i.e., $A[\lambda x + (1-\lambda)y] \leq \lambda A(x) + (1-\lambda)A(y)$.
5. If A_i are dependence functions, so is $\sum_{i=1}^n \alpha_i A_i$, where $\alpha_i \geq 0$ and $\sum_{i=1}^n = 1$.

A may or may not be differentiable. In the former case, H has a joint density everywhere; in the latter, H has a singular component and is not differentiable in a certain region of its support. We shall consider examples of this family of distributions classified as differentiable, nondifferentiable, or Tawn's extension of differentiable. Examples 1–4, 6, and 7 below were discussed by Tawn (1988a).

Nadarajah et al. (2003) studied the local dependance functions for the extreme-value distribution with dependence function A given in (12.29) and (12.30) above.

12.5.3 Differentiable Models

Example 1

The mixed model, also known as Gumbel's type A bivariate extreme-value distribution, sets $A(w) = \theta w^2 - \theta w + 1$ for $0 \leq \theta \leq 1$. Hence

$$\bar{H}(x, y) = \exp\left[-(x+y) + \frac{\theta x y}{x+y}\right]; \tag{12.32}$$

see Gumbel and Mustafi (1967) for further properties.

In the case of marginals with different parameters, Elandt-Johnson (1978) showed that the two crude hazard rates $h_1(x) = \frac{\partial \bar{H}}{\partial x}|_{y=x}$ and $h_2(y) = \frac{\partial \bar{H}}{\partial y}|_{x=y}$ are proportional if and only if the marginal hazard rates $f/\bar{F}$ and $g/\bar{G}$ are proportional.

Example 2

The logistic model, also known as the type B extreme value-distribution, sets $A(w) = [(1-w)^r + w^r]^{1/r}$ for $r \geq 1$. Hence,

$$\bar{H}(x, y) = \exp[-(x^r + y^r)^{1/r}]. \tag{12.33}$$

This is the third type of exponential distribution mentioned, albeit only briefly, by Gumbel (1960); see Gumbel and Mustafi (1967) for further details.

12.5.4 Nondifferentiable Models

Example 3

The biextremal model, also known as the type C bivariate extreme-value distribution, sets $A(w) = \max(w, 1-\theta w)$ for $0 \leq \theta \leq 1$. Hence,

$$\bar{H}(x,y) = \exp\{-\max[x + (1-\theta)y, y]\}\,. \tag{12.34}$$

Example 4

Gumbel's model sets $A(w) = \max[1-\theta w, 1-\theta(1-w)]$ for $0 \leq \theta \leq 1$. Hence,

$$\bar{H}(x,y) = [-(1-\theta)(x+y) - \theta\max(x,y)]\,. \tag{12.35}$$

This is effectively the bivariate exponential distribution of Marshall and Olkin (1967) discussed in Section 10.5.

Example 5

The natural model sets $A(w) = \frac{\beta-1}{\beta-\alpha}\max(1-w,\alpha w) + \frac{1-\alpha}{\beta-\alpha}\max(1-w,\beta w)$ for $0 \leq \alpha \leq 1 \leq \beta < \infty$. Hence,

$$\bar{H}(x,y) = \exp\{-[(\beta-1)\max(x,\alpha y) + (1-\alpha)\max(x,\beta y)]/(\beta-\alpha)\}. \tag{12.36}$$

12.5.5 Tawn's Extension of Differentiable Models

Background

In the dependence functions for the differential models, Tawn (1988a) added an extra parameter ϕ to give further flexibility. This gives us two new models, as follows.

Example 6

The asymmetric mixed model sets $A(w) = \phi w^3 + \theta w^2 - (\theta + \phi)w + 1$ for $\theta \geq 0$, $\theta + \phi \leq 1$, $\theta + 2\phi \leq 1$, $\theta + 3\phi \geq 0$. Hence,

$$\bar{H}(x, y) = \exp\left[-(x + y) + xy\frac{(\theta + \phi)x + (\theta + 2\phi)y}{(x + y)^3}\right]. \tag{12.37}$$

When $\phi = 0$, we get the mixed model presented in Example 1.

Example 7

The asymmetric logistic model sets $A(w) = [\theta^r(1 - w)^r + \phi^r w^r]^{1/r} + (\theta - \phi)w + 1 - \theta$ for $0 \leq \theta \leq 1$, $0 \leq \phi \leq 1$, $r \geq 1$. Hence,

$$\bar{H}(x, y) = \exp[-(1 - \theta)x - (1 - \phi)y - (\theta^r x^r + \phi^r y^r)^{1/r}]. \tag{12.38}$$

When $\theta = \phi = 1$, we get the logistic model presented in Example 2. When $\theta = 1$, we have the biextremal model presented in Example 3, and when $\theta = \phi$ we have Gumbel's model presented in Example 4.

If $r \to \infty$, we get

$$A(w) = \max[1 - \phi w, 1 - \theta(1 - w)], \tag{12.39}$$

a nondifferentiable model with $\Pr(Y = \frac{\theta}{\phi}X) = \frac{\theta\phi}{\theta+\phi-\theta\phi}$. If $\theta = \phi = 1$, (12.39) reduces to the complete dependence model.

12.5.6 Negative Logistic Model of Joe

Joe (1990) generalized the asymmetric logistic model in Example 7 by allowing r to be negative.

Example 8

$$\bar{H}(x, y) = \exp[-(1 - \theta)x - (1 - \phi)y - (\theta^r x^r + \phi^r y^r)^{1/r}], \quad r < 0. \tag{12.40}$$

X and Y are independent if $r \to 0$ and are completely dependent if $r \to \infty$ and $\theta = \phi$. $A(w)$ has the same expression as in Example 7.

12.5.7 Normal-Like Bivariate Extreme-Value Distributions

Example 10

Smith (1991) and Hüsler and Reiss (1989) considered a normal-like bivariate extreme-value distribution with exponential marginals

$$\exp\left[-x\Phi\left(\lambda+\frac{1}{2\lambda}\log\frac{x}{y}\right)-y\Phi\left(\lambda+\frac{1}{2\lambda}\log\frac{x}{y}\right)\right],\quad \lambda\geq 0, \tag{12.41}$$

where $\Phi(x)$ is the standard normal distribution function.

12.5.8 Correlations

X and Y are positively correlated. In fact, as was pointed out by Tawn (1988a), they also have the right-tail increasing (RTI) property; see also Section 3.4.3 for this concept of positive dependence.

Pearson's product-moment correlation may be written as

$$\rho=\int_0^1\frac{dw}{A(w)^2}-1 \tag{12.42}$$

[Tawn (1988a)].

For Example 1 [Tawn (1988a)], we have

$$\rho=\frac{\sin^{-1}(\frac{1}{2}\sqrt{\theta})-\frac{1}{2}\sqrt{\theta(1-\frac{1}{4}\theta)}(1-\frac{1}{2}\theta)}{\sqrt{\theta(1-\frac{1}{4}\theta)^3}}. \tag{12.43}$$

For Example 2 [Tawn (1988a)], we have

$$\rho=\frac{[\Gamma(1/r)]^2}{r\Gamma(2/r)}-1. \tag{12.44}$$

For Example 3, $\operatorname{corr}(-\log X,-\log Y)$ (i.e., the correlation when the marginals are Gumbel's extreme-value distribution) is

$$-6\pi^{-2}\int_0^\theta(1-t)^{-1}\log t\,dt, \tag{12.45}$$

which may also be written as $6\pi^{-2}\mathrm{diln}(\theta)+1$.[1]

[1] This is the nomenclature and notation of Spanier and Oldham (1987, p. 231). Both are different from the usage in the key book on the subject by Lewin (1981). A FORTRAN algorithm for

For Example 4, we have

$$\rho = \theta/(2-\theta). \tag{12.46}$$

See Tiago de Oliveira (1980, 1984) for the correlation coefficients when the marginals are Gumbel's extreme-value distributions in the cases of Examples 1–5; Tiago de Oliveira (1975b) gives the results for the first four, while Section 12.4 gives the first three.

As to Spearman's rho, for Example 1, it is given in (12.11). For Example 4, it is

$$\rho_S = 3\theta/(2+\theta). \tag{12.47}$$

Tiago de Oliveira (1984) gives expressions for Kendall's tau in the case of Examples 2 and 5.

Tawn (1988a) suggests $2[1 - A(\frac{1}{2})]$ as another measure of dependence that is unaffected by the choice of marginals.

12.6 Bivariate Extreme-Value Distributions with Fréchet Marginals

The marginal we considered is the Fréchet distribution with $F(x) = \exp\{-x^{-1}\}$, $x > 0$. A simple transformation $Z = X^{-1}$ yields a unit exponential distribution.

Kotz and Nadarajah (2000) considered a bivariate extreme value distribution with the distribution function written as in (12.29) with Fréchet marginals instead of the exponentials as given by

$$H(x,y) = \exp\left[-\left(\frac{1}{x}+\frac{1}{y}\right) A\left(\frac{x}{x+y}\right)\right], \quad x,y>0, \tag{12.48}$$

where $A(w) = \int_0^1 \max[(1-w)q, w(1-q)] \frac{dB}{dq} dq$, as expressed in (12.30). Instead of using the dependence function A, the bivariate extreme-value distribution is now characterized by $\frac{dB}{dq} = b(q)$.

12.6.1 Bilogistic Distribution

Example 10

Joe et al. (1992) considered

---

calculating the dilogarithm in Lewin's sense has been published by Ginsberg and Zaborowski (1975).

$$H(x,y) = \exp\left[-\int_0^1 \max\left\{\frac{(q_1-1)s^{-1/q_1}}{q_1 x}, \frac{(q_2-1)s^{-1/q_2}}{q_2 y}\right\} ds\right] \tag{12.49}$$

for $q_1 > 0$ and $q_2 > 0$. Here, we have

$$b(w) = \frac{(1-1/q_1)(1-z)z^{1-1/q_1}}{(1-w)w^2\{(1-z)/q_1 + z/q_2\}},$$

where z is the root of the equation

$$(1-1/q_1)(1-w)(1-z)^{1/q_2} - (1-1/q_2)wz^{1/q_1} = 0.$$

12.6.2 Negative Bilogistic Distributions

Example 11

Coles and Tawn (1994) considered a family of distributions having the same distribution function as in Example 8 except that $q_1 < 0$ and $q_2 < 0$ and

$$b(w) = -\frac{(1-1/q_1)(1-z)z^{1-1/q_1}}{(1-w)w^2\{(1-z)/q_1 + z/q_2\}}, \quad q_1 < 0, q_2 < 0.$$

12.6.3 Beta-Like Extreme-Value Distribution

Example 12

Coles and Tawn (1991) considered a beta-like bivariate extreme-value distribution with cumulative distribution function

$$H(x,y) = \exp\left[-\frac{1}{x}\{1 - B_u(q_1+1, q_2)\} - \frac{1}{y}B_v(q_1, q_2+1)\right],$$
$$q_1 > 0,\ q_2 > 0, \tag{12.50}$$

where $u = \frac{q_1 x}{q_1 x + q_2 y}$, $v = \frac{q_1 y}{q_1 x + q_2 y}$, and

$$B_x(a,b) = \frac{\Gamma(a+b)}{\Gamma(a)\Gamma(b)} \int_0^x w^{a-1}(1-w)^{b-1} dw.$$

In this case,

$$b(w) = \frac{q_1^{q_1} q_2^{q_2} \Gamma(q_1+q_2+1)}{\Gamma(q_1)\Gamma(q_2)} \frac{w^{q_1-1}(1-w)^{q_2-1}}{\{q_1 w + q_2(1-w)\}^{1+q_1+q_2}}, \quad w \in (0,1).$$

12.7 Bivariate Extreme-Value Distributions with Weibull Marginals

This distribution was studied by Oakes and Manatunga (1992).

12.7.1 Formula of the Cumulative Distribution Function

The joint distribution function is

$$\bar{H}(x,y) = \exp[-\{(\eta_1^{\kappa_1} x^{\kappa_1})^{\phi} + (\eta_2^{\kappa_2} y^{\kappa_2})^{\phi}\}^{\alpha}. \tag{12.51}$$

Here, the parameter $\alpha = 1/\phi$ represents the degree of dependence between X and Y, and $\alpha = 1 - \tau$ is Kendall's coefficient of concordance. Cases $\alpha = 0$ and $\alpha = 1$ correspond to maximal positive dependence and independence, respectively.

12.7.2 Univariate Properties

The marginal survival functions are

$$\bar{F}(x) = \exp(-\eta_1^{\kappa_1} x^{\kappa_1}), \quad \bar{G}(y) = \exp(-\eta_2^{\kappa_2} y^{\kappa_2}), \quad x, y \geq 0,$$

with scale parameters η_1 and η_2 and shape parameters κ_1 and κ_2, respectively.

12.7.3 Formula of the Joint Density

The joint density function is

$$h(x,y) = \phi\kappa_1\kappa_2\eta_1^{\kappa_1\phi}\eta_2^{\kappa_2\phi}x^{\kappa_1\phi-1}y^{\kappa_2\phi-1}s^{\alpha-2}(1-\alpha+\alpha z)e^{-z}, \tag{12.52}$$

where

$$s = (\eta_1^{\kappa_1} x^{\kappa_1})^{\phi} + (\eta_2^{\kappa_2} y^{\kappa_2})^{\phi}, \quad z = s^{\alpha}.$$

12.7.4 Fisher Information Matrix

Using Lee's (1979) transformation, Oakes and Manatunga (1992) have derived an explicit formula for the elements of the Fisher information matrix for this distribution.

12.7.5 Remarks

- Oakes and Manatunga (1992) numerically calculated the asymptotic variance of the maximum likelihood estimator $\hat{\alpha}$ of α. Calculations reveal that estimators of the scale parameters η_1 and η_2 are almost orthogonal to that of the dependence parameter α.
- By marginal transformation to Gumbel marginals and reparametrizing such that $\tau_1 = \kappa_1^{-1}$ and $\tau_2 = \kappa_2^{-1}$, Shi et al. (2003) have shown that the bivariate Weibull model in (12.51) reduces to type B (the logistic model) with scale parameters τ_1 and τ_2. Thus, testing for $\kappa_1 = \kappa_2$ of the bivariate Weibull becomes testing for the equality of the scale parameters τ_1 and τ_2 of the type B distribution.

12.8 Methods of Derivation

- Bivariate extreme-value distributions arise as the limiting distributions of normalized componentwise maxima. More formally, let (X_i, Y_i), $i = 1, 2, \ldots, n$, be i.i.d. random vectors. Then, $(\max(X_i), \max(Y_i))$, after being suitably normalized, has a bivariate extreme-value distribution.
- (X, Y) has a bivariate extreme-value distribution with unit exponential marginals if and only if the marginals are unit exponentials and, for every $n \geq 1$, $[\bar{H}(x, y)]^n = \bar{H}(nx, ny)$. Pickands (1981) showed that this equation is satisfied if and only if $\bar{H}(x, y)$ can be written as (12.29). For this reason, the dependence function determines the type of bivariate extreme-value distribution; it also expresses the asymptotic connection between two maxima.
- Alternatively, (X, Y) has a bivariate extreme-value distribution with unit exponential marginals if and only if $\min(aX, bY)$ is exponential for all $a, b > 0$.

12.9 Estimation of Parameters

Kotz et al. (2000, Chapter 53) discusses estimation of the parameters of type A, B, and C distributions. Kotz and Nadarajah (2000) have devoted their Section 3.6 to estimation problems for multivariate extreme distributions. Shi (1995a) discussed moment estimation for the logistic model whereas Shi and Feng (1997) considered the maximum likelihood and stepwise method for the parameters of the logistic model.

12.10 References to Illustrations

Plots of the bivariate density along $y = x$ of the mixed and logistic models in Examples 1 and 2, with their marginals being of extreme value of type I form, are given by Gumbel and Mustafi (1967) and Kotz et al. (2000, p. 631). Density and density contour plots of type A and type B (with Gumbel marginals) are given by Arnold et al. (1999, pp. 283–284).

12.11 Generation of Random Variates

Section 3.7 of Kotz and Nadarajah (2000) has given three known methodologies for simulating bivariate extreme-value observations.

12.11.1 Shi et al.'s (1993) Method

Shi et al. (1993) described a scheme for simulating (X, Y) from the bivariate symmetric logistic distribution (type B) as given in (12.12); i.e., $H(x, y) = \exp\left[-(e^{-qx} + e^{-qy})^{1/q}\right]$. Letting $X = Z\cos^{2/q} V$ and $V = Z\sin^{2/q} V$, they observed that the joint density of (U, V) can be factorized as

$$(q^{-1}z + 1 - q^{-1})e^{-z}\sin 2v, \quad 0 < v < \pi/2,\ 0 < z < \infty,$$

which shows that Z and V are independent. It is then shown that V may be represented as $\arcsin U^{1/2}$, where U is uniform on $(0, 1)$, whereas Z is a mixture of two independent exponentials with a ratio $1 - q^{-1} : q^{-1}$. We can now see that (12.12) can be simulated easily.

12.11.2 Ghoudi et al.'s (1998) Method

Ghoudi et al. (1998) described a simulation scheme that is applicable for all bivariate extreme-value distributions. Starting with the expression for the cumulative distribution of the copula associated with the bivariate extreme-value distribution given by (12.1), Ghoudi et al. (1998) first find the joint distribution of $Z = X/(X+Y)$ and $V = A(-X, -Y)$ and then the marginal distribution of Z and the conditional distribution of V given $Z = z$. From these, one can simulate (X, Y), of course!

12.11.3 Nadarajah's (1999) Method

Nadarajah (1999) used the limiting point process result as an approximation to simulate bivariate extreme values.

12.12 Applications

Extreme-value distributions have wide applications in environmental studies (earthquake magnitudes, floods, river flows, storm rainfalls, wind speeds, etc.), insurance and finance, structural design, and telecommunications. There are several books that are devoted to applications of extreme-value distributions; see, for example, Tawn (1994), Embrechts et al. (1997), Kotz and Nadarajah (2000), and Coles (2001). For a more recent survey article, one may refer to Smith (2003).

12.12.1 Applications to Natural Environments

- In the form with extreme-value marginals, the mixed model in Example 1 and the logistic model in Example 2 were both used by Gumbel and Mustafi (1967) to describe the flood of the Fox River at upstream and a downstream gauging stationd. They found the latter fitted better; see Gumbel and Goldstein (1964) for floods of the Ocmulgee River. Tiago de Oliveira (1975b, 1980) mentions that an unpublished paper of Amaral and Gomes in 1975 entitled "The fitting of bivariate extreme models" has reanalyzed these and other datasets.
- The models of Examples 1, 2, 6, and 7 were used by Tawn (1988a) to describe the annual maximum sea levels at Lowestoft and Sheerness.
- Smith (1986) and Tawn (1988b) considered the joint distribution of the r largest observations—they had time series data of sea level, and were

concerned with issues such as the improvement in prediction resulting from making use of the five or ten largest values per year rather than only the largest. Smith's data were from Venice, and Tawn's were from Lowestoft and Great Yarmouth.

- The "station-year" method for the analysis of rainfall or flood maxima is motivated as follows. One may be interested in events with very long return periods (i.e., well out in the tail of the distribution), much larger events than the lengths of the individual rainfall datasets. To make deductions about such rare events, one might wish to combine all datasets from measuring stations in a region to form a single series. The extent to which this is justified depends on the tail of the joint distribution of the rainfall amounts; see Buishand (1984), who considered the ratio $q = \log H(x,x)/\log F(x)$. In the case of independence, this ratio is 2. For annual maximum daily rainfall data from the Netherlands, Buishand plotted q against F for pairs of stations different distances apart. The ratio q increases with both F and distance and seems to be tending to 2. For data restricted to the winter season, the results were more complex.
- Lewis (1975) has briefly mentioned work by himself and Daldry on annual maxima of wind and gust.
- Smith (1991) applied the normal-like bivariate extreme-value distribution to model spatial variations of extreme storms at two locations.
- Coles and Tawn (1994) found the negative bilogistic distribution most suitable for estimating the dependence between the extremes of surge and wave height.
- Yue (2000) used the type A model with Gumbel marginals to model a multivariate storm event, 104-yr daily rainfall data at the Niigata observation station in Japan during 1897–1990.
- Yue (2001) used a type 1 bivariate extreme-value distribution (the logistic model) with Gumbel marginals as a joint distribution of annual maximum storm peaks (maximum rainfall intensities) and the corresponding storm amounts. The model was found to fit well to the rainfall data collected from the Tokushima Meteorological Station of Tokushima Prefecture, Japan.
- In analyzing flood frequency of a region in Northwestern Mexico, Escalante-Sandoval (2007) used a (i) type B bivariate extreme-value two-parameter Weibull distributionas marginals and (ii) type B distribution with mixtures of two Weibull distributions as its marginals. See also Escalante-Sandoval (1998).

A salutary quotation [Klemeš (1987)] is as follows: "The natural frequencies of flood peaks in the historic series are, in fact, almost never analyzed. We do not learn whether there seems to be any pattern such as clustering of high or low peaks, trend, or some other feature, nor any indication of some hydrological, geographical or other context that could shed light on the historic flood record. What happens is that the actual time sequencing is completely ignored and the flood record is declared purely random. The ostensible reason for this is to 'simplify the mathematical treatment.' This, however, is rather amusing when one sees how the laudable resolve to keep things simple is then hastily abandoned and the use of the most advanced theories is advocated for the treatment of this artificially random sample on the pretext that 'greatest amount of information' must be extracted."

12.12.2 Financial Applications

One of the driving forces for the popularity of copulas, especially the extreme-value copulas, is their application in the context of financial risk management. Mikosch (2006, Section 3) explains the reasons why the finance researchers are attracted to copulas. Section 1.15.1 provides a list of applications in this area.

12.12.3 Other Applications

There are many other applications of extreme-value copulas as given in Section 1.15. In addition to what have already been described in Chapter 1, we give the following examples.

- In the form with extreme-value marginals, the mixed model in Example 1 was used by Posner et al. (1969) in their analysis of a spacecraft command receiver.
- The logistic model was also used by Hougaard (1986) to analyze data on tumors in rats.
- For applications to structural design, see Coles and Tawn (1994). Kotz and Nadarajah (2000, p. 145) reanalyzed the Swedish data of ages at death classified according to gender up to year 1997. The result confirms the original finding of independence studied by Gumbel and Goldstein (1964).

12.13 Conditionally Specified Gumbel Distributions

Introducing location and scale parameters, the univariate Gumbel extreme-value distribution (for maxima) has a density of the form

$$f(x) = \frac{1}{\sigma} e^{-(x-\mu)/\sigma} \exp\left(-e^{-(x-\mu)/\sigma}\right), \quad -\infty < x < \infty, \tag{12.53}$$

where μ and σ are, respectively, the location and scale parameters. Chapter 12 of Arnold et al. (1999) considered conditional distributions rather than marginals that are of the Gumbel form.

12.13.1 Bivariate Model Without Having Gumbel Marginals

Section 12.3 of Arnold et al. (1999) considered two conditionally specified Gumbel distributions, neither of them valid bivariate extreme-value distributions. After repametrizations and standardization, we have the following.

Formula of the Joint Density

The joint density function is

$$h(x,y) = k(\theta)\exp[-x-y-e^{-x}-e^{-y}-\theta e^{-x-y}], \tag{12.54}$$

where the normalizing constant is given by

$$k(\theta) = \frac{-}{\theta e^{-1/\theta}} - \mathrm{Ei}(1/\theta), \tag{12.55}$$

in which θ is a dependency parameter and $-\mathrm{Ei}(t) = \int_t^\infty \frac{e^{-u}}{u} du$.

Univariate Properties

The marginal density of X is

$$f(x) = k(\theta)\frac{\exp[-x-e^{-x}]}{1+\theta e^{-x}}, \tag{12.56}$$

and a similar expression holds for $g(y)$.

Conditional Properties

The conditional density of X, given $Y = y$, is

$$f(x|y) = (1+\theta e^{-y})\exp[-x-e^{-x}(1+\theta e^{-y})] \tag{12.57}$$

and

$$g(y|x) = (1+\theta e^{-x})\exp[-y-e^{-y}(1+\theta e^{-x})]. \tag{12.58}$$

Correlations and Dependence

Arnold et al. (1999) have shown that (12.54) is always totally negative of order 2 (also known as RR_2 in Section 3.8), and consequently the correlations are negative.

References to Illustrations

Arnold et al. (1999, p. 285) have presented a density plot and a density contour plot.

12.13.2 Nonbivariate Extreme-Value Distributions with Gumbel Marginals

Arnold et al. (1999) derived another nonvalid bivariate extreme-value distribution by conditional specification as given below (in its standardized form). The specification is through conditional distribution functions rather than conditional densities.

Formula for Cumulative Distribution Function

The joint distribution function is

$$H(x,y) = \exp[-e^{-x} - e^{-y} - \theta e^{-x-y}], \quad 0 < \theta < 1. \tag{12.59}$$

Formula for the Joint Density

The joint density function is

$$\begin{aligned} h(x,y) = &\exp(-e^{-x} - e^{-y} - \theta e^{-x-y} - x - y) \\ &\times[(1+\theta e^{-x})(1+\theta e^{-y}) - \theta]. \end{aligned} \tag{12.60}$$

Univariate Properties

Both marginal distributions are Gumbel distributions.

Conditional Properties

We have

$$\Pr(X \leq x | Y \leq y) = \exp[-e^{-x}(1 + \theta e^{-y})], \qquad (12.61)$$

which is also Gumbel.

Correlations and Dependence

Arnold et al. (1999) have shown that X and Y are NQD and hence have a negative correlation.

References to Illustrations

A density plot and density contour plot of (12.60) are given in Arnold et al. (1999, p. 285).

12.13.3 Positive or Negative Correlation

Tiago de Oliveira (1962) showed that every bivariate extreme model exhibits a non-negative correlation. This result also follows from the fact that X and Y are PQD (see Section 12.2.2), and so they must be positively correlated.

Arnold et al. (1999, p. 282) made the following remark:

> "However, many bivariate data sets are not associated with maxima of sequences of i.i.d. random vectors even though marginally and /or conditionally a Gumbel model may fit quite well.
>
> "Quite often empirical extreme data are associated with dependent bivariate sequences. Unless the dependence is relatively weak, there is no reason to expect the classical bivariate extreme theory will apply in such settings and consequently no a priori argument in favor of non-negative or nonpositive correlation.
>
> "The conditionally specified Gumbel models introduced in this chapter exhibit nonpositive correlations. Thus, the Gumbel–Mustafi models and the conditionally specified models do not compete but, in fact, complement each other. Together they provide us with the ability to fit data sets exhibiting a broad spectrum of correlation structure, both negative and positive."

12.13.4 Fields of Applications

Simiu and Filliben (1975) presented data on annual maximal wind speeds at 21 locations in the United States. About 40% of the 210 pairs of stations in this dataset exhibit negative correlation, so the phenomenon is not an isolated one. Thus, a bivariate extreme-value distribution is not appropriate. Arnold et al. (1999) found that (12.55) and (12.61) provide a good fit to data from two stations, Eastport and North Head.

References

1. Arnold, B.C., Castillo, E., Sarabia, J.M.: Conditional Specification of Statistical Models. Springer-Verlag, New York (1999)
2. Beirlant, J., Goegebeur, Y., Segers, J., Teugels, J.: Statistics of Extreme: Theory and Applications. John Wiley and Sons, Chichester (2004)
3. Buishand, T.A.: Bivariate extreme-value data and the station-year method. Journal of Hydrology **69**, 77–95 (1984)
4. Coles, S.G.: An Introduction to Statistical Modeling of Extreme Values. Springer-Verlag, New York (2001)
5. Coles, S.G., Tawn, J.A.: Modelling extreme multivariate events. Journal of the Royal Statistical Society, Series B **53**, 377–392 (1991)
6. Coles, S.G., Tawn, J.A.: Statistical methods for multivariate extremes: An application to structural design. Applied Statistics **43**, 1–48 (1994)
7. de Haan, L., Resnick, S.I.: Limit theory for multivariate sample extremes. Zeitschrift für Wahrscheinlichkeitstheorie und verwandte Gebiete **40**, 317–337 (1977)
8. Deheuvels, P.: Point processes and multivariate sample extreme values. Journal of Multivariate Analysis **13**, 257–271 (1983)
9. Deheuvels, P.: Probabilistic aspects of multivariate extremes. In: Statistical Extremes and Applications, J. Tiago de Oliveira (ed.) pp.117–130. Reidel, Dordrecht (1984)
10. Deheuvels, P.: Point processes and multivariate extreme values (II). In: Multivariate Analysis-VI, P.R. Krishnaiah (ed.) pp. 145–164. North-Holland, Amsterdam (1985)
11. Elandt-Johnson, R.C.: Some properties of bivariate Gumbel Type A distributions with proportional hazard rates. Journal of Multivariate Analysis **8**, 244–254 (1978)
12. Embrechts, P., Klüppelberg, C., Mikosch, T.: Modeling Extremal Events for Insurance and Finance. Springer-Verlag, New York (1997)
13. Escalante-Sandoval, C.: Multivariate extreme value distribution with mixed Gumbel marginals. Journal of the American Water Resources Association **34**, 321–333 (1998)
14. Escalante-Sandoval, C.: Application of bivariate extreme value distribition to flood frequency analysis: A case study of Northwestern Mexico. Natural Hazards **47**, 37–46 (2007)
15. Galambos, J.: The Asymptotic Theory of Extreme Order Statistics, 2nd edition. Kreiger, Malabar, Florida (1987)
16. Geffroy, J.: Contributions à la théorie des valeurs extrême. Publications de l'Institut de Statistique de l'Université de Paris **7**, 37–121, and **8**, 123–184 (1958/59)
17. Ghoudi, K., Khouddraji, A., Rivest, L.P.: Statistical properties of couples of bivariate extreme-value copulas. Canadian Journal of Statistics **26**, 187–197 (1998)
18. Ginsberg, E.S., Zaborowski, D.: Algorithm 490: The dilogarithm function of a real argument. Communications of the Association for Computing Machinery **18**, 200–202 (Remark, ACM Transactions on Mathematical Software **2**, 112) (1975)
19. Gumbel, E.J.: Distributions à plusieurs variables dont les marges sont données. Comptes Rendus de l'Académie des Sciences, Paris **246**, 2717–2719 (1958)
20. Gumbel, E.J.: Bivariate exponential distributions. Journal of the American Statistical Association **55**, 698–707 (1960)
21. Gumbel, E.J.: Two systems of bivariate extremal distributions. Bulletin of the International Statistical Institute **41**, 749–763 (Discussion, 763) (1965)

22. Gumbel, E.J., Goldstein, N.: Analysis of empirical bivariate extremal distributions. Journal of the American Statistical Association **59**, 794–816 (1964)
23. Gumbel, E.J., Mustafi, C.K.: Some analytical properties of bivariate extremal distributions. Journal of the American Statistical Association **62**, 569–588 (1967)
24. Hougaard, P.: A class of multivariate failure time distributions. Biometrika **73**, 671–678 (Correction **75**, 395) (1986)
25. Hüsler, J., Reiss, R.D.: Maxima of normal random vectors: Between independence and complete dependence. Statistics and Probability Letters **7**, 283–286 (1989)
26. Joe, H.: Families of min-stable multivariate exponential and multivariate extreme value distributions. Statistics and Probability Letters **9**, 75–81 (1990)
27. Joe, H., Smith, R.L., Weissmann, I.: Bivariate threshold method for extremes. Journal of the Royal Statistical Society, Series B **54**, 171–183 (1992)
28. Klemeš, V.: Empirical and causal models in hydrologic reliability analysis. In: Engineering Reliability and Risk in Water Resources, L. Duckstein and E.J. Plate (eds.), pp. 391–403. Nijhoff, Dordrecht (1987)
29. Kotz S., Balakrishnan, N., Johnson N.L.: Continuous Multivariate Distributions, Vol 1: Models and Applications, 2nd edition. John Wiley and Sons, New York (2000)
30. Kotz, K., Nadarajah, D.: Extreme Value Distributions: Theory and Applications. Imperial College Press, London (2000)
31. Lee, L-F.: On comparisons of normal and logistic models in the bivariate dichotomous analysis. Economics Letters **4**, 151–155 (1979)
32. Lewin, L.: Polylogarithms and Associated Functions. North-Holland, New York (1981)
33. Lewis, T.: Contribution to Discussion of paper by Tiago de Oliveira. Bulletin of the International Statistical Institute **46**, 253 (1975)
34. Marshall, A.W., Olkin, I.: A multivariate exponential distribution. Journal of the American Statistical Association **62**, 30–44 (1967)
35. Mikosch, T.: Copulas: Tales and facts. Extremes **9**, 3–20 (2006)
36. Nadarajah, S.: Simulation of multivariate extreme values. Journal of Statistical Computation and Simulation **62**, 395–410 (1999)
37. Nadarajah, S., Mitov, K., Kotz, S.: Local dependence function for extreme value distributions. Journal of Applied Statistics **30**(10), 1081–1100 (2003)
38. Nelsen, R.B.: An Introduction to Copulas, 2nd edition. Springer-Verlag, New York (2006)
39. Oakes, D., Manatunga, A.K.: Fisher Information for a bivariate extreme value distribution. Biometrika **79**, 827–832 (1992)
40. Pickands, J.: Multivariate extreme value distributions. Bulletin of the International Statistical Institute **49**, 859–878 (Discussion, 894–902) (1981)
41. Posner, E.C., Rodemich, E.R., Ashlock, J.C., Lurie, S.: Application of an estimator of high efficiency in bivariate extreme value theory. Journal of the American Statistical Association **64**, 1403–1414 (1969)
42. Resnick, S.: Extreme Values, Regular Variations, and Point Processes. Springer-Verlag, New York (1987)
43. Shi, D-J.: Moment estimation for multivariate extreme value distribution, Applied Mathematics–JCU, **10B**, 61–68 (1995a)
44. Shi D-J.: Multivariate extreme value distribution and its Fisher information matrix. Acta Mathematica Applicatae Sinica **11**, 421–428 (1995b)
45. Shi, D-J.: A property for bivariate extreme value distribution. Chinese Journal of Applied Probability and Statistics **19**, 49–54 (2003)
46. Shi, D-J., Feng, Y-J.: Parametric estimations by maximum likelihood and stepwise method for multivariate extreme value distribution. Journal of System Science and Mathematical Sciences **17**, 243–251 (1997)
47. Shi, D-J., Smith, R.K., Coles, S.G.: Joint versus marginal estimation for bivariate extremes. Technical Report No. 2074, Department of Statistics, University of North Carolina, Chapel Hill (1993)

48. Shi, D-J., Tang, A-L., Wang, L.: Test of shape parameter of bivariate Weibull distribution. Journal of Tianjin University **36**, 68–71 (2003)
49. Simiu, E., Filliben, B.: Structure analysis of extreme winds. Technical Report 868, National Bureau of Standards, Washington, D.C. (1975)
50. Smith, R.L.: Extreme value theory based on the r largest annual events. Journal of Hydrology **86**, 27–43 (1986)
51. Smith, R.L.: Extreme value theory. In: Handbook of Applicable Mathematics, Volume 7, pp. 437–472. John Wiley and Son, New York (1990)
52. Smith, R.L.: Regional estimation from spatially dependent data, Technical Report, Department of Statistics, University of North Carolina, Chapel Hill (1991)
53. Smith, R.L.: Extreme Values. Chapman and Hall, London (1994)
54. Smith, R.L.: Statistics of extremes, with applications in environment, insurance and finance. In: Extreme Values in Finance, Telecommunications, and the Environment, B. Finkenstadt and H. Rootzen (eds.), pp. 1–78. Chapman and Hall/CRC Press, London (2003)
55. Spanier, J., Oldham, K.B.: An Atlas of Functions. Hemisphere, Washington, D.C. and Springer-Verlag, Berlin (1987)
56. Tawn, J.A.: Bivariate extreme value theory: Models and estimation. Biometrika **75**, 397–415 (1988a)
57. Tawn, J.A.: An extreme-value theory for dependent observations. Journal of Hydrology **101**, 227–250 (1988b)
58. Tawn, J.: Applications of multivariate extremes. In: Extreme Value Theory and Applications. Proceedings of the Conference on Extreme Value Theory and Applications, Volume 1, J. Galambos, J. Lechner, and E. Smith (eds.), pp. 249–268. Kluwer Academic Publishers, Boston (1994)
59. Tiago de Oliveira, J.: La représentation des distributions extrêmales bivariées. Bulletin de l'Institut International de Statistique **39**, 477–480 (1961)
60. Tiago de Oliveira, J.: Structure theory of bivariate extremes, extensions. Estudos de Matematica, Estatistica, e Economicos **7**, 165–195 (1962/63)
61. Tiago de Oliveira, J.: Bivariate and multivariate extreme distributions. In: A Modern Course on Distributions in Scientific Work, Volume 1: Models and Structures, G.P. Patil, S. Kotz, and J.K. Ord (eds.), pp. 355–361. Reidel, Dordrecht (1975a)
62. Tiago de Oliveira, J.: Bivariate extremes: Extensions. Bulletin of the International Statistical Institute **46**, 241–252 (Discussion, 253–254) (1975b)
63. Tiago de Oliveira, J.: Bivariate extremes: Foundations and statistics. In: Multivariate Analysis-V, P.R. Krishnaiah (ed.) pp.349–366, North-Holland, Amsterdam (1980)
64. Tiago de Oliveira, J.: Bivariate models for extremes; Statistical decision. In: Statistical Extremes and Applications, J. Tiago de Oliveira (ed.), pp. 131–153. Reidel, Dordrecht (1984)
65. Yue, S.: The Gumbel mixed model applied to storm frequency analysis. Water Resources Management **14**, 377–389 (2000)
66. Yue, S.: The Gumbel logistic model for representing a multivariate storm event. Advances in Water Resources **24**, 179–185 (2001)
67. Yue, S., Ouarda, T.B.M.J., Bobée, B., Legendre, P., Bruneau, P.: The Gumbel mixed model for flood frequency analysis. Journal of Hydrology **226**, 88–100 (1999); see Corrigendum **228**, 283 (2000)
68. Yue, S., Wang, C.Y.: A comparison of two bivariate extreme value distributions. stochastic Environmental Research **18**, 61–66 (2004)
69. Weissman, I.: Multidimensional extreme value theory (Discussion). Bulletin of the International Statistical Institute **51**, 187–190 (1985)

Chapter 13
Elliptically Symmetric Bivariate Distributions and Other Symmetric Distributions

13.1 Introduction

This chapter is devoted to describing a class of bivariate distributions whose contours of probability densities are ellipses; in particular, those ellipses with constant eccentricity. These distributions are generally known as elliptically contoured or elliptically symmetric distributions. A subclass of distributions with contours that are circles are known as spherically symmetric (or simply spherical) distributions. The chapter also includes other symmetric bivariate distributions.

The last 20 years have seen vigorous development of multivariate elliptical distributions as direct generalizations of the multivariate normal distribution that dominated statistical theory and applications for nearly a century. Elliptically contoured distributions retain most of the attractive properties of the multivariate normal distribution. For example, let (X, Y) be an uncorrelated pair from this class. Then, X^2/Y^2 has the usual F-distribution, and $X^2/(X^2 + Y^2)$ has the beta distribution, beta$(\frac{1}{2}, \frac{1}{2})$; see Kelker (1970).

On the application side, members of this class were used to describe the second-order moments of the transformation of a random signal by an instantaneous linear device [McGraw and Wagner (1968)]. Further, van Praag and Wesselman (1989) have shown that many procedures for multivariate analysis in the normal case can be adapted to the elliptical case with the aid of the estimated kurtosis. Bentler and Berkane (1985) went as far as to say, "It is becoming apparent that [elliptical] theory has a potential to displace multivariate normal theory in a variety of applications such as linear structural modelling" (which includes factor analysis and simultaneous equation models). For an early review and bibliography of these distributions, see Chmielewski (1981).

Fang et al. (1990) provided a rather detailed study of these distributions, and their text has now become a standard reference for symmetric multivariate distributions. A more recent review is Fang (1997).

N. Balakrishnan, C.D. Lai, *Continuous Bivariate Distributions*,
DOI 10.1007/b101765_14,

Notation

First, recall some conventions used earlier: Boldface symbols will be used for vectors and matrices; $N(\boldsymbol{\mu}, \boldsymbol{\Sigma})$ denotes the bivariate normal distribution having mean $\boldsymbol{\mu}$ and covariance matrix $\boldsymbol{\Sigma}$; and the transpose of matrix $\mathbf{A}$ is denoted by $\mathbf{A}'$.

There are several ways to describe spherically and elliptically symmetric distributions. In a nutshell, a spherical distribution is an extension of $N(\mathbf{0}, \mathbf{I})$ and an elliptical distribution is an extension of $N(\boldsymbol{\mu}, \boldsymbol{\Sigma})$. Since every circle is also an ellipse, a spherical distribution is an elliptical distribution.

In Section 13.2, we describe the formulation of elliptically contoured bivariate distributions, and then we discuss its properties in Section 13.3. The elliptical compound bivariate normal distribution is discussed in Section 13.4. Next, in Section 13.5, some examples of elliptically and spherically symmetric bivariate distributions are presented. Extremal-type elliptical distributions are discussed in Section 13.6. Tests of spherical and elliptical symmetry and extreme behavior of bivariate elliptical distributions are discussed in Sections 13.7 and 13.8, respectively. Some fields of applications for these distributions are highlighted in Section 13.9. In Sections 13.10 and 13.11, bivariate symmetric stable and generalized bivariate symmetric stable distributions and their properties are discussed. Next, in Sections 13.12 and 13.13, α-symmetric and other symmetric distributions, respectively, are described. Bivariate hyperbolic distributions are outlined in Section 13.14, and finally skew-elliptical distributions are discussed in Section 13.15.

13.2 Elliptically Contoured Bivariate Distributions: Formulations

13.2.1 Formula of the Joint Density

If the probability density $h(x, y)$ is a function only of a positive definite[1] quadratic form

$$(\mathbf{x} - \boldsymbol{\mu})'\boldsymbol{\Sigma}^{-1}(\mathbf{x} - \boldsymbol{\mu}), \tag{13.1}$$

then its contours are ellipses; here, $\mathbf{x}' = (x, y)$, $\boldsymbol{\mu}=(\mu_1, \mu_2)$, and $\boldsymbol{\Sigma}$ is a nonsingular scaling matrix that is determined only up to a multiplicative constant. Its role is like that of the covariance matrix and indeed, when the latter exists, $\boldsymbol{\mu}$ must be proportional to it; see Devlin et al. (1976) for details. More explicitly, the joint density can be expressed as

$$h(x, y) = |\boldsymbol{\Sigma}|^{-1/2} g_c\left((\mathbf{x} - \boldsymbol{\mu})'\boldsymbol{\Sigma}^{-1}(\mathbf{x} - \boldsymbol{\mu})\right), \tag{13.2}$$

[1] An $n \times n$ symmetric matrix $\mathbf{A}$ is positive definite if $\mathbf{x}'\mathbf{A}\mathbf{x} > 0$ for every nonzero $\mathbf{x}$ in R^n.

where $g_c(\cdot)$ is a scalar function referred to as the density generator.

In the special case where $\boldsymbol{\mu} = \mathbf{0}$ and $\boldsymbol{\Sigma} = \mathbf{I}$ (the identity matrix), the distribution is called a *spherically symmetric* (or simply spherical) distribution.

If it is assumed that

$$\boldsymbol{\Sigma} = \begin{pmatrix} 1 & \rho \\ \rho & 1 \end{pmatrix}, \quad -1 < \rho < 1, \tag{13.3}$$

and $\boldsymbol{\mu} = \mathbf{0}$, (13.2) becomes

$$h(x,y) = \frac{1}{\sqrt{1-\rho^2}} g_c\left(\frac{x^2 - 2\rho xy + y^2}{1-\rho^2}\right), \quad -1 < \rho < 1. \tag{13.4}$$

13.2.2 Alternative Definition

If $\mathbf{X}$ has an elliptically contoured bivariate distribution defined in (13.2), it can be written as

$$\mathbf{X} = R\mathbf{L}\mathbf{U}^{(2)} + \boldsymbol{\mu}, \tag{13.5}$$

where R^2 has the same distribution as $(\mathbf{X} - \boldsymbol{\mu})'\boldsymbol{\Sigma}^{-1}(\mathbf{X} - \boldsymbol{\mu})$, $\mathbf{X}' = (X, Y)$, $\boldsymbol{\Sigma} = \mathbf{L}\mathbf{L}'$ (i.e., $\mathbf{L}$ is the lower triangular matrix of the Choleski decomposition of $\boldsymbol{\Sigma}$), and $\mathbf{U}^{(2)}$ is uniformly distributed on the circumference of a unit circle; see, for example, Cambanis et al. (1981) and Johnson et al. (1984). Further, R is independent of $\mathbf{U}^{(2)}$. The stochastic representation in (13.5) may serve as an alternative definition of an elliptically contoured distribution.

Suppose $\mathbf{Y}$ has a spherical distribution. Then the stochastic representation in (13.5) becomes

$$\mathbf{X} = \mathbf{L}\mathbf{Y} + \boldsymbol{\mu}, \tag{13.6}$$

where $\mathbf{L}$ is the lower triangular matrix defined above.

13.2.3 Another Stochastic Representation

Abdous et al. (2005) suggested another stochastic representation for a bivariate elliptical vector. Let X and Y be a pair of random variables with means μ_2, μ_2 and variances σ_1, σ_2, respectively. Then (X, Y) has a bivariate elliptical distribution if

$$(X,Y) = (\mu_1, \mu_2) + \left(\sigma_1 RDU_1, \sigma_2 \rho RDU_1 + \sigma_2\sqrt{1-\rho^2}\, R\sqrt{1-D^2}\, U_2\right), \tag{13.7}$$

where U_1, U_2, R, and D are mutually independent random variables, ρ is Pearson's correlation coefficient, and $\Pr(U_i = -1) = \Pr(U_i = -1) = 1/2$,

$i = 1, 2$. Both D and R are positive random variables and D has probability density function

$$f_D(d) = \frac{2}{\pi\sqrt{1-d^2}}, \quad 0 < d < 1. \tag{13.8}$$

The random variable R is called the generator of the elliptical random vector. Abdous et al. (2005) explained the relationship between this representation with the more classical representation given in (13.5).

Assuming now that X and Y are identically distributed with $\mu_1 = \mu_2 = 0$ and $\sigma_1 = \sigma_2 = 1$, the probability density functions of the generators of some well-known bivariate elliptical distributions (see Section 13.5) are given below.

Example 1. Bivariate Pearson Type VII Distribution

$$f_R(x) = \frac{2(N-1)}{m} x \left(1 + \frac{x^2}{m}\right)^{-N}, \quad x > 0, \ N > 1, \ m > 0.$$

When $m = 1$ and $N = 3/2$, we have the bivariate Cauchy distribution, and when $N = (m+2)/2$, we have the bivariate Student t-distribution.

Example 2. Bivariate Logistic Distribution

$$f_R(x) = 4x \, \frac{\exp\{-x^2\}}{(1+\exp\{-x^2\})^2}, \quad x > 0.$$

Example 3. Kotz-Type Distribution

$$f_R(x) = \frac{2s}{r^{-N/s}\Gamma(N/s)} \, x^{2N-1} \exp\{-rx^{2s}\}, \quad x > 0, \ N, r, s > 0.$$

See Section 13.6.1 below for further information about this distribution.

When $N = 1$, $s = 1$, and $r = 1/2$, we have the bivariate normal distribution.

13.2.4 Formula of the Cumulative Distribution

Naturally, because the only commonality among members of this class is the ellipticity of their contours, there is no single closed form for the distribution function. Except for the bivariate normal, the distribution function $H(x, y)$ is difficult to evaluate in general.

13.2.5 Characteristic Function

The characteristic function φ depends only on a quadratic form $\mathbf{t}'\mathbf{\Sigma}\mathbf{t}$ (assuming $\boldsymbol{\mu} = \mathbf{0}$); see Kelker (1970) and Johnson (1987, p. 107). Here, $\mathbf{t}' = (s, t)$. In general, the characteristic function of an elliptical distribution is given by

$$\varphi(s,t) = e^{it\boldsymbol{\mu}}\phi(\mathbf{t}'\mathbf{\Sigma}\mathbf{t}) \tag{13.9}$$

for some scalar function ϕ, which is called the *characteristic generator* [Fang et al. (1990, p. 32)]. Also,

$$\varphi(s,t) = e^{it\boldsymbol{\mu}}\phi(s^2 + 2\rho st + t^2) \tag{13.10}$$

if $\mathbf{\Sigma}$ is as given in (13.3).

For spherical distributions, we have

$$\varphi(s,t) = \phi(\mathbf{t}'\mathbf{t}) = \phi(s^2 + t^2). \tag{13.11}$$

For any elliptical distribution, the marginal characteristic function is given by

$$\varphi(t) = \phi(t^2). \tag{13.12}$$

13.2.6 Moments

We assume here, without loss of generality, that $\boldsymbol{\mu} = \mathbf{0}$. It follows from Theorems 2.7 and 2.8 of Fang et al. (1990) that the moments associated with (13.4) are

$$E(X) = E(Y) = 0, \qquad \text{var}(X) = \text{var}(Y) = \frac{D_1}{2}, \tag{13.13}$$

and

$$E(X^{2i}Y^{2j}) = \frac{1}{\pi}\, D_{i+j} B\left(\frac{1}{2} + i, \frac{1}{2} + j\right), \tag{13.14}$$

where $i, j \geq 1$ are integers,

$$\text{cov}(X, Y) = \frac{D_1\rho}{2},$$

and

$$D_i = \pi \int_0^\infty x^i g_c(x) dx; \tag{13.15}$$

see also Kotz and Nadarajah (2003).

13.2.7 Conditional Properties

The regression of Y on X is linear. The conditional variance, $\text{var}(Y|X = x)$, is independent of s if and only if X and Y have the bivariate normal distribution; more generally,

$$\text{var}(Y|X = x) = a(x)\sigma_2^2(1 - \rho^2) \tag{13.16}$$

for some function $a(x)$, where σ_2^2 is the variance of Y and ρ is the correlation coefficient between X and Y.

13.2.8 Copulas of Bivariate Elliptical Distributions

Fang et al. (2002) have given a general expression for the copula of an elliptically symmetric bivariate distribution. Explicit expressions are obtained for the Kotz type, bivariate Pearson type VII, bivariate Pearson type II, and symmetric logistic distributions.

13.2.9 Correlation Coefficients

The Pearson product-moment correlation coefficient is ρ if the covariance matrix exists. Fang et al. (2002) pointed out that Spearman's correlation ρ_S is somewhat complicated for elliptically contoured distributions. However, they displayed that Kendall's tau is quite simple and is given by

$$\tau = \frac{2}{\pi} \arcsin(\rho).$$

13.2.10 Fisher Information

The Fisher information matrices for elliptically symmetric Pearson type II and type VII (bivariate Student t, bivariate Cauchy, etc.) distributions were derived by Nadarajah (2006b). Extensive numerical tabulations of the Fisher information matrices were also given for practical purposes.

13.2.11 Local Dependence Functions

Bairamov et al. (2003) introduced a measure of local dependence that is a localized version of the Galton correlation coefficient. Kotz and Nadarajah (2003) provided a motivation for this new measure and derived the exact form of the measure for the class of elliptically symmetric distributions.

13.3 Other Properties

- Any non-negative function $\kappa(\cdot)$ such that $\int_0^\infty \kappa(x)dx < \infty$ can define a density generator g_c for a bivariate elliptical distribution through

$$g_c(x) = \frac{\kappa(x)}{\pi \int_0^\infty \kappa(y)dy}; \tag{13.17}$$

  see Fang et al. (1990, p. 47) and Kotz and Nadarajah (2001).
- All $\alpha X + \beta Y$ with the same variance (if it exists) have the same distribution.
- X and Y are independent if and only if $\mathbf{\Sigma}$ is diagonal and X and Y have a bivariate normal distribution.
- Writing $\mathbf{\Sigma}$ as (c_{ij}), the correlation matrix (assuming it is defined) is given by $\mathbf{\Sigma}/\sqrt{c_{11}c_{22}}$.
- If $\mathbf{X_1}$ and $\mathbf{X_2}$ are members of this class having the same $\mathbf{\Sigma}$ and are independent, then $\mathbf{X_1} + \mathbf{X_2}$ is also a member of this class and has the same $\mathbf{\Sigma}$.
- On the plane $\mathbf{R}^2$, let A_i $(i = 1, 2, 3, 4)$, be the ith quadrant and L_j $(j = 1, \ldots, 6)$ be the ray originating from the origin at an angle of $(j-1)\pi/3$ from the positive directions of the x-axis. Let B_j $(j = 1, \ldots, 6)$ be the region between L_j and L_{j+1}, where we use the convention $L_7 = L_1$. Then, Nomakuchi and Sakata (1988) showed that, for $\boldsymbol{\mu} = \mathbf{0}$, the following two statements are true:

  (i) $\Pr(\mathbf{X} \in A_i) = 1/4$, $i = 1, 2, 3, 4$, if and only if $\mathbf{\Sigma} = \text{diag}\{a, b\}$, where $a, b > 0$.
  (ii) $\Pr(\mathbf{X} \in B_i) = 1/6$, $i = 1, \ldots, 6$, if and only if $\mathbf{\Sigma} = \sigma^2 I = \sigma^2 \text{diag}\{1, 1\}$.

- $h(x, y)$ can be represented as

$$\int_0^\infty \psi(\mathbf{x}; v) \frac{dW}{dv} dv, \tag{13.18}$$

  where $\frac{dW}{dv}$ is a weight function ($\int \frac{dw}{dv} dv = 1$) that may assume negative values and ψ is the density function corresponding to the bivariate distribution $N(\boldsymbol{\mu}, v^{-2}\boldsymbol{\Sigma})$. For an interpretation of (13.18), see Section 13.4 below.

- Slepian's inequality for the bivariate normal distribution (see Section 11.9) also holds for this wider class; see, for example, Gordon (1987). Further probability inequalities applicable to this class have been given by Tong (1980, Section 4.3).
- When the variances of X and Y are equal, consider the probability over the region $\frac{x^2}{a^2} + a^2y^2 \leq 1$ (which is an ellipse of area π). Shaked and Tong (1988, pp. 338–339) showed that the maximum probability content is contained when the rectangle becomes a square.
- For properties concerning the moments, see Berkane and Bentler (1986a,b, 1987).
- Suppose a distribution has a p.d.f. that is constant within the ellipse

$$\frac{1}{1-\rho^2}\left[\frac{(x-\mu_1)^2}{\sigma_1^2} - \frac{2\rho(x-\mu_1)(y-\mu_2)}{\sigma_1\sigma_2} + \frac{(y-\mu_2)^2}{\sigma_2^2}\right] = 4 \qquad (13.19)$$

and is zero outside. Then, this distribution has means μ_1 and μ_2, standard deviations σ_1 and σ_2, and correlation ρ. This ellipse can, of course, be constructed for any distribution with these moments, in which case it is known as the ellipse of concentration. For generation of random variates from such a distribution, see Devroye (1986, p. 567).

13.4 Elliptical Compound Bivariate Normal Distributions

If W in (13.18) is the cumulative distribution function of a positive variable V, then we may say h is an elliptical compound bivariate normal distribution.[2] Fang et al. (1990, p. 48) simply call it a mixture of normal distributions. Obviously, the bivariate normal is itself a member of this class (take V to be a positive constant); other members are longer-tailed [Devlin et al. (1976, p. 369)].

It follows from (13.18) that a special property is that $\mathbf{X} = V^{-1}\mathbf{Z}+\boldsymbol{\mu}$, where $\mathbf{Z} \sim N(\mathbf{0}, \boldsymbol{\Sigma})$ and V and $\mathbf{Z}$ are independent. The question arises as to whether every compound bivariate normal distribution is elliptically contoured. The answer is no, as will be seen from the following:

- If $\mathbf{Z} \sim N(\mathbf{0}, \boldsymbol{\Sigma})$ and V is independent of $\mathbf{Z}$, then it is clear that the density of $V^{-1}\mathbf{Z}$ is a function of a positive definite quadratic form, and hence it is elliptically symmetric. For example, let $V = S/\sqrt{\nu}$, where S has a chi-distribution with ν degrees of freedom; then, $(S/\sqrt{\nu})^{-1}\mathbf{Z}$ has a bivariate t-distribution with ν degrees of freedom.

---

[2] Some authors refer to it as a normal variance mixture.

- Now, if $N(\boldsymbol{\mu}, \boldsymbol{\Sigma})$, with $\boldsymbol{\mu} \neq \mathbf{0}$, then

$$(S/\sqrt{\nu})^{-1}\mathbf{Z} = (S/\sqrt{\nu})^{-1}(\mathbf{Z} - \boldsymbol{\mu}) + \boldsymbol{\mu}(\mathbf{S}/\sqrt{\nu})^{-\mathbf{1}}$$

  which does not have a bivariate t-distribution, and it is not elliptically symmetric since the second term destroys the symmetry.

13.5 Examples of Elliptically and Spherically Symmetric Bivariate Distributions

Table 3.1 of Fang et al. (1990) lists several multivariate spherical distributions together with their densities or characteristic functions. We now select some of them for a brief discussion.

13.5.1 Bivariate Normal Distribution

The bivariate normal distribution plays the central part in the class of elliptically symmetric distributions. Many theoretical results for the bivariate normal distribution also hold in this broader class.

13.5.2 Bivariate t-Distribution

This has already been discussed in Section 9.2. It is also a special case of the bivariate Pearson type VII distribution.

13.5.3 Kotz-Type Distribution

We will discuss this bivariate distribution in Section 13.6 below with details.

13.5.4 Bivariate Cauchy Distribution

This is a special case of the bivariate Pearson type VII distribution; see Section 9.9 for relevant details.

13.5.5 Bivariate Pearson Type II Distribution

This has been discussed in Section 9.11. The special case $\rho = 0$ gives rise to a spherically symmetric distribution whose marginals are symmetric beta (Pearson type II) distributions.

13.5.6 Symmetric Logistic Distribution

The joint density function is

$$h(x, y) = c\frac{\exp - \{x^2 - 2\rho xy + y^2\}}{\sqrt{1-\rho^2}[1 + \exp -\{x^2 - 2\rho xy + y^2\}]^2}. \tag{13.20}$$

This is listed in Table 3.1 of Fang et al. (1990) and studied in Fang et al. (2002). Clearly, it is elliptically symmetric. The density generator g_c is proportional to the density function of a univariate logistic.

13.5.7 Bivariate Laplace Distribution

$$h(x, y) = \frac{1}{16\sigma}|\Sigma|^{-1/2} \exp\left\{-\frac{1}{2}\left[(\boldsymbol{x}-\boldsymbol{\mu})'\Sigma^{-1}(\boldsymbol{x}-\boldsymbol{\mu})\right]^{\frac{1}{2}}\right\}, \tag{13.21}$$

where Σ is the correlation matrix [Ernst (1998) and Lindsey (1999)].

13.5.8 Bivariate Power Exponential Distributions

This family of elliptically contoured distributions was considered by Ernst (1998), Gómez et al. (1998), and Lindsey (1999). This was also called the *bivariate generalized Laplace distribution* by Ernst (1998).

Multivariate p.d.f.

$$h(\boldsymbol{x}; \boldsymbol{\mu}, \boldsymbol{\Sigma}, \beta) = \frac{n\Gamma\left(\frac{n}{2}\right)}{\sigma^{\frac{n}{2}}\sqrt{|\Sigma|}\Gamma\left(1+\frac{n}{2\beta}\right)2^{1+\frac{n}{2\beta}}} \times \exp\left\{-\frac{1}{2}\left[(\boldsymbol{x}-\boldsymbol{\mu})'\Sigma^{-1}(\boldsymbol{x}-\boldsymbol{\mu})\right]^{\beta}\right\}, \tag{13.22}$$

where Σ is the correlation matrix and $\beta > 0$. For $n = 2$, we have a bivariate probability density function:

$$h(x, y) = \frac{\beta}{\sigma\Gamma(\frac{1}{\beta})2^{\frac{\beta+1}{\beta}}} |\Sigma|^{-1/2} \exp\left\{-\frac{1}{2}\left[(\boldsymbol{x}-\boldsymbol{\mu})'\Sigma^{-1}(\boldsymbol{x}-\boldsymbol{\mu})\right]^{\beta}\right\}.$$

Marginal d.f.

$$f(x; \mu, \sigma, \beta) = \frac{1}{\sigma\Gamma\left(1+\frac{1}{2\beta}\right)2^{1+\frac{1}{2\beta}}} \exp\left[-\frac{1}{2}\left|\frac{x-\mu}{\sigma}\right|^{2\beta}\right],$$

$$-\infty < \mu < \infty,\ 0 < \sigma,\ 0 < \beta \leq \infty.$$

Special Cases

When $\beta = 1$, we have a bivariate normal distribution; when $\beta = 1/2$, a bivariate Laplace distribution; and when $\beta \to \infty$, a bivariate uniform distribution. For $\beta < 1$, the distribution has heavier tails than the bivariate normal distribution and can be useful in providing robustness against "outliers."

Moments

Let $\mathbf{X}' = (X, Y)$. Then $E(\mathbf{X}) = \boldsymbol{\mu}$ and $\text{var}(\mathbf{X}) = \frac{2^{\frac{1}{\beta}}\Gamma(\frac{2}{\beta})}{2\Gamma(\frac{1}{\beta})}\Sigma$.

13.6 Extremal Type Elliptical Distributions

The joint density function is

$$h(x, y) = \frac{1}{\sqrt{1-\rho^2}} g_c\left(\frac{x^2 - 2\rho xy + y^2}{1-\rho^2}\right), \tag{13.23}$$

where g_c is a density generator. Recall from (13.17) that

$$g_c(x) = \frac{\kappa(x)}{2\pi\int_0^\infty \kappa(y)dy},$$

where $\kappa(x)$ is one of the three types of univariate extreme-value distributions. Kotz and Nadarajah (2001) obtained the following three extremal type elliptical distributions.

13.6.1 Kotz-Type Elliptical Distribution

This is also called the Weibull-type elliptical distribution in Kotz and Nadarajah (2001). Since 1990, there has been a surge of interest related to this distribution. Nadarajah (2003) provided a comprehensive review of properties and applications of this distribution. Let

$$\kappa(x) = x^{N-1}\exp(-rx^s), \quad r > 0,\ s > 0,\ N > 0, \tag{13.24}$$

which has the form of the type III (Weibull) extreme-value density function. Now, it can be shown that

$$\int_0^\infty \kappa(y)dy = \int_0^\infty y^{N-1}\exp(ry^s)dy = \frac{\Gamma(N/s)}{sr^{N/s}}.$$

It follows from (13.17) that

$$g_c(x) = \frac{sr^{N/s}\kappa(x)}{\pi\Gamma(N/s)},$$

which results in the joint density as presented below.

Formula of the Joint Density

The joint density function is

$$h(x,y) = \frac{sr^{N/s}(x^2-2\rho xy+y^2)^{N-1}}{\pi\Gamma(N/s)(1-\rho^2)^{N-1/2}}\exp\left\{-r\left(\frac{x^2-2\rho xy+y^2}{1-\rho^2}\right)^s\right\}. \tag{13.25}$$

When $N=1, s=1$, and $r=\frac{1}{2}$, this reduces to a bivariate normal. When $s=1$, this is the original Kotz distribution introduced by Kotz (1975). The joint density in (13.25) has been studied by Fang et al. (1990), Iyengar and Tong (1989), Kotz and Ostrovskii (1994), and Streit (1991), among others.

Univariate Properties

Both the marginal p.d.f. and c.d.f. are infinite sums of hypergeometric distributions.

(i) N integer, $s = 1$:

$$f(x) = \frac{r^{N-1/2}x^{2(N-1)}\exp(-rx^2)}{\sqrt{\pi}} \sum_{k=0}^{N-1} \frac{(2k)!}{4^k(N-k-1)!(k!)^2} r^{-k}x^{-2k}.$$

(ii) N integer, $s = \frac{1}{2}$:

$$f(x) = \frac{(N-1)!r^{2N}x^{2(N-1)}}{(2N-1)!\pi} \sum_{k=0}^{N-1} \frac{(2k)!}{2^k(N-k-1)!(k!)^2} r^{-k}x^{-k+1}K_{k+1}(r|x|),$$

where

$$K_\nu(z) = \frac{z^\nu\Gamma(\frac{1}{2})}{2^\nu\Gamma(\nu+\frac{1}{2})} \int_1^\infty \exp(-zy)(y^2-1)^{\nu-1/2}dy$$

is the Bessel function.

(iii) $N = s = \frac{1}{2}$:

$$f(x) = \frac{r}{\pi}K_0(r|x|).$$

Moments

With

$$D_i = r^{i/s}\Gamma\Big(\frac{N}{s}+\frac{i}{s}\Big)\Big/\Gamma\left(\frac{N}{s}\right), \quad i \text{ a positive integer,}$$

it follows from (13.13) that the moments of (13.24) are

$$E(X) = E(Y) = 0, \qquad \text{var}(X) = \text{var}(Y) = \frac{r^{-1/s}}{2}\Gamma\Big(\frac{N}{s}+\frac{1}{s}\Big)\Big/\Gamma\left(\frac{N}{s}\right),$$

$$\text{cov}(X,Y) = \frac{\rho r^{-1/s}}{2}\Gamma\Big(\frac{N}{s}+\frac{1}{s}\Big)\Big/\Gamma\left(\frac{N}{s}\right),$$

and, for $i, j \geq 1$,

$$E(X^{2i}Y^{2j}) = \frac{r^{-(i+j)/s}}{\pi}\Gamma\Big(\frac{N}{s}+\frac{i}{s}+\frac{j}{s}\Big)B\left(\frac{1}{2}+i,\frac{1}{2}+j\right)\Big/\Gamma\left(\frac{N}{s}\right).$$

The Product XY and the Ratio X/Y

The distribution of the product XY was derived by Nadaraja (2005). Nadarajah and Kotz (2005) derived the distribution of the product for the elliptically symmetric Kotz-type distribution. The distributions of the ratio X/Y were derived by Nadarajah (2006a).

Marginal Characteristic Function

The marginal characteristic function turns out to be rather complicated; see Kotz and Nadarajah (2001) for derivations and formulas.

13.6.2 Fréchet-Type Elliptical Distribution

In this case,

$$\kappa(x) = x^{N-1}\exp(-rx^s), \quad r > 0,\ s < 0,\ N < 0,$$

and

$$\int_0^\infty \kappa(y)dy = \int_0^\infty y^{N-1}\exp(ry^s)dy = -\frac{\Gamma(N/s)}{sr^{N/s}},$$

so

$$g_c(x) = -\frac{sr^{N/s}}{\pi\Gamma(N/s)}\kappa(x).$$

This results in the joint density as presented below.

Formula of the Joint Density

The joint density function is

$$h(x,y) = -\frac{sr^{N/s}(x^2-2\rho xy+y^2)^{N-1}}{\pi\Gamma(N/s)(1-\rho^2)^{N-1/2}}\exp\left\{-r\Big(\frac{x^2-2\rho xy+y^2}{1-\rho^2}\Big)^s\right\}. \tag{13.26}$$

Univariate Properties

Both the marginal p.d.f. and c.d.f. are quite complicated, which may be expressed in terms of hypergeometric functions. The expression is simpler if N is an integer.

(i) N an integer, $s = -1$:

$$f(x) = \frac{r^{-2N}|x|^{2N-1}}{2\pi\Gamma(-N)}B\left(\frac{1}{2},\frac{1}{2}-N\right){}_1F_1\left(\frac{1}{2}-N;1-N;-rx^{-2}\right).$$

(ii) N an integer, $s = -\frac{1}{2}$:

$$f(x) = \frac{r^{-2N}x^{2N-2}}{2\pi\Gamma(-2N)}\left\{B\left(\tfrac{1}{2}-N,\tfrac{1}{2}\right)|x|{}_1F_2\left(\tfrac{1}{2}-N;\tfrac{1}{2},1-N;\tfrac{r^2x^{-2}}{4}\right)\right.$$
$$\left.-rB\left(1-N,\tfrac{1}{2}\right){}_1F_2\left(1-N;\tfrac{3}{2},\tfrac{3}{2}-N;\tfrac{r^2x^{-2}}{4}\right)\right\}.$$

Moments

It can be shown that

$$D_i = r^{i/s}\Gamma\left(\frac{N}{s}+\frac{i}{s}\right)\Big/\Gamma\left(\frac{N}{s}\right),$$

provided $i < -N$. Thus, the moments associated with (13.26) are identical to those in Section 13.6.1 except that we must have $N < -1$ in order for the variance and covariance to exit, and we must have $i + j < -N$ in order for the product moment to exit.

Characteristic Function

The marginal characteristic function is quite complex, which can be expressed through a special function called Meijer's G function.

13.6.3 Gumbel-Type Elliptical Distribution

The Gumbel or type I extreme-value distribution has the form

$$\kappa(x) = \exp(-ax)\exp\{-b\exp(-ax)\}, \quad a > 0,\ b > 0.$$

Since

$$\int_0^\infty \kappa(x) = \frac{1-\exp(-b)}{ab},$$

the density generator

$$g_c(x) = \frac{ab\kappa(x)}{\pi\left(1-\exp(-b)\right)},$$

resulting in the following joint density function.

13.6.3.1 Formula of the Joint Density

The joint density function is

$$h(x,y) = \frac{ab(1-\rho^2)^{-1/2}}{\pi\{1-\exp(-b)\}} \exp\left\{-\frac{a(x^2-2\rho xy+y^2)}{1-\rho^2}\right\} \times \exp\left[-b\exp\left\{-\frac{a(x^2-2\rho xy+y^2)}{1-\rho^2}\right\}\right]. \tag{13.27}$$

When $b = 0$, this distribution reduces to a bivariate normal distribution.

Univariate Properties

The marginal density is

$$f(x) = \frac{\sqrt{a}b}{\sqrt{\pi}\{1-\exp(-b)\}} \sum_{k=0}^{\infty} \frac{(-1)^k b^k \exp\{-(k+1)ax^2\}}{k!\sqrt{k+1}},$$

where Φ is the cumulative distribution function of the standard normal distribution, and

$$f(x) = \frac{b}{1-\exp(-b)} \sum_{k=0}^{\infty} \frac{(-1)^k b^k}{(k+1)!} \Phi(\sqrt{2(k+1)}x).$$

Moments

With

$$D_i = \frac{a^i b\Gamma(i+1)}{1-\exp(-b)} \sum_{k=0}^{\infty} \frac{(-1)^k b^k}{k!(k+1)^{i+1}},$$

it follows that

$$E(X) = E(Y) = 0,$$

$$\text{var}(X) = \frac{b}{2a\{1-\exp(-b)\}} \sum_{k=0}^{\infty} \frac{(-1)^k b^k}{k!(k+1)^{i+1}},$$

$$\text{cov}(X,Y) = \frac{b\rho}{2a\{1-\exp(-b)\}} \sum_{k=0}^{\infty} \frac{(-1)^k b^k}{k!(k+1)^{i+1}},$$

and, for $i, j \geq 1$,

$$E(X^{2i}Y^{2j}) = \frac{b\Gamma(i+j+1)}{\pi a^{i+j}\{1-\exp(-b)\}} B\left(\frac{1}{2}+i, \frac{1}{2}+j\right) \sum_{k=0}^{\infty} \frac{(-1)^k b^k}{k!(k+1)^{i+j+1}}.$$

The Characteristic Function

The characteristic function of the marginals is

$$\varphi(u) = \phi(u^2) = \frac{b}{1-\exp(-b)} \sum_{i=0}^{\infty} \frac{(-)^k b^k}{(k+1)!} \exp\left\{-\frac{u^2}{4(k+1)a}\right\}.$$

13.7 Tests of Spherical and Elliptical Symmetry

Elliptical distributions are easily implemented and simulated [Johnson (1987)].

The problem of testing the hypothesis of symmetry of a multivariate distribution has been approached from various points of view. Serfling (2006) gives a review of some of these approaches. Li et al. (1997) introduced some graphical methods by proposing QQ-plots associated with various statistics invariant under orthogonal rotations.

For statistical tests of elliptical symmetry, see, for example, Mardia (1970), Beran (1979), Li et al. (1997), Manzotti et al. (2002), Schott (2002), and Serfling (2006).

Fang and Liang (1999) gave a comprehensive review on tests of spherical and elliptical symmetry.

13.8 Extreme Behavior of Bivariate Elliptical Distributions

The extreme behavior of elliptically distributed random vectors is closely related to the asymptotic behavior of its generator; see, for example, Hashorva (2005). Starting with Sibuya (1960), many authors have studied the subject; see, for example, Hult and Lindskog (2002), Schmidt (2002), Abdous et al. (2005), Demarta and McNeil (2005), Hashorva (2005), and Asimit and Jones (2007).

We note, in particular, that the limiting distribution of the componentwise maxima of i.i.d. elliptical random vectors was discussed in detail by Hashorva (2005) and Asimit and Jones (2007). The latter authors also presented, under certain specified assumptions, the limiting upper copula and a bivariate version of the classical peaks over a high threshold. The research in this area has a potential importance for financial applications.

13.9 Fields of Application

Other members of the class are used as alternatives to the normal when studying the robustness of statistical tests. For instance, Devlin et al. (1976) used them to obtain samples containing outliers and then compared the robustness of two estimators of the correlation; viz., the product-moment correlation r and the quadrant correlation r_q. The latter is defined as $r_q = \sin(\pi q/2)$, where $q = (n_1 + n_3 - n_2 - n_4)/n$, n_i being the number of observations in the ith quadrant using the coordinatewise medians as the origin, and $n = \sum n_i$. Devlin et al. found that r is not robust, while r_q is quite robust.

13.10 Bivariate Symmetric Stable Distributions

13.10.1 Explanations

A univariate d.f. F is "stable" if, for every c_1, c_2, and positive b_1, b_2, there exists c and (positive) b such that

$$F\left(\frac{x - c_1}{b_1}\right) * F\left(\frac{x - c_2}{b_2}\right) = F\left(\frac{x - c}{b}\right), \tag{13.28}$$

where $*$ denotes convolution.[3] By analogy with the univariate case, a bivariate distribution H is said to be *stable* if, for every $b_1 > 0$, $b_2 > 0$, and real $\mathbf{c_1}, \mathbf{c_2}$, there exist $b > 0$ and $\mathbf{c}$ such that

$$H\left(\frac{\mathbf{x} - \mathbf{c_1}}{b_1}\right) * H\left(\frac{\mathbf{x} - \mathbf{c_2}}{b_2}\right) = H\left(\frac{\mathbf{x} - \mathbf{c}}{b}\right). \tag{13.29}$$

$\mathbf{X}$ is said to be symmetric about $\mathbf{a}$ if $\mathbf{X} - \mathbf{a}$ and $-(\mathbf{X} - \mathbf{a})$ have the same distribution.

13.10.2 Characteristic Function

It has been shown by Press (1972a; 1972b, Chapter 6) that a bivariate stable distribution, symmetric about $\mathbf{a}$, and of order m has s characteristic function φ such that

$$\log \varphi(\mathbf{t}) = i\mathbf{a}'\mathbf{t} - \frac{1}{2}\sum_{j=1}^{m}(\mathbf{t}'\mathbf{\Omega}_j\mathbf{t})^{\alpha/2}, \tag{13.30}$$

[3] $F_1 * F_2$ means $\int_{-\infty}^{\infty} F_1(x-t) f_2(t)dt$ (and is the same as $F_2 * F_1$, i.e., $\int_{-\infty}^{\infty} F_2(x-t) f_1(t)dt$).

or equivalently

$$\varphi(\mathbf{t}) = \exp\left\{ i\mathbf{a}'\mathbf{t} - \frac{1}{2}\sum_{j=1}^{m}(\mathbf{t}'\boldsymbol{\Omega}_j\mathbf{t})^{\alpha/2} \right\}, \quad 0 < \alpha \le 2, \tag{13.31}$$

where $\boldsymbol{\Omega}_j$ is a positive semidefinite matrix.[4] It is assumed that no two of the $\boldsymbol{\Omega}_j$'s are proportional and that $\boldsymbol{\Omega} = \sum_{j=1}^{m} \boldsymbol{\Omega}_j$ is positive definite. α is called the *characteristic exponent.*

If $m = 1$, the distribution above is an elliptically symmetric bivariate distribution. When $\alpha = 1$, this gives the log characteristic function of the bivariate Cauchy distribution. When $\alpha = 2$, it becomes that of the bivariate normal even if $m \neq 1$.

13.10.3 Probability Densities

According to Galambos (1985), the only multivariate stable densities known in a closed form, apart from the multivariate normal, are certain Cauchy distributions.

13.10.4 Association Parameter

In bivariate stable distributions with $\alpha < 2$, all second-order moments are infinite, and hence the usual Pearson product-moment correlation coefficient is undefined. We will see that the usual correlation coefficient can be extended below.

13.10.5 Correlation Coefficients

Press (1972a) defined the association parameter ρ for (13.30) as follows. Denote element ij of $\boldsymbol{\Omega}_{\mathbf{k}}$ by $\omega_{ij}(k)$ (where $i, j = 1, 2$ and $k = 1, 2, \ldots, m$). Then

$$\rho = \frac{\sum_{k=1}^{m} \omega_{12}(k)}{[\sum_{k=1}^{m} \omega_{11}(k) \sum_{k=1}^{m} \omega_{22}(k)]^{1/2}}. \tag{13.32}$$

When $\alpha = 2$, then $\boldsymbol{\Sigma} = \sum_{j=1}^{m} \boldsymbol{\Omega}_j$ is the covariance matrix of the bivariate normal distribution, and the defined parameter ρ becomes the ordinary cor-

---

[4] An $n \times n$ symmetric matrix $\mathbf{A}$ is positive semidefinite if $\mathbf{x}'\mathbf{A}\mathbf{x} \ge 0$ for every nonzero $\mathbf{x}$ in R^n.

relation coefficient. Press showed that ρ defined above satisfies $-1 \leq \rho \leq 1$ and that if X and Y are independent, then $\rho = 0$.

13.10.6 Remarks

- The marginals are symmetric stable distributions with characteristic exponent α. But a case of a stable distribution with a vector index—namely, $(\frac{1}{2}, \frac{1}{4})$—arises as a special case of the inverse Gaussian/conditional inverse Gaussian; see Barndorff-Nielsen (1983).
- Every linear combination of X and Y (i.e., $aX + bY$) is symmetric stable [Press (1972a)].
- The distribution is infinitely divisible.

A characterization has been given by Moothathu (1985). For another account of this class of distributions, see Galambos (1985).

13.10.7 Application

Investment economists are concerned with optimal selection of a portfolio of securities. A tradition has been formed in the statistical side of this work of using symmetric stable (univariate) distributions, partly because of empirical evidence and partly because of the theoretical properties of these distributions. Press (1972a; 1972b, Chapter 12) considered an investment portfolio containing two assets whose price changes, X and Y, follow the bivariate symmetric stable distribution in (13.30). Suppose the vector of proportions of resources allocated to the variables-price assets is $\mathbf{c}' = (c_1, c_2)$, so that the return on this allocation of resources is $Q = c_1X + c_2Y$. It is clear from the comments made in Section 13.10.6 above that Q is also symmetric stable. Taking $E(Q) = \mathbf{c}'\mathbf{a}$, the log characteristic function for the return is given by

$$\log \varphi_Q(t) = it(\mathbf{c}'\mathbf{a}) - \frac{1}{2}|t|^\alpha \sum_{j=1}^{n} (\mathbf{c}'\boldsymbol{\Omega}_j\mathbf{c})^{\alpha/2}. \tag{13.33}$$

The "risk" associated with the allocation is taken to be $\frac{1}{2}\sum_{j=1}^{m}(\mathbf{c}'\boldsymbol{\Omega}_j\mathbf{c})$. Define the set of "efficient" portfolios as those for which it is not possible to achieve a greater expected return without increasing risk. Press showed that this set can be obtained as the solution to a programming problem with objective function

$$\lambda(\mathbf{c}'\mathbf{a}) - \frac{1}{2}\sum_{j=1}^{m}(\mathbf{c}'\boldsymbol{\Omega}_j\mathbf{c})^\alpha, \quad 0 < \lambda < \infty. \tag{13.34}$$

That is, for some fixed, preassigned λ, maximize (13.34) with respect to $\mathbf{c}$, subject to $c_i \geq 0$ and $c_1 + c_2 = 1$. The attitude toward avoiding risk-taking is what determines λ; see also Rao (1983).

13.11 Generalized Bivariate Symmetric Stable Distributions

We consider a generalization of (13.30) by de Silva (1978, 1979).

13.11.1 Characteristic Functions

De Silva (1978, 1979) generalized (13.30) to a class of symmetric bivariate stable distributions such that

$$\log \varphi(\mathbf{t}) = i\mathbf{a}'\mathbf{t} - \gamma(\mathbf{t}), \tag{13.35}$$

or equivalently

$$\varphi(\mathbf{t}) = \exp\left\{i\mathbf{a}'\mathbf{t} - \gamma(\mathbf{t})\right\}, \tag{13.36}$$

where

$$\gamma(\mathbf{t}) = \sum_{j=1}^{m} \left(\sum_{k=1}^{r} |c_{kj}s + d_{kj}t|^{\beta} \right)^{\alpha/\beta}, \tag{13.37}$$

c_{kj} and d_{kj} being real constants, $\mathbf{t}' = (s, t)$, and $0 < \alpha < \beta \leq 2$. He showed that, when $\beta = 2$ and $r = 1$, (13.35) is equivalent to (13.30). We consider two special cases here.

13.11.2 de Silva and Griffith's Class

Let us now consider a special case when $\beta = 1$, $m = 1$, $r = 3$. Suppose $c_{11} = \lambda_1 > 0$, $d_{11} = 0$; $c_{21} = d_{21} = \lambda_2 > 0$, and $c_{31} = \lambda_3 > 0$. Then, the characteristic function φ (assuming $\mathbf{a} = \mathbf{0}$) is

$$\varphi(s,t) = \exp(-\lambda_1^{\alpha}|s|^{\alpha} - \lambda_2^{\alpha}|s+t|^{\alpha} - \lambda_3^{\alpha}|t|^{\alpha}. \tag{13.38}$$

This particular class of bivariate symmetric stable distributions has been considered by de Silva and Griffiths (1980). They showed that these can be obtained by the trivariate reduction method (see Section 7.3.4).

13.11.3 A Subclass of de Silva's Stable Distribution

Consider a subclass of de Silva's stable distribution (13.35) such that

$$\log \varphi(x,t) = -\sum_{j=1}^{m} (a_j|s|^\beta + b_j|t|^\beta + c_j|s+\delta t|^\beta)^{\alpha/\beta}, \tag{13.39}$$

where a_j, b_j, and c_j are non-negative constants, $\delta = \pm 1$, and $0 < \alpha < \beta \leq 2$. De Silva (1978) defined an association parameter ρ^* in the following manner.

Let $U = \frac{e^{isX} - \varphi_X(s)}{[1+|\varphi(s)|^2]^{1/2}}$ and $V = \frac{e^{isY} - \varphi_X(t)}{[1+|\varphi(t)|^2]^{1/2}}$, where φ_X and φ_Y are the characteristic functions of X and Y. Define

$$\rho^* = \lim \sup_{s,t \to 0} E(\bar{U}V)|, \tag{13.40}$$

where $\bar{U}$ denotes the complex conjugate of U, and

$$E(\bar{U}V) = \frac{\varphi(-s,t) - \varphi_X(s)\varphi_Y(t)}{[1-|\varphi_X(s)|^2]^{1/2}[1-|\varphi_X(t)|^2]^{1/2}}. \tag{13.41}$$

De Silva (1978) has shown that $\rho^* = \sum_{j=1}^m B_j/2D$, where $D^2 = [\sum_{j=1}^m (a_j + c_j)^{\alpha/\beta} \sum_{j=1}^m (b_j + c_j)^{\alpha/\beta}]$ and $B_j = (a_j + c_j)^{\alpha/\beta} + (b_j + c_j)^{\alpha/\beta} - (a_j + b_j)^{\alpha/\beta}$.

ρ^* reduces to the ordinary correlation coefficient when this class of distributions reduces to the bivariate normal.

Griffiths (1972) [see also de Silva (1978)] has proved that X and Y are independent if and only if $\rho^* = 0$.

A test of independence for the distribution (13.39), based on the empirical characteristic function, has been discussed by de Silva and Griffiths (1980).

13.12 α-Symmetric Distribution

We say that (X, Y) possesses an *α-symmetric bivariate distribution* if its characteristic function can be expressed in the form

$$\varphi(s,t) = \phi(|s|^\alpha + |t|^\alpha), \tag{13.42}$$

where ϕ is a scalar function (known as a "primitive").

For $\alpha = 2$, the 2-symmetric bivariate distribution reduces to a bivariate spherical distribution.

If $\phi(x) = e^{-\lambda x}$, then (13.42) becomes a bivariate stable distribution with independent marginals. Chapter 7 of the book by Fang et al. (1990) provide a detailed discussion on this family of bivariate distributions.

13.13 Other Symmetric Distributions

13.13.1 l_p-Norm Symmetric Distributions

Fang and Fang (1988, 1989) introduced several families of multivariate l_1-norm symmetric distributions, which includes the i.i.d. sample from the exponential as a particular case. A comprehensive treatment of these distributions has been provided in Chapter 5 of Fang et al. (1990).

Yue and Ma (1995) introduced a family of multivariate l_p-norm symmetric distributions that is an extension of the families of l_1-norm distributions. For a bivariate l_p-norm symmetric distribution, its probability density function is given by

$$h(x,y) = \frac{c}{\theta^{2p}} x^{p-1} y^{p-1} g_c\left(\frac{(x^p + y^p)^{1/p}}{\theta}\right), \tag{13.43}$$

where g_c is a density generator and c is the normalizing constant.

Example

Hougaard (1987, 1989) introduced a bivariate Weibull distribution with survival function

$$\bar{H}(x,y) = \exp\{-(\varepsilon_1 x^p + \varepsilon_2 y^p)^\alpha\}. \tag{13.44}$$

Let $\alpha = 1/p$ and $\varepsilon_i = \theta^{-p}$, $i = 1, 2$; then, the distribution above becomes a bivariate l_p-norm symmetric distribution.

13.13.2 Bivariate Liouville Family

This family of bivariate distributions was discussed in Section 9.16. Chapter 6 of Fang et al. (1990) provides a thorough discussion on this subject.

13.13.3 Bivariate Linnik Distribution

Anderson (1992) defined a bivariate Linnik distribution through the joint characteristic function

$$\varphi(s,t) = \frac{1}{1 + (\sum_{i=1}^{m} \mathbf{t}'\mathbf{\Omega}_i\mathbf{t})^\alpha}, \quad 0 < \alpha \leq 2. \tag{13.45}$$

where $\mathbf{\Omega_i}$'s are positive semidefinite matrices with no two $\mathbf{\Omega_i}$'s proportional. This distribution is also closed under geometric compounding. When $m = 1$, it reduces to an elliptically contoured distribution.

13.14 Bivariate Hyperbolic Distribution

This is a bivariate distribution such that the contours of its probability density are ellipses, and yet this is not a member of the family of elliptically symmetric bivariate distributions. This is because the ellipses are not concentric like those of an elliptical distribution as defined earlier in Section 13.2.

13.14.1 Formula of the Joint Density

It is convenient to write the joint density function in multidimensional form as

$$h(x,y) = \frac{\kappa^3 e^{\delta\kappa}}{2\pi\alpha(1+\delta\kappa)} \exp\{-\alpha[\delta^2 + (\mathbf{x}-\boldsymbol{\mu})'\boldsymbol{\Delta}^{-1}(\mathbf{x}-\boldsymbol{\mu})]^{1/2} + \boldsymbol{\beta}'(\mathbf{x}-\boldsymbol{\mu})\} \tag{13.46}$$

for $-\infty < x, y < \infty$, where δ is a scalar parameter, $\boldsymbol{\mu}$ and $\boldsymbol{\beta}$ are vector parameters, $\boldsymbol{\Delta}$ is a symmetric positive definite matrix parameter such that $|\boldsymbol{\Delta}| = 1$, and $\kappa = \alpha - \boldsymbol{\beta}'\boldsymbol{\Delta}\boldsymbol{\beta}$. In the bivariate case, it simplifies to

$$\begin{aligned} h(x,y) = & \frac{\kappa^3 e^{\delta\kappa}}{2\pi\alpha(1+\delta\kappa)} \exp\{-\alpha[\delta^2 + (x-\mu_1)^2\delta_{22} - 2(x-\mu_1)(y-\mu_2)\delta_{12} \\ & -(y-\mu_2)^2\delta_{11}]^{1/2} + \beta_1(x-\mu_1) + \beta_2(y-\mu_2)\}, \end{aligned} \tag{13.47}$$

where $\boldsymbol{\Delta} = \begin{pmatrix} \delta_{11} & \delta_{12} \\ \delta_{21} & \delta_{22} \end{pmatrix}$ and $\boldsymbol{\beta}=(\beta_1, \beta_2)$.

The graph of the log-density is a hyperboloid, which is the reason for the name of this distribution.

13.14.2 Univariate Properties

The marginals are hyperbolic distributions; the p.d.f.'s take the same form as (13.46), except the vectors and matrix become scalars.

The mean and variance of the distribution are $\boldsymbol{\mu} = b_{\kappa\delta}\boldsymbol{\Delta\beta}$ and $c_{\kappa\delta} = \boldsymbol{\Delta\beta}'(\boldsymbol{\Delta\beta}) + b_{\kappa\delta}\boldsymbol{\Delta}$, respectively, where $b_{\kappa\delta} = (\delta^2\kappa^2 + 3\delta\kappa + 3)\kappa^{-2}(1+\delta\kappa)^{-1}$ and $c_{\kappa\delta} = (\delta^3\kappa^3 + 6\delta^2\kappa^2 + 12\delta\kappa + 6)\kappa^{-4}(1+\delta\kappa)^{-2}$.

13.14.3 Derivation

Suppose we have a bivariate normal distribution $N(\boldsymbol{\xi}, \boldsymbol{\Sigma})$ such that the mean $\boldsymbol{\xi}$ and the covariance matrix $\boldsymbol{\Sigma} = \sigma^2\boldsymbol{\Delta}$ are related by $\boldsymbol{\xi} = \boldsymbol{\mu} + \sigma^2\boldsymbol{\beta}$. Here, $\boldsymbol{\Delta}$ is a symmetric positive definite matrix parameter such that the determinant $|\boldsymbol{\Delta}| = \mathbf{1}$. Suppose σ^2 has the generalized inverse Gaussian distribution with p.d.f. given by

$$\frac{(\kappa/\delta)^{\lambda}}{2K_{\lambda}(\delta\kappa)} x^{1/2} \exp[-(\delta^2 x^{-1} + \kappa^2 x)/2] \tag{13.48}$$

for $x > 0$, with $\lambda = 3/2$ and K_{λ} being the modified Bessel function of the third kind with index λ.[5] The bivariate hyperbolic distribution is then obtained by compounding (mixing) the bivariate normal with the generalized inverse Gaussian. In other words, the bivariate hyperbolic distribution is a compound distribution, with the bivariate normal being the compounded distribution and the generalized inverse Gaussian being the compounding distribution.

13.14.4 References to Illustrations

Contours of probability density have been given by Blæsild and Jensen (1981) and Blæsild (1981).

13.14.5 Remarks

- The key references for this distribution are Barndorff-Nielsen (1977, 1978) and Blæsild (1981). Another account has also been given by Barndorff-Nielsen and Blæsild (1983).
- Barndorff-Nielsen proposed using this distribution to represent a two-dimensional Brownian motion with drift $\boldsymbol{\beta}$, starting at $\boldsymbol{\mu}$, and observing at a random time σ^2.
- X and Y cannot be independent [Blæsild (1981, p. 254)]. But if δ_{12} and one of the β's are zero, then X and Y are uncorrelated.
- The bivariate hyperbolic distribution, a normal variance mean mixture, has longer tails than the bivariate normal. (Recall from Section 13.4 that

[5] It is related to the I_{λ} of Section 8.2 by $K_{\lambda}(z) = \frac{\pi}{2}\frac{I_{-\lambda}(z) - I_{\lambda}(z)}{\sin \lambda\pi}$.

a compound bivariate normal, which is a normal variance mixture with $\boldsymbol{\beta} = \mathbf{0}$, also has long tails.)

13.14.6 Fields of Application

Blæsild (1981) fitted this distribution (with one axis log-transformed) to Johannsen's data on the length and width of beans. There are some further remarks in Blæsild and Jensen (1981).

The distribution in three dimensions arises theoretically in statistical physics; see Blæsild and Jensen (1981).

13.15 Skew-Elliptical Distributions

Bivariate and multivariate skew-normal distributions were discussed in Chapter 11. The distribution theory literature on this subject has grown rapidly in recent years, and a number of extensions and alternative formulations have been added. Obviously, there are many similar but not identical proposal coexisting and with unclear connections between them. Recently, Arellano-Valle and Azzalini (2006) have unified these families under a new general formulation, at the same time clarifying their relationships.

Just like the case of skew-normal distributions, we can apply the skewness mechanism to the bivariate t and other elliptical symmetric distributions. There are two approaches to introducing skewness to elliptically symmetric distributions:

(1) the approach introduced by Azzalini and Capitanio (1999) and
(2) the approach by Branco and Dey (2001).

A natural question arises as to how the two approaches are related. The problem has been considered by Azzalini and Capitanio (2003), who found that although a general coincidence could not be established, it is valid for various important cases, notably the multivariate Pearson type II and type VII families; the latter family is of special importance because it includes the t-distribution.

Several other families of skew-elliptical distributions have been defined and studied

- The family studied by Fang (2003, 2004, 2006) includes those of Azzalini and Capitanio (1999) and Branco and Dey (2001).
- The family of Sahu et al. (2003), obtained by using transformation and conditioning, coincides with those of Azzalini and Capitanio (1999) and Branco and Dey (2001) only in the univariate case.

- The generalized skew-elliptical distribution of Genton and Loperfido (2005) includes the multivariate skew-normal, skew t, skew-Cauchy, and skew-elliptical distributions as special cases.
- Arellano-Valle and Genton (2005) also introduced a general class of fundamental skew distributions.
- Ferreira and Steel (2007) considered a class of skewed multivariate distributions. The method is based on a general linear transformation of a multidimensional random variable with independent components, each with a skewed distribution.

Genton (2004) contains 20 chapters, contributed by many authors. This book reviews the state-of-the-art advances in skew-elliptical distributions and provides many new developments, bringing together theoretical results and applications previously scattered throughout the literature. In the editor's words: "The main goal of this research area is to develop flexible parametric classes of distributions beyond the classical normal distribution. The book is divided into two parts. The first part discusses theory and inference for skew-elliptical distribution, the second part presents applications and case studies."

13.15.1 Bivariate Skew-Normal Distributions

This was first studied by Azzalini and Dalla Valle (1996) and Azzalini and Capitanio (1999) by adding an additional parameter that regulates skewness. Consequently, the covariance matrix depends on the mean vector. The distribution was subsequently studied by different authors with various extensions and variants. For the unification of families of skew-normal distributions, see Arellano-Valle and Azzalini (2006).

13.15.2 Bivariate Skew t-Distributions

There are various ways to skew a bivariate t-distribution. Branco and Dey (2001) [see also Azzalini and Capitanio (2003)] constructed a bivariate skew t-distribution in a similar fashion as for the bivariate skew-normal distribution,

$$h(x,y) = h_T(x,y;\nu)T_1\left(\alpha_1 x + \alpha_2 y\left(\frac{\nu+2}{Q+\nu}\right)^{1/2};\nu+2\right), \tag{13.49}$$

where $Q=(x^2-2\rho xy+y^2)/(1-\rho^2)$, $h_T(x,y;\nu)$ is the bivariate t-distribution, and $T_1(x;\nu+2)$ is the cumulative distribution function of the Student t with $\nu+2$ degrees of freedom.

For other bivariate skew t-distributions, see Jones and Faddy (2003), Sahu et al. (2003), and Gupta (2003).

13.15.3 Bivariate Skew-Cauchy Distribution

Consider three independent standard Cauchy random variables W_1, W_2, and U. Let $\boldsymbol{W} = (W_1, W_2)$. Arnold and Beaver (2000) constructed a basic bivariate skew-Cauchy distribution by considering the conditional distribution of $\boldsymbol{W}$ given $\lambda_0 + \lambda_1' \boldsymbol{W} > U$.

13.15.4 Asymmetric Bivariate Laplace Distribution

This is an asymmetric elliptical distribution introduced in Kotz et al. (2001, Chapter 8). The distribution is given by the characteristic function

$$\varphi(s,t) = \left[1 - i\mathbf{t}'\theta + \frac{1}{2}\mathbf{t}'\Sigma\mathbf{t}\right]^{-1}, \tag{13.50}$$

where $\mathbf{t} = (s,t)$ and Σ is the covariance matrix for the symmetrical bivariate Laplace distribution.

When $\theta = 0$, it belongs to the elliptical family of distributions. The mean vector and the covariance matrix are given, respectively, by θ and $\Sigma + \theta\theta'$. The covariance matrix depends on the mean vector, as was the case for the skew normal distribution of the type considered in Azzalini and Dalla Valle (1996) and Azzalini and Capitanio (1999). For estimation and testing of parameters of (13.50), see Kollo and Srivastava (2004).

13.15.5 Applications

The second part of Genton (2004) presents applications and case studies in areas such as biostatistics, finance, oceanography, environmental science, and engineering. For an application to reliability, see Vilca-Labra and Leiva-Sánchez (2006). The skew t-distribution is found to be a sensible parametric distribution applicable for general-purpose robustness study [Azzalini and Genton (2007)]. See also the applications reviewed by Kotz et al. (2001) and Azzalini (2005).

References

1. Abdous, B., Fougères, A.-L., Ghoudi, K.: Extreme behaviour for bivariate elliptical distributions. Canadian Journal of Statistics **33**, 317–334 (2005)
2. Anderson, D.N.: A multivariate Linnik distribution. Statistics and Probability Letters **14**, 333–336 (1992)
3. Arellano-Valle, R.B., Azzalini, A.: On the unification of families of skew-normal distributions. Scandinavian Journal of Statistics **33**, 561–574 (2006)
4. Arnold, B.C., Beaver, R.J.: The skew-Cauchy distribution. Statistics and Probability Letters **49**, 285–290 (2000)
5. Asimit, A.V., Jones, B.L.: Extreme behavior of bivariate elliptical distributions. Insurance: Mathematics and Economics **41**, 53–61 (2007)
6. Azzalini, A.: The skew-normal distribution and related multivariate families. Scandinavian Journal of Statistics **32** 159–188 (2005)
7. Azzalini, A., Capitanio, A.: Statistical applications of multivariate skew normal distribution. Journal of the Royal Statistical Society, Series B **61**, 579–602 (1999)
8. Azzalini, A., Capitanio, A.: Distributions generated by perturbation of symmetry with emphasis on multivariate skew *t* distribution. Journal of the Royal Statistical Society, Series B **65**, 367–390 (2003)
9. Azzalini, A., Dalla Valle, A.: The multivariate skew-normal distribution. Biometrika **83**, 715–726 (1996)
10. Azzalini, A., Genton, M.G.: Robust likelihood methods based on the skew-*t* and related distributions. International Statistical Review **76**, 106–129 (2007)
11. Bairamov, I., Kotz, S., Kozubowski, T.J.: A new measure of linear local dependence. Statistics **37**, 243–258 (2003)
12. Barndorff-Nielsen, O.: Exponentially decreasing distributions for the logarithm of particle size. Proceedings of the Royal Society, Series A **353**, 401–419 (1977)
13. Barndorff-Nielsen, O.: Hyperbolic distributions and distributions on hyperbolae. Scandinavian Journal of Statistics **5**, 151–157 (1978)
14. Barndorff-Nielsen, O.: On a formula for the distribution of the maximum likelihood estimator. Biometrika **70**, 343–365 (1983)
15. Barndorff-Nielsen, O., Blæsild, P.: Hyperbolic distributions. In: Encyclopedia of Statistical Sciences, Volume 3, S. Kotz and N.L. Johnson (eds.), pp. 700–707. John Wiley and Sons, New York (1983)
16. Bentler, P.M., Berkane, M.: Developments in the elliptical theory generalization of normal multivariate analysis. In: American Statistical Association, 1985 Proceedings of the Social Statistics Section, pp. 291–295. American Statistical Association, Alexandria, Virginia (1985)
17. Beran, R.J.: Testing for ellipsoidal symmetry of a multivariate density. Annals of Statistics **7**, 150-162 (1979)
18. Berkane, M., Bentler, P.M.: Moments of elliptically distributed random variates. Statistics and Probability Letters **4**, 333–335 (1986a)
19. Berkane, M., Bentler, P.M.: Characterizing parameters of elliptical distributions. American Statistical Association, 1986 Proceedings of the Social Statistics Section, pp. 278–279. American Statistical Association, Alexandria, Virginia (1986b)
20. Berkane, M., Bentler, P.M.: Characterizing parameters of multivariate elliptical distributions. Communications in Statistics: Simulation and Computation **16**, 193–198 (1987)
21. Blæsild, P.: The two-dimensional hyperbolic distribution and related distributions, with an application to Johannsen's bean data. Biometrika **68**, 251–263 (1981)
22. Blæsild, P., Jensen, J.L.: Multivariate distributions of hyperbolic type. In: Statistical Distributions in Scientific Work, Volume 4: Models, Structures, and Characterizations, C. Taillie, G.P. Patil, and B.A. Baldessari (eds.), pp. 45–66. Reidel, Dordrecht (1981)

23. Branco, M.D., Dey, P.K.: A general class of multivariate skew-ellipitical distributions. Journal of Multivariate Analysis **79**, 99–113 (2001)
24. Cambanis, S., Huang, S., Simons, G.: On the theory of elliptically contoured distributions. Journal of Multivariate Analysis **11**, 368–385 (1981)
25. Chmielewski, M.A.: Elliptically symmetric distributions: A review and bibliography. International Statistical Review **49**, 67–74 (1981)
26. Demarta, S., McNeil, A.J.: The t copula and related copulas. International Statistical Review **73**, 111–129 (2005)
27. de Silva, B.M.: A class of multivariate symmetric stable distributions. Journal of Multivariate Analysis **8**, 335–345 (1978)
28. de Silva, B.M.: The quotient of certain stable random variables. Sankhyā, Series B **40**, 279–281 (1979)
29. de Silva, B.M., Griffiths, R.C.: A test of independence for bivariate symmetric stable distributions. Australian Journal of Statistics **22**, 172–177 (1980)
30. Devlin, S.J., Gnanadesikan, R., Kettenring, J.R.: Some multivariate applications of elliptical distributions. In: Essays in Probability and Statistics, S. Ikeda, T. Hayakawa, H. Hudimoto, M. Okamoto, M. Siotani, and S. Yamamoto (eds.), pp. 365–393. Shinko Tsucho, Tokyo (1976)
31. Devroye, L.: Nonuniform Random Variate Generation. Springer-Verlag, New York (1986)
32. Ernst, M.D.: A multivariate generalized Laplace distribution. Computational Statistics **13**, 227–232 (1998)
33. Fang, B.Q.: The skew elliptical distributions and their quadratic forms. Journal of Multivariate Analysis **87**, 298–314 (2003)
34. Fang, B.Q.: Noncentral quadratic forms of the skew elliptical variables. Journal of Multivariate Analysis **95**, 410-430 (2004)
35. Fang, B.Q.: Sample mean, covariance and T^2 statistic of the skew elliptical model. Journal of Multivariate Analysis **97**, 1675–1690 (2006)
36. Fang, H.B., Fang, K.T., Kotz, S.: The meta-elliptical distributions with given marginals. Journal of Multivariate Analysis **82**, 1–16 (2002)
37. Fang, K.T.: Elliptically contoured distributions. In: Encyclopedia of Statistical Sciences, Updated Volume 1, S. Kotz, C.B. Read, and D. Banks (eds.), pp. 212–218. John Wiley and Sons, New York (1997)
38. Fang, K.T., Fang, B.Q.: Some families of multivariate symmetric distributions related to exponential distributions. Journal of Multivariate Analysis **24**, 109–122 (1988)
39. Fang, K.T., Fang, B.Q.: A characterization of multivariate l_1-norm symmetric distribution. Statistics and Probability Letters **7**, 297–299 (1989)
40. Fang, K.T., Kotz, S., Ng, K.W.: Symmetric Multivariate and Related Distributions. Chapman and Hall, London (1990)
41. Fang, K.T., Liang, J.J.: Spherical and elliptical symmetry, tests of. In: Encyclopedia of Statistical Sciences, Updated Volume 3, S. Kotz, C.B. Read, and D. Banks (eds.), pp. 686–691. John Wiley and Sons, New York (1999)
42. Ferreira, J.T., Steel, M.: A new class of skewed multivariate distributions with applications to regression analysis. Statistica Sinica **17**, 505–529 (2007)
43. Galambos, J.: Multivariate stable distributions. In: Encyclopedia of Statistical Sciences, Volume 6, S. Kotz and N.L. Johnson (eds.), pp. 125–129. John Wiley and Sons, New York (1985)
44. Genton, M.G. (ed.): Skew-Elliptical Distributions and Their Applications: A Journey Beyond Normality. Chapman and Hall/CRC, Boca Raton, Florida (2004)
45. Genton, M.G., Loperfido, N.M.R.: Generalized skew-elliptical distributions and their quadratic forms. Annals of the Institute of Statistical Mathematics **57**, 389–401 (2005)
46. Gómez, E., Gómez-Villegas, M.A., Marin, J. M.: A multivariate generalization of the power exponential family of distributions. Communications in Statistics **27**, 589–600 (1998)

47. Gordon, Y.: Elliptically contoured distributions. Probability Theory and Related Fields **76**, 429–438 (1987)
48. Griffiths, R.C.: Linear dependence in bivariate distributions. Australian Journal of Statistics **14**, 182–187 (1972)
49. Gupta, A.K.: Multivariate skew t-distribution. Statistics **37**, 359–363 (2003)
50. Hashorva, E.: Extremes of asymptotically spherical and elliptical random vectors. Insurance: Mathematics and Economics **36**, 285–302 (2005)
51. Hougaard, P.: Modelling multivariate survival. Scandinavian Journal of Statistics **14**, 291–304 (1987)
52. Hougaard, P.: Fitting a multivariate failure time distribution. IEEE Transactions on Reliability **38**, 444–448 (1989)
53. Hult, H., Lindskog, F.: Multivariate extremes, aggregation and dependence in elliptical distributions. Advances in Applied Probability **34**, 587–608 (2002)
54. Iyengar, S., Tong, Y.L.: Convexity properties of elliptically contoured distributions. Sankhyā, Series A **51**, 13–29 (1989)
55. Johnson, M.E.: Multivariate Statistical Simulation. John Wiley and Sons, New York (1987)
56. Johnson, M.E., Wang, C., Ramberg, J.: Generation of continuous multivariate distributions for statistical applications. American Journal of Mathematical and Management Sciences **4**, 225–248 (1984)
57. Jones, M.C., Faddy, M.J.: A skew extension of the t-distribution, with applications. Journal of the Royal Statistical Society, Series B **65**, 159–174 (2003)
58. Kelker, D.: Distribution theory of spherical distributions and a location-scale parameter generalization. Sankhyā, Series A **32**, 419–430 (1970)
59. Kollo, T., Srivastava, M.S.: Estimation and testing of parameters in multivariate Laplace distribution. Communications in Statistics: Theory and Methods **33**, 2363–2387 (2004)
60. Kotz, S.: Multivariate distributions at a cross road. In: A Modern Course on Distributions in Scientific Work, Volume I: Models and Structures, G.P. Patil, S. Kotz, and J.K. Ord (eds.), pp. 247–270. Reidel, Dordrecht (1975)
61. Kotz, S., Kozubowski, T.J., Podgórski, K.: The Laplace Distribution and Generalizations. Birkhäuser, Boston (2001)
62. Kotz, S., Nadarajah, S.: Some extreme type elliptical distributions. Statistics and Probability Letters **54**, 171–182 (2001)
63. Kotz, S., Nadarajah, S.: Local dependence functions for the elliptically symmetric distributions. Sankhyā, Series A **65**, 207–223 (2003)
64. Kotz, S., Ostrovskii, I.: Characteristic functions of a class of elliptical distributions. Journal of Multivariate Analysis **49**, 164–178 (1994)
65. Li, R.Z., Fang, K.T., Zhu, L.X.: Some Q-Q probability plots to test spherical and elliptical symmetry. Journal of Computational and Graphical Statistics **6**, 435–450 (1997)
66. Lindsey, J.K.: Multivariate elliptically contoured distributions for repeated measurements. Biometrics **55**, 1277–1280 (1999)
67. Manzotti, A., Pérez, F.J., Quiroz, A.J.: A statistic for testing the null hypothesis of elliptical symmetry. Journal of Multivariate Analysis **81**, 274–285 (2002)
68. Mardia, K.V.: Families of Bivariate Distributions. Griffin, London (1970)
69. McGraw, D.K., Wagner, J.F.: Elliptically symmetric distributions. IEEE Transactions on Information Theory **14**, 110–120 (1968)
70. Moothathu, T.S.K.: On a characterization property of multivariate symmetric stable distributions. Journal of the Indian Statistical Association **23**, 83–88 (1985)
71. Nadarajah, S.: The Kotz-type distribution with applications. Statistics **37**, 341–358 (2003)
72. Nadarajah, S.: On the product XY for some elliptically symmetric distributions. Statistics and Probability Letters **75**, 67–75 (2005)

73. Nadarajah, S.: On the ratio X/Y for some elliptically symmetric distributions. Journal of Multivariate Analysis **97**, 342–358 (2006a)
74. Nadarajah, S.: Fisher information for the elliptically symmetric Pearson distributions. Applied Mathematics and Computation **178**, 195–206 (2006b)
75. Nadarajah, S., Kotz, S.: On the product of XY for the elliptically symmetric Kotz type distribution. Statistics **39**, 269–274 (2005)
76. Nomakuchi, K., Sakata, T.: Characterizations of the forms of covariance matrix of an elliptically contoured distributions. Sankhyā, Series A **50**, 205–210 (1988)
77. Press, S.J.: Multivariate stable distributions. Journal of Multivariate Analysis **2**, 444–462 (1972a)
78. Press, S.J.: Applied Multivariate Analysis. Holt, Reinhart and Winston, New York (1972b)
79. Rao, B.R.: A general portfolio model for multivariate symmetric stable distributions. Metron **41**, 29–42 (1983)
80. Sahu, S.K., Dey, D.K., Branco, M.D.: A new class of multivariate skew distributions with applications to Bayesian regression models. Canadian Journal of Statistics **31**, 129–150 (2003)
81. Schmidt, R.: Tail dependence for elliptically contoured distributions. Mathematical Methods of Operations Research **55**, 301–327 (2002)
82. Schott, J.R.: Testing for elliptical symmetry in covariance-matrix-based analyses. Statistics and Probability Letters **60**, 395–404 (2002)
83. Serfling, R.J.: Multivariate symmetry and asymmetry. In: Encyclopedia of Statistical Sciences S. Kotz, N. Balakrishnan, C.B. Read and B. Vidakovic (eds.), pp. 5338–5345. John Wiley and Sons, New York (2006)
84. Shaked, M., Tong, Y.L.: Inequalities for probability contents of convex sets via geometric average. Journal of Multivariate Analysis **24**, 330–340 (1988)
85. Sibuya, M.: Bivariate extreme statistics, I. Annals of the Institute of Statistical Mathematics **11**, 195–210 (1960)
86. Streit, F.: On the characteristic functions of the Kotz type distributions. Comptes Rendus Mathématiques de l'Académie des Sciences **13**, 121–124 (1991)
87. Tong, Y.L.: Probability Inequalities in Multivariate Distributions. Academic Press, New York (1980)
88. van Praag, B.M.S., Wesselman, B.M.: Elliptical multivariate analysis. Journal of Econometrics **41**, 189–203 (1989)
89. Vilca-Labra, F., Leiva-Sánchez, V.: A new fatigue life model based on the family of skew-elliptical distributions. Communications in Statistics: Theory and Methods **35**, 229–244 (2006)
90. Yue, X., Ma, C.: Multivariate l_p-norm symmetric distributions. Statistics and Probability Letters **24**, 281–288 (1995)

Chapter 14
Simulation of Bivariate Observations

14.1 Introduction

Devroye (1986) has provided an exhaustive treatment on the generation of random variates. Gentle (2003) has also recently provided a state-of-the-art treatise on random number generation and Monte Carlo methods. For this reason, we provide here a brief review of this subject and refer readers to these two references for a comprehensive treatment. In view of the importance of simulation as a tool while analyzing practical data using different parametric statistical models as well as while examining the properties and performance of estimators and hypothesis tests, we feel that it is very important for a reader of this book to know at least some essential details about the simulation of observations from a specified bivariate probability function.

We referred to published algorithms (some coded in a language such as FORTRAN, while some are not coded) at several points in all the preceding chapters, and here we present a concise review.

In Section 14.2, we detail some of the common approaches for simulation in the univariate case, while in Section 14.3 simulation methods for some specific univariate distributions are described. In Section 14.4, some available software for simulation in the univariate case is listed. Some general approaches for simulation in the bivariate case are presented in Section 14.5. In Sections 14.6 and 14.7, simulations from bivariate normal distributions and copulas are detailed. Some methods of simulating observations from some specific variate distributions with simple forms are described in Section 14.8. Simulations from bivariate exponential and bivariate gamma distributions are explained in Sections 14.9 and 14.10, respectively. In Section 14.11, simulation methods for conditionally specified bivariate distributions are detailed. In Sections 14.12 and 14.13, simulation methods for elliptically contoured bivariate distributions and bivariate extreme-value distributions are presented. In Sections 14.14 and 14.15, generation of bivariate and multivariate skewed distributions and generation methods for bivariate distributions with given

N. Balakrishnan, C.D. Lai, *Continuous Bivariate Distributions*,
DOI 10.1007/b101765_15,

marginals are described. Finally, in Section 14.16, simulaton of bivariate distributions with specified correlations is presented.

14.2 Common Approaches in the Univariate Case

14.2.1 Introduction

Our starting point is that we assume we can easily and efficiently generate independent uniform random numbers on a computer; see, for example, Cohen (1986). We then wish to use these in some way to obtain random variates with specific distributional properties of interest.

Valuable references include Bratley et al. (1983, Chapter 5), Devroye (1981, 1986, especially Chapters VII–IX), Fishman (1978, Chapters 8 and 9), Hoaglin (1983), Kennedy and Gentle (1980, Chapter 6), Knuth (1981, especially Section 3.4.1), Law and Kelton (1982), Leemis and Schmeiser (1985), Morgan (1984, Chapters 4 and 5), Ripley (1983, 1987, Section 3.4), and Rubinstein (1981, Chapter 3).

We will describe here the following common approaches for generating univariate random variates:

- Inverse probability integral transform.
- Composition.
- Acceptance/rejection.
- Ratio of uniform variates.
- Transformation.
- Markov Chain Monte Carlo—MCMC.

We note that the best method of random variate generation for a given distribution may depend on the value of some parameter of that distribution, such as its shape parameter. In fact, there are many other things that may affect which method is the "best" one, including the number of variates to be generated, the availability of its generator or otherwise of fast generators for some related distribution; whether the algorithm is to be coded in a high or low language, the criteria that we use to make the assessment, such as speed, portability, or simplicity; and so on.

The notation $U(a, b)$ is used to denote the uniform variate over $[a, b]$. Sometimes $U(a, b)$ will be written simply as U if it does not cause any confusion, and sometimes a subscript will be affixed to denote the first, second, etc.; as usual, u will be used for a particular value of U.

14.2.2 Inverse Probability Integral Transform

The use of the inverse of the cumulative distribution function is of wide applicability in simulation algorithms; in principle, it can be used whenever the distribution function is known.[1] It is based on the following well-known result. If X has a continuous distribution function F, then $U = F(X)$ is distributed with uniform density over $[0, 1]$; conversely,[2] $X = F^{-1}(U)$ has distribution function $F(x)$. We can then express the method simply as follows:

1. Generate u from U.
2. Set $x = F^{-1}(u)$, or the modified expression[2] if F^{-1} does not exist uniquely.

For discrete distributions, this method is easy to apply, but straightforward application to continuous distributions is limited to variates having F^{-1} in a simple closed-form. For example, we can obtain exponential distribution from the transformation $X = -\log(1 - U)$ or $X = -\log U$. For other members of the gamma family, i.e., for arbitrary shape parameter, this approach is inefficient since the evaluation of F^{-1} must be done iteratively.[3]

We note that, according to Hoaglin (1983), this method is generally slower than other, seemingly more complicated, methods. However, Schmeiser (1980) has given three reasons for its use: easy generation of order statistics, easy generation from truncated distribution, and easy implementation of "variance reduction" technique in simulation models.

If only the density f is known explicitly, and not the distribution function. F, then the additional step of numerical integration becomes an added burden. The work of Ulrich and Watson (1987) is directed towards this issue.

14.2.3 Composition

The mixture method or composition technique is based on representing the density f from which variates are to be generated as $f(x) = \sum_{i=1}^{n} p_i f_i(x)$, where $\sum_{i=1}^{n} p_i = 1$. The mixture algorithm then simply generates variates from each f_i with probability p_i; see Peterson and Kronmal (1982, 1985) for a detailed discussion on this method.

[1] There are times when the distribution is defined through the characteristic function φ and it is difficult or impossible to get F directly. In this situation, with some conditions about the integrability and continuity of φ and its first two derivatives, one may generate uniform variates by means of the acceptance/rejection method of Section 14.2.4 [Devroye (1986, pp. 695–716)]. Similarly, sometimes only a sequence of moments or Fourier coefficients may be known; see Devroye (1989).

[2] If F is not continuous or not strictly increasing (or both), then the inverse does not exist. In this case, the definition $X = \inf\{x : F(x) \geq U\}$ becomes useful, which simply means that x assumes the infimum (or smallest value) for which $F(x)$ is at least u.

[3] There are popular algorithms for numerical inversion if this should be needed to obtain $F^{-1}(u)$, such as the bisection method, the secant method, and the Newton–Raphson method; see Devroye (1986, pp. 32–33).

Composition must not be confused with convolution. The linear combination of random variables such as $X = \sum_{i=1}^{n} a_i X_i$ is a convolution and one can easily produce such an r.v. X directly from the X_i's; but, this is quite different from the composition construction.

14.2.4 Acceptance/Rejection

The acceptance/rejection method[4] has been a useful approach for developing new algorithms for generating univariate observations; see Tadikamalla (1978a). The basic idea of this method is to generate a variate from a density function that somewhat resembles the desired density function f. First, select a function t such that $t(x) \geq f(x)$ for all values of x, and t is called a *dominating or majorizing function.* Let $g(x) = t(x)/c$, where $c = \int_{-\infty}^{\infty} t(x)dx$, so that g becomes a probability density function.

The algorithm works as follows:

1. Generate x from the density function $g(x)$.
2. Generate u from U.
3. If $u \leq f(x)/t(x)$, accept x; otherwise go to Step 1.

The number c is then the number of "trials" (or iterations) until an acceptance. The value $1/c$ is generally referred to as the "efficiency" of the procedure. The factors to be considered in selecting the density g are:

- Step 1 of the algorithm should be executed quickly.
- c should be close to 1.
- The acceptance/rejection test should be simple; i.e., $f(x)/t(x)$ should be easy to evaluate.

Kronmal et al. (1978) and Kronmal and Peterson (1979) proposed what they called the *alias-rejection-mixture* method. It is based on two methods: (i) Walker's (1974a,b) alias method and (ii) the rejection-mixture method, which is a combination of the mixture and acceptance/rejection methods. Kronmal and Peterson (1984) have also proposed an *acceptance-complement method.*

14.2.5 Ratio of Uniform Variates

For a density f, if (U, V) is uniformly distributed over the region $0 \leq u \leq \sqrt{f(v/u)}$, then $X = V/U$ has the desired density f; see Kinderman and Monahan (1977), Hoaglin (1983), and Devroye (1986, Section IV.7).

---

[4] Referred to in the literature sometimes as simply the rejection method.

14.2.6 Transformations

Many methods for simulating observations are based on first generating some intermediate nonuniform random variates $Y_1, Y_2, \ldots, Y_n$, and then setting $X = T(Y_1, Y_2, \ldots, Y_n)$.

Among the transformations of a single variate (i.e., $n = 1$), familiar examples include nonstandard normal variates by $X = \mu + \sigma Z$, where Z is the standard normal variate; $U(a, b)$ by $X = a+(b-a)U$; and lognormal variates by $X = \exp(Y)$, where Y is the appropriate normal variate.

As for generating one random variate from two or more nonuniform variates, some well-known examples include gamma variates (having an integer shape parameter k) as the sum of k exponential variates, beta variates as a ratio of gammas, t from the standard normal and chi-squared variates, F from chi-squared variates, chi-squared variates from the normal, and so on.

14.2.7 Markov Chain Monte Carlo—MCMC

There are various ways of using a Markov chain to generate random variates from some distribution related to the chain. Such methods are called *Markov Chain Monte Carlo* or simply MCMC.

The Markov chain Monte Carlo method has become one of the most important tools in recent years, particularly in Bayesian analysis and simulation. An algorithm based on a stationary distribution of a Markov chain is an iterative method because a sequence of operations must be performed until they converge. The stationary distribution is chosen to correspond to the distribution of interest (called the target distribution).

The techniques for generating random numbers based on Markov chains are generally known as "samplers."

Two prominent samplers are (i) the Metropolis–Hastings and (ii) the Gibbs samplers. These algorithms are obtained in Sections 4.10 and 4.11 of Gentle (2003). There are several variations of the basic Metropolis–Hastings algorithm as well; see Gentle (2003, pp. 143–146) for pertinent details. Gentle (2003, pp. 157–158) has also described another method, called the *hit-and-run sampler.*

The Markov chain samplers generally require a "burn-in" period (that is, a number of iterations before a stationary distribution is achieved). In practice, the variates generated during the burn-in periods are therefore discarded. The number of iterations needed varies with the distribution and can be quite large sometimes, even in the thousands. We also note that a sequence of observations generated by a sampler is autocorrelated, and thus variance estimation must be performed with care since the estimated variance may be biased. The method of batch means [see Section 7.4 of Gentle (2003)] or some other method that accounts for autocorrelation should be used.

A computer program known as BUGS (Bayesian Inference Using Gibbs Sampling) designed for MCMC methods is widely used in this regard. Information on BUGS can be obtained at the site

http://www.mrc-bsu.cam.ac.uk/bugs/

14.3 Simulation from Some Specific Univariate Distributions

14.3.1 Normal Distribution

A well-known exact method for generating normal variates is that of Box and Muller (1958). It gives two independent standard normal variates X_1 and X_2, $X_1 = R\sin\alpha$ and $X_2 = R\cos\alpha$, where $R = \sqrt{-2\log(U_1)}$ and $\alpha = 2\pi U_2$. A change of variables argument demonstrates the validity of this algorithm. Alternatively, observe that (R, α) are the polar coordinates of (X_1, X_2). Let X_1 and X_2 be two independent standard normal variates. Then, their density is symmetric about the origin. So, α is uniform over $(0, 2\pi)$, and $R^2 = X_1^2 + X_2^2$ has a chi-squared distribution with two degrees of freedom, which is the exponential distribution with mean 2.

A modified polar method, due to Marsaglia and Bray (1964), avoids the use of trigonometric functions. From variates V_1 and V_2 that are uniformly distributed over $[-1, 1]$, $W = V_1^2 + V_2^2$ is calculated. If $W > 1$, the pair (V_1, V_2) is rejected. With an acceptable pair, we then calculate the normal variates as

$$X = \left(\frac{-2\log W}{W}\right)^{1/2} V_1, \qquad Y = \left(\frac{-2\log W}{W}\right)^{1/2} V_2. \tag{14.1}$$

Subsequent algorithms have been primarily based on the composition and acceptance/rejection techniques. One example is the rectangle-wedge-tail method of Marsaglia et al. (1964); for example, the normal distribution is seen as made up from rectangles, wedges between rectangles and the true density, and tails.

Schmeiser (1980) has provided a detailed list of references; see also Devroye (1986, especially Section IX.1), Rubinstein (1981, Section 3.6.4), and Ripley (1987, especially pp. 82–87). FORTRAN codes for the Box and Muller (1958) and Ahrens and Dieter (1972) methods are given in Bratley et al. (1983, p. 297 and p. 318). There are two FORTRAN programs in Best (1978b). A comparison of the algorithms made by Kinderman and Ramage (1976) is noteworthy in this regard for paying attention to user-oriented features such as machine independence, brevity, and implementation in high-level language rather than being confined to speed and accuracy.[5]

14.3.2 Gamma Distribution

It is sufficient to generate random variates from the standard gamma distribution with density

$$f(x) = x^{\alpha-1} \exp(-x)/\Gamma(\alpha), \qquad x \geq 0; \tag{14.2}$$

if the scale parameter is other than 1 and/or the lower end of the distribution is other than 0, a linear transformation can be applied easily.

For the case $\alpha = 1$ (exponential variates), as mentioned earlier, the usual method is the inverse transformation $x = -\log(u)$. For other methods, see Schmeiser (1980, pp. 84–85).

For the case $\alpha = K$ an integer (Erlang variates), the variates may be generated as $x = -\log(\prod_{i=1}^{k} u_i)$; however, the execution time grows linearly with k.

According to Schmeiser (1980), the easiest exact method for generating a gamma variate for any $\alpha > 0$ is due to Jöhnk (1964). However, this is based on the method for Erlang variates and so has the same disadvantage. Since the mid-1970s, many algorithms have been developed. Among them are those of Ahrens and Dieter (1974), Atkinson and Pearce (1976), Cheng (1977), Best (1978a,b), Tadikamalla (1978a,b), Tadikamalla and Johnson (1978), and Schmeiser and Lal (1980). Most of these algorithms are listed in Fishman (1978, pp. 422–429). For extensive surveys on generating gamma variates, see Schmeiser (1980) and Ripley (1987, pp. 88–90). FORTRAN listings of some leading algorithms are given in the second of these,[6] and there are several in Văduva (1977) and two more in Bratley et al. (1983, pp. 312–313). Other works in this direction include those of Barbu (1987), Monahan (1987), and Minh (1988).

---

[5] In passing, we note that Kinderman and Ramage stated, “The algorithms discussed in this paper were coded in as comparable manner as the authors could manage... We have experimented with several versions of the coding.” This throws light on something that concerned us from time to time: How can we know that a purported comparison of algorithms really is that, rather than a comparison of their coding? The tone of the quotation above suggests there is not—or was not in the mid-1970s—any great science in the step from algorithm to FORTRAN code.

[6] Minh (1988) notes a couple of misprints in the algorithm G4PE.

14.3.3 Beta Distribution

We may obtain beta variates as $X = W/(W+Y)$, where W and Y are gamma variates with shape parameters a and b, respectively.

A method due to Jöhnk (1964) uses the relationship between beta and uniform variates. Set $Y = U_1^{1/a}$, $Z = U_2^{1/b}$. If $Y + Z \leq 1$, calculate $X = Y/(Y+Z)$. Then, X has a beta distribution with parameters a and b. Several other methods are discussed in the references cited in Section 14.2.1.

Unfortunately, the execution time for Jöhnk's algorithm grows indefinitely with increasing a and/or b because $U_1^{1/a} + U_2^{1/b}$ will end up being greater than 1. Cheng (1978) described an algorithm, denoted by BB, whose execution time becomes constant as a or b increases. It involves using the acceptance/rejection method to generate an observation from the beta distribution of the second kind. According to Schmeiser (1980), algorithm B4PE developed in Schmeiser and Babu (1980) executes in about half the time of BB, but the setup time is longer and it requires more lines of code. Cheng's method is coded in FORTRAN in Bratley et al. (1983, p. 295). Algorithms devised by Sakasegawa (1983) have been shown by that author to be very fast in execution, albeit with considerable setup time (as with B4PE).

A s stated by Devroye (1986, p. 433), "The bottom line is that the choice of a method depends upon the user: if he is not willing to invest a lot of time, he should use the ratio of gamma variates. If he does not mind coding short programs, and a and/or b vary frequently, one of the rejection methods based upon analysis of beta density or upon universal inequalities can be used. The method of Cheng is very robust. For special cases, such as symmetric beta densities, rejection from the normal density is very competitive. If the user does not foresee frequent changes in a and b, a strip table method or the algorithm of Schmeiser and Babu (1980) are recommended. Finally, when both parameters are smaller than one, it is possible to use rejection from polynomial densities or to apply Jöhnk method."

14.3.4 t-Distribution

Random variates from the t-distribution with ν degrees of freedom may be generated by the transformation method as the ratio of a normal variate and the square root of an independent gamma variate having shape parameter $\alpha = \nu/2$ divided by α. For other methods, see Devroye (1986, pp. 445–454).

Best (1978b) showed that the t_ν-variate can be generated by the acceptance/rejection method. The density function of t_ν for $\nu \geq 3$ is dominated by a multiple of density t_3. Hence, the algorithm involves the following steps: (i) generating a t_3 variate by a ratio-of-uniform method, for which one may refer to Section 14.2.5 and Devroye (1986, pp. 194–203), and (ii) generating t_3 (for

$\nu > 3$) by the acceptance/rejection method based on t_3. Best (1978b) has provided a FORTRAN program, and this algorithm has been summarized and slightly modified by Devroye (1986, pp. 449–451). Several other diverse methods have been listed by Hoaglin (1983).

14.3.5 Weibull Distribution

The Weibull distribution with parameters $\alpha > 0$ and $\beta > 0$ has probability density function

$$f(x) = \frac{\alpha}{\beta} x^{\alpha-1} e^{-x^{\alpha}/\beta}, \quad 0 \le x < \infty. \tag{14.3}$$

The simple inverse probability integral transform method applied to the standard Weibull distribution (i.e., $\beta = 1$) is quite efficient. The formula is simply $x = (-\log u)^{\frac{1}{\alpha}}$. Of course, an acceptance/rejection method could also be used to avoid the evaluation of the logarithmic function. The standard Weibull variate is finally scaled by $\beta^{1/\alpha}$ to obtain variates from (14.3).

14.3.6 Some Other Distributions

Tadikamalla (1984) has discussed the simulation from normal, gamma, beta, and t-distributions, as well as the inverse Gaussian and exponential power distributions, and has given references for the stable distribution and some others. Section IX.2 of Devroye (1986) is on the exponential distribution, Section IX.3 on the gamma distributions, Section IX.6 on stable densities, and Section IX.7.5 on the generalized inverse Gaussian distributions. There is a FORTRAN routine in Bratley et al. (1983, pp. 314–315) for generating a stable variate. For "phase-type" distributions, one may refer to Neuts and Pagano (1981).

14.4 Software for Random Number Generation

Chapter 4 of Gentle (2003) has listed software for random number generation. Monte Carlo simulation often involves many hours of computer time, and so computational efficiency is very important in software for random number generation.

Implementing one of the simple methods to convert a uniform variate to that of another distribution may not be as efficient as a special method specifically oriented toward the target distribution. The IMSL Libraries and

S-Plus have a number of modules that use efficient methods to generate variates from several common distributions.

14.4.1 Random Number Generation in IMSL Libraries

The well-known IMSL Libraries contain a large number of routines to generate random variates from many continuous univariate distributions[7] and multivariate normal distribution. All the IMSL routines for random number generation are available in both the FORTRAN and C programming languages. Morgan (1984, Appendix 1) has described the programs more fully. Lewis (1980) and Gentle (1986) have discussed the use of IMSL in simulation and statistical analysis. A package by Lewis et al. (1986) [reviewed by Burn (1987)] includes routines for generating random variates from normal, Laplace, Cauchy, gamma, Pareto, and beta distributions.

14.4.2 Random Number Generation in S-Plus and R

The software system S was developed at Bell Laboratories in the mid-1970s and has evolved considerably since the early versions. S is both a data analysis system and an object-oriented programming language.

S-Plus is an enhancement of S developed by StatSci, Inc. (now a part of Insightful Corporation). The enhancements made include graphical interfaces, more statistical analysis functionality, and support.

There is a freely available package, called R, that provides generally the same functionality in the same language as S; see Gentleman and Ihaka (1997). The R programming system is available at the site

http:/www.r-project.org/

Like the IMSL Libraries, S-Plus and R have a number of modules that use efficient methods to generate variates from several common distributions as listed in Table 8.2 of Gentle (2003). It is also pointed out there that S-Plus and R do not use the same random number generators.

14.5 General Approaches in the Bivariate Case

[7] Beta, Cauchy, chi-squared, exponential mixture, F, gamma, inverted beta, logistic, lognormal, normal, stable, t, triangular, and Weibull variates can be obtained from IMSL or NAG or both of these libraries.

14.5.1 Setting

We shall now turn our attention to the random number generation of variates from bivariate continuous distributions. Much of our discussion applies to the multivariate case as well.

To establish a framework for specific generation algorithms detailed below in Sections 14.6–14.15, we will first describe two general methods—the conditional distribution and transformation techniques. The usual context for application of the former is where the conditional distribution function is explicitly known, while the latter does not need this and is usually met in the trivariate reduction context.

The acceptance/rejection method, which is used extensively in the univariate situation, is not discussed in detail in this section. Although, in principle, the method applies to multivariate situations, practical difficulties have stifled its use. According to Johnson et al. (1984), these difficulties include

- a lack of suitable dominating functions,
- complications in optimizing the choice of parameters in the dominating function, and
- low efficiencies.

For a discussion of these difficulties and an idea about how to overcome them, see Johnson (1987, pp. 46–48).

As in the univariate situation, the composition (or probability mixing) method is also used to generate bivariate random vectors. Schmeiser and Lal (1982) have used this method to generate bivariate gamma random variables.

In what follows, we shall use (X_1, X_2), rather than (X, Y), to denote the pair of variates that we wish to generate.

14.5.2 Conditional Distribution Method

This idea, usually attributed to Rosenblatt (1952), is as follows:

1. Generate x_1 from the marginal distribution of X_1.
2. Generate x_2 from the conditional distribution of X_2, given $X_1 = x_1$.

The suitability of this method for a given bivariate distribution depends on there being an efficient method for generating from the required univariate distributions.

One might ask which variate ought to be X_1 and which one should be X_2. Rubinstein (1981, p. 61) has stated bluntly, "Unfortunately, there is no way to find a priori the optimal order of representing the variates in the vector to minimize the CPU time."

14.5.3 Transformation Method

This method is best presented in the multivariate context. The idea is that we let $\mathbf{X} = (X_1, X_2, \ldots, X_p)'$ be the p-dimensional random vector that we want, whose distribution may be difficult to generate directly, and we let $\mathbf{Y} = (Y_1, Y_2, \ldots, Y_q)'$ be a q-dimensional ($q \geq p$) random vector having a distribution that is easier to generate from. Then, if there exists a function $\mathbf{a}(\mathbf{Y}) = (a_1(\mathbf{Y}), \ldots, a_p(\mathbf{Y}))$ such that $\mathbf{a}(\mathbf{Y})$ has the same distribution as $\mathbf{X}$, we can get a random realization of $\mathbf{X}$ by first generating $\mathbf{Y}$ and then evaluating $\mathbf{a}(\mathbf{Y})$.

This method is most appealing when the specific transformation is already available. For some distributions, this is indeed the case.[8] However, for some arbitrary multivariate density $h(\mathbf{x})$, it is seldom obvious what transformation of what vectors (that are themselves easy to generate) will give rise to $\mathbf{X}$. The following advice [Johnson et al. (1984) and Johnson (1987, p. 46)] may help:

- Carry out a thorough search of the literature for an appropriate construction scheme. Rarely does a distribution emerge from a vacuum; usually there is some derivation, possibly by compounding, convolution, or transformation.
- Attempt (invertible) transformation of $\mathbf{X}$. Is there a recognizable result? Start with a component transformation (and compare the resulting expression with known bivariate uniform distributions) or a transformation to exponential marginals (and compare the result with known bivariate exponential distributions).
- Perhaps h can be recognized as a mixture (viz., $ph_1 + (1-p)h_2$, with h_1 and h_2 being well known).
- Check if the p.d.f. can be written as a function of the quadratic form $ax_1^2 + bx_1x_2 + cx^2$. Then, generation of random variates is easy, as the distribution would then belong to the elliptical class discussed in Chapter 13.

14.5.4 Gibbs' Method

The Gibbs sampler is one of the MCMC methods. It also uses the conditional distribution approach. Suppose X and Y have a joint density $h(x, y)$ and conditional densities $f(x|y)$ and $g(y|x)$. Let X_i (Y_i) be a sequence of observations from X (Y). Then, observations on X and Y can be generated as a Markov chain with elements having densities

[8] For example, we need to look no further than bivariate t, which can be obtained as $X_i = Z_i/\sqrt{W/\nu}$ for $(i = 1, 2)$, where (Z_1, Z_2) has a standardized bivariate normal distribution, and W, independent of the Z's, has a chi-squared distribution with ν degrees of freedom.

$$g(y_i|x_{i-1}),\ f(x_i|y_i),\ g(y_{i+1}|x_i),\ f(x_{i+1}|y_{i+1}),\ldots .$$

For multivariate distributions, Gibbs' algorithm may be given by the following steps [see Algorithm 4.20 of Gentle (2003)].

Algorithm

0. Set $k = 0$.
1. Choose $\mathbf{x}^{(k)} \in S \subseteq R^p$.
2. Generate $x_1^{(k+1)}$ conditionally on $x_2^{(k)}, x_3^{(k)}, \ldots, x_p^{(k)}$.
   Generate $x_2^{(k+1)}$ conditionally on $x_1^{(k+1)}, x_3^{(k)}, \ldots, x_p^{(k)}$.
   $\vdots$
   Generate $x_{p-1}^{(k+1)}$ conditionally on $x_1^{(k+1)}, x_2^{(k+1)}, \ldots, x_p^{(k)}$.
   Generate $x_p^{(k+1)}$ conditionally on $x_1^{(k+1)}, x_2^{(k+1)}, \ldots, x_{p-1}^{(k+1)}$.
3. If convergence has occurred, then deliver $\mathbf{x} = \mathbf{x}^{(k+1)}$; otherwise, set $k = k + 1$ and go to Step 2.

Casella and George (1992) have presented a simple proof that this iterative algorithm converges, but to determine whether the convergence has occurred or not is not a simple matter.

Another type of Metropolis method is the "hit-and-run" sampler. In this method, all components of the vector are updated at once. The method has been presented in Algorithm 4.21 of Gentle (2003) in its general version as described by Chen and Schmeiser (1996).

14.5.5 Methods Reflecting the Distribution's Construction

Some bivariate distributions are more easily thought of in terms of how they are constructed rather than in terms of a formula for a c.d.f. or p.d.f. It may happen in such cases that the method of construction can be directly adapted to random variate generations, as in the case of a trivariate reduction or other form of transformation.

14.6 Bivariate Normal Distribution

The conditional distribution and transformation methods for generating bivariate and multivariate normal random variates have been available for some time now; see, for example, Scheuer and Stoller (1962) and Hurst and Knop (1972).

Let $(X_1, X_2)'$ denote the bivariate normal vector with covariance matrix $\mathbf{\Sigma}$. Define

$$\begin{aligned} Y_1 &= (X_1 - \mu_1)/\sigma_1, \\ Y_2 &= \frac{(X_2 - \mu_2) - \frac{\sigma_2}{\sigma_1}(X_1 - \mu_1)\rho}{\sigma_2(1-\rho^2)^{1/2}}. \end{aligned} \tag{14.4}$$

Then, Y_1 and Y_2 are two independent standard normal variables, and we can now express

$$\begin{aligned} X_1 &= \sigma_1 Y_1 + \mu_1, \\ X_2 &= \sigma_2 \rho Y_1 + \sigma_2(1-\rho^2)^{1/2} Y_1 + \mu_2. \end{aligned} \tag{14.5}$$

Univariate standard generators are widely available for this purpose; see Section 14.3.1.

More generally, let $\mathbf{X} \sim N(\boldsymbol{\mu}, \mathbf{\Sigma})$, i.e., $\mathbf{X}$ is a p-dimensional multivariate normal random vector with mean vector $\boldsymbol{\mu}$ and covariance matrix $\mathbf{\Sigma}$. Let $\mathbf{L}$ be the lower triangular matrix of the Choleski decomposition of $\mathbf{\Sigma}$, i.e., a matrix such that $\mathbf{\Sigma} = \mathbf{LL}'$. (Routines for computing $\mathbf{L}$ are available in many computer software packages.) Given p independent univariate standard variates, $\mathbf{Y}' = (Y_1, \ldots, Y_p)$, transform them via

$$\mathbf{X} = \mathbf{LY} + \boldsymbol{\mu} \tag{14.6}$$

to achieve $N_p(\boldsymbol{\mu}, \mathbf{\Sigma})$ distribution.

In two or three dimensions, $\mathbf{L}$ can be expressed easily. For example, if

$$\mathbf{\Sigma} = \begin{pmatrix} \sigma_1^2 & \rho_{12}\sigma_1\sigma_2 & \rho_{13}\sigma_1\sigma_3 \\ \rho_{12}\sigma_1\sigma_2 & \sigma_2^2 & \rho_{23}\sigma_2\sigma_3 \\ \rho_{13}\sigma_1\sigma_3 & \rho_{23}\sigma_2\sigma_3 & \sigma_3^2 \end{pmatrix},$$

we have

$$\mathbf{L} = \begin{pmatrix} \sigma_1 & 0 & 0 \\ \sigma_2\rho_{12} & \sigma_2(1-\rho_{12}^2)^{1/2} & 0 \\ \sigma_3\rho_{13} & \frac{\sigma_3(\rho_{23}-\rho_{12}\rho_{13})}{(1-\rho_{12}^2)^{1/2}} & \sigma_3[(1-\rho_{12}^2)(1-\rho_{13}^2) - (\rho_{23}-\rho_{12}\rho_{13})^2]^{1/2} \end{pmatrix}. \tag{14.7}$$

For the bivariate case, simply delete the third row and column of $\mathbf{L}$ in (14.7).

The conditional distribution method is almost the same. In the first term in (14.4), we would write $(X_1 - \mu_1)/\sigma_1$ and then the distribution of X_2 conditional upon the known value of X_1.

The trivariate reduction method is also simple, albeit at the expense of using three independent normal variates for the pair (X_1, X_2) generated.

A FORTRAN program that includes the Choleski decomposition procedure has been published by Bedall and Zimmermann (1976). Routines for generating vectors from multivariate normal distribution are in both the IMSL and NAG collections (see Section 14.4.1 above). A routine for the bivariate normal is present in STATLIB [Brelsford and Relies (1981, p. 375)]. Ghosh and Kulatilake (1987) have published a FORTRAN program listing to generate random variates from the multivariate normal distribution. An APL listing for the bivariate case has been given by Bouver and Bargmann (1981).

For more details, interested readers may refer to Devroye (1986, Section XI.2), Gentle (2003, pp. 197–198), Johnson (1987, pp. 52–54), Kennedy and Gentle (1980, Section 6.5.9), Ripley (1987, pp. 98–99), Rubinstein (1981, Section 3.5.3), and Văduva (1985).

14.7 Simulation of Copulas

Section 4.10 of Drouet-Mari and Kotz (2001) presents simulation procedures for copulas. This section has been subdivided into the following cases:

- The general cases.
- The Archimedean copulas.
- Archimax distributions.
- Marshall and Olkin's mixture of distributions.
- Three-dimensional copulas with truncation invariance.

The first two items were described in Section 1.13.

Using a copula, a data analyst can construct a bivariate (multivariate) distribution by specifying marginal univariate distributions and choosing a particular copula to provide a correlation structure between variables.

Yan (2007) presented the design, features, and some implementation details of the R package copula, which contains codes to generate commonly used copulas, including the elliptical, Archimedean, extreme value, and Farlie–Gumbel–Morgenstern families.

14.8 Simulating Bivariate Distributions with Simple Forms

14.8.1 Bivariate Beta Distribution

Recall that the bivariate beta distribution has a density given by

$$h(x_1, x_2) = \frac{\Gamma(\theta_1 + \theta_2 + \theta_3)}{\Gamma(\theta_1)\Gamma(\theta_2)\Gamma(\theta_3)} x_1^{\theta_1 - 1} x_2^{\theta_2 - 1} (1 - x_1 - x_2)^{\theta_3 - 1}$$
$$x_1, x_2 > 0, \ x_1 + x_2 < 1.$$

Arnason and Baniuk (1978) have described several ways to generate variates from the Dirichlet distribution (the bivariate beta above is a Dirichlet), including a sequence of conditional betas and the use of the relationship of order statistics from a uniform distribution to a Dirichlet. The most efficient method seems to be the one using the relationship between independent gamma variates and a Dirichlet. If Y_1, Y_2, Y_3 are independently distributed gamma random variables with shape parameters $\theta_1, \theta_2, \theta_3$, respectively, (X_1, X_2) with

$$X_j = \frac{Y_j}{Y_1 + Y_2 + Y_3}, \qquad i = 1, 2,$$

has a bivariate beta distribution with parameters θ_1, θ_2, and θ_3. This relationship yields a straightforward method of generating bivariate betas through generating independent gammas.

Loukas (1984) presented five methods for generating bivariate beta observations (X_1, X_2):

- the bivariate version of Jöhnk's rejection method based on Jöhnk's (1964) rejection method for simulating univariate beta variates;
- the bivariate version of Jöhnk's transformation method;
- the bivariate rejection method, which is a an extension of the simple rejection technique;
- the conditional method that is based on the property that $X_2|(1-X_1)/X_1$ also has a beta distribution; and
- The gamma method discussed in the preceding paragraph.

14.9 Bivariate Exponential Distributions

14.9.1 Marshall and Olkin's Bivariate Exponential Distribution

This bivariate distribution may be generated either through the trivariate reduction of three independent exponential variates or by the generation of univariate Poisson variates because the distribution may be derived in terms of Poisson shocks; see Dagpunar (1988).

14.9.2 Gumbel's Type I Bivariate Exponential Distribution

The marginals in this case are exponential, and the conditional distribution of X_2, given $X_1 = x_1$, has density

$$g(x_2|x_1) = [(1 + \theta x_1)(1 + \theta x_2) - \theta] \exp[-x_2(1 + \theta x_1)],$$

which can be rewritten as

$$g(x_2|x_1) = p\beta \exp(-\beta x_2) + (1 - p)\beta^2 x_2 \exp(-\beta x_2), \tag{14.8}$$

where $\beta = 1 + \theta x_1$ and $p = (\beta - \theta)/\beta$. The form in (14.8) is a mixture density arising by a mechanism that with probability p generates an exponential variate with mean β^{-1} and with probability $1 - p$ generates the sum of two independent exponential variates each having mean β^{-1}. Generation in this mixture form is therefore straightforward, which is evidently an example of composition (see Section 14.2.3).

14.10 Bivariate Gamma Distributions and Their Extensions

14.10.1 Cherian's Bivariate Gamma Distribution

Cherian's bivariate gamma distribution is discussed in Section 8.10. Let $Y_i \sim \text{gamma}(\theta_i)$ (for $i = 1, 2, 3$) be three independent gamma variates. Define $X_1 = Y_1 + Y_3, X_2 = Y_2 + Y_3$. Then (X_1, X_2) have Cherian's bivariate gamma distribution. One can see that generation of this joint distribution is easy.

14.10.2 Kibble's Bivariate Gamma Distribution

Kibble's bivariate gamma is discussed in Section 8.2. The marginal distributions have a gamma distribution with shape parameter α. For an arbitrary value of $\alpha > 0$, the simulation does not appear to be easy. However, when 2α is a positive integer, then the pair (X_1, X_2) can be easily generated through the bivariate normal distributions. See Section 8.2.7 for details.

14.10.3 Becker and Roux's Bivariate Gamma

Becker and Roux (1981) defined a bivariate extension gamma distribution that serves as a useful model for failure times of two dependent components in a system. The joint density is given in Section 8.20. Gentle (2003, pp. 122–125) has used this example to illustrate how the acceptance/rejection method can be applied to multivariate distributions. A simulation procedure is listed with a majorizing density that is composed of two densities, a bivariate exponential and a bivariate uniform. The example also serves to illustrate the difficulty in using the acceptance/rejection method in higher dimensions.

14.10.4 Bivariate Gamma Mixture of Jones et al.

Jones et al. (2000) considered a bivariate gamma mixture distribution by assuming two independent gammas with the scale parameters having a generalized Bernoulli distribution. The joint density was presented in Chapter 8. Simulation from this distribution consists of the following two steps:

(i) Simulate a pair of scale parameters (γ, β) from the probability matrix of $p_{\gamma_i \beta_j}$, $i, j = 1, 2$.
(ii) Simulate two independent gammas, each with the scale parameter obtained from the first step.

14.11 Simulation from Conditionally Specified Distributions

Appendix A in Arnold et al. (1999, pp. 371–380) presents an overview on generating observations from conditionally specified bivariate distributions. Arnold et al. (1999) have commented, "Despite the fact that we often lack analytical expressions for the densities, it turns out to be quite easy to devise relatively efficient simulation schemes." Two methods, including their simulation algorithms, have been described there:

(i) The acceptance/rejection method will often accomplish the goal.
(ii) Alternatively, the importance sampling simulation scheme also allows us to forget about the normalizing constant problem.

The second scheme involves a score function that is the ratio of the population density function and the simulation density function. This scheme has been described in Section A.3 of Arnold et al. (1999). More details and examples of the importance method can be found in Castillo et al. (1997), Salmerón (1998), and Hernández et al. (1998).

14.12 Simulation from Elliptically Contoured Bivariate Distributions

We are concerned here with the class of elliptically symmetric distributions and not with all distributions whose contours are ellipses. The class, as mentioned in Chapter 13, includes the bivariate normal, Cauchy, and t distributions. Define $R^2 = (X_1^2 - 2\rho X_1 X_2 + X_2^2)/(1-\rho^2)$, where ρ is the off-diagonal entry in the scaling matrix $\mathbf{\Sigma} = \begin{pmatrix} 1 & \rho \\ \rho & 1 \end{pmatrix}$. Let $\mathbf{L}$ be the lower triangle (Choleski) decomposition of $\mathbf{\Sigma}$. Then, from (13.5), $\mathbf{X}$ may be represented as

$$(X_1, X_2)' = R\mathbf{L}\mathbf{U}^{(2)} + \boldsymbol{\mu}, \tag{14.9}$$

where $U^{(2)}$ is uniformly distributed on the circumference of a unit circle and is independent of R.

Generation of $U^{(2)}$ is naturally easy. The expression of $\mathbf{L}$ in the bivariate case is simple. The choice of a particular member of this class of distributions determines the distribution of R. Generation of the vector $\mathbf{X}$ is therefore just as easy or just as difficult as generation of the single variate R.

For the p-dimensional case, an explicit expression for $\mathbf{L}$ would be complicated, and we would need a random vector with a uniform distribution on the surface of the p-dimensional unit hypersphere instead of on the circumference of the unit circle, but except for this, the method would be the same.

Ernst (1998) has described a multivariate generalized Laplace distribution that includes the multivariate normal and Laplace distributions. The joint density for (X_1, X_2) is given by

$$h(x_1, x_2) = \frac{\lambda}{2\pi}\Gamma(2/\lambda)|\mathbf{\Sigma}|^{1/2}\exp\left\{-[(\mathbf{x}-\boldsymbol{\mu})'\mathbf{\Sigma}^{1/2}(\mathbf{x}-\boldsymbol{\mu})]^{\lambda/2}\right\}. \tag{14.10}$$

Ernst (1998) has shown further that the density of R is simply the Stacy distribution. Recall that if X has a gamma distribution, then $X^{1/\lambda}$ has a Stacy distribution. Thus, random variate generation from this bivariate distribution becomes quite easy.

14.13 Simulation of Bivariate Extreme-Value Distributions

Section 3.7 of Kotz and Nadarajah (2000) describes three known methods for simulating bivariate extreme-value observations.

14.13.1 Method of Shi et al.

See Section 12.11.1 for details.

14.13.2 Method of Ghoudi et al.

Ghoudi et al. (1998) described a simulation scheme for (X_1, X_2) that is applicable for all bivariate extreme-value distributions. They first obtained the joint distribution of $Z = X_1/(X_1 + X_2)$ and $V = C(\exp(-X_1), \exp(-X_2))$, where C is the cumulative distribution function of the copula that is associated with the bivariate extreme-value distribution:

$$C(u, v) = \exp[\log(uv)A\{\log(u)/\log(uv)\}].$$

Then, from the joint distribution of Z and V, they obtain as the marginal distribution function of Z

$$G_Z(z) = z + z(1-z)\frac{A'(z)}{A(z)} \tag{14.11}$$

and as the conditional distribution function of V, given $Z = z$,

$$vp(z) + (v \log v)\{1 - p(z)\},$$

where

$$p(z) = \frac{z(1-z)A''(z)}{A(z)g_Z(z)}$$

and $g_Z(z)$ is the p.d.f. of Z.

Thus, given $Z = z$, the distribution of V is uniform over $[0, 1]$ with probability $p(z)$ and is the distribution of the product of two independent uniform variables over $[0,1]$ with probability $1-p(z)$. Thus, to simulate (X_1, X_2) from a bivariate extreme-value distribution, we can use the following procedure:

- Simulate Z according to the distribution in (14.11).
- Having generated Z, take $V = U_1$ with probability $p(Z)$ and $V = U_1U_2$ with probability $1-p(Z)$, where U_1 and U_2 are independent uniform variables over $[0,1]$.
- Finally, set $X_1 = V^{Z/A(Z)}$ and $X_2 = V^{(1-Z)/A(Z)}$.

14.13.3 Method of Nadarajah

Nadarajah's (1999) scheme differs from the two methods above in that it does not simply simulate from a bivariate extreme-value distribution directly. Instead, it uses the limiting point process result as an approximation to simulate bivariate extreme values. The procedure is described in detail in Kotz and Nadarajah (2000, pp. 143–144).

14.14 Generation of Bivariate and Multivariate Skewed Distributions

R has many functions for simulating univariate and multivariate observations with specified distributions.

For generating multivariate skew-normal and multivariate skew t-distributions using R, see the documentation at the library

http://pbil.univ-lyon1.fr/library/sn/html/

written by Professor Adelchi Azzalini, Dipart. Scienze Statistiche, Universitá di Padova, Italy.

14.15 Generation of Bivariate Distributions with Given Marginals

14.15.1 Background

Many simulation methods require the specification of joint bivariate distributions as input. If there is adequate theory or sufficient data on which to choose a specific bivariate distribution, the problem is well defined and can usually be solved by one of the methods discussed above.

In many situations, there is really no adequate theory or sufficient data to be able to specify a unique bivariate distribution. However, it may be realistic to specify the marginal distribution of the random variables and a measure of dependence between them. For example, Johnson and Tenenbein (1978, 1981) stated that in the context of an investment portfolio simulation, a joint distribution of stock and bond returns may have to be specified. Because of a lack of data, it may be difficult to specify the joint distribution of stock and bond returns completely, but it would be possible to specify the marginal distributions and some measures of dependence between the variables concerned.

We shall denote the marginal d.f.'s of X_1 and X_2 by F_1 and F_2, respectively.

14.15.2 Weighted Linear Combination and Trivariate Reduction

Johnson and Tenenbein (1978, 1979, 1981) considered two procedures to generate bivariate random variables when the marginals are specified and their measures of dependence, Kendall's τ or Spearman's ρ_S, can be specified.

Let W_1 and W_2 be i.i.d. r.v.'s with common density function g (uniform, Laplace, or exponential). Set

$$\begin{aligned} Y_1 &= W_1, \\ Y_2 &= cW_1 + (1-c)W_2 \end{aligned} \tag{14.12}$$

$(0 < c < 1)$, and then find the marginals of Y_1 and Y_2. Suppose F_1 and F_2 are the marginals that we require Then obtain X_1 and X_2 by appropriate univariate transformations, $X_1 = F^{-1}[G_1(Y_1)]$ and $X_2 = F_2^{-1}[G_2(Y_2)]$, where G_1 and G_2 are the distribution functions of Y_1 and Y_2, respectively; these are determined by c, which in turn is determined by the chosen value of τ or ρ_S. (Notice that this method is not applicable if the product-moment correlation is specified, as it would get changed by the univariate transformations of Y_i to get X_i.)

Trivariate reduction is similar to the weighted linear combination method, except that Y_1 and Y_2 are now defined as

$$\begin{aligned} Y_1 &= W_1 + \beta W_3, \\ Y_2 &= W_2 + \beta W_3, \end{aligned} \tag{14.13}$$

$(0 < \beta < \infty)$, where W_3, independent of W_1 and W_2, also has density g.

14.15.3 Schmeiser and Lal's Methods

Schmeiser and Lal (1982) pointed out that many random phenomena may be modeled by dependent gamma variates with correlation coefficient ρ and then went on to show how (i) the trivariate reduction algorithm (i.e., Cherian's method) and (ii) the composition algorithm made up from the independence case and the upper Fréchet bound can be used to generate a pair of gamma variates. Moreover, they developed a family of algorithms that can produce bivariate gamma vectors having any parameters α_1, α_2 (shape parameters), β_1, β_2 (scale parameters), and ρ (Pearson's product-moment correlation coefficient).

Let $Z \sim \text{gamma}(\gamma, 1)$ and $W_i \sim \text{gamma}(\delta_i, 1)$ be mutually independent. Then, because of the reproducibility of the gamma distribution,

$$X_1 = (\text{Ga}_1^{-1}(U) + Z + W_1)\beta_1 \tag{14.14}$$

and

$$X_2 = (\text{Ga}_2^{-1}(V) + Z + W_2)\beta_2 \tag{14.15}$$

are both gamma variates, with shape parameters $\alpha_1 = \lambda_1 + \gamma + \delta_1$ and $\alpha_2 = \lambda_2 + \gamma + \delta_2$ and scale parameters β_1 and β_2, respectively, and either $V = U$ or $V = 1 - U$. In the above, Ga_i denotes the distribution function of $\text{gamma}(\lambda_i, 1)$, $i = 1, 2$. Pearson's product-moment correlation is then given by

$$\rho = \frac{E[\text{Ga}_1^{-1}(U)\text{Ga}_2^{-1}(V)] - \lambda_1\lambda_2 + \gamma}{\sqrt{\alpha_1\alpha_2}}. \tag{14.16}$$

Given the values of α_1, α_2, β_1, β_2, and ρ, we now wish to select values of λ_1, λ_2, γ, δ_1, δ_2, β_1, and β_2 such that

$$\left.\begin{array}{ll} \lambda_1 + \gamma + \delta_1 & = \alpha_1 \\ \lambda_2 + \gamma + \delta_2 & = \alpha_2 \\ E[\text{Ga}_1^{-1}(U)\text{Ga}_2^{-1}(V)] - \lambda_1\lambda_2 + \gamma & = \rho\sqrt{\alpha_1\alpha_2} \\ \lambda_1, \lambda_2, \gamma, \delta_1, \delta_2 & \geq 0 \end{array}\right\}. \tag{14.17}$$

(β_1 and β_2 do not appear here; they can be set directly.) As we are using five variables to satisfy three equations, finding a set of parameter values corresponds to finding a feasible solution, rather than an optimal solution, to a linear programming problem. Schmeiser and Lal (1982) have given guidelines for an efficient solution and developed an algorithm called GBIV, which determines the parameter values as well as generating the random vector (X_1, X_2). It takes $\gamma = \delta_2 = 0$. For some scatterplots of data generated by this algorithm, see Schmeiser and Lal (1982) and Hsu and Nelson (1987).

14.15.4 Cubic Transformation of Normals

Fleishman (1978) proposed generating a random variable with given mean, standard deviation, skewness, and kurtosis by taking a standard random variate Z and forming a new r.v. X from a cubic expression, $\sum_{i=0}^{3} a_i Z^i$. (The a_i's can be obtained by solving a system of equations involving the desired values of the first four moments.) Although the method was criticized by Tadikamalla (1980), its simplicity makes it attractive, and it has been extended to the multivariate situation by Vale and Maurelli (1983). Their method is to set $X_1 = \sum_{i=0}^{3} Z_1^i$ and $X_2 = \sum_{i=0}^{3} b_i Z_2^i$, where (i) the a_i's and b_i's are determined by the desired univariate moments and (ii) the correlation between Z_1 and Z_2 is determined by the desired correlation between X_1 and X_2—solution of the cubic equation $\rho_{X_1 X_2} = \sum_{i=0}^{3} c_i \rho_{Z_1 Z_2}^i$ is required, where the c_i's are given in terms of the a_i's and b_i's.

14.15.5 Parrish's Method

Parrish (1990) has presented a method for generating variates from a multivariate Pearson family of distributions. A member of the Pearson family is specified by the first four moments, which of course includes the covariances.

14.16 Simulating Bivariate Distributions with Specified Correlations

14.16.1 Li and Hammond's Method for Distributions with Specified Correlations

Li and Hammond (1975) propose a method for a p-variate distribution with specified marginals and covariance matrix. This method uses the inverse probability integral transform method to transform a p-variate normal into a multivariate distribution with specified marginals. The covariance matrix of the multivariate normal is chosen to yield the specified covariance matrix of the target distribution. The determination of the covariance matrix for the multivariate normal to yield the desired target distribution is difficult, however, and does not always yield a positive definite covariance matrix.

Lurie and Goldberg (1998) modified the Li–Hammond approach by iteratively refining the correlation matrix of the underlying normal using the sample correlation matrix of the transformed variates.

14.16.2 Generating Bivariate Uniform Distributions with Prescribed Correlation Coefficients

The simple method described below was due to Falk (1999). The approach here is in the same spirit as that of Li and Hammond (1975) given above.

Suppose we wish to generate a pair of uniform variates with correlation coefficient ρ, which is also Spearman's rho. Let (X_1, X_2) have a standard bivariate normal distribution with correlation coefficient $\rho' = 2\sin(\rho\pi/6)$. Such bivariate normal distributions can be easily generated.

Consider now the marginal transformation $(U, V) = (\Phi(X_1), \Phi(X_2))$. Then (U, V) will have a bivariate uniform distribution with a grade correlation (Spearman's rho) ρ_S. It is well known (see Section 4.7.2, for example) that

$$\rho_S = \frac{6}{\pi}\sin^{-1}\frac{\rho'}{2}.$$

Since we set $\rho' = 2\sin(\rho\pi/6)$, it is clear that $\rho_S = \rho$. We note that for any bivariate uniform distributions, Spearman's rho is simply Pearson's product-moment correlation coefficient.

14.16.3 The Mixture Approach for Simulating Bivariate Distributions with Specified Correlations

This method, proposed by Michael and Schucany (2002), was inspired by concepts found in Bayesian inference. The theory of the mixture approach is as follows. Let the random variable X_1 have a prior represented by the p.d.f.

$$f(x_1, \theta), \tag{14.18}$$

where the parameter θ may be multidimensional. Next, conditioning on $X_1 = x_1$, let the random variable Z for the data have a likelihood with probability function or probability density function

$$g(z|x_1; \eta), \tag{14.19}$$

where the parameter η may also be multidimensional. Multiplying (14.18) and (14.19) yields the joint distribution of X_1 and Z:

$$j(x_1, z; \theta, \eta) = f(x_1; \theta)g(z|x_1; \eta). \tag{14.20}$$

Integrating out x_1 yields the marginal of Z,

$$m(z;\theta,\eta) = \int j(u,z;\theta,\eta)du. \tag{14.21}$$

Dividing (14.20) by (14.21) yields the posterior of X_1 given $Z = z$, from which we define the p.d.f. of a new random variable X_2:

$$p(x_2|z;\theta,\eta) = j(x_2,z;\theta,\eta)/m(z;\theta,\eta). \tag{14.22}$$

Finally, multiplying (14.22) and (14.20), one obtains the trivariate distribution of X_1, X_2, and Z with the density

$$h(x_1,x_2,z;\theta,\eta) = j(x_1,z;\theta,\eta)j(x_2,z;\theta,\eta)/m(z;\theta,\eta) \tag{14.23}$$

It was pointed out that the bivariate distribution of (X_1, X_2) does not depend on Z. The parameter η in the likelihood (14.19) plays a critical role such that the choice of its value precisely controls the correlation between X_1 and X_2 because the outcome of Z defines the mixture of posteriors that will be used to simulate X_2.

Steps in Mixture Simulation

The proposed mixture method involves three successive steps to generate the desired pair (x_1, x_2):

1. Simulate x_1 from a specified prior.
2. Simulate z from a specified likelihood given x_1.
3. Simulate x_2 from the derived posterior given z.

Examples

Three examples were given to illustrate the mixture method:

- a new bivariate beta family;
- a new bivariate gamma family; and
- a bivariate unform family.

References

1. Ahrens, J.H., Dieter, U.: Computer methods for sampling from the exponential and normal distributions. Communications of the Association for Computing Machinery **15**, 873–882 (1972)
2. Ahrens, J.H., Dieter, U.: Computer methods for sampling from gamma, beta, Poisson, and binomial distributions. Computing **12**, 223–246 (1974)

3. Arnason, A.N., Baniuk, L.: A computer generation of Dirichlet variates. In: Proceedings of the Eighth Manitoba Conference on Numerical Mathematics and Computing, Utilitas Mathematica, pp. 97–105 (1978)
4. Arnold, B.C., Castillo, E., Sarabia, J.M.: Conditional Specification of Statistical Models. Springer-Verlag, New York (1999)
5. Atkinson, A.C., Pearce, M.C.: The computer generation of beta, gamma and normal random variables. Journal of the Royal Statistical Society, Series A **139**, 431–448 (Discussion, 448–461) (1976)
6. Barbu, Gh.: A new fast method for computer generation of gamma and beta random variables by transformations of uniform variables. Statistics **18**, 453–464 (1987)
7. Becker, P.J., Roux, J.J.J.: A bivariate extension of the gamma distribution. South African Statistical Journal **15**, 1–12 (1981)
8. Bedall, F.K., Zimmermann, H.: On the generation of $N\ (\mu, \Sigma)$-distributed random vectors by $N(0, 1)$-distributed random numbers. Biometrische Zeitschrift **18**, 467–471 (1976)
9. Best, D.J.: Letter to the editor. Applied Statistics **27**, 181 (1978a)
10. Best, D.J.: A simple algorithm for the computer generation of random samples from a Student's t or symmetric beta distribution. In: COMPSTAT 1978. Proceedings in Computational Statistics, L.C.A. Corsten and J. Hermans (eds.), pp. 341–347. Physica-Verlag, Heidelberg (1978b)
11. Box, G.E.P., Muller, M.E.: A note on the generation of random normal deviates. Annals of Mathematical Statistics **29**, 610–611 (1958)
12. Bouver, H., Bargmann, R.E.: Evaluation and graphical application of probability contours for the bivariate normal distribution. In: American Statistical Association, 1981 Proceedings of the Statistical Computing Section, pp. 272–277. American Statistical Association, Alexandria, Virginia (1981)
13. Bratley, P., Fox, B.L., Schrage, L.E.: A Guide to Simulation. Springer-Verlag, New York (1983)
14. Brelsford, W.M., Relies, D.A.: STATLIB: A Statistical Computing Library. Prentice-Hall, Englewood Cliffs, New Jersey (1981)
15. Burn, D.A.: Software review of "Advanced Simulation and Statistics Package: IBM Professional FORTRAN Version" by P.A.W. Lewis, E.J. Orav, and L. Uribe, 1986. The American Statistician **41**, 324–327 (1987)
16. Casella, G., George, E.I.: Explaining the Gibbs sampler. The American Statistician **46**, 167–174 (1992)
17. Castillo, E., Gutiérrez, J.M., Hadi, A.S.: Expert Systems and Probabilistic Network Models. Springer-Verlag, New York (1997)
18. Chen, M.H., Schmeiser, B.W.: General hit-and-run Monte Carlo sampling for evaluating multidimensional integrals. Operations Research Letters **19**, 161–169 (1996)
19. Cheng, R.C.H.: The generation of gamma variables with noninteger shape parameter. Applied Statistics **26**, 71–75 (1977)
20. Cheng, R.C.H.: Generating beta variates with nonintegral shape parameters. Communications of the Association for Computing Machinery **21**, 317–322 (1978)
21. Cohen, M-D.: Pseudo-random number generators. In: Encyclopedia of Statistical Sciences, Volume 7, S. Kotz and N.L. Johnson (eds.), pp. 327–333. John Wiley and Sons, New York (1986)
22. Dagpunar, J.: Principles of Random Variate Generation. Clarendon Press, Oxford (1988)
23. Devroye, L.: Recent results in nonuniform random variate generation. In: 1981 Winter Simulation Conference Proceedings, Volume 2, T.I. Ören, C.M. Delfosse, and C.M. Shub (eds.), pp. 517–521. Piscataway, Institute of Electrical and Electronics Engineers, New Jersey (1981)
24. Devroye, L.: Nonuniform Random Variate Generation. Springer-Verlag, New York (1986)

25. Drouet-Mari, D., Kotz, S.: Correlation and Dependence. Imperial College Press, London (2001)
26. Ernst, M.D.: A multivariate generalized Laplace distribution. Computational Statistics **13**, 227–232 (1998)
27. Falk, M.: A simple approach to the generation of uniformly distributed random variables with prescribed correlations. Communications in Statistics: Simulation and Computation **28**, 785–791 (1999)
28. Fishman, G.S.: Principles of Discrete Event Simulation. John Wiley and Sons, New York (1978)
29. Fleishman, A.I.: A method for simulating non-normal distributions. Psychometrika **43**, 521–532 (1978)
30. Gentle, J.E.: Simulation and analysis with IMSL routines. In: 1986 Winter Simulation Conference Proceedings, J.R. Wilson, J.O. Henriksen, and S.D. Roberts (eds.), pp. 223–226. Institute of Electrical and Electronics Engineers, Piscataway, New Jersey (1986)
31. Gentle, J.E.: Random Number Generation and Monte Carlo Methods. Springer-Verlag, New York (2003)
32. Gentleman, R., Ihaka, R.: The R language. Computing Science and Statistics **28**, 326–330 (1997)
33. Ghosh, A., Kulatilake, P.H.S.W.: A FORTRAN program for generation of multivariate normally distributed random variables. Computers and Geosciences **13**, 221–233 (1987)
34. Ghoudi, K., Khouddraji, A., Rivest, L.P.: Statistical properties of couples of bivariate extreme-value copulas. Canadian Journal of Statistics **26**, 187–197 (1998)
35. Hernández, L.D., Moral, S., Salmerón, A.: A Monte Carlo algorithm for probabilistic propagation based on importance sampling and stratified simulation techniques. International Journal of Approximate Reasoning **18**, 53–92 (1998)
36. Hoaglin, D.C.: Generation of random variables. In: Encyclopedia of Statistical Sciences, Volume 3, S. Kotz and N.L. Johnson (eds.), pp. 376–382. John Wiley and Sons, New York (1983)
37. Hsu, J.C., Nelson, B.L.: Control variates for quantile estimation. In: 1987 Winter Simulation Conference Proceedings, A. Thesen, H. Grant, and W.D. Kelton (eds.), pp. 434–444. Institute of Electrical and Electronics Engineers, Piscataway, New Jersey (1987)
38. Hurst, R.L., Knop, R.E.: Algorithm 425: Generation of random correlated normal variables. Communications of the Association for Computing Machinery **15**, 355–357 (1972)
39. Jöhnk, M.D.: Erzeugung von betaverteilten und gammaverteilten Zufallszahlen. Metrika **8**, 5–15 (1964)
40. Johnson, M.E.: Multivariate Statistical Simulation. John Wiley and Sons, New York (1987)
41. Johnson, M.E., Tenenbein, A.: Bivariate distributions with given marginals and fixed measures of dependence. In: American Statistical Association, 1978 Proceedings of the Statistical Computing Section, pp. 261–263. American Statistical Association, Alexandria, Virginia (1978)
42. Johnson, M.E., Tenenbein, A.: Bivariate distributions with given marginals and fixed measures of dependence. Informal Report LA-7700-MS, Los Alamos Scientific Laboratory, Los Alamos, New Mexico (1979)
43. Johnson, M.E., Tenenbein, A.: A bivariate distribution family with specified marginals. Journal of the American Statistical Association **76**, 198–201 (1981)
44. Johnson, M.E., Wang, C., Ramberg, J.: Generation of continuous multivariate distributions for statistical applications. American Journal of Mathematical and Management Sciences **4**, 225–248 (1984)
45. Jones, G., Lai, C.D., Rayner, J.C.W.: A bivariate gamma mixture distribution. Communications in Statistics: Theory and Methods **29**, 2775–2790 (2000)

46. Kennedy, W.J., Gentle, J.E.: Statistical Computing. Marcel Dekker, New York (1980)
47. Kinderman, A.J., Monahan, J.F.: Computer generation of random variables using the ratio of uniform deviates. ACM Transactions on Mathematical Software **3**, 257–260 (1977)
48. Kinderman, A.J., Ramage, J.G.: Computer generation of normal random variables. Journal of the American Statistical Association **71**, 893–896 (1976)
49. Knuth, D. E.: The Art of Computer Programming, Volume 2, Seminumerical Algorithms, 2nd edition. Addison-Wesley, Reading, Massachusetts (1981)
50. Kotz, K., Nadarajah, S.: Extreme Value Distributions: Theory and Applications. Imperial College Press, London (2000)
51. Kronmal, R.A., Peterson, A.V.: The alias and alias-rejection-mixture methods for generating random variables from probability distributions. In: 1979 Winter Simulation Conference, Volume 1, H.J. Highland, M.G. Spiegel, and R. Shannon (eds.), pp. 269–280. Institute of Electrical and Electronics Engineers, New York (1979)
52. Kronmal, R.A., Peterson, A.V.: An acceptance-complement analogue of the mixture-plus-acceptance-rejection method for generating random variables. ACM Transactions on Mathematical Software **10**, 271–281 (1984)
53. Kronmal, R.A., Peterson, A.V., Lundberg, E.D.: The alias-rejection-mixture method for generating random variables from continuous distributions. In: American Statistical Association, 1978 Proceedings of the Statistical Computing Section, pp. 106–110. American Statistical Association, Alexandria, Virginia (1978)
54. Law, A.M., Kelton, W.D.: Simulation Modeling and Analysis. McGraw-Hill, New York (1982)
55. Leemis, L., Schmeiser, B.: Random variate generation for Monte Carlo experiments. IEEE Transactions on Reliability **34**, 81–85 (1985)
56. Lewis, P.A.W.: Chapter G of the IMSL library: Generation and testing of random deviates: Simulation. In: Proceedings of the 1980 Winter Simulation Conference, T.I. Ören, C.M. Shub, and P.F. Roth (eds.), pp. 357–360. Institute of Electrical and Electronics Engineers, New York (1980)
57. Lewis, P.A.W., Orav, E.J., Uribe, L.: Advanced Simulation and Statistics Package: IBM Professional FORTRAN Version (Software). Wadsworth and Brooks/Cole, Monterey, California (1986)
58. Li, S.T., Hammond, J.L.: Generation of pseudo-random numbers with specified univariate distributions and correlation coefficients. IEEE Transactions on Systems, Man, and Cybernetics **5**, 557–560 (1975)
59. Loukas, S.: Simple methods for computer generation of bivariate beta random variables. Journal of Statistical Computation and Simulation **20**, 145–152 (1984)
60. Lurie, P.M., Goldberg, M.S.: An approximation method for sampling correlated random variables from partially specified distributions. Management Sciences **44**, 203–218 (1998)
61. Marsaglia, G., Bray, T.A.: A convenient method for generating normal variables. SIAM Review **6**, 260–264 (1964)
62. Marsaglia, G., MacLaren, M.D., Bray, T.A.: A fast procedure for generating normal random variables. Communications of the Association for Computing Machinery **7**, 4–10 (1964)
63. Michael, J.R., Schucany, W.R.: The mixture approach for simulating bivariate distributions with specified correlations. The American Statistician **56**, 48–54 (2002)
64. Minh, D.L.: Generating gamma variates. ACM Transactions on Mathematical Software **14**, 261–266 (1988)
65. Monahan, J.F.: An algorithm for generating chi random variables. ACM Transactions on Mathematical Software **13**, 168–172 (Correction, 320) (1987)
66. Morgan, B.J.T.: Elements of Simulation. Chapman and Hall, London (1984)
67. Nadarajah, S.: Simulation of multivariate extreme values. Journal of Statistical Computation and Simulation **62**, 395–410 (1999)

68. Neuts, M.F., Pagano, M.E.: Generating random variates from a distribution of phase type. In: 1981 Winter Simulation Conference Proceedings, Volume 2, T.I. Ören, C.M. Delfosse, and C.M. Shub (eds.), pp. 381–387. Institute of Electrical and Electronics Engineers, Piscataway, New Jersey (1981)
69. Parrish, R.S.: Generating random deviates from multivariate Pearson distributions. Computational Statistics and Data Analysis **9**, 283–295 (1990)
70. Peterson, A.V., Kronmal, R.A.: On mixture methods for the computer generation of random variables. The American Statistician **36**, 184–191 (1982)
71. Peterson, A.V., Kronmal, R.A.: Mixture method. In: Encyclopedia of Statistical Sciences, Volume 5, S. Kotz and N.L. Johnson (eds.), pp. 579–583. John Wiley and Sons, New York (1985)
72. Ripley, B.D.: Computer generation of random variables: A tutorial. International Statistical Review **51**, 301–319 (1983)
73. Ripley, B.D.: Stochastic Simulation. John Wiley and Sons, New York (1987)
74. Rosenblatt, M.: Remarks on a multivariate transformation. Annals of Mathematical Statistics **23**, 470–472 (1952)
75. Rubinstein, R.Y.: Simulation and the Monte Carlo Method. John Wiley and Sons, New York (1981)
76. Sakasegawa, H.: Stratified rejection and squeeze method for generating beta random numbers. Annals of the Institute of Statistical Mathematics **35**, 291–302 (1983)
77. Salmerón, A.: Algoritmos de Propagación II. Métodos de Monte Carlo. In: Sistemas Expertos Probabilísticos, J.A. Gámez and J.M. Puerta (eds.), pp. 65–88. Ediciones de la Universidad de Castilla-La Mancha, Cuenca (1998)
78. Scheuer, E.M., Stoller, D.S.: On the generation of normal random vectors. Technometrics **4**, 278–281 (1962)
79. Schmeiser, B.W.: Random variate generation: A survey. In: Simulation with Discrete Models: A State-of-the-Art View, T.I. Ören, C.M. Shub, and F. Roth (eds.), pp. 79–104. Institute of Electrical and Electronics Engineers, New York (1980) [Updated in T.I. Ören, C.M. Delfosse, and C.M. Shub (eds.), 1981 Winter Simulation Conference Proceedings, Vol. 1, pp. 227–242, Institute of Electrical and Electronics Engineers, Piscataway, New Jersey (1981)]
80. Schmeiser, B.W., Babu, A.J.G.: Beta variate generation via exponential majorizing functions. Operations Research **28**, 917–926 (1980)
81. Schmeiser, B.W., Lal, R.:, Squeeze methods for generating gamma variates. Journal of the American Statistical Association **75**, 679–682 (1980)
82. Schmeiser, B.W., Lal, R.: Bivariate gamma random vectors. Operations Research **30**, 355–374 (1982)
83. Tadikamalla, P.R.: Computer generation of gamma variables. Communications of the Association for Computing Machinery **21**, 419–422 (1978a)
84. Tadikamalla, P.R.: Computer generation of gamma random variables-II. Communications of the Association for Computing Machinery **21**, 925–928 (1978b)
85. Tadikamalla, P.R.: On simulating non-normal distributions. Psychometrika **45**, 273–279 (1980)
86. Tadikamalla, P.R.: Modeling and generating stochastic inputs for simulation studies. American Journal of Mathematical and Management Sciences **4**, 203–223 (1984)
87. Tadikamalla, P.R., Johnson, M.E.: A survey of methods for sampling from the gamma distribution. In: 1978 Winter Simulation Conference, H.J. Highland, N.R. Nielsen, and L.G. Hull (eds.), pp. 130–134. Institute of Electrical and Electronics Engineers, New York (1978)
88. Ulrich, G., Watson, L.T.: A method for computer generation of variates from arbitrary continuous distributions. SIAM Journal on Scientific and Statistical Computing **8**, 185–197 (1987)
89. Vădiva, I.: On computer generation of gamma random variables by rejection and composition procedures. Mathematische Operationsforschungund Statistik, Series Statistics **8**, 545–576 (1977)

90. Văduva, I.: Computer generation of random vectors based on transformation of uniformly distributed vectors. In: Proceedings of the Seventh Conference on Probability Theory, pp. 589–598. VNU Science Press, Bucharest and Editura Academiei, and Utrecht (1985)
91. Vale, C.D., Maurelli, V.A.: Simulating multivariate nonnormal distributions. Psychometrika **48**, 465–471 (1983)
92. Walker, A.J.: New fast method for generating discrete random numbers with arbitrary frequency distributions. Electronics Letters **10**, 127–128 (1974a)
93. Walker, A.J.: Fast generation of uniformly distributed pseudo-random numbers with floating point representation. Electronics Letters **10**, 553–554 (1974b)
94. Yan, J.: Enjoy the joy of Copulas: With a package copula. Journal of Statistical Software **21**, 21 pages. http://www.jstatsoft.org/ (2007)

Author Index

Subject Index

MIX
Papier aus verantwortungsvollen Quellen
Paper from responsible sources
FSC® C105338

If you have any concerns about our products,
you can contact us on
ProductSafety@springernature.com

In case Publisher is established outside the EU,
the EU authorized representative is:
Springer Nature Customer Service Center GmbH
Europaplatz 3, 69115 Heidelberg, Germany

Printed by Libri Plureos GmbH
in Hamburg, Germany